Combustion Physics

Michael A. Liberman

Combustion Physics

Flames, Detonations, Explosions, Astrophysical Combustion and Inertial Confinement Fusion

Springer

Michael A. Liberman
NORDITA, KTH Royal Institute
of Technology
Stockholm University
Stockholm, Sweden

ISBN 978-3-030-85138-5 ISBN 978-3-030-85139-2 (eBook)
https://doi.org/10.1007/978-3-030-85139-2

This Springer imprint is published by the registered company Springer Nature Switzerland AG
The registered company address is: Gewerbestrasse 11, 6330 Cham, Switzerland

Preface

In 1825, Michael Faraday initiated popular Christmas lectures in London, which remain the flagship scientific lectures in the U.K. until today. Eventually, Faraday presented a series of six lectures centered around flame, "The Chemical History of a Candle",[1] which proved to be hugely popular—and for a good reason: the quest for fire illuminated human history for millennia but only at the time of the Industrial Revolution combustion emerged as a distinct area of science, setting out to analyze the chemical process of fuel oxidation. In the past one hundred and fifty years, this field experienced explosive growth alongside the appearance of the multitude of uses, from combustion engines to gas turbines to jet engines and space rockets! At last, combustion became a scientific field due to the combined progress in chemical kinetics and fluid mechanics. It was possible to explain various combustion modes, such as flame, explosions, and detonation, which all make a strong impact on our lives.

Today, combustion remains a truly interdisciplinary subject, and as such it requires merging of various fields of science, from chemical kinetics and thermodynamics to fluid mechanics, nonlinear phenomena, and turbulence. Because of this complexity, the in-depth scientific investigation of combustion is a relatively recent phenomenon, despite its long-standing impact on all types of human activities.

Delivering his lectures on The Chemical History of a Candle, Faraday justified his choice by claiming that *there is no better, there is no more open door by which you can enter into the study of science than by considering the physical phenomena of a candle*. Nowadays, combustion remains a fascinating and fast-developing field of physics. Faraday's words about an *open door by which you can enter into the study of science* can be just as well applied to combustion physics. While we often hear predictions about the end of new discoveries in physics, the actual end of new discoveries never seems to come. The elements of combustion, such as a flame, explosion, and detonation, are known for a long time, but somewhat surprisingly

[1] The Chemical History of a Candle: A Course of Lectures Delivered Before A Juvenile Audience at the Royal Institution by Michael Faraday, D. C. L., F. R. S. Edited by William Crookes, F. C. S. CHATTO AND WINDUS, PUBLISHERS, London.

their first principal description was obtained not such a long time ago and for a limited range of conditions. A more coherent picture emerged in recent times, as a result of many experimental and theoretical studies, sponsored by energy production and industrial needs.

Combustion is a process of heat release in exothermal reactions, which is accompanied by mass and heat transfer. The chemical combustion involves a chemical transformation between fuel substances and oxidizers. In these processes, nuclei of the entering substances are not altered but formed new molecules and atoms, which are combustion products. Heat release in the process of bonding is most interesting since this energy can be captured and usefully exploited. Combustion may involve all phases of matter—solid, liquid, and gas: solid rocket propellants, liquid droplets burning in diesels, and gaseous combustion in SI engines. The principal difficulties in understanding combustion processes are the huge differences of time and space scales involved, chemical complexity, and multidimensional nature of the flow configuration. In turbulent combustion, the strong nonlinear coupling of the turbulence and the chemistry bring additional complications. Chemical reactions' rates are coupled to molecular diffusion at the smallest scales of the turbulence. Furthermore, the heat release associated with combustion affects the turbulent flow, both from variations in the mean density field and from the effects of local dilatation. Because of its importance in many areas of physics and technology, combustion science is developing rapidly and there is an extensive volume of literature on the subject. A fast progress has been achieved in combustion science during a little more than 10 years since the publication of my book "Introduction to Physics and Chemistry of Combustion" in 2008. Because of this progress, it was clear that a new book is necessary to bring readers to the frontiers of the subject.

Combustion processes surround us in our everyday lives. Chemical combustion is used for energy production in power plants, gas turbines, and engines. Processes of thermonuclear combustion are the heat source for the Sun and stars. A special interest to the thermonuclear explosions of type Ia supernovas is due to the possibility to use the exploding supernovas as cosmic standard light markers to analyze the fate of the universe and the universe geometry. Using type Ia supernovas as distance indicators led to the discovery of the accelerated expansion of the universe and changed our understanding of the universe geometry to a nearly flat universe that is expanding at an accelerated rate. It was found that the universe expansion is speeding up rather than slowing down, whereas previously it was thought that gravity should slow down the universe expansion.

This book provides a comprehensive presentation of state of the art of the combustion theory, focusing on the combustion fundamentals. The book contains the most recent scientific results in various fields related to combustion. In each chapter of the book, and on each issue, an exhaustive list of references on the original research is provided, including the most recent publications. The presented material is aimed at illustrating the physics of the phenomena under consideration and its quantitative description obtained using analytical methods and multidimensional computer modeling, as well as comparison with experimental results. The level of the book is suitable for graduate university courses in combustion and for an advanced combustion course. In general, it would require a solid background in thermodynamics and fluid mechanics, but I tried to present the material in a way suitable for 2nd–3rd years and Ph.D. students of physical, mathematical, chemical, and engineering specialties. The last two chapters of the book describe applications of theoretical methods developed in combustion to study type Ia supernova explosions and stability problems in Inertial Confinement Fusion. Therefore, the book can be useful for astrophysicists and plasma physicists.

I am indebted to many people who helped in the preparation of this book. This especially applies to my colleagues at Nordic Institute of Theoretical Physics (NORDITA). In particular, I want to thank Axel Brandenburg for continued collaboration; Gregory Sivashinsky for carefully reading many chapters of the book and giving useful comments on the book's layout; and to Chengeng Qian for his meticulous help with preparing the illustrations.

Uppsala, Sweden
June 2021

Michael A. Liberman

Contents

Chapter 1
Combustion Chemistry

1.1 Chemical Reactions

From the very beginning of combustion research, it was understood that the chemical reactions of the combustion process consist of the oxidation of fuel. For example, it was clear that when hydrogen–oxygen is burned, water is formed, and when methane–oxygen is burned, water and carbon dioxide are formed:

$$2H_2 + O_2 \rightarrow 2H_2O,\ \ CH_4 + 2O_2 \rightarrow 2H_2O + CO_2.$$

However, it was only at the beginning of the twentieth century that it became clear that combustion reactions never proceed in a simple way as they are written above. The true path of products' creation in a burning process usually consists of many intermediate elementary reactions at the molecular or atomic level. While considering the chemical equilibrium states, we can calculate the final state of the system from the known initial state of the reactive mixture if chemical and thermal equilibria have been established. However, the equilibrium calculations cannot answer the question about how the mixture transits from the initial state to the final state, how much time it takes, etc. In a chemical transformation, the molecular bonds holding atoms in one kind of molecule are broken, and new bonds are formed with other molecules which appear during the reaction. Obviously, the number of atoms does not change in a chemical reaction; they are only regrouped in accordance with a specific release of energy. The conservation of the elementary composition can be written down in the form of a balance equation for the number of atoms of each element

$$\sum_i n_{ik} N_i = A_k, \tag{1.1.1}$$

where n_{ik} is the number of atoms of the kth element in the ith molecule, N_i is the number of molecules of type i in the system, and A_k is the total number of atoms

M. A. Liberman, *Combustion Physics*,
https://doi.org/10.1007/978-3-030-85139-2_1

of the given element in the mixture. The number of Eq. (1.1.1) corresponds to the number of elements present in the initial state of the reactive mixture. The summation is carried out over all types of molecules. Equation (1.1.1) is satisfied for any state of the reacting mixture: before the onset of the reaction when the excess energy is stored in chemical bonds, after the reaction is finished and complete chemical equilibrium has been achieved, and at any intermediate time during the reaction. The numbers A_k are the same for all states of the reacting mixture.

The rate of the reaction may be expressed in terms of the concentration of any reacting species or any product species of the reaction as the rate of decrease in the concentration of the reactive species or as the rate of increase of the reaction product. If only one reaction occurs in a system, or a complex reaction can be described as a one-step chemical reaction, then Eq. (1.1.1) for the initial and final states can be written in the form of the stoichiometric equation (2.6.1) with the stoichiometric coefficients of the reaction:

$$\sum_{i=1}^{M} n_i A_i = \sum_{i=1}^{M'} n_j A_j,$$

where A_i and A_j are the chemical symbols of the initial reagent i and of the reaction product j; n_i and n_j are the stoichiometric coefficients, which are interrelated by conditions of the form

$$\sum_{i=1}^{M} n_i n_{ik} = \sum_{i=1}^{M'} n_j n_{jk}.$$

If the initial substances are taken in the proportion such that the chemical transformation can, in principle, convert all of them fully into the reaction products, then the mixture is referred to as *stoichiometric*. If one of the substances appears in the initial mixture in an amount smaller than that required according to the stoichiometric equation, then this substance is *deficient*.

From the standpoint of thermodynamics, it does not matter what is the true reaction mechanism, because in thermodynamics only the initial and final equilibrium states are considered. But often we need the true reaction path to study combustion phenomena, especially in the case of a transient and nonstationary combustion process. Also, it can be interesting information on the rate in proceeding from the initial state to the final state. Such rates of chemical reactions are provided by chemical kinetics. The determination of actual reaction paths involves very complex inference from often incomplete data and theoretical arguments. Reliable detailed chemical kinetic models for practical fuels still remain among the major challenges. The availability of such models is essential for the design of advanced combustion engines and for gaining scientific insight into important combustion phenomena such as combustion and knocking in spark-ignition engines, stability gas turbine combustion, emissions of pollutants, efficiency of energy conversion, and device reliability.

From an engineering viewpoint, complete knowledge of a reaction scheme is often not necessary, and different useful approximations can be made and used in engineering practice. Often, researchers use the approximate approach of the one-step Arrhenius law for reaction, which allows obtaining simple analytical solutions and yields valuable qualitative insight into flame behavior.

The transition from the initial to the final state is characterized by the chemical reaction rates for the various components of the reacting mixture. The reaction rate for one of the components is defined as

$$\dot{\omega}_i = -\frac{1}{V}\frac{dN_i}{dt}, \quad \dot{\omega'}_j = \frac{1}{V}\frac{dN'_j}{dt}, \tag{1.1.2}$$

where N_i and N'_j are the number of molecules (atoms) of the ith reactant and the jth product in volume V at time t.

The relationship between the reaction rates for different species is determined by the conservation of the elementary composition given by Eq. (1.1.1), which implies that

$$\frac{\dot{\omega}_i}{n_i} = \frac{\dot{\omega'}_j}{n_j} = \dot{\omega}. \tag{1.1.3}$$

If the volume is constant in the course of reaction, then the reaction rates are defined as the time derivatives of the volume concentrations, $c_i = N_i/V$ and $c'_i = N'_j/V$:

$$\dot{\omega}_i = -\frac{dc_i}{dt}, \quad \dot{\omega'}_j = \frac{1}{V}\frac{dc'_j}{dt}. \tag{1.1.4}$$

The concentration of molecules in reacting gaseous mixtures, however, may also change, for example, because of thermal expansion of the gas or gas dynamic processes. In these cases, the reaction rate must be calculated according to Eq. (1.1.2).

Atoms or molecules can react only when they collide. The rate of atomic or molecular collisions is proportional to the number densities (number per unit volume) of the colliding species. If a reaction takes place as a result of the collision of two molecules (bimolecular reaction) of substances A_1 and A_2, then the reaction rate is

$$-\frac{dc_1}{dt} = -\frac{dc_2}{dt} = kc_1c_2.$$

For example, the elementary reaction between substances A and B can be described by the general expression

$$\mathrm{a \cdot A + b \cdot B \rightarrow c \cdot C + d \cdot D},$$

where a, b, c, d are stoichiometric coefficients. The corresponding rate of reaction progress can be expressed as (referred to as the law of mass action)

$$\dot{\omega} = k[\mathrm{A}]^{\mathrm{a}}[\mathrm{B}]^{\mathrm{b}}.$$

This expression for the reaction rate states that the reaction rate is proportional to the concentration of reactants. Since usually not more than two molecules are involved in elementary reactions, the values of a, b, c, d are one or can be two.

For example, in the reaction

$$\mathrm{O} + \mathrm{H} \rightarrow \mathrm{OH}$$

the rate of reaction for the reactants must be proportional to $c_{\mathrm{H}}c_{\mathrm{O}}$. We read the reaction above as "one atom of oxygen collides with one atom of hydrogen to form one molecule of radical OH." We define the reaction rate $\dot{\omega}$ through the coefficient, k, called the specific reaction rate constant, by

$$\dot{\omega} = kc_{\mathrm{H}}c_{\mathrm{O}}.$$

In all cases, the reaction rate is proportional to the product of the reacting species concentrations with the coefficient of proportionality called the Arrhenius rate constant k. For example, for the reaction of recombination of hydrogen atoms in a hydrogen molecule

$$\mathrm{H} + \mathrm{H} + \mathrm{H} \rightarrow \mathrm{H}_2 + \mathrm{H} \tag{1.1.5}$$

$\dot{\omega} = kc_{\mathrm{H}}^3$. The reaction rate tells how often a reaction occurs and is used to tell the rate of concentration change of the reacting species. For example, in the reaction $\mathrm{O} + \mathrm{H} \rightarrow \mathrm{OH}$, the units of k are given by the equality:

$$\frac{dc_{\mathrm{O}}}{dt} = -(1)kc_{\mathrm{O}}c_{\mathrm{H}} = -\dot{\omega},$$

where the minus sign is chosen because a free oxygen atom disappears in the reaction, forming the radical OH. In the example of reaction (1.1.5), because both hydrogen atoms disappear in the reaction forming the hydrogen molecule, the change in concentration of H-atoms is

$$\frac{dc_{\mathrm{H}}}{dt} = -3\dot{\omega} + \dot{\omega} = -2\dot{\omega} = -2kc_{\mathrm{a}}^3. \tag{1.1.5a}$$

The above two cases are examples of two-body and three-body reactions, indicating, respectively, a collision between either two or three atoms that must take place before a reaction can occur. Notice, in order for the units to be correct, the units of k

depend upon the number of colliding bodies. The most common are two-body and three-body reactions. The probability that four or more bodies could get together is extremely low, especially in ideal gas systems at not too high pressures, and they never been observed in experiments.

The law of acting mass states that the rate of disappearance of the reacting particles is proportional to the concentration of the reacting particles, each concentration being raised to a power equal to the corresponding stoichiometric coefficient [1]:

$$\dot{\omega}_i = k \prod_{j=1}^{M} (n_j)^{\nu_j}, \tag{1.1.6}$$

where the factor k is the specific reaction rate constant, which is a function of the gas temperature. The stoichiometric coefficient ν_j is called the order of the reaction with respect to species j. In practice, the overall order of the reaction is often used, which is defined as $\sum_j \nu_j$. Thus, the rate of change of the concentration of the given species i in a reacting system is given by the equation

$$\frac{dN_i}{dt} = (\nu_i' - \nu_i) k \prod_{j=1}^{M} (N_j)^{\nu}, \tag{1.1.7}$$

which means that ν' moles of N_i are formed for every ν moles of consumed N_j.

For the above example (1.1.5), this will be Eq. (1.1.5a), or $d[\mathrm{H}]/dt = -2k[\mathrm{H}]^3$. For a single-step reaction $\sum_j \nu_j$ gives the overall order of the reaction.

The sum of the concentration's exponents, or the number of atoms or molecules that interact with each other at the molecular level, leading to the completion of the reaction is called the molecularity of the reaction. It indicates the number of particles entering the reaction. In general, the molecularity of majority reactions is 2 or 3, however, for complex reactions the concept of molecularity has no sense, and the overall order of the reaction can be fractional. For two-body reactions, the molecularity is two.

Since $c_i = P_i/RT = X_i P/RT$, then for fixed composition and temperature, c_i is proportional to pressure. Whenever we discuss the pressure dependence of the reaction rate, it is assumed that values of T and X_i are fixed. Consequently, two-body reactions have a rate proportional to P^2, and three-body reactions have a rate proportional to P^3. Therefore, as pressure increases, the rate of three-body reactions increases faster than that of two-body reactions. In general, reactions involving the combination of active radicals, such particles as ions, atoms, and excited molecules, which do not require that the activation energy barrier be overcome, and which release a considerable amount of energy, cannot take place in bimolecular reactions and require the participation of a third particle to remove this excess of energy.

Majority of elementary reactions have overall molecularity of two, representing two-body collisions. However, reactions with overall molecularity of three are important ones for the recombination of two colliding particles. When two active particles collide, the resulting complex contains a significant excess of energy, which causes it to break down rapidly. In order for the complex to survive as a stable chemical compound, the excess energy must be removed. Radiation does not make a significant contribution in this case because the probability of photon emission is rather small (corresponding to the small parameters $e^2/\hbar c = 1/137$, which appear in the emission probability and the ratio of the size of the molecule to the wavelength of the emitted light). Therefore, the only possibility remains that the energy is removed by a third particle, which may have low chemical activity. For example, this kind of process occurs, in hydrogen–oxygen flames in which active radicals (hydrogen atoms) recombine in the three-body collision according to reaction $H + O_2 + M \rightarrow H_2O + M$. Reactions involving stable chemical compounds may also occur in three-body molecular collisions. However, since the three-body collision rate is substantially below the binary collision rate, even at high pressures, the main change will take place in bimolecular reactions. For large molecules, the excess energy can be transferred to the various vibrational modes of the complex molecules, and a third body may not be needed. Reactions with the overall molecularity of one are also possible. For example, a molecule can become excited when irradiated by a light source of a particular frequency and subsequently dissociate into two atoms or molecules. A chemically active molecule can also become deactivated through collisions with solid walls.

The temperature dependence of k follows from the fact that only a very few collisions between reactive molecules A and B lead to a chemical reaction that is the formation of a new molecule. The collision frequency of the molecule A with another molecule B in the gas with molecular number densities n_A and n_B is

$$Z = \sqrt{2}\sigma_{AB}n_A n_B \bar{c} = \sigma_{AB}n_A n_B\sqrt{\frac{8}{\pi}\frac{k_B T}{\mu_{AB}}},$$

where σ_{AB} is the effective collision cross section of molecule A with molecule B, $\mu_{AB} = m_A m_B/(m_A + m_B)$ is the reduced mass, $\bar{c}$ is mean speed of molecules given by the Boltzmann distribution (the thermal velocity of molecules), and $k_B = 1.38 \cdot 10^{-23}$ J/(K mol) is Boltzmann constant. The collision frequency of molecule A with molecule B depends on the product of their concentrations c_A and c_B, and on the mutual collision cross section (roughly equals the square of the sum of molecules radii) [2]

$$Z_{AB} = \sigma_{AB}c_A c_B\sqrt{\frac{8}{\pi}\frac{k_B T}{\mu}}.$$

In order for a reaction to occur, the colliding molecules must have a sufficient kinetic energy in order to overcome a potential barrier and to destroy or change the chemical bonds and electronic structure of the reactants. This potential barrier is characterized by the activation energy E_a and the temperature dependence of the specific reaction rate k owing to the existence of the activation energy. It is described by the *Arrhenius equation* [3]

$$k = k_0 \exp(-E_a/RT). \quad (1.1.8)$$

Here, k_0 is called the *pre-exponential factor*, $R = 1.986$ cal/(mol K) is the universal gas constant, and T is the temperature in Kelvin (°K), which is 273.15+°C. The units of the pre-exponential factor k_0 are identical to those of the rate constant and will vary depending on the order of the reaction. The activation energy, E_a, may be considered as the minimum energy required in a collision for the reaction to occur, or as the height of potential energy barrier that molecules have to overcome to be able to react. The pre-exponential factor may have a weak compared to the exponential temperature dependence. The law expressed by Eq. (1.1.8) is called the *Arrhenius law*.[1]

The extremely strong temperature dependence of the reaction rate in the Arrhenius equation Eq. (1.1.8) is a fundamental characteristic of the chemical processes under non-isothermal conditions, such as explosions and combustion. In gases, the Arrhenius law reflects the fact that according to a Maxwell–Boltzmann energy distribution, only the exponentially small fraction of the total number of reactive molecules belonging to the high energy "tail" of the distribution function can react. On the other hand, those molecules, which are moving at the same speed in the same direction, will not participate in a chemical reaction even though they may have a large kinetic energy. Thus, even for highly energetic molecules, a chemical reaction is a rare process.

The exponential term in the Arrhenius equation Eq. (1.1.8) can be interpreted as the probability of a successful collision leading to products. In combustion chemistry, reactions often have high activation energy such that the reaction rates are very sensitive to temperature. On the other hand, recombination reactions usually have either low or no activation energies so that the forward rate constants are insensitive to temperature. Because recombination reactions require three-body collisions to occur, the overall forward rate scales as P^3. As the pressure increases, the molecules are forced closer together so that the likelihood of three-body collisions increases. Therefore, the forward rate of a recombination step increases more rapidly with pressure than two-body reaction steps that scale as P^2. The values of k_0 and E_a are determined experimentally using shock tubes or flow reactors. The Arrhenius rate constant is calculated from the rate of progress of the experimental data, and the values of k_0 and E_a are found by plotting $\ln k = \ln k_0 - E_a/RT$ versus $1/T$. The

[1] Temperature dependence of the rate coefficient (3.1.8) had been suggested by Jacobus Van't Hoff (the first winner of Nobel Prize in Chemistry, 1901). Svante Arrhenius (received Nobel Prize in Chemistry, 1903) has given theoretical explanation and formulated the concept of activation energy.

rate of the reaction is then expressed as

$$\dot{\omega} = k_0[\mathrm{A}]^{\mathrm{a}}[\mathrm{B}]^{\mathrm{b}} \exp(-E_{\mathrm{a}}/RT). \tag{1.1.9}$$

Most chemical reactions are of second order, taking place so that their rates are dominated by collisions of two species. However, there are reactions that are dominated by a decomposition processes, e.g. by loose bond braking step. These reactions as well as isomerization reactions are the first-order reactions. The first-order reactions occur as a result of a two-step process, and therefore their rates also obey the Arrhenius law.

For arbitrary second-order reaction is written as $\mathrm{A} + \mathrm{B} \rightarrow \mathrm{C} + \mathrm{D}$, for example, $\mathrm{H} + \mathrm{O}_2 \rightarrow \mathrm{OH} + \mathrm{O}$, the rate of reaction for the reactants is proportional to $c_{\mathrm{A}}c_{\mathrm{B}}$ ($c_{\mathrm{H}}c_{\mathrm{O}_2}$):

$$\omega = \frac{d[\mathrm{A}]}{dt} = -kc_{\mathrm{A}}c_{\mathrm{B}} = -\frac{d[\mathrm{C}]}{dt}. \tag{1.1.10}$$

The pre-exponential factor k_0 in Eq. (1.1.8), which is the rate of collisions, may have weak temperature dependence, which can be estimated from simple kinetic theory as

$$[\mathrm{A}] \cdot [\mathrm{B}]\, \sigma_{\mathrm{AB}} (8\pi k_{\mathrm{B}} T/\mu)^{1/2}, \tag{1.1.11}$$

where σ_{AB}, the reduced mass μ, and k_{B} were defined above. The last term in (1.1.11) is the thermal velocity of molecules. The pre-exponential factor has a mild temperature dependency that generally can be ignored compared to the exponential dependence, or sometimes is taken as being proportional to T^m, where m is an empirical value.

Most elementary binary reactions exhibit Arrhenius behavior over modest ranges of temperature. However, for the large temperature ranges "non-Arrhenius" behavior is more likely, especially for the processes with a small energy barrier. Finally, it should be noticed that the units for the reaction rate constant k_0 when the reaction order is n will be $\{[(\text{concentration})^{n-1}][\text{time}]\}^{-1}$. For example, for the first-order reaction the units of the reaction rate constant are $[\mathrm{s}^{-1}]$, and for a second-order reaction $[\mathrm{cm}^3][\mathrm{mol}]^{-1}[\mathrm{s}^{-1}]$.

One other type of reaction occurs with highly reactive species in enclosures. The walls of the vessel may absorb some species. In order for this to occur, the species must diffuse to the proximity of the wall and then collided with it. The time to get near a wall is proportional to the gas density ρ, which, in turn, is proportional to pressure, P. Alternatively, the rate at which molecules appear near the wall is proportional to $1/P$. These arguments support the contention that wall reaction rates are dependent upon pressure to no larger than the first power. The consequence is that three reaction types (wall, two-body, and three-body) depend upon pressure in different ways.

The reactions written above indicated only forward reactions: the left to right reactions. A backward reaction can also occur. We denote the rate constants as $k_{\rm f}$ and $k_{\rm b}$, for the forward and reverse reactions, respectively. For example,

$$\mathrm{H} + \mathrm{H} + \mathrm{H} \leftrightarrow^{k_{\rm f}}_{k_{\rm b}} \mathrm{H}_2 + \mathrm{H}$$

with

$$\frac{dc_{\rm H}}{dt} = -2k_{\rm f}c_{\rm H}^3 + 2k_{\rm b}c_{\rm H_2}c_{\rm H}. \tag{1.1.12}$$

Note, that for the case of equilibrium, which corresponds to $dc_{\rm H}/dt = 0$, from (1.1.12) steams

$$\frac{c_{\rm H_2}}{c_{\rm H}^2} = \frac{k_{\rm f}(T)}{k_{\rm b}(T)} = \mathrm{RT}\frac{P_{\rm H_2}}{P_{\rm H}^2} = RTK_{\rm p}(T) = \frac{RT}{K_{\rm f f,H}^2}. \tag{1.1.13}$$

That is, since the functions of temperature are related through the equilibrium constant based on concentrations and the functions cannot change even when out of equilibrium, only one of the reaction rate constants is independent. If either the forward or the reverse reaction rate constant is known, the other is determined through thermodynamic relation.

In the general case, the net reaction rate of the forward and the reversed reactions is (see 1.1.6)

$$\dot{\omega} = k_{\rm f}\prod_{i=1}^{M}(n_i)^{\nu_i} - k_b\prod_{j=1}^{M}(n_j)^{\nu_j}. \tag{1.1.14}$$

At equilibrium $\dot{\omega} = 0$, and the ratio of k_f and k_b is the equilibrium constant

$$\frac{k_{\rm f}}{k_{\rm b}} = \prod_{i=1}^{M}(n_i)^{\nu_i-\nu_j} = K_{\rm c}. \tag{1.1.15}$$

Practically, all or majority of the reactions are initiated by bimolecular collisions. However, many of the bimolecular reactions exhibit first-order kinetics. As was mentioned earlier, this is a result of the two-step reaction sequence in which the reacting molecule is activated by collision processes and then the activated species decomposes to products. Another situation when a pseudo-first-order reaction arises is one when one of the reactants (usually the oxidizer in a fuel lean combustion with O_2 being in large excess) is in large excess. In this case, the concentration of another reagent does not change appreciably, and with this accuracy the reaction appears as the overall first order.

In many situations, the reverse reaction can be much slower than the forward reaction. For example, if products have a large temperature the concentration of radicals is much lower than it is for the starting reactants, and correspondingly the collision frequency for the reverse reaction is much lower.

1.2 Non-branching Chain Reaction: Hydrogen Chlorine

Most chemical reactions are complex and include a series of elementary steps. A complete solution requires the integration of an entire system of kinetic equations, which consists of the kinetic equations for each type of molecule or atom that participates in the reaction. Each of these equations must take into consideration the loss and production of the given type of particle in all the elementary steps of the chemical reaction. An analytical solution of the system of kinetic equations for complex reactions can be obtained only for some simplest cases. The situation is even more complicated for non-isothermal processes when the system of kinetic equations must be solved along with the energy balance equation.

The widely used approximation technique is the method of stationary concentrations. The method assumes that concentrations of the active intermediate species (radicals) are constant; that is, their time derivatives in the kinetic equations are equal to zero and the corresponding differential equations are replaced by algebraic equations, so that the overall order of the system is reduced. This is equivalent to the assumption that a small concentration of highly reactive intermediate species that participate in the reaction is regenerated in a sequence of reactions. This means an assumption that the intermediate highly reactive species react so rapidly with other substances that over the time during which the concentrations of the principal substances (reagents and reaction products) change by a small amount, the difference in the rates of formation and consumption of the active intermediate particles remains small compared with these rates. In these reactions the active radicals are chain carriers, and correspondingly the reactions are called chain reactions. A typical example of such reaction is the hydrogen–halogen reaction. Chain reactions are further classified into straight chain, or simply chain reactions, and chain-branching reactions.

In chain reactions active species (radicals) react with reagents or intermediate species to form both combustion products and new active radicals. This means that active radicals are "regenerated"—reproduce themselves during the reactions. As an example of a chain reaction we consider the reaction of hydrogen with chlorine. In principle, the reaction of hydrogen with chlorine could proceed via simple binary collisions

$$H_2 + Cl_2 = 2HCl. \tag{1.2.1}$$

However, a chain process is much faster so that reaction (1.2.1) is of no importance whatsoever. In fact, H_2 and Cl_2 do not directly react with each other at all; breaking both H–H and Cl–Cl bonds simultaneously during a single molecular collision is much less probable than other chemical pathways. The end product, hydrogen chloride, is formed as a result of a sequential alternation of reactions involving atomic chlorine and atomic hydrogen as active centers. Equation (1.2.1) is only a "global" reaction; the reaction of hydrogen with chlorine involves many "elementary reactions", each containing two or three species. The first active centers are chlorine atoms formed by dissociation of molecular chlorine by collisions of the type

$$Cl_2 + M = Cl + Cl + M \quad (k_1), \tag{1.2.2}$$

where k_1 is the rate constant and the notation "M" denotes any molecules that collide with H_2 or Cl_2, and are referred to as the third body particle. The third body particles serve as energy carriers. The formation of atomic hydrogen in the reaction $H_2 + M = H + H + M$ analogous to (1.2.2) can be neglected in view of the much greater stability of the H_2 molecular bond (104 kcal/mole for H_2 molecule compared to 57.9 kcal/mole for Cl_2). This is quite similar for all hydrogen–halogen reactions. For example, in the reaction $H_2 + B_2 \rightarrow 2HBr$ dissociation of the Br_2 during a collision with the third particle accompanied by the formation of two active radicals of Br is the chain initiation reaction since the Br_2 molecular bond energy 46 kcal/mole is considerably less than that of the H_2 molecular bond. In the theory of chain reactions, processes of the type (1.2.2), in which active centers are formed, are called *chain initiation reactions*. The third particle M, which participates in the initiation reaction, is needed to transfer energy to the Cl_2 molecule to break the molecular Cl–Cl bond.

Reaction (1.2.2) is reversible; that is, two chlorine atoms can recombine into a molecule in a three-body collision:

$$Cl + Cl + M = Cl_2 + M \quad (k_1'). \tag{1.2.3}$$

A reaction of this type, accompanied by the loss of free valence (*recombination*), is a *chain termination reaction*. In the example of the H_2–Br_2 reaction, the corresponding chain termination reaction is $2Br + M = Br_2 + M$. In the presence of molecular hydrogen, however, the chlorine atoms formed in chain initiation reaction may react in another way:

$$Cl + H_2 = HCl + H, \quad (k_2). \tag{1.2.4}$$

Although the concentration of Cl atoms is very small and the number of $Cl + H_2$ collisions is much smaller than the number of collisions between chlorine molecules and hydrogen molecules, reaction (1.2.4) takes place much more rapidly than the bimolecular reaction $H_2 + Cl_2$ because of its *very low activation energy* ($E \simeq$ 5 kcal/mole compared to roughly 40 kcal/mole for the reaction $H_2 + Cl_2$). Note also that according to the Arrhenius equation, the rate of a reaction with a lower activation energy is exponentially larger than that with higher activation energy. However, even this fact is insufficient to explain the rapidity of the entire process.

Although the chlorine atoms react rapidly, they are formed very slowly. Reaction (1.2.4) shows that the formation of the end product, HCl, is accompanied by the production of a new active center—a hydrogen atom, which reacts in the process

$$H + Cl_2 = HCl + Cl, \quad (k_3) \tag{1.2.5}$$

even more rapidly than a chlorine atom reacts with hydrogen. In this way, any reaction of the chlorine atom that participates in reaction (1.2.4) will be inevitably followed by reaction (1.2.5), which is recovered after the formation of two HCl molecules. This is shown in the overall balance equation for the reactions (1.2.4) and (1.2.5):

$$Cl + H_2 + H + Cl_2 = 2HCl + H + Cl. \tag{1.2.6}$$

The regenerated chlorine atom again participates in reaction (1.2.4) and so on. A small amount of chlorine atoms keeps the reaction going constantly and, since these atoms are not in effect consumed, their number is roughly constant. This mechanism ensures that the rate for the whole process is large. Reactions such as (1.2.4) and (1.2.5), which conserve free valency, are known as *chain propagation reactions.*

We shall now determine the concentration of chlorine atoms in the system and derive a formula for the reaction rate using the method of stationary concentrations by setting

$$\frac{d[H]}{dt} = k_2[Cl][H_2] - k_3[H][Cl_2] = 0 \tag{1.2.7}$$

$$\frac{d[Cl]}{dt} = 2k_1[Cl_2][M] - 2k_1'[Cl]^2[M] = 0. \tag{1.2.8}$$

Here, symbol [...] as usually means concentration of species. From (1.2.7) and (1.2.8), the steady-state concentrations of the active centers (hydrogen and chlorine atoms) are

$$[H] = \frac{k_2}{k_3}[H_2]\left(\frac{k_1}{k_1'}\right)^{1/2}[Cl_2]^{-1/2} \text{ and } [Cl] = \left(\frac{k_1}{k_1'}\right)^{1/2}[Cl_2]^{1/2}. \tag{1.2.9}$$

The overall rate of the process now can be written (see (1.2.4) and (1.2.5)) as

$$\dot{\omega} = \frac{d[HCl]}{dt} = k_2[Cl][H_2] + k_3[H][Cl_2]. \tag{1.2.10}$$

After substitution of the steady-state concentrations for [H] and [Cl] from (1.2.9) in (1.2.10), we obtain for the overall rate

$$\dot{\omega} = 2k_2\left(\frac{k_1}{k_1'}\right)^{1/2}[Cl_2]^{1/2}[H_2]. \tag{1.2.11}$$

This equation shows that the reaction is of first order in hydrogen and of the order 1/2 in chlorine, so that the overall order of the reaction is 1.5. A fractional power of one of the reagent concentrations in the reaction rate formula indicates that the reaction is complex, usually a chain reaction. In this particular case, this means that the reaction proceeds via a chlorine atom.

To evaluate the magnitude of the rate of hydrogen chloride formation, we notice that the ratio k_1/k_1' is the equilibrium constant for the reaction $2Cl \leftrightarrow Cl_2Cl \leftrightarrow Cl_2$. Using JANNAF tables we find

$$K_1 = 2.5 \cdot 10^{25} e^{-56800/RT} cm^{-3},$$

and k_2 is the constant for the bimolecular elementary process

$$k_2 = 10^{-10} e^{-6000/RT} cm^3/s.$$

Substituting these values in Eq. (1.2.11), we obtain

$$\dot{\omega} = 10^3 e^{-34400/RT} [Cl_2]^{1/2} [H_2]. \tag{1.2.12}$$

For a temperature $T = 600$ K and the initial concentrations of chlorine and hydrogen $[H_2]_0 = [Cl_2]_0 = 0.7 \cdot 10^{19} cm^{-3}$ (we assume $P = 1$ bar), this equation gives an initial maximum reaction rate of $\dot{\omega}_0 \approx 6.5 \cdot 10^{18} s^{-1} cm^{-3}$ which is about 10^6 times greater than the rate of the simple bimolecular reaction $H_2 + Cl_2$ which is of the order $1 s^{-1}$. If a hydrogen–chlorine mixture were kept at a constant temperature and there are no hydrogen and chlorine atoms at the initial moment, then the reaction rate would first rise to a value close to $\dot{\omega}_0$ and then fall slowly, approaching zero as the reagents were consumed. The initial rise in the reaction rate takes place over the time required for the production of a sufficient equilibrium quantity of chlorine atoms.

1.3 Formation Mechanisms of Nitrogen Oxides

Another example of the non-branching chain reaction is the oxidation of nitrogen, accompanied by the formation of nitrogen oxide NO The chemical reactions of nitrogen compounds that occur in combustion processes attracted special attention because of the harmful effects of nitric oxides even at very small concentrations. The formation of nitric oxides has been the subject of intensive study for more than 50 years. Much of this research has been motivated by the impact of the nitrogen compounds emitted from combustion sources on the environment and by the role that these species play in the combustion of energetic materials (explosives). At the present time, many of the complex reaction routes responsible for the formation and destruction of nitrogen species in combustion have been established, and the reaction rates of the elementary reactions have been determined.

Two of the gaseous oxides of nitrogen, nitric oxide NO, and nitrogen dioxide NO_2, collectively known as NO_x, are emitted from combustion processes. In nature, the normal concentration of NO_x is very low. It mainly results from forest fires and is present in the atmosphere in low concentrations. However, the high local concentrations of NO_x generated in man-made combustions (combustion in engines, industrial furnaces etc.) have a variety of harmful effects on the environment. During combustion, most of the nitrogen oxides are produced as NO with a minor (1–3%) concentration of nitrogen dioxide NO_2. The kinetics of nitrogen oxide formation in fuel combustion processes is a very complex issue. The knowledge in this area is constantly enriched due to experimental studies combined with numerical simulations and the increased knowledge on combustion chemical kinetics. Considerable progress in the understanding of NO_x reactions has been achieved in the last years due to the increased performance of computers and comprehensive modeling of NO_x formation and destruction [5–8]. This progress helps to develop technologies which have the potential to enhance the application of various combustion techniques used to reduce NO_x emissions from practical combustion systems.

Nitrogen oxide from combustion systems originates from three main processes: *thermal* NO, *prompt* NO, and *fuel* NO. Various details of these basic processes have been discussed in reviews [4–8]. The first fundamental research on the kinetics of NO_x has been performed by Ya. B. Zel'dovich in 1946 [9, 10] and is known as the Zel'dovich thermal mechanism [11]. In 1971, C. P. Fenimore [12] proposed the *prompt* NO mechanism to explain the excess of NO above the Zel'dovich thermal mechanism observed from gaseous pyridine and ammonia flames. The additional, *fuel-bound route* of NO formation was studied by Pershing and Wendt [13]. *Thermal* NO is formed from oxidation of atmospheric nitrogen, *prompt* NO is formed in reactions of atmospheric nitrogen with hydrocarbon radicals in fuel-rich regions of flames, and *fuel* NO is formed from the oxidation of nitrogen bound in the fuel. The importance of understanding the mechanisms of the formation of nitrogen oxides stems from the fact that nitrogen oxides are one of the principal pollutants emitted by combustion processes. Because of stringent laws to control the emissions of pollutants, suitable schemes must be developed for controlling the amount of NO_x Therefore, the understanding of mechanisms of NO_x formation is essential for developing clean combustion technologies. Below, we consider the main mechanisms of NO_x formation.

1.3.1 Thermal NO Mechanism

Thermal NO is the main source of NO in gaseous combustion systems. Thermal NO is formed from the oxidation of atmospheric nitrogen at relatively high temperatures in fuel-lean combustion, and has a strong temperature dependence. The formation of nitric oxide is unavoidable when air, which contains 78% nitrogen, is used as an oxidant. Moreover, in order to ensure complete combustion of fuels, the fuel-to-air ratio is adjusted so that a slight excess of oxygen remains in the combustion products. At high concentrations, NO enters further in the reaction $2NO + O_2 = 2NO_2$. The presence of NO_2 is easily observed by the yellowish color it imparts to smoke. Nitrogen dioxide NO_2 is easily absorbed by water or alkaline solutions. However, the oxidation reaction $2NO + O_2NO + O_2$ is a trimolecular reaction, which proceeds very slow at low concentrations of NO. Therefore, the oxidation of nitrogen and the understanding conditions under which this reaction is suppressed are of particular importance. It is now generally accepted that in the combustion of lean and near-stoichiometric fuel–air mixtures, the principal reactions governing formation of NO from molecular nitrogen are those originally proposed by Zel'dovich [9].

Zel'dovich, Frank-Kamenetskii, and Sadovnikov have shown for the first time that NO is produced in the non-branching chain reactions involving free oxygen and nitrogen atoms as active centers:

$$O + N_2 \leftrightarrow NO + N - 75\ \text{kcal/mole}, \quad (k_1 k_2') \tag{1.3.1}$$

and

$$N + O_2 \leftrightarrow NO + O + 32\,\text{kcal/mole} \quad (k_2 k_2') \tag{1.3.2}$$

Here, the heats of the reaction have been calculated from the known dissociation energies of N_2 (226 kcal/mole), NO (151 kcal/mole), and O_2 (118 kcal/mole). The rate of production of NO is limited by the first, endothermic reaction (1.3.1) because it has very high activation energy due to the strong bound of the N_2 molecule. The consumption of the oxygen atom in reaction (1.3.1) is compensated by its regeneration in reaction (1.3.2). The N atom formed in the first reaction immediately reacts with the oxygen molecule to yield another active oxygen atom. If it would be assumed that the reaction occurs through collisions of N_2 and O_2 molecules, then the calculated reaction rate would be a thousand times smaller than the experimentally observed value. The experimental heat of activation is roughly 135 kcal/mole.

Using the method of stationary concentrations, we obtain

$$-\frac{d[\mathrm{N}]}{dt} = -k_1[\mathrm{O}][\mathrm{N}_2] + k_1'[\mathrm{NO}][\mathrm{N}] + k_2[\mathrm{N}][\mathrm{O}_2] - k_2'[\mathrm{NO}][\mathrm{O}] = 0 \tag{1.3.3}$$

so that the concentration of N atoms can be expressed in terms of the steady-state concentration of oxygen atoms as

$$[N] = \frac{k_1[N_2] + k_2'[NO]}{k_1'[NO] + k_2[O_2]}[\mathrm{O}]. \tag{1.3.4}$$

Substituting this expression in the kinetic equation for nitric oxide

$$\frac{d[\mathrm{NO}]}{dt} = k_1[\mathrm{O}][\mathrm{N}_2] + k_2[\mathrm{N}][\mathrm{O}_2] - k_1'[\mathrm{NO}][\mathrm{N}] - k_2'[\mathrm{NO}][\mathrm{O}], \tag{1.3.5}$$

we obtain

$$\frac{d[\mathrm{NO}]}{dt} = 2\frac{k_1k_2[\mathrm{N}_2][\mathrm{O}_2] - k_1'k_2'[\mathrm{NO}]^2}{k_2[\mathrm{O}_2] + k_1'[\mathrm{NO}]}[\mathrm{O}] \approx 2\left(k_1[\mathrm{N}_2] - \frac{k_1'k_2'[\mathrm{NO}]^2}{k_2[\mathrm{O}_2]}\right)[\mathrm{O}]. \tag{1.3.6}$$

Since the constants k_1' and k_2 which determine the exothermic reaction rate of atoms with molecules in reactions (1.3.1) and (1.3.2) are of the same order of magnitude and since the concentration of nitric oxide is much lower than the concentration of oxygen molecules (the equilibrium concentrations of nitric oxide at temperatures from 3000 K up to 10,000 K and $P = 1\,atm$ do not exceed a few percent), we can neglect the term $k_1'[\mathrm{NO}]$ in the denominator of Eq. (1.3.6).

Using the principle of detailed balance for the reactions (1.3.1), (1.3.2), and for the reaction $O_2 \leftrightarrow 2O$ (k_3, k_3'), we can write

$$K_1 = \frac{k_1}{k_1'} = \frac{(\mathrm{NO})(\mathrm{N})}{(\mathrm{O})(\mathrm{N}_2)}, \quad K_2 = \frac{k_2}{k_2'} = \frac{(\mathrm{NO})(\mathrm{O})}{(\mathrm{N})(\mathrm{O}_2)}, \quad K_3 = \frac{k_3}{k_3'} = \frac{[\mathrm{O}]^2}{[\mathrm{O}_2]}. \tag{1.3.7}$$

Here, parentheses denote equilibrium concentrations. With the help of (1.3.7), Eq. (1.3.6) is reduced to

$$\frac{d[\mathrm{NO}]}{dt} = k'[\mathrm{N}_2][\mathrm{O}_2] - [\mathrm{NO}]^2 = k\{(\mathrm{NO})^2 - [\mathrm{NO}]^2\}, \tag{1.3.8}$$

where experimentally measured in [10] values of reaction rates k and k' (in $\mathrm{cm}^3/\mathrm{sec}$) are

$$k = \frac{k'}{K_1K_2} = \frac{2k_1'k_2'K_3^{1/2}}{k_2[\mathrm{O}_2]^{1/2}} = \frac{53}{[\mathrm{O}_2]^{1/2}}\exp\{-92000/RT\} \tag{1.3.9}$$

$$k' = 2\frac{k_1K_3^{1/2}}{[\mathrm{O}_2]^{1/2}} = \frac{1.1\cdot 10^3}{[\mathrm{O}_2]^{1/2}}\exp\{-135000/RT\}. \tag{1.3.10}$$

Equation (1.3.8) for the reaction rate of NO differs from the usual kinetic equation for the bimolecular reaction of N_2 with O_2. In Eq. (1.3.8), the reaction rate constants depend on the concentration of only oxygen molecules, which is the consequence of the chain nature of the reaction.

The experimental value of the activation energy for the production of nitric oxide is 135 kcal/mole. From Eq. (1.3.10) for k', the temperature dependence of this reaction is determined by the temperature dependence of the equilibrium constant, which is $K_3 \propto \exp(-61000/RT)$. This follows from the fact that 61 kcal/mole is the effective energy of formation of the oxygen atom at temperatures of order 2000–5000 K, and of the value of reaction rate k_1. Thus, we obtain that the activation energy of the first step reaction (1.3.1) of the oxidation of nitrogen $E_1 = 135 - 61 = 74$ kcal/mole, which is almost the same as the heat of the endothermic reaction. This means that the reverse of (1.3.1) reaction proceeds with practically zero activation energy.

In some cases when NO formation from the original Zeldovich mechanism is underestimated [14, 15], an additional elementary reaction should be added to the thermal mechanism. Lavoie, Heywood, and Keck [16] have suggested that the reaction

$$N + OH = NO + H \tag{1.3.11}$$

can contribute to NO production, especially in near-stoichiometric and rich fuel–air mixtures. The extended Zeldovich mechanism, consisting of the reactions (1.3.1), (1.3.2), and (1.3.11), considers the effect of oxygen and hydrogen radicals on NO formation. Typically, these radicals are calculated assuming the equilibrium state. However, with this assumption, the predicted thermal NO concentrations are always lower than the experimentally observed concentrations [5, 14, 15]. All the same, the use of measured concentrations of the oxygen radical gives a good agreement [14, 15] with the predicted thermal nitric oxide concentrations.

The thermal NO reactions proceed in a few tens of microseconds, and are highly dependent on temperature, residence time, and atomic oxygen concentration. The first reaction (1.3.1) is usually accepted as being the rate-determining step due to its high activation energy, and results in the Zel'dovich thermal mechanism being temperature-sensitive. The temperature dependence also results from the temperature sensitivity of the oxygen atom equilibrium concentration. The last reaction (1.3.11) is usually negligible except in fuel-rich flames [5]. The overall rate for the three reversible, thermal NO reactions (1.3.1, 1.3.2, 1.3.11) is [15]

$$\frac{d[\mathrm{NO}]}{dt} = 2[\mathrm{O}]\left\{\frac{k_1[N_2] - \frac{k_1' k_2' [\mathrm{NO}]^2}{k_2[\mathrm{O}_2]}}{1 + \frac{k_1'[\mathrm{NO}]}{k_2[\mathrm{O}_2] + k_3[\mathrm{OH}]}}\right\} \mathrm{g\,mol/cm^3\,s}. \tag{1.3.12}$$

Assuming a steady-state condition for N atoms, and using the method of stationary concentrations, we obtain instead of (1.3.12) a simplified expression in a way similar to (1.3.6). This assumption means that the concentration of NO is much less than the equilibrium value, so that the backward reaction can be neglected, and only the forward rates of the Zel'dovich thermal mechanism are significant:

$$\frac{d[\mathrm{NO}]}{dt} = 2k_1[\mathrm{O}][\mathrm{N}_2]\ \mathrm{g\,mol/cm^3\,s}. \tag{1.3.13}$$

Kinetic Eqs. (1.3.12) and (1.3.13) are coupled to the fuel oxidation process through competition for the oxygen atom, whose local concentration must be estimated when a comprehensive kinetic scheme is not used to compute the fuel oxidation chemistry.

Accurate estimation of the nonequilibrium oxygen concentration is critical for the correct prediction of thermal NO reactions [17]. The commonly used reaction rate constants for the extended Zel'dovich thermal mechanism reactions are taken in the form [15]:

$$k = AT^{\beta} \exp(-E/RT) \text{ cm}^3/(\text{g mol s K}),$$

with the following parameters: $A = 1.36 \cdot 10^{14}$, $\beta = 0$, $E = 315.9$ for the reaction (1.3.1); $A = 6.4 \cdot 10^9$, $\beta = 1$, $E = 26.3$ for (1.3.2); $A = 3.28 \cdot 10^{13}$, $\beta = 0$, $E = 0$ for (1.3.11), where dimensions are A, $\text{cm}^3/(\text{g mol s K})$, E, (J/(g mol).

At relatively low temperatures below 1800 K, thermal NO formation by the Zel'dovich mechanism is noticeably reduced. It is also considerably less in fuel-rich combustions because of the low concentration of oxygen atoms. As a result of the lower temperatures (<1800 K) and for the fuel-rich combustions (which lack oxygen atoms), the Zel'dovich mechanism does not appear to be a major source of NO.

1.3.2 Prompt NO Mechanism

Prompt mechanism of NO formation is an important, and perhaps dominating mechanism initiated by the reaction of atmospheric nitrogen with hydrocarbon radicals, CH_n in fuel-rich regions of flames in turbulent hydrocarbon/air diffusion flames. It was first proposed by Fenimore [12] who studied NO formation in burners from gaseous pyridine and ammonia. He observed a higher NO formation rate in the fuel-rich regions of hydrocarbon flames. The main reactions proposed to describe this process in hydrocarbon flames are [18]

$$CH + N_2 \rightleftarrows HCN + N, \tag{1.3.14}$$

$$N_2 + CH_2 \rightleftarrows HCN + HN + N + ..., \tag{1.3.15}$$

$$C_2 + N_2 \rightleftarrows 2CN, \tag{1.3.16}$$

$$N + OH \rightleftarrows NO + H. \tag{1.3.17}$$

Reaction (1.3.14) is the primary route and the limiting reaction with the rate constant, $k_{f,14} = 4.4 \cdot 1012 \exp(-11060K/T)$ $\text{cm}^3/(\text{mol s})$. The activation energy for this reaction is approximately 75–92 kJ/mole, which is much lower than 319 kJ/mole for the formation of thermal NO. It was believed for a long time that the primary route reaction proceeds as (1.3.14) to form HCN + N. However, it was shown by Moskaleva and Lin [19] that the kinetics is more complex: this reaction proceeds via insertion of CH into the N_2 triple bond, yielding NCH + H:

$$CH + N_2 \rightleftarrows NCN + H. \tag{1.3.18}$$

Prompt NO formation occurs in fuel-rich regions where hydrocarbon radicals increase the formation of HCN through the reactions producing HCN and NO. Some of them are

$$NCN + H \rightleftarrows HCN + N, \tag{1.3.19}$$

$$NCN + O \rightleftarrows CN + NO, \tag{1.3.20}$$

$$NCN + OH \rightleftarrows HCN + NO, \tag{1.3.21}$$

$$NCN + O_2 \rightleftarrows NCO + NO, \tag{1.3.22}$$

$$CH_2 + N_2 \rightleftarrows HCN + NH, \tag{1.3.23}$$

$$CH + N_2 \rightleftarrows HCN + N. \tag{1.3.24}$$

Since the prompt NO mechanism requires a hydrocarbon to initiate the reaction with nitrogen, this mechanism is more prevalent in fuel-rich flames than in fuel-lean hydrocarbon flames. In practical combustion systems which are usually operated with fuel-lean or stoichiometric mixtures, the contribution of prompt NO to the total NO formation is small. Therefore, prompt NO is more prevalent in cooler, hydrocarbon flames of "clean" fuels (containing no fuel–nitrogen). Prompt NO reactions are neglected in many NO models due to the complexity of nitrogen chemistry. The modeling of industrial flames is often restricted to the thermal-NO formation, the rate of which can be approximated throughout the analytical function of oxygen and nitrogen concentrations and temperature. However, a reliable predictive model requires considering a rather detailed chemical model [20].

Cui et al. [21] have pointed that the calculated rate constant of reaction (1.3.14) with the product channel HCN + N is about two orders of magnitude lower than experimental measurements of the reactant consumption. Williams and Fleming [22] proposed modifications to the Konnov detailed reaction mechanism for small hydrocarbon combustion [23] related to the prompt-NO route and concluded that their modeling supports the inclusion of the reactions of CH_2 and C_2O with N_2 into NO formation schemes. The role of C_2O radicals in the prompt-NO mechanism was re-evaluated by Konnov [24]. The measured NO concentrations in the post-flame zone of different hydrocarbon + O_2 + N_2 flames at standard temperature and atmospheric pressure available in the literature were compared with predictions of the original Konnov reaction mechanism [23] and with the same mechanism extended by the reaction of C_2O with N_2. Modified prompt NO sub-mechanism of [25] allows more accurate prediction of NO formation in lean and rich flames of methane, ethylene, ethane, and propane. Nowadays, the availability of detailed chemical schemes and the reaction rate constants [26–32], at least in principle, allows accurate prediction of NO with a satisfactory agreement with experimentally measured values.

1.3.3 Fuel NO

Fuel NO is formed from nitrogen bound in the fuel, and is usually assumed to proceed through the formation of HCN or NH_3 which are oxidized to NO while being competitively reduced to N_2 according to the overall reactions [34]:

$$HCN/NH_3 + O_2 \rightarrow NO + \ldots \tag{1.3.25}$$

$$NO + HCN/NH_3 \rightarrow N_2 + \ldots \tag{1.3.26}$$

Nitrogen bound in the coal is released during the devolatilization process. A fraction of the nitrogen is rapidly converted to HCN, and the remaining portion of the fuel nitrogen reacts to form NH_3. The formed NO can be reduced by a heterogeneous reaction with char particles. Various global reaction rates have been proposed, e.g. [8, 33]. The global reaction rates are typically measured under controlled conditions in the temperature range of interest, and apply to both fuel-rich and fuel-lean conditions. When modeling the fuel NO reaction process, fuel nitrogen is assumed to evolve from the fuel as HCN or NH_3, which further react to form either NO or N_2 in reactions (1.3.25) and (1.3.26) depending on the local conditions. In fuel-rich regions, the nitrogen-containing species will typically be reduced to N_2, and in fuel-lean regions they are generally oxidized to form NO. Controlling the local conditions in which nitrogen is released from the fuel is a primary means of controlling NO emissions.

Fuel NO is the dominant NO formation mechanism in flames which contain nitrogen in the fuel, and typically accounts for more than 80% of the NO formed. Fuel NO is formed more readily than thermal NO because the N–H and N–C bonds most common in fuel-bound nitrogen are much weaker than the triple bond in molecular nitrogen which must be broken for thermal NO formation. The conversion of nitrogen species to NO is only weakly dependent on the specific nitrogen compound.

Experiments indicate that HCN is normally the first measurable volatile nitrogen species. The appearance of fuel nitrogen as NH_3 in the gas phase may be important during char oxidation or during high-temperature gasification of low-rank coals. HCN may consist of only a small portion of the nitrogen compounds initially evolved from the coal, but the remainder of these nitrogen compounds is rapidly converted to HCN in the absence of oxygen.

For fuel-lean systems, NO is the main nitrogen pollutant, and low amounts of HCN and NH_3 are observed. As the system becomes more fuel-rich, less NO is produced, but higher amounts of HCN and NH_3 are observed. The relative amounts of formed HCN and NH_3 appear to be functions of coal type with bituminous coals producing higher quantities of HCN and lignite coals producing higher quantities of NH_3. Lower temperatures also result in decreased NO emissions but increased HCN and NH_3 emissions. The presence of CO influences reduction of NO which tends to create additional radicals, such as OH, H, and O, by the branching CO oxidation which then participate in the reduction process. A shift to lower temperatures for NO reduction has been observed when CO is present [34].

1.3.4 Flames of the First-Order Reaction

Nitric oxide and nitrogen dioxide are acid rain precursors and participate in the generation of photochemical smog, while nitrous oxide is a greenhouse gas. Nitrogen dioxide is formed through the oxidation of nitric oxide at temperatures lower than the typical temperature in flames [35]. Nitrogen dioxide is formed through the sequence of reversible reactions [8, 35, 36]:

$$NNH + O_2 \rightleftarrows N_2 + HO_2, \tag{1.3.27}$$

$$NO + HO_2 \rightleftarrows NO_2 + OH, \tag{1.3.28}$$

$$NH_2 + NO_2 \rightleftarrows H_2NO + NO, \tag{1.3.29}$$

$$H_2NO + O_2 \rightleftarrows HNO + HO_2, \tag{1.3.30}$$

$$HNO + O_2 \rightleftarrows NO + HO_2. \tag{1.3.31}$$

The formation of N_2O is described by complex chemistry and depends on temperature and oxygen concentration [8]. There are two main reactions producing N_2O [8, 36, 37]:

$$NH_2 + NO_2 \rightleftarrows N_2O + H_2O. \tag{1.3.32}$$

This reaction is active predominantly at lower temperatures and larger $[O_2]$ where NO_2 is readily available. The reaction

$$NH + NO \rightleftarrows N_2O + H \tag{1.3.33}$$

competes with reaction $NH + O_2$ at higher temperatures and lower values of $[O_2]$. The predictions obtained with the present model agree qualitatively with the measured concentrations [8], but the model overpredicts N_2O under conditions where the NH + NO reaction is the main source.

1.4 Chain-Branching Reactions: Explosions

Compounds, which are highly stable at normal temperatures, usually mimic the noble gases (helium, argon, etc.) in the configuration of their electrons. That is, these highly inert gases have outer electron shells empirically filled by 2 and 8 electrons in the case of He and Ne, respectively. Such compounds are usually quite stable and hard to break apart. An example is a hydrogen molecule, H_2 which looks like He in the sense that the hydrogen molecule has two electrons in the shell of which each hydrogen atom shares its electron with other hydrogen atom. Similarly, a water molecule, H_2O, looks like Ne where there are total 8 electrons being shared, 6 from the outer shell in oxygen and 2 from each of the hydrogen atoms.

Any atom or compound with the unfilled outer electron configuration is highly reactive. Examples are the hydroxyl radical, OH, and the oxygen and hydrogen atoms, O and H. Species such as H, O, OH and HO_2 are called *radicals*. Radicals have unpaired valence electrons, which make them very reactive and short-lived. They try to partner with other radicals to form covalent bonds. While an equilibrium calculation at usual temperatures shows that radicals are only present in small amounts, they play a central role in chemical kinetics reaction schemes. In *chain-branching reactions* there is a net production of radicals. Chain-branching reactions lead to rapid production of radicals, which causes the overall reaction to proceed very fast—explosively. The reaction comes to completion through chain termination reactions, where the radicals recombine to form final products.

Radicals usually have positive heats of formation, which means that energy is required for their formation from the standard state elements. They soak up collision energy in their formation and wish to give up this energy in attacking other species in the path toward reaction completion. Hungry for electrons, radicals are efficient intermediates in reaction schemes for taking the reaction path forward. It should be pointed out that the above reasoning concerning the electron configuration is not always straightforward. For example, a complex electron sharing occurs with CO, which is not a radical in the sense described above, although it would appear to be one. Radicals are extremely important in the kinetic schemes relevant to atmospheric pollution. Examples of reactions producing nitric oxide as a harmful air pollutant through Zel'dovich's thermal mechanism discussed previously are

$$O + N_2 \rightarrow NO + N, \quad N + O_2 \rightarrow NO + O.$$

Some chemical reactions develop in time smoothly, others develop "explosively" and are usually accompanied by some phenomena such as a flash, sounds, and shock waves. The important features of explosions are that the reaction rate varies extremely rapidly with temperature. For example, at room temperature and atmospheric pressure hydrogen and oxygen hardly react, and their mixture can remain unchanged over a period of many years. However, at sufficiently high temperatures, if the mixture is kept at a temperature slightly higher 450C and the pressure in the vessel at a few torrs, or at high pressures, a violent exothermic reaction will be initiated with water formed as the product. This kind of very fast exothermic reaction is called an explosion. As the temperature is raised, the reaction rate remains extremely small up to some critical value, which depends on the experimental conditions. For a stoichiometric mixture of hydrogen/oxygen, the so-called "detonating mixture", the critical temperature at atmospheric pressure is about 550 C. At higher temperatures, even those only a few degrees above critical, the detonating mixture reacts very rapidly, the pressure increases explosively fast, and the vessel may be broken apart. Unlike an ordinary reaction, an explosive reaction is characterized by the existence of a certain temperature at which the reaction rate changes very rapidly, almost discontinuously. This temperature is called the *ignition* (or *self-ignition*) *temperature*. An equally rapid change in the reaction rate can be obtained by changing the pressure at a given temperature. At some pressures the reaction either does not proceed or it proceeds very slowly, but at some critical pressures even a very small pressure change can be sufficient to cause a very fast reaction taking place over a short time. In observing explosive reactions, the first impression is that below critical values of temperatures and pressures nothing is happening in the mixture, but when the critical parameters are reached everything reacts all at once. This is the reason for calling such reactions explosive.

1.4.1 Hydrogen–Oxygen Explosions: Explosion Limits

Explosions are conventionally classified into two different categories: *chain-branching explosions*, in which the reaction rate increases spontaneously without limit due to chain branching, and *thermal explosions* in which the reaction rate increases exponentially due to exothermic chemical reaction resulting in heating of the reactants and an increase of the reaction rate constant. A general theory of chain reactions branched and non-branched has been developed by N. N. Semenov [38] and C. N. Hinshelwood [39]. (In 1956, Sir Cyril Norman Hinshelwood (Oxford, UK) and Nikolay Nikolaevich Semenov (Moscow, Russia) shared the Nobel Prize in Chemistry "for their researches into the mechanism of chemical reactions".).

To what extend a reaction is accelerated by the branching process can be demonstrated by a numerical example. Let a 1 cm^3 container contain a mixture of reactant molecules at normal conditions and one active radical. The number density of reactant molecules is $n = 10^{19}\,cm^{-3}$. The average collision frequency of molecules is $n\sigma\overline{v}$, where σ is the cross section of collision, and $\overline{v}$ is the average thermal velocity of molecules. For normal conditions, $n = 10^{19}\,\text{cm}^{-3}$, $\sigma \approx 2 \cdot 10^{-15}\text{cm}^2$, and $\overline{v} \approx 5 \cdot 10^4\,\text{cm/s}$, and we have $n\sigma\overline{v} \approx 10^9\text{s}^{-1}$. If the reaction in the volume is chain-carrying, i.e. one active radical generates another active radical, then the time required for all the molecules to react will be $10^{19}/10^9 = 10^{10}\text{s} \approx 3.000$ years. If the reaction is chain-branching such that one active radical generates two new chain radicals, then the number of generations needed for reproduction of the active radical up to 10^{19} will be defined by the geometric progression

$$\sum_{i=0}^{N} 2^i = \frac{2^{N+1} - 1}{2 - 1} = 10^{19}.$$

Thus, the number of generations is $N = 62$, and the time of the reaction becomes $62 \cdot 10^{-9}\text{s} \approx 0.6\,\mu\text{s}$.

In the mixture of hydrogen and oxygen, the explosion can be spontaneous, without requiring any ignition source, such as a spark. Though one might think it reasonable that the reaction scheme

$$H_2 + H_2 + O_2 = H_2O + H_2O$$

would explain the formation of water, it never occurs in reality. First of all, it would require a three-body collision that would require three bodies to be perfectly aligned in the proper orientation in order to form two water molecules. The probability of this is extremely small. Second, it would require that all diatomic bonds be broken. Third, since this is a highly exothermic reaction, where would the energy go? Note that though the total heat of the reaction $2H_2 + O_2 = 2H_2OH_2 + O_2 = 2H_2O$ is 116 kcal/mol, it requires to break bonds of two hydrogen molecules (206 kcal/mole) and one oxygen molecule (118 kcal/mol), which means that the activation energy for such reaction should be extremely high, and it will never proceed in this way. It is presumed plausible that in a mixture of hydrogen and oxygen, the bulk reaction $2H_2 + O_2 = 2H_2O$ contains a chain of elementary intermediate steps involving free valence radicals—active centers, such as hydrogen and oxygen atoms and OH radical. The minimum number of individual reactions, which can explain realistic reaction of hydrogen–oxygen, depends on temperature and pressure.

The global hydrogen–oxygen reaction proceeds via the following elementary reactions, known as a reaction mechanism:

$$\begin{aligned}
&H_2 + M \rightarrow H + H + M && \} \text{ chain initiation} \\
&H + O_2 + M \rightarrow HO_2 + M && \\
&HO_2 + H_2 \rightarrow H_2O + OH && \text{chain propagation} \\
&OH + H_2 \rightarrow H_2O + H && \\
&H + O_2 \rightarrow OH + O && \\
&H_2 + O \rightarrow OH + H && \text{chain branching} \\
&H + OH + M \rightarrow H_2O + M && \\
&H + H + M \rightarrow H_2 + M && \text{chain termination.} \\
&O + O + M \rightarrow O_2 + M &&
\end{aligned}$$

Some of these reactions are bimolecular reactions, and their rates will scale with pressure as the second power of pressure. Some of these reactions involve three particle collision and will increase its rate in proportion to the third power of pressure. Besides, at very low pressure a radical formation will be terminated by wall destruction, and no explosions may occur. The wall reactions scale with pressure as approximately the first power of pressure.

In experimental studies of the reaction of hydrogen with oxygen, three explosion limits have been observed. These limits can be conveniently represented in the form of a "pressure–temperature" diagram. The explosion limits of a stoichiometric hydrogen–oxygen mixture are plotted in Fig. 1.1. If the initial pressure and temperature of the mixture correspond to the point lying to the right from the curve ABCD, then ignition occurs. The segment AB corresponds to the first explosion limit, the segment BC to the second, and the segment CD to the third explosion limit. The kinetic curves for the hydrogen–oxygen in the explosion region have the

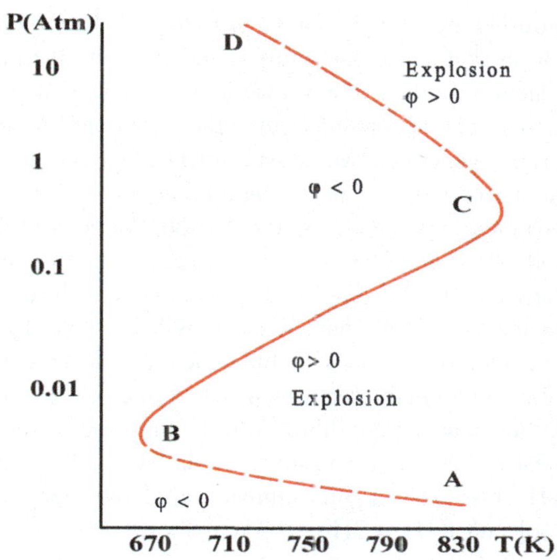

Fig. 1.1 *P–T* diagram of kinetic curves for a stoichiometric hydrogen–oxygen mixture

following appearances: over a certain time t_i, known as the *induction period* (or *ignition delay time*), there is practically no reaction (the reaction is so slow that it cannot be observed) and the concentration is practically constant; then the reaction takes place rapidly and ends up in a very short time. Using the reaction of hydrogen with oxygen as an example, we can distinguish two characteristic features of explosive reactions: the induction period and the existence of critical conditions, which separate the explosion region from the non-explosion region in which the reaction hardly occurs (more exactly, the reaction is going extremely slow).

Let us now consider the oxidation of hydrogen, which is one of the simplest and most widely studied chain reactions. The bulk reaction $2H_2 + O_2 = 2H_2O$ contains a series of elementary reaction steps, which are chain initiation, chain propagation, chain branching, and chain termination, listed above. The chain initiation reaction is

$$H_2 + M = H + H + M. \tag{1.4.1}$$

Consider the following reactions involving active radicals (OH, H, and O), which play an important role:

$$OH + H_2 = H_2O + H \quad (6\ \text{kcal/mol}) \tag{1.4.2}$$

$$H + O_2 = OH + O \quad (17\ \text{kcal/mol}) \tag{1.4.3}$$

$$O + H_2 = OH + H \quad (10\,\text{kcal/mol}). \tag{1.4.4}$$

Reactions (1.4.2) and (1.4.4) have activation energies of roughly 6 and 10 kcal/mol, respectively, and therefore proceed much more rapidly than reaction (1.4.3) whose activation energy is 17 kcal/mol. Therefore, reaction (1.4.3) is the limiting step for the entire process and determines its overall rate. The hydrogen atoms that participate in these reactions are the main active radicals governing the reaction. In the cycle of reactions (1.4.2–1.4.4), a hydrogen atom entering reaction (1.4.3) is regenerated in the other reactions. A single H atom that enters reaction (1.4.3) generates two active radicals, OH and O, which in its turn regenerates two hydrogen atoms, that is, chain branching occurs. Since the rate of reaction (1.4.3) is much larger than the rate of the reactions (1.4.2) and (1.4.4), in which active radicals H and OH are created (the dissociation of hydrogen molecules, $H_2 \to 2H$ requires 103 kcal/mol and can be neglected), the concentration of active radicals in the branching chain reaction will be many times larger than their equilibrium concentration. This is the fundamental difference between chain-branching and non-branching chain reactions in which the concentration of active radicals cannot be larger than the equilibrium level. In terms of reaction energy this means that the heat released in the basic reaction $2H_2 + O_2 = 2H_2O + 116\,\text{kcal/mol}H_2 + O_2 = 2H_2O + 116\,\text{kcal/mol}$ almost completely goes into the dissociation of hydrogen molecules, $H_2 = 2H - 103\,\text{kcal/mol}$.

If the initial temperature of the mixture corresponds to a point to the right of the curve ABCD in Fig. 1.1, then the ignition that occurs leads to a thermal explosion. The self-acceleration of a branching chain reaction occurs because the rate of loss of active radicals is lower than their rate of formation. At lower temperatures and pressures below the pressure, corresponding to point B (below the first explosion limit in Fig. 1.1, corresponding to the AB branch of ABCD), active radicals can diffuse to and be destroyed at the vessel walls. In order to describe this process, reactions (1.4.2–1.4.4) must be supplemented by equations describing diffusion, absorption, and distortion of H, O, and OH radicals. In this process (diffusion), the rate of the "loss" of radicals is proportional to the first power of pressure. As the pressure increases, the diffusion rate decreases and critical conditions for the explosion may be again reached for the temperatures and pressures above the AB and below the BC branches. When the pressure is raised further, the rate of production of active radicals exceeds their rate of loss owing to diffusion to the walls, providing again conditions for explosion (above the branch AB). In the initial stage of the reaction, when the concentration of H atoms is rising, only a little amount of heat is released. The thermal yield of the reaction is released only when the concentration of hydrogen atoms is close to the maximum and they begin to recombine with a large thermal release. In the initial stage of the reaction, the thermal energy of the reagents goes mainly into the chemical energy of active radicals, thereby causing the rapid progress of this reaction.

At still higher pressures, three-body collisions became sufficiently frequent and three-body reactions such as

$$H + O_2 + M = HO_2 + M \tag{1.4.5}$$

begin to play an important role. Reaction (1.4.5) competes with reaction $H + O_2 = OH + O$. Reaction (1.4.5) leads to the formation of HO_2 radicals, which is very inert from a chemical standpoint and is able to diffuse to the vessel walls before it reacts. Thus, this reaction begins to play as a chain termination reaction. It should be noted that the loss of the HO_2 radical depends strongly on the material of which the wall is made. With some materials, all radicals that reach the wall are lost, while with others a substantial fraction is reflected back into the vessel volume without change. With increasing pressure, the frequency of three-body collisions according to reaction (1.4.5) increases more rapidly than the frequency of binary collisions yielding reactions (1.4.2–1.4.4), and the rate of loss of active centers exceeds their rate of production after a certain pressure. There is no explosion to the left from curve BCD and this explains the existence of the second explosion limit in Fig. 1.1.

At even higher pressures, the diffusion becomes less effective and before radicals reach the walls, the HO_2 radicals begin to participate in another phase of chain propagation and branching reactions of the following types:

$$HO_2 + H_2 = H_2O_2 + H - 15\,\text{kcal/mol} \tag{1.4.6}$$

$$HO_2 + H = OH + OH + 37\,\text{kcal/mol} \tag{1.4.7}$$

$$HO_2 + H = H_2O + O + 54\,\text{kcal/mol}. \tag{1.4.8}$$

A substantial amount of heat is released in reactions (1.4.7) and (1.4.8) which also leads to an increase in the reaction rate that, in turn, causes further heating of the mixture. If the rate of heat release exceeds the rate of cooling owing to heat transfer to the vessel walls, then a thermal explosion takes place. This is the origin of the third explosion limit above the branch CD of the curve ABCD.

These are the basic concepts involved in the H_2–O_2 reaction, which can be viewed as a qualitative model of the chain-branching model reaction and explosion. At present, the mechanism of this reaction has been studied quite thoroughly for a wide range of reaction conditions, using a much more detailed chemical kinetic models with detailed discussions given in a number of textbooks. Most fuel and oxidizer combinations have more complicated chemical kinetics compared with the H_2–O_2 mixture.

For the first time, explosion limits for hydrogen–oxygen have been calculated by Maas and Warnatz [40] for 1D geometries using a detailed reaction mechanism of 37 elementary reactions. The calculated three explosion limits of the hydrogen–oxygen mixture and the ignition delay time were found in a satisfactory good agreement with experimentally measured values. Up to this point in explaining the existence of critical conditions for explosive reactions, we mainly proceeded from the idea of branching chain reactions. However, the process that occurs at the third explosion limit of a hydrogen–oxygen explosive mixture is associated with the heating of the reacting mixture by the energy released as a result of chemical change and corresponds to thermal explosion.

1.4.2 Oxidation and Explosions of Hydrocarbons

N.·N. Semenov for the first time proposed the general mechanism [38] that became the basis for the further research of explosions and explosion limits. The simplest case of hydrogen oxidation and the importance of the high-temperature chain mechanism were discussed in detail in 1.4.1. The electronic structure of hydrocarbons and, accordingly, the detailed chemistry of hydrocarbon oxidation are much more complex than that for hydrogen–oxygen. Moreover, since some details of hydrocarbon chemistry are still unclear, determining the explosion limits of hydrocarbons is a much more complicated task. Nevertheless, a qualitative understanding can be obtained by exploring some simplifications.

Structures of several simplest and most widely used in experimental studies hydrocarbons are shown in Figs. 1.2, 1.3, and 1.4.

Figure 1.2 Paraffins are single-bond compounds with chemical formula C_nH_{2n+2}. These are methane CH_4, ethane C_2H_6, propane C_3H_8, etc.

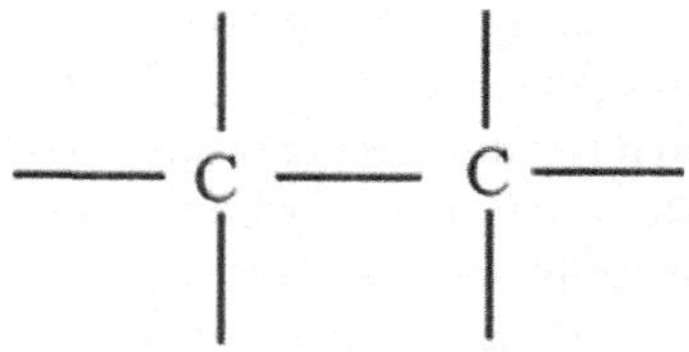

Fig. 1.2 Paraffins

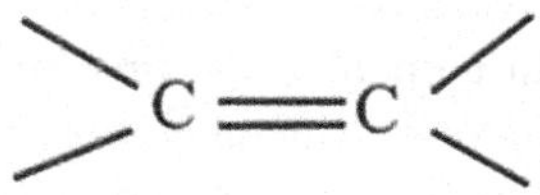

Fig. 1.3 Olefins

Fig. 1.4 Acetylenes

Figure 1.3 Olefins are compounds containing two double bonds with chemical formula C_nH_{2n-2}. These are ethylene C_2H_4, propylene C_3H_6, butylene C_4H_8, etc.

Figure 1.4 Acetylenes are compounds containing two double bonds with chemical formula C_nH_{2n-2}. These are acetylene C_2H_2, methyl acetylene C_3H_4, ethyl acetylene C_4H_6, etc.

At temperatures below 200 °C, the hydrocarbon–oxygen mixtures react very slowly. At higher temperatures, a variety of oxygen-containing compounds begin to form. As the temperature increases further, CO and H_2O dominate in the products and simultaneously such compounds as hydrogen peroxide (H_2O_2), formaldehyde (CH_2O), and CO_2 appear. Since almost all carbon atoms are converted to CO and because the oxidation of CO into CO_2 is the highly exothermic part of hydrocarbons oxidation, it requires a special analysis. At temperatures around 400 °C, a faint blue light often appears, which indicates the appearance of cool flames that could be followed by the explosion.

The mechanisms explaining the cool flame phenomenon are referred to as the cool flame chemistry [41]. Depending on the specific experimental conditions and mixtures, many different types of cool flames propagating through a given mixture have been observed. Also, cool flames have been observed in flow systems [42]. It is essential to establish the specific mechanisms that explain the cool flame phenomenon, as well as the hydrocarbon combustion characteristics. The higher order hydrocarbons, such as propane and above, oxidize much more slowly than hydrogen and are known to form metastable molecules that are important in explaining the explosion limits of hydrogen and carbon monoxide. In the theory of the explosion limits N. Semenov [38] explained the long induction period by hypothesizing unstable, but long-lived species that form as intermediates and then undergo different reactions at the higher temperature. The existence of the metastable molecules makes it possible to explain qualitatively the explosion limits of the complex hydrocarbons. It is not surprising that the character of chain-branching mechanisms for the complex hydrocarbons is different from that of the hydrogen–oxygen reaction since the larger hydrocarbon molecules and their intermediates tend to break down more readily to form radicals that initiate fast reactions. Experimental studies show that hydrocarbons exhibit certain characteristics that are consistent with the explosion limit curves.

It should be noted that the combustion mechanism of methane and its specific oxidation mechanism were better understood only in recent years. Methane exhibits oxidation characteristics that are different from those of all other hydrocarbons. In particular, the first broken C–H bond in methane takes about 9.6 kcal, that is, more considerably more than the C–H bonds in longer chain hydrocarbons. Therefore, it is not surprising that various experiments show that ignition is more difficult with methane/air mixtures than it is with other larger hydrocarbons. Many extensive models of the high-temperature oxidation of methane [43–46] and for other higher order hydrocarbons [46, 47] have been published. Such models are quite complex, often including hundreds of reactions and intermediate species, and they are continuously improved and modified using data obtained experimentally. More detailed information can be found in [47–54].

References

1. L.D. Landau, E.M. Lifshitz, *Course of Theoretical Physics*, vol. 5 (Pergamon Press, Oxford, Statistical Physics, 1989)
2. J.O. Hirschfelder, C.F. Gurtiss, R.B. Bird, *Molecular Theory of Gases and Liquids* (Wiley, NewYork, 1964)
3. S.A. Arrhenius, Über die Reaktionsgeschwindigkeit bei der Inversion von Rohrzucker durch Säuren. Z. Phys. Chem. **4**, 226–248 (1989)
4. A.N. Hayhurst, I.M. Vince, Nitric oxide formation from N_2 in flames: The importance of "prompt" NO. Prog. Energy Combust. Sci. **6**, 35–51 (1980)
5. J.A. Miller, C.T. Bowman, Mechanism and modeling of nitrogen chemistry in combustion. Prog. Energy Combust. Sci. **15**, 287–338 (1989)
6. J.C. Kramlich, W.P. Linak, Nitrous oxide behavior in the atmosphere, and in combustion and industrial systems. Prog. Energy Combust. Sci. **20**, 149–202 (1994)
7. S.C. Hill, L.D. Smoot, Modeling of nitrogen oxides formation and destruction in combustion systems. Prog. Energy Combust. Sci. **26**, 417–458 (2000). https://doi.org/10.1016/S0360-1285(00)00011-3
8. P. Glarborga, J.A. Miller, B. Ruscic, S.J. Klippenstein, Modeling nitrogen chemistry in combustion. Prog. Energy Combust. Sci. **67**, 31–68 (2018)
9. Ya. B. Zel'dovich, The oxidation of nitrogen in combustion explosions. Acta Physicochimica U.S.S.R. **21**, 577–628 (1946)
10. Ya.B. Zeldovich, D.A. Frank-Kamenetskii, P. Sadovnikov, oxidation of nitrogen in combustion. Published. Acad. Sci. USSR (1947)
11. Y.B. Zeldovich, G.I. Barenblatt, V.B. Librovich, G.M. Makhviladze, The Mathematical theory of combustion and explosions. Science Publ. Moscow, 1980 (in Russian). English translation: Consultants Bureau, New York (1985).
12. C.P. Fenimore, Formation of nitric oxide in premixed hydrocarbon flames, in *Symposium (International) on Combustion*, vol 13, pp. 373–380 (1971).

13. D.W. Pershing, O.L. Wendt, Relative contributions of volatile nitrogen and char nitrogen to NO_x emissions from pulverized coal flames. Ind. Eng. Chem. Process. Des. Dev. **18**, 60–67 (1979)
14. C.T. Bowman, Kinetics of nitric oxide formation in combustion processes, in *Proceedings of Fourteenth Symposium (International) on Combustion, The Combustion Institute, Pittsburgh, PA* (1973). pp. 729–738.
15. C.T. Bowman, Kinetics of pollutant formation and destruction in combustion. Prog. Energy Combust. Sci. **1**, 33–45 (1975)
16. G.A. Lavoie, J.B. Heywood, J.C. Keck, Experimental and theoretical study of nitric oxide formation in internal combustion engines. Combust. Sci. Technol. **1**, 313–326 (1970)
17. J.F. Driscoll, R.H. Chen, Y. Yoon, Nitric oxide levels of turbulent jet diffusion flames: effects of varying residence time and Damköhler number Combust. Flame **88**, 37–49 (1992)
18. C. Morely, The formation and destruction of hydrogen cyanide from atmospheric and fuel nitrogen in rich atmospheric-pressure flames Combust. Flame **27**, 189–204 (1976)
19. L.V. Moskaleva, M.C. Lin, The spin-conserved reaction CH: a major pathway to prompt NO studied by quantum/statistical theory calculations and kinetic modeling of rate constant. Proc. Combust. Inst. **28**, 2393–2401 (2000)
20. A.A. Konnov, On the relative importance of different routes forming in hydrogen flames. Combust Flame **134**, 421–424 (2003)
21. Q. Cui, K. Morokuma, J.M. Bowman, S.J. Klippenstein, The spin-forbidden reaction $CH(2\Pi)+N_2 \rightarrow HCN+N(4S)$ revisited. II. Nonadiabatic transition state theory and application. J. Chem. Phys. 110 (1999) pp. 9469–9482
22. B.A. Williams, J.W. Fleming, Experimental and modeling study of formation in 10 torr methane and propane flames: evidence for additional prompt- precursors. Proc. Combust. Inst. **31**, 1109–1117 (2007)
23. F.H.V. Coppens, J. De Ruyck, A.A. Konnov, The effects of composition on burning velocity and nitric oxide formation in laminar premixed flames of + + +. Combust. Flame **149**, 409–417 (2007)
24. A.A. Konnov, On the role of radicals in the prompt- mechanism. Combust. Expl. Shock Waves **44**, 497–501 (2008)
25. A.A. Konnov, Implementation of the NCN pathway of prompt-NO formation in the detailed reaction mechanism. Combust. Flame **156**, 2093–2105 (2009)
26. D.L. Baulch, D.D. Drysdale, D.G. Home, A.C. Lloyd, Evaluated kinetic data for high temperature reactions. Butterworths, London **1**, 2 (1973)
27. D.L. Baulch, C.G. Cobos, R.A. Cox, P. Frank, G. Hayman, T.H. Just, J.A. Kerr, T. Murrells, M.J. Pilling, J. Troe, R.W. Walker, J. Warnatz, Summary table of evaluated kinetic data for combustion modeling. Combust. Flame **98**, 59–79 (1994)
28. D.L. Baulch, C.T. Bowman, C.G. Cobos, R.A. Cox, T.H. Just, J.A. Kerr, M.J. Pilling, D. Stocker, J. Troe, W. Tsang, R.W. Walker, J. Warnatz, Evaluated kinetic data for combustion modeling: supplement II. J. Phys. Chem. Ref. Data **34**, 757–1397 (2005)
29. R.K. Hanson, S. Salimian, Survey of rate constants in the N-H-O system, in *Combustion Chemistry*, ed. by W.C. Gardiner, Jr (Springer-Verlag, NY, 1984).
30. W. Tsang, J.T. Herron, Chemical Kinetic Data Base for Propellant Combustion. I. Reactions involving NO, NO_2, HNO, HNO_2, HCN and N_2O. J. Phys. Chem. Ref. Data **20**, 609–663 (1991)
31. J. Warnatz, Rate coefficients in the C-H-O system, in *Combustion Chemistry*, ed. by W.C. Gardiner Jr (Springer, NY, 1984)
32. R.J. Kee, E.M. Rupley, J.A. Miller, Chemkin-II: a fortan chemical kinetics package for the analysis of gas-phase chemical kinetics. Sandia National Laboratories Report. SAND898009 (1990)
33. D. De Soete, Overall reaction rate of and formation from fuel nitrogen, in *Fifteenth Symposium (International) on Combustion. The Combustion Institute, Pittsburgh, PA* (1975). pp. 1093–1102

34. P. Glarborg, P.G. Kristensen, K. Dam-Johansen, J.A. Miller, Branching fraction of the reaction between 1210 and 1370 K. J. Phys. Chem. A **101**, 3741–3745 (1997)
35. P. Glarborg, K. Dam-Johansen, and J.A. Miller. The reaction of ammonia with nitrogen dioxide in a flow reactor: implications for the reaction. Int. J. Chem. Kinet. **27**, 1207–1220 (1995)
36. J.A. Miller, P. Glarborg, Modeling the formation of and in the Thermal De-NOx process. Springer Ser. Chem. Phys. **61**, 318–333 (1996)
37. J. A. Miller and P. Glarborg. Modeling the Thermal DeNOx process: Closing in on a final solution. Int. J. Chem. Kinet. **31**, 757765 (1999)
38. N.N. Semenov, *Chain Reactions, Goskhimtekhizdat, Leningrad, 1934* (Chemical Kinetics and Chain Reactions, Oxford University Press, Oxford, English Translation, 1935)
39. C.N. Hinshelwood, *The Kinetics of Chemical Change* (Oxford University Press, Oxford, 1940)
40. U. Maas, J. Warnatz, Ignition processes in hydrogen-oxygen mixtures. Combust. Flame **74**, 53–69 (1988)
41. G.J. Minkoff, C.F.H. Tipper, *Chemistry of Combustion Reactions* (Butterworth, London, 1962)
42. F.W. Williams, R.S. Sheinson, Manipulation of cool and blue flames in the winged vertical tube reactor. Combust. Sci. Technol. **7**, 85–92 (1973)
43. J. Warnatz, in *Combustion Chemistry*, ed. by W. C. Gardiner , Jr (Springer, New York, 1984).
44. M. Frenklach, H. Wang, M.J. Rabinowit, Optimization and analysis of large chemical kinetic mechanisms using the solution mapping method—combustion of methane. Prog. Energy Combust. Sci. **18**, 47–73 (1992)
45. T.B. Hunter, H. Wang, T.A. Litzinger, M. Frenklach, The oxidation of methane at elevated pressures: experiments and modeling. Combust. Flame **97**, 201–224 (1994)
46. P. Dagaut, J.C. Boettner, M. Cathonnet, Methane oxidation: experimental and kinetic modeling study. Combust. Sci. Technol. **77**, 127–148 (1991)
47. P. Dagaut, M. Cathonnet, J.C. Boettner, F. Guillard, Kinetic modeling of propane oxidation. Combust. Sci. Technol. **56**, 23–63 (1987)
48. M. Cathonnet, Chemical kinetic modeling of combustion from 1969 to 2019, Combust. Sci. Technol. **98**, 265–279 (1994)
49. M. Nehse, J.Warnatz, C. Chevalier, Kinetic modeling of the oxidation of large aliphatic hydrocarbons, in *Symposium (International) on Combustion*, vol 26 (1996). pp. 773–780
50. J.A. Sutton, B.A. Williams, J.W. Fleming, Investigation of NCN and prompt-NO formation in low-pressure C1–C4 alkane flames. Combust. Flame **159**, 562–576 (2012)
51. J.A. Miller, M.J. Pilling, J. Troe, Unravelling combustion mechanisms through a quantitative understanding of elementary reactions. Proc. Combust. Inst. **30**, 43–88 (2005)
52. C. K. Westbrook, W. J. Pitz, O.r Herbinet, H. J. Curran, E. J. Silke. A comprehensive detailed chemical kinetic reaction mechanism for combustion of n-alkane hydrocarbons from n-octane to n-hexadecane. Combust. Flame, Elsevier **156**, 181–199 (2008)
53. I. Glassman, R.A. Yetter, *Combustion*, 4th edn (Elsevier, 2008)
54. H.J. Curran, Developing detailed chemical kinetic mechanisms for fuel combustion. Proceed. Combust. Institute **37**, 57–81 (2019)

Chapter 2
Self-Accelerating Reactions

When the combustible mixture ignites, the temperature of the mixture rises rapidly, and is accompanied by a rapid increase in pressure as well. Understanding ignition processes is important across a wide range of combustion processes, from the design and development of practical combustion devices to preventing catastrophic explosions and fires. Although the term explosion is commonly used, it is vague and rather uncertain. An explosion can be defined as a scenario in which, due to the energy released in a system the pressure increases much faster than it can be smoothly equalized through the system by sound waves. Ignition of the combustible mixture can occur as a forced ignition, which is triggered by an external energy source, such as an electric spark, hot wire, pilot flame, and a laser beam, that locally heats and increases the temperature of the reactants initiating combustion. A spontaneous ignition, also known as autoignition, occurs through the self-heating of the reactants. Since the rate of a chemical reaction is a strong function of temperature, if the heat generated by the reaction is greater than the heat losses to the surroundings, then the temperature of the reactants increases until they spontaneously ignite. The conditions under which a mixture can be ignited or an explosion may occur are of great importance in various areas. From the point of view of safety problems, it is necessary to understand the conditions under which ignition can lead to disastrous explosions, such as coal mine explosions and dust explosions in the chemical industry. To design combustion devices we need to know how much energy is required for ignition and what should be the rate of energy input for ignition of a certain combustion mode. Finally, one of the most important problems in combustion research is what kind of combustion mode occurs as the result of ignition. It can be either slow combustion (deflagration) or a detonation accompanied by strong and destructive shock waves produced by detonation. Ignition is an unsteady, time-dependent transient combustion process controlled by chemical kinetics. It is not surprising therefore that the correct description of ignition requires knowledge of the detailed chemical reaction mechanisms for reproducing combustion parameters in a wide range of pressures and temperatures. Knowledge of the detailed reaction mechanisms is of special importance for understanding unsteady combustion processes, which are controlled by

M. A. Liberman, *Combustion Physics*,
https://doi.org/10.1007/978-3-030-85139-2_2

chemical kinetics, such as ignition and self-ignition processes (e.g. engine knock), flame acceleration, and the deflagration-to-detonation transition [1]. All the same, a qualitative understanding can be obtained using a chemical reaction in the form of a one-step Arrhenius model to solve the problem in question in explicit analytical form.

2.1 A One-Step Chemical Reaction Model

As mentioned earlier, the complex chemical kinetics of hydrocarbon fuels is described by chemical mechanisms, which may consist of many hundreds or thousands of chemical species. The number of species and reaction steps is tremendously growing with the number of carbon atoms in the fuel. Such reaction schemes are too complex for modeling even for the most powerful computers, and it is impossible to comprehend physical significance from such large chemical mechanisms. Engineers often want to know the behavior of practical, multidimensional systems, and for this purpose use computers for modeling combustion kinetics in simplified chemical models. A simplified description of chemical kinetics is useful for practical applications of engineering problems. A one-step global reaction is often used in practical simulations due to its simplicity.

For a hydrocarbon fuel with an overall combustion stoichiometry the corresponding global rate of the reaction progress can be expressed as

$$\dot{\omega} = A_0[fuel]^{\alpha}[O_2]^{\beta} \exp(-E_a/RT). \tag{2.1.1}$$

The pre-exponential factor, A_0, activation energy, E_a, and exponents α and β are either obtained experimentally in flow reactors [2], or can be tuned so that a one-step reaction reproduces correctly the normal velocity of deflagration. Note that units for A_0 and the fuel and oxidizer concentrations are chosen in a way that the rate of progress has units of mol/cm^{3}s. Because of the high activation energy, the reaction rate is highly dependent on temperature and therefore can be sensitive to the equivalence ratio of the mixture due to the change in flame temperature. For the one-step reaction model (2.1.1), the consumption rate of reactant A is expressed by

$$\frac{d[A]}{dt} = -\alpha\dot{\omega}. \tag{2.1.2}$$

The reaction rate depends on pressure through the concentrations of fuel and oxygen. Assuming the ideal equation of state ($P = RT$) and introducing concentrations X_{fuel} and X_{O_2} in Eq. (2.1.1), we obtain

$$\dot{\omega} = A_0 X_{fuel}^{\alpha} X_{O_2}^{\beta} \exp(-E_a/RT)\left(\frac{P}{RT}\right)^{\alpha+\beta}. \tag{2.1.3}$$

It is seen that the one-step model shows that with the pressure raise the reaction rate increases since $\alpha + \beta > 0$. Using the expression of the consumption rate (2.1.2), we can introduce the heat release rate in the reaction as

$$Q = -\frac{d[A]}{dt} q m_{fuel} = -\alpha \dot{\omega} q m_{fuel}, \tag{2.1.4}$$

where q_r is specific heat of the reaction.

2.2 Elementary Theory of Thermal Explosion

Some chemical reactions develop in time slowly while others develop extremely fast—"explosively", and they can be accompanied by such phenomena as a flash, sound, and shocks. The important feature of explosions is that the rate of the reaction varies extremely rapidly with temperature. As it was discussed in the previous chapter, very often an explosive path of the reaction is explained by its chain character, that is, by the dependence of its rate on the concentration of active radicals formed during the reaction.

Chain reactions take place through the formation of active radicals. During a reaction, the amount of these particles may increase for two reasons. First, active radicals may be formed independently of the chain reaction as a result of the increased temperature and the corresponding increase in the thermal velocity of particles. Reagent molecules may dissociate upon more frequent collisions with one another, but the rate for this process can be sufficiently high only at high temperatures. Second, branching of the chain may occur, that is, an elementary process involving one active atom or radical that yields two atoms (or free radicals). Branching always occurs due to the reaction energy. For example, from the point of view of the material balance the reaction $H + H_2 = 3H$ is possible, but it is energetically impossible. On the other hand, the reaction $H + 3H_2 + O_2 = 3H + 2H_2O$ is possible. The rate of formation of active particles in this case is proportional to their concentration. The presence of a source of active particles is extremely important and determines the main features of chain kinetics. The appearance of active particles can initiate branching that is the formation of multiple numbers of active particles if it is possible energetically. Besides, there are always processes, which cause the loss of active particles. The rate of removal of active particles from the reaction by collisions with stable molecules or by diffusion to the vessel walls is proportional to the concentration of active particles. If the rate of chain branching becomes greater than the rate of chain termination, then explosion conditions are achieved. When conditions for the fast loss of active particles exist, the explosion may be impossible. Essentially, the same concept holds for thermal explosion (or thermal ignition). When the rate of thermal energy release is greater than the rate of thermal energy loss, an explosive condition exists. But if the rate of thermal energy loss exceeds the rate of thermal energy release, the explosion is impossible.

Considering a simple model [3], we can write down the equation for the time variation in the concentration of active particles in the form

$$\frac{dn}{dt} = W_0 + fn - gn, \tag{2.2.1}$$

where W_0 is the rate of creation of active particles, and f and g are the rate constants for chain branching and chain termination. Designating $f - g \equiv \varphi$, we obtain

$$\frac{dn}{dt} = W_0 + \varphi n. \tag{2.2.2}$$

Note that the rate of creation of active particles, W_0, is very small, almost equal to zero, such that in the experiment we can't measure either change in the concentration of active particles or heat of the reaction. A change in the external conditions (temperature, pressure) causes changes in both values of f and g. Since no activation energy is required for loss of active particles, the constant characterizing the formation of new active particles by the chain branching typically has a much stronger temperature dependence than the rate of chain termination. Thus, the difference $f - g = \varphi$ changes sign as the temperature increases: at low temperatures it is negative, and at high temperatures it is positive. For the initial condition $n(t = 0) = 0$ one can obtain that when $\varphi < 0$ (at low temperatures), the concentration of active particles approaches the limit $W_0/(-\varphi)$, still a small quantity, and $dn/dt \to 0$. On the other hand, when $\varphi > 0$, the rate of formation and the concentration of active particles are increasing continuously. Thus, depending on the sign of φ, the type of solution changes drastically. The temperature at which φ equals zero is the critical temperature; below this temperature, an explosion is impossible. These qualitative considerations explain the existence of explosion conditions for explosive mixtures at the fixed initial pressure when the initial temperature passes through the critical value corresponding to $\varphi = 0$.

The solution of Eq. (2.2.2) satisfying the initial condition $n(t = 0) = 0$ is

$$n = \frac{W_0}{\varphi}(e^{\varphi t} - 1). \tag{2.2.3}$$

The rate of formation of the end products, that is the reaction rate, is

$$W = \nu \cdot (f \cdot n) = \frac{\nu \cdot f \cdot W_0}{\varphi}(e^{\varphi t} - 1), \tag{2.2.4}$$

where ν is an integer of the order unity that indicates how many molecules of the end product are formed as the result of the reaction of one active particle. When $\varphi > 0$, i.e. chain branching occurs more rapidly than the chain termination, the concentration of active particles increases with time exponentially as does the reaction rate. In the early stages of the reaction, while there is a buildup of active particles, the rate remains lower than the limit of measurability since W_0 is extremely small for explosive reactions and the change hardly can be noticed experimentally. After a certain time, the rate reaches a measurable value and continues to rise without limit. This time corresponds to the *ignition delay* also called the *induction time* t_i. Since W_0 is always much smaller than W_{min} for an explosive reaction, $\exp(\varphi t)$ is much greater than unity for $t > t_i$ and with sufficiently good accuracy we can set

$$W \approx \nu \frac{W_0 \cdot f}{\varphi} e^{\varphi t}. \tag{2.2.5}$$

Therefore, the induction time equals

$$t_i = \frac{1}{\varphi} \ln\left\{\frac{W_{min}\varphi}{\nu W_0 f}\right\}. \tag{2.2.6}$$

The variation in the logarithmic term can be neglected, so that

$$t_i \approx const/\varphi.$$

For large φ, when $f \gg g, \varphi = f - g \approx f$, and

$$W \approx \nu \cdot W_0 \exp(\varphi),\ and\ t_i = \frac{1}{\varphi} \ln\left\{\frac{W_{min}}{\nu \cdot W_0}\right\} \approx const_1/\varphi. \tag{2.2.7}$$

A change in φ for constant W_0 causes a sharp change in the induction time and reaction rate. Thus, when φ is reduced there is a noticeable increase in t_i, and the reaction rate curve is expanded in time. As φ decreases, the self-acceleration of the reaction becomes less and less distinct and, finally, when $\varphi = 0$, the kinetic curve becomes a straight line given by the equation

$$W = \nu \cdot W_0 \cdot f \cdot t. \tag{2.2.8}$$

The value of the induction time corresponding to detectable rate W_{min} is reached in time:

$$t_i = \frac{W_{min}}{\nu \cdot f \cdot W_0}. \tag{2.2.9}$$

In this limiting case, the reaction rate increases without a limit until the burnout of reagents begins to have an effect. If φ is still further decreased and becomes negative $\varphi < 0$, then the character of the reaction changes substantially. From Eq. (2.2.4), it is seen that for $\varphi < 0$, the reaction rate approaches the limit $\nu \cdot f \cdot W_0/|\varphi|$. Since this quantity is less than W_{min}, being extremely small, we conclude that when $\varphi < 0$ the reaction does not proceed. For very small negative φ, only a small change in the external conditions (pressure or temperature) leading to a very small increase in f is needed to cause φ to change sign and to make the reaction self-accelerating. Finally, it should be noticed that during the ignition delay (induction) time, the concentration of active radicals increases exponentially, but the fuel consumption and the energy released in the reaction is still small. This means that in chain-branching reactions the temperature increase during induction time also remains small. Due to the strong exponential temperature dependence of elementary reactions, the induction time also exponentially depends on the temperature as $\tau_{ind} \propto \exp(E_a/RT)$.

2.3 Thermal Self-Ignition

The self-acceleration of the chemical reaction may be associated with the buildup of heat in the system. The concept of thermal ignition of gaseous explosions is crucial to the concept of ignition and it plays a key role in consideration of safety problems. The explosion limits discussed in Chap. 1 exist for particular conditions of pressure and temperature. If the thermal conditions become suitable for the reaction to begin, then some heat will be produced in the reaction. Depending on the experimental conditions, there can be different scenarios. If the heat of the reaction is dissipated at the walls of the vessel containing the gas mixture sufficiently fast to retain the temperature at the initial value, then a steady reaction prevails and a slight pressure rise is observed. But if conditions are such that the heat of reaction is dissipated too slow, then the rate of chain carrier generation can exceed the rate of chain termination. In this case, the reaction rate grows progressively and subsequently an explosion occurs and a considerable pressure rise can be observed.

Two conditions must be met for the thermal explosion to occur: first, the mixture must release a certain amount of heat during the reaction and, second, the reaction rate must rise with temperature. The overwhelming majority of reactions between stable molecules satisfy the second condition. These reactions do not proceed at low temperatures while at high temperatures they proceed rapidly. If the first condition is satisfied, that is, the reactions are exothermic, then the heat of the reaction alone may cause the reaction to speed up and lead to an explosion, unlike chain reactions in which the rate of the initial thermal reaction W_0 is very small.

Let us consider the simple case when the reaction rate depends on the concentration of only one reacting component. The kinetic equation is

$$\frac{da}{dt} = -W(a, t) = -k_0 a^n e^{-E_a/RT}, \tag{2.3.1}$$

where n is the order of the reaction, for example, for bimolecular reactions $n = 2$. Let Q be the heat of the reaction, and T_0 be the initial temperature of the mixture consisting of N_0 moles. From the first law of thermodynamics we can write down conditions for the enthalpy balance for the adiabatic process

$$c_P N_0(T - T_0) = Q(N_0 - N), \ or \ c_P(T - T_0) = Q(a_0 - a), \tag{2.3.2}$$

where N is the number of reacted moles, and $a = N/N_0$ is the concentration.

For the highest temperature T_b reached during the complete combustion of the reagent under adiabatic conditions in a vessel, we obtain

$$c_P(T_b - T_0) = Qa_0. \tag{2.3.3}$$

If the reaction proceeds in a vessel of constant volume, then we obtain formulas similar to (2.3.2) and (2.3.3) but with the specific heat at constant volume c_V instead of the specific heat at constant pressure, c_P.

Using (2.3.2) and (2.3.3), we can express concentration via temperature as

$$a = a_0 \frac{T_b - T}{T_b - T_0}, \tag{2.3.4}$$

and rewrite the equation for (2.3.1) for the only variable, for example, for concentration

$$\frac{da}{dt} = -k_0 a^n \exp\left\{-\frac{E_a/RT_0}{1 + (Qa_0/c_p\rho_0 T_0)(1 - a/a_0)}\right\} = -W_1(a), \tag{2.3.5}$$

where we used the relation between adiabatic temperature and concentration for the isobaric process $T_b - T_0 = Qa_0/c_P\rho_0$.

The corresponding equation for temperature is

$$\frac{dT}{dt} = k_0 a_0^{n-1}(T_b - T_0)^{1-n}(T_b - T)^n e^{-E_a/RT} = W_2(T). \tag{2.3.6}$$

Equations (2.3.5) and (2.3.6) determine the time variations of the concentration and temperature and the reaction rate. The reaction rate increases very rapidly with the reduction in the number of reactants that have been consumed, and the reaction rate reaches a maximum and begins to fall only when very little reactants are left. Such dependence of the reaction rate is the typical feature of explosive reactions that are catalyzed by the heat released during the chemical change with the rate given by the Arrhenius law with large activation energy E_a.

For typical reactions, the activation energy is of the order of several tens of kilocalories so that $E_a/RT \gg 1$ over the entire temperature range of interest (300–4000)K. Calculating the maximum of functions $W_1(a)$ and $W_2(T)$ for $RT_b/E_a \ll 1$, we find expressions for T_* and a_*, corresponding to the maximum of the reaction rate

$$T_*/T_b = 1 - n\,(RT_b/E) \tag{2.3.7}$$

$$a_*/a_0 = \frac{n\,R\,T_b^2}{E\,(T_b - T_0)}. \tag{2.3.8}$$

For $RT_b/E_a \ll 1$, the maximum of the functions $W_1(a)$ and $W_2(T)$ at the points for a_* and T_* is very sharp. Using (2.3.7) and (2.3.8), we obtain

$$\frac{W(a_*, T_*)}{W(a_0, T_0)} = \frac{n}{e}\left(\frac{RT_b^2}{(E(T_b - T_0)}\right)^n e^{\frac{E_a}{RT_0}\left(1-\frac{T_0}{T_b}\right)}. \tag{2.3.9}$$

Let us consider the effect of the reactant burnup on the peak value of the reaction rate. For this purpose, we use the ratio

$$\frac{W(a_*, T_*)}{W(a_0, T_*)} = \left(\frac{a_*}{a_0}\right)^n = n\left(\frac{RT_b^2}{E_a(T_b - T_0)}\right)^n. \tag{2.3.10}$$

It is easy to verify that for the reaction where $E_a/RT \gg 1$ the maximum rate of the reaction corresponds to temperatures very close to T_b and to very small concentrations. Figure 2.1 shows the result of numerical solutions [3] of Eqs. (2.3.5) and (2.3.6): the time variations of the temperature, concentration, and reaction rate for $E_a/RT_0 = 12.5, T_b = 2.5T_0, T_0 = 800\ K$, and $n = 1$. It is seen that the entire process has two characteristic stages, which are similar to the case of chain-branching reactions. From the beginning, the temperature of the mixture and the concentration of reagents change slowly over a relatively long time interval—the *induction period*.

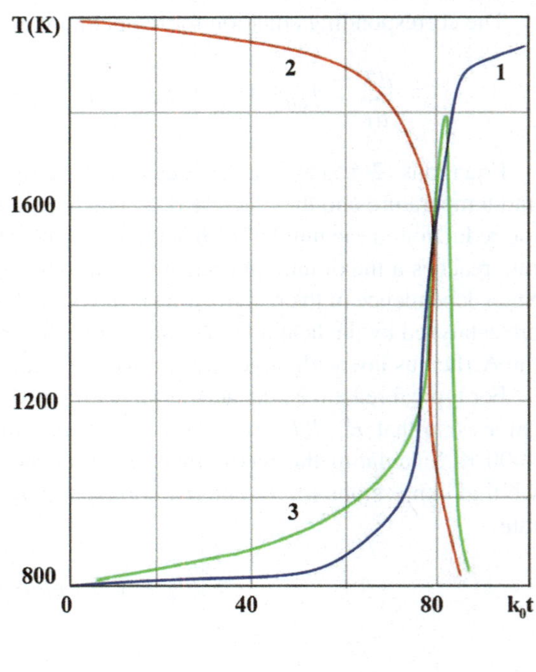

Fig. 2.1 The time variations of temperature (1), concentration (2), and reaction rate (3) in the adiabatic thermal explosion, from [3]

The second stage is a rapid heating of the mixture accompanied by fast burnup of the reacting species. If we estimate the induction period for this example to be the time over which the reaction reaches 5% of its maximum value, then the induction period occupies approximately 85% of the entire duration of the process. As higher activation energy as more distinct the difference in duration of the induction stage, during which a small amount of the reactant is consumed, followed by practically instantaneous burnup.

Taking this into account, we can for high activation energies use a model with the time dependence of the reaction rate being approximated by a delta function:

$$W(t) = A_0 \delta(t - t_i) \tag{2.3.11}$$

where the normalizing constant A_0 is determined from the condition that all the fuel be used up in the reaction,

$$\int_0^{\infty} W(t)dt = A_0, \tag{2.3.12}$$

and the induction period t_i can be defined as the time over which the reaction rate reaches its maximum value.

The induction period can be calculated by integrating Eq. (2.3.5) with the initial condition $T(0) = T_0$

$$k_0 a_0^{n-1} (T_b - T_0)^{1-n} t = \int_{T_0}^{T_b} \frac{\exp(E_a/RT)dT}{(T_b - T)^n}. \tag{2.3.13}$$

For the reaction orders $n \geq 1$, the integral in (2.3.13) diverges as $T \to T_b$. This means that the final state, characterized by complete consumption of the reagent, was attained asymptotically over an infinite time. Therefore, the time for completion of the reaction is usually estimated as the time over which the concentration of the reagent falls to the order of 1% of its initial value a_0. The integral in (2.3.13) can be calculated analytically using the expression for the exponential integral and its asymptotic for large argument [3]. For the adiabatic induction time, we obtain

$$t_{ia} = \frac{1}{k_0} \frac{RT_0^2}{E_a(T_b - T_0)} \exp(E_a/RT_0), \quad for\ n = 1, \tag{2.3.14}$$

$$t_{ia} = \frac{1}{k_0 a_0} \frac{RT_0^2}{E_a(T_b - T_0)} \exp(E_a/RT_0), \quad for\ n = 2. \tag{2.3.15}$$

2.4 Frank-Kamenetskii Transformation

The fact that the rate of a chemical reaction depends strongly on the temperature according to the Arrhenius law for $E_a/RT \gg 1$ allows to simplify calculations considerably, and to investigate analytically the basic properties of explosions and combustion. For this purpose, we will use approximation for the temperature dependence of the reaction rate in the Arrhenius law, which was proposed by D. A. Frank-Kamenetskii [4], and is widely used in combustion theory to obtain analytical solutions. The purpose of the Frank-Kamenetskii transformation is to replace the function $W(T) = C\exp(-E/RT)$ by the function $W'(T) = C'\exp\{(T - T_1)/\Theta\}$ which is very close to the Arrhenius temperature dependence over a temperature interval in the vicinity of some temperature T_1 for suitable values of the constants C' and Θ. To make this replacement, we expand the exponent in the Arrhenius formula in a Taylor series. With accuracy up to the first linear term in the vicinity of T_1, we obtain

$$W(T) = C\,e^{-\frac{E}{RT}} \approx C\,e^{-\frac{E}{RT_1}}\,e^{\frac{E}{RT_1^2}(T-T_1)}, \tag{2.4.1}$$

with

$$C' = C\,e^{-\frac{E}{RT_1}}\ and\ \Theta = RT_1^2/E. \tag{2.4.2}$$

The functions W(T) and W'(T) and their first derivatives are equal at $T = T_1$. Since the basic equations of the temperature dependence of the reaction rate are nonlinear, it is much easier to work with the function $\exp(x)$ than with $\exp(1/x)$, explicitly introducing into the analysis the characteristic temperature interval $\Theta = RT_1^2/E$ and representing small parameters in asymptotic solutions with good accuracy. Formally, the characteristic temperature interval is the temperature range over which the rate of a chemical reaction varies by a factor of *e*. The table below shows values of the Arrhenius law temperature dependence of the reaction rate, W(T), and the corresponding approximation W'(T) for $E/RT_1 = 2000\,K$ at various temperatures. For $T_1 = 2000\,K$ and $E = 80000\,kcal/mole$, Eq. (2.4.2) gives $\Theta = 100\,K$.

T	1000	1800	1900	2000	2100	2200	4000
T/T_1	0.5	$1-2\Theta/T_1$	$1-\Theta/T_1$	1	$1+\Theta/T_1$	$1+2\Theta/T_1$	2
$W(T)/W(T_1)$	$2\ 10^{-9}$	$1.1\ 10^{-1}$	$3.49\ 10^{-1}$	1	2.59	6.16	$2.2\ 10^{4}$
$W'(T)/W(T_1)$	$4.5\ 10^{-5}$	$1.35\ 10^{-1}$	$3.67\ 10^{-1}$	1	2.72	7.39	$4.85\ 10^{4}$

It is seen from the table that within a characteristic interval of T_1, the error in using the approximation is less than 5%. Though the relative error increases rapidly for large differences between T and T_1, this difference is not important since the reaction rate at temperatures far from T_1 is very small and may be neglected compared with the rates at temperatures near T_1. The peak temperature, which typically is the adiabatic flame temperature, is usually well known. The use of the Frank-Kamenetskii transformation makes the analytic analysis significantly easier [3].

2.5 Semenov Theory of Thermal Explosions

The development of the reaction in a combustible gaseous mixture with the initial temperature T_0 in a vessel, whose walls are kept at the same fixed temperature T_0, throughout the process have been first considered by N. N. Semenov [5–7] and C. N. Hinshelwood [8].

If at the beginning the gas temperature is T_0, then the amount of heat released due to chemical reaction per unit time (q_+)can be represented in a simplified form as

$$q_+ = Q\,W = Q\,a^n A \exp(-E/RT), \tag{2.5.1}$$

where $\dot{\omega} = A \exp(-E/RT)$ is the reaction rate, Q is the thermal energy release of the reactions, a is the concentration, n is the reaction order and T is the temperature of the gaseous mixture after the reaction. The rate of heat transfer to the vessel walls can be estimated, taking into account that the heat flux through the surface element dS is $dS\frac{\partial T}{\partial n}$, and the total heat flux from the vessel of volume V to the surface area S of the walls of the containing vessel is

$$q_- = \kappa \frac{S}{V}(T - T_0), \tag{2.5.2}$$

where κ is the thermal conduction coefficient of the gaseous mixture.

The temperature change for the system with simultaneous heat generation and heat losses is defined by the overall energy conservation

$$\rho c_V \frac{dT}{dt} = (q_+ - q_-), \tag{2.5.3}$$

where c_V is the specific heat of the mixture at constant volume.

Neglecting by the burning up of the initial reactants, we obtain the condition for stationary state of the reaction by equating the heat production to the heat loss ($q_+ - q_-$) and determine the value of the steady-state temperature in the vessel. The corresponding solutions are shown in Fig. 2.2, where q_+ and q_- are plotted as functions of the temperature. Note that q_+ rises with the temperature in the same way as the reaction rate in accordance with the Arrhenius law (curve 1), while the rate of heat loss rises linearly (curve 2).

Let the heat transfer be large enough so that the heat loss line intersects the heat release curve. Since the temperature of the mixture at the initial time equals the wall temperature T_0, there is no heat loss from the beginning. As soon as heat is released in

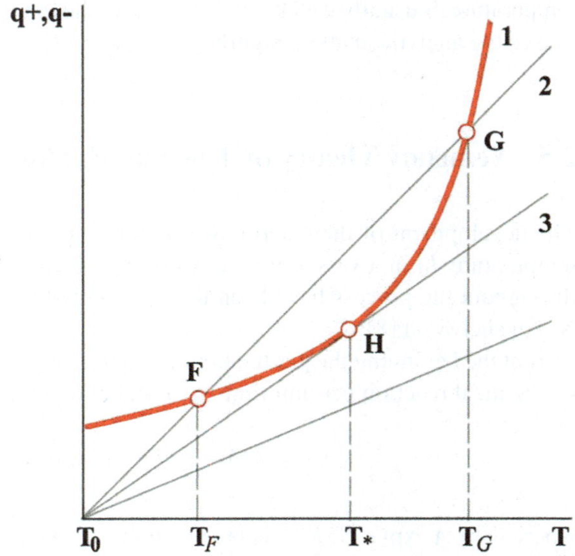

Fig. 2.2 The rate of heat release (1), heat losses (2, 3) depending on temperature

the reaction, the temperature of the mixture begins to rise. The temperature difference $T - T_0$ between the gas and vessel walls rises and the rate of heat loss increases. At some gas temperature $T = T_F$, the rates of heat production and heat loss become equal (point F in Fig. 2.2), and the gas temperature will not increase further. If for some reason the temperature becomes slightly greater than T_F, then the temperature falls to T_F since at that point the heat loss is more intense than the heat release. This means that the steady-state regime corresponding to point F is stable. Therefore, if the heat transfer is so large that the heat loss line intersects the heat production curve, then the gas initially at temperature T_0 will be heated to some temperature T_F slightly higher than the initial value T_0, and the reaction continues for a long time at this temperature.

The second intersection point, G, existing at the temperature $T = T_G$, corresponds to the unstable regime. If the temperature drops below T_G, the rate of heat loss below $T = T_G$ exceeds the rate of heat production and the gas temperature falls to the value $T = T_F$. If the gas rises to $T > T_G$, then the heat release will rise further since the rate of the reaction heat release exceeds the rate of heat loss for temperatures $T > T_G$. For an initial gas temperature below T_G, the gas cools to T_F and the reaction proceeds at the constant rate, while for an initial gas temperature above T_G the mixture ignites and the temperature will rise further. Ignition occurs only if we heat the mixture to a temperature $T > T_G$, much higher than the vessel temperature. A mixture with an initial temperature $T < T_G$ cannot ignite, which means that at temperatures $T < T_G$ there is no self-ignition.

Let us assume that the heat losses decreased, for example, owing to the decrease in S/V for a larger vessel, while the wall temperature T_0 remains the same. The heat release curve is the same, but the slope of the heat loss line is smaller. If the q_+

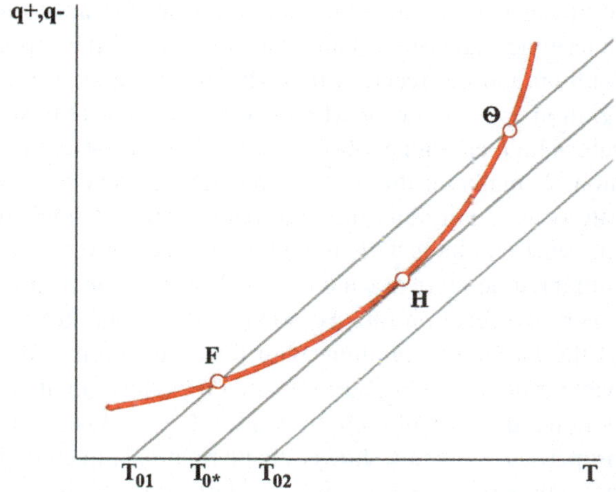

Fig. 2.3 Heat release and heat loss at different initial temperatures

curve is above q_- for all temperatures (the critical slope corresponds to curve 3 in Fig. 2.2), then the temperature of the reacting mixture will increase unlimitedly from the initial temperature T_0. The reaction rate will also rise unlimitedly, which means that the stable exothermal reaction regime does not exist and self-ignition will occur.

The heat release q_+ is a function of pressure through the density term, which is proportional to $\rho^n \propto P^n$ according to Eq. (2.5.1), while the pressure dependence of the heat loss q_- through the thermal conduction coefficient of the gaseous mixture is a less sensitive function of pressure. Therefore, at the higher initial pressure, the heat release curve shifts to higher values, and if the value of pressure is such that q_+ is everywhere greater than q_- then self-ignition will occur.

The same conclusion can be obtained by changing the wall temperature T_0 for the same vessel. In this case, the slope of the heat loss line is unchanged but the q_- line shifted parallel to itself in the direction of the temperature change (Fig. 2.3). A steady regime corresponding to temperature T_F exists for a wall temperature $T_{01} < T_{0*}$, but for a wall temperature $T_{02} > T_{0*}$, the lines q_+ and q_- do not intersect and an explosion occurs (Fig. 2.1).

The highest wall temperature at which a stationary regime is possible corresponds to $T_0 = T_{0*}$. At this temperature, the heat production curve is tangent to the heat loss curve. From these considerations, it is clear that the ability of a fuel mixture to explode, i.e. a self-ignition of the mixture to occur, depends on the shape and dimensions of the vessel, the thermal conductivity of the gas, as well as the rate of the chemical reaction. The point H in both Figs. 2.2 and 2.3 represents the ignition temperature $T_0 = T_{0*}$.

A fundamental feature of Semenov's theory is that a steady-state approach is used in the analysis of what might seem to be an essentially unsteady process. The condition of an explosion (nonstationary regime) is formulated as the loss of stability of the steady-state solution, that is, as the impossibility of a balance between heat production and heat losses. Semenov showed that although the reaction takes place at higher temperatures than the initial wall temperature, the wall temperature does determine many important characteristics of the process, such as the condition for ignition and the reaction time. If, for example, at the temperature of 580 °C self-ignition does not occur while at 600 °C it does, this does not mean that the reaction rate is high enough at 600 °C to produce the observable manifestations of an explosion. Actually, at this initial temperature, the conditions are such that the reaction rate rises rapidly and the high temperatures (above 2000 K) are reached at which the reaction rate is high enough to produce effects accompanied by explosions. The initial temperature at which an explosion occurs is not some kind of a singular point where the reaction rate experiences a discontinuity, but ignition is a consequence of the fact that, beginning with this temperature, the heat loss cannot balance the heat production. Contrary for a chain explosion with chain termination in the vessel volume, the critical ignition condition $\varphi = 0$ is a consequence only of the mixture properties. Semenov showed that a relationship could be obtained between the ignition temperature and the initial temperature of the mixture, that is, the temperature of the wall with the initial temperature of the mixture and the temperature at which the vessel's wall is maintained are the same (T_0). It should be emphasized that T_0 is a wall temperature that may cause a mixture to ignite.

2.6 Frank-Kamenetskii Theory of Thermal Explosion

An understanding of ignition is important in a wide range of combustion processes. Usually, two different processes of a combustible mixture ignition are distinguished: spontaneous ignition known also as autoignition and artificial or forced ignition. Spontaneous ignition occurs through the self-heating of the combustible mixture and can be considered as thermal instability resulting in an unwanted explosion. A forced ignition occurs with the help of an external ignition source with the heat generated in the combustible mixture by means of an electric spark, hot wire, or by focusing radiation capable of heating locally a gas mixture. The conditions under which a mixture can be ignited and/or an explosion may occur are of interest in various fields of human activity. From the point of view of safety problems, it is necessary to understand the conditions under which ignition of a mixture can lead to an explosion to develop methods preventing unwanted explosions, such as coal mine explosions, dust explosions in chemical industry, and many others. To design practical combustion devices, we need to know how much energy is required for ignition, what should be the rate of energy input, and in what volume for the desired combustion mode. For example, what is the rate of fuel ignition in the engine cylinder of a car at which efficiency is maximal but there is no "knocking"? Finally, one of the most important problems is what kind of combustion mode occurs as the result of ignition. It can be either slow combustion (deflagration) or explosion accompanied by destructive shock waves produced by a detonation.

Frank-Kamenetskii considered the non-steady heat conduction equation with a chemical reaction rate term that is included as a source of heat produced by the gaseous fuel–oxidizer mixture fuel which undergoes exothermic transformations in a vessel. The heat conduction equation for gas mixture in the vessel is

$$\rho c_V \frac{\partial T}{\partial t} = div(\kappa\, grad\, T) + \dot{\omega} Q, \tag{2.6.1}$$

where the last term is the heat release rate density.

We consider here a simplified Eq. (4.5.3) instead of (2.6.1) to obtain critical conditions for thermal explosion depending on parameters characterizing the problem. Let us apply the Frank-Kamenetskii transformation to analyze the critical conditions for a thermal explosion [3, 9–11]. Consider the system parameters at the boundary separating two possible regimes of a stationary slow reaction and an explosion. Substituting the expressions for q_- and q_+, and W in the heat balance Eq. (4.5.3) and applying the Frank-Kamenetskii transformation near the wall temperature $T = T_0$, which equals the initial gas temperature, we obtain

$$\frac{d\theta}{d\tau} = e^{\theta} - \gamma\theta, \tag{2.6.2}$$

where we introduced the dimensionless variables

$$\theta = \frac{T - T_0}{\Theta}, \; \tau = \frac{t}{t_{ia}}, \; t_{ia} = \frac{\Theta\, c_V\, \rho}{Q\, k_0\, a_0^n} e^{E/RT_0} \tag{2.6.3}$$

and the parameter

$$\gamma = t_{ia}/t_a, \; t_a = c_V \rho V/\kappa S, \tag{2.6.4}$$

which is the ratio of the adiabatic induction time to the characteristic time of the heat loss. The magnitude $\gamma = t_{ia}/t_a$ depends on the volume and surface area of the vessel, the thermal conductivity of the mixture, the rate of the chemical heat release, and the wall temperature T_0. If the heat losses are small, so that $t_{ia} \ll t_a$, then $\gamma = t_{ia}/t_a \ll 1$ and we can neglect by the second term on the right side of Eq. (2.6.2). For small γ, the characteristic heat transfer time is much greater than the characteristic heat production time; the heat cannot be removed to the vessel walls and, therefore, an explosion must occur. This means that for sufficiently small γ, the right-hand side of Eq. (2.6.2) must be positive for arbitrary θ; that is, $d\theta/dt > 0$ for all t, and the equation

$$e^{\theta} - \gamma\theta = 0 \tag{2.6.5}$$

has no roots, so that the temperature increases unlimitedly. If γ is such that the roots of Eq. (2.6.5) are real, then a stationary regime is possible: the temperature initially rises to a value such that $e^{\theta} - \gamma\theta$ goes to zero and then the reaction proceeds at a constant temperature. The condition for the transition from a stationary burning regime to the explosion is the same as it was obtained previously in the Semenov theory (see Fig. 2.2-3). This is the tangency of the curves $\exp\theta$ and $\gamma\theta$. At the point separating the two reaction regimes,

$$e^{\theta} = \gamma\theta, \; and \; \frac{d}{d\theta}e^{\theta} = \frac{d}{d\theta}\gamma\theta. \tag{2.6.6}$$

The compatibility of these two conditions gives the ignition condition in the form

$$\gamma = e, \; and \; \theta = 1. \tag{2.6.7}$$

The equality $\theta = 1$ can be written in the form

$$T_* - T_0 = \Theta, \tag{2.6.8}$$

which defines the maximum temperature of the mixture just before the explosion begins, the value of the maximum temperature a mixture can be heated without explosion.

If we use the condition of tangency of $q_+(T)$ and $q_-(T)$ without the Frank-Kamenetskii transformation, we obtain for the pre-explosion temperature $T_* = \frac{E}{2R}\left[1 - \sqrt{1 - 4\frac{RT_0}{E}}\right]$, which coincides with (2.6.8) for $RT_0 \ll E$. Substituting the expression for γ from Eq. (2.6.4) in the first of conditions (2.6.7), and using (2.6.3) for t_{ia}, we obtain

$$\kappa\frac{S}{V}\frac{\Theta\exp(E/RT_0)}{k_0 a_0^n Q} = e. \tag{2.6.9}$$

This equation gives the relation between the initial temperature T_0, the reaction energy release Q, the concentration of reagents a_0, the volume and shape of the vessel (V and S), the heat conductivity coefficient of the mixture, and the initial pressure of the fuel mixture at the explosion limit ($a_0^n \propto \rho^n \propto P^n$).

The induction time of a thermal explosion can be obtained integrating Eq. (2.6.2) and neglecting by the burnout of reagents (the quantity of products at the beginning of the reaction is very small):

$$\tau_i = \int_0^{\infty} \frac{d\theta}{e^{\theta} - \gamma\theta}. \tag{2.6.10}$$

The upper limit in the integral (2.6.10) can be taken equal to infinity because $E(T_b - T_0)/RT_0^2 \gg 1$. If in addition it is possible to neglect by the heat losses, then the induction time calculated from (2.6.10) will coincide with the adiabatic induction time, which was calculated in Sect. 2.3 and given by Eqs. (2.3.14) and (2.3.15):

$$t_{ia} = \frac{1}{k_0 a_0} \frac{RT_0^2}{E_a(T_b - T_0)} \exp(E_a/RT_0), \quad for\ n = 2. \tag{2.6.11}$$

The reagent consumption can be neglected for high activation energies, if $Qa_0/\rho c_V \gg \Theta = T_* - T_0$. However, the classical theory developed by Semenov and Frank-Kamenetskii does not consider the effects of the heating rate in the transition from slow to ignition events and the reagent consumption. An extension of the classical Semenov theory with an account of the competition between the characteristic time scales: the chemical energy release time scale, the wall heat transfer time, and ramp heating time was considered in [12]. Those authors explore the competition between the physical processes represented by these time scales and the effect of reaction consumption. The effect of the walls' heating and competition between the rate of chemical energy release and heat loss and the rate of increasing wall temperature on the ignition delay time, which can be important for the systems with low rate of the chemical heat release was also analyzed in [12].

If the heat release time scale is much shorter than the heat transfer time scale, then the ignition delay time can be found assuming a one-step chemical model, and for large activation energy the energy equation (2.6.2) reduces to the following form:

$$\frac{d\theta}{d\tau} = e^{\theta}. \tag{2.6.12}$$

This equation can be integrated, and we obtain

$$\theta = -\ln(1 - \tau). \tag{2.6.13}$$

It is seen from (2.6.13) that the temperature tends to infinity when $\tau = t/\tau_{in} \to 1$, which is the induction time $\tau = t/\tau_{in}$. The nonphysical infinite temperature is a consequence of neglecting reagent consumption.

2.7 Spark Ignition and Minimum Ignition Energy

If the combustion is initiated by an external energy source which locally heats the mixture to a high temperature, the process is called the forced ignition. Burning is then sustained once the ignition source is removed. The forced ignition is essentially associated with gaseous mixtures. To ignite a solid fuel the energy input first must vaporize the material and then ignite the vaporized gas-phase reaction. There are many means to achieve the forced ignition of a reactive gaseous mixture; the ignition can be achieved using an electric spark, a hot wire heated by electric current, adiabatic compression of the mixture, glow plug, shock wave, and any sufficiently hot source capable to locally heat the mixture to sufficiently high temperature. The most commonly studied concepts are those associated with processes that take place in the spark ignition, automotive engine.

Let's consider the case of a spark generated with a spark plug, such as in a car engine. A high voltage, which is applied to the electrodes spaced apart a distance Δ, creates an electric arc across the gap between the electrodes, heating the combustible mixture in between. The energy required for igniting the mixture is important for both engineering applications and explosion safety. The most commonly used spark systems are those developed from electric discharge. In the automobile cylinder, the spark initiates a flame that propagates through the engine cylinder. Typically, the spark is fired before the piston reaches the top dead center, and as the flame is traveling the combustible mixture ahead of the flame is compressed. Under certain circumstances, the mixture ahead of the flame can explode, in which case the phenomenon of knock occurs, which can destroy the cylinder walls and therefore should be avoided.

Let us estimate the minimum energy of a spark required for ignition. We assume that the spark can be modeled as a point heat source, which releases a quantity of heat. In this model, the distribution of the heat is given by equation

$$c_P \rho \frac{\partial T}{\partial t} = \kappa \, \Delta T \tag{2.7.1}$$

with the boundary conditions $T(r = \infty) = T_0$.

The input energy at any time must obey the equality

$$Q/c_P \rho = \int T(x, y, z) \, d^3x = 4\pi \int_0^\infty (T(r) - T_0) r^2 dr, \tag{2.7.2}$$

where Q is the energy released either at constant pressure, or at constant volume.

Equations (2.7.1) and (2.7.2) are equivalent to Eq. (2.7.1) with the sources of energy in the form of delta function, $q = Q \, \delta(r) \, \delta(t)$. The solution of Eq. (2.7.1) can be written as

$$T(\mathbf{r}, t) - T_0 = \frac{Q}{8(\pi \chi t)^{3/2}} \exp\left(-\frac{r^2}{4\chi t}\right), \; \chi = \kappa / \rho c_P. \tag{2.7.3}$$

It is seen that the maximum temperature is at the point $r = 0$ and it decreases with time as $1/t^{3/2}$. Simultaneously, the temperature of the surrounding gas is gradually increasing. The condition for the mixture ignition can be specified assuming that the cooling time associated with $T(0)$ is greater than the reaction duration time. A simple estimate of the ignition condition can be obtained considering the stability of the solutions of stationary equations associated with Eq. (2.7.1) with the energy source in the right-hand part. If the right-hand part is positive, then positive is the time derivation of the temperature, which can be view as instability and ignition. In such consideration, cooling is associated with energy loss due to thermal conduction.

Table 2.1 Minimum ignition energies of stoichiometric fuel/air mixtures (1 atm and 20 °C)

Fuel	Energy (mJ)
Hydrogen	0.02
Acetylene	0.03
Methanol	0.21
n-Hexane	0.29
Methane	0.30
Propane	0.40

The larger the temperature gradient is, the larger the thermal losses are. This leads to the condition for ignition as a condition for the size of the spark where energy was released

$$r_0 > 4.2L_f, \tag{2.7.4}$$

where $L_f = T/\max\{|\nabla T|\}$ is the width of the laminar flame

Due to the importance of understanding ignition processes for practical applications such as ignition in the SI-engine cylinder [13] and for safety problems, there is a vast number of works in which the ignition processes are studied for different mixtures, using the detailed chemistry and 2D and 3D numerical simulations [14–20]. Minimum ignition energy for some mixtures is shown in Table 2.1.

Ignition of a combustible mixture requires the mixture to be heated to a sufficiently high temperature. Ignition caused by a spark occurs in a local region and spreads by a flame in the surrounding space. Ignition can be achieved by adiabatic compression of the gaseous fuel, or by a strong shock wave propagating through the mixture, which is rapidly compressed and heated by the shock. Ignition by the adiabatic compression is similar to the conditions in Diesel engines. For an ideal gaseous mixture, the temperature in adiabatic compression rises as $T^{\gamma} P^{\gamma-1}$ = constant or $TV^{\gamma-1}$ = constant, where $\gamma = C_p/C_v$. The temperature increase in a shock wave is $T/T_0 = \frac{(\gamma+1)P_0+(\gamma-1)P_0}{(\gamma-1)P_0+(\gamma+1)P_0}(P/P_0)$. In a strong shock wave, with Mach number $M \gg 1$: $T/T_0 = \frac{2\gamma(\gamma-1)}{(\gamma+1)^2}M^2$ and $P/P_0 = \frac{2\gamma}{\gamma+1}M^2$.

While conventional knock in spark-ignited (SI) engines has direct relevance to the end gas autoignition [13, 21–24], super-knock is found to be closely associated with detonation development induced by randomly localized hot spot [25–31], and ignition by the shock compression is similar to the conditions that generate knock in a high compression spark-ignited (SI) engine. Fuel ignition delay time (induction time), the kinetics data, and ignition delays for the fuels are often studied using shock tubes [32–38].

2.8 Induction Time: One-Step and Detailed Chemical Models

While models with fairly simplified gas dynamics and a one-step Arrhenius kinetics in many cases make it possible to solve the problem in question in explicit analytical form, many important features of combustion cannot be explained without account of the reactions' chain nature, describing qualitatively a few major properties of the phenomena in question with some poor accuracy if any, often rendering misinterpretation of a variety of combustion phenomena. However, many important features of combustion cannot be explained without account of the reaction chain nature. The range of validity of simplified schemes is necessary, very limited, and their applicability to the understanding and modeling of combustion phenomena must be examined with great care. Furthermore, it became clear that the use of a one-step Arrhenius model has not always been interpreted properly, rendering a misinterpretation of a variety of combustion phenomena. A one-step model does not reproduce two distinct stages of the reaction: the induction stage and the exothermal one. Another fundamental difference between a one-step Arrhenius model and the chain-branching reaction is an avalanche multiplication of intermediate species, atoms and radicals, and their repetitive reactions constituting reaction chains.

A correct description of unsteady, transient combustion processes controlled by chemical kinetics requires knowledge of the detailed chemical reaction mechanisms for reproducing combustion parameters in a wide range of pressures and temperatures. Knowledge of the detailed reaction mechanisms is of special importance for understanding unsteady and transient combustion processes, which are controlled by chemical kinetics, such as ignition and self-ignition processes (e.g. engine knock), flame acceleration, and the deflagration-to-detonation transition (DDT). The use of the detailed reaction mechanisms is absolutely necessary for a correct description of kinetically controlled pollutant formation. Development and exploitation of reliable detailed chemical kinetic models, identification of the important kinetic pathways, and accurate kinetic transport models remain among the major challenges in combustion science and technology being essential for the design of efficient and reliable engines and for controlling emissions. The availability of such models is essential for gaining scientific insight into the most fundamental combustion phenomena. The detailed reaction mechanisms were developed for different combustible mixtures, and today they are available and can be used in the simulation of spatially homogeneous reaction systems. If, however, real 3D and often turbulent flows are considered, we have to use reduced chemical schemes, since the use of detailed reaction mechanisms involves massive computing times which can be difficult and even impossible to implement. Even in the combustion of simple hydrocarbons, the chemical kinetics is reproduced by relatively large reaction mechanisms, and for complex hydrocarbons the number of chemical species can be up to several hundreds and the number of elementary reactions up to several thousands. Therefore, there is great interest in reduced reliable chemical reaction schemes consisting of a not too large number of elementary reactions.

A quintessential example of chain mechanisms in chemical kinetics and combustion science is the H_2/O_2 mechanism, which has been a major topic of research for many decades. The number of available mechanisms is significant, yet there are uncertainties related to the elementary reactions involved [39] and available for validation experimental data is scarce, especially for low-temperature kinetics (T < 1000 K). Usually, the verification of the reduced kinetic schemes covers the data on induction periods, equilibrium composition, temperature of the products, and the normal laminar flame velocity. The latter are pure thermodynamic characteristics and are more or less in good agreement with experimental data at normal pressure. To study unsteady combustion processes, such as flame acceleration and ignition, which are accompanied by compression and shock waves, it is necessary to take into account parameters such as the induction period and the period of exothermal reaction, which determine the chemical time scales competing with the transport time scales in establishing the zone of energy release, which in turn determines the evolution of the flame.

We compare the one-step chemical models, which have been widely used for 2D simulations of the flame acceleration and DDT in a highly reactive hydrogen/air mixture and in a low reactive methane/air with the detailed chemical mechanisms. It was shown [40] that to be able to predict correctly the ignition delay times, the reduced mechanisms must consist of at least ten reactions. The requirements to the chemical schemes imply their capability to reproduce correctly laminar flame speeds and structure, the ignition delay time calculated with a detailed chemical model GRI Mech 3.0 [55], and agreement with experimental measurements. These requirements stem from the need of ensuring the controllable capability of characterizing ignition for transient combustion processes evolving over wide ranges of temperature and pressure during the flame acceleration and DDT. Finally, computational costs to use chemical schemes with DNS or LES solvers must be reasonable for the feasibility of multidimensional simulations.

2.8.1 Hydrogen–Air: A Single-Step and Detailed Chemical Models

A single-step chemistry approach suggests that the complex set of reactions can be modeled by a one-step Arrhenius reaction:

$$W = A\rho Y \exp(-E_a/RT), \tag{2.8.1}$$

where $A = 6.85 \cdot 10^{12} \mathrm{cm}^3/(g\,s)$ is the pre-exponential factor, $E_a = 46.37\ RT_0$ is the activation energy of the reaction, and R is the universal gas constant. The detailed mechanism chosen to model the hydrogen–air chemistry is the mechanism developed by Kéromnès et al. [41], which consists of 19 reactions and 9 species. An advantage of this mechanism is that it was extensively validated over a large number of experimental conditions, especially focused on high pressures [42], and demonstrated an excellent agreement between the modeling and experimental measurements of the flame velocity–pressure dependence over a wide range of pressures and equivalence ratios.

Figures 2.4a, b from [56] show the induction times versus temperature for hydrogen–air at pressures (a) 1 bar and (b) 10 bar, computed using the one-step model Eq. (2.8.1) and the induction times computed with the detailed chemical model [41]; empty symbols show experimental results. It is seen that the induction times predicted by the detailed chemical model [41] are in good agreement with the experimental results but differ up to three orders of magnitude from that predicted by the one-step model. Another feature of the "real" induction time is an abrupt change of $(d\tau_{ind}/dT)$ at the crossover temperatures, which corresponds to the transition from the endothermal induction stage to the exothermal stage.

2.8.2 Methane–Air: A Single-Step and Detailed Chemical Models

We compare a one-step Arrhenius model, which was calibrated to give a reasonable approximation of the key properties of methane/air which are the laminar flame speed, the adiabatic flame temperature and the speed of sound

$$W = A\rho Y \exp(-E_a/RT), \qquad (2.8.2)$$

where for methane/air: $E_a = 67.55\, RT_0$, $A = 1.64 \cdot 10^{13}\mathrm{cm}^3/(g\, s)$.

In addition, we compare the induction times calculated for more advanced two-step or four-step models, which to some extent mimic the chain-branching kinetics. As a representative example, we consider the widely used versions of two-step models: 2S-CH_4 mechanism developed by Westbrook et al. [47] and 2S-CH_4-BFER developed by Franzelli et al. [48]. The 2S-CH4-BFER model is an improved version of the classical two-step mechanism [47], and was proved to be valid for lean and stoichiometric mixtures, providing an accurate description of the chemical flame structure and velocity. The 2S-CH_4-BFER mechanism uses 6 species (CH_4, O_2, CO_2, CO, H_2O, and N_2) for 2 global reactions and correctly predicts the chemical flame structure and velocity of a premixed laminar methane/air flame for a wide range of equivalence ratio and pressures ($1bar < P_0 < 10bar$) and for fresh mixture temperatures ($300K \leq T_0 \leq 800K$). A further improvement of the reaction mechanism into several steps is the four-step reaction scheme developed by Jones and Lindstedt [49]. The scheme consists of four global reactions describing chain branching, chain breaking, and CO oxidation: (I) $CH_4 + 2H + H_2O \Leftrightarrow CO + 4H_2$, (II) $CO + H_2O \Leftrightarrow CO_2 + H_2$, (III) $H + H + M \Leftrightarrow H_2 + M$, and (IV) $3H_2 + O_2 \Leftrightarrow 2H + 2H_2O$. The four-step reaction scheme makes the predicted product

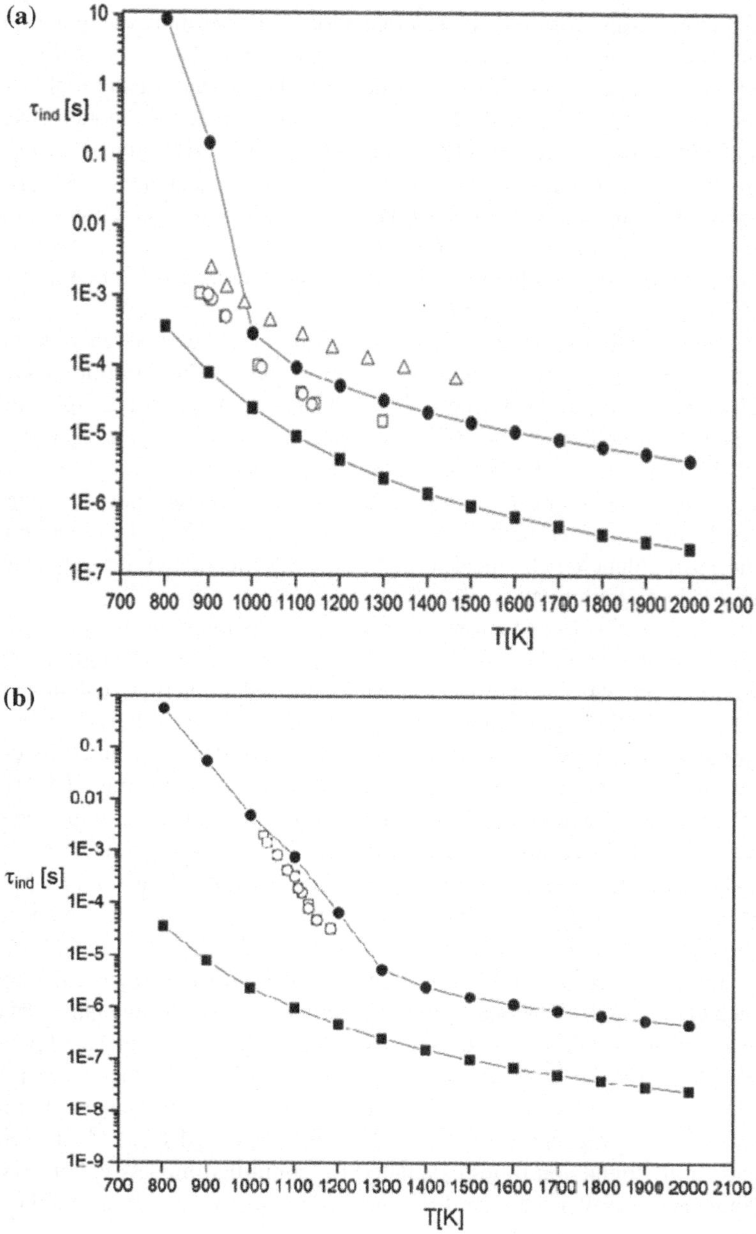

Fig. 2.4 **a, b** Induction times for hydrogen–air mixture at pressure: $P = 1atm$: (■)—one-step model Eq. (2.8.1), (•)—detailed model [41]. Experiments: □- [44]; ○- [45]; △—[45]. **b** Induction times for stoichiometric hydrogen–air mixture at pressures $P = 10atm$ calculated for one-step (■) and detailed (•) chemical models [41]. Experimental data are □- [46], ○- [46, 47]

temperature and composition more accurate, though it costs more computation time. The multi-step detailed mechanisms chosen to model methane–air chemistry is the reduced detailed reaction model DRM-19 developed by Kazakov and Frenklach [50], which consists of 19 species and 84 reactions and the 35 reactions model, developed by Smooke [51]. The reaction mechanism DRM-19 has been chosen as the main one for simulating as it was extensively validated by many researchers for combustion characteristics related to ignition delay times and laminar flame velocities over a wide range of pressures, temperatures, and equivalence ratios [52–54]. The results for all chemical models were compared with the standard detailed reference mechanism GRI 3.0 Mech [55] and experimental data.

The ignition delay times were calculated for different chemical reaction schemes using the standard constant volume adiabatic model. The ignition delay time can be defined as the time during which the maximum rate of temperature rise ($Max\{dT/dt\}$) is achieved, which is close to the time of exothermic reactions activation. While one-step models allow to reproduce the laminar flame speed U_L and the adiabatic flame temperature T_b with satisfactory good accuracy, the induction times predicted by all simplified models are significantly shorter than the induction times calculated using detailed chemical models, GRI 3.0 Mech, and the experimentally measured induction times.

Figure 2.5a shows the induction times versus temperature at the initial pressure $P_0 = 1atm$ computed for simplified (one-step, two-step, and four-step) models and for detailed chemical reaction models, as well as the induction times calculated using GRI 3.0 Mech [55]. Empty symbols in Fig. 2.5a are the induction times experimentally measured in [57, 58]. It is seen that the induction times predicted by detailed chemical models are in good agreement with those calculated using GRI 3.0 Mech and with the experimental data. The induction times predicted by simplified chemical models is up to three orders of magnitude smaller than that predicted by detailed chemical models. The value of $\tau_{ind}(T)$ calculated for DRM-19 and GRI 3.0 Mech for temperatures in the range 800–1000 K is shown by dashed line because of uncertainty of the “cool flames” contribution.

This trend of a very large difference between the induction times calculated for simplified models and the “actual” ignition times measured experimentally and calculated using detailed chemical models remains almost the same at elevated pressures. Figure 2.5b shows the induction times versus temperature at the initial pressure $P_0 = 10\,atm$ computed for simplified models and for detailed chemical reaction models, as well as the induction times calculated using GRI 3.0 Mech [55]. The difference in the induction times calculated for simplified models is not very large and significantly smaller than the difference from experimental data and the values calculated for detailed models.

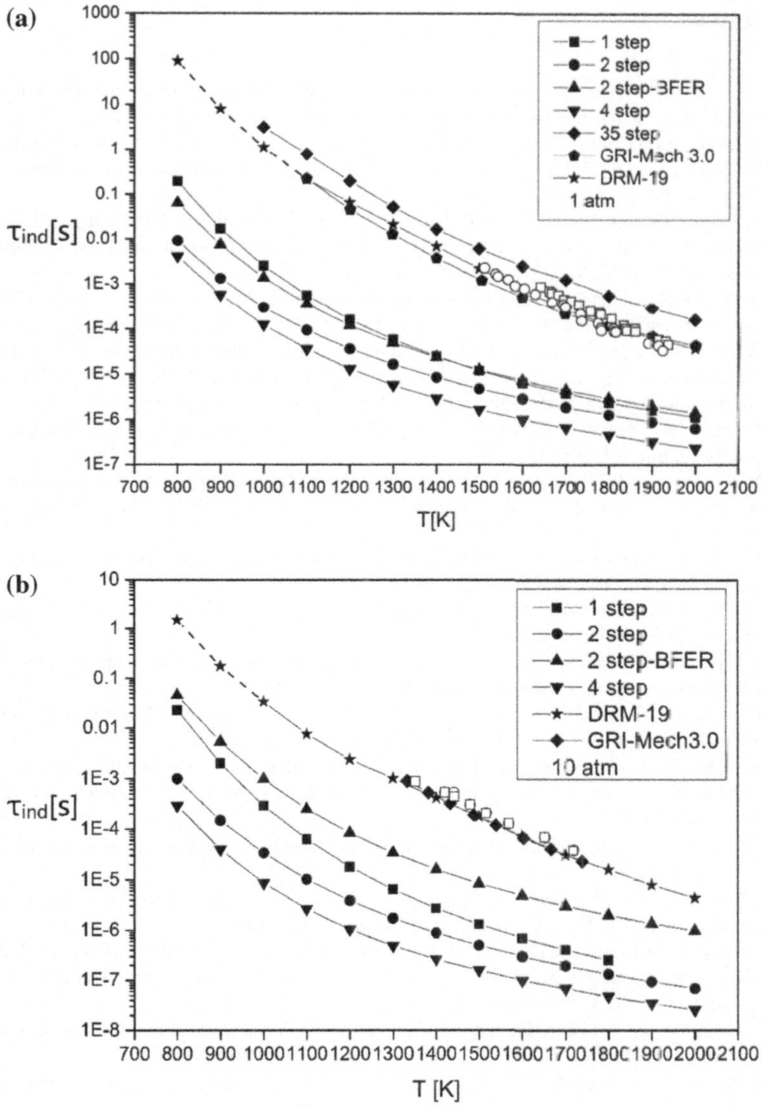

Fig. 2.5 **a** Induction times for stoichiometric methane–air mixture at pressure $P_0 = 1bar$ calculated for simplified and detailed chemical kinetics schemes. Empty circles and squares are experimental data from [57, 58]. **b** Induction times for stoichiometric methane–air mixture at pressure $P_0 = 10atm$ calculated for simplified and detailed chemical kinetics schemes. Empty circles and squares are experimental data from [57, 58]

References

1. M.A. Liberman, Unsteady combustion processes controlled by detailed chemical kinetics, in *Notes on Numerical Fluid Mechanics and Multidisciplinary Design*, ed. by R. King, Active Flow and Combustion Control, vol. 127 (Springer, 2014), pp. 317–341. ISBN: 978-3-319-11966-3
2. C.K. Westbrook, F.L. Dryer, Chemical kinetic modeling of hydrocarbon combustion. Prog. Energy Comb. Sci. **10**, 1–57 (1984)
3. Y.B. Zeldovich, G.I. Barenblatt, V.B. Librovich, G.M. Makhviladze, The Mathematical theory of combustion and explosions. Science Publ. Moscow, 1980 (in Russian). English translation (Consultants Bureau, New York, 1985)
4. D.A. Frank-Kamenetskii, Temperature distributions in a reaction vessel and the stationary theory of thermal explosion. J. Phys. Chem. **13**, 738–755 (1939). ((in Russian))
5. N.N. Semenov, Zur Theorie des Verbrennungsprozesses. Z. Phys. Chemie **48**, 571–582 (1928)
6. N.N. Semenov, *Chain Reactions, Goskhimtekhizdat, Leningrad, 1934* (Chemical Kinetics and Chain Reactions, Oxford University Press, Oxford, English Translation, 1935)
7. N.N. Semenov, *Some Problems in Chemical Kinetics and Reactivity* (Princeton University Press, Princeton, NJ, 1958)
8. C.N. Hinshelwood, *The Kinetics of Chemical Change* (Oxford University Press, Oxford, 1940)
9. D.A. Frank-Kamenetskii, Calculation of thermal explosion limits. Acta Physicochimica U.R.S.S. 10 (1939), pp. 365–370
10. D.A. Frank-Kamenetskii, On the mathematical theory of thermal explosions. Acta Physicochimica U.R.S.S. 16 (1942), pp. 357–361
11. D.A. Frank-Kamenetskii, *Diffusion and Heat Exchange in Chemical Kinetics* (Princeton University Press, Princeton, New Jersey, 1955)
12. P.A. Boettcher, R. Mevel, V. Thomas, J.E. Shepherd, The effect of heating rates on low temperature hexane air combustion. Fuel **96**, 392–403 (2013)
13. J.B. Heywood, *Internal Combustion Engine Fundamentals*, 2nd edn. (McGraw-Hill Education, New York-Chicago-San Francisco, 2018). ISBN: 9781260116106
14. J. Warnatz, in "Combustion Chemistry", ed. by W.C. Gardiner, Jr. (Springer, New York, 1984)
15. B. Deshaies, G. Joulin, On the initiation of a spherical flame kernel. J. Combust. Sci. Technol. **37**, 99–116 (1984)
16. U. Maas, J. Warnatz, Ignition processes in hydrogen-oxygen mixtures. Combust. Flame **74**, 53–69 (1988)
17. M. Thiele, S. Selle, U. Riedel, J. Warnatz, U. Maas, Numerical simulation of spark ignition including ionization. Proc. Combust. Institute **28**, 1177–1185 (2000)
18. M. Thiele, J. Warnatz, A. Dreizler, S. Lindenmaier, R. Schieß, U. Maas, A. Grant, P. Ewart, Spark ignited hydrogen/air mixtures: two-dimensional detailed modeling and laser based diagnostics. Combust. Flame **128**, 74–87 (2002)
19. C. Vázquez-espí, A. Liñán, Fast, non-diffusive ignition of a gaseous reacting mixture subject to a point energy source. Combust. Theory Model. **5**, 485–498 (2001)
20. A.P. Kelley, G. Jomaas, C.K. Law, Critical radius for sustained propagation of spark-ignited spherical flames. Combust. Flame **156**, 1006–1013 (2009)
21. M.A. Liberman, M.F. Ivanov, O. Peil, D. Valiev, L.-E. Ericsson, Self-acceleration and fractal structure of outward radially propagating wrinkling flames. Phys. Fluids **16**(7), 2476–2482 (2004)
22. M.A. Liberman, L.-E. Eriksson, M.F. Ivanov, Hot spot formation by the propagating flame and the influence of EGR on knock occurrence in SI-engines. Comb. Sci. Tech. **178**, 1613–1647 (2006)
23. Z. Wang, H. Liu, R.D. Reitz, Knocking combustion in spark-ignition engines. Prog. Energy Combust. Sci. **61**, 78–112 (2017)
24. L. Bates, D. Bradley, G. Paczko, N. Peters, Engine hot spots: modes of auto-ignition and reaction propagation. Combust. Flame **166**, 80–85 (2016)
25. G. Kalghatgi, Knock onset, knock intensity, super-knock and preignition in spark ignition engines. Int. J. Engine Res. **19**, 7–20 (2018)

26. G.T. Kalghatgi, D. Bradley, Pre-ignition and “super-knock” in turbo-charged spark-ignition engines. Int. J. Engine Res. **13**, 399–414 (2012)
27. Z. Wang, Y. Qi, X. He, J. Wang, S. Shuai, C.K. Law, Analysis of pre-ignition to super-knock: hotspot-induced deflagration to detonation. Fuel **144**, 222–227 (2015)
28. Z. Wang, H. Liu, T. Song, Y.L. Qi, X. He, S.J. Shuai, J.X. Wang, Relationship between super-knock and pre-ignition. Int. J. Engine Res. **16**, 166–180 (2015)
29. P. Dai, Z. Chen, Supersonic reaction front propagation initiated by a hot spot in n-heptane/air mixture with multistage ignition. Combust. Flame **162**, 4183–4193 (2015)
30. P. Dai, C.K. Qi, Z. Chen, Effects of initial temperature on autoignition and detonation development in dimethyl ether/air mixtures with temperature gradient. Proc. Combust. Inst. **36**, 3643–3650 (2017)
31. P. Dai, Z. Chen, X. Gan, M.A. Liberman, Autoignition and detonation development from a hot spot inside a closed chamber: effects of end wall reflection. Proc. Combust. Inst. (2020). https://doi.org/10.1016/j.proci.2020.09.025
32. W. Tsang, A. Lifshitz, Shock tube techniques in chemical kinetics. Ann. Rev. Phys. Chem. **41**, 559–599 (1990)
33. D.F. Cooke, A. Williams, Shock tube studies of methane and ethane oxidation. Combust. Flame **24**, 245–256 (1975)
34. A.A. Borisov, B.E. Gelfand, V.N. Zamanskiy, V.V. Lisyanskiy, G.I. Skachkov, K.Y. Troshin, Ignition of combustible gaseous mixtures in conditions of reflected shock wave focusing. Sov. J. Chem. Phys. **7**, 1387–1402 (1988)
35. J. DeVries, J.M. Hall, S.L. Simmons, M.J.A. Rickard, D.M. Kalitan, E.L. Petersen, Ethane ignition and oxidation behind reflected shock waves. Combust. Flame **150**, 137–150 (2007)
36. M.O. Kapp, E.L. Petersen, Shock tube study of ignition of multi-component syngas mixture with and without ammonia impurities. Proc. Combust. Inst. **34**, 3211–3218 (2013)
37. V.P. Zhukov, V.A. Sechenov, A.Y. Starikovski, Spontaneous ignition of methane air mixtures in a wide range of pressure. Combust. Explosion and Shock Waves, **39**, 487–495 (2003)
38. S.M. Frolov, S.N. Medvedev, V.Y. Basevich, F.S. Frolov, Self-ignition of hydrocarbon-hydrogen-air mixtures. Intern. J. Hydrogen energy **38**, 4177–4184 (2013)
39. A.A. Konnov, Remaining uncertainties in the kinetic mechanism of hydrogen combustion. Combust. Flame **152**, 507–528 (2008)
40. A.C. Zambon, H.K. Chelliah, Explicit reduced reaction models for ignition, flame propagation, and extinction of C_2H_4/ CH_4/H_2 and air systems. Combust. Flame **150**, 71–91 (2007)
41. A. Kéromnès, W.K. Metcalfe, K.A. Heufer, N. Donohoe, A.K. Das, C.-J. Sung, J. Herzler, C. Naumann, P. Griebel, O. Mathieu, M.C. Krejci, E.L. Petersen, W.J. Pitz, H.J. Curran, Combust. Flame **160**, 995–1011 (2013)
42. G. Dayma, F. Halter, P. Dagaut, New insights into the peculiar behavior of laminar burning velocities of hydrogen–air flames according to pressure and equivalence ratio. Combust. Flame **161**, 2235–2241 (2014)
43. A.D. Snyder, J. Robertson, D.L. Zanders, G.B. Skinner, Technical Report AFAPL-TR- (1965), pp. 65–93.
44. M. Slack, A. Grillo, Investigation of hydrogen-air ignition sensitized by nitric oxide and by nitrogen dioxide, Tech. Rep. RE-537, Grumman Research Department (1977).
45. E. Hu, L. Pan, Z. Gao, Shock tube study on ignition delay of hydrogen and evaluation of various kinetic models. Int. J. Hydrogen Energy **41**, 13261–13280 (2016)
46. L. Pan, E. Hu, F. Deng, Z. Zhang, Z. Huang, Effect of pressure and equivalence ratio on the ignition characteristics of dimethyl ether-hydrogen mixtures. Int. J. Hydrogen Energy **39**, 19212–19223 (2014)
47. C.K. Westbrook, F.L. Dryer, Simplified reaction mechanisms for the oxidation of hydrocarbon fuels in flames. Combust. Sci. Technol. **27**, 31–43 (1981)
48. B. Franzelli, E. Riber, L.Y. Gicquel, T. Poinsot, Large eddy simulation of combustion instabilities in a lean partially premixed swirled flame. Combust. Flame **159**, 621–637 (2012)
49. W. Jones, R. Lindstedt, Global reaction schemes for hydrocarbon combustion. Combust. Flame **73**, 233–249 (1988)

50. A. Kazakov, M. Frenklach, Reduced Reaction Sets based on GRI–Mech 1.2: 19–species reaction set. http://www.me.berkeley.edu/drm/
51. M.D. Smooke, *Reduced Kinetic Mechanisms and Asymptotic Approximations for Methane-Air Flames, Series: Lecture Notes in Physics, 384* (Springer, Berlin- New York, 1991)
52. M. Goswami, S.C.R. Derks, Kris Coumans, W.J. Slikker, M.H. de Andrade Oliveira, R.J.M. Bastiaans, C.C.M. Luijten, L. Philipus, H. de Goey, A.A. Konnov, The effect of elevated pressures on the laminar burning velocity of methane + air mixtures. Combust. Flame **160**, 1627–1635 (2013)
53. M. Goswami, K. Coumans, R.J.M. Bastiaans, A.A. Konnov, L.P.H. de Goey, Numerical simulations of flat laminar premixed methane-air flames at elevated pressure. Combust. Sci. Technol. **186**, 1447–1459 (2014)
54. H. Yu, E. Hu, Y. Cheng, X. Zhang, Z. Huang, Experimental and numerical study of laminar premixed dimethyl ether/methane-air flame. Fuel **136**, 37–45 (2014)
55. M. Frenklach, N.W. Moriarty, B. Eiteneer, M. Goldenberg, C.T. Bowman, R.K. Hanson, S. Song, W.C. Gardiner, Jr. V.V. Lissianski, G.P. Smith, D.M. Golden, Z. Qin, GRI-mech 3.0 http://www.me.berkeley.edu/gri-mech/
56. C. Wang, C. Qian, J. Liu, M.A. Liberman, Influence of chemical kinetics on detonation initiating by temperature gradients in methane/air. Combust. Flame **197**, 400–415 (2018)
57. W. Zeng, H. Ma, Y. Liang, E. Hu, Experimental and modeling study on effects of N_2 and CO_2 on ignition characteristics of methane/air mixture. J. Adv. Res. **6**, 189–201 (2015)
58. E. Hu, X. Li, X. Meng, Y. Chen, Y. Cheng, Y. Liang, X.Z. Huang, Laminar flame speeds and ignition delay times of methane–air mixtures at elevated temperatures and pressures. Fuel **158**, 1–10 (2015)

Chapter 3
Laminar Flames

3.1 Reaction Waves

In *premixed flames* the reactants are perfectly mixed with the oxidizer before the start of the reaction. Contrary to premixed flames, in *non-premixed combustion* the fuel and oxidizer are initially spatially separated and brought together through the molecular diffusion and the bulk convective motion to the region where mixing and subsequently the reaction take place. It should be emphasized that in premixed combustion, diffusion is also important to transport the combustible mixture to the reaction zone and the combustion products away from the reaction zone. Under ordinary conditions, i.e. atmospheric pressure and room temperature, the rate of chemical reactions in most combustible mixtures is negligibly small. A perfectly premixed fuel–oxidant mixture ignites only under conditions when either there is a spark, which can heat considerably the mixture, or the hot wall of the vessel can heat the mixture or some active centers initiating the reaction are introduced. If these conditions are realized at some place, then a strongly exothermic reaction, such that a quantity of the released heat sufficiently raises the temperature, the reaction which started at some location raises the temperature of the surrounding mixture and once having begun, it will extend to the surrounding gaseous mixture. This process is called the combustion of a gas mixture. The combustion of a gas mixture is inevitably accompanied by gas motion. Therefore, combustion is the process representing a chemical phenomenon coupled with gas dynamics. This means that the nature of the combustion process is described by a solution of chemical kinetics equations together with equations of gas dynamics.

A self-sustained wave of the exothermic chemical reaction spreading through a homogeneous combustible mixture is known to occur either as a subsonic *deflagration* (flame) or supersonic *detonation* wave. From the mathematical point of view, since both deflagration and detonation are stationary solutions of the complete system of equations, they appear to be stable attractors each being linked to its own base of initial data. The deflagrative mode of a chemical reaction wave propagates at a velocity substantially below that of sound and is associated with molecular thermal

M. A. Liberman, *Combustion Physics*,
https://doi.org/10.1007/978-3-030-85139-2_3

conductivity or diffusion. During thermal propagation of a deflagration wave, the heat released in the chemical reaction is transmitted to neighboring parts of the unburned reactant gas by thermal conductivity, heats the unreacted mixture, and initiates chemical reactions in the neighboring layer of the unburned mixture. In chain propagation, the reaction in neighboring layers is driven by the diffusion of active radicals. Also, it is possible that diffusion and thermal conductivity may act together. A detonation wave is caused by rapid compression and heating of the unreacted mixture in a shock wave, so the reaction can proceed at a sufficiently high rate. In turn, the heat released in the chemical reaction supports the intensity of the shock wave, thereby ensuring that it will propagate over large distances.

The activation energy of many reactions is so large that the reaction rate at the room temperature may be taken as zero. On the contrary, the increase of the fuel temperature even by a factor of 2 may lead to the increase of the reaction rate by many orders of magnitude. In the case of a strongly exothermic reaction when a considerable energy release is involved, a relatively slight increase of the temperature at some region ignites the reaction, which eventually extends over the whole gas. Once a reaction is ignited, it can propagate in a self-supporting regime.

Typically, the burning proceeds in a layer, which is very thin compared to length scales of the problem, for example, compared with the size of the combustion chamber. This layer is the flame front consisting of the reaction zone and the preheated layer. The thickness of the flame front is by the order of the distance which the heat released in the chemical reaction propagates during the characteristic reaction time. The typical width of the laminar flame front is $L_f \approx 10^{-2} - 10^{-4}$cm. In the first approximation, the propagation of a flame can be treated as a propagation of a geometrical surface of zero thickness, which separates the unburned mixture and combustion products, and which is moving relative to the unburned gas with a known velocity. This velocity is called the *normal propagation velocity of the laminar flame* or is also referred to as the *laminar flame speed*. The thermal theory of flame propagation was first suggested by Mallard and Le Chatelier [1], who postulated that the heat transfer controls the flame propagation, and the flame consists of two zones separated at the surface where the next layer of combustible mixture ignites. Futher development of the theory of laminar flame propagation is due to Lewis and von Elbe [2, 3] and Zel'dovich and Frank-Kamentskii [4–6].

The normal propagation velocity of a flame determines the volume of fuel mixture consumed per unit time per unit surface has the dimensionality of linear velocity (cm/s). For a curved flame front, the normal combustion velocity characterizes the velocity of the flame front relative to the unburned mixture in the direction normal to the surface of the flame front. The magnitude of the normal propagation velocity of a flame is determined by the kinetics of chemical reaction and by transport processes inside the flame front: thermal conductivity and diffusion. Because of poor knowledge of the reaction kinetic coefficients, for practical applications usually the magnitude of the normal flame velocity is taken from experimental measurements rather than calculated theoretically. Such experimental measurements can be easily performed with high precision.

Considering a flame front as a geometrical surface of zero thickness, it is convenient to choose a coordinate system co-moving with the flame front, so the flame front is at rest in this coordinate system. The unburned fuel mixture moves into the flame front with velocity u_n (upstream flow), and the combustion products flow out of the flame front with velocity u_b (downstream flow). Clearly, the upstream velocity u_n is equal to the normal velocity of the flame relative to the motionless gas, $u_n = U_f$. The relationship between u_n and u_b for a plane flame front steams from the condition of conservation of the total mass flux. The total mass of gas mixture entering the flame per unit area of the flame front must be equal to the mass of combustion products leaving this surface downstream:

$$\rho_u u_n = \rho_b u_b, \tag{3.1.1}$$

where ρ_u and ρ_b are the densities of the initial unburned mixture and combustion products. Thus,

$$u_b = \frac{\rho_u}{\rho_b} u_n. \tag{3.1.2}$$

Obviously, the velocity u_b of the burned products is greater than u_n since the gas is heated and expands during combustion, which means that the flame front propagating in the motionless gas mixture in a laboratory coordinate system will inevitably involve gas in the motion. If the gas mixture is at rest ahead of the flame front, the gas of the burned products will be involved in a motion and products move with velocity $u_b = U_f(\rho_u/\rho_b)$ in the coordinate system co-moving with the flame, and with velocity $u_{bL} = U_f\left(\frac{\rho_u}{\rho_b} - 1\right)$ in the laboratory coordinate system. The overall picture of the gas dynamics ahead and behind the flame depends on the boundary conditions of the particular problem. For example, since the burning mixture undergoes expansion ($\rho_b < \rho_u$), the flame propagating from the closed end of a tube will move in the laboratory system, relative to the tube walls, with velocity $U_{fl} = U_f(\rho_u/\rho_b)$, while the gas mixture ahead of the flame moves with velocity $u_u = U_f\left(\frac{\rho_u}{\rho_b} - 1\right)$ relative to the tube walls. Thus, the flame itself moves relative to the unburned gas with velocity U_f. Because of the expansion, the burning process becomes strongly coupled to the gaseous flow; which leads to different flame instabilities, generation of pressure waves, etc.

As we shall see, the pressure varies as $\Delta P \approx P(U_f/a_s)^2$, where $M = u/a_s \ll 1$ is the Mach number. Since the normal velocity of a flame, U_f, is much smaller than sound speed, a_s, the pressure is practically constant over the gas volume, $\Delta P/P \sim M^2 \ll 1$. This means that, with high precision, the pressure is constant and we can rewrite (3.1.1) as

$$\frac{u_b}{u_n} = \frac{\rho_u}{\rho_b} = \frac{\mu_u}{\mu_b}\frac{T_b}{T_u}, \tag{3.1.3}$$

where T_u and T_b are the temperatures of the initial unburned mixture and the temperature of combustion products (the adiabatic flame temperature); μ_u and μ_b are the molecular weight of the unburned mixture and combustion products. For typical combustion reactions $\mu_u \approx \mu_b$, and (3.1.3) can be written as

$$\frac{u_b}{u_n} = \frac{\rho_u}{\rho_b} = \frac{T_b}{T_u}. \tag{3.1.4}$$

In the flame, the temperature of combustion products increases by a factor of 6–12, so that $\rho_u/\rho_b = u_b/u_n \approx 6-12$. The expansion coefficient, $\Theta = \rho_u/\rho_b$, is one of the most important flame parameters. It is convenient to introduce the combustion mass flux $\rho_u u_n$, which is the product of the gas density and velocity. This quantity has the dimensions g/cm^2 s and represents the mass of material burned per square centimeter per second in the flame front. It is also convenient to refer the rate of heat release to the unit surface area of the flame. The corresponding quantity has the dimensions cal/cm^2 s.

We consider the simplest case of the steady propagation of a 1D, planar flame relative to a combustible mixture. Following Mallard and Le Chatelier, the flame can be in two zones. In the first, the preheat zone, the gas mixture is heated by thermal conduction and the temperature rises up to ignition level. From this point is much narrower the chemical reaction zone, where chemical energy is released and temperature reaches the final value $T = T_f$ in the combustion products. Let δ_T and δ_{ch} be widths of the preheat and chemical zones, T_0 be the initial temperature of the mixture, and T_{ig} be the ignition temperature at the front edge of the chemical zone. Then the energy balance equation can be written as

$$U_f \rho_u C_P (T_{ig} - T_0) \approx \lambda (T_f - T_{ig})/\delta_{ch}. \tag{3.1.5}$$

From (3.1.5), the flame speed is

$$U_f \approx \frac{\chi (T_f - T_{ig})}{(T_{ig} - T_0)\delta_{ch}}, \tag{3.1.6}$$

where $\chi = \lambda/\rho C_P$ is the coefficient of thermal diffusivity. Introducing the reaction time as $\tau_{ch} \approx \delta_{ch}/U_f$, we obtain for the flame velocity

$$U_f = \sqrt{\frac{\chi (T_f - T_{ig})}{\tau_{ch}(T_{ig} - T)_0}}. \tag{3.1.7}$$

Though this formula gives only a qualitatively correct value for the laminar flame speed, it demonstrates important functional dependences. The laminar flame speed is proportional to $\sqrt{\chi}$, and the smaller the τ_{ch} (higher highly reactive mixtures), the larger the flame speed τ_{ch}.

3.2 Velocity and Thickness of Laminar Flames

In this section, we shall discuss the physical basis of the flame propagation, which serves as the foundation of the concept of a reaction wave propagation.

For the sake of simplicity, we shall consider premixed flames propagating in a gaseous combustible mixture, which is the opposite case to diffusion flames. This means that all components necessary for the reaction are present in the fuel mixture from the very beginning and in order to start the reaction one has only to heat the mixture. In the case of premixed flames, the burning rate is not controlled by the diffusion process, still even for premixed flames the fuel diffusion is an important factor of propagation together with thermal conduction. Particularly, when the fuel diffusion is stronger than the thermal conduction, then the thermal-diffusion instability of a planar stationary flame may develop. We shall consider mostly the situations of equally strong fuel diffusion and thermal conduction, when the hydrodynamic flame properties can be elucidated.

A remarkable feature of premixed combustion is a very strong dependence of the reaction rate on temperature expressed by the Arrhenius law for the reaction rate, $\dot{\omega} \propto \exp(-E_a/RT)$, for high activation energy, $E_a/RT \gg 1$. The activation energy of many reactions is so large that the reaction rate at the room temperature may be taken equal to zero. On the contrary, the increase of fuel temperature even by a factor of 2 may lead to the increase of the reaction rate by 10–12 orders of magnitude. In the case of a strongly exothermic reaction when a considerable amount of energy releases, relatively slight increase of the temperature at some region ignites the reaction, which eventually extends over the whole gas. It is obvious that because of the strong dependence of the reaction rate on the temperature, the chemical reaction proceeds nonuniformly inside a flame front, and the reaction will be mainly concentrated in the portion of the flame front adjacent to the hot combustion products. Where the temperature is lower, the reaction proceeds at a much lower rate and can generally be neglected.

The transport processes dominate because of the rapid changes in the temperature and concentrations of the reactants through the flame front. Once a reaction is ignited, it can propagate in a self-sustained regime. The burned matter has a larger temperature, and thermal conduction transfers energy from the hot burned matter to the cold fuel: the temperature of the fuel close to the burned matter increases, and the reaction in this fuel develops faster until another portion of the fuel is burned. The released energy is transferred by thermal conduction to the next fuel layer and so on, resulting in propagation of the reaction front.

Flame velocity and thickness may be estimated on the basis of a simple dimensional analysis [7]. Let the burning process be characterized by some characteristic chemical reaction time τ_{ch}, which is the time of the heat release in a chemical reaction. This value is much larger than the average time between molecule collisions because only an exponentially small number of molecules have sufficiently large energy necessary to overcome the potential barrier to participate in the reaction. Then the only combination of velocity dimensionality that may be constructed out of the thermal diffusivity coefficient $\chi = \lambda/\rho_u C_P$ and the reaction time τ_{ch} is

$$U_f \propto \sqrt{\lambda/\rho_u C_P \tau_{ch}} = \sqrt{\chi/\tau_{ch}}. \tag{3.2.1}$$

This is consistent with the previously obtained formula (3.1.7). Using the dimensional analysis, we can also estimate the flame thickness. The distance heated by thermal conduction can be evaluated as

$$L_f = U_f \tau_{ch} = \sqrt{\chi \tau_{ch}}. \tag{3.2.2}$$

Combining Eqs. (3.2.1) and (3.2.2), we obtain

$$L_f U_f \approx \chi. \tag{3.2.3}$$

The exact magnitude in these formulas depends on the definition since both a thermal conduction and density depend on the gas temperature. We can choose the value $\lambda = \lambda_u(T_f)$ as the definition for L_f. The important conclusion from these simple estimates is that the shorter the reaction time, τ_{ch}, and the stronger the thermal conduction, the faster the flame propagates and the thinner the flame front.

Let us compare the laminar flame velocity with the sound speed. From kinetic theory, the thermal diffusivity coefficient can be expressed through the sound speed a_s (that is about the thermal velocity of molecules) and the mean free time τ_{coll}, which is the average time between molecule collisions, $\chi \approx a_s^2 \tau_{coll}$. Then we obtain $U_f/a_s \approx \sqrt{\tau_{coll}/\tau_{ch}}$. $\tau_{coll} \ll \tau_{ch}$ since only a very small fraction of colliding molecules reacts because of the large potential barrier of a reaction (large activation energy). This means that the flame velocity is much smaller than the sound speed, $U_f/a_s \ll 1$. Typical velocities of the flame at the normal laboratory conditions ($P = 1$ bar, $T = 300$ K) and sound speeds for some mixtures are shown in Table 3.1.

Let us designate ℓ to be the average molecule mean free path. The reaction time τ_{ch} is the product of the average number Z of molecule collisions resulting in the reaction and the mean collision time ℓ/a_s between two collisions:

$$\tau_{ch} \approx Z(\ell/a_s). \tag{3.2.4}$$

Substituting this dependence in the expression for flame velocity, we obtain

$$U_f \approx a_s/\sqrt{Z}, \tag{3.2.5}$$

and for the characteristic width of the flame

$$L_f \approx \ell\sqrt{Z}. \tag{3.2.6}$$

Equation (3.2.5) is similar to Einstein's expression for the mean diffusion velocity in a given direction with Z being the total number of collisions experienced by the

Table 3.1 Velocities of flames at normal conditions ($P = 1$ atm, $T_u = 300$ K)

Fuel	Combustible mixture	U_f (cm/s)	a_s (m/s)	T_b (K)	$\Theta = \rho_u/\rho_b$
Methane CH_4	6% CH_4 + 94%Air	5	353.9	1850	6.3
Methane CH_4	6.84%CH_4–Air	15		1900	6.3
Methane CH_4	7.76%CH_4–Air	27		2000	6.7
Methane CH_4	9.5%CH_4–Air	43		2250	7.5
Methane CH_4	11.6%CH_4–Air	25		2100	7.5
Methane CH_4	10%CH_4–O_2	80		2200	7.3
Methane CH_4	25%CH_4–O_2	304		3000	10
Methane CH_4	40%CH_4–O_2	305		3000	12.7
Methane CH_4	50%CH_4–O_2	112		2650	13.3
Propane C_3H_8	2.86%C_3H_8–Air	28		1870	6.4
Propane C_3H_8	3.64%C_3H_8–Air	35		2170	7.5
Propane C_3H_8	4.02%C_3H_8–Air	40		2240	7.8
Propane C_3H_8	5.5%C_3H_8–Air	17		2030	7.5
Propane C_3H_8	5.9%C_3H_8–Air	5.9		1830	6.9
Propane C_3H_8	7.4%C_3H_8–O_2	240		2500	9
Propane C_3H_8	19.3%C_3H_8–O_2	320		3000	13.6
Propane C_3H_8	21.9%C_3H_8–O_2	235		2300	14.3
Propane C_3H_8	22.7%C_3H_8–O_2	190		2750	14.5
Ethane C_2H_6	$C_2H_6 + 3.5\ O_2$	44	316	3086	10.3
Butane C_4H_{10}	$C_4H_{10} + 6.5\ O_2$	45		3101	10.3
Benzene C_6H_6	$C_6H_6 + 7.5\ O_2$	48		3136	
Hydrogen H_2	20%H_2–Air	100	408.7	1910	5.7
Hydrogen H_2	30% H_2–Air	195		2500	6.5
Hydrogen H_2	40%H_2–Air	265		2240	6.5
Hydrogen H_2	57%H_2–Air	190		1850	5.5
Hydrogen H_2	$H_2 + 0.5O_2$	900	539.5	3072	10
Hydrogen H_2	$H_2 + 0.5O_2 + 0.5N_2$	520	408	2350	7.8
Acetylene C_2H_2	$C_2H_2 + 2.5O_2$	1330	331	3342	11.1
Carbon Monoxide	$CO + 0.5O_2$	29	337	2977	9.9
Ethylene/Oxygen	$C_2H_4 + 3O_2$	486	328.2	3112	10.6

particle during the diffusion time and Eq. (3.2.6) is similar to Einstein's formula for the mean distance covered by a molecule after Z collisions. It is clear from the molecular picture of flame propagation: a fuel molecule undergoes Z collisions in the combustion zone and during this time it covers a distance L_f at mean velocity U_f. The difference between a diffusion process and the propagation of a flame is that the mean velocity of diffusing particles falls with time (as $l/\sqrt{t}$) while the velocity of a flame is constant in time. The point is that the motion of a flame does not involve displacement of the molecules along the entire path followed by the flame front. A given molecule moves within the combustion zone for only an extremely short time and undergoes Z collisions on the average, after which it reacts and, thereby, creates the conditions necessary for the next molecule to pass the same way. Since according to (3.2.6) $L_f \gg \ell$, the flam propagation and combustion dynamics can be described by the hydrodynamic equations, including thermal conduction and diffusion.

Let us consider qualitatively the propagation of the reaction wave from another point of view. In the co-moving coordinate system, the flame is at rest and we consider the reaction zone as a surface located at $x = 0$ with fixed temperature T_a. Then, the temperature distribution is defined by the equation of thermal conduction

$$\rho u C_p \frac{dT}{dx} = \lambda \frac{d^2T}{dx^2}, \tag{3.2.7}$$

with the boundary conditions $T(-\infty) = T_0$ and $T(0) = T_f$. For simplicity, all the material properties are assumed to be constant and independent of the temperature. Obviously, Eq. (3.2.7) is applicable only to the region $x < 0$ where there is no source of heat and the reaction is very slow due to strong dependence on temperature. For $x > 0$, the combustion products are at a constant temperature T_f. Equation (3.2.7) can be readily integrated by introducing a new variable $z = dT/dx$:

$$z = \frac{\lambda}{\rho u C_p} \frac{dz}{dx}, \tag{3.2.8}$$

which is easily integrated, and the solution of (3.2.7) satisfying the boundary conditions is

$$T = T_0 + (T_f - T_0)e^{ux/\chi}. \tag{3.2.9}$$

The solution (3.2.7) shows that we can choose the distance over which the heating rises by a factor of e as the scale length for the heating zone with the velocity of the flow entering the flame front being equal to the flame velocity U_f. Then we come to the already known formula (3.2.3)

$$L_f U_f = \chi. \tag{3.2.10}$$

Let us estimate the flame width for typical laboratory flames: a slow-reactive mixture (9% methane in air) and the fast-reactive hydrogen–oxygen flame ($2H_2+O_2$). Taking velocities from Table 3.1, and setting $\chi = 0.85\,\mathrm{cm^2/s}$, we obtain $L_f = 0.0009\,cm$ for the hydrogen–oxygen flame (experimental value is $L_f = 0.0027$ cm) and $L_f = 0.027$ cmfor methane–air flame. In both cases, the width of the heating zone, which can be regarded as the thickness of the flame, is rather small. The typical thickness of the combustion zone ranges by the order of magnitude from 10^{-2} for the flame in highly reactive mixtures to 10^{-4} cm for low reactive mixtures.

The reaction time is of the same order as the residence time of the reactive mixture in the preheat zone. An order of magnitude of the residence time of the reactive mixture in the flame can be evaluated by dividing the width of the preheat zone by the flame speed: $t_{L_f} \approx 10^{-3}$ s for CH_4 and $t_{L_f} = 10^{-7}$ s for $2H_2 + O_2$. This time is many times greater than the mean collision time of the molecules. Because of the large activation energy, a very large number of molecular collisions takes place in a flame for each effective collision that leads to the chemical reaction. It should be emphasized that the temperature and composition of the mixture vary in a flame for other reasons besides the chemical reaction. They also vary in the preheat zone where there is no chemical reaction: the temperature changes because of the thermal conductivity of the gas and the composition may change because of diffusion. In a flame, the chemical reaction takes place in a mixture which has already been heated and whose composition has already been changed. The typical structure of a premixed flame is shown in Fig. 3.1, and a structure of a diffusion flame is shown in Fig. 3.2. For diffusion flames, the chemical transformation is focused in a narrow zone near the peak temperature. Diffusion continuously supplies reagents to the reaction zone and redistributes them there. Completion of the reaction and thermodynamic equilibrium is achieved asymptotically, where the diffusive fluxes practically become equal to zero.

The flame is not the only possible self-sustained regime of reaction propagation. A reaction can propagate also in a fast-supersonic regime of detonation, which will be discussed later. In case of a detonation, the reaction is induced by a shock wave compressing and heating the fuel. The burning mixture expands and acts like a piston pushing a leading shock and supporting the detonation. The transitions from the slow deflagration regime of flame propagation to a detonation regime are observed quite often in experiments. In industrial combustion and for safety reasons, this can be a very undesirable phenomenon.

We can equally consider another type of flame instead of a chemical one, where energy release may be supplied from other sources. An example is the thermonuclear burning of white dwarfs, which results in a Supernova Type Ia explosion. In this case, the thermonuclear flame expands from the center of the star, and the energy

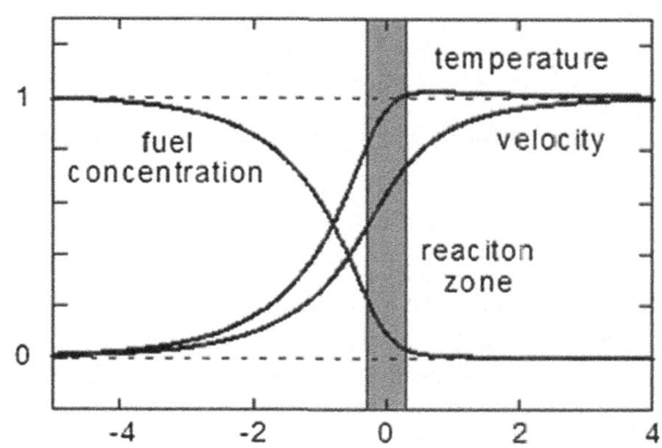

Fig. 3.1 Structure of the flame front in premixed combustion

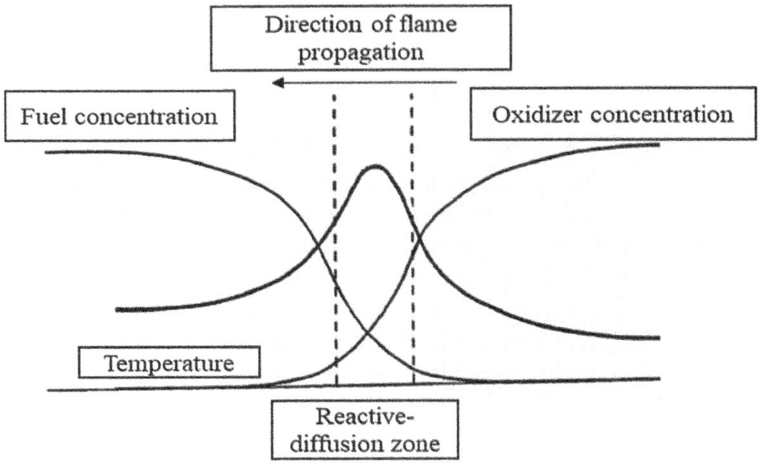

Fig. 3.2 Structure of a diffusion flame

source is thermonuclear reactions. Another example is the laser radiation absorbed by plasma layers close to the critical surface of a target used for experiments on inertially confined fusion. For all these examples, the released energy is transported by thermal conduction, and the flame propagates relatively slow in comparison with the sound speed. Therefore, a flame is defined as a subsonic regime of reaction wave propagation, called a *deflagration*.

3.3 Temperature and Concentration Distributions

We saw that since the flame velocity is much smaller than sound speed, the conservation of momentum flux yields pressure to be nearly constant. Also, conservation of the energy yields constancy of the enthalpy. Below, we will use subscripts u and b for variables in the unburned fresh mixture and for products. Thus, we can write

$$P_u = P_b, \quad H_u = H_b. \tag{3.3.1}$$

Assuming for simplicity constant specific heat we obtain

$$H_u = H_{u0} + c_{pu}T_u = H_b = H_{b0} + c_{pb}T_b, \tag{3.3.2}$$

where c_{pu} and c_{pb} are specific heats of the fuel and combustion product at constant pressure. Since $Q = H_{u0} - H_{b0}$ is a heat of the reaction, this implies that the temperature of the combustion products is determined only by the initial enthalpy of the fuel mixture ():

$$T_b = \frac{c_{pu}}{c_{pb}} T_u + \frac{1}{c_{pb}} Q, \quad Q = H_{u0} - H_{b0}. \tag{3.3.3}$$

With an account of (3.1.3), we also obtain

$$\frac{\rho_u}{\rho_b} = \frac{c_{pu}}{c_{pb}} \left(1 + \frac{Q}{c_{pu} T_u}\right). \tag{3.3.4}$$

The structure of a laminar flame is described by a system of thermal conductivity and diffusion equations. We shall examine the simplest case of a plane flame front and a chemical reaction described by a one-step Arrhenius model for a stoichiometric mixture. In the coordinate system co-moving with the flame front, the change of the heat flux, J_T, and of the mass flux, J_i, are given by equations:

$$\frac{d}{dx} J_T = Q W(a_i, T), \tag{3.3.5}$$

$$\frac{d}{dx} J_i = -\frac{\nu_i}{\nu_1} W(a_i, T). \tag{3.3.6}$$

Here, Q is the heat of reaction, a_i is the mass concentrations of the reagents, and $W(a_i, T)$ is the rate of the reaction. The chemical reaction rate is calculated relative to one mole of the substance denoted by the subscript 1. We shall assume that this substance is deficient in the initial mixture. The fluxes of heat and mass are made up of convective and molecular transport terms:

$$J_T = -\lambda \frac{dT}{dx} + \rho u c_p T, \tag{3.3.7}$$

$$J_i = -\rho D_i \frac{da_i}{dx} + \rho u a_i, \tag{3.3.8}$$

where D_i is the coefficient of diffusion. Using the equation of continuity of the mass flux, $\rho u = \rho_0 u_u = \rho_b u_b = const$, Eqs. (3.3.7) and (3.3.8) can be rewritten as

$$\rho u c_p \frac{dT}{dx} = \frac{d}{dx} \lambda \frac{dT}{dx} + Q W(a_i, T), \tag{3.3.9}$$

$$\rho u \frac{da_i}{dx} = \frac{d}{dx} \rho D_i \frac{da_i}{dx} - \frac{\nu_i}{\nu_1} W(a_i, T), \tag{3.3.10}$$

where ν_i is stoichiometric coefficients.

We assume for simplicity that the diffusion coefficients are the same in the unburned mixture and in the products that is true if the molecular weights of the reactants and products are almost the same. In addition, we assume that the Lewis number

$Le = D_i/\chi = 1$ (the ratio of the thermal diffusivity to the diffusion coefficients) over the entire temperature interval, which is true for many gases. Multiplying the diffusion equation (3.3.10) for the concentration a_i by $1/\nu_i$, and the diffusion equation for a_j by $1/\nu_j$, and subtracting one from the other, we obtain a linear equation of diffusion (without the term with the reaction rate function) for the linear combination $(a_i/\nu_i) - (a_j/\nu_j)$ of the densities. The solution of this equation, which is finite for all $+\infty > x > -\infty$, with the conditions of stoichiometric mixture for a one-step chemical model, gives the linear relation between concentrations:

$$\frac{a_{i0} - a_i}{\nu_i} = \frac{a_{j0} - a_j}{\nu_j}, \tag{3.3.11}$$

where the subscript 0 denotes the concentrations of the components in the initial mixture.

Thus, the concentrations of all the reagents can be calculated using (3.3.11) if the concentration of one of them is known. Therefore, Eqs. (3.3.9) and (3.10) can be written as two equations for only one reagent (we omit the subscript),

$$\rho u c_p \frac{dT}{dx} = \frac{d}{dx}\lambda\frac{dT}{dx} + QW(a, T), \tag{3.3.12}$$

$$\rho u \frac{da}{dx} = \frac{d}{dx}\rho D\frac{da}{dx} - W(a, T). \tag{3.3.13}$$

With the help of Eq. (3.3.11), in these equations the reaction rate is written as a function of the temperature and of only one of the concentrations. Multiplying (3.3.13) by Q, adding it to the thermal conductivity equation, for $\chi = D$, we obtain the equation for enthalpy, which does not include the reaction rate:

$$\rho u \frac{dH}{dx} = \frac{d}{dx}\left(\frac{\lambda}{c_p}\frac{dH}{dx}\right). \tag{3.3.14}$$

The total enthalpy of the system is the sum of the chemical and thermal energies per unit mass of gas for constant pressure. The only restricted for all x solutions of Eq. (3.3.14) is

$$H = H_0 = Qa_0 + \int_0^{T_0} c_p T = \text{const} \tag{3.3.15}$$

where the constant is defined by conditions for the initial state of the fuel mixture:

$$H_0 = Qa_0 + \int_0^{T_0} c_p T. \tag{3.3.16}$$

The idea that the sum of the chemical and thermal energies is constant in a flame was first proposed as a hypothesis by Lewis and von Elbe [2] for a chain reaction initiated by the diffusion of active centers. However, as we have shown, the condition that the total enthalpy is constant in a stationary flow depends on the relationship between the coefficients of diffusion and thermal conductivity in the medium, rather than on the mechanism of the reaction.

For a constant specific heat, the constancy of the total enthalpy within the flame can be written in the form

$$Qa + c_p T = Qa_0 + c_p T_0 = c_p T_b, \tag{3.3.17}$$

that is, the temperature distribution and the distribution of concentrations are similar:

$$\frac{a_0 - a}{a_0} = \frac{T - T_0}{T_b - T_0}. \tag{3.3.18}$$

The temperature distribution and the concentration distribution of other reagents, as well as concentrations of combustion products, are similar in exactly the same way since the concentration of one of the reagents is linearly related to others according to Eq. (3.3.11). The similarity also exists when the rate of a simple reaction depends on the concentrations of the end products as in the reversible reaction or autocatalysis by the end products. When the reaction is complex, stoichiometric relations do not relate to the concentrations of intermediate products and these similarities do not exist. However, even in this case, when all the transport coefficients are equal ($D_i = \chi$), the total enthalpy of the system is still constant, but this fact cannot be used to derive a unique temperature dependence for each of the substances involved in the reaction.

The simplest case, in which the propagation of a flame is described by the system of Eqs. (3.3.9) and (3.3.10), is that when the reaction rate depends on the concentration of only one of the reacting substances. This situation is often encountered during the propagation of flames in fuel mixtures whose composition deviates strongly from being stoichiometric. In such mixtures, the reaction rate is limited by the concentration of one substance that is deficient. When the diffusion and thermal conductivity coefficients for this substance are equal, its concentration and the temperature have similar distributions. The deficient substance may be viewed as the carrier of chemical energy since the ultimate temperature reached by the mixture during combustion depends on its concentration.

As was shown before, the conservation of enthalpy

$$H_0 = Qa_0 + \int_0^{T_0} c_p T = \int_0^{T_b} c_p T \tag{3.3.19}$$

implies that the final temperature of the combustion products is determined only by the initial enthalpy of the fuel mixture; see Eq. (3.3.3). When $D \neq \chi$ within the flame the sum of the chemical and heat energies is not conserved, but we can predict the direction of the shift for the distribution of temperature and concentrations. For example, in the lean mixture of hydrogen with air, the diffusion coefficient of hydrogen molecules (the carrier of chemical energy) is on the order of magnitude greater than the thermal conductivity of the mixture. Therefore, hydrogen diffuses from the preheat zone more rapidly than the mixture is heated by molecular thermal conductivity. As a result, for a hydrogen–air mixture (when $D > \chi$) the total enthalpy has a minimum in the preheat zone. On the other hand, in a lean mixture of a high-molecular-weight hydrocarbon (for example, benzene–air), the thermal conductivity exceeds the diffusion coefficient of benzene vapor ($\chi > D$) and heating of the mixture in the preheat zone is more intense than the diffusion of benzene into the reaction zone. Therefore, the total enthalpy has a maximum in the preheat zone. Typical distributions of reagent concentration, temperature, and total enthalpy in a flame for various relationships between the coefficients of diffusion and thermal conductivity are shown in Fig. 3.3.

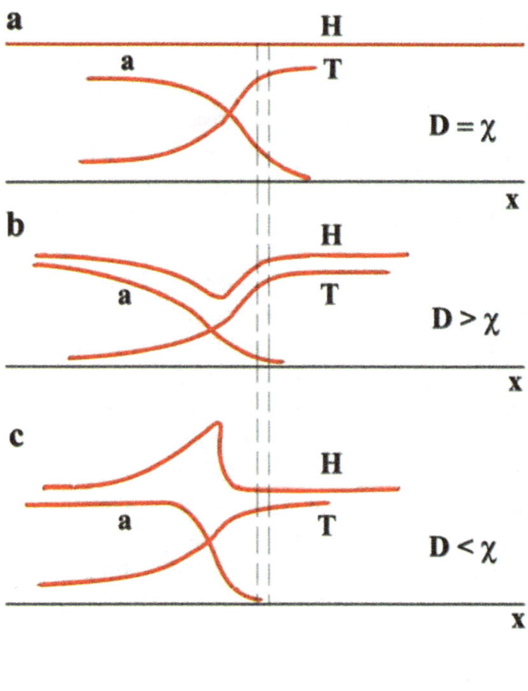

Fig. 3.3 Schematic distributions of reagent concentration, temperature, and total enthalpy in a flame front for various relationships between the diffusion coefficient and the thermal conductivity

3.4 Zel'dovich–Frank-Kamentskii Theory: Laminar Flame Speed

To derive the formula for the laminar flame speed, Zel'dovich and Frank-Kamentskii [2–4] adopted the flame structure of the flame consisting of the preheat and reaction zones proposed by Mallard and LeChatelier [1]. However, in addition to the energy conservation, they used also the species-conservation equation. In a simple case, when the diffusion and thermal conductivity coefficients are equal and the concentration and temperature are expressed by the linear relation between them, we can reduce the two differential equations for the fuel balance and energy to a single differential equation [4–6]. To do this, let us express the concentration in terms of the temperature in the formula for the chemical reaction rate and write the reaction rate $W(a, T)$ as a function of temperature only: $W[a(T), T] = W(T)$. Then we arrive at the equation for temperature:

$$\rho u c_{\mathrm{p}} \frac{dT}{dx} = \frac{d}{dx}\left(\lambda \frac{dT}{dx}\right) + QW(T). \tag{3.4.1}$$

This is the second-order equation for function $T(x)$ containing the mass flux ρu. This means that the normal flame velocity is an *eigenvalue* of the problem, which must be found in solving the problem.

Here, we shall obtain an approximate solution for the flame (asymptotic solution) using the fact of the very strong dependence of the reaction rate on temperature, that is, a rapid chemical reaction takes place over a narrow range of temperatures close to the final temperature achieved in the flame. The heat released in the reaction is mainly carried from the reaction zone to the preheat zone by thermal conduction for heating the unburned mixture from the initial temperature to the temperature of the reaction zone. The temperature derivative in the narrow chemical reaction zone changes from a finite value on the side facing the reactive mixture to zero on the side facing the combustion products where thermal and chemical equilibria are reached. This rapid change in the temperature gradient over a short distance means that the contribution of the thermal conductivity is large. Because of this, the contribution of the convective heat flux may be neglected in the reaction zone compared to that of the thermal conductivity in the heat conduction equation. Hence, in the reaction zone, we can take the "truncated" equation without the convective heat flux

$$\frac{d}{dx}\left(\lambda \frac{dT}{dx}\right) + QW(T) = 0. \tag{3.4.2}$$

This equation can be integrated using the independent variable T and setting $\lambda dT/dx = z$,

$$\frac{d}{dx}\lambda \frac{dT}{dx} = \frac{z}{\lambda}\frac{dz}{dT} \tag{3.4.3}$$

$$z\frac{dz}{dT} + \lambda Q W(T) = 0. \tag{3.4.4}$$

In the reaction products $a = 0$ and the temperature is nearly constant, being equal to the combustion temperature T_b, so that the boundary conditions behind the flame are

$$T = T_b, \quad \lambda\frac{dT}{dx} = z = 0. \tag{3.4.5}$$

Integrating Eq. (3.4.4) with respect to the variable temperature T up to T_b (the thermal conductivity is constant over the reaction zone, $\lambda = \lambda_b = \text{const}$), we obtain the thermal flux from the reaction zone into the heating zone

$$\lambda_b \frac{dT}{dx}\bigg|_{T_b} = \sqrt{2\lambda_b Q \int_{T}^{T_b} W(T)dT}. \tag{3.4.6}$$

The integration in Eq. (3.4.6) can be extended to the initial temperature $T_0 = T_u$ since at low temperatures the reaction rate is negligibly small, and the integral can be regarded as equal to zero in the low-temperature region: the heat release function will contribute to the integral only at high temperatures. The actual integration region is on the order of the characteristic temperature interval (for an Arrhenius heat release function). The function $W(T)$ includes the concentration dependence $a(T)$, so that $W(T_b) = 0$. The amount of heat released per unit time by the chemical reaction and removed by thermal conductivity is equal to the chemical energy stored in the flowing fuel mixture (in co-moving with the flame coordinate system, the flame is at rest while the fuel mixture moves at a speed equal to the normal propagation velocity of the flame). Thus, we can write

$$\sqrt{2\lambda_b Q \int_{T_0}^{T_b} W(T)dT} = \rho_u c_p u_n Q a_0. \tag{3.4.7}$$

This equation gives the formula for the normal propagation velocity of a laminar flame

$$U_f = u_n = \frac{1}{\rho_0 c_p a_0}\sqrt{\frac{2\lambda_b}{Q}\int_{T_0}^{T_b} W(T)dT}. \tag{3.4.8}$$

We used here subscript "0" for the variables in the unburned mixture.

Formula (3.4.8) shows that the flame speed depends on the integral of the heat release function and, therefore, it must be related to the order of the chemical reaction, its activation energy, and other kinetic and physical–chemical characteristics of the fuel mixture. The only assumption used in deriving (3.4.8) was that about a strong temperature dependence of the rate of heat release, which is true for large activation energy.

Since $W(T)$ depends strongly on the temperature according to the Arrhenius law, the integral in Eq. (3.4.8) can be estimated as the product of the maximum value $W = W_{max}$ and the effective width of the temperature interval equal to $RT_b^2/E \equiv ZT_b$, where $Z = RT_b/E \ll 1$ is called the Zel'dovich number. Putting $a_0 = 1$, we obtain

$$U_f = u_n \approx \frac{1}{\rho_0 c_p}\sqrt{\frac{2\lambda_b W_{max} Z T_b}{Q}}. \tag{3.4.9}$$

Introducing the chemical reaction time τ_{ch} as the time over which all the heat contained in the initial mixture is released at the maximum heat production rate, and using the heat balance equation

$$Q W_{max} \tau_{ch} \simeq \rho_0 c_p T_b, \tag{3.4.10}$$

we can write the expression for the flame velocity as

$$U_f \approx \sqrt{\frac{2\lambda_b Z}{\rho_0 c_p T_b}\frac{1}{\tau_{ch}}} = \sqrt{\frac{2Z}{T_b}} \cdot \sqrt{\frac{\chi_b}{\tau_{ch}}}. \tag{3.4.11}$$

The relationship $U_f \propto \sqrt{\chi_b/\tau_{ch}}$ was obtained earlier (Eq. (3.2.1)) from dimensional considerations. It is useful to obtain the expression for the normal velocity of a laminar flame in a slightly different form. Integrating (3.4.2), we obtain similar to Eq. (3.4.6)

$$\lambda_b \frac{dT}{dx}\bigg|_{T_b} = \sqrt{2\lambda_b Q \int_T^{T_b} W(T) dT}. \tag{3.4.12}$$

At the same time, integrating Eq. (3.4.1) in the thermal conduction zone, where we can omit the reaction term, we consider the integral of the equation

$$\rho u c_p \frac{dT}{dx} = \frac{d}{dx}\left(\lambda \frac{dT}{dx}\right), \tag{3.4.13}$$

which is

$$\lambda \frac{dT}{dx} = \rho c_P u T + const. \tag{3.4.14}$$

With the boundary condition $T = T_0$ and $dT/dx = 0$ at $x = -\infty$, we obtain

$$\lambda \frac{dT}{dx} = \rho c_P u(T - T_0). \tag{3.4.15}$$

The continuity of the heat flux leads to the equation similar to (3.4.12)

$$\rho c_P u(T - T_0) = \sqrt{2\lambda Q \int_T^{T_b} W(T)dT}, \tag{3.4.16}$$

where T is the temperature at the end of the thermal conduction zone.

Considering that for large activation energy the maximum reaction is very close to $T = T_b$ and the reaction zone is very narrow, we can take $T = T_b$ on the left-hand side of (3.4.16). To estimate the integral in the right-hand side of (3.4.16), we expand the argument at the Arrhenius exponent in the expression for $W(T) = A\exp(-E/RT)$ in the power series:

$$\exp(-E/RT) \simeq \exp\left[-\frac{E}{RT_b}\left(1 + \frac{T_b - T}{T_b}\right)\right] = \exp\left(-\frac{E}{RT_b}\right) \cdot \exp\left(-\frac{E}{RT_b}(T_b - T)\right).$$

Then integral on the right-hand side can be easily calculated, and the result is

$$\int_T^{T_b} W(T)dT \simeq A\frac{RT_b^2}{E}\exp\left(-\frac{E}{RT_b}\right). \tag{3.4.17}$$

Finally, substituting (3.4.17) with (3.4.16) we obtain for the normal flame velocity in the form

$$U_f = u_n \approx \sqrt{\frac{2\lambda}{\rho_0 c_P}\frac{A \cdot RT_b^2}{E(T_b - T_0)}} \cdot e^{-E/2RT_b}. \tag{3.4.18}$$

If we allow removing the restricting assumptions for the Lewis number not to be equal to unity and allow an arbitrary ratio of the number of moles of reactant and products (n_r/n_p), then the result for a first-order reaction becomes

$$U_f = \sqrt{\frac{2\lambda c_{Pb}}{\rho_0 \langle c_P \rangle^2}\frac{T_0}{T_b}\frac{n_r}{n_p}(Le)\frac{A \cdot R^2 T_b^2}{E^2(T_b - T_0)^2}} \cdot e^{-E/2RT_b}, \tag{3.4.19}$$

and for a second-order reaction

$$U_f = \sqrt{\frac{2\lambda c^2 \rho_b}{\rho_0 \langle c_P \rangle^3} \frac{T_0}{T_b} \left(\frac{n_r}{n_p}\right)^2 (Le)^2 \frac{A \cdot R^3 T_b^3}{E^3 (T_b - T_0)^3}} \cdot e^{-E/2RT_b}, \tag{3.4.20}$$

where c_{Pb} is the specific heat at $T = T_b$, and $\langle c_P \rangle$ is the average specific heat in the interval between T_0 and T_b.

The structure of the reaction zone is a consequence of the nonlinearity of the rate of heat release, which is zero everywhere except a narrow region near the final temperature. This is exactly what makes it possible to find the temperature distribution inside the flame and to calculate the flame velocity without solving the eigenvalue problem of Eq. (3.4.1).

The constant characteristic velocity of the normal flame propagation is associated with the structure of the reaction zone in which the chemical transformation takes place. The temperature distribution within this zone in the used asymptotic approximation and the resulting rate of change of the fuel mixture into combustion products are independent of the flame velocity. We can say that the reaction zone "dictates" the condition: the magnitude of the heat flux that goes into heating the fuel mixture, which in turn determines the flame propagation velocity.

3.5 Consequences of the Formula for Normal Flame Velocity

It may seem surprising that the heat of the reaction Q of the mixture appears in the denominator of the formula (3.4.8) for the flame velocity. This seems to be obvious that as the heat of the reaction is increased, the flame velocity should increase too. The paradox is explained by the fact that a rise in Q also causes a rise in the combustion temperature, so that the increase in the integral of the reaction rate is even larger than the increase in Q in the denominator. It should be noticed also that the initial fuel concentration enters formula (3.4.8) both in the denominator as well as through the fact that as a_0 increases, T_b increases and the rate of the reaction also becomes larger in the integration region.

Since the density is proportional to the pressure and the reaction rate depends on the pressure exponentially as P^n (where n is the order of the reaction), we find that

$$U_f \propto P^{\frac{n}{2}-1}. \tag{3.5.1}$$

Thus, for bimolecular reactions the flame velocity is independent of the pressure, and it rises with pressure for trimolecular reactions, and for the first-order reactions it falls with pressure.

For the second-order reactions, the rate of chemical change is determined by the number of binary collisions between reactive molecules, which is independent of pressure. The mean free path and the collision time decrease in inverse proportion to pressure, while the speed of the molecules is independent of pressure. This conclusion stems also from Eq. (3.2.3), which says that the flame velocity is proportional to a_s and, thus, is independent of pressure. Furthermore, it can be stated that the propagation of a flame in which the chemical change involves an arbitrary number of bimolecular reactions will be self-similar as the pressure is changed. This is because as the pressure increases, all the spatial scales (the dimensions of the preheat and chemical reaction zones) and temporal scales (mean collision times of the molecule, chemical reaction time) decrease in proportion to $1/P$, while the number of collisions, the thermal velocity of the molecules, and the flame velocity remain unchanged. This similarity is destroyed in the case of monomolecular or three-molecular reactions for which it is necessary to consider the characteristic decay time of the molecule or the lifetimes of metastable chemical complexes. On the other hand, the combustion mass flux $\rho_0 u_n$ increases with pressure. Since $\rho_0 u_\mathrm{n} \propto P^{n/2}$, this means that the reaction rate increases with pressure. According to experimental data, the dependence on the pressure of the flame velocity varies from $u_n \propto P^{-0.5}$ ($\rho_0 u_\mathrm{n} \propto P^{0.5}$) for slowly burning mixtures (hydrocarbons) to $u_n \propto P^{0.5}$ ($\rho_0 u_n \propto P^{3/2}$) for fast-burning mixtures (rich carbon monoxide–oxygen mixtures, hydrogen–oxygen mixtures, etc.).

3.6 Laminar Non-premixed Flames

Contrary to premixed flames discussed previously, where the fuel and oxidizer are initially perfectly premixed, there are combustion systems where the fuel and oxidizer are initially spatially separated. Examples of such systems are a lighted candle, burning wood in a fireplace, droplets spray burning in diesel engines or in industrial furnaces, all producing *diffusion or non-premixed* flames. In these and many other practical systems, a fuel and oxidizer come together in a reaction zone through molecular and/or turbulent diffusion and convection if any. The fuel may be in the form of a gaseous fuel jet or a solid or liquid, and the oxidizer may be a flowing oxygen or air stream or the quiescent atmosphere. The distinctive characteristic of a diffusion flame is that the burning rate is determined by the rate at which the fuel and oxidizer are brought together. The mixing rate caused by diffusion is much lower than the rate of chemical reaction. Correspondingly, the characteristic mixing time is much longer than the chemical reaction times. The dimensionless number, the ratio of mixing (diffusion) time, t_{dif}, to chemical reaction time, t_ch, is called the *Karlovitz number* (B. Karlovitz, Hungarian physicist), $Ka = t_\mathrm{dif}/t_\mathrm{ch}$. Commonly used in combustion is the *Damköhler number* (Gerhard Damköhler, German chemist), which is related to the Karlovitz number, $Da = 1/Ka$. Typically, in such systems the mixing and reaction will take place only in thin reaction zones separated by fuel and oxidizer. Although we will consider mainly premixed flames, in this section we briefly describe the distinctive properties of diffusion combustion. A detailed description of the various aspects of diffusion combustion can be found in books and reviews [8–12].

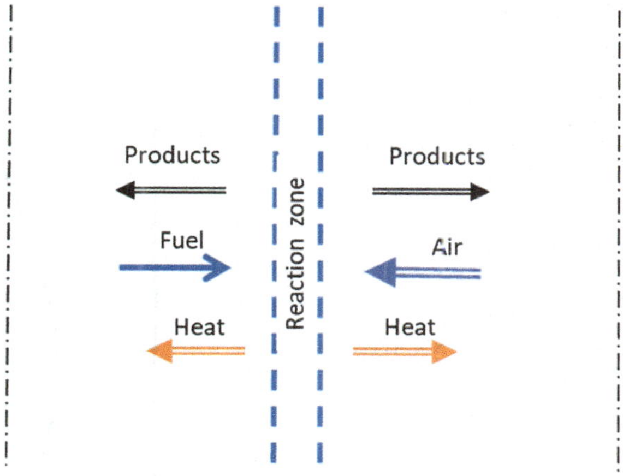

Fig. 3.4 Configuration of a 1D non-premixed flame

The structure of a non-premixed flame consists of three zones, with a reaction zone (flame) separating a fuel zone and an oxidizer (air/oxygen) zone which is shown in Fig. 3.4.

As the fuel and oxidizer are transported toward each other, through diffusion as well as whatever convective motion the system may have, they are heated by the heat transferred from the high-temperature reaction zone, and eventually meet and mix within the reaction zone. The reaction between the fuel and oxidizer takes place much faster than the characteristic times of diffusion or convection. The combustion products together with the heat of reaction are in turn transported away from the reaction zone in both directions.

3.6.1 Non-premixed Laminar Flames

The non-premixed laminar jet flame is the classical simplified representation of jet combustion, which has been first described quantitatively by Burke and Schumann [13]. Non-premixed jet flames are well characterized and are very helpful in understanding the important features of a typical non-premixed flame including its structure, flame location, flame temperature, and the flame height. In a concentric configuration, which is shown in Fig. 3.5, the air is flowing through the outer annulus and the fuel is coming from a nozzle. They meet and mix within the reaction zone where the flame that develops takes a closed, elongated form. A similar flame forms when a fuel jet issues into the quiescent atmosphere. In a concentric configuration, which is shown in Fig. 3.5, the air flowing through the outer annulus and the fuel are coming from a nozzle. They meet and mix within the reaction zone where the flame that develops takes a closed, elongated form. A similar flame forms when a fuel jet issues into the quiescent atmosphere. In general, the shape and height of the flame vary with the flow conditions: the volumetric flow rate of air in the outer annulus can

Fig. 3.5 Schematic of a simple jet flame structure

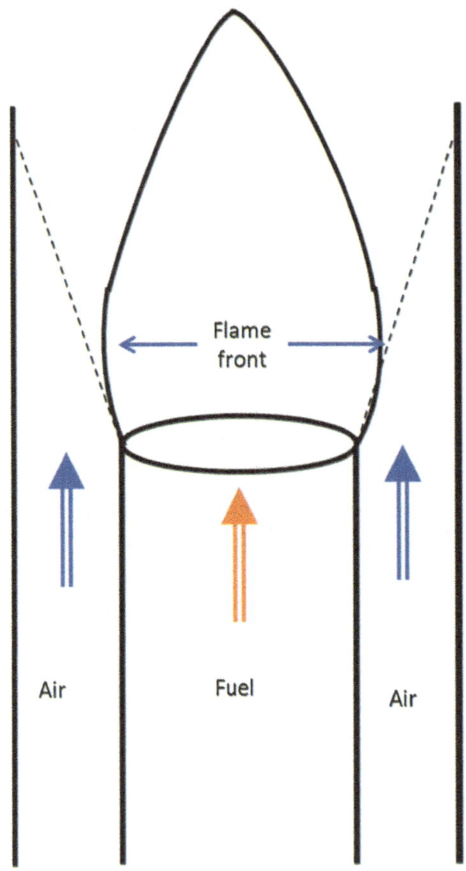

be in excess of the stoichiometric amount required for the volumetric flow rate of the inner fuel jet—overventilated flames, and underventilated flame is the air supply in the outer annulus which is below that corresponding to the stoichiometric required amount.

Unlike premixed flames, which have a very narrow reaction zone, diffusion flames have a considerably wider reaction zone due to interdiffusion of reactants, oxidizer, and products where chemical reactions occur. The flame front is located at the point of maximum temperature, which corresponds to that at which the maximum concentrations of major products exist. The maxima arise due to diffusion of reactants in the direction normal to the flowing air and fuel because, for the concentric configuration, the molecular diffusion establishes a bulk velocity component in the normal direction.

The primary processes for diffusion flames are diffusion and mixing processes of fuel and oxidizer. For a simple free jet flame configuration in Fig. 3.5, the problem can be solved assuming steady-state conditions [13], in particular for calculation of the height and the shape of a non-premixed flame, which are important parameters indicating the size of a flame. Readers can find detailed solutions, which are expressed in the Bessel functions of the first kind of order 0 and 1 [8–12]. Lewis and von Elbe [9] have shown that the flame height and shape predicted by such solutions are in good agreement with experimental results, even so, the basic assumptions seem to be oversimplified. Current computer simulations accurately predict diffusion flame structure and behavior. We do not consider here these solutions, but instead show that the parameters controlling the behavior of non-premixed jet flames can be determined from dimensional analysis, using various scales characteristic of the flame. From the time scale of the oxidizer diffusion to the jet centerline, $\tau_{\mathrm{diff}} = r_{\mathrm{jet}}^2/D$ and from the fuel jet velocity, V_{jet}, the only one function of length dimensionality characterizing the flame can be constructed: $h_{\mathrm{f}} = \tau_{\mathrm{diff}} V_{\mathrm{jet}} = r_{\mathrm{jet}}^2 V_{\mathrm{jet}}/D$. This formula implies that the flame height increases linearly with the fuel flow speed (V_{jet}). Indeed, such a linear dependence is observed in experiments.

3.6.2 Combustion of Liquid Fuel Droplets

Liquid fuel droplet combustion involves sequential vaporization and combustion of vapors in the gas phase. Since the fuel vapor and oxidizer (oxygen or air) are separated, they burn in the form of a diffusion flame as it shown in Fig. 3.6. The heat from the combustion and surrounding hot gases is transferred to the droplet to vaporize liquid fuel from the droplet surface. During the droplet combustion, the liquid fuel droplet will be vaporized layer by layer and finally will be consumed. The fuel evaporates from the droplet surface and diffuses to the flame front as the oxygen moves here from the surrounding atmosphere. Within the reaction zone, the fuel vapor is heated, mixed with oxygen, and "feeds" the ongoing reaction. Thus, the flame front in the process of liquid droplet combustion acts as a source of heat for droplet vaporization. Otherwise, the mechanism is the same as pure vaporization. First of all, of interest is the mass consumption rate, or the rate of a liquid fuel droplet vaporization. The similarity to the case of diffusion jet flames arises from the fact that the liquid phase must be gasified, and consequently, there must be an energy input into the liquid droplet material. Since droplet heating involves the change of a liquid phase, and therefore occurs over a longer period than that of gas-phase transport, the problem can be considered as quasi-steady. What determines the rate of droplet evaporation is the heat flux at the surface. The time required to completely consume a droplet of a given diameter d can be computed if we know the burning rate and the amount of heat from the flame and surrounding hot gases, which are transferred to the droplet surface.

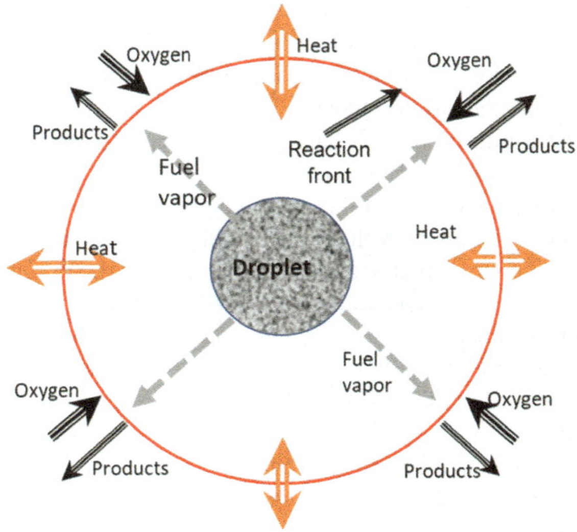

Fig. 3.6 Schematic model of liquid fuel droplet combustion

In the case of vaporization of an isolated spherical droplet of diameter d in a uniform, quiescent atmosphere, the rate of droplet vaporization can be written in the form of the d^2 law of droplet vaporization, introduced by C. K. Law [14–16]:

$$d_d^2 = d_0^2 - K_{\mathrm{vp}} t. \tag{3.6.1}$$

Here, d_0 is the initial drop diameter, and K_{vp} is the evaporation rate constant [14]

$$K_{\mathrm{vp}} = 8\chi \cdot \ln\left(1 + \frac{c_{\mathrm{p}}(T_\infty - T_{\mathrm{s}})}{q_{\mathrm{v}}}\right), \tag{3.6.2}$$

where q_{v} is the specific heat of vaporization, T_∞ is the temperature of the ambient gas far away from the droplet, and T_{s} is the temperature at the droplet surface. Equation (3.6.2) shows that the square of the droplet diameter decreases linearly with time. This is called the d^2 law of droplet vaporization. Note that the time of complete burnout of the droplet is $\tau_0 = r_0^2/K_{\mathrm{vp}}$. It can be readily seen $K_{\mathrm{vp}} \approx D\,(\rho/\rho_{\mathrm{d}})$, so that $\tau_0 = (r_0^2/D)(\rho_{\mathrm{d}}/\rho)$, where D and ρ are coefficient of diffusion and density in the gas phase.

For steady combustion of the spherical symmetric liquid droplet, Eqs. (3.6.1 and 3.6.2) must be supplemented by the equations of thermal conduction and diffusion, describing the heat origin in the flame and the heat transfer to the droplet surface from the flame front. Assuming for simplicity the coefficient of thermal conduction, λ, and the coefficient of diffusion, D, to be constant, and taking $r = 0$ at the center of the droplet, we obtain equations, which are similar to Eqs. (3.3.9 and 3.3.10):

$$\lambda\left(\frac{d^2T}{dr^2}+\frac{2}{r}\frac{dT}{dr}\right)+Q\,\dot{\omega}\,(a,T)=0, \tag{3.6.3}$$

$$\rho D\left(\frac{d^2a}{dr^2}+\frac{2}{r}\frac{da}{dr}\right)-\dot{\omega}\,(a,T)=0. \tag{3.6.4}$$

Since the chemical reaction time is the smallest time scale, the width of the reaction zone and flame is much smaller than the droplet radius and the width of the preheat zone (boundary layer). Therefore, Eqs. (3.6.3) and (3.6.4) in the space between the flame and the droplet can be solved without the term with chemical reaction. The steady solutions for the temperature and concentration are

$$T=T_\infty+(T_\mathrm{s}-T_\infty)\frac{r_\mathrm{d}}{r}, \tag{3.6.5}$$

$$a=a_\infty(1-r_d/r), \tag{3.6.6}$$

where T_s is the temperature at the droplet surface.

Conclusions obtained from this simple description of the liquid fuel droplet combustion agree at least qualitatively with experimental observations [15–19]. Combustion of liquid fuels is employed in many techniques such as industrial furnaces, boilers, gas turbines, diesel, spark ignition, and rocket engines. These varied applications have led to numerous experimental and numerical studies of both spray combustion and associated processes, such as droplet evaporation and combustion [20–27].

References

1. E. Mallard, H. Le Chatelier, Recherches Experimentales et Theoriques sur la Combustion des Melanges Gaseoux Explosifs. Ann. Mines **8**, 274–568 (1883)
2. B. Lewis, G. von Elbe, On the theory of flame propagation. J. Chem. Phys. **2**, 537–561 (1934)
3. B. Lewis, G. von Elbe, *Combustion, Flames, and Explosions of Gases* (Cambridge Press, 1938)
4. Ya.B. Zel'dovich, D.A. Frank-Kamenetski, A theory of thermal propagation of flame. Acta Physicochimica. 9, 341–350 (1938)
5. Ya.B. Zel'dovich "K Teorii Rasprostranenia Plameni." Zhurnal Fizicheskoi Khimii (USSR), T. 22, 1948, pp. 27–49 in Russian). Translated: Ya.B. Zel'dovich, Theory of flame propagation, NASA, Technical Memorandum 1282, Washington 1951.
6. Ya.B. Zel'dovich, G. I. Barenblatt, V. B. Librovich, G. M. Makhviladze, The Mathematical theory of combustion and explosions. Science Publ. Moscow, 1980 (in Russian). English translation: Consultants Bureau, New York (1985).
7. V.V. Bychkov, M.A. Liberman, Dynamics and stability of premixed flames. Phys. Rep. **325**, 115–237 (2000)
8. I. Glassman, R.A. Yetter, *Combustion*, 4th edn. (Elsevier, New York-Oxford, 2008)
9. B. Lewis, G. von Elbe, *Combustion* (Third edition, Academic Press, New York, Flames and Explosion of Gases, 1987)
10. K.K. Kuo, *Principles of Combustion*, 2nd edn. (John Wiley, New York, 2005)

11. C.K. Law, *Combustion Physics* (Cambridge University Press, 2006).
12. T. Poinsot, D. Veynante, *Theoretical and Numerical Combustion* (R.T. Edwards Inc., Philadelphia, 2005)
13. S.P. Burke, T.E.W. Schumann, Diffusion flames. Ind. Eng. Chem. **20**, 998–1004 (1928)
14. C.K. Law, H.K. Law, A d^2-law for multicomponent droplet vaporization and combustion. AIAA J. **20**, 522–527 (1982)
15. C.K. Law, H.K. Law, A theoretical study of ignition in the laminar mixing layer. J. Heat Transfer **104**, 329–337 (1982)
16. C.K. Law, Prog. Recent advances in droplet vaporization and combustion. Energy Combust. Sci. 8, 171–201 (1982).
17. A. Lee, C.K. Law, Gasification and shell characteristics in slurry droplet burning. Combust. Flame **85**, 77–93 (1991)
18. A. Lee, C.K. Law, An experimental investigation on the vaporization and combustion of methanol and ethanol droplets. Combust. Sci. Technol. **86**, 253–265 (1992)
19. C.R. Shaddix, D.R. Hardesty, Combustion properties of biomass flash pyrolysis oils: final project report. Sandia National Laboratory Report, SAND99-8238 (1999).
20. G. Blanquart, H. Pitsch, Modeling autoignition in non-premixed turbulent combustion using a stochastic flamelet approach. Proc. Combust. Inst. **30**, 2745–2753 (2005)
21. N. Peters, Laminar diffusion flamelets in non-premixed turbulent combustion. Prog. Energy Combust. Sci. **3**, 319–339 (1984)
22. D. Bradley, P.H. Gaskell, X.J. Gu, The mathematical modeling of liftoff and blowoff of turbulent non-premixed methane jet flames at high strain rates. Proc. Combust. Inst. **27**, 1199–1206 (1998)
23. E. Mastorakos, Ignition of turbulent non-premixed flames. Prog. Energy Combust. Sci. **35**, 57–97 (2009)
24. S.K. Aggarwal, Single droplet ignition: theoretical analyses and experimental findings. Prog. Energy Combust. Sci. **45**, 79–107 (2014)
25. W.A. Sirignan, Advances in droplet array combustion theory and modeling. Prog. Energy Combust. Sci. **42**, 54–86 (2014)
26. E.D. Gonzalez-Juez, A.R. Kerstein, R. Ranjan, S. Menon, Advances and challenges in modeling high-speed turbulent combustion in propulsion systems. Prog. Energy Combust. Sci. **60**, 26–67 (2017)
27. K. Luo, C. Shao, M. Chai, J. Fan, Level set method for atomization and evaporation simulations. Prog. Energy Combust. Sci. **73**, 65–94 (2019)

Chapter 4
Hydrodynamics of Premixed Laminar Flames

4.1 Flame Hydrodynamics: Combustion Regimes

In many situations, a detailed scheme of chemical reactions are unnecessary and the hydrodynamics of the flame can be described satisfactory well using an accurate hydrodynamic model coupled to a one-step simplified irreversible reaction that converts the fresh fuel mixture to combustion products. When we talk about a flame, we mean, first of all, an ordinary chemical flame, a flame that can be observed in a laboratory, or fuel combustion in a car engine, combustion processes in industry, etc. Still, there are many other physical phenomena that possess very similar properties and may be called flames in a sense. These are, for example, the laser ablation in the laser-induced acceleration of thin foils in inertial confinement fusion [1–10], thermonuclear flames in Supernova events [11–21], ionizing waves in gas discharges [22–25], evaporation wave in an overheated fluid [26, 27], and many other phenomena.

When we consider a flame in a gaseous fuel mixture, we assume that the burning process does not involve a phase transition: both the matter before the burning and the matter after the burning is in the gaseous phase. This is a typical situation that takes place in car engines or aircraft gas turbines. From the theoretical point of view, the absence of phase transitions makes easier analysis of the flames, besides, the equation of state of a perfect gas is applicable in most cases. The other important point inherent to the gaseous burning is that the burning mixture undergoes strong expansion. Because of the expansion, the burning process becomes strongly coupled to the fuel flow; the coupling leads to different flame instabilities, generation of pressure waves, etc. Particularly, one of the most important flame parameters is the expansion coefficient, $\Theta = \rho_f/\rho_b$, defined as the ratio of the fuel density ρ_f and the density of the burned matter ρ_b. For most of the laboratory flames, the expansion coefficient is about $\Theta = 5 - 12$.

When a flame is premixed (which is the opposite case to diffusion flames), then it implies that all components necessary for the reaction are present in the fuel mixture from the very beginning and in order to start the reaction one has only to

M. A. Liberman, *Combustion Physics*,
https://doi.org/10.1007/978-3-030-85139-2_4

heat the mixture. In the case of premixed flames, the burning rate is not controlled by the diffusion process, still even for premixed flames the fuel diffusion (together with thermal conduction) is an important factor influencing the flame propagation and cannot be neglected. Particularly, when the fuel diffusion is stronger than the thermal conduction, the thermal-diffusion instability of a planar stationary flame may develop. In the opposite case of zero diffusion (which is not typical for a gaseous fuel), a planar stationary flame cannot propagate either: the domination of thermal conduction over diffusion leads to pulsations of the reaction zone [28]. Since we are interested in flame hydrodynamics, we consider now the situation of equally strong fuel diffusion and thermal conduction.

A remarkable feature of premixed combustion is very strong dependence of the reaction rate on temperature expressed by the Arrhenius law for the reaction rate $\propto \exp(-E/T)$, where E is the activation energy measured in temperature units. The activation energy of many reactions is so large that the reaction rate at room temperature may be taken as zero. On the contrary, the increase of the fuel temperature even by a factor of 2 may lead to the increase of the reaction rate by 10–12 orders of magnitude. In the case of a strongly exothermic reaction when a considerable energy release is involved, a relatively slight increase of the temperature at some region ignites the reaction, which eventually extends over the whole gas. Once a reaction is ignited, it can propagate in a self-supporting regime. The physical mechanism of flame propagation may be described as follows. Thermal conduction transfers heat from the hot combustion products behind the flame front to the cold fuel–oxygen mixture ahead of the flame front. The temperature of the unburned mixture close to the flame increases, the reaction in this location accelerates until another portion of the fuel is burned, and some additional energy is released. The released energy is transported by thermal conduction to the next fuel layer and so on, resulting in propagation of the reaction front. The order of magnitude of the distance over which heat is transferred by thermal conductivity is $\mathrm{L_f} \sim \sqrt{\chi \tau_\mathrm{R}}$, where $\chi = \kappa/\rho_f C_P$ is the thermal diffusivity. The time τ_R is some characteristic time of the reaction, which depends only on the particular combustible mixture. Next, we suppose that the width of the combustion zone is negligibly small compared to the characteristic dimensions of the problem (for example, the width of the hydrogen–oxygen flame at normal conditions is $L_f = 0.0027cm$; the width of the methane–air flame is $L_f = 0.0451cm$). Therefore, we can consider the combustion zone as a discontinuity surface, which separates the unburned gas mixture from the combustion products. The flame velocity U_f is defined by the condition that during time τ_R, the flame front propagates a distance of the order $\mathrm{L_f}$:

$$U_f \sim L_f/\tau_R = \sqrt{\frac{\kappa}{\rho_f C_P \tau_R}}. \tag{4.1.1}$$

Since the value of the flame thickness depends on the definition, we can adopt the equation

$$L_f \approx \sqrt{\chi \tau_R} \tag{4.1.2}$$

as the definition of the flame thickness, taking the value of the thermal conduction coefficient in the fuel $\kappa = \kappa_f$ for the definition.

Using the estimate for thermal diffusivity coefficient, $\chi \approx c_s^2 \tau_{coll}$, expressed through the sound speed a_s (that is about the thermal velocity of molecules) and the mean free time between molecules collision, τ_{coll}, we obtain $\mathrm{U_f/c_S} \sim \sqrt{\tau_{\mathrm{coll}}/\tau_{\mathrm{R}}}$. Since only a very small fraction of colliding molecules react, roughly one reaction over thousand of more collisions because of the large potential barrier of a reaction, then $\tau_{coll} \ll \tau_R$. This means that the flame velocity, U_f, which is called a normal deflagration speed, is much smaller than the sound speed.

A flame considered as a discontinuity surface can be regarded as the phenomenon that is overall of fluid-mechanical nature. The necessary condition for the existence and uniqueness of the solution in the form of a discontinuity surface is the evolutionarity condition (see details in the book [29]). The evolutionarity condition requires that the number of possible outgoing simple waves, which can propagate away from both sides of the discontinuity surface, $s_1 + s_2$, be equal to one less than the number, r, of boundary conditions, e.g.

$$s_1 + s_2 = r - 1. \tag{4.1.3}$$

The evolution of small perturbations of the discontinuity surface is defined by the set of linearized boundary conditions which must be satisfied at the discontinuity surface. The stability condition is satisfied if the number of these equations is the same as the number of independent parameters in these equations. A small perturbation of the discontinuity surface consisting of an infinitesimal displacement in the direction of the normal to the surface can be considered as a superposition of small perturbations of the gas velocity, pressure and temperature on both sides of the discontinuity surface. The perturbation originated at the discontinuity surface then propagates away from both sides of the discontinuity surface. Thus, a perturbation can be considered as a superposition of sound waves propagating with the local sound speed (pressure, density and velocity) and perturbation of entropy, which is carried by the flow.

Let us consider the coordinate system co-moving with the flame. In this coordinate system, a flux of fresh fuel flows in the flame front with velocity $u_0 = U_f$, and the combustion products left the flame front with velocity u_2. The boundary conditions at the flame front are given by the continuity of mass, momentum, and energy fluxes.

$$\rho_0 u_0 = \rho_2 u_2, \tag{4.1.4}$$

$$P_0 + \rho_0 u_0^2 = P_2 + \rho_0 u_2^2, \tag{4.1.5}$$

$$\tfrac{1}{2}\rho_0 u_0^2 + H_0 = \tfrac{1}{2}\rho_2 u_2^2 + H_2. \tag{4.1.6}$$

For simplicity, we consider a perfect gas with the specific heats assumed to be constant and the same for unburned and burned gases. The enthalpy for a perfect gas is

$$H = c_P T + H_{00} = \frac{\gamma}{\gamma - 1}\frac{P}{\rho} + H_{00}. \tag{4.1.7}$$

First of all, let us verify that the evolutionarity conditions for the flame considered as a discontinuity surface are satisfied. There are three simple waves (characteristics) which can propagate away from the flame front. Behind the flame, in the downstream flow these are sound waves, propagating with velocity $-(u_2+a_{s2})$, and the entropy is carried by the downstream flow with velocity $-u_2$. Since a deflagration is a subsonic combustion mode, in the upstream flow the sound wave propagates away from the flame front with velocity $(a_{s0} - u_0) > 0$. There are three boundary conditions, Eqs. (4.1.4)–(4.1.6), given by the continuity of the mass, momentum, and energy fluxes. Since there are three simple waves propagating away from the flame front, an additional condition is required to satisfy the evolutionarity condition. This additional condition is a certain value of the flame velocity that cannot be determined if we consider the flame front as a discontinuity surface. To determine the flame velocity, we must consider the structure of the front of finite width, including, for example, thermal conductivity.

Since the flame velocity is much smaller than the sound speed, from (4.1.5) and (4.1.6) follows the continuity of pressure and enthalpy

$$P_0 = P_2,\ H_0 = H_2. \tag{4.1.8}$$

The continuity of enthalpy can be written as

$$c_P T_0 + H_{00} = c_P T_2 + H_{02} \ \Rightarrow\ c_P T_0 + Q = c_P T_2, \tag{4.1.9}$$

where the chemical energy released is $Q = H_{00} - H_{02}$. Then we obtain the following relations between the quantities characterizing the unburned mixture and combustion products:

$$P_0 = P_2,\ T_2 = T_0 + Q/c_P,\ \frac{u_2}{u_0} = \frac{\rho_0}{\rho_2} = 1 + \frac{Q}{c_P T_0}, \tag{4.1.10}$$

It can be readily verified that in the case of different specific heats of unburned and burned gases, the result is qualitatively similar to (4.1.10):

$$T_2 = \frac{c_{P0}}{c_{P2}} T_0 + Q/c_{P2},\ \frac{u_2}{u_0} = \frac{\rho_0}{\rho_2} = \left(1 + \frac{Q}{c_{P0} T_0}\right)\frac{\gamma_0(\gamma_2 - 1)}{\gamma_2(\gamma_0 - 1)}. \tag{4.1.11}$$

In the laboratory coordinate system, the velocity ahead of the flame is zero. The flame velocity relative to the unburned gas is $u_0 = U_f$. The velocity of the burned products in the laboratory coordinate system is $u_2 - U_f$. Using the continuity of mass flux $\rho_0 u_0 = \rho_2 u_2$, we obtain the velocity of burned products in a laboratory coordinate system: $u_{2lab} = U_f(\Theta - 1)$, where $\Theta = \rho_0/\rho_2 \sim 5-12$ is dimensionless compression.

It is convenient to analyze possible regimes of combustion waves considered as a discontinuity surface on the plot P_2 as a function of $V_2 = 1/\rho_2$ which we will call the *combustion adiabatic* by analogy with shock adiabatic. Similar to the derivation of shock adiabatic (Appendix: Sect. Shock adiabatic), P_2 as a function of $V_2 = 1/\rho_2$, we introduce the mass flux density at the discontinuity surface, $j \equiv \rho_0 u_0 = \rho_2 u_2$. From (4.1.4) and (4.1.5) can be readily obtained the following relations:

$$u_0 = j/\rho_0 = jV_0, \quad u_2 = j/\rho_2 = jV_2, \tag{4.1.12}$$

$$j^2 = (P_2 - P_0)/(V_0 - V_2). \tag{4.1.13}$$

Using (4.1.13) and (4.1.5), we obtain the equation for the combustion adiabatic

$$H_0 - H_2 + \frac{1}{2}(P_2 - P_0)(V_0 + V_2) = 0. \tag{4.1.14}$$

The difference between the combustion adiabatic (4.1.14) and the shock adiabatic is clearly seen if we assume a perfect gas equation for unburned and burned gases with a constant and the same specific heats. Then, (4.1.14) becomes

$$P_2\left(\frac{\gamma+1}{\gamma-1}V_2 - V_0\right) - P_1\left(\frac{\gamma+1}{\gamma-1}V_1 - V_2\right) = 2Q. \tag{4.1.15}$$

The left-hand part of (4.1.15) is exactly the same as for a shock adiabatic. The combustion adiabatic is shown in Fig. 4.1. Unlike the shock adiabatic, the Rankine–Hugoniot curve does not pass through the point of initial state, (V_0, P_0), because of the difference of enthalpy of the unburned and burned gases. Equation (4.1.15) for $Q = 0$ gives the shock adiabatic passing through the point of initial state, (V_0, P_0). Since formula (4.1.13) is a consequence of the continuity of mass and momentum fluxes, so similar to the shock adiabatic, the value of j^2(called the *Rayleigh line*) is the slope of the line connecting the point of initial state, (V_0, P_0), with any point on the Rankine–Hugoniot curve. It is seen from Fig. 4.1 that the value of j^2 is limited by slopes of the two tangent lines (V_0, P_0) – (CJ - detonation) and (V_0, P_0) – (CJ - deflagration). The states corresponding to these two points are called the Chapman–Jouguet (CJ) detonation (upper point) and CJ deflagration, respectively. In these two points, the straight lines down from the point (V_0, P_0) are tangent lines to the Rankine–Hugoniot curve. At the tangent points of the Rankine–Hugoniot curve, the slope of the Rayleigh line (4.1.13) is

$$\left(\frac{dP_2}{dV_2}\right)_{CJ} = \frac{P_2 - P_0}{V_2 - V_0}. \tag{4.1.16}$$

Differentiating (4.1.15) with respect to V_2

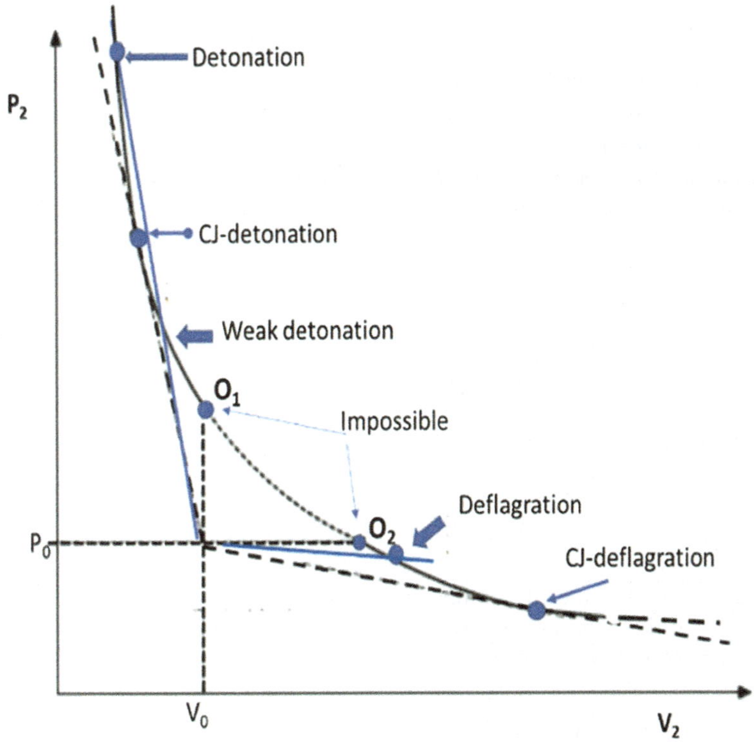

Fig. 4.1 Schematic combustion adiabatic; the Rankine–Hugoniot curve

$$\frac{dP_2}{dV_2} = -\frac{P_2(\gamma+1)+P_0(\gamma-1)}{V_2(\gamma+1)-V_0(\gamma-1)}, \tag{4.1.17}$$

and equating the right-hand parts of (4.1.16) and (4.1.17), we obtain

$$j_{CJ}^2 = \frac{P_2 - P_0}{V_2 - V_0} = \gamma \rho_2 P_2 = \rho_2^2 u_2^2. \tag{4.1.18}$$

According to (4.1.18), the velocity of the flow behind combustion waves equals the local sound speed:

$$u_2^2 = a_s^2 = \gamma \frac{P_2}{\rho_2}. \tag{4.1.19}$$

At the Rankine–Hugoniot curve above the point O_1, $V_2 < V_0$ and consequently $\rho_2 > \rho_0$, which means that these points correspond to compression waves. On the contrary for the points below O_2, $V_2 > V_0$ and $\rho_2 < \rho_0$. In the part of the Rankine–Hugoniot curve between points O_1 and O_2, the value of j^2 is negative. The imaginary value of j (which is a mass flux) has no sense, and therefore this part of the curve does not correspond to any solution.

Let us consider possible regimes of the combustion waves shown in Fig. 4.1. Below, we will follow the analysis of evolutionary conditions presented in the book [29]. It is convenient to use dimensionless variables normalized on their values ahead of the wave:

$$u = u_2/u_0,\ \ \rho = \rho_2/\rho_0,\ \ T = T_2/T_0,\ \ P = P_2/P_0,\ \ M_0 = u_0/a_{s0},\\ q = Q/a_{s0}. \tag{4.1.20}$$

Here, $q = Q/a_{s0} \gg 1$ is dimensionless energy release. To make the results more transparent, we again assume that the specific heats are the same for unburned and burned gases, so $\gamma_1 = \gamma_2 \equiv \gamma$. Equation (4.1.13) in dimensionless variables is

$$\gamma M_0^2 = \rho(P-1)/(\rho-1). \tag{4.1.21}$$

Excluding density using Eq. (4.1.4), after a simple algebra we obtain Eqs. (4.1.5)–(4.1.6) for dimensionless variables

$$u - 1 + \frac{1}{\gamma M_0^2}\left(\frac{T}{u} - 1\right) = 0, \tag{4.1.22}$$

$$u^2 - 1 + \frac{2}{(\gamma-1)M_0^2}[T-1] + \alpha\frac{q}{M_0^2} = 0. \tag{4.1.23}$$

Here, $\alpha = 1$ ahead of the reaction front and $\alpha = 0$ in the combustion products. These two equations can be reduced to one equation for dimensionless velocity

$$(1-u_k)(u_k - u_{GD}) = \frac{2(\gamma-1)}{(\gamma+1)}\frac{q}{M_0^2}. \tag{4.1.24}$$

Here, subscripts $k = 1,\ 2$ correspond to two real roots of Eq. (4.1.24). Supersonic waves of detonation have subscript $k = 1$; subsonic waves of deflagration have subscript $k = 2$; $u_{GD} = \frac{2+(\gamma-1)M_0^2}{(\gamma+1)M_0^2}$. The root u_1 of (4.1.24) corresponds to supersonic waves, $M_0 \geq M_{CJ}^{(1)} = \sqrt{2q(\gamma^2-1)}$, and the root u_2 corresponds to subsonic waves, $M_0 \leq M_{CJ}^{(2)} = 1/\sqrt{2q(\gamma^2-1)}$. The equality sign corresponds to the CJ detonation

$$u_{CJ}^{(1)} = \sqrt{2q(\gamma^2-1)}, \tag{4.1.25}$$

and to the CJ deflagration

$$u_{CJ}^{(2)} = 1/\sqrt{2q(\gamma^2-1)}. \tag{4.1.26}$$

Next, we consider the number of boundary conditions required to satisfy the evolutionarity condition for possible combustion modes [29]. There are four different combinations of speeds ahead and behind the discontinuity surface. For supersonic detonation wave:

$$u_0 > a_{s0}, \ \ u_1 \leq a_{s1}, \tag{4.1.27}$$

$$u_0 > a_{s0}, \ \ u_1 > a_{s1}. \tag{4.1.28}$$

For subsonic deflagration wave:

$$u_0 < a_{s0}, \ \ u_2 < a_{s2}, \tag{4.1.29}$$

$$u_0 < a_{s0}, \ \ u_2 > a_{s2}. \tag{4.1.30}$$

There are three boundary conditions given by the continuous fluxes of mass, momentum, and energy. The case (4.1.27) is similar to the shock wave, therefore, for this detonation mode the number of boundary conditions and the evolutionarity condition remain the same as for an ordinary shock wave: the boundary conditions given by the continuity of mass, momentum, and energy flux satisfy the evolutionarity condition.

The case (4.1.28) corresponding to the part of the Rankine–Hugoniot curve between points of CJ detonation and O_1 is called a weak detonation. The evolutionarity condition in this case requires one additional condition because behind the discontinuity surface, a sonic wave propagating to the discontinuity surface from infinity with velocity $+a_{s1}$ will be carried away from the discontinuity with velocity $(a_{s1} - u_1) < 0$. Since velocity is proportional to the slope of j, velocity of the "weak detonation" increases down from the CJ detonation point and tends to infinity at the point O_1. Modes of the "weak detonation" require an additional boundary condition, which is the certain velocity of this mode. While the detonation velocity (above CJ detonation point) is determined by the shock wave velocity, the propagation velocity of "weak detonation" is determined, for example, by artificial (external) ignition of the chain of chemical reactions ahead of the weak detonation. The velocity of a detonation wave above the CJ detonation point is also larger than that of the CJ detonation. Thus, the velocity of CJ detonation is the lowest velocity of detonation modes.

The subsonic mode, $u_0 < a_{s0}, \ u_2 < a_{s2}$, corresponds to normal deflagration. The evolutionarity condition in this case requires one additional "boundary" condition, which is a certain value of the normal velocity of the deflagration. The evolutionarity of the subsonic mode, $u_0 < a_{s0}, \ u_2 > a_{s2}$, corresponding to the part of the Rankine–Hugoniot curve below the CJ deflagration point, requires two additional conditions, and therefore such regime does not exist. The temperature of products in detonation wave

$$T_{CJ}^{(1)} = 2q(\gamma^2 - 1). \tag{4.1.31}$$

Remember, that we used the same value of γ for the fresh gas and combustion products and $q \gg 1$. It should be noted that in majority cases a detonation propagates in the ChapmanJouguet regime. The detailed structure of detonation waves and possibility of CJ deflagration will be considered in Chap. 12.

4.2 Complete System of Equations for Propagating Flame

The simplest configuration of a slow combustion wave is a planar stationary flame. The dynamics of a flame is described by the hydrodynamic equations of mass, momentum, and energy conservation with the account of reaction kinetics and transport processes of thermal conduction, fuel diffusion, and viscosity. For simplicity, a single irreversible reaction will be admitted, such that the governing equations are the following:

$$\frac{\partial}{\partial t}\rho + \frac{\partial}{\partial x_i}(\rho v_i) = 0, \tag{4.2.1}$$

$$\frac{\partial}{\partial t}(\rho v_i) + \frac{\partial}{\partial x_j}\left(\rho v_i v_j + \delta_{ij} P - \tau_{ij}\right) = \rho g_i, \tag{4.2.2}$$

$$\frac{\partial}{\partial t}\left(\rho e + \frac{1}{2}\rho v_i v_i\right) + \frac{\partial}{\partial x_i}\left(\rho v_i h + \frac{1}{2}\rho v_i v_j v_j + q_i - v_j \tau_{ij}\right) = \rho g_i v_i, \tag{4.2.3}$$

$$\frac{\partial}{\partial t}(\rho Y) + \frac{\partial}{\partial x_i}\left(\rho v_i Y - \kappa \frac{\mathrm{Le}}{C_P}\frac{\partial Y}{\partial x_i}\right) = -\frac{\rho^n Y^n}{\rho_R^{n-1}\tau_R}\exp\left(-\frac{E}{T}\right). \tag{4.2.4}$$

where Y is the fuel mass fraction, $e = QY + C_V T$ is the internal energy, $h = QY + C_P T$ is the enthalpy, C_P and C_V are specific heats per unit mass at constant pressure and volume, respectively, their ratio determines the adiabatic exponent, and $\mathbf{g}$ is the gravity acceleration. In the development of the reaction the fuel fraction Y changes from 1 to 0. We consider a reaction of the order n with the energy release Q; the Arrhenius law gives the temperature dependence of the reaction rate with the activation energy E (taken in temperature units) and with constants of time dimension τ_R and density dimension ρ_R. Usually, the reaction order is of no importance for hydrodynamic properties of a flame with the exception of the problem of flame dynamics in a closed burning chamber, where flames with the first-, second-, and third-order reactions behave in a different way [30]. The stress tensor and the energy diffusion vector are given by the formulas

$$\tau_{ij} = \kappa \frac{\mathrm{Pr}}{C_P}\left(\frac{\partial v_i}{\partial x_j} + \frac{\partial v_j}{\partial x_i} - \frac{2}{3}\frac{\partial v_k}{\partial x_k}\delta_{ij}\right), \tag{4.2.5}$$

$$q_i = -\kappa\left(\frac{\partial T}{\partial x_i} - \mathrm{Le}\frac{Q}{C_P}\frac{\partial Y}{\partial x_i}\right), \tag{4.2.6}$$

where Pr is the Prandtl number characterizing the relative strength of viscosity and thermal conduction and *Le* is the Lewis number that shows the relative role of fuel diffusion and thermal conduction. We assume that the combustible mixture is a perfect gas:

$$P = \frac{\gamma - 1}{\gamma} C_P \rho T. \tag{4.2.7}$$

The set of Eqs. (4.2.1)–(4.2.7) is typically used in direct numerical simulations of flame dynamics. For a perfect gas, the transport coefficients of thermal conduction, fuel diffusion, and viscosity are proportional to each other. For example, the Prandtl number of a perfect gas depends on the adiabatic exponent $\mathrm{Pr} = 4\gamma/(9\gamma - 5)$ and changes within the limits $2/3 < \mathrm{Pr} < 1$. For simplicity, we assume also that the adiabatic exponent and the specific heats do not change in the reaction.

4.3 Isobaric Approximation

The theoretical approach to flame dynamics based on isobaric approximation considerably simplifies the analysis. It was pointed out earlier that flames propagate with velocities much smaller than the sound speed. The Mach number, which is the ratio of the velocity of a plane stationary flame U_f and the sound speed in the fuel a_s, characterizes the compressibility effects. Even for the fastest flames in a highly reactive fuel, the Mach number is very small. For example, the flame velocity in the hydrogen–oxygen mixture at normal conditions is $U_f = 9.4m/s$, the sound speed is $a_s = 539.5m/s$, and $\mathrm{M} = U_f/a_s \approx 0.02$. Since the change in pressure is $\Delta \mathrm{P} \sim \rho_\mathrm{f} \mathrm{U}_\mathrm{f}^2$, the relative change of pressure in a flame-generated flow is small. At the same time, the changes of temperature and density of a burning matter in a flame are about the temperature and the density themselves. We can use isobaric approximation

$$\rho T = \rho_u T_u. \tag{4.3.1}$$

The equation of energy can be also simplified in the isobaric approximation. Using (4.2.1) and (4.2.2) and the equation of state (4.2.7), we can rewrite (4.2.3) in the form

$$\begin{aligned} &\rho\left(\frac{\partial}{\partial t} + \mathbf{v}\cdot\nabla\right)(C_P T + QY) - \frac{\partial P}{\partial t} - \mathbf{v}\cdot\nabla P - \frac{\partial}{\partial x_i}(v_j \tau_{ij}) \\ &= \nabla\cdot[\kappa\,(\nabla T + \mathrm{Le}\frac{Q}{C_P}\nabla Y)]. \end{aligned} \tag{4.3.2}$$

In the isobaric approximation, the terms related to the pressure derivatives in (4.3.2) and to the viscous stress are small as $M^2 \ll 1$ and may be neglected compared to the terms including changes of temperature and the fuel fraction. Therefore, the equation of energy is reduced to the equation of thermal conduction

$$\rho\left(\frac{\partial}{\partial t}+\mathbf{v}\cdot\nabla\right)(C_P T+QY)=\nabla\cdot\left[\kappa\left(\nabla T+\mathrm{Le}\frac{Q}{C_P}\nabla Y\right)\right]. \tag{4.3.3}$$

Thus, in the isobaric approximation, flame dynamics is described by the hydrodynamic Eqs. (4.1.1) and (4.1.2), by the equation of thermal conduction (4.3.3), by the equation of chemical kinetics (4.1.4), and by the equation of state in the form (4.3.1).

In the case of equally strong thermal conduction and fuel diffusion (when the Lewis number is Le=1), further simplification is possible [28]. In this case, the internal enthalpy is conserved, $h = C_P T + QY = C_P T_u + Q$, and the fuel fraction is related to temperature as

$$Y = 1 - C_P(T - T_u)/Q. \tag{4.3.4}$$

In the fuel (unburned gaseous mixture) one has $Y = 1$, while in the combustion products $Y = 0$ and the final temperature of the burned products can be calculated as $T_b = T_u + Q/C_P$. Taking into account Eq. (4.3.1), we find out that in the isobaric approximation the ratio of the temperature of the burned matter and the temperature of the fuel as well as the ratio of densities are determined by the expansion coefficient, $\Theta \equiv \rho_u/\rho_b = T_b/T_u$. Therefore, Θ can be expressed, using $T_b = T_u + Q/C_P$, through the energy released in the reaction

$$\Theta = 1 + Q/C_P T_u. \tag{4.3.5}$$

It should be pointed out that the relation between the expansion coefficient Θ and the energy release Q given by (4.3.5) is valid for any isobaric flame independent of the value of the Lewis number. Due to the enthalpy conservation, the equation of thermal conduction (4.3.3) and the equation of chemical kinetics (4.1.4) may be replaced by a single equation of thermal conduction with the account of the energy released in the reaction expressed in terms of temperature only

$$\rho\frac{\partial T}{\partial t}+\rho\mathbf{v}\cdot\nabla T=\nabla\cdot\left(\frac{\kappa}{C_P}\nabla T\right)+W_R(T) \tag{4.3.6}$$

with the reaction rate given by the formula

$$W_R(T)=\frac{\rho_f^n T_u}{(\Theta-1)^{n-1}\rho_R^{n-1}\tau_R}\left(\frac{T_b-T}{T}\right)^n\exp(-E/T). \tag{4.3.7}$$

Besides, it was found that in a majority of practically interesting cases, viscosity is of no importance for flame dynamics. For this reason in many problems, one can use simplified equations of hydrodynamics for an ideal gas for analysis of the flame dynamics

$$\frac{\partial \rho}{\partial t} + \nabla \cdot (\rho \mathbf{v}) = 0, \tag{4.3.8}$$

$$\rho \frac{\partial \mathbf{v}}{\partial t} + \rho(\mathbf{v} \cdot \nabla)\mathbf{v} = -\nabla P + \rho \mathbf{g}. \tag{4.3.9}$$

Quite often, the simplified set of Eqs. (4.3.6)–(4.3.9) and the isobaric approximation are used for the analytical investigation of flame dynamics instead of the complete system of hydrodynamic Eqs. (4.2.1)–(4.2.7).

In the case of slow combustion (deflagration), the isobaric approximation means that conservation of the momentum flux and energy flux, and continuity of the pressure and enthalpy, $P_1 = P_2$, $H_1 = H_2$. Thus, we can write

$$H_1 = H_{01} + C_{P1}T_1 = H_2 = H_{02} + C_{P2}T_2, \tag{4.3.10}$$

where $H_{01} - H_{02} = Q$ is the heat released in the reaction, and we have the following relations between thermodynamic variables of the unburned and burned flow on both sides of the propagating flame:

$$P_1 = P_2, \quad T_2 = (Q + C_{P1}T_1)/C_{P2}, \quad \frac{\rho_1}{\rho_2} = \frac{u_2}{u_1} = \frac{\gamma_1(\gamma_2 - 1)}{\gamma_2(\gamma_1 - 1)}\left(1 + \frac{Q}{C_{P1}T_1}\right). \tag{4.3.11}$$

4.4 Theory of Planar Flames

Let us consider a planar stationary flame propagating with the velocity U_f in the negative direction along the z-axis. In the reference frame co-moving with the flame, the flow consists of a uniform flow of the fresh fuel mixture moving with the velocity $u_z = U_f$ toward the flame front, of the transition region where heating and the reaction are in progress (which is the flame front itself) and of a uniform flow of the burned gas flowing away from the flame front. Our purpose is to find the structure of the heating and reaction region and the velocity of the planar flame front as a function of the thermal and chemical fuel parameters.

In co-moving with the planar stationary flame coordinate system, the hydrodynamic Eqs. (4.3.8) and (4.3.9) take the form

$$\frac{d}{dz}(\rho u_z) = 0, \tag{4.4.1}$$

$$\frac{d}{dz}\left(P + \rho u_z^2\right) = 0. \tag{4.4.2}$$

The corresponding integrals across the flame front are

$$\rho u_z = \rho_u U_f, \tag{4.4.4}$$

$$P + \rho u_z^2 = P_u + \rho_u U_f^2. \tag{4.4.5}$$

Using Eq. (4.4.4), the thermal conduction equation takes the form

$$\rho_u U_f \frac{dT}{dz} = \frac{d}{dz}\left(\frac{\kappa}{C_P}\frac{dT}{dz}\right) + \frac{\rho_u T_u}{\tau_R}\frac{T_b - T}{T}\exp(-E/T) \tag{4.4.6}$$

with the boundary conditions in the fuel ahead of the flame and in the burned matter:

$$T(z = -\infty) = T_u; \quad T(z = \infty) = T_b; \ (notice\, that\ u_u \equiv U_f). \tag{4.4.7}$$

The burned matter temperature depends only on the fuel temperature and the energy released in the reaction and may be expressed through the expansion coefficient as $T_b = \Theta T_u$. Equation (4.4.6) together with the boundary conditions (4.4.7) constitutes an eigenvalue problem, where the flame velocity U_f is the eigenvalue and the flame internal structure T=T(z) is the eigenfunction.

The Zel'dovichFrank-Kamenetskii theory of a planar flame [28, 31] is based on the fact that the activation energy of the reaction is usually very large, $E/T > E/T_b \gg 1$. Because of the strong Arrhenius dependence, the last term in (4.4.6) is exponentially small and differs from zero only in a thin reaction zone (that is as it was assumed by Mallard and LeChatelier [32]), where temperature is nearly equal to the final value, $(T_b - T)/T_b \sim T_b/E$. Outside the reaction zone, the equation of thermal conduction becomes

$$\rho_u U_f \frac{dT}{dz} = \frac{d}{dz}\left(\frac{\kappa}{C_P}\frac{dT}{dz}\right) \tag{4.4.8}$$

and after straightforward integration, we obtain the temperature distribution in the flame outside the reaction zone

$$\frac{T(z)}{T_u} = \begin{cases} 1 + (\Theta - 1)\exp\left(z/L_f\right), & z < 0 \\ \Theta, & z > 0 \end{cases} \tag{4.4.9}$$

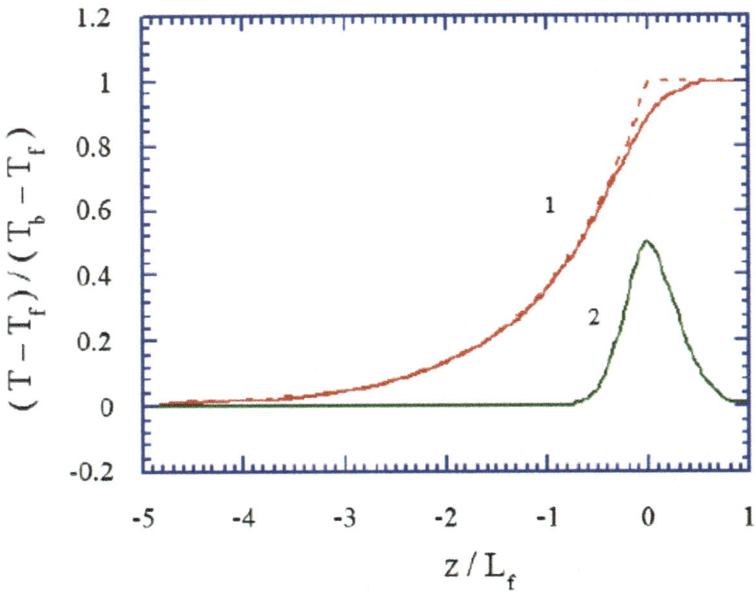

Fig. 4.2 Distribution of the scaled temperature (curve 1) and reaction rate (curve 2) scaled by the double maximum value in a planar flame with the expansion coefficient $\Theta = 8$ and the activation energy $E/RT_b = 8.75$. The dashed line corresponds to the analytical solution (4.4.9)

with other hydrodynamic variables depending on the temperature as $\rho = \rho_u T_u/T$, $u_z = U_f T/T_u$, $P = P_u + \rho_u U_f^2(1 - T/T_u)$. The solution (4.4.9) for the temperature distribution in a planar flame is shown in Fig. 4.2 by the dashed curve. The solid curve in Fig. 4.2 depicts the numerical solution of the eigenvalue problem (4.4.6)–(4.4.7) for the case of the expansion coefficient $\Theta = 8$ and the activation energy of the reaction $E/T_b = 8.75$. One can see that the analytical solution is in good agreement with the numerical one everywhere except the thin zone of active reaction, where the discontinuity of the temperature derivative of (4.4.9) is smoothed. The distributions of density, velocity, and concentration for the same flame parameters are shown in Fig. 4.3.

The characteristic thickness of the reaction zone L_R where temperature is close to the final value, $(T_b - T)/T_b \sim T_b/E$, is much smaller than the total thickness of the flame front, $L_R \sim L_f T_b/E \ll L_f$. For this reason, the term with the first derivative of temperature in Eq. (4.4.6) may be neglected compared to the term with the second derivative, and the equation of thermal conduction takes the form

$$\frac{\kappa_b}{C_f}\frac{d^2T}{dz^2} + Q\frac{\rho_u T_u}{\tau_R}\frac{T_b - T}{T_b}\exp(-E/T) = 0, \tag{4.4.10}$$

where $\kappa_b = \kappa(T_b)$, and the boundary conditions outside the reaction zone can be written as

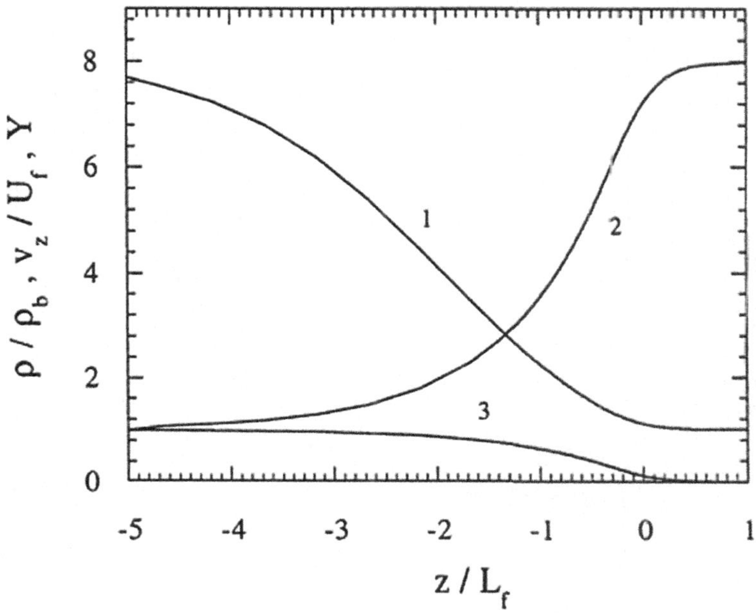

Fig. 4.3 Distribution of the scaled density (curve 1) velocity (curve 2) and the fuel fraction in a planar flame with the expansion coefficient $\Theta = 8$ and the activation energy $E/RT_b = 8.75$

$$\frac{dT}{dz} = (\Theta - 1)\frac{T_u}{L_f}, \ for \ \frac{T_b - T}{T_b} \gg \frac{T_b}{E} \tag{4.4.11}$$

ahead of the reaction zone, and

$$\frac{dT}{dz}(T = T_b) = 0 \tag{4.4.12}$$

behind the reaction zone. In the solution (4.4.9), the eigenvalue U_f is "hidden" in the definition for the flame thickness $L_f = \kappa_b/\rho_u C_p U_f$ in the boundary condition (4.4.11).

Figure 4.4 shows the calculated in [33] profiles of the scaled temperature, $\Theta = T/T_u$, for different values of the activation energy, $\varepsilon = E/T_u$ and thermal expansion, $\Theta_2 = T_b/T_u$ in comparison with the Zel'dovich and Frank-Kamenetskii analytical solution (4.4.9). It is seen from Fig. 4.4 that the analytical solution (4.4.9) shown by curve (4) is a very good approximation for a wide range of activation energies and thermal expansion.

Multiplying (4.4.10) by the temperature derivative and using the Frank-Kamenetskii approximation for the Arrhenius function in the reaction zone

$$\exp(-E/T) \approx \exp(-E/T_b)\exp\left(E\frac{T - T_b}{T_b^2}\right), \tag{4.4.13}$$

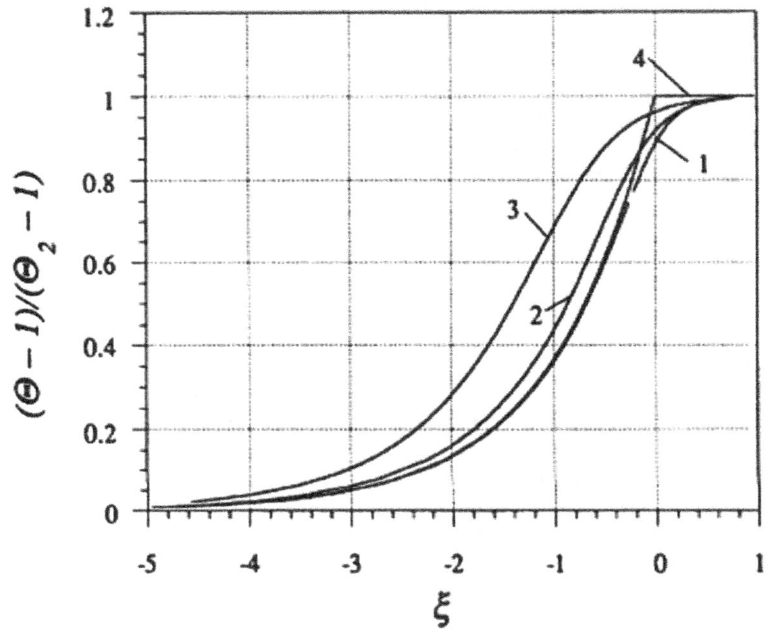

Fig. 4.4 Calculated scaled profiles (from [33]) of the dimensionless temperature in a deflagration wave. (1) $\varepsilon = 70$, $\Theta_2 = 8$, $\ln \Lambda = 14.3$; (2) $\varepsilon = 42$, $\Theta_2 = 6$, $\ln \Lambda = 11.7$; (3) $\varepsilon = 21$, $\Theta_2 = 3$, $\ln \Lambda = 10.4$. (4) corresponds to analytical solution (4.4.9). Here, $\Lambda = L_f/U_f\tau_R$ is dimensionless parameter, pre-exponential factor in (4.3.7); $\xi \equiv z/L_f$

and after integration across the reaction zone we obtain the equation for the flame thickness

$$\frac{\kappa_b}{2C_P}\frac{T_u^2(\Theta-1)^2}{L_f^2} = \frac{\rho_u T_u}{\tau_R T_b}\frac{T_b^4}{E^2}\exp(-E/T_b). \tag{4.4.14}$$

Using the definition for the flame thickness, $L_f = \kappa_b/\rho_u C_p U_f$, we obtain from (4.4.14) for the flame velocity

$$U_f = \left(\frac{2\kappa_b}{C_P\rho_u\tau_R}\right)^{1/2}\frac{\sqrt{\Theta}}{(\Theta-1)}\frac{T_b}{E}\exp\left(-\frac{E}{2T_b}\right). \tag{4.4.15}$$

In agreement with the estimate for the flame speed obtained in Chap. 3 from the dimensional analysis, the flame velocity is proportional to the square root of thermal conduction and inversely proportional to the square root of the characteristic reaction time $\tau_b^{1/2} \sim \tau_b^{1/2}\exp(E/2T_b)$:

$$U_f \propto \left(\frac{\kappa_b}{C_P\rho_u\tau_R}\right)^{1/2}\exp\left(-\frac{E}{2T_b}\right). \tag{4.4.16}$$

The Zel'dovich–Frank-Kamenetskii formula (4.4.15) is in good agreement with the numerical solution of the eigenvalue problem (4.4.6)–(4.4.7). Particularly, for the flame with the expansion coefficient $\Theta = 8$ and the activation energy $E/T_b = 8.75$ the analytical formula (4.4.15) gives the dimensionless value $\ln(L_f/U_f\tau_R) = 14.2$, while the result of numerical calculations [33] is $\ln(L_f/U_f\tau_R) = 14.3$.

In a general case of an arbitrary reaction order, n, and an arbitrary Lewis number different from unity, the expression for the velocity of a planar flame becomes

$$U_f = \left(\frac{2n!}{(\text{Le})^n\Theta}\frac{\kappa_b}{C_P\rho_u\tau_R}\right)^{1/2}\left(\frac{\rho_b}{\rho_R}\right)^{\frac{n-1}{2}}\left(\frac{\Theta T_b}{[\Theta-1]E}\right)^{\frac{n+1}{2}}\exp\left(-\frac{E}{2T_b}\right). \quad (4.4.17)$$

Formula (4.4.17) can be derived using the set of isobaric hydrodynamic equations and the equation of state using the same method as was used for derivation (4.4.15).

According to (4.4.17), the flame velocity depends on the pressure as $U_f \propto P^{(n/2)-1}$: the flame propagates slower at higher pressures in the case of the first-order reaction, while flame velocity increases with pressure for the third-order reaction. For flames with second-order reactions, the flame velocity is independent of the fuel pressure. Also, the flames of higher order reactions are more sensitive to the increase of the initial temperature.

Strictly speaking, the one-step reaction takes place at any finite temperature $T_u \neq 0$ and has a finite rate ahead of the flame, at $z = -\infty$. Since it proceeds an infinite time while it flows toward the flame, the fuel is completely consumed before it reached the flame front. This means that the governing Eqs. (4.4.4)–(4.4.7) are actually ill-defined mathematically. According to Ya. Zel'dovich, we should think that there is an imaginary transparent net, which cools the mixture to zero temperature and stops the reaction at some distance ahead of the flame. This also can be interpreted as an ignition temperature, T_{ig}, such that the reaction rate vanishes at $T < T_{ig}$. This assumption is physically reasonable because at large activation energy the reaction rate is negligibly (exponentially) small everywhere except the reaction zone.

References

1. W.H. Manheimer, D.G. Colombant, G.H. Gardner, Steady-state planar ablative flow. Phys. Fluids **25**, 1644–1652 (1982)
2. A.B. Bud'ko, M.A. Liberman, Suppression of the Rayleigh-Taylor instability by convection in ablatively accelerated laser target. Phys. Rev. Lett. **68**, 178–181 (1992)
3. V.V. Bychkov, S.M. Golberg, M.A. Liberman, Rayleigh-Taylor instability of combustion and laser produced ablation fronts. Phys. Fluids B **5**, 3822–3824 (1993)
4. V.V. Bychkov, S.M. Golberg, M.A. Liberman, Self-consistent theory of the Rayleigh-Taylor instability in ablatively accelerated laser plasma. Phys. Plasmas **1**, 2976–2986 (1994)
5. V.V. Bychkov, S.M. Golberg, M.A. Liberman, Growth of the Rayleigh-Taylor instabilities in an inhomogeneous ablative accelerated laser plasma. Sov. Phys. JETP **99**, 1162–1185 (1991)
6. P. Clavin, L. Masse, Instabilities of ablation fronts in inertial confinement fusion: a comparison with flames. Phys. Plasmas **11**, 690–705 (2004)

7. R. Betti, V. Goncharov, R. McCrory, C. Verdon, Self-consistent stability analysis of ablation fronts with small Froude numbers. Phys. Plasmas **3**, 4665–4676 (1996)
8. R. Betti, V. Goncharov, R. McCrory, C. Verdon, Growth rates of the ablative Rayleigh-Taylor instability in inertial confinement fusion. **5**, 1446–1454 (1998)
9. J. Sanz, L. Masse, P. Clavin, The linear Darrieus-Landau and Rayleigh-Taylor instabilities in inertial confinement fusion revisited. Phys. Plasmas **13**, 102702 (2006)
10. A. Piriz, R. Portugues, Landau-Darrieus instability in an ablation front. Phys. Plasmas **10**, 2449–2456 (2003)
11. S. Woosley, T. Weaver, The physics of supernova explosions. Annu. Rev. Astron. Astrophys. **24**, 205–253 (1986)
12. J.C. Niemeyer, W. Hillebrandt, Microscopic instabilities of nuclear flames in type Ia supernovae. Astrophys. J. **452**, 769–784 (1995)
13. V.V. Bychkov, M.A. Liberman, Thermal instability and pulsations of the flame front in white dwarfs. Astrophys. J. **451**, 711–716 (1995)
14. V.V. Bychkov, M.A. Liberman, Self-consistent theory of white dwarf burning in the supernova Ia events. Astrophys. Space Sci. **233**, 287–292 (1995)
15. K. Nomoto, K. Iwamoto, N. Kishimoto, Type Ia supernovae: their origin and possible applications in cosmology. Science **276**, 1378–1382 (1997)
16. V.V. Bychkov, M.A. Liberman, Flame instabilities and models of white dwarf burning, in "*Thermonuclear Supernovae*", ed. by P. Ruiz-Lapuente, R. Canal, & J. Isern (Dordrecht: Kluwer, 1996), Conference on Thermonuclear Supernovae, NATO Institute for Advanced Science, (Aiguablava, Spain, June 20–30, 1995), pp. 379–388
17. V. Gamezo, A. Khokhlov, E. Oran, A. Chtchelkanova, R. Rosenberg, Thermonuclear supernovae: simulations of the deflagration stage and their implications. Science **299**, 77–81 (2003)
18. S.E. Woosley, Type la supernovae: burning and detonation in the distributed regime. Astrophys. J. **668**, 1109–1117 (2007)
19. F.K. Röpke, J.C. Niemeyer, Delayed detonations in full-star models of type Ia supernova explosions. Astron. & Astrophys. **464**, 683–686 (2007). https://doi.org/10.1051/0004-6361:20066585
20. M. Fink, M. Kromer, W. Hillebrandt, F.K. Röpke, R. Pakmor, I.R. Seitenzahl, S.A. Sim, Thermonuclear explosions of rapidly differentially rotating white dwarfs: candidates for superluminous type Ia supernovae? Astron. & Astrophys. **618**, A124 (2018)
21. A. Tanikawa, K. Nomoto, N. Nakasato, Three-dimensional simulation of double detonations in the double-degenerate model for type Ia supernovae and interaction of ejecta with a surviving white dwarf companion. Astrophys. J. **868**, 12–90 (2018)
22. M.A. Liberman, A.T. Rakhimov, Hydrodynamic instability of the high frequency gas discharge. Phys. Lett. **38A**, 61–63 (1972)
23. D.L. Turcotte, R.S.B. Ong, The structure and propagation of ionizing wave fronts. J. Plasma Phys. **2**, 145–155 (1968)
24. A.N. Lagarkov, I.M. Rutkevich, *Ionization Waves in Electrical Breakdown of Gases* (Springer-Verlag, New York Berlin Heidelberg, 1994)
25. I. Rutkevich, Two-dimensional instability of fast ionization waves propagating in an external electric field. Phys. Plasmas **5**, 3054–3064 (1998)
26. U. Ebert, W. van Saarloos, C. Caroli, Propagation and structure of planar streamer fronts. Phys. Rev. E **55**, 1530–1549 (1997)
27. P. Rodin, U. Ebert, A. Minarsky, I. Grekhov, Theory of superfast fronts of impact ionization in semiconductor structures. J. Appl. Phys. **102**, 034508 (2007)
28. Y.B. Zeldovich, G.I. Barenblatt, V.B. Librovich, G.M. Makhviladze, *The Mathematical Theory of Combustion and Explosion* (Consultants Bureau, New York, 1985)
29. M. Liberman, A. Velikovich, *Physics of Shock Waves in Gases and Plasmas* (Springer-Verlag, Berlin - New York, 1985)
30. V.V. Bychkov, M.A. Liberman, On the stability of a flame in a closed chamber. Phys. Rev. Lett. **78**, 1371–1734 (1997)

31. Y.B. Zel'dovich, D.A. Frank-Kamenetski, A theory of thermal propagation of flame. Acta Physicochimica. **9**, 341–350 (1938)
32. E. Mallard, H. Le Chatelier, Recherches Experimentales et Theoriques sur la Combustion des Melanges Gaseoux Explosifs. Ann. Mines **8**, 274–568 (1883)
33. M.A. Liberman, V.V. Bychkov, S.M. Golberg, D. Book, Stability of a planar flame front in the slow-combustion regime. Phys. Rev. E **49**, 445–457 (1994)

Chapter 5
Flame Instabilities

5.1 Darrieus-Landau Instability of Zero Thickness Flame

Combustion is characterized by various instabilities, which are of fundamental and practical interest, since they affect the dynamics of combustion and the operation of industrial combustible systems. The simplest scenario of a premixed combustion is the propagation of a planar flame in a homogeneous combustible mixture. However, in real life, flame fronts are rarely planar. In a quiescent combustible mixture, the flame propagates at a constant speed, which is determined by the thermodynamic and mechanical properties of the mixture. A flame propagating in a tube acquires a curved shape and almost never propagates as a planar stationary front. The flame front is stretched along the tube walls due to no-slip boundary conditions. In the case of rough tube walls, a transition to a turbulent flame can occur, which is accompanied by a considerable amplification of the flame velocity. Besides, the flame front may become spontaneously wrinkled because of the hydrodynamic instability, which was independently discovered in the theoretical studies by G. Darrieus [1] and L. Landau [2, 3]. Known a as the Darrieus–Landau (DL) instability, it is a consequence of the gas expansion caused by the heat released in combustion. Due to the Darrieus—Landau instability, the initial small perturbations of the flame front grow exponentially with time.

To begin with, we will consider a flame as a zero-thickness surface, which is a discontinuity surface separating fresh combustible mixture from burned gas products. It is known that a necessary condition for the existence a solution to a flow with a discontinuity is the "evolutionarity condition" [4, 5]. The evolutionarity condition is satisfied if the number of boundary conditions at the discontinuity is one more than the total number of simple waves that can run away of both sides of the discontinuity surface. For example, the direction of quantities variation in the shock wave, and that the shock velocity is greater than sound speed is known in hydrodynamics as the Zemplen theorem [3]. Let us show that this immediately follows from the evolutionarity condition. In this case there are 3 boundary conditions, which means that only two simple waves can run away from the shock front. Behind the shock

M. A. Liberman, *Combustion Physics*,
https://doi.org/10.1007/978-3-030-85139-2_5

there are two such waves: sound wave and entropy wave going along with the flow. This means that in order to satisfy the evolutionarity condition, a sound wave cannot leave the shock wave upstream, which is possible only if the shock velocity is greater than the sound speed.

In general, the boundary conditions at the discontinuity surface consist of the 'basic' boundary conditions that follow from the conservation laws, which are conservation laws at the discontinuity surface, and an additional "boundary" condition that does not follow from the conservation laws [4]. This additional boundary condition can be obtained only by solving the complete problem, considering a front of a small but finite thickness. For example, considering the flame front as a discontinuity surface, this additional condition is the certain value of the flame velocity. The flame velocity can be found in turn as the eigenvalue of the complete system of equations, including thermal conduction, diffusion, the energy release in chemical reactions, etc. It is obvious that, if the additional condition is required for the unperturbed problem, then the stability problem also requires an additional condition. Sometimes to justify the deficit of the condition at the discontinuity, an additional condition is introduced as a "physically reasonable" one. The sensitivity of the solution obtained in this way and the accuracy of the assumed boundary condition can be verified only by solving the complete problem with finite thickness and real structure of the fronts. The most prominent example of such a problem is the intrinsic hydrodynamic, Darrieus– Landau instability of the planar flame front. Another problem, where similar difficulties arise due to the lack of the boundary condition, is, for example, stability of ablation wave in inertial confined fusion in the approximation when the ablation front is considered as a discontinuity surface. Thus formulated, this problem leads to a similar difficulty: the number of 'basic' boundary conditions at the ablation surface is insufficient to determine the solution [6, 7].

The linear stability analyses of Darrieus and Landau raised a number of questions, primarily about their assumption that the normal propagation velocity of the flame relative to the unburned gas remains unchanged in the presence of perturbations. Fortunately, the additional condition proposed by Landau turned out to be correct, but, as will be shown below, it is valid only in the long-wavelength approximation, such that $L_f/\lambda \ll 1$. If we consider a flame as a discontinuity surface, the only dimensional variables that can be used to construct the dimension of the growth rate of instability are the flame velocity, U_f and the wave number $k = 2\pi/\lambda$, so that in linear approximation the growth rate of instability should be $\sigma = kU_f\Gamma(\Theta)$. Here, $\Gamma(\Theta)$ is some function of the expansion coefficient $\Theta = \rho_u/\rho_b$ of the order of unity. The particular value of λ for which the asymptotic solution is applicable cannot be found using the discontinuity model, and requires to solve the problem for the finite thickness of the flame front.

The DL instability is inherent to all flames in gaseous mixtures, since the instability is related to the gas expansion in exothermal reactions and, it appears to be universal, explaining the instability of the flame fronts, ionizing waves, ablation wave in laser inertial fusion, etc. In the approach used by Darrieus and Landau a flame front is treated as a surface of zero thickness separating the fresh fuel and the burnt matter and propagating with a constant velocity U_f with respect to the fuel, which is considered as an external parameter.

We consider a planar flame propagating in the positive direction of the z-axis normal to the flame front. In the reference frame co-moving with the flame front, the flow consists of a discontinuity surface being at rest at $z = 0$. The uniform upstream flow enters the flame front with velocity U_f and flows out of the flame with velocity $u_2 = \Theta U_f$. Since any perturbation can be represented as a Fourier series, we can look for perturbations of the unperturbed stationary solution in the form of a periodic function of time and coordinate:

$$F'(y, z, t) = F'(z) \exp(\sigma t + iky), \tag{5.1.1}$$

where $F'(z)$ is the amplitude of the perturbed variable, y is the coordinate in plane of the flame front, $k = 2\pi/\lambda$ is the perturbation wave-number, λ is the wavelength of the perturbation, σ is the growth rate of instabilities. Our purpose is to find the dispersion relation that is the growth rate as a function of the wave number $\sigma = \sigma(k)$. Flame instability means that the real part of complex σ is positive, $\mathrm{Re}[\sigma] > 0$. In general, the perturbations (5.1.1) should be taken in a three-dimensional (3D) form $F' \exp(\sigma t + i\mathbf{kr})$, but there is no difference between 2 and 3D configurations since we always can take $\mathbf{k}$ along the y-axis.

Following the method, used by Landau, we consider the flame front as a discontinuity surface separating the flow of the unburned gas upstream and of the burned gas downstream. Since the flow is essentially subsonic, it can be treated as isobaric and incompressible and in addition, the upstream flow is potential. The linearized equations for small perturbations of the velocity, $u' \ll u$ and pressure, $P' \ll P$ have the form:

$$\nabla \cdot \mathbf{u}' = 0, \quad \frac{\partial \mathbf{u}'}{\partial t} + (\mathbf{u}\nabla)\mathbf{u}' = -\frac{1}{\rho}\nabla P'. \tag{5.1.2}$$

Considering the flow velocity is constant on the both sides of the flame front, and applying the operator ∇ to both sides of the second equation in (5.1.2), we obtain

$$\nabla^2 P' = \Delta P' = \frac{\partial^2 P'}{\partial z^2} + \frac{\partial^2 P'}{\partial y^2} = 0. \tag{5.1.3}$$

Substituting $P_1' = f(z) \exp(-i\omega t + iky)$ into (5.1.3), we obtain solution for $f(z)$

$$f(z) = Ae^{kz} + Be^{-kz}. \tag{5.1.4}$$

Since all the perturbed quantities must vanish for $z \to -\infty$, we take $B = 0$, so that

$$P_1' = Ae^{-i\omega t + iky + kz}. \tag{5.1.5}$$

Substituting (5.1.5) into Eqs. (5.1.2) and (5.1.3) we obtain

$$-i\omega u'_{1z} + ku_1 u'_{1z} = -A\frac{k}{\rho_1}e^{-i\omega t+ky+kz}, \quad u'_{1y} = iu'_{1z}, \tag{5.1.6}$$

Thus, from (5.1.6) we obtain

$$u'_{z1} = A\frac{k}{\rho_1(ku_1 - i\omega)}\mathrm{e}^{-i\omega t+kz+iky}. \tag{5.1.7}$$

Since we are interested in unstable solutions when $\mathrm{Im}\{\omega\} > 0$, we introduce $\sigma \equiv -i\omega$. The unstable solutions then correspond to real positive value of σ. Note that there is another solution of the Euler equation (5.1.2) corresponding to the uniform equation with $P' = 0$ in the second Eq. (5.1.2):

$$\frac{\partial \mathbf{u}'}{\partial t} + (\mathbf{u}\nabla)\mathbf{u}' = 0. \tag{5.1.8}$$

However, solutions of (5.1.8) for real positive σ

$$u'_z = C\,e^{\sigma t+iky-\frac{\sigma}{u_1}z} \quad and \quad u'_y = C\,e^{\sigma t+iky-\frac{\sigma}{u_1}z}$$

tends to infinity at $z \to \infty$, so that they must be considered only at $z > 0$. Thus, at $z < 0$ we have

$$P'_1 = A\mathrm{e}^{-i\omega t+iky+kz}, \tag{5.1.9}$$

$$u'_{z1} = -A\frac{k}{\rho_1(\sigma + ku_1)}\mathrm{e}^{\sigma t+kz+iky}, \tag{5.1.10}$$

$$u'_{y1} = -A\frac{ik}{\rho_1(\sigma + ku_1)}\mathrm{e}^{\sigma t+kz+iky}. \tag{5.1.11}$$

In a similar way, at $z > 0$, we obtain solutions

$$P'_2 = B\mathrm{e}^{-i\omega t+iky-kz}, \tag{5.1.12}$$

$$u'_{z2} = \left(B\frac{k}{\rho_2(\sigma - ku_2)}\mathrm{e}^{-kz} + C\mathrm{e}^{-\frac{\sigma}{u_2}z}\right)\mathrm{e}^{\sigma t+iky}, \tag{5.1.13}$$

$$u'_{y2} = \left(B\frac{-ik}{\rho_2(\sigma - ku_2)}\mathrm{e}^{-kz} + C\frac{i\sigma}{ku_2}\mathrm{e}^{-\frac{\sigma}{u_2}z}\right)\mathrm{e}^{\sigma t+iky}, \tag{5.1.14}$$

where the solution of the uniform Eq. (5.1.8) was include in expressions (5.1.13–5.1.14).

Let $\zeta = \zeta(y, t) = D \exp(\sigma t + iky)$ be the displacement of the flame surface caused by small perturbations. On the surface of the flame front considered as a discontinuity surface must be satisfied the following conditions at $z = 0$. Continuity of pressure:

$$P_1'(0) = P_2'(0). \tag{5.1.15}$$

Continuity of the tangential velocity component

$$u_{y1}' + u_1 \frac{\partial \zeta}{\partial y} = u_{y2}' + u_2 \frac{\partial \zeta}{\partial y}. \tag{5.1.16}$$

Finally, it is assumed that the flame velocity in the direction normal to the flame surface remains unchanged, which means that

$$u_{z1}' - \frac{\partial \zeta}{\partial t} = u_{z2}' - \frac{\partial \zeta}{\partial t} = 0. \tag{5.1.17}$$

Substituting solutions (5.1.9)–(5.1.11) and (5.1.12)–(5.1.14) into the boundary conditions (5.1.15)–(5.1.17) at $z = 0$, we obtain a system of four uniform equations for the coefficients A, B, C, D, which using equation $\rho_1 u_1 = \rho_2 u_2$, is written in the form

$$A = B = -\frac{\sigma}{k} \rho_1 (\sigma + k u_1) D, \tag{5.1.18}$$

$$C = \sigma^2 \frac{u_1 + u_2}{u_1(\sigma - k u_2)} D. \tag{5.1.19}$$

After a simple algebra the compatibility condition (the existence of a nontrivial solution) of the equations for the coefficients A, B, C, D is reduced to

$$\sigma^2 (u_1 + u_2) + 2\sigma u_1 u_2 + k^2 u_1 u_2 (u_1 - u_2) = 0. \tag{5.1.20}$$

Since $u_1 = U_f$, and as a consequence of the flux of mass continuity, $u_2 = \Theta U_f$, we can rewrite (5.1.20) as

$$\sigma^2 + \frac{2\Theta}{\Theta + 1} U_f k \sigma - \Theta \frac{\Theta - 1}{\Theta + 1} U_f^2 k^2 = 0. \tag{5.1.21}$$

Since $\Theta = \rho_1/\rho_2 = u_2/u_1 > 1$, both roots of the quadratic Eq. (5.1.21) are real and the positive root corresponds to instability:

$$\sigma = U_f k \Gamma(\Theta) = U_f k \frac{\Theta}{\Theta + 1} \left(\sqrt{\Theta + 1 - 1/\Theta} - 1 \right). \tag{5.1.22}$$

The positive root (5.1.22) means that the perturbations of the flame front grow exponentially in time. Pure imaginary ω means that perturbations in the flame front grow locally, they do not propagate along the flame front. Formally, the obtained solution means that the flame front is absolutely unstable against perturbations of any wavelength: the larger is the wavenumber (shorter the wavelength), the larger is σ and the faster is growing perturbations.

From Eq. (5.1.21) it is seen that there is no instability if there is no gas expansion across the flame front, since $\Gamma(\Theta = 1) = 0$. The factor Γ increases with the increase of the expansion coefficient Θ, so that the larger the difference between the fuel density and the density of the burnt products, the stronger the DL instability. If the density of the burning matter changes only slightly, $\Theta - 1 \ll 1$, then the instability growth rate becomes as small as $\sigma/U_f k = (\Theta - 1)/2$. For the value of Θ corresponding to real flames the growth rate only slightly changes: from $\Gamma(\Theta = 6) = 1.437$, till $\Gamma = 2.092$ for $\Theta = 10$. Though formula (5.1.22) was obtained for 2D configuration, it also holds for a 3D case with the only replacement of k by the absolute value of the wavenumber vector $|\mathbf{k}|$.

Note, that the DL theory predicts the growth of perturbations for all wavenumbers: the shorter the wavelength of the perturbation, the faster it grows. Since the solution was obtained assuming a zero thickness of the flame, it holds only for the wavelength exceeding the flame width, $\lambda \gg L_f$, or $kL_f \ll 1$. Hence, flame instability can be expected to lead to the formation of 2D or 3D structures at the flame front with length scales ranging from flame thickness to some larger size. In scope of the discontinuity model we do not know how strong is inequality $\lambda \gg L_f$. Experimental studies of the DL instability have shown that the characteristic length scale of the instability is at least two orders of magnitude larger than the flame thickness, which cannot be explained by the theory with a discontinuity surface.

In the derivation of the DL instability we assumed that the flow is potential ahead of the flame front. According to Kelvin's theorem of the conservation of circulation, the flow remains potential everywhere ahead of the flame, at $x < 0$. Using formulas (5.1.10), (5.1.11), (5.1.13), (5.1.14) and (5.1.18), (5.1.19) we obtain for the velocity components

$$u'_{z1} = \sigma D e^{-i\omega t + kz + ikx}, \tag{5.1.22}$$

$$u'_{y1} = i\sigma D e^{-i\omega t + kz + ikx}, \tag{5.1.23}$$

$$u'_{z2} = \left(\frac{\sigma^2 (u_1 + u_2)}{u_1(\sigma - ku_2)} e^{-\frac{\sigma}{u_2} z} - \frac{\sigma u_2(\sigma + ku_1)}{u_1(\sigma - ku_2)} e^{-kz} \right) D e^{\sigma t + ikx}, \tag{5.1.24}$$

$$u'_{x2} = i \left(\frac{\sigma u_2(\sigma + ku_1)}{u_1(\sigma - ku_2)} e^{-kz} - \frac{\sigma^3 (u_1 + u_2)}{u_1 u_2 k(\sigma - ku_2)} e^{-\frac{\sigma}{u_2} z} \right) D e^{\sigma t + ikx}. \tag{5.1.25}$$

Calculating the vorticity, $\boldsymbol{\omega} = \nabla \times \mathbf{u}$, we find that ahead of the flame front vorticity is zero

$$\omega_1 = \frac{\partial u'_{z1}}{\partial x} - \frac{\partial u'_{x1}}{\partial z} = 0, \tag{5.1.26}$$

however, behind the flame, in the combustion products, vorticity is not zero

$$\omega_2 = \frac{\partial u'_{z2}}{\partial x} - \frac{\partial u'_{x2}}{\partial z} \propto i\frac{\sigma^3(\sigma + ku_1)(u_1 + u_2)}{ku_1u_2^2(\sigma - ku_2)} De^{\sigma t - \frac{\sigma}{u_2}z + ikx} \neq 0. \tag{5.1.27}$$

Thus, the flame instability generates vorticities in the flow behind the flame.

5.2 Hydrodynamic Instability; Flames of Finite Thickness

Since many experimental results have shown a spontaneous transition from a laminar regime of flame propagation to a turbulent flame, it was believed at the beginning that the DL instability leads to turbulization of the flame. Later, Markstein [8] and Zel'dovich [9] proposed phenomenological models accounting the stabilizing role of curvature and strain (through diffusion) that can be parameterized in linear stability analysis by one or two length scales called the Markstein length and therefore leading to the cut-off wavelength depending on the curvature of the flame of finite thickness. Markstein [8] introduced the modified DL model, where the correction to the flame speed arising from the speed dependence on the local stretch rate was mimics by the Markstein length L

$$U_f = U_{f0}(1 - \mathbf{K}\mathrm{L}), \tag{5.2.1}$$

where U_{f0} is a normal velocity of a plane deflagration, and **K** is the local curvature of the flame front. The dispersion relation with this "stabilizing" term becomes

$$\sigma^2(\Theta + 1) + 2\sigma\Theta kU_f(1 + k\mathrm{L}) + k^2U_f^2\Theta(1 + 2k\Theta\mathrm{L} - \Theta) = 0. \tag{5.2.2}$$

For L $= 0$ Eq. (5.2.2) reduces to (5.1.20) with the growth rate $\sigma_{DL} = \Gamma(\Theta)U_f k$, given by Eq. (5.1.22), and for L $\neq 0$, the growth rate is

$$\sigma = kU_f\frac{\Theta}{\Theta + 1}\left(\sqrt{\Theta + 1 - 1/\Theta - k\mathrm{L}(2\Theta - k\mathrm{L})} - \frac{\Theta}{\Theta + 1}(1 + k\mathrm{L})\right). \tag{5.2.3}$$

According to (5.2.3), short-wavelength perturbations are stabilized with the cut-off wavelength $\lambda_c = \frac{4\pi\Theta}{\Theta - 1}\mathrm{L}$. These phenomenological models led to the conclusion that the self-turbulization of a flame is unlikely. As an alternative, it has been suggested that the development of perturbations leads to a smoothly curved and, possibly, stationary flame shape. In this sense, the question about the nonlinear stage of DL instability can be formulated as the problem of a flame propagating in a tube with ideally slip and adiabatic walls. The configuration of a flame in an ideal tube is equivalent to the configuration of a freely propagating flame, and the ideal tube walls can be considered as symmetry planes.

To avoid the assumption used in DL theory that the normal velocity of flame propagation relative to the unburned gas remains unchanged, and to determine the cutoff wavelength, it is necessary to consider the internal structure of the flame front, which is formed due to dissipations and the chemical energy release. Such studies were carried out by Joulin and Klavin [10], Pelche and Klavin [11], Frankel and Sivashinsky [12], Matalon and Matkovsky [13, 14] in the early 1980s using a multiscale asymptotic analysis of the complete system of hydrodynamic equations. These studies led to the correction term to the relation (5.1.22), which is the next term in the Taylor expansion in small wavenumbers, $kL_f \ll 1$

$$\sigma = \Gamma U_f k(1 - k/k_c), \tag{5.2.4}$$

where k_c and $\lambda_c = 2\pi/k_c$ are a cut-off wavenumber and cut-off wavelength. For the dispersion relation in the form (5.2.4) a planar flame propagating in a channel of width $D > \lambda_c$ is unconditionally unstable, and it is stable if $D < \lambda_c$ because all linearly unstable modes do not fit the width of the channel.[1]

In the previous section, it was assumed that instability of a flame does not change the flame velocity in the direction normal to the flame surface. A self-consistent approach to the problem must include the processes, that determine velocity of the flame. Since the complete solution is cumbersome, for simplicity, we will consider a flame propagating in an inviscid fuel with a constant coefficient of thermal conduction, unit Lewis number and zero gravitational acceleration. We consider a deflagration wave, taking into account the structure of the transition region in the case most often encountered in practice, in which the Lewis number $Le = \rho C_p D/\kappa = 1$, where D and κ are the diffusivity and thermal conductivity of the gas with the specific heat at constant pressure, C_p and the gas density ρ. The equality of the diffusivity and thermal conductivity, $Le = 1$, implies that for a steady flow the temperature and the concentration of the fuel mixture have similar profiles, so that the thermal conduction and diffusion equations are reduce to a single equation. The flow consists of homogeneous upstream and downstream flows separated by the flame front of finite thickness, consisting of a heating zone and a reaction zone. If, for example, we consider flame in a tube, then the tube diameter gives the typical hydrodynamic length scale D. The typical length scale of the thermal conduction zone is of the order of the flame thickness L_f, while the width of the reaction zone is much smaller, $L_R \ll L_f$.

Solutions in the hydrodynamic zone were obtained in the Sect. 5.1. From Eqs. (5.1.9)–(5.1.19) the potential mode in the upstream flow ($z < 0$) is

$$u'_{x-} = iu'_{z-}, \tag{5.2.5}$$

[1] We mean here an idealized situation of a channel with slip walls.

$$P'_{-} = -\rho_f U_f \left(1 + \frac{\sigma}{U_f k}\right) u'_{z-}. \tag{5.2.6}$$

In the downstream flow ($z > 0$) there are two potential modes ($u'_{zp} \propto \exp(-kz)$)

$$u'_{xp} = -i u'_{zp}, \tag{5.2.7}$$

$$P'_p = -\rho_f U_f \left(1 - \frac{\sigma}{U_f k}\right) u'_{zp}, \tag{5.2.8}$$

and a vorticity mode ($u'_{zp} \propto \exp(-\sigma z/\Theta U_f)$),

$$P' = 0, \quad u'_{xv} = -i \frac{\sigma}{\Theta U_f k} u'_{zv}. \tag{5.2.9}$$

To solve the problem for the case of a finite thickness of the flame we need to integrate the linearized hydrodynamic Eqs. (6.2.1)–(6.2.4) across the flame front, which are

$$\sigma \rho' + \frac{d}{dz}(\rho' u_z + \rho u'_z) + ik\rho u'_x = 0, \tag{5.2.10}$$

$$\sigma \rho u'_x + \rho u_z \frac{d}{dz} u'_x = -ikP', \tag{5.2.11}$$

$$\sigma \rho u'_z + \sigma u_z \rho' + ik\rho u_z u'_x + \frac{d}{dz}(P' + 2\rho u_z u'_z + u_z^2 \rho') = 0, \tag{5.2.9}$$

$$\frac{\kappa}{C_p}\frac{d^2 T'}{dz^2} - \rho_f U_f \frac{dT'}{dz} + \frac{dW_R}{dT} T' = (\rho u'_z + \rho' u_z)\frac{dT}{dz} + \rho \sigma T' + \frac{\kappa}{C_p} k^2 T'. \tag{5.2.12}$$

Perturbations of density are related to temperature perturbations according to the isobaric equation of state

$$\rho' = -\rho_u \frac{T_u}{T^2} T'. \tag{5.2.13}$$

It is convenient to use dimensionless variables: $\zeta = z/L_f$, $\theta = T/T_u$, $\theta' = T'/T_u$, and to introduce perturbations of the mass flux and two components of the momentum flux $j' = \frac{\rho u'_z + \rho' u_z}{\rho_f U_f}$, $w' = iu'_x/U_f$, $p' = \frac{P' + 2\rho u_z u'_z + \rho' u_z^2}{\rho_f U_f^2}$, and the scaled parameters $K = kL_f$, $S = \frac{\sigma}{U_f k}$. In the dimensionless variables the equation of thermal conduction for the stationary flame becomes

$$\frac{d^2\theta}{d\zeta^2} - \frac{d\theta}{d\zeta} + W_R(\theta) = 0. \tag{5.2.14}$$

The Zel'dovich-Frank-Kamenetsky solution for the temperature distribution in the dimensionless variables is reduced to

$$\theta = \begin{cases} 1 + (\Theta - 1)\exp\zeta, & \zeta < 0 \\ \Theta, & \zeta > 0 \end{cases}. \tag{5.2.15}$$

The linearized equations take the form

$$\frac{dj'}{d\zeta} = SK\frac{\theta'}{\theta^2} - K\frac{w'}{\theta}, \tag{5.2.16}$$

$$\frac{dw'}{d\zeta} = -SK\frac{w'}{\theta} + Kp' - 2K\theta j' - K\theta', \tag{5.2.17}$$

$$\frac{dp'}{d\zeta} = -SKj' - Kw', \tag{5.2.18}$$

$$\frac{d^2\theta'}{d\zeta^2} - \frac{d\theta'}{d\zeta} - \frac{dW_R}{d\theta}\theta' = j'\frac{d\theta}{d\zeta} + SK\frac{\theta'}{\theta} + K^2\theta', \tag{5.2.19}$$

where $W_R = W_R(\theta)$ is the scaled reaction.

The boundary conditions for Eqs. (5.2.16)–(5.2.19) require that all perturbations vanish at infinity, which can be written in the form of perturbation modes in the uniform flows far ahead of the flame front and far behind the flame front. In addition to the modes (5.2.9)–(5.2.12), which we described previously, the set of Eqs. (5.2.16)–(5.2.19) has two more independent modes related to the equation of thermal conduction (5.2.19). The thermal conduction modes go to zero on small length scales of about or less than the flame thickness, therefore in the hydrodynamic zone these modes correspond to the condition $\theta' = 0$. The last condition written in the dimensional variables gives $\rho' = 0$. The other boundary conditions in the hydrodynamic zone in the dimensionless variables in the upstream flow are

$$w'_- = -j'_-, \quad p'_- = -(S-1)j'_-, \tag{5.2.20}$$

and in the downstream flow

$$w'_+ + p'_+ - (S + 2\Theta)j'_+ = 0 \tag{5.2.21}$$

We will consider the solution of Eqs. (5.2.16)–(5.2.19) in the long-wavelength approximation, $K \ll 1$. To do this we first write down Eq. (5.2.19) in the form

$$\hat{W}_R[\theta'] = j'\frac{d\theta}{d\zeta} + KS\frac{\theta'}{\theta} + K^2\theta', \tag{5.2.22}$$

where the operator $\hat{W}_R$ is defined as

$$\hat{W}_R[\theta'] \equiv \frac{d^2\theta'}{d\zeta^2} - \frac{d\theta'}{d\zeta} - \frac{d\hat{W}_R}{d\theta}\theta', \tag{5.2.23}$$

so that $\hat{W}_R\left[\frac{d\theta'}{d\zeta}\right] = 0$. It was shown by Liberman et al. [15] that the first and second terms in the right-hand side of (5.2.22) are of the same order, so that in the limit $K \ll 1$ the solution of (5.2.22) in zero order of K is

$$\theta' = \zeta_T\frac{d\theta}{d\zeta}, \tag{5.2.24}$$

which means that the flame is shifted in ζ direction on the distance ζ_T. Using formula (5.2.24) for the temperature perturbation, the integrals of Eqs. (5.2.16)–(5.2.19) with the accuracy of the first order terms in $K \ll 1$ will be:

$$j' - j'_- = KS\zeta_T\frac{\theta - 1}{\theta}, \tag{5.2.25}$$

$$w' - w'_- = -K\zeta_T(\theta - 1), \tag{5.2.26}$$

$$p' = p'_-. \tag{5.2.27}$$

Integrals (5.2.25)–(5.2.27) relating to the downstream flow coincide with the linearized conservation laws with $F' = -\zeta_T L_f$, and can be written as

$$\left(\rho u'_z - \sigma\rho F'\right)\Big|_-^+ = 0, \tag{5.2.28}$$

$$\left(u'_x + ikv_z F'\right)\Big|_-^+ = 0, \tag{5.2.29}$$

$$\left(P' + 2\rho u'_z u_z\right)\Big|_-^+ = 0. \tag{5.2.30}$$

Recall that $F'(x,t) = F'\exp(\sigma t + ikx)$ and the formulas (5.2.24), (5.2.28)–(5.2.30) give the inner structure of perturbations inside the flame front.

To obtain the supplementary boundary condition on the perturbed flame velocity $\partial F'/\partial t$ with the same accuracy we have to obtain the relation between the perturbed flux of mass j'_- and the flame front shift ζ_T. Substituting j' from (5.2.25) in the right-hand side of (5.2.22), we obtain with accuracy of the first order in K

$$\hat{W}_R[\theta'] = (KS + j'_-/\zeta_T)\theta'. \tag{5.2.31}$$

The function $\theta' = \zeta_T \frac{d\theta}{d\zeta}$ is the eigenfunction of Eq. (5.2.31) for the eigenvalue $KS + j'_-/\zeta_T = 0$, and it is the unique eigenvalue for which $\mathrm{Re}[\sigma] > 0$, when the instability can develop. Thus, the supplementary condition is

$$KS\zeta_T + j'_- = 0. \tag{5.2.32}$$

The condition (5.2.32) written for the dimensional variables gives the condition of constant flame velocity with respect to the fuel ($u'_z = \sigma F'$), which was assumed in derivation of the DL instability in previous section.

The integrals (5.2.25)–(5.2.27) related to the downstream flow together with the boundary conditions (5.2.20) and (5.2.21) and the supplementary condition (5.2.32) give the dispersion relation for the growth rate of perturbations within the accuracy of the first order terms in the expansion in K

$$S = \frac{\sigma}{kU_f} = \Gamma(\Theta) = \frac{\Theta}{\Theta + 1}\left(\sqrt{\Theta + 1 - 1/\Theta} - 1\right). \tag{5.2.33}$$

Thus, the first order term in the expansion in K corresponds to the DL instability of the infinitely thin flame (5.1.22), and this is the first order term in the long wavelength approximation, $L_f/\lambda \ll 1$. The cut-off wavenumber corresponding to the thermal stabilization of the DL instability, can be found from the solution obtained with the second order in the expansion in K accuracy.

To do this we substitute the first order solution (5.2.25)–(5.2.27) into Eqs. (5.2.16)–(5.2.18) and perform the straightforward integration

$$j' = j'_- + KS\zeta_T \frac{\theta - 1}{\theta} - K \int_{\zeta_-}^{\zeta} \frac{w'_0}{\theta} dy, \tag{5.2.34}$$

$$w' = w'_- - K\zeta_T(\theta - 1) - KS \int_{\zeta_-}^{\zeta} \frac{w'_0}{\theta} dy + K \int_{\zeta_-}^{\zeta} p'_0 dy - 2K \int_{\zeta_-}^{\zeta} \theta j'_0 dy, \tag{5.2.35}$$

$$p' = -(S - 1) j'_- - KS \int_{\zeta_-}^{\zeta} j'_0 dy - K \int_{\zeta_-}^{\zeta} w'_0 dy. \tag{5.2.36}$$

Here we used the following designations: ζ_- is some point in the hydrodynamic zone ahead of the flame front, j'_0, w'_0, p'_0 are the first order terms in the expansion of j', w', p' in the expansion of K, defined by (5.2.25)–(5.2.27) with account the boundary conditions (5.2.20). If the temperature distribution for the planar flame is known $\theta = \theta(\zeta)$, then integrals in (5.2.34)–(5.2.36) give the linearized conservation laws at the perturbed flame front of finite thickness.

The supplementary condition for a flame of finite thickness is calculated as an eigenvalue of the linearized equation of thermal conduction similar to the problem of planar flame propagation. Substituting j' from (5.2.34) into the right-hand side of (5.2.22) we find

$$\hat{W}_R[\theta'] = \left[(1 - K\zeta_-)\frac{j_-}{\zeta_T} + KS\right]\theta' + K^2 H(\zeta)\theta', \tag{5.2.37}$$

where $H(\zeta) = 1 - S\zeta - (S+1)J_1(\zeta)$ and

$$J_1(\zeta) = \int_{\zeta_-}^{\zeta} (1-\theta)/\theta \, dy, \tag{5.2.38}$$

According to [16], Eq. (5.2.37) can be reduced to the equation with a Hermitian operator by means of the substitution $\theta' = \theta'_H \exp(\zeta/2)$. Considering the term $K^2 H(\zeta)\theta'$ as a small correction and using the standard quantum mechanical technique of perturbation theory [17], the supplementary boundary condition with accuracy of the order K^2, can be written as

$$(1 - K\zeta_-)j'_- + \left(KS + K^2\langle H\rangle\right)\zeta_T = 0, \tag{5.2.39}$$

where

$$\langle H\rangle = \frac{\int_{-\infty}^{\infty} H(\zeta)\left(\frac{d\theta}{d\zeta}\right)^2 \exp(-\zeta)d\zeta}{\int_{-\infty}^{\infty} \left(\frac{d\theta}{d\zeta}\right)^2 \exp(-\zeta)d\zeta}. \tag{5.2.40}$$

Substituting integrals across the flame front, Eqs. (5.2.34)–(5.2.36), into equations in the hydrodynamic zone, (5.2.20) and (5.2.21), and, using the supplementary in the form (5.2.39), we can obtain the dispersion relation of the DL instability for a flame of finite thickness, if the temperature distribution for the planar stationary flame $\theta = \theta(\zeta)$ is known. Using the Zel'dovich-Frank-Kamenetsky temperature distribution (6.4.9) for the case of a flame with the unit Lewis number and assuming that the coefficient of thermal conduction does not depend on temperature, we come to the dispersion relation of the DL instability for a flame of finite thickness

$$S^2 + \frac{2\Theta}{\Theta+1}\left(1 + \frac{\Theta\ln\Theta}{\Theta-1}K\right)S - \Theta\frac{\Theta-1}{\Theta+1}K^2\left[1 - K\left(1 + \Theta\ln\Theta\frac{\Theta+1}{(\Theta-1)^2}\right)\right] = 0. \tag{5.2.41}$$

In the dimensional variables the dispersion relation (5.2.41) takes the form

$$\begin{aligned}&\sigma^2+\sigma\frac{2\Theta}{\Theta+1}\left(1+\frac{\Theta\ln\Theta}{\Theta-1}kL_f\right)U_fk-\\&-\Theta\frac{\Theta-1}{\Theta+1}\left[1-kL_f\left(1+\Theta\ln\Theta\frac{\Theta+1}{(\Theta-1)^2}\right)\right]U_f^2k^2=0.\end{aligned}\tag{5.2.42}$$

Equation (5.2.42) for the growth rate of instabilities, which was first derived by Pelce and Clavin [11], can be written as

$$\sigma=\Gamma U_fk\left(1-\frac{k\lambda_c}{2\pi}\right),\tag{5.2.43}$$

where Γ is the coefficient found in the Darrieus-Landau theory (5.1.22), and λ_c is the cut-off wavelength for which the instability is suppressed by thermal conduction. For the case of a unit Lewis number and the constant thermal conduction coefficient, the cut-off wavelength is given by the formula

$$\lambda_c=\frac{\pi L_f(\Theta-1)}{\Gamma\sqrt{\Theta+1-1/\Theta}}\left(1+\Theta\ln\Theta\frac{\Theta+1+2\Gamma}{(\Theta-1)^2}\right).\tag{5.2.44}$$

For the expansion coefficients typical for laboratory flames, $\Theta=6\div10$, the cut-off wavelength is $\lambda_c\approx20L_f$, and it depends only slightly on thermal expansion. The wavelength of the fastest growing perturbations is approximately two times larger than the cut-off wavelength, $\lambda=2\lambda_c\approx40L_f$, and much greater than the flame thickness.

For the long wavelength perturbations, $kL_f\ll1$, the analytical expression for the cut-off wavelength obtained from the dispersion relation (5.2.39) differs from (5.2.44):

$$\lambda_c=\frac{2\pi L_f}{\Theta-1}\left(\Theta-1+\Theta\ln\Theta\frac{\Theta+1}{\Theta-1}\right).\tag{5.2.45}$$

However, the numerical values of analytical formulas (5.2.44) and (5.2.45) for the expansion coefficients in the range $\Theta=5-10$ differ insignificantly [18]. For small expansion coefficients, $\Theta-1\ll1$, we come to the asymptotic value of the cut-off wavelength, $\lambda_c=4\pi L_f/(\Theta-1)\to\infty$ for $\Theta\to1$. The analytical formula (5.2.44) for the growth rate is in a good agreement with the numerical solution of the eigenvalue problem Eqs. (5.2.15)–(5.2.19) with the boundary conditions (5.2.20)–(5.2.21) obtained in [15]. The results of the numerical solution are shown in Figs. 5.1 and 5.2. Figure 5.1 shows the calculated ratio of the instability growth rate for a finite flame thickness and for an infinitely thin flame front, $\sigma/\Gamma(\Theta)U_fk$ versus the scaled wave number kL_f.

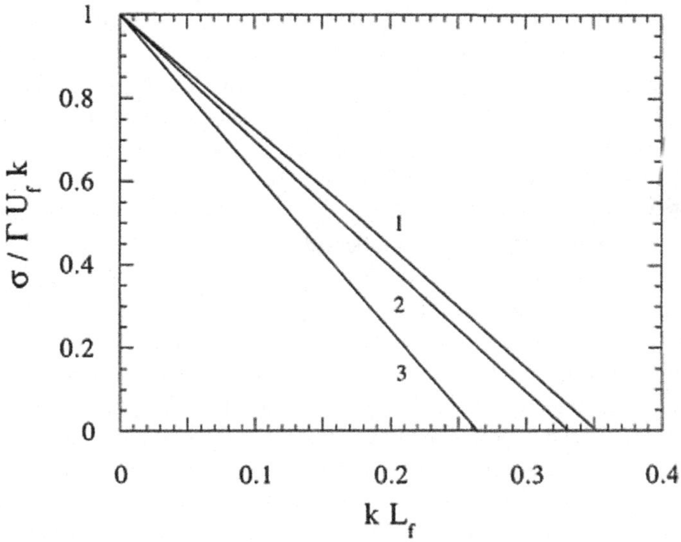

Fig. 5.1 Scaled growth rate $\sigma/\Gamma(\Theta)U_f k$ for the flame instability as a function of the dimensionless wave numbers kL_f. Lines 1, 2, 3 correspond to the expansion coefficients $\Theta = 8,\ 6,\ 3$

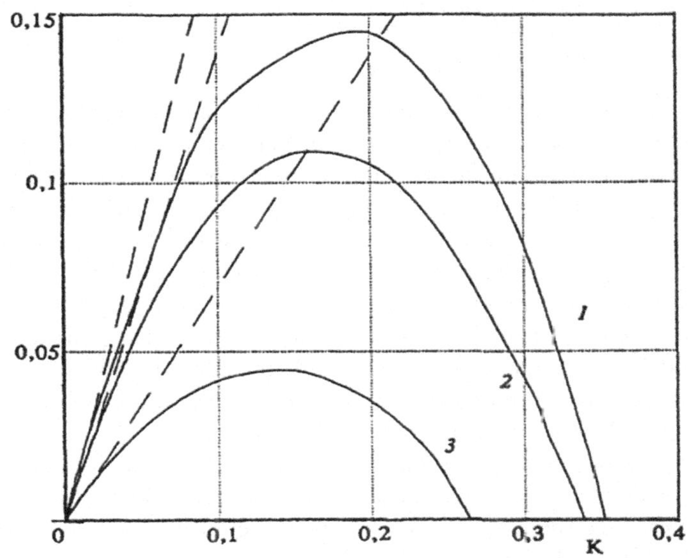

Fig. 5.2 The scaled growth rates of the DL instability versus the scaled wave number for the expansion coefficients $\Theta = 8,\ 6,\ 3$ shown by curves 1, 2, 3 respectively, and activation energies used in Fig. 4.4.3. Dashed lines show corresponding asymptotic solutions for the infinitely thin flame front

The analytical formulas (5.2.43–5.2.44) for the growth rate is in a good agreement with the numerical solution of the eigenvalue problem Eqs. (5.2.15)–(5.2.19) with the boundary conditions (5.2.20)–(5.2.21) obtained in [15]. The results of the numerical solution are shown in Figs. 5.1 and 5.2. Figure 5.1 shows the calculated ratio of the instability growth rate for a finite flame thickness and for an infinitely thin flame front, $\sigma/\Gamma(\Theta)U_f k$ versus the scaled wave number kL_f.

Figure 5.2 shows the scaled growth rate $\sigma/\Gamma(\Theta)U_f k$ for the flame instability as a function of the dimensionless wave numbers kL_f for different expansion coefficients and activation energies: (1) $\varepsilon = 70$, $\Theta_2 = 8$; (2) $\varepsilon = 42$, $\Theta_2 = 6$; (3) $\varepsilon = 21$, $\Theta_2 = 3$. In the approximation of an incompressible flow ($M \ll 1$) and for the coefficient of thermal conduction independent on temperature, the calculated cut-off wavelength λ_c in Eq. (5.2.44) depends on the expansion coefficient Θ and on the type of thermal conduction only. Matalon and Matkowsky [14] generalized their hydrodynamic model developed in [13], to take into account temperature-dependent transport and preferential diffusion, in particular, for non-unity Lewis numbers.

Based on the generalized hydrodynamic model the effect of stoichiometry and the reaction order was incorporated in the model [19]. The cut-off wavelength for the incompressible flow and for the thermal conduction coefficient depending on temperature as $\kappa \propto T^{\mu}$ was calculated in [20, 21]:

$$\frac{\lambda_c}{L_f} = \frac{2\pi\Theta}{\mu(\Theta-1)}\left(\frac{\Theta+1}{\Theta-1}(1-\Theta^{-\mu}) + \frac{1-\Theta^{-(\mu+1)}}{1-\mu^{-1}}\right). \tag{5.2.46}$$

Compared to the cut-off wavelength (5.2.44) for the temperature independent thermal conduction coefficient, $\lambda_c \approx 20L_f$, a strong temperature dependence ($\mu > 1$) of the thermal conduction coefficient may result in a considerably smaller value, $\lambda_c \approx (3-5)L_f$, of the cut-off wavelength compared with that given by Eq. (5.2.44), which is of interest in situations when a flame propagates due to electron thermal conduction ($\kappa \propto T^{5/2}$) or due to radiation heat transfer ($\kappa \propto T^{13/2}$).

The effect of the shear on DL instability was studied analytically in [22]. It was shown that at low shear, the behaviour of instabilities is analogous to Darrieus–Landau model, the increasing shear has a stabilizing effect, but for sufficiently high shear the flame becomes unstable and its most unstable wavelength can be much smaller than the Markstein length of the zero-shear flame.

While there is a good progress in understanding a flame instability, and analytical theories provide a qualitative understanding of the development of hydrodynamic instability, a common drawback is the use of a one-step chemical model. All the same in spite detailed chemical models are available nowadays, the perturbed problem of Navier- Stokes equations coupled with a detailed chemical model is a formidable task for analytical analysis.

5.3 Flame in a Gravitational Field; Rayleigh–Taylor Instability

If the flame propagates upward in a gravitational field, then the DL instability is amplified by the Rayleigh–Taylor (RT) instability. The RT instability develops at the interface between a heavy and a light fluid, when a heavier fluid (gas) is supported by the fluid with lower density in a gravitational field. In the case of a flame propagating upward, unburned fuel plays the role of a heavy gas, and the combustion products act as a lighter gas, and the DL instability is enhanced by the RT instability. In contrast, if the flame propagates downward, the DL instability is attenuated and can be completely suppressed. Formula (5.1.22) for the Darrieus-Landau instability increment can be easily generalized to take into account gravitational forces [2]. We first consider the increment of instability of an infinitely thin flame propagating upward in a gravitational field. The problem is similar to the problem considered in Sect. 5.1, including the linearized equations of an ideal fluid. The boundary conditions at the flame front are more convenient in the form (5.2.28)–(5.2.29), but the linearized equation for the normal component of the momentum flux (5.2.30), has an additional term due to gravity acceleration g

$$\left(P' + 2\rho u_x u_x' - \rho g F'\right)\Big|_-^+ = 0. \tag{5.3.1}$$

Since the linearized equations of an ideal incompressible gas do not change in the presence of gravity, the relations in the uniform flows ahead of the flame (5.2.5)–(5.2.6) and behind the flame (5.2.8) and (5.2.9) remain the same. Resolving the relations in the uniform flows together with the boundary conditions at the flame front we obtain the dispersion relation

$$\sigma^2 + \frac{2\Theta}{\Theta+1}U_f k\sigma - \Theta\frac{\Theta-1}{\Theta+1}U_f^2k^2 - \frac{\Theta-1}{\Theta+1}gk = 0, \tag{5.3.2}$$

and the instability growth rate of an infinitely thin flame propagating upwards is

$$\sigma = \sqrt{\frac{\Theta-1}{\Theta+1}gk + \frac{\Theta^2+\Theta-1}{(\Theta+1)^2}\Theta U_f^2k^2} - \frac{\Theta}{\Theta+1}kU_f \tag{5.3.3}$$

It can be written as

$$\sigma_0 = \frac{\sigma}{kU_f} = \frac{\Theta}{\Theta+1}\left(\sqrt{\Theta + 1 - \frac{1}{\Theta} + \frac{\Theta^2-1}{\Theta^2}\frac{1}{KFr}} - 1\right) \tag{5.3.4}$$

where, $K \equiv kL_f$ and the Froude number, $Fr = U_f^2/gL_f$ shows the relative contribution between the gravitational field and the mass flow. The dispersion relation (5.3.4) coincides with the DL instability growth rate if the gravity acceleration tends to zero, $Fr \to \infty$. In the opposite case of a large acceleration, $Fr \to 0$, or for perturbations of a very long wavelength ($k \to 0$), the instability growth rate goes over to the expression for the RT instability

$$\sigma_{RT} = \sqrt{\frac{\Theta - 1}{\Theta + 1} gk} = \sqrt{\frac{\rho_2 - \rho_1}{\rho_2 + \rho_1} gk}. \tag{5.3.5}$$

From (5.3.4) follows that the flame instability in a gravitational field is stronger than both the DL instability and the RT instability.

To find the dispersion relation for a flame of finite thickness in a gravitational field, one has to solve the set of linearized hydrodynamic equations similar to Sect. 5.2 with the one extra term in Eq. (5.2.9) related to the gravitational field

$$\sigma \rho u_z' + \sigma u_z \rho' + ik\rho u_z u_x' + \frac{d}{dz}\left(P' + 2\rho u_z u_z' + u_z^2 \rho'\right) = g\rho', \tag{5.3.6}$$

Corresponding changes should also be made to the dimensionless Eq. (5.2.18)

$$\frac{dp'}{d\zeta} = -SKj' - Kw' - Fr^{-1}\frac{\theta'}{\theta^2}. \tag{5.3.7}$$

Remember that here $K = kL_f$ is dimensionless wave number.

The integration of (5.3.7) within the first order terms in dimensionless $K = kL_f$ leads to the variations of the perturbed normal momentum flux inside the flame front

$$p' - p'_- = \zeta_T \frac{\theta - 1}{\theta\, Fr}. \tag{5.3.8}$$

Integration similar to that performed in the previous section leads to the dispersion relation, which have been derived first by Pelce and Clavin [11]

$$\begin{aligned} &\sigma^2 + \frac{2\Theta}{\Theta + 1}\left(1 + \frac{\Theta \ln \Theta}{\Theta - 1} kL_f\right) U_f k\sigma - \\ &\Theta \frac{\Theta - 1}{\Theta + 1} U_f^2 k^2 \left[1 - kL_f\left(1 + \Theta \ln \Theta \frac{\Theta + 1}{(\Theta - 1)^2}\right)\right] - \frac{\Theta - 1}{\Theta + 1} gk = 0 \end{aligned} \tag{5.3.9}$$

with the instability growth rate

$$\frac{\sigma}{kU_f} = \frac{\sigma_0}{kU_f} - \frac{1}{2}\frac{\Theta k L_f}{(\Theta + 1)\sigma_0 + \Theta}\left[\frac{2\Theta \ln \Theta}{\Theta - 1}\sigma_0 + \left(\frac{\Theta + 1}{\Theta - 1}\Theta \ln \Theta + \Theta - 1\right)\right], \tag{5.3.10}$$

where σ_0 is the instability growth rate (5.3.4) of an infinitely thin flame.

5.3.1 Comparison with Numerical Simulations

The dispersion relation (5.3.9) describes the RT instability of an upward propagating flame of finite thickness for the case of a unit Lewis number and a constant coefficient of thermal conduction. The growth rates of the upward propagating flame according to the analytical solution (5.3.10) and obtained from the numerical solution of the stability problem [23–25] are shown in Figs. 5.3 and 5.4. Figure 5.3 shows that with the increase of gravitational acceleration, the maximum growth rate also increases, while the cut-off wavenumber λ_{RT} shifts toward a shorter wavelength (larger wavenumbers $K = kL_f$). At the same time, the maximum value of the growth rate is shifted towards long wavelength perturbations as compared to $\lambda_{RT}/2$. Figure 5.4 shows the dependence of the scaled maximal growth rate of instability (5.3.10) on the scaled gravitational acceleration, $1/Fr$ compared with the results of numerical simulations. It is seen that the numerical solutions [23–25] confirm the good accuracy of the analytical formula (5.3.10) in the domain of long-wavelength perturbations up to the fastest perturbations with the maximum growth rate.

In the opposite case of a downward propagating flame the gravity acceleration plays stabilizing role. Unlike the thermal stabilization inherent to the short wavelength perturbations, the stabilizing effect of gravitational acceleration concerns the

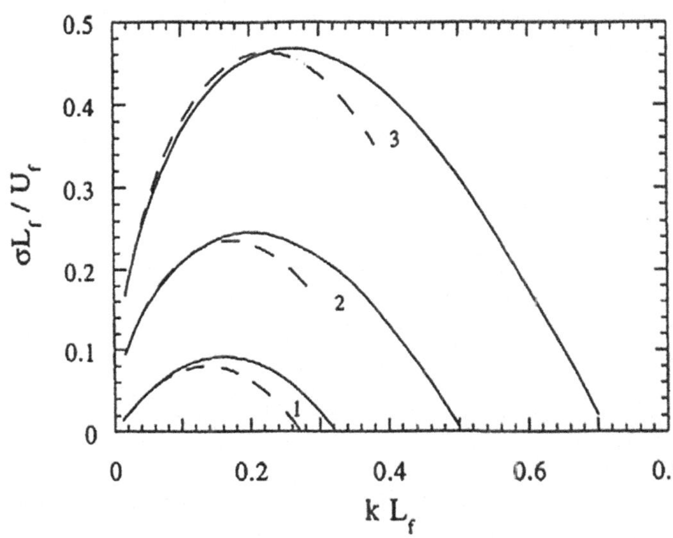

Fig. 5.3 The scaled growth rates (5.3.10) of the DL instability of a flame propagating upwards in a gravitational field for $1/Fr = 0,\ 1,\ 3$ and $\Theta = 5$ are shown by dashed curves 1, 2, 3 respectively. Solid lines are from numerical solutions [25]

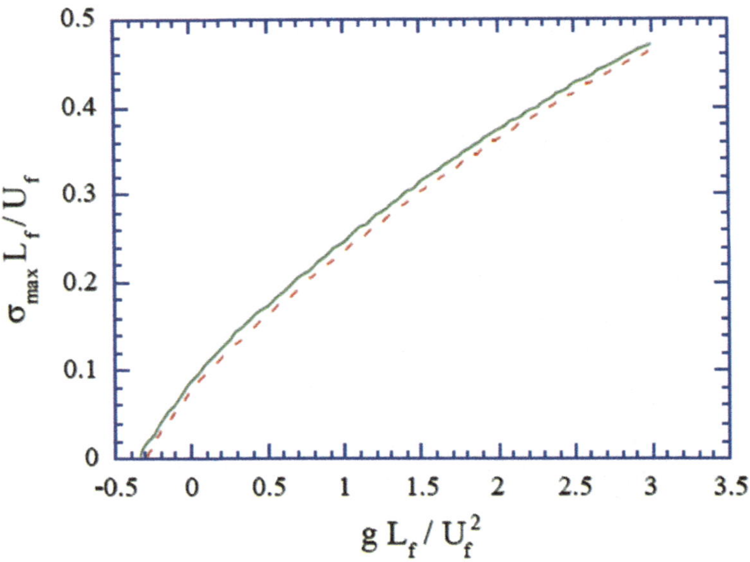

Fig. 5.4 The scaled maximum critical instability growth rate for the flame ($\Theta = 5$) propagating upwards versus the scaled gravity acceleration; dashed line is according to formula (5.3.10); solid line is the numerical solution [25]

long wavelength perturbations. The joint effect of the stabilizing gravity and thermal conduction may lead to complete suppression of the flame instability. To investigate the stabilizing effect of gravitational acceleration it is convenient to rewrite the dispersion relation (5.3.9) in the form [11]

$$\sigma^2 + \frac{2\Theta}{\Theta+1}\left(1 + \frac{\Theta \ln \Theta}{\Theta - 1} k L_f\right) U_f k \sigma + \Theta \frac{\Theta - 1}{\Theta + 1} \frac{U_f^2 k}{k_c} (k - k_1)(k - k_2) = 0, \tag{5.3.11}$$

where $k_c = 2\pi/\lambda_c$ is the cut-off wave number in the absence of gravity with the cut-off wavelength determined by (5.2.45), and the marginal wave numbers k_1 and k_2 show the stability limits in the case of stabilizing gravitational acceleration

$$k_1 = \frac{k_c}{2}\left(1 - \sqrt{1 - \frac{4|g|}{\Theta U_f^2 k_c}}\right), \quad k_2 = \frac{k_c}{2}\left(1 + \sqrt{1 - \frac{4|g|}{\Theta U_f^2 k_c}}\right) \tag{5.3.12}$$

Complete stabilization of the DL instability takes place when the marginal wave numbers coincide, $k_1 = k_2$, which holds for the gravitational acceleration

$$|g_c| = \frac{\Theta}{4} U_f^2 k_c. \tag{5.3.13}$$

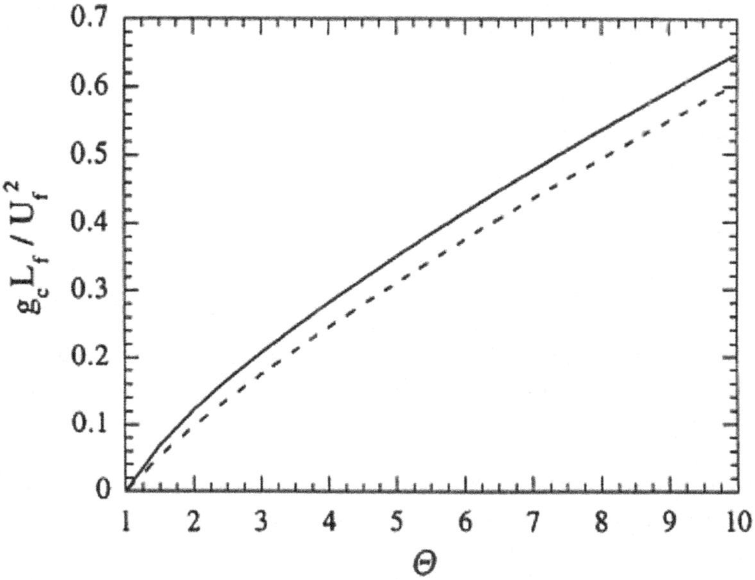

Fig. 5.5 Critical gravitational acceleration suppressing completely the hydrodynamical instability in a downward propagating flame versus the flame expansion coefficient. Solid line—the numerical solution; dashed line—the Eq. (5.3.14)

Using (5.2.45) for $k_c = 2\pi/\lambda_c$, this condition for the Froude number is

$$Fr_c = \frac{4\Theta}{(\Theta - 1)}\left[\Theta - 1 + \frac{\Theta \ln \Theta}{\Theta - 1}\right]. \tag{5.3.14}$$

The dependence of the critical gravitational acceleration ($1/Fr_c$) on the expansion coefficient for the flame with a unit Lewis number and a constant coefficient of thermal conduction is shown in Fig. 5.5.

When the gravity acceleration is close to the critical value $|g/g_c - 1| \ll 1$ the instability growth rate can be written in the form

$$\sigma = \sigma_m\left(1 - \frac{(2k - k_c)^2}{|g/g_c - 1|k_c}\right), \tag{5.3.15}$$

where the maximal value of the increment σ_m is

$$\sigma_m = \frac{1}{4}\frac{(\Theta - 1)^2 U_f k_c}{2(\Theta - 1) + \Theta \ln \Theta k_c L_f}|g/g_c - 1| \tag{5.3.16}$$

When the gravity acceleration is close to the critical value $|g/g_c - 1| \ll 1$ the instability growth rate can be written in the form

$$\sigma = \sigma_m\left(1 - \frac{(2k - k_c)^2}{|g/g_c - 1|k_c}\right), \tag{5.3.15}$$

where the maximal value of the increment σ_m is

$$\sigma_m = \frac{1}{4}\frac{(\Theta - 1)^2 U_f k_c}{2(\Theta - 1) + \Theta \ln \Theta k_c L_f}|g/g_c - 1| \tag{5.3.16}$$

In this case, the instability is localized within a small interval of wavenumbers near the wavenumber of the maximal increment $k_c/2$. The calculated instability growth rates for the case of a downward propagating flame with the expansion coefficient $\Theta = 5$ are shown in Fig. 5.6. The value of the critical gravitational field $g_c L_f/U_f^2 = -0.35$, obtained from the numerical solution is close to the analytical estimate $g_c L_f/U_f^2 = -0.31$.

The marginal perturbation wavenumbers k_1 and k_2 found numerically for different values of the gravitational acceleration are compared to analytical formulas (5.3.12) in Fig. 5.7.

For arbitrary Lewis numbers and for transport coefficients varying with temperature, the calculation of the dispersion relation of the flame instability becomes more tedious and cumbersome. Such dispersion relation was derived in a closed form by Searby and Rochwerger [26] under the assumption that the Lewis number is close to unity, as $(\mathrm{Le} - 1) \sim 1/Z$, and for a large activation energy $Z = E(T_b - T_u)/T_b^2 \gg 1$, which is typical for a gaseous fuel

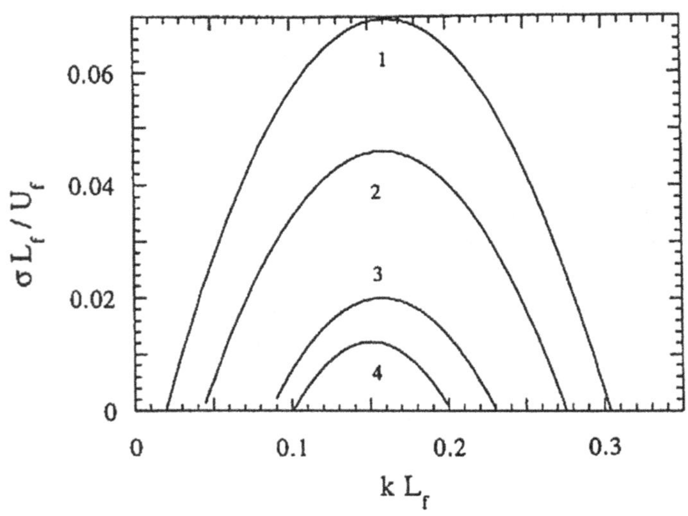

Fig. 5.6 The scaled instability growth rate of a downward propagating flame ($\Theta = 5$) for dimensionless gravitational acceleration $gL_f/U_f^2 = 0.1, 0.2, 0.3, 0.33$, shown by curves 1, 2, 3, 4 respectively

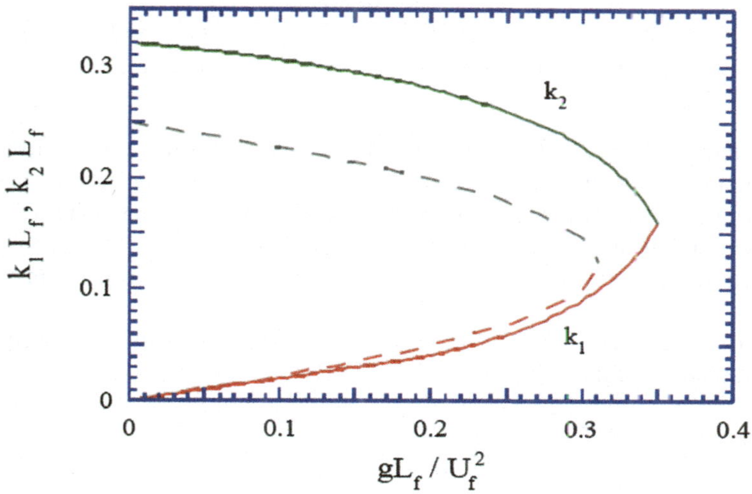

Fig. 5.7 The marginal wave numbers for a downward propagating flame ($\Theta = 5$) versus the scaled gravitational acceleration; dashed line -formulas (5.3.12), solid line—numerical solution

$$A\sigma^2 + BU_f k\sigma - CU_f^2 k^2 - C_1 gk = 0. \tag{5.3.17}$$

The dimensionless coefficients A, B, C, C_1 in (5.3.17) depend on the flame parameters as

$$A = 1 + \frac{\Theta - 1}{\Theta + 1} kL_f \left(\mathrm{Ma} - \frac{\Theta}{\Theta - 1} J_2 \right), \tag{5.3.18}$$

$$B = \frac{2\Theta}{\Theta + 1} \left[1 + \Theta k L_f (\mathrm{Ma} - J_2) \right], \tag{5.3.19}$$

$$C = \Theta \frac{\Theta - 1}{\Theta + 1} \left(1 - \frac{kL_f}{\Theta - 1} \left[J_3 + (3\Theta - 1)\mathrm{Ma} - 2\Theta J_2 + 2\,\mathrm{Pr}(\Theta - 1)\frac{\kappa_b}{\kappa_u} - 2\,\mathrm{Pr}\, J_3 \right] \right), \tag{5.3.20}$$

$$C_1 = \frac{\Theta - 1}{\Theta + 1} \left[1 - kL_f \left(\mathrm{Ma} - \frac{\Theta J_2}{\Theta - 1} \right) \right]. \tag{5.3.21}$$

The integrals J_2 and J_3 in Eqs. (5.3.18)–(5.3.21) are equal

$$J_2 = \int_{T_u}^{T_b} \frac{\kappa_b}{\kappa_u} \frac{dT}{T}, \quad J_3 = \int_{T_u}^{T_b} \frac{\kappa_b}{\kappa_u} \frac{dT}{T_u}. \tag{5.3.22}$$

The parameter Ma is the Markstein number, which is a measure of the sensitivity of the local flame velocity to curvature and stretch, and for the case of a simple one-step Arrhenius reaction with a large activation energy, this parameter is

$$\mathrm{Ma} = \frac{\Theta}{\Theta - 1} J_2 + \frac{\mathrm{Le} - 1}{\Theta - 1} \frac{E(T_b - T_u)}{2T_b^2} \int_{T_u}^{T_b} \ln\left(\frac{T - T_u}{T_b - T_u}\right) \frac{k_b dT}{k_u T}. \tag{5.3.23}$$

The integral in (5.3.23) is negative and in the case of a fuel diffusion stronger than thermal conduction ($\mathrm{Le} > 1$) the Markstein number may become negative $\mathrm{Ma} < 0$ and the stabilizing terms proportional to the flame thickness in (5.3.20) may become destabilising. In latter case, the hydrodynamic instability of the flame front is enhanced for short wavelength perturbations due to thermal-diffusion instability [16].

The dispersion relation (5.3.17) can be written in the form of differential equation for perturbations of the flame front

$$A \frac{d^2 F'}{dt^2} + B U_f k \frac{dF'}{dt} - C U_f^2 k^2 F' - C_1 g k F' = 0 \tag{5.3.24}$$

The dispersion relation in this form can be used to study evolution of the flame perturbations under the action of time dependent effective acceleration that is, for example, the acceleration in shocks or acoustic waves [26]. In a modified form, this equation was also used to study both hydrodynamic and thermo-diffusive stability of an outward propagating spherical flame in [27]. Studies on the stability of spherical flames have shown that time dependent stretch of the expanding flame strongly influences the development of small perturbations. Wavelength of any perturbations at the spherical flame increases linearly in time together with the increase of the flame radius. Therefore, any perturbation of a given angular shape first attenuates in amplitude relative to the flame radius, and then eventually grows as t^{σ}, where σ is a dimensionless exponent characterising the growth rate of perturbations [16, 27, 28]. For the given parameters of the flame, one can also identify a particular angular mode, the relative amplitude of which has the shortest period of decay. It is if practical interest that in the presence of acoustic or shock waves a flame can be either stabilized or destabilized, since the acoustic and shock waves produce an effective "gravitational" acceleration.

5.4 Thermal-Diffusive Instability

In premixed flames, the primary instability is the Darrieus-Landau instability, which results from thermal expansion across the flame front. In non-premixed flames, thermo-diffusive instabilities may be predominant, while the hydrodynamic instability plays a minor role, and there is a lot of experimental evidence of instabilities in diffusion flames. Another form of instability in non-premixed systems is spontaneous flame oscillations. In the previous section we considered the DL instability of a planar flame using models, which are valid for the Lewis number close to unity.

5.4.1 *Thermal Diffusive Instability* ($Le \neq 1$)

The thermal diffusive instability is associated with a disparity of molecular and thermal diffusivity for the non-unity Lewis number. The consequence of $Le \neq 1$ is that the temperature and concentration profiles are not similar, and we must consider separately the equation of thermal conduction and the diffusion equation for concentration. The physically non-unity Lewis numbers represent the competition between the heat transferred away from the flame front and the diffusion of fuel and oxidant toward the reaction zone. While the DL instability is an inherent feature of any flame, the occurrence of diffusive instability depends on the mixture composition. Unlike DL instability, which is characterized by relatively large scales and can be suppressed in small-scale systems, the diffusive modes of instability manifest themselves in the emergence of a small-scale irregular cellular structure of the flame front, which can be observed in laboratory experiments. The hydrodynamic and thermal diffusive instabilities may well coexist, yet in theoretical studies they are analysed separately. The most elaborate results were obtained under the assumption that the density is constant in order to filter out hydrodynamic instabilities.

For the first time, such instabilities were considered by Ya. B. Zel'dovich and co-authors[2] [16, 29, 30] within the framework of a simplified combustion model, assuming a large activation energy in a one-step chemical model, expressed by a large value of the Zeldovich number, $Ze = E(T_b - T_u)/2RT_b^2 \gg 1$. In this model the reaction zone is very thin compared with the thermal zone of the flame, $L_R \sim L_f/Ze$. The unburned mixture is heated by thermal conduction in the thermal zone and simultaneously diffuses toward the reaction zone where it is consumed. A flame stabilization due to thermal conduction and destabilization due to diffusion can be qualitatively understood as follows. If diffusion is the dominant process, for example, $Le = \chi/D < 1$, then the diffusion of fresh fuel is larger at the convex parts, increasing the reaction rate, and enhancing the bending of the flame front. On the contrary if $Le = \chi/D > 1$, then thermal conductivity is the dominant process. In this case when the flame front is slightly bended, the convex parts of the front loss more heat than the concave parts, so that the convex parts will be slackened while the concave parts accelerated, thus smoothing the flame surface. In [16, 29, 30], the dispersion relation was obtained in the case of a one-step reaction model with a high activation energy, $Ze \gg 1$, considering the reaction zone as a discontinuity surface, neglecting by thermal expansion and considering only transport processes. Then,

[2] Probably Lewis and Elbe [31] have been the first who assumed that flame becomes unstable if thermal diffusivity exceeds mass diffusivity. However, a qualitative theory [16, 29, 30] was developed for the opposite case, Le > 1.

the equations for the adiabatic flame will be reduced to the equations of thermal conduction and diffusion equation for the concentration of a deficient component, which in a co-moving with the flame co-ordinate system are

$$\frac{\partial T}{\partial t} + u_1 \frac{\partial T}{\partial z} = \chi \left(\frac{\partial^2 T}{\partial z^2} + \frac{\partial^2 T}{\partial y^2} \right), \tag{5.4.1}$$

$$\frac{\partial a}{\partial t} + u_1 \frac{\partial a}{\partial z} = D \left(\frac{\partial^2 a}{\partial z^2} + \frac{\partial^2 a}{\partial y^2} \right). \tag{5.4.2}$$

The unperturbed solution for the temperature profile and concentration are

$$T^0 = T_u + (T_b - T_u)\, e^{u_1 z/\chi}, \quad a^0 = a_u (1 - e^{u_1 z/D}), \quad for\ z < 0, \tag{5.4.3}$$

$$T = T_b, \quad a^0 = 0, \quad for\ z > 0. \tag{5.4.4}$$

In the reference frame comoving with the flame, the unperturbed solution is a stationary planar flame front, z is the coordinate along the normal to the unperturbed flame surface.

The boundary conditions at the surface of the chemical reaction are

$$T_{z-} = T_{z+}, \quad a_{z-} = a_{z+}, \tag{5.4.5}$$

$$\lambda \left(\left.\frac{\partial T}{\partial n}\right|_{z+} - \left.\frac{\partial T}{\partial n}\right|_{z-} \right) + Q\rho D \left(\left.\frac{\partial a}{\partial n}\right|_{z+} - \left.\frac{\partial a}{\partial n}\right|_{z-} \right) = 0, \tag{5.4.6}$$

where $\lambda = \rho C_P \chi$ is the coefficient of thermal conduction, Q is the reaction heat release, $\partial/\partial n$ is derivative along the normal to the surface of reaction zone, $z = z_f(t, y)$. Since the concentration of the deficient reagent, $a \equiv 0$ for $z > z_f$, then

$$a|_{z+} = a|_{z-} = 0 \ and \ \left.\frac{\partial a}{\partial n}\right|_{z+} = 0. \tag{5.4.7}$$

For the case $Ze \to \infty$, the width of the reaction zone tends to zero simultaneously with the average concentration of reagents within the reaction zone, but the amount of fuel consumed per unit time remains finite

$$j = -\rho D \left.\frac{\partial a}{\partial n}\right|_{z-}. \tag{5.4.8}$$

The magnitude of j, which is defined by solution of (5.4.2) with boundary conditions (5.4.7) depends on the temperature in the reaction zone

$$T_R = T|_{z+} = T|_{z-}. \tag{5.4.9}$$

The amount of the consumed fuel is defined by the Arrhenius law

$$j \propto \exp(-E/2RT_R). \tag{5.4.10}$$

The boundary condition (5.4.7) required that for the unperturbed solution

$$C_P(T_b - T_u) = Qa_0. \tag{5.4.11}$$

While for the case $Le = 1$, chemical reaction completes at the adiabatic flame temperature T_b, in the case of perturbed flames with $\chi \neq D$ temperature at the surface of chemical reaction zone can differ from T_b. Small perturbations of temperature, concentration and the flame surface in the upstream (subscript 1) and downstream (subscript 2) flows are

$$T_1' = \varepsilon_1(T_b - T_u)\exp(\sigma t + k_1 z + iky), \tag{5.4.12}$$

$$T_2' = \varepsilon_2(T_b - T_u)\exp(\sigma t - k_2 z + iky), \tag{5.4.13}$$

$$a_1' = \varepsilon_3 a_u \exp(\sigma t + k_3 z + iky). \tag{5.4.14}$$

Perturbations of a planar flame front against small disturbances bending the front are

$$\varsigma(y, t) = \varepsilon_4 \exp(\sigma t + iky), \tag{5.4.15}$$

where z and y are the coordinates along the normal to the unperturbed flame surface, and along the flame surface, respectively, $k = 2\pi/\lambda$ is the wave-number of perturbation, λ is the wavelength of the perturbation, σ is the perturbation growth rate, and $\varepsilon_1, \varepsilon_2, \varepsilon_3, \varepsilon_4$ are the same order small amplitudes of perturbations.

The linearized equations for perturbations are

$$\frac{\partial T_1'}{\partial t} + u_1\frac{\partial T_1'}{\partial z} = \chi\left(\frac{\partial^2 T_1'}{\partial z^2} + \frac{\partial^2 T_1'}{\partial y^2}\right), \tag{5.4.16}$$

$$\frac{\partial T_2'}{\partial t} + u_1\frac{\partial T_2'}{\partial z} = \chi\left(\frac{\partial^2 T_2'}{\partial z^2} + \frac{\partial^2 T_2'}{\partial y^2}\right), \tag{5.4.17}$$

$$\frac{\partial a_1'}{\partial t} + u_1\frac{\partial a_1'}{\partial z} = D\left(\frac{\partial^2 a_1'}{\partial z^2} + \frac{\partial^2 a_1'}{\partial y^2}\right). \tag{5.4.18}$$

Reagents are fully consumed in the reaction zone, and therefore, $a_z \equiv 0$.

Substituting (5.4.12)–(5.4.14) in (5.4.16)–(5.4.18), we obtain for k_1, k_2, k_3

$$k_1 = k_2 = \pm\frac{u_1}{2\chi} + \sqrt{\frac{u_1^2}{4\chi^2} + \frac{\sigma}{\chi} + k^2}, \quad k_3 = \frac{u_1}{2D} + \sqrt{\frac{u_1^2}{4D^2} + \frac{\sigma}{D} + k^2}. \quad (5.4.19)$$

Linearized conditions at the flame surface with accuracy up to the first order terms, $O(\varepsilon^2)$, are

$$T|_{z=z_f} = T^0(0) + \left.\frac{\partial T^0}{\partial z}\right|_0 \cdot z_f + T'(0), \quad (5.4.20)$$

$$\left.\frac{\partial T}{\partial n}\right|_{z=z_f} \approx \left.\frac{\partial T}{\partial z}\right|_{z=z_f} = \left.\frac{\partial T^0}{\partial z}\right|_0 + \left.\frac{\partial^2 T^0}{\partial z^2}\right|_0 \cdot z_f + \left.\frac{\partial T'}{\partial z}\right|_0, \quad (5.4.21)$$

$$a|_{z=z_f} = \left.\frac{\partial a^0}{\partial z}\right|_0 + \left.\frac{\partial^2 a^0}{\partial z^2}\right|_0 \cdot z_f + \left.\frac{\partial a'}{\partial z}\right|_0. \quad (5.4.22)$$

With help of Eqs. (5.4.20)–(5.4.22) the boundary conditions (5.4.6)–(5.4.7) can be written as

$$\left.\frac{\partial T_1^0}{\partial z}\right|_0 \cdot z_f + T_1'(0) = T_2'(0) = T_R - T_b, \quad and \quad \left.\frac{\partial a_1^0}{\partial z}\right|_0 \cdot z_f + a_1'(0) = 0, \quad (5.4.23)$$

$$\rho C_P \chi\left(\left.\frac{\partial T_2'}{\partial z}\right|_0 - \left.\frac{\partial^2 T_1^0}{\partial z^2}\right|_0 \cdot z_f - \left.\frac{\partial T_1'}{\partial z}\right|_0 \right) - Q\rho D\left(\left.\frac{\partial^2 a_1^0}{\partial z^2}\right|_0 \cdot z_f + \left.\frac{\partial a_1'}{\partial z}\right|_0 \right) = 0. \quad (5.4.24)$$

Equating perturbations of the fuel flux obtained from (5.4.8)

$$\frac{j'}{j_0} = \left(\left.\frac{\partial^2 a_1^0}{\partial z^2}\right|_0 \cdot z_f + \left.\frac{\partial a_1'}{\partial z}\right|_0 \right) \Big/ \left(\left.\frac{\partial a_1^0}{\partial z}\right|_0 \right), \quad (5.4.25)$$

and taking into account that (5.4.10) implies

$$\frac{j'}{j_0} = \frac{E T_2'(0)}{2RT_b^2}, \quad (5.4.26)$$

we obtain equation for small perturbations of the reagent concentration on the flame surface

$$\left(\left.\frac{\partial^2 a_1^0}{\partial z^2}\right|_0 \cdot z_f + \left.\frac{\partial a_1'}{\partial z}\right|_0 \right) \Big/ \left(\left.\frac{\partial a_1^0}{\partial z}\right|_0 \right) = Ze\left(\frac{T_R - T_b}{T_b - T_u} \right). \quad (5.4.27)$$

Substituting (5.4.12)–(5.4.15) in (5.4.23) and (5.4.27) we obtain system of linear uniform equations for unknown $\varepsilon_1, \varepsilon_2, \varepsilon_3, \varepsilon_4$

$$\frac{u_1}{D}\varepsilon_4 - \varepsilon_3 = 0, \quad \frac{u_1}{\chi}\varepsilon_4 + \varepsilon_1 - \varepsilon_2 = 0, \tag{5.4.28}$$

$$\frac{u_1^2}{\chi^2}\left(1 - \frac{\chi}{D}\right)\varepsilon_4 + k_3\frac{D}{\chi}\varepsilon_3 + k_1\varepsilon_1 + k_2\varepsilon_2 = 0, \quad \frac{u_1}{D}\varepsilon_4 - k_3\frac{D}{u_1}\varepsilon_3 - \varepsilon_2 Ze = 0. \tag{5.4.29}$$

It is convenient to introduce dimensionless variables for the growth rate, $\Sigma = 4\frac{L_f}{U_f}\sigma$, and for the wave vector, $K = 2L_f k$, taking into account that $u_1 = U_f$ is normal velocity of the flame and the width of the flame (defined by the thermal zone) is $L_f U_f \approx \chi$. The condition of compatibility of the system of four Eqs. (5.4.28)–(5.4.29) written in dimensionless variables is

$$\begin{aligned} &\sqrt{1+\Sigma+K^2}\left(1 - \frac{Ze}{Le} - \sqrt{1+\Sigma Le + K^2 Le^{-2}}\right) + \\ &+ Ze\left(\frac{1}{Le} - 1 + \sqrt{1+\Sigma/Le + K^2 Le^{-2}}\right) = 0. \end{aligned} \tag{5.4.30}$$

For a long wavelength perturbation, $\lambda \gg L_f \rightarrow K \ll 1$, and, taking into account that we consider a very high activation energy, $Ze \gg 1$, we obtain from (5.4.31)

$$\Sigma = K^2 Ze(\frac{1}{Le} - 1). \tag{5.4.32}$$

The r.h.s. of (5.4.32) is positive for $Le = \chi/D < 1$, which means that the long wavelength perturbations, $\lambda \gg L_f$ exponentially grow with time.

The theory of thermal-diffusive instability developed in [16, 29, 30] does not include thermal expansion in order to decouple the hydrodynamics from the transport equations. The assumption of the infinitely large activation energy is another limitation of the theory. Nevertheless, the Zel'dovich theory has been the first successful approach elucidating the relative role of the thermal conduction and mass diffusion on the flame stability. Further development in understanding of the thermo-diffusive flame instability was due to Sivashinsky [32, 33], who has pointed that a flame, though possibly unstable for long-wave perturbations, is nevertheless always stable for short-wave perturbations. He also introduced the critical the Lewis number, such that the flame is stable if $Le > Le_c = 1 - 2/\beta$, where $\beta = E(T_b - T_u)/RT_b^2$, and derived the dispersion relation at $Le \simeq Le_c$, incorporating the relaxation effect of short-wave perturbations

$$\sigma = \chi\left\{[\beta(1 - Le)/2]k^2 - 4L_f^2 k^4\right\}. \tag{5.4.33}$$

In subsequent fundamental works, a general asymptotic formulation was obtained for the Lewis numbers slightly different from unity, and an equation was derived [34], which describes a nonlinear development of instabilities. Numerical simulations of this equation for outward expanding spherical flame [35] was apparently one of the first attempts to explain the experimentally observed acceleration of a spherically expanding flame [36].

Most of the models [10–14, 37–40] used the assumption of high activation energies, asymptotic analysis and exploit the multi-scale nature of the problem characterized, for example, by a diffusion length scale representing the flame thickness and the hydrodynamic length scale associated, with the vessel dimension, or with the size of the wrinkles on the flame front. The most elaborated models were obtained resolving the internal structure of the reaction zone, using the matching conditions and assuming that the density is constant. The dispersion relation can be derived in a closed explicit form, but due to its transcendental nature, roots for the growth rate can be found only with the help of numerical calculations [41, 42]. The effect of thermal expansion on the stability of planar diffusion flames has been considered in [43]. Unlike premixed flames, the burning rate of diffusive flames is determined by the rate at which the fuel and oxidizer are transported from opposite sides in the certain proportions to sustain the chemical reaction. The model of the flame instability derived in [37] includes the effects of thermal expansion and assumption that a thin flame interacts with the flow on a larger scale, so that the flame curvature is weak compared to the flame front thickness. It does capture the pulsating instability, but does not exhibit its high-wavenumber cut off. On the contrary, the Matalon-Matkowsky model [13] exhibits a high-wavenumber cut-off for the DL instability, but does not capture the pulsating instability because of restriction $Le = 1 + 0(1)$. In [40] the model was derived for an arbitrary Lewis number, large Peclet numbers and weakly curved flame, and it captures the pulsating instability and exhibits its high-wavenumber cut-off for all instabilities. The analytical models of the flame instability derived in [11, 13, 21, 39] used multiscale asymptotic analysis and expansion in powers of the inverse Peclet number, $1/Pe \ll 1$. As was pointed in [42], the planar pulsation instability is associated with relatively large Lewis numbers, $Le \approx 1.2 - 2$. A specific manifestation of thermal-diffusive instability is spontaneous flame oscillations, which were experimentally observed in candles, suspended fuel droplets and jet diffusion flames [44, 45].

An accurate investigation of the interactions between flow, chemistry and transport processes at the linear stage, as well as during the nonlinear evolution of the flame front, must consider the finite thickness of the flame front, and can only be achieved by numerical solution the complete system of the reactive Navier–Stokes equations. The initial linear growth of perturbations in premixed flame domains were studied in two dimensional numerical simulations using a one-step chemical model for hydrogen/air [46]. Both unity Lewis number, where only hydrodynamic instability appears, and subunit Lewis number, where the flame propagation is strongly affected by the combined effect of hydrodynamic and thermo diffusive instabilities were considered in [47, 48]. In the second case the mixture was characterized by an effective Lewis number lower than unity ($Le_{eff} = 0.404$), so that the perturbed planar flame is subjected to thermo diffusive and hydrodynamic instability. The growth rates during the initial linear stage were computed and compared with analytical results. In agreement with the theory, the hydrodynamically unstable flames with $Le = 1$ acquire a single cell structure which propagates with a constant velocity. It was found that the flame speed is modified due to a strong dependence on the flame curvature. The important consequence is that the flame speed definition as the local consumption speed, which is based on reactivity is not appropriate. For thermo diffusively unstable flames, both reactivity and heat conduction vary locally and their relative contribution to the local displacement speed depends on the flame curvature. The increased speed in convex regions is caused by the enhanced local reactivity. In concave regions reactivity is lower by an order of magnitude in comparison to the convex regions, nevertheless the local flame speed is higher due to the dominant influence of the tangential heat conduction. The increase of the global speed of the propagating flame over the laminar flame speed is always equal to the increase of the flame area for the case $Le = 1$, and it well agrees with the hydrodynamic theory. In contrast with the case $Le = 1$, the global flame speed for the flames with $Le < 1$ is significantly larger than the corresponding flame area increase.

5.4.2 *Thermal Diffusive Instability of Solid Propellant*

The problem of a supplementary boundary condition arises also in the theory of flames in solid propellants if one replaces the reaction zone by a discontinuity surface [16, 29]. We consider an exact analytical solution for the stability problem of gasless combustion propagating in solid propellant, which was obtained in [48]. It is known, that combustion in a solid propellant is unstable and the fastest (1D) instability results in a pulsating regime of the combustion front. For the case of large activation energy $E \gg T_2^2/(T_2 - T_1)$, where T_1 and T_2 are temperatures of the fuel and the combustion products respectively, the chemical reaction zone is thin in comparison with the preheat zone. The solution of the stability problem obtained in [29], considered the reaction zone as the inner discontinuity surface assuming that the temperature perturbation is continuous at the discontinuity surface. The growth rate for the fastest perturbations was found to be

$$\sigma = \frac{u_f^2}{\chi} \frac{E^2(T_2 - T_1)^2}{16T_2^4}, \tag{5.4.2.1}$$

where u_f is the velocity of the combustion wave and χ is thermal diffusivity.

It is easily seen that there is a contradiction between this solution and the basic assumption, since the temperature perturbations change on the length scale $\sim \sqrt{\chi/\sigma}$, which is of the same order as the thickness of the reaction zone $\delta_{ch} \approx \chi T_2/u_f E$. This means that perturbations are localized inside the chemical reaction zone, which therefore cannot be treated as a surface of zero thickness.

We consider the spectral problem of the flame stability in a condensed matter, taking into account the finite thickness of the reaction zone [48]. The diffusion is negligible and the flame propagation is described by the equations

$$\frac{\partial T}{\partial t} + \mathbf{u}_f \nabla T = \nabla(\chi \nabla T) + \frac{aQ}{C\tau} \exp(-E/T), \tag{5.4.2.2}$$

$$\frac{\partial a}{\partial t} + \mathbf{u}_f \nabla a = -\frac{a}{\tau} \exp(-E/T), \tag{5.4.2.3}$$

where a is the fuel concentration, Q is the heat of the reaction, which is taken as the first order Arrhenius reaction; τ is a constant with dimension of time, C is the specific heat. We assume a steady planar flame propagating along the z- axis. Using dimensionless variables:

$$\Theta = T/T_2, \quad \alpha = aQ/CT_2, \quad \kappa = \chi/\chi_2, \quad \xi = zu_f/\chi_2,$$

$$\tau = tu_f^2/\chi_2, \quad \mathfrak{A} = E/T_2, \quad \Lambda = \frac{\chi_2}{u_f^2 \tau}, \tag{5.4.2.4}$$

the velocity of a steady unperturbed flame is determined by the eigenvalue Λ of the system of Eqs. (5.4.2.2)–(5.4.2.3) written in dimensionless variables

$$\frac{d}{d\xi}\left(\kappa \frac{d\Theta}{d\xi}\right) - \frac{d\Theta}{d\xi} + \alpha\Lambda \exp(-\mathfrak{A}/\Theta) = 0, \tag{5.4.2.5}$$

$$\frac{d\alpha}{d\xi} + \alpha\Lambda \exp(-\mathfrak{A}/\Theta) = 0, \tag{5.4.2.6}$$

with the boundary conditions

$$\Theta = \Theta_1, \quad \alpha = \alpha_1 \quad for \quad \xi \to -\infty, \tag{5.4.2.7}$$

$$\frac{d\Theta}{d\xi} = 0, \quad \alpha = 0 \quad for \quad \xi \to +\infty. \tag{5.4.2.8}$$

The solution of the problem (5.4.2.5)–(5.4.2.8) exists as an intermediate asymptotic in the limit $\Lambda \exp(-\mathfrak{A}/\Theta_1) \approx 0$, which means the reaction time of the fuel in the initial state is much larger any other time scales of the problem, so that with a high precision we can consider that there is no reaction in the initial state.

In the limit $\mathfrak{A}(1 - \Theta_1) \gg 1$ the fuel concentration is constant and equal to the initial value α_1 everywhere up to the thin chemical reaction zone of thickness $\Delta_R \sim 1/\mathfrak{A} \ll 1$. For the steady flame the temperature changes on a length scale of order unity in dimensionless variables, while inside the reaction zone the temperature changes slowly as $1/\mathfrak{M}$. Thus, inside the reaction zone Eqs. (5.4.2.5) and (5.4.2.6) to within terms of order $1/\mathfrak{M}$ can be written as

$$\frac{d\Theta}{d\xi} = \alpha, \tag{5.4.2.9}$$

$$\frac{d^2\Theta}{d\xi^2} - \frac{d\Theta}{d\xi}\Lambda \exp(-\mathfrak{A}) \exp[\mathfrak{A}(\Theta - 1)] = 0. \tag{5.4.2.10}$$

The solution to Eq. (5.4.2.10) with boundary conditions: $d\Theta/d\xi = 1 - \Theta_1$ for $\mathfrak{A}(1 - \Theta) \gg 1$, and $d\Theta/d\xi = 0$ for $\Theta = 1$, is

$$\mathfrak{A}(\Theta - 1) = \ln[\mathfrak{A}(\Theta - 1)] = \mathfrak{A}(1 - \Theta_1)\xi. \tag{5.4.2.11}$$

The corresponding eigenvalue is

$$\Lambda = \mathfrak{A}(1 - \Theta_1)\exp(\mathfrak{A}). \tag{5.4.2.12}$$

For planar unperturbed flow, the small perturbations can be taken in the form

$$\tilde{\varphi}(\xi, \zeta, \tau) = \tilde{\varphi}(\xi)\exp(S\tau + iK\zeta), \tag{5.4.2.13}$$

where, $\zeta = xu_f/\chi_2$, $S = \sigma\chi_2/u_f^2$, and $K = k\chi_2/u_f$ are the dimensionless coordinate, instability growth rate, and the perturbation wave number, respectively.

System of Eqs. (5.4.2.2)–(5.4.2.3) linearized for the small perturbations is

$$\frac{d^2}{d\xi^2}\left(\kappa\tilde{\Theta}\right) - \frac{d\tilde{\Theta}}{d\xi} - (S - \kappa K^2)\tilde{\Theta} + \Lambda(\tilde{\alpha} + \alpha\mathfrak{A}\tilde{\Theta}/\Theta^2)\exp(-\mathfrak{A}/\Theta) = 0 \tag{5.4.2.14}$$

$$\frac{d\tilde{\alpha}}{d\xi} - S\tilde{\alpha} + \Lambda(\tilde{\alpha} + \alpha\mathfrak{A}\tilde{\Theta}/\Theta^2)\exp(-\mathfrak{A}/\Theta) = 0. \tag{5.4.2.15}$$

The boundary conditions for Eqs. (5.4.2.14) and (5.4.2.15) are that the perturbations vanish at $\xi \to \pm\infty$.

Since inside the reaction zone $\Theta \approx 1$, we obtain from (5.4.2.15)

$$\tilde{\alpha} = \mathfrak{A}\alpha \exp(-S\xi)\int_{-\infty}^{\xi}\tilde{\Theta}\frac{d(\ln(\alpha))}{d\eta}\exp(S\eta)d\eta. \tag{5.4.2.16}$$

Assuming that $S \sim \mathfrak{A}^2$, and perturbations of the temperature $\tilde{\Theta}$ change on the length scale $1/\sqrt{S} \sim 1/\mathfrak{A}$, which is comparable to the thickness of the reaction zone, we obtain from (5.4.2.16) to within terms of order $1/\mathfrak{A} \ll 1$ '

$$\tilde{\alpha} = \frac{\mathfrak{A}}{S}\frac{d\alpha}{d\xi}\tilde{\Theta}. \tag{5.4.2.17}$$

Substituting this expression for $\tilde{\alpha}$ into Eq. (5.4.2.14) and considering only the zero-order terms of the expansion in $1/\mathfrak{A} \ll 1$, we obtain equation for the temperature perturbation

$$\frac{d^2\tilde{\Theta}}{d\xi^2} - \tilde{\Theta}\left(S + K^2 + \mathfrak{A}\frac{d\alpha}{d\xi}\right) = 0. \tag{5.4.2.18}$$

Taking into account (5.4.2.9) and (5.4.2.11) for the structure of the reaction zone we obtain

$$\frac{d\alpha}{d\xi} = -\frac{\mathfrak{A}}{4}(1 - \Theta_1)^2 \cosh^{-2}[\mathfrak{A}(1 - \Theta_1)\xi/2]. \tag{5.4.2.19}$$

Substituting (5.4.2.19) into Eq. (5.4.2.18) and introducing the variables

$$\eta = \mathfrak{A}(1 - \Theta_1)\xi/2, \quad S_\eta = 4\frac{S + K^2}{\mathfrak{A}^2(1 - \Theta_1)^2}, \tag{5.4.2.20}$$

we obtain

$$\frac{d^2\tilde{\Theta}}{d\xi^2} - \left(S_\eta - \cosh^{-2}\eta\right)\tilde{\Theta} = 0. \tag{5.4.2.21}$$

Equation (5.4.2.21) is the Schrodinger equation for the one-dimensional motion of a particle of unit mass in the potential well $\cosh^{-2}\eta$. This equation has a well-known exact solution [17] with the following eigenvalues

$$S_{\eta,n} = \frac{1}{4}(-1 - 2n + \sqrt{5}), \tag{5.4.2.22}$$

where $n = 0, 1, 2, \ldots$ Only the mode $n = 0$ corresponds to growing perturbations ($S > 0$), so the instability growth rate is

$$S = \frac{\sqrt{5} - 1}{16}\mathfrak{A}^2(1 - \Theta_1)^2 - K^2. \tag{5.4.2.23}$$

In the dimensional variables the instability growth rate is

$$\sigma = \frac{\sqrt{5} - 1}{16}\frac{u_f^2}{\chi_2}\frac{E^2(T_2 - T_1)^2}{T_2^4} - \chi_2 k^2. \tag{5.4.2.24}$$

It should be noted that the exact solution (5.4.2.24) for the instability growth rate differs from the solution obtained for the discontinuity model and for the limiting case of large activation energy in [16] by the factor $\sqrt{5} - 1$. It is very likely that for the case of a finite activation energy the difference between the exact solution and the solution obtained for the discontinuity model will be even more pronounced.

5.5 Flame Dynamics; Evolution Equation

As was shown in previous sections of this chapter, the analysis of a flame stability is greatly simplified if we consider a flame as a discontinuity surface. At the same time, in order to understand how is limited the growth rate of instabilities and to find the cut-off wave lengths, we must consider structure of the flame front, which means that we have to integrate complete system of the Navier–Stokes equations with the energy release in chemical reactions to account dissipations caused by thermal conduction and viscosity, which is a rather complicated problem. The problem can be simplified considerably if the region of thermal conduction and energy release can be replaced by a surface of discontinuity, separating the fresh mixture and the combustion products. Practically this means that we have to specify the evolution equation for the deflagration front and the boundary conditions at the flame front. It was shown in previous sections that the structure of the flame front remains important even on the scales exceeding significantly the flame thickness. At the same time, the flame front can be considered as a discontinuity surface, but this does not necessarily mean the introduction of a discontinuity surface of zero thickness: the structure of the flame front can be considered in the boundary conditions and in the evolution equation as an external parameter that depends on the fuel properties. In this case we must specify the evolution equation for the deflagration front and the boundary conditions at the front. The evolution equation for a flame front of finite thickness and the boundary conditions at the front for the case of an ideal gas equation of state and a one-step chemical reaction with high activation energy were derived first by Matalon and Matkowsky [13] and by Clavin and Joulin [47]. The evolution equation for a curved deflagration front of finite thickness for the case of an arbitrary equation of states of the combustible mixture and for arbitrary types of dissipations and energy release was derived in [49].

5.5.1 Evolution Equation; Infinitely Thin Flame Front

Since a deflagration propagates with essentially subsonic velocity, we can consider the flow as incompressible. We consider the isobaric approximation and the unit Lewis number, Le $=$ 1. In the reference frame comoving with the deflagration wave propagating in the direction of z-axis, the stationary solution consists of the upstream homogeneous flow of the fresh combustible mixture, a narrow transition zone of the flame structure, which here we consider as a surface of zero thickness, and the downstream homogeneous flow of the combustion products. We will use the same designations as in the derivation of Eq. (5.2.14). In dimensionless variables: $\xi = z/L_f$, $u = u_z/U_f$, $\mathbf{w} = \mathbf{u}_\perp/U_f$, $\theta = T/T_-$, $\Pi = (P - P_-)/\rho_- U_f^2$, the equation of thermal conduction reads

$$\frac{d^2\theta}{d\xi^2} - \frac{d\theta}{d\xi} + W_R(\theta) = 0, \tag{5.5.1.1}$$

with boundary conditions $\theta = 1$ for $\xi \to -\infty$, and $\theta = \Theta$ for $\xi \to \infty$.

Let us assume that the flame generated flow can be characterized by some length scale R, e.g. size of combustion chamber or a tube radius, and the scaled position of the flame $f(\mathbf{x}, t)$ is defined as $F(z, \mathbf{x}, t) = z - Rf(\mathbf{x}, t) = 0$. Typically, the hydrodynamic length scale is always much larger than the flame thickness. It is convenient to introduce the dimensionless time $\tau = U_f t/R$, the dimensionless transverse coordinate, $\boldsymbol{\eta} = \mathbf{x}/R$, and the small parameter, $\varepsilon = L_f/R \ll 1$, characterizing the small thickness of the flame front. In the coordinate system co-moving with the flame front

$$\zeta = \xi - f(\boldsymbol{\eta}, \tau)/\varepsilon, \quad \eta = \eta, \quad \tau = \tau. \tag{5.5.1.2}$$

The hydrodynamic equations for the isobaric condition take the form

$$-\frac{\varepsilon}{\theta^2}\frac{\partial\theta}{\partial\tau} + \frac{\partial\mathbf{j}}{\partial\zeta} + \varepsilon\nabla_\perp \cdot \left(\frac{\mathbf{j}}{\theta}\right) = 0, \tag{5.5.1.3}$$

$$\frac{\varepsilon}{\theta}\frac{\partial\mathbf{w}}{\partial\tau} + j\frac{\partial\mathbf{w}}{\partial\zeta} + \frac{\varepsilon}{\theta}(\mathbf{w}\cdot\nabla_\perp)\mathbf{w} = -\varepsilon\nabla_\perp\boldsymbol{\Pi} + \nabla_\perp f\frac{\partial\boldsymbol{\Pi}}{\partial\zeta}, \tag{5.5.1.4}$$

$$\frac{\varepsilon}{\theta}\frac{\partial\mathbf{u}}{\partial\tau} + \mathbf{j}\frac{\partial\mathbf{u}}{\partial\zeta} + \frac{\varepsilon}{\theta}(\mathbf{w}\cdot\nabla_\perp)\mathbf{u} = -\frac{\partial\boldsymbol{\Pi}}{\partial\zeta}, \tag{5.5.1.5}$$

$$\begin{aligned}\frac{\varepsilon}{\theta}\frac{\partial\theta}{\partial\tau} + \mathbf{j}\frac{\partial\theta}{\partial\zeta} + \frac{\varepsilon}{\theta}\mathbf{w}\cdot\nabla_\perp\theta = \mathbf{N}^2\frac{\partial^2\theta}{\partial\zeta^2} + \varepsilon^2\nabla_\perp^2\theta + \\ +\varepsilon\nabla_\perp\cdot\left(\nabla_\perp f\frac{\partial\theta}{\partial\zeta}\right) - \varepsilon\nabla_\perp f\cdot\nabla_\perp\frac{\partial\theta}{\partial\zeta} + \Theta_R(\theta).\end{aligned} \tag{5.5.1.6}$$

Here the mass flux j is given by the expression

$$\mathbf{j} = \frac{1}{\theta}\left(u - \mathbf{w}\cdot\nabla_\perp f - \frac{\partial f}{\partial\tau}\right), \tag{5.5.1.7}$$

and $\nabla_\perp$ is the "transverse" part of the operator ∇ related to the coordinate in the flame front surface. The Laplace operator in these variables takes the form

$$\nabla^2 = N^2\frac{\partial^2}{\partial\zeta^2} - \varepsilon\nabla_\perp^2 f\frac{\partial}{\partial\zeta} - 2\varepsilon\frac{\partial}{\partial\zeta}(\nabla_\perp f\cdot\nabla_\perp) + \varepsilon^2\nabla_\perp^2, \tag{5.5.1.8}$$

where $N = \sqrt{1+(\nabla_\perp f)^2}$.

It is convenient to rewrite Eqs. (5.5.1.4) and (5.5.1.5) in another form, extracting the expression for the momentum flux in the left-hand side

$$\begin{aligned} j\frac{\partial}{\partial\zeta}(\mathbf{w}+u\nabla_\perp f) = -\frac{\varepsilon}{\theta}\left(\frac{\partial}{\partial\tau}+\mathbf{w}\cdot\nabla_\perp\right)(\mathbf{w}+u\nabla_\perp f)+ \\ \varepsilon\frac{u}{\theta}\left(\frac{\partial}{\partial\tau}+\mathbf{w}\cdot\nabla_\perp\right)\nabla_\perp f - \varepsilon\nabla_\perp\Pi, \end{aligned} \tag{5.5.1.9}$$

$$\begin{aligned} N^2\frac{\partial}{\partial\zeta}\left(\Pi+\frac{\theta j^2}{N^2}\right) = -\varepsilon\frac{\partial j}{\partial\tau} - \varepsilon\nabla_\perp\cdot(j\mathbf{w}) - \frac{\varepsilon}{\theta}\frac{\partial^2 f}{\partial\tau^2} - \\ -2\frac{\varepsilon}{\theta}\mathbf{w}\cdot\nabla_\perp\frac{\partial f}{\partial\tau} - \frac{\varepsilon}{\theta}\mathbf{w}\cdot(\mathbf{w}\cdot\nabla_\perp)\nabla_\perp f + \varepsilon\nabla_\perp f\cdot\nabla_\perp\Pi. \end{aligned} \tag{5.5.1.10}$$

Expanding j and θ in (5.5.1.3)–(5.5.1.6) in power series of the small parameter ε: $j = j_0+\varepsilon j_1+O(\varepsilon^2)$ and $\theta = \theta_0+\varepsilon\theta_1+O(\varepsilon^2)$, we obtain a solution of Eqs. (5.5.1.3), (5.5.1.9), (5.5.1.10) with the accuracy $O(\varepsilon)$

$$j = j_0(\eta,\tau) + O(\varepsilon), \tag{5.5.1.11}$$

$$\mathbf{w} + u\nabla_\perp f = \mathbf{w}_- + u_-\nabla_\perp f + O(\varepsilon), \tag{5.5.1.12}$$

$$\Pi + \frac{\theta j^2}{N^2} = \Pi_- + \frac{j_-^2}{N^2} + O(\varepsilon), \tag{5.5.1.13}$$

where label "–" refers to the values just ahead of the flame front.

Written in dimensional variables integrals (5.5.1.11)–(5.5.1.13) are the conservation laws of mass and momentum across the flame front of zero thickness:

$$\rho\left(\mathbf{n}\cdot\mathbf{v}+\frac{1}{N}\frac{\partial F}{\partial t}\right)\Big|_-^+ = 0, \tag{5.5.1.11a}$$

$$\mathbf{n}\times\mathbf{v}|_-^+ = 0, \tag{5.5.1.12a}$$

$$\left[P+\rho\left(\mathbf{n}\cdot\mathbf{v}+\frac{1}{N}\frac{\partial F}{\partial t}\right)^2\right]\Big|_-^+ = 0, \tag{5.5.1.13a}$$

where $(...)|_-^+$ represents the change of a quantity across the flame front from the fresh mixture (-) to the combustion products (+), for a planar stationary flame moving in the negative direction $F(\mathbf{x}, t) = z + U_f t = 0$, $\mathbf{n} = \nabla F/|\nabla F|$ is the unit normal vector directed towards the combustion products and $N = |\nabla F|$. Equation of thermal conduction with the same accuracy becomes

$$N^2 \frac{\partial^2 \theta}{\partial \zeta^2} - j \frac{\partial \theta}{\partial \zeta} + W_R(\theta) = 0. \tag{5.5.1.14}$$

Since Eq. (5.5.1.14) is similar to the equation of thermal conduction for a planar stationary flame, the solution of Eq. (5.5.1.14) can be written in the form:

$$\theta_0 = \theta_p(\zeta/N), \quad j_0(\eta, \tau) = N \equiv \sqrt{1 + (\nabla_\perp f)^2}. \tag{5.5.1.15}$$

Here $\theta = \theta_p(\xi)$ is the temperature profile for a planar flame, which is an unperturbed solution of the ordinary differential equation (5.5.1.1), which, for example, can be taken as the Zeldovich-Frank-Kamenetskii solution for the temperature distribution for a stationary flame. The obtained integrals give the solution for the velocity components and pressure inside the curved flame

$$u = u_- + \frac{\theta - 1}{N} + O(\varepsilon), \tag{5.5.1.16}$$

$$\mathbf{w} = \mathbf{w}_- - \frac{\theta - 1}{N} \nabla_\perp f + O(\varepsilon), \tag{5.5.1.17}$$

$$\Pi = \Pi_- - \theta + 1 + O(\varepsilon). \tag{5.5.1.18}$$

Equations (5.5.1.11) and (5.5.1.15) determine the evolution equation for an infinitely thin deflagration front

$$\frac{\partial f}{\partial \tau} + \mathbf{w}_- \cdot \nabla_\perp f + \sqrt{1 + (\nabla_\perp f)^2} - u_- = O(\varepsilon). \tag{5.5.1.19}$$

Evolution Eq. (5.5.1.19) together with Eqs. (5.5.1.11)–(5.5.1.13) describe the dynamics of the infinitely thin flame front and correspond to the Darrieus-Landau model of the infinitely thin deflagration front.

5.5.2 *Evolution Equation; Influence of Finite Flame Thickness*

Evolution of a flame front depends essentially upon the finite thickness of the flame front even on the length scales exceeding considerably the flame front thickness. To consider a finite thickness of the flame front, we must leave the second-order terms in the expansion in ε. Integral of the Eq. (5.5.1.3) for the mass flux with the accuracy ε^2 is

$$
\begin{aligned}
j_1 = j_{1-} - \zeta \nabla_\perp \cdot \mathbf{w}_- - \frac{\theta_0 - 1}{\theta_0}\left(\frac{\partial N}{\partial \tau} + \mathbf{w}_- \cdot \nabla_\perp N + \frac{\nabla_\perp f \cdot \nabla_\perp N}{N}\right)\frac{\zeta}{N} - \\
-\left(\frac{\partial N}{\partial \tau} + \nabla_\perp \cdot (N\mathbf{w}_-) + \nabla_\perp^2 f\right) J_1(\zeta/N),
\end{aligned}
\tag{5.5.2.1}
$$

where $J_1(\zeta)$ is defined by Eq. (5.2.38). The conservation law for the mass flux across the flame of a finite thickness follows from Eq. (5.5.2.1), where we substitute the Zeldovich-Frank-Kamenetskii expression for the temperature profile

$$
\frac{1}{\theta}\left(u - \mathbf{w} \cdot \nabla_\perp f - \frac{\partial f}{\partial \tau}\right)\Big|_-^+ = \varepsilon \Theta \ln \Theta \left(\frac{\partial N}{\partial \tau} + \nabla_\perp \cdot (N\mathbf{w}_-) + \nabla_\perp^2 f\right). \tag{5.5.2.2}
$$

Integrating (5.5.1.9) and (5.5.1.10) and, taking into account asymptotic behavior outside the flame front, we obtain for the jump of the of momentum flux components across the flame front of finite thickness

$$
(\mathbf{w} + u \nabla_\perp f)|_-^+ = \varepsilon \Theta \ln \Theta \left(\hat{D}\mathbf{w}_- + \nabla_\perp f \hat{D} u_- + N^{-1} \hat{D} \nabla_\perp f\right), \tag{5.5.2.3}
$$

$$
\begin{aligned}
\left(\Pi + \frac{\theta j^2}{N^2}\right)\Big|_-^+ = \varepsilon(\Theta - 1)\nabla_\perp \cdot \left(N^{-1}\nabla_\perp f\right) + \varepsilon \frac{\Theta \ln \Theta}{N} \times \\
\times \left(\frac{\partial^2 f}{\partial \tau^2} + 2\mathbf{w}_- \cdot \nabla_\perp \frac{\partial f}{\partial \tau} + \mathbf{w}_- \cdot (\mathbf{w}_- \cdot \nabla_\perp)\nabla_\perp f + 2\hat{D}N - \frac{\nabla_\perp f \cdot \nabla_\perp N}{N}\right),
\end{aligned}
\tag{5.5.2.4}
$$

where $\hat{D}$ is the differential operator, which is defined as

$$
\hat{D} = \frac{\partial}{\partial \tau} + \mathbf{w}_- \cdot \nabla_\perp + \frac{1}{N}\nabla_\perp f \cdot \nabla_\perp. \tag{5.5.2.5}
$$

To find the evolution equation with the accuracy up to terms of order ε^2, we need to obtain the relation for the next order term of the mass flux just ahead of the flame front j_{1-} as an eigenvalue of the equation of thermal conduction (5.5.1.6). Substituting expression for the mass flux (5.5.2.1) into Eq. (5.5.1.6) and extracting the terms of order $O(\varepsilon)$

$$
\begin{aligned}
& j_{1-}\frac{\partial\theta_0}{\partial\zeta} - \left(\frac{\partial N}{\partial\tau} + \nabla_\perp\cdot(N\mathbf{w}_-)\right)\frac{\zeta}{N}\frac{\partial\theta_0}{\partial\zeta} - \\
& -\left(\frac{\partial N}{\partial\tau} + \nabla_\perp\cdot(N\mathbf{w}_-) + \nabla_\perp^2 f\right)J_1(\zeta/N)\frac{\partial\theta_0}{\partial\zeta} + \\
& +\nabla_\perp^2 f\frac{\partial\theta_0}{\partial\zeta} - 2\nabla_\perp f\cdot\frac{\nabla_\perp N}{N}\frac{\partial}{\partial\zeta}\left(\zeta\frac{\partial\theta_0}{\partial\zeta}\right) = \\
& = N^2\frac{\partial^2\theta_1}{\partial\zeta^2} - N\frac{\partial\theta_1}{\partial\zeta} + \frac{dW_R}{d\theta}\theta_1,
\end{aligned}
\tag{5.5.2.6}
$$

we can see that the right-hand side of Eq. (5.5.2.6) is zero for $\theta_1 = \partial\theta_0/\partial\zeta$:

$$
\hat{W}_R\left(\frac{\partial\theta_0}{\partial\zeta}\right) \equiv N^2\frac{\partial^3\theta_0}{\partial\zeta^3} - N\frac{\partial^2\theta_0}{\partial\zeta^2} + \frac{dW_R}{d\theta}\frac{\partial\theta_0}{\partial\zeta} = 0. \tag{5.5.2.7}
$$

Just as it was used in the derivation of (5.2.41–5.2.42), we can consider the left-hand side of Eq. (5.5.2.6) as small corrections to the expression in the right-hand side. Then, reducing the operator $\hat{W}_R$ to a Hermitian form and employing the perturbation technique [17], we obtain the relation for the next term in ε of the mass flux just ahead of the flame front

$$
\begin{aligned}
j_{1-} = & \left(\frac{\partial N}{\partial\tau} + \nabla_\perp\cdot(N\mathbf{w}_-) + \nabla_\perp^2 f\right)\left\langle(\xi + J_1)\frac{d\theta_p}{d\xi}\right\rangle - \\
& -\nabla_\perp^2 f\left\langle(\xi+1)\frac{d\theta_p}{d\xi}\right\rangle + 2\nabla_\perp f\cdot\frac{\nabla_\perp N}{N}\left\langle\frac{d}{d\xi}\left(\xi\frac{d\theta_p}{d\xi}\right)\right\rangle.
\end{aligned}
\tag{5.5.2.8}
$$

Here the averaging $\langle...\rangle$ is defined in the same way as in Sect. 5.2. Calculating coefficients in Eq. (5.5.2.8) using the Zeldovich-Frank-Kamenetskii profile for the temperature, we can write the evolution equation for the flame front in the form

$$
\begin{aligned}
& \frac{\partial f}{\partial\tau} + \mathbf{w}_-\cdot\nabla_\perp f + \sqrt{1 + (\nabla_\perp f)^2} - u_- = \\
& = \varepsilon\frac{\Theta\ln\Theta}{\Theta - 1}\left(\frac{\partial N}{\partial\tau} + \nabla_\perp\cdot(N\mathbf{w}_-) + \nabla_\perp^2 f\right)
\end{aligned}
\tag{5.5.2.9}
$$

Equation (5.5.2.9) provides a relation between the finite thickness flame stretching and the flame velocity. Together with the conservation laws (5.5.2.2)–(5.5.2.4) the evolution Eq. (5.5.2.9) specifies the discontinuity model for a flame of small but finite thickness.

The evolution Eq. (5.5.2.9) can be written in dimensional variables in a coordinate free form for an arbitrary Lewis number and an arbitrary coefficient of thermal conduction depending on temperature. For a flame front described by a function $F(z, \mathbf{x}, t) = 0$, for example, $F = z - Rf(\mathbf{x}, t) = 0$, we have $N = |\nabla F|$ and the evolution of a flame front is described by the equation

$$\mathbf{n} \cdot \mathbf{v}_- + \frac{1}{N} \frac{\partial F}{\partial \tau} = U_f - \mathrm{Ma} L_f \Lambda, \tag{5.5.2.10}$$

where $\mathbf{v}_-$ is the flow velocity in the flow ahead of the flame front, $\mathbf{n} = \nabla F / |\nabla F|$ is the unit normal vector directed towards the combustion products, Λ is the flame front stretch, which can be written in a coordinate free form according to [13] as

$$\Lambda = \frac{1}{N} \frac{\partial N}{\partial \tau} + \frac{1}{N} \nabla \cdot \left[N \left(\mathbf{v}_- - U_f \mathbf{n} \right) \right] \tag{5.5.2.11}$$

where Ma is the Markstein number defined by (5.3.23):

$$\mathrm{Ma} = \frac{\Theta}{\Theta - 1} J_2 + \frac{\mathrm{Le} - 1}{\Theta - 1} \frac{E(T_b - T_u)}{2T_b^2} \int\limits_{T_u}^{T_b} \ln\left(\frac{T - T_u}{T_b - T_u} \right) \frac{k_b dT}{k_u T}.$$

The conservation laws for the mass and momentum flux also can be written in the coordinate free form as:

$$\rho \left(\mathbf{n} \cdot \mathbf{v} + \frac{1}{N} \frac{\partial F}{\partial t} \right) \Big|_-^+ = \rho_f L_f J_2 \Lambda, \tag{5.5.2.12}$$

$$(\mathbf{n} \times \mathbf{v})|_-^+ = \frac{L_f}{U_f} [J_2 + (\Theta - 1)\,\mathrm{Pr}] \Lambda \mathbf{n} \times \hat{D} \left(\mathbf{v}_- - U_f \mathbf{n} \right), \tag{5.5.2.13}$$

$$\begin{gathered} \left(P + \rho \left[\mathbf{n} \cdot \mathbf{v} + \frac{1}{N} \frac{\partial F}{\partial t} \right]^2 \right) \Big|_-^+ = -\rho_f U_f^2 L_f J_3 \nabla \cdot \mathbf{n} + \\ + \rho_f L_f J_2 \left(\frac{U_f}{N} \hat{D} N + \mathbf{n} \cdot \frac{\partial \mathbf{v}_-}{\partial t} + \mathbf{n} \cdot (\mathbf{v}_- \cdot \nabla) \mathbf{v}_- \right) \end{gathered} \tag{5.5.2.14}$$

where the integrals J_2 and J_3 were defined earlier by Eq. (5.3.22), and the operator $\hat{D}$ in the dimensional variables is

$$\hat{D} = \frac{\partial}{\partial \tau} + \mathbf{v}_- \cdot \nabla - U_f \mathbf{n} \cdot \nabla. \tag{5.5.2.15}$$

It should be emphasized that the hydrodynamic equations of flame dynamics quite complex even being integrated across a curved flame front. An analytical solution of the eigenvalue problem for the curved stationary flames propagating in the tube with adiabatic and slip walls, considering also nonlinear terms related to the flame front evolution, was not obtained yet even for an infinitely thin flame front [16].

5.5.3 Evolution Equation; Arbitrary Equation of State

The evolution equation for a curved deflagration front of finite thickness and an arbitrary equation of state of the combustible mixture was derived in [49]. In this case, the equation of the energy transfer must be written for enthalpy per unit mass. We consider a subsonic deflagration wave assuming an incompressible deflagration-induced flow. Then, the deflagration wave is described by equations

$$\frac{\partial \rho}{\partial t} + \nabla \cdot (\rho \mathbf{v}) = 0\,, \tag{5.5.3.1}$$

$$\rho \frac{\partial \mathbf{v}}{\partial t} + (\rho \mathbf{v} \cdot \nabla)\mathbf{v} + \nabla P = \mu_F \Delta \mathbf{v} + \left(\mu_S + \frac{1}{3}\mu_F\right)\nabla \cdot (\nabla \mathbf{v}), \tag{5.5.3.2}$$

$$\rho \frac{\partial H}{\partial t} + \rho \mathbf{v} \cdot \nabla H = \nabla \cdot (\kappa \nabla T) + \frac{Q\rho_-}{\tau_R} W(T). \tag{5.5.3.3}$$

Here H is the enthalpy per unit mass, κ is the coefficient of thermal conduction, Q is the energy release per unit mass, τ_R is the preexponential factor with dimension time in the rate of the energy release $W(T)$, μ_F and μ_S are the coefficients of the first and second viscosity respectively, which are taken constant. For incompressible flow the enthalpy and thermal conduction depend only on temperature. We will use a subscript "-" for variables ahead of the flame in the fresh mixture, and a subscript "+" for variables in combustion products. For a given enthalpy of the fresh mixture, H_-, the enthalpy of combustion products, H_+, is computed from the conservation of energy, $H_+ = H_- + Q$. The temperature and the density in the combustion products follows from the equation of state: $T_+ = T(H_+)$, $\rho_+ = \rho(H_+)$. The dimensionless function $W(T)$ gives the temperature dependence of the reaction rate. It vanishes in the fresh mixture, ahead of the flame, $W(T_-) = 0$ and in the combustion products, $W(T_+) = 0$.

A stationary solution of Eqs. (5.5.3.1)–(5.5.3.3) is a planar deflagration front propagating along z-axis. In the reference frame comoving with the deflagration front the stationary solution consists of the upstream homogeneous flow of the fresh mixture, the transition zone of heating and energy release, and the homogeneous downstream flow of the combustion products. In the stationary flow $\rho u = \rho_- U_f = const$, and the speed of the stationary deflagration wave U_f is an eigenvalue of the energy transfer equation

$$\rho_- U_f \frac{dH}{dz} = \frac{d}{dz}\left(\kappa \frac{dT}{dz}\right) + \frac{Q\rho_-}{\tau_R} W(T), \tag{5.5.3.4}$$

with the boundary conditions $T = T_-$ in the upstream flow ahead of the flame and $T = T_+$ in the downstream flow behind the flame.

We will use dimensionless variables: $\xi = z/L$, $\Theta = \rho/\rho_-$, $\theta = T/T_-$, $h = H/H_-$, $\Pi = (P - P_-)/\rho_- U_f^2$, $\chi = \kappa/\kappa_-$, $\Pr_{F,S} = \mu_{F,S} H_-/T_- \kappa_-$, $\Lambda = QL/\tau_R H_- U_f$, where $L_f = \kappa_- T_-/H_- \rho_- U_f$ is the width of the planar stationary deflagration front propagating with velocity U_f. The parameter Λ is the eigenvalue of the energy equation (5.5.3.3), which in dimensionless variables is

$$\frac{d}{d\xi}\left(\chi \frac{d\theta}{d\xi}\right) - \frac{dh}{d\xi} + \Lambda W(\theta) = 0, \qquad (5.5.3.5)$$

with the boundary conditions $\theta = 1$ for $\xi \to -\infty$, and $\theta = \theta_+$ for $\xi \to \infty$.

Using the Zel'dovich-Frank-Kamenetskii method we obtain the analytical solution of Eq. (5.5.3.5) for the case of negligible thickness of the zone of energy release

$$\Lambda \approx \frac{(h_+ - 1)^2}{2\chi_+}\left(\int_1^{\theta} W(\theta)d\theta\right)^{-1}, \qquad (5.5.3.6)$$

and the deflagration velocity depending on the specific function of the energy release

$$U_f = \sqrt{Q\kappa_- T_-/\Lambda H_-^2 \rho_- \tau_R}. \qquad (5.5.3.7)$$

We derive the evolution equation for the curved deflagration front of zero thickness described by the function $z = Rf(\mathbf{x}, t)$, where the scale $R \gg L_f$ characterizes the evolution of the flame front. It is convenient to introduce dimensionless variables: time $\tau = U_f t/R$, the transverse coordinate $\boldsymbol{\eta} = \mathbf{x}/R$, and the small parameter $\varepsilon = L_f/R$. In the coordinate system comoving with the flame front:

$$\zeta = \xi - f(\boldsymbol{\eta}, \tau)/\varepsilon, \quad \boldsymbol{\eta} = \boldsymbol{\eta}, \quad \tau = \tau \qquad (5.5.3.8)$$

and $u = v_z/U_f$, $\mathbf{v} = v_x/U_f$.

Then the hydrodynamic equations take the form

$$\varepsilon\frac{\partial \Theta}{\partial \tau} + \frac{\partial j}{\partial \zeta} + \varepsilon \nabla_\perp(\Theta \mathbf{u}) = 0, \qquad (5.5.3.9)$$

$$\varepsilon\Theta\frac{\partial \mathbf{v}}{\partial \tau} + j\frac{\partial \mathbf{v}}{\partial \zeta} + \varepsilon\Theta(\mathbf{v}\nabla_\perp)\mathbf{v} = -\varepsilon\nabla_\perp\Pi + \nabla_\perp f\frac{\partial \Pi}{\partial \zeta} + \Pr_F \Delta_\zeta \mathbf{v} +$$
$$\left(\Pr_S + \frac{1}{3}\Pr_F\right)\left(\varepsilon\nabla_\perp - \nabla_\perp f\frac{\partial}{\partial \zeta}\right)\left(\frac{\partial}{\partial \zeta}\left(\frac{j}{\Theta}\right) + \varepsilon\nabla_\perp \mathbf{v}\right) \qquad (5.5.3.10)$$

$$\varepsilon\Theta\frac{\partial u}{\partial \tau}+j\frac{\partial u}{\partial \zeta}+\varepsilon\Theta(\mathbf{v}\nabla_\perp)u=-\frac{\partial \Pi}{\partial \zeta}+\Pr_F\Delta_\zeta u+ \left(\Pr_S+\frac{1}{3}\Pr_F\right)\frac{\partial}{\partial \zeta}\left(\frac{\partial}{\partial \zeta}\left(\frac{j}{\Theta}\right)+\varepsilon\nabla_\perp\mathbf{v}\right), \tag{5.5.3.11}$$

$$\varepsilon\Theta\frac{\partial h}{\partial \tau}+j\frac{\partial h}{\partial \zeta}+\varepsilon\Theta\mathbf{v}\nabla_\perp h=N^2\frac{\partial}{\partial \zeta}\left(\chi\frac{\partial \theta}{\partial \zeta}\right)+ \varepsilon\nabla_\perp\left[\chi\left(\varepsilon\nabla_\perp\theta-\nabla_\perp f\frac{\partial \theta}{\partial \zeta}\right)\right]-\varepsilon\nabla_\perp f\frac{\partial}{\partial \zeta}(\chi\nabla_\perp\theta)+W(\theta), \tag{5.5.3.12}$$

where $\nabla_\perp$ is the part of the operator ∇ related to the coordinate η along the front surface, and the mass flux is

$$j=\Theta u-\Theta\frac{\partial f}{\partial \tau}-\Theta\mathbf{v}\nabla_\perp f. \tag{5.5.3.13}$$

The Laplace operator in these variables takes the form

$$\Delta_\zeta=N^2\frac{\partial^2}{\partial \zeta^2}-\varepsilon\nabla_\perp^2 f\frac{\partial}{\partial \zeta}-2\varepsilon\frac{\partial}{\partial \zeta}\nabla_\perp f\nabla_\perp+\varepsilon^2\nabla_\perp^2, \tag{5.5.3.14}$$

where $N=\sqrt{1+(\nabla_\perp f)^2}$.

Let us rewrite Eqs. (5.5.3.10) and (5.5.3.11) with the momentum flux in the left-hand side

$$j\frac{\partial}{\partial \zeta}(\mathbf{v}+u\nabla_\perp f)=-\varepsilon\Theta\left(\frac{\partial}{\partial \tau}+\mathbf{v}\nabla_\perp\right)(\mathbf{v}+u\nabla_\perp f)+ \varepsilon\Theta u\left(\frac{\partial}{\partial \tau}+\mathbf{v}\nabla_\perp\right)\nabla_\perp f-\varepsilon\nabla_\perp\Pi+\Pr_F(\nabla_\zeta\mathbf{v}+\nabla_\perp f\nabla_\zeta u)+ \varepsilon\left(\Pr_S+\frac{1}{3}\Pr_F\right)\nabla_\perp\frac{\partial}{\partial \zeta}\left(\frac{\partial}{\partial \zeta}\left(\frac{j}{\Theta}\right)+\varepsilon\nabla_\perp\mathbf{v}\right), \tag{5.5.3.15}$$

$$N^2\frac{\partial}{\partial \zeta}\left(\Pi+\frac{j^2}{N^2\Theta}\right)=-\varepsilon\frac{\partial j}{\partial \tau}+\varepsilon\nabla_\perp(j\mathbf{v})-\varepsilon\Theta\frac{\partial^2 f}{\partial \tau^2}-2\varepsilon\Theta\mathbf{v}\nabla_\perp\frac{\partial f}{\partial \tau} -\varepsilon\Theta\mathbf{v}(\mathbf{v}\nabla_\perp)\nabla_\perp f+\varepsilon\nabla_\perp f\nabla_\perp\Pi+\Pr_F\Delta_\zeta\left(\frac{j}{\Theta}\right)+\Pr_F\mathbf{v}\Delta_\zeta\nabla_\perp f +\Pr_F\Delta_\zeta\frac{\partial f}{\partial \tau}+\left(\Pr_S+\frac{\Pr_F}{3}\right)\left(N^2\frac{\partial}{\partial \zeta}-\varepsilon\nabla_\perp f\nabla_\perp\right)\left(\frac{\partial}{\partial \zeta}\left(\frac{j}{\Theta}\right)+\varepsilon\nabla_\perp\mathbf{v}\right). \tag{5.5.3.16}$$

Expanding the solution of Eqs. (5.5.3.9–5.5.3.12) in power series of a small parameter ε: $j=j_0+\varepsilon j_1+O(\varepsilon^2)$, $\theta=\theta_0+\varepsilon\theta_1+O(\varepsilon^2)$ etc., we obtain with accuracy $O(\varepsilon^2)$ Eqs. (5.5.3.9), (5.5.3.12), (5.5.3.15), (5.5.3.16) in the form

$$\frac{\partial j}{\partial \zeta} = 0, \tag{5.5.3.17}$$

$$j\frac{\partial}{\partial \zeta}(\mathbf{v} + u\nabla_{\perp} f) = \underset{F}{\Pr} N^2 \frac{\partial^2}{\partial \zeta^2}(\mathbf{v} + u\nabla_{\perp} f), \tag{5.5.3.18}$$

$$N^2 \frac{\partial}{\partial \zeta}(\Pi + \frac{j^2}{N^2\Theta}) = \left(\frac{4}{3}\underset{F}{\Pr} + \underset{S}{\Pr}\right) N^2 \frac{\partial^2}{\partial \zeta^2}\left(\frac{j}{\Theta}\right), \tag{5.5.3.19}$$

$$j\frac{\partial h}{\partial \zeta} = N^2 \frac{\partial}{\partial \zeta}\left(\chi \frac{\partial \theta}{\partial \zeta}\right) + W(\theta). \tag{5.5.3.20}$$

From Eq. (5.5.3.17) it follows that $j = j_0(\eta, \tau) + O(\varepsilon)$. Similar to (5.5.3.5) the solution of Eq. (5.5.3.20) can be presented in the form:

$$\theta_0 = \theta_p(\zeta/N), \quad j_0(\eta, \tau) = N = \sqrt{1 + (\nabla_{\perp} f)^2}. \tag{5.5.3.21}$$

Here $\theta = \theta_p(\xi)$ is the temperature profile for a planar deflagration front, which is a solution of the ordinary differential equation (5.5.3.5). From the equation of state of the fresh mixture we obtain

$$\Theta_0 = \Theta(\theta_0) = \Theta_p(\zeta/N), \quad h_0 = h(\theta_0) = h_p(\zeta/N). \tag{5.5.3.22}$$

Equations (5.5.3.18) and (5.5.3.19) integrated across the deflagration front give the conservation laws for the momentum flux at the infinitely thin surface of discontinuity. In particular, from Eq. (5.5.3.18) we find

$$\mathbf{v} + u\nabla_{\perp} f = \mathbf{v}_- + u_- \nabla_{\perp} f + O(\varepsilon). \tag{5.5.3.23}$$

Equations (5.5.3.13) and (5.5.3.23) give the solution for the velocity components

$$u = u_- - \frac{1}{N}\frac{\Theta - 1}{\Theta} + O(\varepsilon), \tag{5.5.3.24}$$

$$\mathbf{v} = \mathbf{v}_- - \frac{\nabla_{\perp} f}{N}\frac{\Theta - 1}{\Theta} + O(\varepsilon). \tag{5.5.3.25}$$

From Eq. (5.5.3.19) we obtain the following expression for pressure

$$\Pi = \Pi_- + \frac{\Theta - 1}{\Theta} + \left(\frac{4}{3}\underset{F}{\Pr} + \underset{S}{\Pr}\right) N \frac{\partial}{\partial \zeta}\left(\frac{1}{\Theta}\right) + O(\varepsilon). \tag{5.5.3.26}$$

The evolution equation for an infinitely thin deflagration front can be obtained from Eqs. (5.5.3.13) and (5.5.3.21):

$$\frac{\partial f}{\partial \tau} + \mathbf{v}_{-}\nabla_{\perp} f + \sqrt{1 + (\nabla_{\perp} f)^2} - u_{-} = O(\varepsilon). \quad (5.5.3.27)$$

The evolution Eq. (5.5.3.27) together with (5.5.3.24)–(5.5.3.26) describes the dynamics of an infinitely thin deflagration front, which, in fact, is the Darrieus-Landau model of a zero-thickness deflagration front instability.

The next terms of the expansion in ε describe the influence of the finite thickness of the deflagration front. The evolution equation for a deflagration front of finite thickness with the accuracy $O(\varepsilon^2)$ was derived in [49]. An advantage of this model is that it determines the dynamics of a deflagration front in a mixture with an arbitrary equation of state and arbitrary type of thermal conduction and energy release. Therefore, it is applicable for physical problems of different nature, e.g. different kind of chemical flames or thermonuclear flames in thermonuclear Supernovae Ia, propagation of the ablation front in the laser irradiated targets, etc. The simplest result, which readily follows from the model, is the expression for the growth rate of the instability of a plane flame propagating in the mixture with a high activation energy. Taking into account the explicit expressions for the perturbation modes we obtain the instability growth rate from the consistency condition of the linearized equations:

$$\sigma = \Gamma(\Theta) U_f k (1 - k\lambda_c / 2\pi), \quad (5.5.3.28)$$

where $\Gamma(\Theta) = \frac{\Theta^+}{\Theta^+ + 1}\left(\sqrt{\Theta^+ + 1 - 1/\Theta^+} - 1\right)$ is the coefficient in the DL theory of the flame instability, and $\Theta^+ = \rho_- / \rho_+$. The instability is suppressed by thermal conduction for the perturbations with wavelength shorter than

$$\lambda_c = \frac{\pi L_f}{2\Gamma[\Gamma(\Theta^+ + 1) + 1]}\left[\frac{\Theta^+ - 1}{\Theta^+} + \frac{\Theta^+ \ln \Theta^+}{\Theta^+ - 1}\left(\frac{2\Gamma}{\Theta^+} + \frac{1}{\Theta^+} + 1\right)\right]. \quad (5.5.3.29)$$

For typical laboratory flames $\lambda_c \approx 20 L_f$, and this is the reason why the DL instability does not result in strongly corrugated flames.

Example of different kind of combustion is a thermonuclear deflagration in supernova. The hydrodynamic length scale of the deflagration wave in supernovae is about the white dwarf radius $R_W \approx 100 km$, which by many orders of magnitude exceeds the thermonuclear flame thickness $L_{fw} \sim 10^{-3} cm$. One of the possible scenarios of the flame dynamics in white dwarfs is the formation of a fractal structure of the flame front during its expansion outward from the star center [51]. This scenario means the development of many cascades of cells of different sizes growing because of the DL instability. The velocity of a fractal flame increases with the flame radius as

$$\lambda_c = \frac{2\pi L_f \Theta^+}{2\Gamma} \frac{(b_1 + b_2 \Gamma)\Theta^+}{(\Theta^+ + 1)\Gamma + 1}. \quad (5.5.3.30)$$

Here $2+d$ is the fractal dimension, n_c is the critical number of a spherical functions for the fractal accelerating flame [52], and the coefficients b_1 and b_2 depending upon the ratio of densities Θ^+ were calculated in [52]. The cut-off wavelength λ_c defines the low limit of the flame front fractal structure in white dwarfs. This value can be calculated considering that the density near the center of a white dwarf is $\rho_c \approx 3 \cdot 10^9 g/cm^3$. For such densities the equation of state is determined by the Fermi degenerate ultra-relativistic electrons [53]

$$H = 4P/\rho, \quad P = 0.77\hbar c\left(\frac{\rho Z}{M_i}\right)^{4/3}\left(1 + 2.06\left(\frac{M_i}{\rho Z}\right)^{2/3}\frac{T^2}{\hbar^2 c^2}\right), \tag{5.5.3.31}$$

where eZ and M_i are the average electrical charge and the average mass of ions. For such a huge density in a white dwarf, thermal conduction of degenerate electrons is dominant

$$\kappa_e = \frac{0.81\hbar c^2}{e^4 Z\Lambda_{ei}}\left(\frac{\rho Z}{M_i}\right)^{1/3} T, \tag{5.5.3.32}$$

where eZ and M_i are the average electrical charge and the average mass of the ions.

Using the equation of state and the thermal conduction (5.5.3.31–5.5.3.32), in a way similar to the case of a flame in an ideal gaseous fuel, we find the cut-off wavelength of the DL instability for the deflagration wave in white dwarfs:

$$\lambda_c = \frac{2\pi L_f \Theta^+}{2\Gamma}\frac{(b_1 + b_2\Gamma)\Theta^+}{(\Theta^+ + 1)\Gamma + 1}. \tag{5.5.3.33}$$

The coefficients b_1 and b_2 in (5.5.3.33) are functions of the density expansion in the front of the thermonuclear wave. The density decrease in the thermonuclear deflagration near the white dwarf center is $\Theta^+ = \rho_-/\rho_+ \approx 1.1 - 1.2$, which is much smaller then it is in terrestrial flames. The DL instability for such small expansion is quite weak, while the thermal stabilization is strong and the cut-off wavelength exceeds the flame front thickness by two–three orders of magnitude, $\lambda_{cW} \approx 400L_f$, and at least by one order of magnitude exceeds the cut-off wavelength of usual terrestrial flames $\lambda_c \approx 20L_f$.

5.5.4 Influence of Compressibility

In previous sections we considered the Darrieus Landau instability of an infinitesimally thin deflagration front under the condition of an incompressible flow. This is a reasonable approach because of flames propagate with essentially subsonic speed.

Recall that the existence of a solution in the form of a discontinuity surface of zero-thickness is possible only if the evolutionarity condition is satisfied. The evolutionarity condition requires that the number of boundary conditions on the discontinuity surface exceeds by one the total number of simple waves (the number of characteristics) outgoing from the discontinuity surface. Considering the flame as a discontinuity surface, we find the "deficit" of boundary conditions, since the conservation laws of fluxes mass, momentum and energy give only three boundary conditions. To satisfy the evolutionarity condition for the existence of the solution in the form of a discontinuity surface, an additional condition is required. This additional condition is a prescribed velocity of the flame propagation. Obviously, the same is true for the stability problem if the flame is considered as a discontinuity surface. Indeed, in the stability problem, we have for 4 unknown independent amplitudes of perturbations of the velocities and the flame front position, but only 3 relations, which are given by the conservation laws at the discontinuity. Within the model of an incompressible fluid, an additional condition was the condition that the flame velocity remains constant with respect to the fresh combustible mixture. The validity of this condition was proved in previous sections, where the stability problem was solved taking into account of the finite thickness of the flame.

Since the relative difference in pressures is about the square of the Mach number, $\delta P/P \sim M^2 \ll 1$, the isobaric approximation and the model of incompressible fluid are satisfactory good in the theory of DL instability. For example, for the fast hydrogen–oxygen flames, $U_f = 9.4m/s$, sound speed $a_s = 539.5m/s$, $M = 0.02$ and $M^2 = 0.0003$, and for a slow methane/air flame $U_f = 0.38m/s$ and $a_s = 354m/s$, and the Mach numbers are r $M = 0.001$ and $M^2 = 0.000001$, respectively. Thus, the isobaric approximation is satisfied with high accuracy, although the compressibility effect can be noticeable, at least for fast flames. The concept of the fast flame propagating due to thermal conduction and fuel diffusion should not be confused with the convection-driven fast flames. For example, if a flame propagates in a channel with non-slip or rough walls or in the obstructed channel, it can accelerate up to the velocity close to the CJ deflagration speed, and can transit to detonation. Another example where the influence of compressibility can be important is the laser ablation of a small targets in the inertial thermonuclear fusion (ICF) [6, 7, 54].

In the ICF approach, a spherical capsule with thermonuclear fuel is spherically symmetric irradiated by the beams of high-power laser or X-rays. Ablation of the outer surface of the capsule causes the expansion of the hot plasma, resulting in compression of the thermonuclear target due to the reaction forces. In the adiabatic spherically symmetric compression, the temperature increases as $T_2/T_0 = (R_0/R_2)^{3(\gamma-1)}$. The isothermal CJ regime is the ablated hot plasma expands into vacuum, so that the head of the rarefaction propagates with isothermal sound speed. Since instabilities destroy the spherical symmetry, a high degree of compression, which is necessary to ignite the thermonuclear reaction cannot be reached.

Importance of the compressibility effects for DL instability has been pointed out first by Kadowaki [55]. The development of DL instability in a compressible flow was studied in [55] by solving the reactive Navier–Stokes equations with a finite activation energy. The linear stage of the DL instability in a compressible flow was studied in [56], taking into account finite flame thickness and transport processes. The dynamics of curved flames in compressible flows has been investigated in [57] using 2D numerical simulations. It was shown that the growth rate of the DL instability and the velocity amplification of curved flames are very sensitive to the effects of compressibility. The influence of compressibility on the DL instability of deflagration waves considered as a discontinuity surface has been studied in numerous analytical works and by means of numerical simulation [60–58]. Similar problem of the DL and the Rayleigh–Taylor (RT) instability of an ablation front has attracted much attention because the ablation front instability presents the most difficult obstacle on the way to achieve the high compression required for the target ignition in the inertial confinement fusion [64–68]. The DL and RT instabilities of an ablation front will be discussed in Chap. 18.

A flame propagating in a compressible flow is characterized by the Mach number in the fresh mixture $\mathrm{M}_0 = U_f/c_0$, where the sound speed in a fresh mixture is $c_0^2 = \gamma P_0/\rho_0$. In the approximation of the infinitely small Mach number the change in pressure is negligibly small for the isobaric flame, while for a finite Mach number the pressure drops across the flame front, though the change in the pressure is quite small. The relative change in pressure across a planar flame from the fresh mixture to the combustion products is calculated from the conservation of momentum flux as

$$P_b/P_0 = 1 - \gamma \mathrm{M}_0^2(\Theta - 1). \qquad (5.5.4.1)$$

For example, the change in pressure for the hydrogen/oxygen flame $P_b/P_0 \approx 9.996$, is quite small. However, due to pressure changes, the expansion coefficient, defined as the ratio of the unburned mixture density and the density of the combustion products $\Theta = \rho_0/\rho_b$, does not coincide with the ratio of temperatures, $\theta_b = T_b/T_0$, as it was for isobaric flames. The relation between the thermal expansion coefficient $\theta_b = T_b/T_0$, and the expansion coefficient, $\Theta = \rho_u/\rho_b$, follows from Eq. (5.5.4.1)

$$\theta_b = \Theta - \gamma \mathrm{M}_0^2 \Theta(\Theta - 1). \qquad (5.5.4.2)$$

For example, for the Mach number $M_0 = 0.17$, the ratio of densities is $\Theta = 8.7$, while the thermal expansion coefficient is $\theta_b = 6$. At the same time both the expansion coefficient and the thermal expansion coefficient depend on the scaled heat release in the reaction $Q/C_P T_0$ through the conservation law of energy flux, which can be written in the form

$$\frac{Q}{C_P T_0} = \frac{\gamma - 1}{2} \mathrm{M}_0^2 (\Theta^2 - 1) + \theta_b - 1. \qquad (5.5.4.3)$$

Besides, it should be remembered that the Mach number is not a completely free parameter of the flow, which can take any values. Since the drifting velocity of the combustion products with respect to the flame front cannot be larger than the isothermal sound speed in the burnt mixture, $\sqrt{P_b/\rho_b}$, it can be shown [56] that the Mach number cannot exceed the maximum value, which is

$$M_{max}^2 = \frac{1}{\gamma(2\Theta - 1)}. \tag{5.5.4.4}$$

The maximum Mach number corresponds to the isothermal CJ deflagration regime. For $\gamma = 1.4$ and for the expansion coefficient $\Theta = 9$, the Mach number cannot exceed $M_{max} = 0.2$. Besides, the maximum value of the Mach number decreases with an increase of the expansion coefficient Θ, as well as with an increase of the thermal coefficient θ_b.

Figure 5.8 shows the dependence of the expansion coefficient on the Mach number for different thermal expansion coefficients [56]. As it is seen in Fig. 5.8, the ratio of densities increases with the increase of the Mach number and for the Mach numbers 0.13–0.18 it can almost two times exceed the respective thermal expansion coefficients. Figure 5.8 shows that the difference between the expansion coefficient and the thermal expansion coefficient increases when the Mach number approaches the critical value. Still the internal structure of a flame in a compressible flow remains qualitatively similar to the structure of isobaric flames [51].

As we have already said, to solve the stability problem of a flame considered as a discontinuity surface, an additional condition is required. For any discontinuous model there is a deficit of matching conditions at the discontinuity surface. In [55] this deficit of the boundary conditions was resolved assuming the same as in the DL

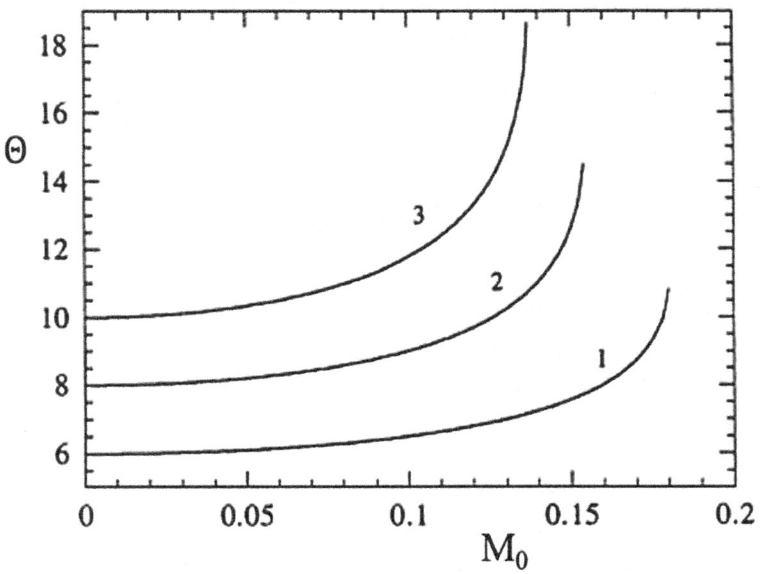

Fig. 5.8 Dependence of the expansion coefficient in a flame versus the Mach number of the upstream flow for different thermal coefficients [56]: $\theta_b = 6, 8, 10$ (curves 1, 2, 3, respectively)

theory, it was suggested that perturbations do not influence the normal velocity of a flame front, and the effect of compressibility was considered as a small correction.

In the DL theory the additional condition, required to solve the problem, was introduced for an incompressible flow, assuming that the flame velocity remains constant relative to the unburned mixture: $u'_{z1} - \frac{\partial\zeta}{\partial t} = 0$. In the model of incompressible flow this condition also leads to Eq. (5.1.17):

$$u'_{z1} - \frac{\partial\zeta}{\partial t} = 0 = u'_{z2} - \frac{\partial\zeta}{\partial t}. \tag{5.5.4.5}$$

In Sect. 5.2 the stability problem was solved using the complete system of equations for incompressible flow. This solution showed that the DL model is the correct long-wavelength asymptotic of the solution, obtained without any additional assumptions. However, the assumption made in [55] that the additional condition in the form of vanishing either left-hand or right-hand sides of (5.1.17) in the case of a compressible model was not justified. In the attempt to resolve the problem of the additional condition for the discontinuity model, in [58] the system of the perturbed equations was complemented by the perturbed equation of energy conservation. In this case there is an additional fourth equation however, there is also the additional unknown: either the perturbation of entropy or temperature, which was not considered in [58], and which is an additional assumption. If the equation of the energy transfer is included in the system of equations for the discontinuity model, the perturbation of density (we consider a compressible model) can be expressed via perturbation of the flame front as

$$\rho'_1 u_{z1} + \rho_1 (u'_{z1} - \frac{\partial\zeta}{\partial t}) = 0. \tag{5.5.4.6}$$

For the incompressible flow the density perturbation vanishes and (5.5.4.6) becomes the additional condition in the DL model.

As a matter of fact, it is not obvious that the supplementary boundary condition at a perturbed flame front in a compressible flow is the Darrieus - Landau condition of constant normal flame velocity as it was assumed in [55] and [58]. The only way to check if such supplementary condition is correct is to solve the eigenvalue problem of the flame stability considering the finite flame thickness and the internal structure of the flame front, as it was done in [56, 57].

Propagation and stability of a planar flame front in a compressible flow is described by the equations:

$$\frac{\partial\rho}{\partial t} + \frac{\partial}{\partial x_i}(\rho u_i) = 0, \tag{5.5.4.7}$$

$$\rho\frac{\partial u_i}{\partial t} + \rho u_k \frac{\partial u_i}{\partial x_k} = -\frac{\partial P}{\partial x_i} + \frac{\partial\sigma_{ik}}{\partial x_k}, \tag{5.5.4.8}$$

$$\frac{\partial}{\partial t}\left(\frac{\rho u^2}{2}+\frac{C_P\rho T}{\gamma}\right)+\frac{\partial}{\partial x_k}\left[\rho u_k\left(\frac{u^2}{2}+C_P T\right)-u_i\sigma_{ik}-\kappa\frac{\partial T}{\partial x_i}\right]=QW, \tag{5.5.4.9}$$

$$\rho\frac{\partial a}{\partial t}+\rho u_k\frac{\partial a}{\partial x_k}=\frac{\partial}{\partial x_k}\left(\rho D\frac{\partial a}{\partial x_k}\right)-W. \tag{5.5.4.10}$$

Here, $P=\frac{(\gamma-1)}{\gamma}C_P\rho T$ is the gas pressure; $\sigma_{ik}=\eta\left(\frac{\partial u_i}{\partial x_k}+\frac{\partial u_k}{\partial x_i}-\frac{2}{3}\delta_{ik}\frac{\partial u_l}{\partial x_l}\right)$ is the viscosity stress tensor; C_P is the specific heat at constant pressure; γ is the adiabatic exponent; Q is the heat release of a chemical reaction with the reaction rate described by a first-order Arrhenius law $W=\frac{a\rho}{\tau_R}\exp(-E/R_B T)$, a is the fuel concentration, R_B is the universal gas constant, E is the activation energy and τ_R is the constant of time dimension.

We consider a planar flame front propagating along the z-axis with the velocity u_{z1}. In the coordinate system where that flame is at rest, the stationary Eqs. (5.5.4.7)–(5.5.4.10) constitute the eigenvalue problem with the flame velocity being the eigenvalue. Taking into account conservation of mass in the stationary flow we introduce the dimensionless variables: $\Theta=\rho_1/\rho$, $\theta=T/T_1$, $Y=Qa/(C_{|P}T_1)$, dimensionless activation energy $\varepsilon=E/R_B T_2$, and the dimensionless coordinate $\xi=z/L_f$, where $L_f=\kappa/(\rho_1 C_P u_{z1})$ is the flame front thickness, and the subscript 1 denotes quantities in the upstream flow of the unburned mixture. In terms of the dimensionless variables the stationary Eqs. (5.5.4.7)–(5.5.4.10) take the form:

$$\frac{4}{3}\Pr\gamma M_1^2\frac{\partial\Theta}{\partial t}=\frac{\theta}{\Theta}-1+\gamma M_1^2(\Theta-1), \tag{5.5.4.11}$$

$$\frac{\partial\theta}{\partial\xi}+Le\frac{\partial Y}{\partial\xi}=\frac{\gamma-1}{\gamma}\left[\Theta-\theta-\frac{\gamma M_1^2}{2}(\Theta-1)^2\right]+\theta-1+Y-Y_1, \tag{5.5.4.12}$$

$$Le\frac{d^2Y}{d\xi^2}-\frac{dY}{d\xi}-\Lambda\frac{Y}{\Theta}\exp(-\varepsilon\theta_2/\theta)=0. \tag{5.5.4.13}$$

Here, $M=u_z/\sqrt{\gamma P/\rho}$, $\Pr=\eta C_P/\kappa$, $Le=C_P\rho D/\kappa$ are the Mach, Prandtl, and Lewis numbers, respectively; $\Lambda=L_f/u_{z1}\tau_R$ is the reduced eigenvalue of the problem. Since we are interested in the DL flame instability, we take the Lewis number equal to unity. The boundary conditions for Eqs. (5.5.4.11)–(5.5.4.13) are

$$\Theta=1,\ \ \theta=1,\ \ Y=Y_1,\ \ for\ \xi\to-\infty, \tag{5.5.4.14}$$

$$\Theta=\Theta_2,\ \ \theta=\theta_2,\ \ Y=0,\ \ for\ \xi\to+\infty, \tag{5.5.4.15}$$

where the quantities Y_1, Θ_2, and θ_2 are related by the following equations

$$Y_1 = \frac{\gamma - 1}{2} M_1^2 (\Theta_2^2 - 1) + \theta_2 - 1, \tag{5.5.4.16}$$

$$\theta_2 = \Theta_2 [1 - \gamma M_1^2 (\Theta_2 - 1)]. \tag{5.5.4.17}$$

As it was mentioned before, for a given thermal expansion θ_2 the Mach number M_1 cannot exceed some critical value defined by inequality: $\gamma M_1^2 \leq 2[\theta_2 - \sqrt{\theta_2(\theta_2 - 1)}] - 1$.

The eigenvalue problem determined by (5.5.4.11)–(5.5.4.13) with the boundary conditions (5.5.4.14)–(5.5.4.15) has been solved for different values of the activation energy, the thermal expansion, and the Mach number in the fresh mixture M_1, which was treated as the free parameters. The influence of the compressibility effects on the instability growth rate becomes significant even for rather small Mach numbers. Slight increase of the Mach number leads to strong increase of the instability growth rate, so that the maximum growth rate for a flame with a rather small Mach number can be about three times larger than the maximum growth rate for a flame in the incompressible flow. Figure 5.9 shows the dependence of the maximum growth rate on the Mach number for different heating coefficients.

It is seen in Fig. 5.9 that development of the DL instability becomes more sensitive to the compressibility effects for flames with a larger expansion coefficient. Figure 5.10 shows the calculated scaled cut-off wave number, $K \equiv k_c L_f$, on the Mach number, for which thermal stabilization of the DL instability takes place. It is

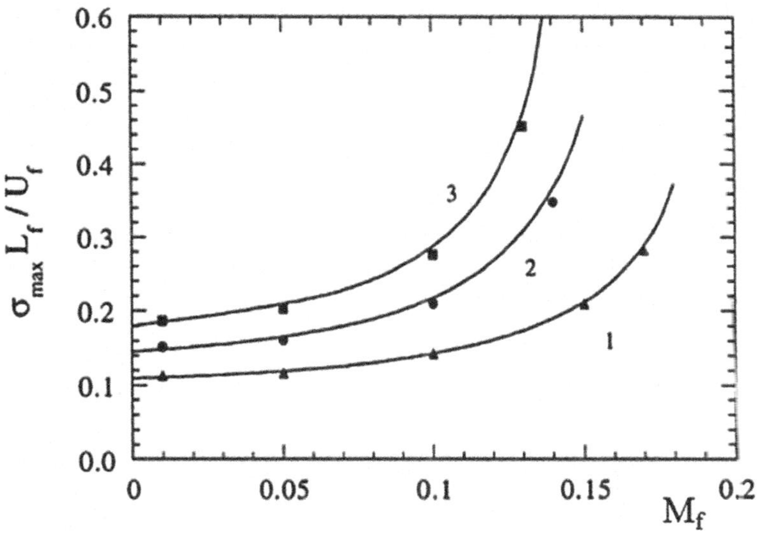

Fig. 5.9 The scaled maximum instability growth rate as a function of the Mach number for different heating coefficients $\theta_2 = 6, 8, 10$ (curves 1, 2, 3, respectively). Markers show the results of 2D simulations from [56]

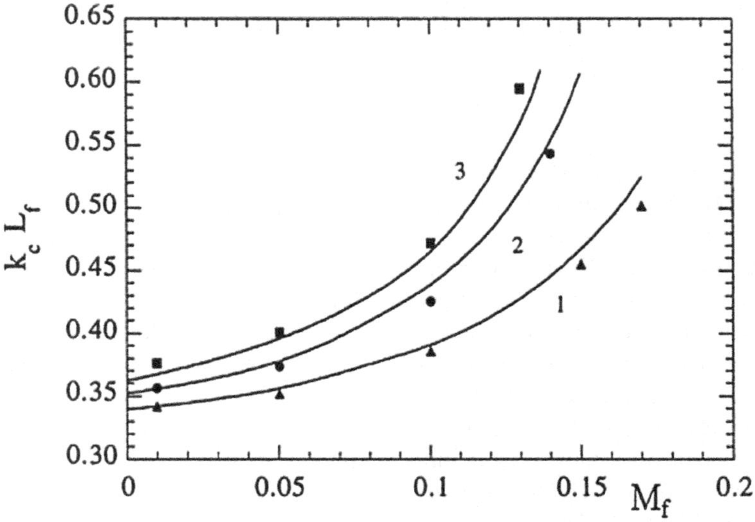

Fig. 5.10 Dependence of the scaled cut-off wave number on the Mach number for different heating coefficients $\theta_2 = 6,\ 8,\ 10$ (curves 1, 2, 3, respectively). Markers show the results of 2D simulations [56]

seen that the larger the Mach number, the larger the cut-off wave number (the shorter the cut-off wavelength).

Dynamics of curved flames during the nonlinear stage of the DL instability in a compressible flow has been considered using 2D numerical simulations [57]. When the amplitude of perturbations becomes comparable to the respective unperturbed value, then the instability growth slows down because of nonlinear effects and the flame front acquires a stationary curved shape. Due to increase of the flame surface area curved flames travel at speeds exceeding the respective velocities of planar flames.

Consider now infinitesimally small perturbations of the flame front in the form $\varphi' = \varphi'(z)\exp(\sigma t + ikx)$, and introduce the dimensionless combinations of perturbations similar to Sect. 5.2:

$$j' = \frac{\rho u_z' + \rho' u_z}{\rho_1 u_{z1}}, \quad \Theta' = \frac{u_z'}{u_{z1}}, \quad V' = i\frac{u_x'}{u_{z1}}, \quad \theta' = \frac{T'}{T_1},$$

$$P' = \frac{p'}{\rho_1 u_{z1}^2}, \quad Y' = \frac{Qa'}{C_P T_1}, \quad K = kL_f, \quad S = \sigma/u_{z1}k. \tag{5.5.4.18}$$

The linearized Eqs. (5.5.4.7)–(5.5.4.10) in dimensionless perturbations (5.5.4.18) are:

$$\frac{dj'}{d\xi} = \frac{SK}{\Theta}\left(\frac{\Theta'}{\Theta} - j'\right) - K\frac{V'}{\Theta}, \tag{5.5.4.19}$$

$$\Pr\frac{d^2V'}{d\xi^2}-\frac{dV'}{d\xi}=K\left(\frac{4}{3}K\,\Pr+\frac{S}{\Theta}\right)V'-KP'+K\frac{\Pr}{3}\frac{d\Theta'}{d\xi}, \qquad (5.5.4.20)$$

$$\frac{4}{3}\Pr\frac{d^2\Theta'}{d\xi^2}-\frac{d\Theta'}{d\xi}=K\left(\mathrm{KPr}+\frac{S}{\Theta}\right)\Theta'+j'\frac{d\Theta}{d\xi}+\frac{dP'}{d\xi}-\frac{\Pr}{3}K\frac{dV'}{d\xi}, \qquad (5.5.4.21)$$

$$\frac{\kappa}{C_p}\frac{d^2\theta'}{d\xi^2}-\frac{d\theta'}{d\xi}=K\left(K+\frac{S}{\Theta}\right)\theta'+j'\frac{d\theta}{d\xi}-\frac{\gamma-1}{\gamma}\Theta'\frac{d}{d\xi}\left(\frac{\theta}{\Theta}\right)-$$
$$(\gamma-1)M_1^2\left[SKP'+\Theta\frac{dP'}{d\xi}+\frac{4}{3}\Pr\left(2\frac{d\Theta'}{d\xi}-KV'\right)\frac{d\Theta}{d\xi}\right]-W'. \qquad (5.5.4.22)$$

$$Le\frac{d^2Y'}{d\xi^2}-\frac{dY'}{d\xi}=K\left(Le+\frac{S}{\Theta}\right)Y'+j'\frac{dY'}{d\xi}+W', \qquad (5.5.4.23)$$

where dimensionless perturbations of pressure and the chemical reaction rate are:

$$(\gamma-1)M_1^2P'=\frac{\theta'}{\Theta}+\frac{\theta}{\Theta}\left(j'-\frac{\Theta'}{\Theta}\right), \qquad (5.5.4.24)$$

$$W'=\Lambda\left[\frac{Y'}{\Theta}-\frac{Y}{\Theta^2}\Theta'+\frac{Y}{\Theta}\left(j'+\theta'\frac{\varepsilon}{\theta^2}\right)\right]\exp\left(-\frac{\varepsilon\theta_2}{\theta}\right). \qquad (5.5.4.25)$$

The boundary conditions for the system (5.5.4.19)–(5.5.4.23) are that all perturbations vanish at infinity in the uniform upstream $\xi\to-\infty$ and downstream $\xi\to+\infty$ flows. In the uniform flows from the both sides from the flame any solution of Eqs. (5.5.4.19)–(5.5.4.23) can be presented as a superposition of the modes $\varphi'(\xi)=\sum\varphi_i'\exp(\mu_i\xi)$ with $\mu_i>0$ in the unburned mixture and $\mu_i<0$ in the combustion products.

The flame instabilities simulations [56] were initiated by superimposing small perturbations on the longitudinal velocity $u_z=u_{z1}+u_z'(z,x)$, where

$$u_z'(z,x)=A_0u_z(z)\cos(\pi x/L)\exp(-\pi^2z^2/L^2). \qquad (5.5.4.26)$$

The initial dimensionless amplitude of perturbation $A_0=10^{-3}$. The point $z=0$ corresponds to the position of the flame front, which was calculated as a position of the maximum of the reaction rate. The velocity perturbation induces perturbations of other parameters of the flow and results in the flame front bending. To study development of the instability the flame front was expanded into series of Fourier harmonics $f(x,t)=\sum f_n(t)\cos(n\pi x/L)$. and the time history of amplitude of the first harmonic with the wavelength equal to the doubled width of the tube. The perturbation wave numbers permitted by the boundary conditions at the tube walls are $k=n\pi/L, n=1,2,3....$ On the linear stage, the first harmonic grows exponentially with the growth rate depending on the wavelength, the Mach number, and the thermal

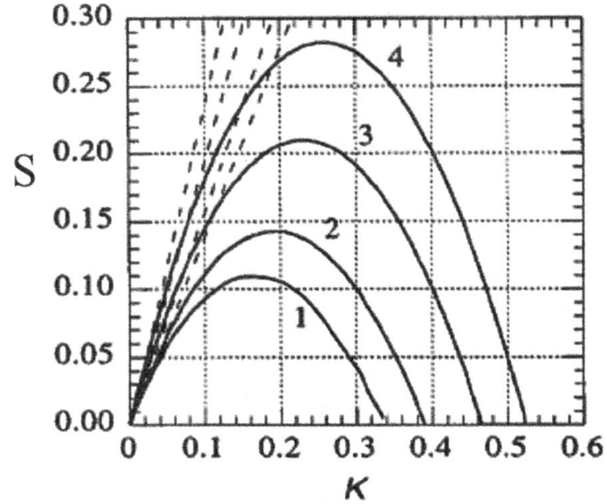

Fig. 5.11 Dimensionless growth rate $S = \sigma(L_f/U_f)$ versus the perturbation wave number for: $M_1 = 0$; 0.1; 0.15; 0.17, curves 1, 2, 3, 4, respectively. The scaled activation energy and the thermal expansion coefficient are $\varepsilon = 7$ and $\theta_2 = 6$. Dashed lines correspond to the solution in the approximation of a discontinuous flame front [55]

coefficient. When the amplitude of the growing harmonic became considerably large, its growing rate slows down because of the nonlinear effects.

It follows from Eqs. (5.5.4.19)–(5.5.4.23) that there are four independent modes in the upstream flow and five modes in the downstream flow. The structure of the modes determines the boundary conditions for the eigenvalue problem Eqs. (5.5.4.19)–(5.5.4.23) with the parameter S being the eigenvalue. The eigenvalue problem was solved numerically in [56, 57] using the method described in details in [15]. The numerical solution of (5.5.4.19)–(5.5.4.23) determines the dispersion relation $S = S(K)$ for different values of the Mach number M_1 and the expansion coefficient θ_2. Figure 5.11 shows the dimensionless growth rate of instability as a function of the wave number K for different values of the Mach number for the flame with the thermal expansion coefficient $\theta_2 = 6$, the adiabatic exponent $\gamma = 1.4$. The instability increment grows linearly with the increase of the wave number for long wavelength perturbations, while the perturbations of shorter wavelength are stabilized. Similar to the incompressible case, the growth rate reaches its maximal value for perturbations of rather small wave numbers $K = 0.15 - 0.3$ (wavelengths $(20-40)L_f$). It was found that the growth rate is practically independent of viscosity and for the case of Le $= 1$ the growth rate is independent of the activation energy. The dashed lines in Fig. 5.11 show the growth rate of instability according to the solution obtained in [55]. Assuming that the same supplementary condition is that perturbations do not influence the normal velocity of a flame similar to the incompressible case, the obtained in [55] solution is

$$\frac{\sigma}{U_f k} = \Gamma_0 + \Gamma_1 \mathrm{M}_1^2 \tag{5.5.4.27}$$

where Γ_0 is the coefficient that appears in the DL theory and the factor Γ_1 is:

$$\Gamma_1 = \frac{\Theta}{2}\frac{\Gamma_0(\Gamma_0+1)}{\Gamma_0(\Theta+1)+\Theta}\left[\frac{1}{2}(\Gamma_0+1)^3 + \frac{1}{2\Theta}\frac{\Theta^2-\Gamma_0^2}{1-M_1^2(\Theta-1)} + \Theta - 1\right]. \tag{5.5.4.28}$$

It is seen that the growth rate of instability obtained in [55] is close to the numerical solution [56] only asymptotically, for the long wavelength perturbations $K \ll 1$. Since the factor Γ_1 is always positive, from Eq. (5.5.4.27) follows that compressibility enhances the instability and the effect as larger as larger the Mach number M_1. In the opposite case of infinitely small Mach numbers, Eq. (5.5.4.26) goes over to the classical DL theory.

The effect of compressibility increases the growth rate of the instability for premixed flames, and it also increases the range of unstable wavelengths. The analytical and numerical studies demonstrated this for a flame of finite thickness with the first order Arrhenius reaction. Still it is not totally clear if this result remains valid for the second order reaction. In general, we can expect that the growth rate of instability depends on the Mach number, on expansion or thermal expansion coefficients, and on the factor Γ_0 in the DL theory, $\Gamma \equiv \sigma/U_f k = \Gamma(M_1, \theta, \Gamma_0)$. The instability growth rate increases with the increase in the Mach number for a fast flame However, in practice the flame velocity increases, for example, due to stretching of the flame front when the flame propagates in a channel with no-slip walls. In this case bending of the flame front caused by the DL instability has a negligible effect compared to the increase of the flame front surface due to stretching along the channel wall. The development of local curvature of the flame also can lead to a stabilizing effect. The DL stability of CJ deflagration is to some extent only of academic interest, since the existence of a planar flame front in the CJ-deflagration mode is rather unrealistic. Nevertheless, from the theoretical point of view, the CJ-deflagration front should be stable similarly to the CJ-shock wave, although this is not completely clear.

The nonlinear stage of the DL instability achieved when the amplitude of perturbations becomes comparable with the respective unperturbed value. Then the instability growth slows down because of nonlinear effects and the flame front acquires a stationary curved shape. Due to the increase of the flame surface, curved flames travel at speeds exceeding the respective velocities of planar flames. The acceleration of a flame due to the development of the curved shape makes the flame act like a piston driving a weak shock wave ahead of the flame front. A shock of noticeable intensity generated by a curved flame is one of the main differences between slow isobaric curved flames and the fast flames in a compressible fuel mixture. Typical distributions of temperature, velocity, density, and pressure in a curved flame and in a flame generated weak shock are shown in Figs. 5.12 and 5.13, obtained in [57].

Figure 5.12 shows the scaled temperature distribution at the time $t = 5L/U_f$ after small perturbations have been imposed at a planar flame front propagating in a tube of width $R = 2R_c$ ($R_c = \pi/k_c$) with the expansion coefficient $\Theta_b = 6$ and the Mach number $M_1 = U_f/c_s = 0.17$, which is close to $\gamma M_{max}^2 = 2\left[\Theta_b - \sqrt{\Theta_b(\Theta_b - 1)}\right] - 1$, which is the maximum possible value. The time instant was chosen in such a way that the flame has almost acquired the final stationary curved shape, while the flame

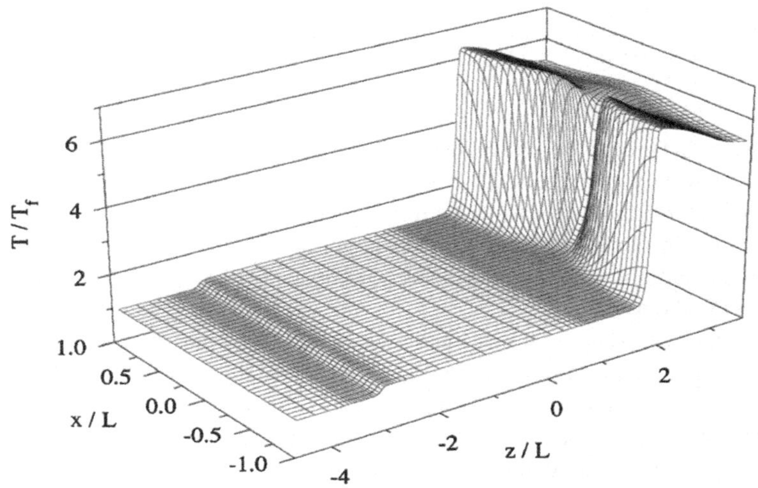

Fig. 5.12 The scaled temperature distribution in the flow with the flame with the heating coefficient $\Theta_b = 6$, the initial Mach number $\mathrm{M_1} = 0.17$, the width of the tube $R = 2R_c$. The time instant is $t = 5L/U_f$, from [57]

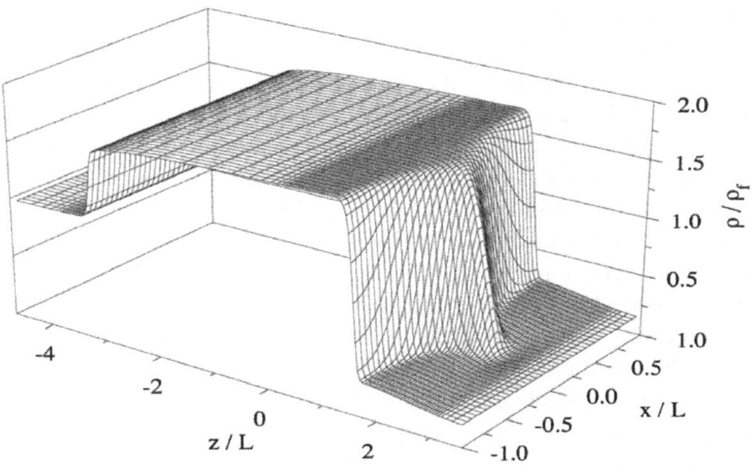

Fig. 5.13 The scaled density distribution in the flow with the flame for the same parameters as in Fig. 5.12, from [57]

generated shock is still close to the flame. This choice is possible since the flame acceleration happens mostly in a short time $\sim L/U_f$, just before the flame settles down to the stationary shape. To make the figures more illustrative it is shown not only the flow in the domain of calculations $[0, R]$, but also the flow in the domain $[-R, 0]$. In a compressible gas instead of the density ratio $\Theta = \rho_u/\rho_b$, was used the ratio of temperatures $\theta_b = T_u/T_b$.

The combustible mixture flows from the region of negative z to meet the weak shock wave moving ahead of the curved stationary flame. As it is seen in Fig. 5.12, the heating of the fuel in the shock wave is small compared to the temperature rise at the flame front, while the changes in flow velocity, density, and pressure are considerable. Particularly, it is seen in Fig. 5.13 that for the used parameters the shock compresses the fresh fuel by the factor 1.7. As a consequence of the density change, the parameters of a flame propagating in the compressed fuel change too, including the planar flame velocity and thickness, the heating coefficient and the cutoff wave number of the DL instability, etc.

As a consequence of the density change, the parameters of a flame propagating in the compressed fuel change too, which includes the planar flame velocity and thickness, the heating coefficient, the cutoff wave number of the DL instability, etc. The stronger the shock wave, the larger the difference between the initial flame parameters in the fuel prior to the shock compression used for the scaling of variables and the actual parameters of a planar flame corresponding to the matter behind the shock. Typically, the change of the flame velocity due to the fuel compression implies some acceleration of the flame front with respect to the fuel just ahead of the flame and reduction of the flame thickness. Besides, the shock involves the fuel mixture to flow in the direction of the shock, which can be observed in Fig. 5.14 as the velocity drop from the initial velocity U_f to the negative value $-2.3U_f$. The latter implies that the weak shock induces the drift of the flame together with the shocked matter. Figure 5.15 shows the scaled pressure distribution. The velocity jump in the flame generated by a weak shock wave scaled by the flame velocity can be used as a characteristic of the shock intensity. In general, the strength of a flame generated shock depends on initial and boundary conditions. For example, a shock strength

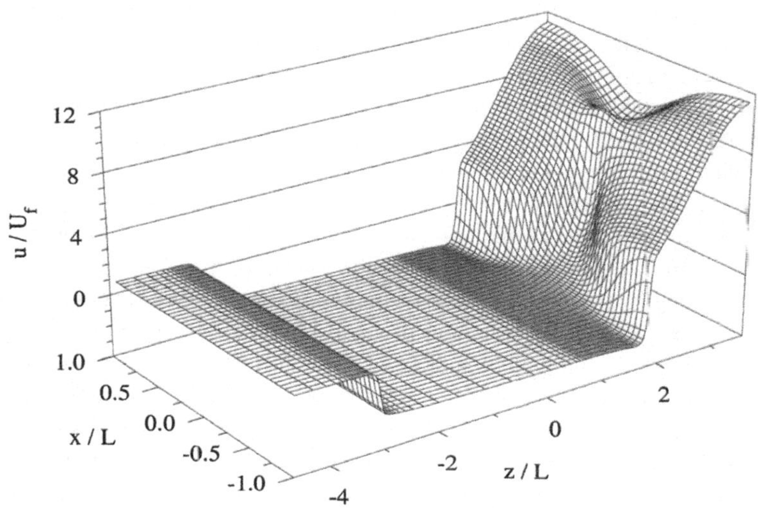

Fig. 5.14 Distribution of the scaled longitudinal flow velocity in the flame for the same parameters as in Fig. 5.12, from [57]

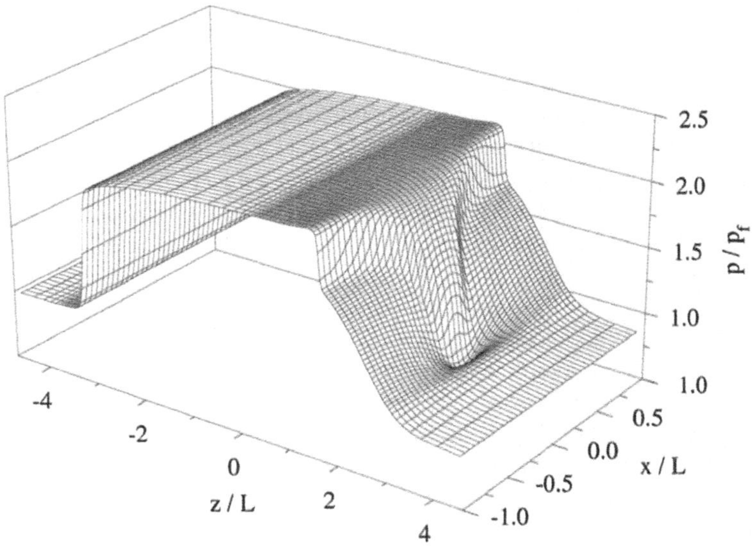

Fig. 5.15 The scaled pressure distribution in the flow of the flame with the same parameters as in Fig. 5.12, from [57]

is different for flames propagating in open tubes and closed tubes. Shock waves generated by the flame are typically weak. The superposition of shock intensities takes place when one shock catches up to another shock, but usually the shock intensity (Mach number) $M < 2.5 - 3$.

The strength of the shock wave produced by the flame depends both on the velocity of the driving flame and on the thermal expansion of the burnt products. The leading shock travels with the speed larger than the sound speed in the fresh fuel, and much larger than the flame velocity.

When the distance between the curved flame and the shock exceeds considerably the tube width, the flame shape settles down to some stationary configuration and the flame front propagates at a constant velocity larger than the velocity of the planar flame. The velocity of a curved flame in a tube depends strongly on the tube width, especially for tubes of moderate width comparable to the cutoff wavelength. It should be reminded that the tube width specifies a half of the largest possible perturbation wavelength on the linear stage of the DL instability. If a tube is sufficiently narrow $L < R_c/2$, then development of small perturbations in the linear stage of the DL instability is completely suppressed by thermal conduction and the flame front remains planar (this is different for the flame in a tube with no-slip walls). The nonlinear instability implies that for narrow tubes all perturbations of small amplitudes vanish and the flame is linearly stable, but perturbations of a sufficiently large amplitudes can develop into curved stationary structures. In wider tubes, the tube width determines the wavelength of the most rapidly growing harmonic, the shape of the resulting flame front, and finally the flame velocity.

References

1. G. Darrieus, Propagation d'un front de flamme. Unpublished work presented at La Technique Moderne, and at Le Congrés de Mécanique Appliqueée (1945), 1938
2. L. Landau, On the theory of slow combustion. Acta Phys. Chem. **19**, 77–85 (1945)
3. L.D. Landau, E.M. Lifshitz, *Fluid Mechanics* (Pergamon Press, Oxford, 1989)
4. M. Liberman, A. Velikovich, *Physics of Shock Waves in Gases and Plasmas* (Springer-Verlag, Berlin, New York, 1985)
5. A. Jeffrey, T. Taniuti, *Nonlinear Wave Propagation* (Academic, New York, 1964)
6. S.E. Bodner, Rayleigh-taylor instability and laser-pellet fusion. Phys. Rev. Lett. **33**, 761–765 (1974)
7. W.H. Manheimer, D.G. Colombant, G.H. Gardner, Steady-state planar ablative flow. Phys. Fluids **25**, 1644–1652 (1982)
8. G.H. Markstein, *Nonsteady Flame Propagation* (Pergamon Press, Oxford, 1964)
9. Ya. B. Zel'dovich, An effect stabilizing a curved front of a laminar flame. Prikl. Mekh. Tekh. Fiz. **1**. 102–118 (1966)
10. G. Joulin, P. Clavin, Linear stability analysis of nonadiabatic flames: diffusional-thermal model. Combust. Flame **35**, 139–153 (1979)
11. P. Pelce, P. Clavin, Influence of hydrodynamics and diffusion upon the stability limits of laminar premixed flames. J. Fluid Mech. **124**, 219–237 (1982)
12. M.L. Frankel, G.I. Sivashinsky, The effect of viscosity on the hydrodynamic stability of a plane flame front. Combustion Sci. Technol. **29**, 207–224 (1982)
13. M. Matalon, B.J. Matkowsky, Flame as gasdynamic discontinuities. J. Fluid Mech. **124**, 239–259 (1982)
14. M. Matalon, B.J. Matkowsky, Flames in fluids: their interaction and stability. Combust. Sci. Techn. **34**, 295–316 (1983)
15. M.A. Liberman, V.V. Bychkov, S.M. Golberg, D. Book, Stability of a planar flame front in the slow-combustion regime. Phys. Rev. E **49**, 445–457 (1994)
16. Y.B. Zeldovich, G.I. Barenblatt, V.B. Librovich, G.M. Makhviladze, *The Mathematical Theory of Combustion and Explosion* (Consultants Bureau, New York, 1985)
17. L.D. Landau, E.M. Lifshitz, *Quantum Mechanics* (Pergamon Press, Oxford, 1976)
18. V.V. Bychkov, M.A. Liberman, Dynamics and stability of premixed flames. Phys. Rep. **325**, 115–237 (2000)
19. M. Matalon, C. Cui, J.K. Bechtold, Hydrodynamic theory of premixed flames: effects of stoichiometry, variable transport coefficients and arbitrary reaction orders. J. Fluid Mech. **487**, 179–210 (2003)
20. P. Clavin, P. Garcia, The influence of the temperature dependence of diffusivities on the dynamics of flame fronts. J. Mech. Theor. Appl. **2**, 245–263 (1983)
21. P. Clavin, Dynamic behavior of premixed flame fronts in laminar and turbulent flows. Prog Energy Combust Sci. **11**, 1–59 (1985)
22. X. Lu, C. Pantano, Linear stability analysis of a premixed flame with transverse shear. J. Fluid Mech. **765**, 150–166 (2015)
23. M.A. Liberman, V.V. Bychkov, S.M. Golberg, On the stability of a flame in the gravitational field. Sov. Phys. JETP **77**, 227–236 (1993)
24. V.V. Bychkov, S.M. Golberg, M.A. Liberman, Rayleigh-Taylor instability of combustion and laser produced ablation fronts. Phys. Fluids B **5**, 3822–3824 (1993)
25. V.V. Bychkov, S.M. Golberg, M.A. Liberman, Self-consistent theory of the rayleigh-taylor instability in ablatively accelerated laser plasma. Phys. Plasmas **1**, 2976–2986 (1994)
26. G. Searby, D. Rochwerger, A parametric acoustic instability in premixed flames. J. Fluid Mech. **231**, 529–543 (1991)
27. J.K. Bechtold, M. Matalon, Hydrodynamics and diffusion effects on the stability of spherically expanding flames. Combust. Flame **67**, 77–90 (1987)
28. A.G. Istratov, V.B. Librovich, The effect of transport processes on the stability of a plane flame front. Astronautica Acta **14**, 453–467 (1969)

29. Г.И. Баренблат, Я.Б. Зельдович, А.Г. Истратов, О диффузионно тепловой устойчивоти ламинарных пламен. Журнал прикладной механики и технической физики. Том 2, №4 (1962) стр. 21–26 (in Russian) [G. I. Barenblatt, Y. B. Zeldovich, and A. G. Istratov, On diffusive thermal stability of a laminar flame. J. Appl. Mech. Tech. Phys. **4**, 21–26 (1962)
30. Y.B. Zel'dovitch, A.G. Istratov, N.I. Kidin, V.B. Librovitch, Flame propagation in tubes: hydrodynamics and stability. Combust. Sci. Tech. **24**, 1–13 (1980)
31. B. Lewis, G. Elbe, On the theory of flame propagation. J. Chem. Phys. **2**, 537–546 (1934)
32. G.I. Sivashinsky, Diffusional-thermal theory of cellular flames. Combust. Sci. Technol. **15**, 137–145 (1977)
33. G.I. Sivashinsky, Instabilities, pattern formation, and turbulence in flames. Ann. Rev. Fluid Mech. **15**, 179–199 (1983)
34. G.I. Sivashinsky, Nonlinear analysis of hydrodynamic instability in laminar flames-I. Derivation of basic equations. Acta Astronaut. **4**, 1177–1206 (1977)
35. L. Filyand, G.I. Sivashinsky, M.L. Frankel, On self-acceleration of outward propagating wrinkled flames. Physica D **72**, 110–118 (1994)
36. Yu. Gostintsev, A. Istratov, Yu. Shulenin, Self-similar propagation of a free turbulent flame in mixed gas mixtures. Combust. Expl. Shock Waves 24, 563–569 (1988)
37. A. Linan, The asymptotic structures of counter-flow diffusion flames for large activation energies. Acta Astronaut. **1**(1974), 1007–1039 (1974)
38. G.I. Sivashinsky, On a distorted flame front as a hydrodynamic discontinuity. Acta Astronaut. **3**, 889–918 (1976)
39. P. Clavin, F.A. Williams, Effects of molecular diffusion and of thermal expansion on the structure and dynamics of premixed flames in turbulent flows of large scale and low intensity. J. Fluid Mech. **116**, 251–282 (1982)
40. A.G. Class, B.J. Matkowsky, A.Y. Klimenko, Stability of planar flames as gasdynamic discontinuities. J. Fluid Mech. **491**, 11–49 (2003)
41. S. Cheathamand, M. Matalon, A general asymptotic theory of diffusion flames with application to cellular instability. J. Fluid Mech. **414**, 105–144 (2000)
42. P. Metzener, M. Matalon, Diffusive-thermal instabilities of diffusion flames: onset of cells and oscillations. Combust. Theory Model. **10**, 701–725 (2006)
43. M. Matalon, P. Metzener, The effect of thermal expansion on diffusion flame instabilities. J. Fluid Mech. **647**, 453–472 (2010)
44. H.D. Ross, R.G. Sotos, J.S. T'ien, Observations of candle flames under various atmospheres in microgravity. Combust. Sci. Technol. 75, 155–160 (1991)
45. M. Füri, P. Papas, P.A. Monkewitz, Non-premixed jet flame pulsations near extinction. Proc. Combust. Inst. **28**, 831–838 (2000)
46. C. Altantzis, C.E. Frouzakis, A.G. Tomboulides, M. Matalon, Hydrodynamic and thermo diffusive instability effects on the evolution of laminar planar lean premixed hydrogen flames. J. Fluid Mech. **700**, 329–361 (2012)
47. P. Clavin, G. Joulin, Premixed flames in large scale and high intensity turbulent flow. J. Physique Lett. **44**, 1–12 (1983)
48. V.V. Bychkov, M.A. Liberman, Stability of solid propellant combustion. Phys. Rev. Letters **73**, 1998–2000 (1994)
49. V.V. Bychkov, M.A. Liberman, On the dynamics of a curved deflagration front. J. Exp. Theor. Phys. (JETP) **84**, 281–288 (1997)
50. B.J. Matkowsky, G.I. Sivashinsky, Propagation of a pulsating reaction front in solid fuel combustion, S.I.A.M. J. Applied Math. **35**, 465–478 (1978)
51. S.I. Blinnikov, P.V. Sasorov, Landau-Darrieus instability and the fractal dimension of flame fronts. Phys. Rev. E **53**, 4827–4841 (1996)
52. V.V. Bychkov, M.A. Liberman, On the stability and fractal structure of spherical flame in a self-similar regime. Phys. Rev. Lett. **76**, 2814–2818 (1996)
53. L.D. Landau, E.M. Lifshitz, Vol. 5 Statistical Physics, Pergamon, Oxford (1980)

54. D. John, Lindl, inertial confinement fusion: the quest for ignition and energy gain using indirect drive. Amer. Inst. of Physics (1998)
55. S. Kadowaki, Instability of a deflagration wave propagating with finite Mach number. Phys. Fluids **7**, 220–222 (1995)
56. O.Y. Travnikov, M.A. Liberman, V.V. Bychkov, Stability of a planar flame front in a compressible flow. Phys. Fluids **9**, 3935–3937 (1997)
57. O.Yu. Travnikov, V.V. Bychkov, M. A. Liberman, Influence of compressibility on propagation of curved flames, Phys. Fluids 11, 2657–2666 (1999)
58. L. He, Analysis of compressibility effects on Darrieus-Landau instability of deflagration wave. Europhys. Lett. **49**, 576–583 (2000)
59. S. Kadowaki, Numerical study on lateral movements of cellular flames. Phys. Rev. E **56**, 2966–2971 (1997)
60. G. Sharpe, Linear stability of planar premixed flames: reactive Navier-Stokes equations with finite activation energy and arbitrary Lewis number. Combust. Theory Modelling **7**, 45–65 (2003)
61. S. Kadowaki, Numerical study on the formation of cellular premixed flames at high Lewis numbers. Phys. Fluids **12**, 2352–2359 (2000)
62. S. Kadowaki, T. Hasegawa, Numerical simulation of dynamics of premixed flames: flame instability and vortex–flame interaction. Prog. Energy Combust. Sci. **31**, 193–241 (2005)
63. A. Fecteau, J. McDonald, A. Sow, M. Radulescu, The effects of compressibility on the propagation of premixed deflagration. 27th International Colloquium on the Dynamics of Explosions, Beijing, Chains (2019)
64. V. Bychkov, S. Golberg, M. Liberman, Self-consistent model of the Rayleigh-Taylor instability in ablatively accelerated laser plasma. Phys. Plasmas **1**, 2976–2986 (1994)
65. A.R. Piriz, J. Sanz, L.F. Ibanez, Rayleigh-Taylor instability of steady ablation fronts: the discontinuity model revisited. Phys. Plasmas **4**, 1117–1126 (1997)
66. P. Clavin, L. Masse, Instabilities of ablation fronts in inertial confinement fusion: a comparison with flames. Phys. Plasmas **11**, 690–705 (2004)
67. J. Sanz, L. Masse, P. Clavin, The linear Darrieus-Landau and Rayleigh-Taylor instabilities in inertial confinement fusion revisited. Phys. Plasmas **13**(1–14), 102702 (2006)
68. A.R. Piriz, N.A. Tahir, Physics of ablative Rayleigh–Taylor and Landau–Darrieus instabilities. New J. Phys. **15**(1–14), 015013 (2013)

Chapter 6
Flame–Acoustic Interaction: Thermoacoustic and Parametric Instabilities

6.1 Early Experimental Research on the Flame–Acoustic Interaction

In previous chapters, we considered flame dynamics in homogeneous unbounded fuel mixtures. A more common situation is a flame propagating in a closed combustion chamber. The spontaneous radiation of sound waves by a flame and their impact on the flame were already observed in the nineteenth century by Mallard and Le Châtelier in their pioneering experiments to measure the laminar flame speed [1]. If a flame propagates in a closed chamber, then there are many additional effects which influence flame dynamics and the development of the DL instability [2–7]. In particular, when a propagating or accelerating flame produces pressure waves and as a consequence, the flame interacts with these waves. The generated pressure waves can be considered as acoustic waves if the time necessary for a sound wave to pass through the burning chamber is much less than the inverse growth rate of the DL instability. In the opposite case, if the characteristic time needed for sound to pass through the burning chamber is longer than the characteristic time of flame dynamics (e.g. the inverse growth rate of the DL instability), the pressure waves generated by the flame can be considered as weak shocks.

From the beginning, the interaction of flames with acoustic waves has been studied on the basis of the phenomenological model for a flame [8] using the Markstein number which describes the details of the internal structure of the flame and can be measured experimentally [9]. From the very beginning, even in the studies based on a phenomenological approach, two important effects related to the flame–acoustic interaction have been pointed out. These are stabilization of the DL instability by sound waves of small amplitudes and the parametric instability of a flame front induced by sound waves of sufficiently large amplitudes. The quantitative investigation of the flame interaction with sound waves began after the linear equation for a flame instability in a gravitational field has been derived (Ref. [10] in Sect. 3.2).

A fundamental investigation of the propane flame–acoustic interaction performed by Searby and Rochwerger [11] and Searby [10], where stability limits for a flame in

M. A. Liberman, *Combustion Physics*,
https://doi.org/10.1007/978-3-030-85139-2_6

the oscillating acoustic field have been calculated numerically and measured experimentally, initiated a series of more detailed investigations of the flame dynamics in an acoustic wave [12–16], and finally led to the first direct experimental measurement and visualization of the growth rate of the DL instability of a planar laminar premixed flame by Clanet and Searby [17]. All the previous studies on planar flames only measured cellular structures that is nonlinearly saturation stage, which provides only rough estimates.

The dynamics of propane–air flames propagating from the open top end to the closed bottom of a vertical tube 120 cm long and 10 cm wide has been studied in [10]. The experiments were mainly concerned lean mixtures without spontaneous cellular structures of a diffusive-thermal nature. The flame was ignited at the top and initially takes the form of a smooth curved front, resulting from the Darrieus–Landau hydrodynamical instability. The regime of the flame propagation strongly depends on the flame velocity. The laminar flame velocity was a sensitive parameter in the experiment [10]. For sufficiently lean mixtures with a laminar flame velocity below 16 cm/s, the flame propagated from the top to the bottom of the tube producing no sound waves and the flame shape remained almost planar due to the stabilizing effect of gravitation on downward propagating flames. For flames with velocities 16 cm/s $< U_f <$ 25 cm/s, the DL instability prevails over the gravitational stabilization and results in curved shapes of the flame front at the upper half of the tube. The curved flames propagated with about doubled velocity compared to the velocity of a planar flame. The development of the DL instability caused the generation of sound waves in the tube with a linear growth rate of about 10 s^{-1}, which was called a primary acoustic instability in [10]. For flames with velocities in the range 16 cm/s $< U_f <$ 25 cm/s, the growth of the sound wave amplitude was saturated after about 0.25 s, when the amplitude of acoustic waves reached about 0.005 bar, and when the flame was close to the center of the tube. This saturation was accompanied by the stabilization of the DL instability after about 0.25 s, and transition from a curved flame front to the almost planar flame propagating with the normal slightly oscillating at the acoustic frequency flame velocity. The sound intensity continued to increase but at a much smaller rate of about 0.6 s^{-1} compatible with theoretical predictions for the flame–acoustic coupling in the case of planar flames [17]. For faster flames, 25 cm/s $< U_f <$ 40 cm/s, there is a third stage, which begins at the lower half of the tube where a violent secondary instability occurs on the planar front. Small cellular structures with the size of a few millimeters appeared at the flame front and grow rapidly in amplitude oscillating in time with about twice the acoustic period, while the average velocity of the flame propagation can reach 170 cm/s [10]. At the same time amplitudes of the sound increase with the growth rate about 30 s^{-1} and reach $0.05 - 0.2$bar, depending on the flame velocity. The oscillation period of these structures, $14rmms$, that is twice of the acoustic period, indicates that the cellular structure develops as a result of parametric instability of the flame front in a periodic effective gravitational field created by the acoustic waves [7]. For sufficiently fast flames (of about 40 cm/s), the described growth of the flame cellular structure and the sound wave amplitude develops directly from the primary DL instability without intermediate flame stabilization. Finally, for fast flames the described instability may result in turbulization of the flame leading to the increase of the flame velocity up to 750 cm/s. As soon as the breakdown to turbulent flame occurs, the amplitude of sound waves drops abruptly, which implies that the acoustic energy is converted to the energy of the turbulent motion.

6.2 Flame Stabilization by Acoustic Waves

The theory describing the linear stage of the DL instability stabilization by small amplitude acoustic waves and the parametric instability induced at a flame front by sound waves of the sufficiently large amplitudes was developed in [10, 11, 15, 18–20]. The acoustic instability induced by a slightly curved flame close to the threshold of the DL instability for a downward propagating flame was considered in [10, 12]. Using the linear theory (Sect. 5.3) of a perturbed flame in a gravitational field, a quantitative investigation of the flame–acoustic interaction for propane flames has been performed in [10], where stability limits for a flame in an oscillating acoustic field have been calculated numerically and measured experimentally.

Let us consider a slightly perturbed almost planar flame front propagating along the z-axis and planar acoustic waves propagating to the flame. We choose the reference frame co-moving with the flame front. The corresponding experimental configuration is the flame which is forced to keep a certain position in an open tube by the inflow of the fresh mixture and the sound wave produced by a loudspeaker. Let us consider the flame response to the sound wave with the linear equation for perturbations of a small amplitude imposed at a planar flame front in an arbitrary gravitational field, Eqs. (5.3.17)–(5.3.20). The action of the sound wave, $u_z = -U_a \sin(\omega t)$, on the flame can be considered as an effective oscillating gravitational field

$$g_a = -\omega U_a \cos(\omega t). \tag{6.2.1}$$

In addition, there is a constant terrestrial gravitation creating acceleration g. The acceleration produced by sound waves can be considered as pressure waves propagating along the normal to the flame front. With the account of the constant terrestrial–gravitational acceleration g, the equation for small perturbations $f'(t, x)$ of a planar flame front (7.3.17) can be written as

$$A\frac{d^2 f'}{dt^2} + BU_f k\frac{df'}{dt} - CU_f^2 k^2 f' - C_1 \omega U_a k \cos(\omega t) f' = 0. \tag{6.2.2}$$

The linear equation implies that the flame front propagates upwards or downwards. For flames propagating downwards the acceleration in (6.2.2) is positive, while for upward propagating flames it is negative. The coefficients in Eq. (6.2.2), depending on the flame parameters are given by the formulas (7.3.18)–(7.3.20):

$$A = 1 + \frac{\Theta - 1}{\Theta + 1} k L_f \left(\mathrm{Ma} - \frac{\Theta}{\Theta - 1} J_2 \right), \tag{6.2.3}$$

$$B = \frac{2\Theta}{\Theta + 1} \left[1 + \Theta k L_f (\mathrm{Ma} - J_2) \right] \tag{6.2.4}$$

$$C = \Theta \frac{\Theta - 1}{\Theta + 1} \left(1 - \frac{k L_f}{\Theta - 1} \begin{bmatrix} J_3 + (3\Theta - 1)\mathrm{Ma} - 2\Theta J_2 + \\ 2\,\mathrm{Pr}(\Theta - 1)\frac{\kappa_b}{\kappa_u} - 2\,\mathrm{Pr}\, J_3 \end{bmatrix} \right) \tag{6.2.5}$$

$$C_1 = \frac{\Theta - 1}{\Theta + 1} \left[1 - k L_f \left(\mathrm{Ma} - \frac{\Theta J_2}{\Theta - 1} \right) \right]. \tag{6.2.6}$$

The effect of flame stabilization by acoustic waves is similar to the well-known stabilization of an inverted pendulum (know as the Kapitza pendulum) in a vertical position by the vertical vibration of the pendulum suspension at the appropriate amplitude and frequency [21]. We will use the same method as was used for the pendulum problem. We look for the solution in the form of a superposition of a slow component $f_1'(t)$ that changes on a typical time $\propto 1/\sqrt{C} U_f k$ and the fast oscillating component $f_2'(t) = f_2' \cos(\omega t)$:

$$f' = f_1'(t) + f_2' \cos(\omega t). \tag{6.2.7}$$

In the limit of high frequencies of the acoustic oscillations $\sqrt{C} U_f k/\omega << 1$, the function f_1' is almost a constant on the oscillation period $2\pi/\omega$ and the following conditions hold:

$$\frac{d^2 \tilde{F}_2}{dt^2} \propto \omega^2 \tilde{F}_2 >> B U_f k \frac{d\tilde{F}_2}{dt} >> C U_f^2 k^2 \tilde{F}_2. \tag{6.2.8}$$

Substituting (6.2.7) into linear Eq. (6.2.2) for the flame front perturbations, we obtain

$$\begin{aligned} & A \frac{d^2 f_1'}{dt^2} - A\omega^2 f_2' \cos(\omega t) + B U_f k \frac{d f_1'}{dt} - \left(C U_f^2 k^2 - C_1 g k \right) f_1' \\ & - C_1 \omega U_a k f_1' \cos(\omega t) - \frac{1}{2} C_1 \omega U_a k f_2' [1 + \cos(2\omega t)] = 0. \end{aligned} \tag{6.2.9}$$

Separating in (6.2.9) terms with the frequency ω, we obtain the equation which expresses the amplitude f_2' via f_1'

$$f_2' = -\frac{C_1 U_a k}{A\omega} f_1'. \tag{6.2.10}$$

With the help of (6.2.10) and neglecting by the term containing $\cos(2\omega t)$ which is the high frequency additional term in (6.2.7) which can be neglected because of its small amplitude, we obtain the equation for slow-changing terms

$$A\frac{d^2 f_1'}{dt^2} + BU_f k\frac{df_1'}{dt} - (CU_f^2 k^2 - C_1 gk) f_1' + \frac{C_1^2}{2A} U_a^2 k^2 f_1' = 0. \tag{6.2.11}$$

For the perturbation in the form $f_1'(t) = f_1' \exp(\sigma t)$ the solution of Eq. (6.2.11) corresponding to the growth rate of the unstable mode is:

$$\frac{\sigma}{U_f k} = -\frac{B}{2} + \frac{1}{2}\left(B^2 + 4C - 4C_1 \frac{g}{U_f^2 k} - \frac{2C_1^2 U_a^2}{AU_f^2}\right)^{1/2}. \tag{6.2.12}$$

From (6.2.12) it follows that the larger the amplitude of the acoustic oscillations U_a, the smaller the growth rate of instability. The stability limit, that is the amplitude of the acoustic oscillations when the DL instability is fully suppressed ($\sigma = 0$), is

$$\frac{U_a}{U_f} = \frac{\sqrt{2CA}}{C_1}\left(1 - \frac{C_1}{C}\frac{g}{U_f^2 k}\right)^{1/2}. \tag{6.2.13}$$

One should remember that the coefficients A, C, and C_1 are functions of the perturbation wave number, therefore, Eq. (6.2.13) determines the critical amplitude of the acoustic oscillations necessary to suppress the growth of perturbations with a fixed wave number k.

Figure 6.1 shows the dependence of the critical amplitude of the acoustic oscillations U_a on the perturbation wave number k for the unit Lewis number $\mathrm{Le} = 1$, zero gravitational acceleration, and constant thermal conduction for different expansion coefficients Θ.

Region (I) in Fig. 6.1 corresponds to the development of the DL instability. It is seen that the maximal critical amplitude U_a corresponds to the small wave number $k \to 0$. Acoustic oscillations with such amplitude completely suppress the DL instability for all wavelengths. Using expressions (7.3.18)–(7.3.20) for the coefficients $A,\ C,\ C_1$, we find the amplitude of sound waves providing a complete suppression of the DL instability

$$\max\{U_a/U_f\} = \sqrt{\left(\Theta\frac{\Theta+1}{\Theta-1}\right)}. \tag{6.2.14}$$

Note that in the case of zero gravity, the critical amplitude depends only on the expansion coefficient of the flame and does not depend on the parameters describing the internal flame structure such as the scaled thermal conduction and the Markstein number. Besides, for flames with realistic expansion coefficients $\Theta = 7 - 10$, the critical amplitude of sound waves depends only slightly on the expansion coefficient, and the numerical value of (6.2.14) becomes approximately $\max\{U_a/U_f\} = 4 - 5$. When the amplitudes of acoustic waves in experiments [11] exceed the critical values,

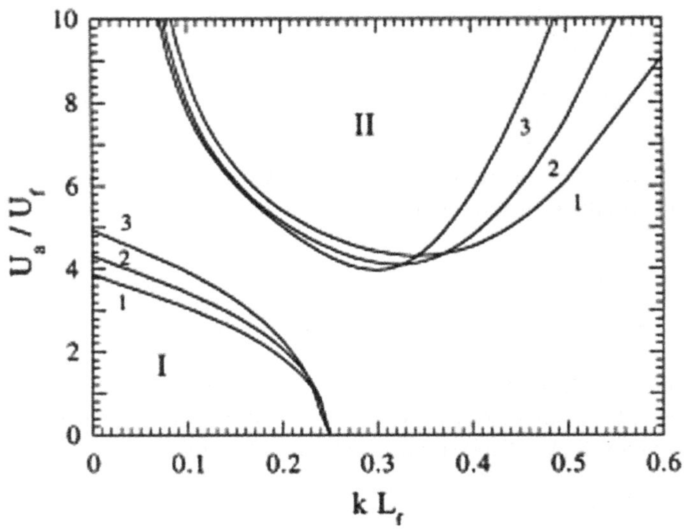

Fig. 6.1 Stability limits (amplitudes of acoustic waves) for a flame with $Le = 1$ in an acoustic wave with frequency $\omega L_f / U_f = 1$ for the DL instability (region I) and the parametric instability (region II). Curves 1, 2, and 3 correspond to the flame expansion coefficients $\Theta = 5,\ 7,\ 9$; adopted from [16]

the complete stabilization can be observed provided that these acoustic waves are still sufficiently weak and the parametric instability is not induced yet.

Figure 6.2 shows the stabilization of the DL instability by acoustic waves for the case of a nonzero gravitational field. For a flame propagating downwards with a stabilizing gravitational field, the critical amplitude is somewhat reduced in comparison with (6.2.14), and the maximum of the curve bounding the region I is achieved at a smaller but finite nonzero scaled perturbation wave number $K = kL_f$. On the contrary, the gravity destabilizes flames propagating upwards and the DL instability is enhanced by the Rayleigh–Taylor instability.

As one can see in Fig. 6.2 and Eq. (6.2.12), the complete stabilization of an upward propagating flame by an acoustic wave is impossible, since the Rayleigh–Taylor instability is always stronger for perturbations of a sufficiently long wavelength than the stabilizing effect of acoustic waves. Still, for a particular experimental installation, one can observe stabilization of the Rayleigh–Taylor instability by an acoustic wave, since the perturbation wavelength for a flame in a tube is limited from above by the tube diameter. However, since the wavelengths of perturbations for a flame propagating in a tube are restricted from above by the tube diameter, stabilization of the Rayleigh–Taylor instability by an acoustic wave is possible for a specially designed experiment.

Aldredge [22] considered the influence of the Saffman–Taylor instability on flame propagating between two closely separated walls, resulting from the disparity between the viscosities of the reactant and product mixtures. The stability limits

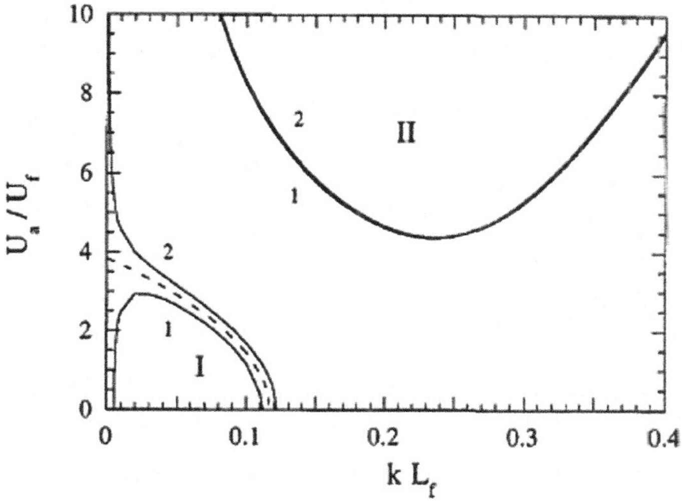

Fig. 6.2 Stability limits for the flame with $\Theta = 5$, $Le = 1$, and $\Pr = 1$ for the acoustic wave with frequency $\omega L_f / U_f = 1$. Region (I) is for the DL instability and region (II) for the parametric instability. Curves 1 and 2 correspond to the flames propagating downwards and upwards with velocity $U_f^2/|g|L_f = 40$. The dashed line corresponds to zero gravitation: from [16]

diagram calculated taking into account the influence of the Saffman–Taylor instability is similar to that shown in Fig. 6.2.

6.3 Parametric Flame Instability and Sound Waves

Another effect of the flame interaction with sound waves is the parametric instability of the flame front. Similar to the parametric instability of the oscillator in classical mechanics [20], we consider the solution of Eq. (6.2.2) in the form

$$f' = a \exp(\sigma t) \cos\left(\frac{\omega t}{2}\right) + b \exp(\sigma t) \sin\left(\frac{\omega t}{2}\right). \qquad (6.3.1)$$

We omit the higher frequency terms, which is justified if the amplitude of the last acoustic term in Eq. (6.2.2) is small

$$A\frac{d^2 f'}{dt^2} >> [C_1 \omega U_a k \cos(\omega t)] f'. \qquad (6.3.2)$$

The condition (6.3.2) means that $\omega >> C_1 U_a k$, which is a rigorous approach in the case of high-frequency oscillations and for long-wavelength perturbations. Otherwise, this approach provides an evaluation of the stability limits for the parametric instability. Substituting (6.3.1) into Eq. (6.2.2), we come to the linear equations for the amplitudes a and b, which give us the dispersion relation for the parametric instability

$$\begin{aligned} & a\left(A\sigma^2 - \frac{1}{4}A\omega^2 + BU_f k\sigma - CU_f^2 k^2 + C_1 gk - \frac{1}{2}C_1\omega k U_a\right) + \\ & b\left(A\sigma\omega + \frac{1}{2}BU_f k\omega\right) = 0 \end{aligned} \tag{6.3.3}$$

$$\begin{aligned} & a\left(A\sigma\omega + \frac{1}{2}BU_f k\omega\right) + \\ & b\left(A\sigma^2 - \frac{1}{4}A\omega^2 + BU_f k\sigma - CU_f^2 k^2 + C_1 gk + \frac{1}{2}C_1\omega k U_a\right) = 0. \end{aligned} \tag{6.3.4}$$

From the condition that Eqs. (6.3.3) and (6.3.4) are compatible, we obtain the dispersion relation for the parametric instability

$$\begin{aligned} & \left(A\sigma^2 - \frac{1}{4}A\omega^2 + BU_f k\sigma - CU_f^2 k^2 + C_1 gk\right)^2 + \\ & \omega^2\left(A\sigma + \frac{B}{2}U_f k\right)^2 - \frac{\omega^2}{4}C_1^2 U_a^2 k^2 = 0. \end{aligned} \tag{6.3.5}$$

The stability limits for the parametric instability correspond to $\sigma = 0$, and from Eq. (6.3.5) this gives the expression for the critical amplitude of the acoustic oscillations

$$\frac{U_a^2}{U_f^2} = \frac{B^2}{C_1^2} + \frac{\left(A\omega^2 + 4CU_f^2 k^2 - 4C_1 gk\right)^2}{4C_1^2\omega^2 U_f^2 k^2}. \tag{6.3.6}$$

With the explicit form (6.2.3)–(6.2.6) for coefficients A, B, C, and C_1 as functions of the perturbation wave number k, Eq. (6.3.6) defines the critical amplitude of the sound wave capable to initiate the parametric instability for perturbations with the specific wave number k. The dependence of the critical amplitude U_a on the perturbation wave number is shown in Fig. 6.1 for Le $= 1$, zero gravitational acceleration, thermal conductivity of an ideal gas, $\chi \propto \sqrt{T}$, and different expansion coefficients Θ, where the development of the parametric instability corresponds to the region II. It is seen that there is a minimal amplitude of the sound wave needed to induce the parametric instability at any wave number, which slightly depends on the expansion coefficient. For the scaled frequency $\omega L_f/U_f = 1$ and for $\Theta = 5 - 10$, the minimal amplitude of the sound wave capable to induce the parametric instability is $U_a/U_f \approx 4 - 5$. The minimal critical amplitude corresponds to a certain finite wave number of the perturbation, which in turn characterizes the inverse length scale of the structures developing at the flame front as a result of the parametric instability.

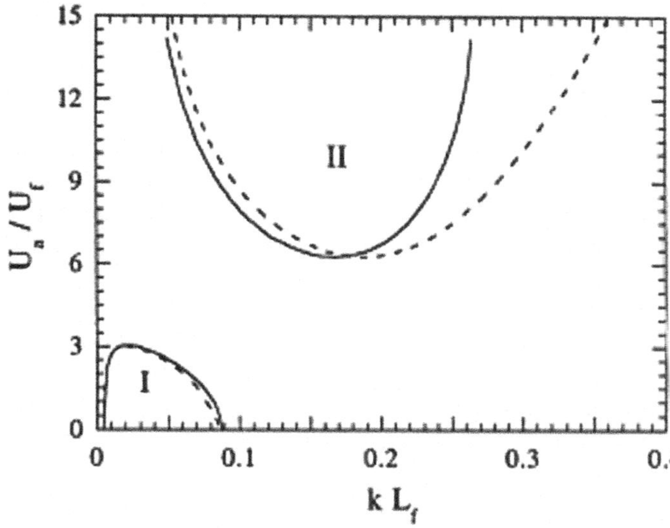

Fig. 6.3 Stability limits (acoustic wave amplitude versus perturbation wave number) for propane flames in the acoustic wave with frequency $\omega L_f/U_f = 1$ for the DL instability (I) and for the parametric instability (II). The solid lines are the numerical results [9]; the dashed lines are Eqs. (6.2.13) and (6.3.6). The scaled acceleration $gL_f/U_f^2 = 0.025$ corresponds to the Froude number $U_f^2/gL_f = 40$; adopted from [16]

As one can see in Fig. 6.1, the length scale of the parametric instability is typically smaller than the length scale of the DL instability, if the latter is not suppressed yet by the sound wave oscillations. Still the perturbation wavelength of the parametric instability is sufficiently long compared to the flame thickness, and this justifies the applicability of the long-wavelength perturbations [19] to use Eq. (6.2.2). For the fixed frequency of the acoustic wave $\omega L_f/U_f = 1$, the parametric instability starts with perturbations of wave numbers $kL_f = 0.3 - 0.4$, for $\Theta = 5 - 10$. For a flame with $\Theta = 10$, the amplitude of the acoustic waves needed to initiate the parametric instability is smaller than the amplitude necessary to suppress the DL instability. Such a flame cannot be completely stabilized by acoustic waves since with increasing amplitude of the acoustic wave, the parametric instability starts before the DL instability was suppressed. This is similar to the case of fast flames 25 cm/s $< U_f <$ 40 cm/s in Searby's experiments [11] where the primary acoustic instability is developed directly into the parametric instability without intermediate smoothing of the flame front.

The analytical solutions are in good agreement with the numerical and experimental results on propane flames [9]. For the values characterizing propane flames taken from [9], $\Theta = 5.62$, Ma $= 4.5$, Pr $= 0.69$, and $\kappa_b/\kappa_f = 3.2$, $J_2 = 3.3$, $J_3 = 10.3$ for (6.2.3)–(6.2.6), we can evaluate using Eqs. (6.2.13) and (6.3.6) the stability limits of the DL instability and for the parametric instability of a propane flame in an acoustic wave. The results are shown in Fig. 6.3 together with the numerical results obtained in [9] for the scaled frequency of the acoustic wave $\omega L_f / U_f = 1$ acting on a downward propagating flame with the scaled gravitational acceleration $g L_f / U_f^2 = 0.025$.

It is seen in Fig. 6.3 that the DL instability stabilization by the acoustic wave obtained from Eq. (6.2.13) is in good agreement with the numerical results obtained in [9]. For the parametric instability, the agreement with the numerical results [9] is satisfactorily good for the long-wavelength perturbations: from Eq. (6.2.13), the minimum of the critical amplitude of the acoustic wave is at $kL_f = 0.19$, while according to [9] the minimum is achieved at $kL_f = 0.16$. Besides, Eq. (6.2.13) and numerical calculations [9] predict almost the same value of the minimal amplitude $\min\{U_a/U_f\} = 6.3$. The agreement with numerical results [9] is not so good for short-wavelength perturbations. However, one should remember that the basic Eq. (6.2.2) for both numerical and analytical calculations has been derived in the limit of long-wavelength perturbations.

Figure 6.4 shows the minimal amplitude of acoustic oscillations necessary to initiate the parametric instability for propane flames as a function of the oscillation frequency calculated in using Eq. (6.3.6) and obtained experimentally in [10]. Different markers in Fig. 6.4 correspond to different frequencies from 57 to 265 Hz used in experiments [9]. All the experimental points in Fig. 6.4 are close to the curve from small frequencies $\omega L_f / U_f = 0.1$ to high frequencies $\omega L_f / U_f = 10$. The minimal amplitude of the sound wave needed to initiate the parametric instability obtained from Eq. (6.3.6) is in good agreement with the experimental data. Both the experiment [10] and the analytical theory [16, 19] show the general tendency of a

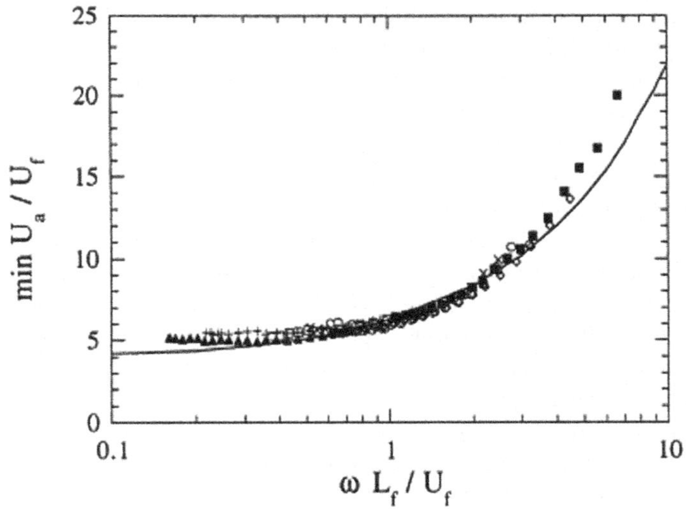

Fig. 6.4 The minimum amplitude of the acoustic wave exciting the parametric instability for a propane flame versus the acoustic oscillation frequency. The solid line is calculated using Eq. (6.3.6); the markers are the experimental measurements [10]

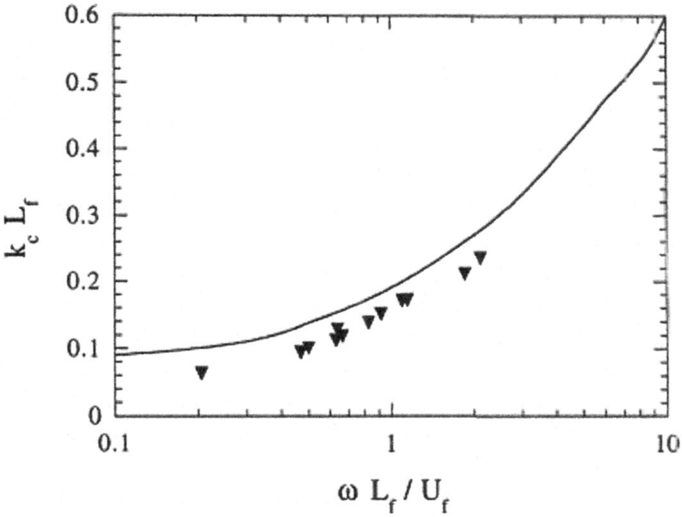

Fig. 6.5 The characteristic wave number of the parametric instability for a propane flame versus the acoustic oscillation frequency. The solid line shows the analytical results (6.3.6); the markers are from the experiments [10]

significant increase in the critical acoustic amplitude from $U_a/U_f = 5$ typical for small oscillation frequencies up to $U_a/U_f = 20$ for high frequencies $\omega L_f/U_f = 10$.

The typical length scale of the parametric instability for propane flames versus the oscillation frequency of the acoustic waves calculated from Eq. (6.3.6) and measured experimentally in [10] is shown in Fig. 6.5. It is seen that with the increase of the oscillation frequency, the domain of parametric instability obtained from Eq. (6.3.6) and from the experiment is shifted to the shorter wavelength perturbations.

Numerical simulations [23, 24] confirmed qualitatively theoretical and experimental results. It was found that the moderate amplitude acoustic oscillations stabilize a plane flame, which normally should be unstable against the DL instability. For higher amplitudes of acoustic oscillations, the flame becomes unstable with the cells oscillating with double acoustic frequency, which is qualitatively consistent with the stability limits in Figs. 6.1 and 6.3.

When the flame is ignited and begins propagating from the closed end of a tube, the initial increase in pressure at the nonstationary stage causes the pressure wave to run forward creating the upstream flow ahead of the flame front. The characteristic time for which a stationary flow is established and pressure is equalized on the order of the time for which the sound passes the characteristic scale of the tube. A stationary flow is established with the upstream flow ahead of the flame in a short time. From the conservation of the mass flux and the boundary condition of zeroth velocity at the solid wall, the velocity the upstream flow is $U_a = (\Theta - 1)U_f$. For rough estimates on the development of flame instabilities we can take this value as an amplitude of acoustic waves generated by the flame under confinement and compare with 2D numerical simulations [25]. For this purpose, we take the same parameters as were used in [25]: the flame expansion coefficient $\Theta = 5$, the amplitude of the sound waves $U_a = 4U_f$, the frequency of the acoustic waves $\omega \approx \pi a_s/L$, where a_s is the sound speed, and L is the tube length. For the flame with the Mach number $\mathrm{M} = U_f/a_s = 10^{-2}$, the Lewis number $Le = 1$, and constant thermal conduction taken in numerical simulations [24], the scaled acoustic frequency $\omega\, L_f/U_f = 1$ shown in Fig. 6.1 corresponds to the tube length $L \approx \pi L_f/\mathrm{M} \approx 300L_f$. For these parameters, Eq. (6.2.13) predicts the stabilization of the DL instability by sound waves. At the same time, the amplitude of the sound waves $U_a = 4U_f$ is close but nevertheless smaller than the critical value needed to excite the parametric instability (Figs. 6.1 and 6.3). Therefore, we do not expect the development of parametric instabilities. Indeed, in the numerical simulations of flames with the above parameters in closed tubes of the lengths $L/L_f = 120, 200, 500$, the observed stabilization of the DL instability was stronger than it could be without the stabilizing influence of the acoustic waves [25]. Still, one should remember that the flame dynamics in a closed chamber is complicated due to involving pre-compression of the combustible mixture and gradual change of the flame parameters. For this reason, the estimate $U_a = (\Theta - 1)U_f$ for the amplitude of the flame-generated acoustic waves is rather crude and provides only qualitative estimates of the flame dynamics.

6.4 Experimental Study of the Darrieus–Landau Instability

A long time ago, it was established theoretically that the planar flame propagating in a gaseous combustible mixture is intrinsically unstable against small perturbations bending the flame front. The Darrieus–Landau instability bending the flame front is inherent to all flames in gaseous mixtures since it is related to the gas expansion in exothermal reactions. On the linear stage of the DL instability, small perturbations of a planar flame front grow exponentially with the instability growth rate depending on the perturbation wave number. Due to the DL instability, a planar flame front becomes spontaneously curved, wrinkled by small perturbations, and on subsequent nonlinear stages acquires a curved or corrugated shape. The saturation of the growth of perturbations is due to nonlinear effects, which lead to the formation of stationary curved flames. Since the surface of wrinkled or corrugated flame grows, the flame consumes more fuel per unit time, and the flame speed undergoes a noticeable amplification.

The outcome of the DL instability at the nonlinear stage may lead to the cellular structure of the freely propagating flames shown in Fig. 6.6, or to a smooth curved shape of the flame. If the cell size is large enough compared to the cutoff wavelength, then the cellular flame in turn may become unstable on a smaller scale. As a result of this secondary DL instability, a fine structure arises on the larger cells.

Although the DL instability has been thoroughly studied theoretically, including the nonlinear stage, the experimental verification of the theoretical results for the growth rate of the DL instability has been obtained after about 50 years after the discovery of the DL instability. The main difficulty for the experimental study of

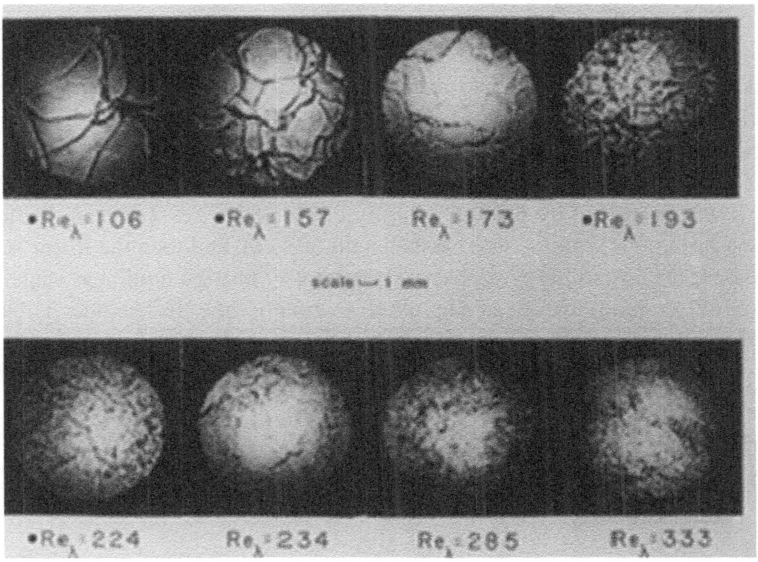

Fig. 6.6 Cellular structure of the flame propagating in a tube

the DL instability is the creation of an initially planar flame and to impose on the flame front a perturbation with a given wavelength. The first direct experimental measurement of the growth rate of the DL instability of a planar laminar premixed flame was reported by Clanet and Searby [16]. All the previous studies of planar flames' instability only measured cellular structures that were already at nonlinear stage. The acoustic stabilization of a freely propagating premixed laminar flame by imposed acoustic field elaborated by Searby [9, 11] has been used to convert an intrinsically unstable flame in an almost perfectly planar laminar propane–air flames propagating from the open top end to the closed bottom of a vertical tube 120 cm long and 10 cm wide prior to observing the growth of the DL instability. In fact, the effect of gravity together with the effect of flame curvature explains why it is possible to experimentally implement stable planar flame fronts. However, the experimental investigations of such flames are not simple since they require an active stabilization [9, 17] and elaborate ignition methods. Heat losses to the walls may seriously complicate the analysis of the results.

The dynamics of a flame propagating from the open top end to the closed bottom of a vertical tube was described in Sect. 6.1. From the map of the flame front stability limits shown in Fig. 6.1, it is seen that one can choose an appropriate amplitude and wave number of acoustic oscillations to be fitted in the stability "widow", where the DL instability for perturbations of all wavelengths is completely suppressed. The stability limits were calculated in [16] using numerical solution of Eq. (6.2.2) for the conditions used in experiments [16]: the lean propane–air flame with velocity $U_f = 19\,\mathrm{cm}/s$, acoustic wave of frequency 230 Hz, and amplitude in the range $3.1 < U_a/U_f < 3.6$. The stability limits in Fig. 6.1 are qualitatively very similar to those used in [17]. For the lean propane–air flame used in [17], the characteristic time of the DL instability is 3–20 ms for the wavelength $\lambda = 2\,\text{cm}$. The flame ignited at the top of the vertical tube initially takes the form of a smooth curved front, resulting from the Darrieus–Landau hydrodynamical instability. The imposed acoustic field almost fully stabilized the flame at about half-length from the top of the tube and was removed in a short time of about 1 ms. Simultaneously, the upstream gas flow was perturbed by placing an array of parallel wires 2 mm in diameter to excite perturbations with the fixed wavelength of 2 cm, and then the linear stage of the growing DL instability was observed. After 140 ms, the nonlinear stage of the DL instability saturation indicated by the formation of the cells became visible. The experimentally measured growth rate of the DL instability as a function of laminar flame speed for wavelength 2 cm was in excellent agreement with the theoretical predictions Eq. (6.2.2).

Figure 6.7 shows the photo of luminous emission from the flame recorded by a high-speed camera during the growth of the DL instability [17]. Figure 6.7 adopted from [17] shows the experimentally measured growth rate of the DL instability as a

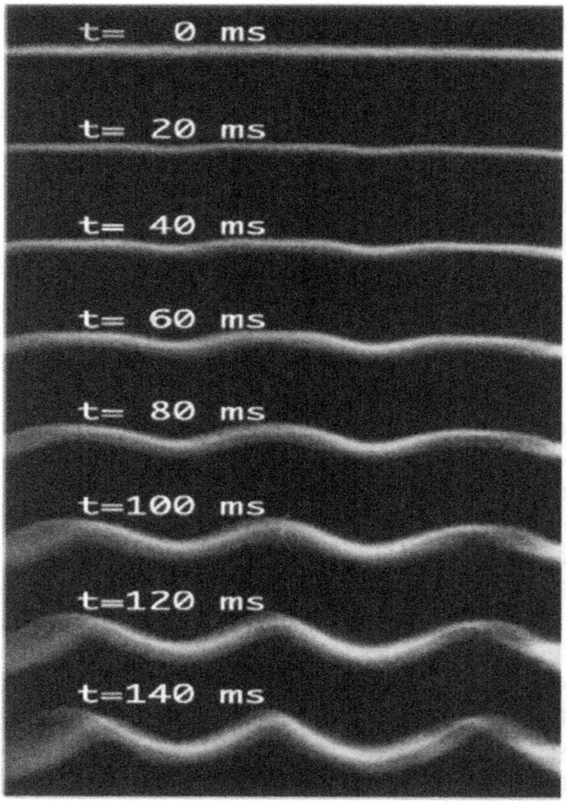

Fig. 6.7 Images of the growth of the DL instability obtained by Clanet and Searby [17]

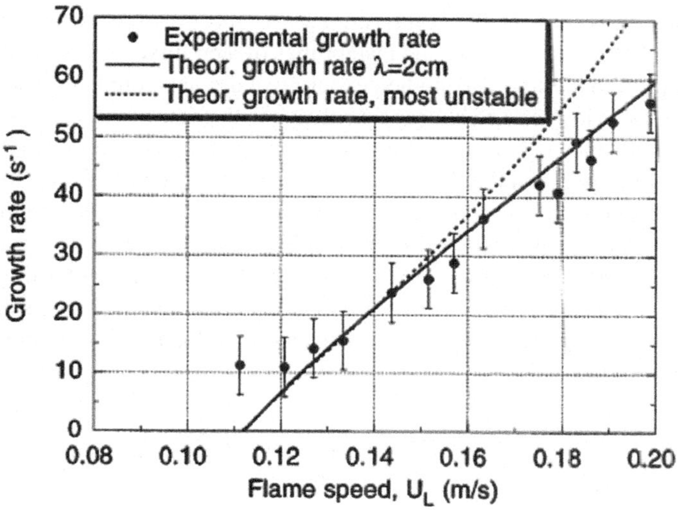

Fig. 6.8 Comparison of experimental [17] and theoretical growth rates adopted from [17]

function of laminar flame speed together with theoretical values of the growth rates for the mode $\lambda = 2\,\text{cm}$ and the most unstable mode (Fig. 6.8).

Since the experimental technique [17] was limited to the relatively slow flames, further progress in the experimental verification of the DL instability was associated with the expansion of the flame velocities and verification of theoretical predictions regarding the dependence of the growth rate of instability on the perturbation wavelength and the existence of the perturbation with a wavelength having the highest growth rate [26, 27]. For this purpose, the authors of [26, 27] used a special construction of burner to produce a 2D inverted "V" premixed flame and an electrostatic deflection system which can impose a periodic sinusoidal 2D wrinkle on the flame front with a certain wavelength. The measured [26, 27] growth rate of the Darrieus–Landau instability for propane flames as a function of wave number is well consistent with the theoretical formula (7.2.43) at least within the experimental uncertainty of about 10%.

6.5 Flame–Acoustic Interaction: Thermoacoustic Instabilities

In [28], thermoacoustic instabilities of a flame were studied experimentally, and their nature was analyzed theoretically. Thermoacoustic instabilities result from a coupling between local heat release fluctuations in the flame front and pressure oscillations associated with the sound waves in the combustion chamber. A strong, resonant amplification of instabilities can occur if feedback is formed between fluctuations in the heat release rate of the combustion wave and pressure fluctuations in the sound wave. In [28], thermoacoustic instabilities of a premixed spray combustion of decane droplets in air was also studied experimentally and compared with that of a premixed flame. The measured linear growth rate in the case of the spray combustion was found to be on order of magnitude higher than that for the premixed flame. The strong difference between the linear growth rate of acoustic instabilities for gaseous and spray combustion can apparently be explained by the fact that in the case of the gaseous flame fluctuations of the heat release rate of the combustion wave is caused by pressure fluctuations, but in the case of the spray combustion the heat release rate is caused by the spray particles velocity fluctuations in the sound wave, not by adiabatic temperature variations. The droplet velocity v is determined by the equation of motion:

$$\frac{dv}{dt} = \left[U_f - v + u_s \cos(\omega_a t)\right]/\tau_{st}. \tag{6.5.1}$$

Here, U_f is the velocity of upstream flow (in the coordinate system co-moving with the flame front), u_s is the amplitude of the sound wave, $\tau_{st} = 10^{-3} s m_{dr}/3\pi\rho\nu d \approx (1-3)\,\mu\text{m}$ is the Stokes time, ν is the kinematic viscosity of gaseous phase of density ρ, and $m_{dr} = (\pi/6)\rho_{dr} d^3$ is the droplet mass, $\rho_{dr} >> \rho$. For the size of spray particles $d \approx (1-3)\mu\text{m}$, the Stokes time is by order $\tau_{st} = 10^{-3}s$, which is comparable with the period of acoustic modes with the frequencies in the range 700–750 Hz. For $\tau_{st} \propto \omega_a^{-1}$, the velocity fluctuations of liquid droplets produce oscillations in the flux of fuel entering into the reaction zone and modulation of the heat release with the frequency $\approx \omega_a$. Since the wavelength of the acoustic waves is much larger than the tube diameter, acoustic waves can be considered as the planar pressure waves. A resonant character of the thermoacoustic instability is defined by the condition $\omega_a \tau_t = 1$, where $\tau_t = \chi / U_f^2$ is a characteristic time of changing the flame structure. Since $\omega_a \propto a_s/L$ is gradually changing with changing the distance L from the flame to the closed end, the condition $\omega_a \tau_t = 1$ will be always reached. For the spray combustion studied in [28], the parametric instability was found to be much stronger than the primary instability leading to pressure peaks about the order of magnitude higher.

Seminal works done by Searby and his group [9–11, 17, 26–28] have been extended and refined by many authors and included as experimental studies of the DL instability and the influence of acoustic waves for laboratory conditions and application to engineering problems. Research on such thermoacoustic instability has been conducted for a long time in various ways; experiments on a flame propagating in a tube have been used due to their simplicity. The thermoacoustic instabilities manifest themselves as pressure fluctuations. The two-way interaction between the flame and the acoustics can lead to strong self-sustained severe sound emissions, which may have disastrous effects on the components of an engine, for example, by causing vibrations and structural fatigue. Interactions between combustion and acoustic oscillations are of fundamental importance for industrial applications such as combustion in gas turbines and rocket engines. A variety of thermoacoustic combustion instabilities limit the operational ranges of gas turbines and rocket engines. For gas turbines, the frequency range can roughly be specified to 100–500 Hz. At these frequencies, the acoustic field inside the combustor can be approximated by planar acoustic waves. Combustion instabilities in gas turbines, rocket engines are much more complex compared to those investigated in laboratory experiments. In real engines, there are specific unstable modes such as azimuthal instabilities in gas turbines and transverse modes in rocket chambers, not to mention that engine combustion is almost always turbulent. In order to control these instabilities, first of all it is necessary to understand the condition for the Rayleigh criterion to be satisfied [29], which requires that pressure and heat release variations must be in phase for the acoustic instability to develop. A flame captures the pressure of a sound wave at a specific position, amplifies and phase-changes the signals, and emits them at a different location. In this way, the heat source establishes a feedback cycle which may become unstable and lead to a self-oscillatory state. All this leads to a new and complex physics, whose understanding required more detailed experimental studies and advanced numerical simulations. Some progress was achieved in the last years using advanced simulations, in particular, LES methods [30–34] and DNS simulations with detailed chemical models for large hydrocarbon fuels [35]. In [31], it was concluded that because the high complexity of these systems, theory plays an important role for understanding both experimental and simulation results and for identification of mitigation techniques.

Further progress in laboratory experimental studies of the DL and thermoacoustic instabilities was performed using different constructions: flame propagating downward in tubes [20, 36–40], anchored ducted inverted V-flame [26, 41], 2D flames in Hele-Shaw cell of two closely spaced parallel plates [36, 42–49]. Many experimental studies were focused on different configurations such as spherical bombs [50–52]. Here, a mixture of fuel and oxidizer is ignited by a spark that causes a radially expanding deflagration front. In order to ensure a negligible increase of the pressure in the combustor during the combustion process, the volume of the combustion chamber has to be large compared to the one occupied by the burning mixture. Advantages of such constructions is the relatively simple adjustment of experimental parameters such as pressure and temperature of the mixture, the absence of initial turbulence in the unburned fluid, and a well-defined and uniformly distributed flame stretch. In particular, the latter property qualifies these configurations for measurements of Markstein numbers and laminar flame speeds [53].

6.6 Applicability of Analytical Models

Since the pioneering experimental work of Searby [10], several studies have been performed to achieve a better understanding of the thermo-acoustic instability in downward propagating flames. Numerous works were devoted to the further improvement of experimental studies of a flame interaction with sound waves, the primary and parametric instability of premixed flames [22, 37–40, 54, 55], and comparisons of experimental measurements with predictions of theoretical models, but any quantitative comparison with experiments was not reported. It should be emphasized that in order to obtain analytical solutions, an oversimplified one-step Arrhenius model for chemical reactions is inevitably used. A one-step Arrhenius kinetics model in many cases makes it possible to solve the problem in question in explicit analytical form and gives at least a qualitatively correct description of stationary processes. However, the range of validity of a one-step chemical model is very limited, and its applicability to the understanding and modeling of combustion phenomena must be examined with great care. The use of a one-step chemical model often gives a rather good qualitative description of stationary processes, and sometimes useful functional dependencies on the problem parameters. However, in the case of nonstationary processes, such as flame instability, the interaction of a flame with acoustic waves, and ignition, analytical models can at best give only a qualitative description of the processes, if at all possible. Many important features of combustion cannot be explained without considering a detailed mechanism of chemical reactions. A one-step model does not reproduce even two distinct stages of the combustion: the induction stage and the exothermal one. Another fundamental difference between a one-step Arrhenius model and the chain-branching reaction is an avalanche multiplication of intermediate species, atoms, and radicals, and their repetitive reactions constituting reaction chains. To study unsteady combustion processes, such as flame acceleration, instabilities, and ignition, which are accompanied by compression and shock waves, it is necessary to account correctly such processes as the flame velocity-pressure dependence, the induction period and the period of exothermal reaction, which determine the chemical time scales competing with the transport time scales in establishing zone of energy release, and which in turn determines the flame evolution. The limited applicability of analytical models for quantitative, and sometimes even for a qualitative description of unsteady combustion processes, was obvious at the very beginning of theoretical studies of combustion, when as the first attempt to mimic chain-branching character of the reaction Zel'dovich proposed to use two-step models [55–58]. Two-step models contain only reactions of hydrocarbon oxidation, which are most important for the heat release. Later it became clear that only detailed chemical models can give a quantitatively correct understanding of unsteady processes.

A similar situation occurred when Clavin and Searby [59] found that the experimentally measured in [60] frequency dependence of the unsteady 1D premixed flame to the acoustic pressure wave does not agree with the predictions obtained from analytical models that used the Zel'dovich–Frank-Kamenetskii model with a one-step high activation energy Arrhenius kinetics [12, 16, 18, 19, 28] as well as with numerical simulations, which also used a one-step Arrhenius model [61–64]. In this approximation, a premixed flame response to the pressure oscillations is mainly due to the temperature change associated with adiabatic compressions. These models predict that the flame response increases as the root square of the frequency with the burning velocity and decreases with the burnt temperature, but the experimental results [60] were found to be independent of frequency. In fact, already earlier experiments [10, 28] indicated that the acoustic instability of flames propagating in tubes is more sensitive to the velocity coupling and the corresponding changing in the geometry of the curved flame front. Discrepancies between the experimental results and predictions of analytical models led to the conclusion that the standard analytical models do not provide an adequate description of the unsteady response of real flames and therefore must be modified with an account of more realistic chemical kinetics. It should be noted that up to the time of experiments [60], there was only a qualitative agreement of theoretical models and experiments, but the models have never been validated by a detailed comparison with experiments. The combined analysis [59] with experiments [60] are the first accurate comparisons of the experimentally measured frequency dependence of a premixed flame response with the prediction of analytical models [12, 18, 28]. The numerical study [65] has been probably the only attempt to consider the premixed flame response to a fast pressure drop using a more realistic two-step model, and considering flames of different reaction order. Clavin and Searby [59] used a slightly modified two-step Zel'dovich–Liñan model [5758] to develop the 1D analytical model of the lean methane–air flame– response to acoustic waves.

Using detailed chemistry (GRI 3.0), Schmidt and Jiménez [66] performed numerical studies of the unsteady response of the 1D premixed methane–air flame to acoustic pressure waves for the frequencies below and above the inverse flame transit time and compared the results of their simulations with simulations performed with one-step and two-step chemical models and with experimental results of [61]. They found that the results of simulations with detailed chemical models agree satisfactory well with the experimental results for the amplitudes and for the fuel consumption rate response. However, the simulations did not reproduce either the experimental trend for the lower frequencies or the negative phase shift for the higher frequencies found in [60]. The reason for these discrepancies between the experimental results and the numerical simulations is not totally clear. In subsequent studies, the experimental results for hydrogen flames, in which chemical kinetics is much simpler than that of methane–air, were compared with the simulations employing detailed chemical kinetics. Relatively good agreement was found between experiments and numerical simulations similar to that found in [66], and a similar disagreement with the difference between the heat release and OH* chemiluminescence fluctuations. Therefore, it was concluded that the use of OH* emission as a diagnostic tool can overestimate the source term in investigations of thermoacoustic instability of hydrogen combustion. In any case, the experiments [58, 65] confirmed that models using a one-step chemical kinetics cannot reproduce the experimentally observed trends of hydrogen and methane–air flames response to pressure fluctuations in sound waves.

References

1. E. Mallard, H.L. Le Chatelier, Combustion des mélanges gazeux explosifs. Ann. Mines Paris, Series **8**(4), 274–388 (1883)
2. M. Gonzalez, R. Borghi, A. Saouab, Interaction of a flame front with its self-generated flow in an enclosure: the 'tulip flame' phenomenon. Combust. Flame **88**, 201–220 (1992)
3. B. N'konga, G. Fernandez, H. Guillard, B. Larroutrou, Numerical investigations of the tulip flame instability. Comparisons with experimental results. Combust. Sci. Technol. **87**, 68–89 (1993)
4. J.L. McGrevy, M. Matalon, Hydrodynamic instability of a premixed flame under confinement. Combust. Sci. Technol. **100**, 75–94 (1994)
5. R.W. Schcefer, F.M. Namazian, J. Kelly, M. Perrin, Effect of confinement on bluff-body burner recirculation zone characteristics and flame stability. Combust. Sci. Technol. **120**, 185–211 (1996)
6. V.V. Bychkov, M.A. Liberman, Stability of a flame in a closed chamber. Phys. Rev. Lett. **78**, 1371–1734 (1997)
7. M.A. Liberman, V.V. Bychkov, S.M. Golberg, L.E. Eriksson, Numerical study of curved flames under confinement. Combust. Sci. Technol. **130**, 1–31 (1998)
8. G. H. Markstein, Nonsteady Flame Propagation ~Pergamon, Oxford, 1964.
9. G. Searby, J. Quinard, Direct and indirect measurements of Markstein numbers of premixed flames. Combust. Flame **82**, 298–311 (1990)
10. G. Searby, Acoustic instability in premixed flames. Combust. Sci. Technol. **81**, 221–231 (1992)
11. G. Searby, D. Rochwerger, A parametric acoustic instability in premixed flames. J. Fluid Mech. **231**, 529–543 (1991)
12. P. Pelce, D. Rochwerger, Vibratory instability of cellular flames propagating in tubes. J. Fluid Mech. **239**, 293–307 (1992)
13. T. Poinsot, B. Yip, D. Veynante, A. Trouve, J.M. Sarmaniego, S. Candel, Active control: an investigation method for combustion instabilities. J. Phys. (Paris) **2**, 1331–1357 (1992)
14. P. Clavin, Premixed combustion and gasdynamics. Ann. Rev. Fluid Mech. **26**, 321–352 (1994)
15. O.Y. Travnikov, V.V. Bychkov, M.A. Liberman, Numerical studies of flames in wide tubes: stability limits of curved stationary flames. Phys. Rev. E **61**, 468–474 (2000)
16. V.V. Bychkov, M.A. Liberman, Dynamics and stability of premixed flames. Phys. Reports **325**, 115–237 (2000)
17. C. Clanet, G. Searby, First experimental study of the Darrieus-Landau instability. Phys. Rev. Lett. **27**, 3867–3870 (1998)
18. P. Clavin, P. Pelce, L. He, One-dimensional vibratory instability of planar flames propagating in tubes. J. Fluid Mech. **216**, 299–322 (1990)
19. V.V. Bychkov, Analytical scalings for flame interaction with sound waves. Phys. Fluids **11**, 3168–3173 (1999)
20. P. Pelce, P. Clavin, Influence of hydrodynamics and diffusion upon the stability limits of laminar premixed flames. J. Fluid Mech. **124**, 219–237 (1982)
21. L.D. Landau, E.M. Lifshitz, *Mechanics* (Pergamon, Oxford, 1960)
22. R.C. Aldredge, Saffman-Taylor Influence On Flame Propagation In Thermo-acoustically Excited Flow **177**, 53–73 (2004)
23. B. Denet, A. Toma, Numerical study of premixed flames parametric acoustic instability. Combust. Sci. Tech. **109**, 23–33 (1995)
24. M. Gonzalez, Acoustic instability of a premixed flame propagating in a tube. Combust. Flame **107**, 245–259 (1996)
25. M.A. Liberman, V.V. Bychkov, S.M. Golberg, L.E. Eriksson, Numerical study of curved flames under confinement. Combust. Sci. Tech. **130**, 1–31 (1998)
26. J.-M. Truffaut, G. Searby, Experimental study of the darrieus-landau instability on an inverted-V flame, and measurement of the markstein number. Combust. Sci. Technol. **149**, 35–52 (1999)
27. G. Searby, J.-M. Truffaut, G. Joulin, Comparison of experiments and a nonlinear model equation for spatially developing flame instability. Phys. Fluids **13**, 3270–3276 (2001)

28. C. Clanet, G. Searby, P. Clavin, Primary acoustic instability of flames propagating in tubes: cases of spray and premixed gas combustion. J. Fluid Mech. **385**, 157–197 (1999)
29. J. Rayleigh, The explanation of certain acoustical phenomena. Nature **18**, 319–321 (1878)
30. T. Lieuwen, Modeling premixed combustion—acoustic wave interactions: a review. J. Propul. Power **19**, 765–781 (2003)
31. T. Poinsot, Prediction and control of combustion instabilities in real engines. Proc. Combust. Inst. **36**, 1–28 (2017)
32. S. Bomberg, T. Emmert, W. Polifke, Thermal versus acoustic response of velocity sensitive premixed flames. Proc. Combust. Inst. **35**, 3185–3192 (2015)
33. T. Emmert, S. Bomberg, S. Jaensch, W. Polifke, Acoustic and intrinsic thermoacoustic modes of a premixed combustor. Proc. Combust. Inst. **36**, 3835–3842 (2017)
34. S. Candel, Combustion dynamics and control: progress and challenges. Proc. Combust. Institute. **29**, 1–28 (2002)
35. G. Beardsell, G. Blanquart, Impact of pressure fluctuations on the dynamics of laminar premixed flames. Proc. Combust. Inst. **37**, 1895–1902 (2019)
36. C. Almarcha, B. Denet, J. Quinard, Premixed flames propagating freely in tubes. Combust. Flame **162**, 1225–1233 (2015)
37. J. Yanez, M. Kuznetsov, J. Grune, Flame instability of lean hydrogen-air mixtures in a smooth open-ended vertical channel. Combust. Flame **162**, 2830–2839 (2015)
38. R.C. Aldredge, N.J. Killingsworth, Experimental evaluation of Markstein-number influence on thermoacoustic instability. Combust. Flame **137**, 178–197 (2004)
39. S.H. Yoon, T.J. Noh, O. Fujita, Onset mechanism of primary acoustic instability in downward-propagating flames. Combust. Flame **170**, 1–11 (2016)
40. S.H. Yoon, L. Hu, O. Fujita, Experimental observation of pulsating instability under acoustic field in downward-propagating flames at large Lewis number. Combust. Flame **188**, 1–4 (2018)
41. C.M. Luzzato, A.S. Morgans, The effect of a laminar moving flame front on thermoacoustic oscillations of an anchored ducted V-flame. Combust. Sci. Technol. **187**, 410–427 (2015)
42. C. Almarcha, J. Quinard, B. Denet, E. Al-Sarraf, J. M. Laugier, E. Villermaux, Experimental two dimensional cellular flames. Phys. Fluids **27**, 091110 (2015)
43. E. Al Sarraf, C. Almarcha, J. Quinard, B. Radisson, B. Denet, P. Garcia-Ybarra, Darrieus–Landau instability and Markstein numbers of premixed flames in a Hele-Shaw cell, Proc. Combust. Inst. **37**, 1783–1789 (2019)
44. E. Al Sarraf, C. Almarcha, J. Quinard, B. Radisson, B. Denet, quantitative analysis of flame instabilities in a hele-shaw burner. Flow Turbul. Combust. **101**, 851–868 (2018)
45. F. Veiga-López, D. Martínez-Ruiz, E. Fernández-Tarrazo, M. Sánchez-Sanz, Experimental analysis of oscillatory premixed flames in a Hele-Shaw cell propagating towards a closed end. Combust. Flame **201**, 1–11 (2019)
46. D. Fernandez-Galisteo, V.N. Kurdyumov, P.D. Ronney, Analysis of premixed flame propagation between two closely-spaced parallel plates. Combust. Flame **190**, 133–145 (2018)
47. D. Fernandez-Galisteo, V.N. Kurdyumov, Impact of the gravity field on stability of premixed flames propagating between two closely spaced parallel plates. Proc. Combust. Institute **37**, 1937–1943 (2019)
48. M.M. Alexeev, O. Yu Semenov, S.E. Yakush, Experimental study on cellular premixed propane flames in a narrow gap between parallel plates. Combust. Sci. Technol. **191**, 1256–1275 (2019)
49. B. Radisson, J. Piketty-Moine, C. Almarcha, Coupling of vibro-acoustic waves with premixed flame. Phys. Rev. Fluids **4**, 121201(1–9) (2019)
50. D. Bradley, P.H. Gaskell, X.J. Gu, Burning velocities, markstein lengths, and flame quenching for spherical methane-air flames: a computational study. Combust. Flame **104**, 176–198 (1996)
51. A.S. Al-Shahrany, D. Bradley, M. Lawes, R. Woolley, Darrieus-Landau and thermo-acoustic instabilities in closed vessel explosions. Combust. Sci. Tech. **178**, 1771–1802 (2006)
52. D. Bradley, M. Lawes, R. Mumby, P. Ahmed, The stability of laminar explosion flames. Proc. Combust. Institute **37**, 1807–1813 (2019)
53. A.K. Dubey, Y. Koyama, N. Hashimoto, O. Fujita, Experimental and theoretical study of secondary acoustic instability of downward propagating flames: Higher modes and growth rates. Combust. Flame **205**, 316–326 (2019)

54. A.K. Dubey, Y. Koyama, S.H. Yoon, N. Hashimoto, O. Fujita, Range of "complete" instability of flat flames propagating downward in the acoustic field in combustion tube: Lewis number effect. Combust. Flame **216**, 326–337 (2020)
55. Ya. B. Zel'dovich, On the Theory of flame propagation, Zhurnal Fizicheskoi Khimii **22**, 27–49 (1948) (in Russian). Translated: NACA, Technical memorandum 1282, Washington, 1951
56. Ya. B. Zel'dovich, Chain reactions in hot flames—an approximate theory for flame velocity. Kinetika I Katalis **2**, 305–313 (1961)
57. A. Liñan, Tehnical Report no. 1. Afosr contract no. e00ar68-0031, INTA Madrid (1971)
58. T. Mantel, F.N. Egolfopoulos, C.T. Bowman, A new methodology to determine kinetic parameters for one- and two-step chemical models. Center for Turbulence Research, Proceedings of the Summer Program (1996), pp. 149–166
59. P. Clavin, G. Searby, Unsteady response of chain-branching premixed-flames to pressure waves. Combust. Theor. Model. **12**, 545–567 (2008)
60. A. Wangher, G. Searby, J. Quinard, Experimental investigation of the unsteady response of a flame front to pressure waves. Combust. Flame **154**, 310–318 (2008)
61. A.C. McIntosh, The linearized response of the mass burning rate of a premixed flame to rapid pressure change. Combust. Sci. Technol. **91**, 329–346 (1993)
62. O.J. Teerling, A.C. McIntosh, J. Brindley, Pressure wave excitation of natural flame frequencies. Combust. Theory Model **11**, 147–164 (2007)
63. A. Petchenko, V. Bychkov, V. Akkerman, L.E. Eriksson, Flame–sound interaction in tubes with nonslip walls. Combust. Flame **149**, 418–434 (2007)
64. H. Schmidt, C. Jiménez, Numerical study of the direct pressure effect of acoustic waves in planar premixed flames. Combust. Flame **157**, 1610–1619 (2010)
65. C. Jiménez, J. Quinard, J. Graña-Otero, H. Schmidt, G. Searby, Unsteady response of hydrogen and methane flames to pressure waves. Combust. Flame **159**, 1894–1908 (2012)
66. E. Motheau, F. Nicoud, T. Poinsot, Mixed acoustic–entropy combustion instabilities in gas turbines. J. Fluid Mech. **749**, 542–576 (2014)

Chapter 7
Interaction of Flames with Weak Shocks

7.1 Linear Theory of Flame–Shock Interaction

A flame under confinement (flames propagating in a tube or combustion chamber) is often subject to processes leading to the flame acceleration. The accelerating flame acts as a semitransparent piston producing pressure waves. The pressure wave ahead of the flame may steepen into a weak shock, which after reflection from the end wall, returns back toward the flame. These weak shock waves influence the flame shape and velocity, which may lead under certain conditions to considerable flame acceleration, or trigger additional instability of the flame. In addition, the shock waves heat and compress the unburned mixture, which can even cause an explosion or facilitate triggering detonation ahead of the flame front. If the typical time scale of flame dynamics in a closed chamber is larger than the time needed for a sound to traverse the chamber, then the pressure waves can be considered as acoustic waves. The flame-acoustic wave interaction was considered in Chap. 6. However, for certain conditions, the characteristic time of a flame propagation can be sufficiently large to allow the pressure wave to steepen into a weak shock but less than the time during which the shock wave is attenuated significantly before it the reflects from the wall and returns to the flame front. When a shock propagates in a gas mixture, it involves the gas behind the shock front into the motion together with the shock, so that the gas that was initially at rest starts moving with some velocity ΔU in the direction of shock propagation. If the shock velocity generated by a flame is D_{Sh}, and the sound speed is a_s, then for a weak shock $(D_{Sh} - c_s)/D_{Sh} = \varepsilon \ll 1$. The shock wave involves the gas flowing behind the shock with velocity $\Delta U = \varepsilon D_{sh} \ll D_{sh}$. The changes in density and pressure behind a weak shock are also small: $\Delta\rho/\rho \propto \Delta P/P \propto \varepsilon \ll 1$ in comparison with the initial density and pressure of the gas mixture, while in the laboratory flames with the expansion coefficients $\Theta = 5$–10, the density variations in a flame are about the density itself. For a weak shock, a small change in pressure behind the shock wave can be estimated in the same way as the pressure change in a sound wave, $\Delta P \approx \rho c_s \Delta U$. Notice, that the pressure increase in a weak shock $\Delta P = \rho_f c_f \Delta U$ exceeds considerably the pressure change at the flame front, $\propto$

M. A. Liberman, *Combustion Physics*,
https://doi.org/10.1007/978-3-030-85139-2_7

$\rho_f U_f^2$, while the density change in the weak shock $\Delta\rho/\rho \propto \varepsilon \ll 1$ is negligible compared with the density change in the flame, $(\rho_f - \rho_b)/\rho_f = 1 - \Theta^{-1} \approx 1$. In what follows, it is convenient to characterize the intensity of the shock interacting with the flame by the ratio $\Delta U/U_f$, where U_f is a normal laminar flame velocity, and we will neglect by the change of density in the weak shocks, which are small as $\Delta U/c_f$.

Interaction of a curved flame with a weak shock is one of the classical problems in combustion science. First experiments on the flame–shock interaction by Markstein [1] demonstrated that shocks colliding with the flame may lead to a significant increase in the flame surface area, can increase the flame velocity, and sometimes result in flame turbulization. For the shock intensity used in the experiments [1] a head-on collision of the flame and a shock caused reversion of the flame shape, when the initially convex parts of the flame front became concave and vice versa. Similar effects have been observed in numerical experiments on the shock collision with cylindrical flames [2, 3] and in numerical simulations of curved flames under confinement [4]. The first theoretical analysis of flame–shock interaction has been undertaken by Markstein [1], where the linear response of a slightly perturbed almost planar flame to a passing shock has been considered on the basis of a phenomenological linear equation for a flame front. In particular, it was concluded in [1] that any head-on collision of a shock with a curved flame front causes inversion of the initial flame shape. On the contrary, numerical simulations of flame dynamics in closed tubes [4] have shown that flame inversion is only one of the possible outcomes of the flame–shock collision. It was found that the shock–flame collisions may lead to stabilization of the flame front without inversion, and sometimes flame inversion happens with stabilization or additional destabilization of the flame front depending on the shock intensity and on the flame parameters.

The flame velocity is changing during the flame collision with a weak shock, which means that the flame accelerates during the collision. We consider stabilization of the DL instability caused by the flame–shock collision [5–7] on the basis of the linear theory of the flame front in a gravitational field (Sect. 5.2). Figure 7.1 shows the density distribution before and after the collision of a flame with the expansion coefficient $\Theta = 8$ with a weak shock wave of intensity $\Delta U = 3.8U_f$ obtained in numerical calculations [7].

We consider a head-on collision of a perturbed flame front and a weak shock in a tube with open ends. The flame in Fig. 7.1a propagates to the left and a weak shock moves from the left to the right. The density of the unburned mixture ahead of the flame is ρ_f, and the density of the combustion products $\rho_b = \rho_f/\Theta$. The temperature and pressure behind the flame in the isobaric approximation are: $T_b = \Theta T_f$ and $P_b = P_f - (\Theta - 1)\rho_f U_f^2$. In the reference frame co-moving with the flame front, the fresh mixture flows into the flame front with velocity U_f and the burned products flow away with velocity ΘU_f. The sound speeds in the unburned mixture and in the burned products are: c_f and $c_b = c_f\sqrt{\Theta}$. The weak shock compresses the unburned mixture slightly and increases its velocity from U_f to $U_f + \Delta U_0$ as it is seen in Fig. 7.1a. After the flame–shock collision, a rarefaction wave propagates away from

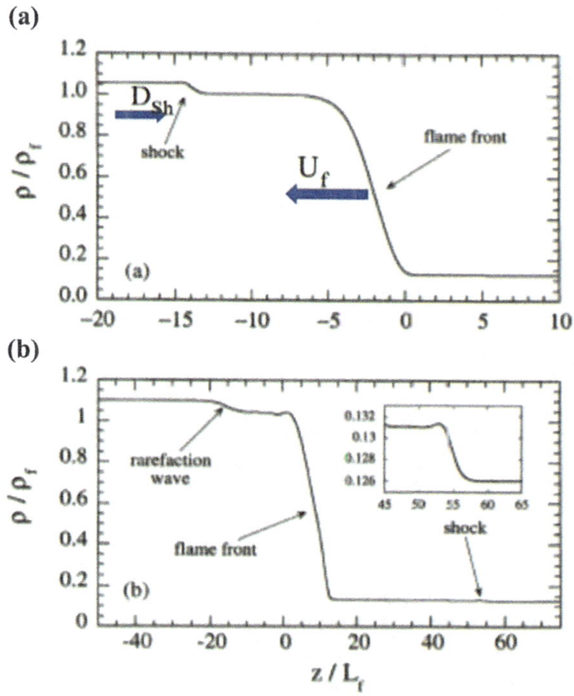

Fig. 7.1 Density distributions for the flame–shock collision obtained in numerical simulations [7]: **a** before the collision, **b** after the collision. The expansion coefficient of the flame is $\Theta = 8$, the shock intensity is $\Delta U = 3.8 U_f$. The inset in figure (**b**) shows the secondary shock formed after the collision; from [7]

the flame front to the left, while in the burned gas the secondary weak shock shown in Fig. 7.1b is formed.

Since properties of a weak shock are similar to that of a sound wave, the velocity change in a weak shock can be calculated similar to the classical problem of the change in the amplitude of a sound wave which crosses a surface separating two gases with different sound speeds [8]. The pressure in the fresh mixture just ahead of the flame front consists of the sum of the initial pressure P_f, additional pressure caused by the shock $\Delta P_0 = \rho_f c_f \Delta U_0$, and the negative contribution to the pressure due to the weak rarefaction wave, $\Delta P_{rfw} = \rho_f c_f \Delta U_{rfw}$

$$P_{f1} = P_f + \rho_f c_f \Delta U_0 - \rho_f c_f \Delta U_{rfw}. \quad (7.1.1)$$

The velocity of the fresh mixture at the same point is

$$u_{f1} = U_f + \Delta U_0 + \Delta U_{rfw}. \quad (7.1.2)$$

The velocity of combustion products behind the flame and the secondary shock wave is

$$u_{b1} = \Theta U_f + \Delta U_1. \quad (7.1.3)$$

The pressure increase caused by the secondary shock can be written in the same way as before, $\Delta P_1 = \rho_b c_b \Delta U_1$, thus, the pressure increase in the combustion products is

$$P_{b1} = P_b + \rho_b c_b \Delta U_1. \tag{7.1.4}$$

From the conservation of mass flux and the balance of pressure

$$u_{b1} - u_{f1} = (\Theta - 1)U_f, \tag{7.1.5}$$

$$P_{b1} - P_{f1} = -(\Theta - 1)\rho_f U_f^2, \tag{7.1.6}$$

we obtain the velocity increase, ΔU_1, the intensity of the secondary weak shock expressed via the intensity of the primary shock ΔU_0

$$\Delta U_1 = \frac{2\sqrt{\Theta}}{\sqrt{\Theta} + 1} \Delta U_0. \tag{7.1.7}$$

According to Markstein [1], the collision of the shock with the flame front can be interpreted as a pulse acceleration $g(t) = \Delta U_1 \delta(t)$ in with $t = 0$ chosen as the instant of the flame–shock collision. Similar to Sec. 6.2, where a flame stabilization by acoustic waves was considered, we can use Eq. (6.2.2) for small perturbations of the flame front $f'(t, x)$ of small but finite thickness, and a gravity acceleration $g(t) = \Delta U_1 \delta(t)$

$$A\frac{d^2 f'}{dt^2} + BU_f k\frac{df'}{dt} - CU_f^2 k^2 f' - C_1 \Delta U_1 \delta(t) k f' = 0, \tag{7.1.8}$$

with coefficients A, B, C, C_1 in (7.1.8) given by Eqs. (5.3.18)–(5.3.20).

For simplicity, we consider the case of unity Lewis number and the constant thermal conduction. Then the coefficients in (7.1.8) become

$$A = 1,\ B = \frac{2\Theta}{\Theta + 1}\left(1 + \frac{\Theta \ln \Theta}{\Theta - 1} k L_f\right)$$

$$C = \Theta\frac{\Theta - 1}{\Theta + 1}\left(1 - \frac{k\lambda_c}{2\pi}\right),\ C_1 = \frac{\Theta - 1}{\Theta + 1}. \tag{7.1.9}$$

In the absence of a shock wave, small perturbations at the flame front grow exponentially in time as $f'(t) = f_0' \exp(\sigma_g t)$ with the instability growth rate σ_g given by the formula (5.2.43) $\sigma = \Gamma(\Theta) U_f k\left(1 - \frac{k\lambda_c}{2\pi}\right)$, where $\lambda_c = \frac{\pi L_f(\Theta - 1)}{\Gamma\sqrt{\Theta + 1 - 1/\Theta}}\left(1 + \Theta \ln \Theta \frac{\Theta + 1 + 2\Gamma}{(\Theta - 1)^2}\right)$, and the expansion coefficient, $\Gamma(\Theta) = \frac{\Theta}{\Theta + 1}\left\{\sqrt{\Theta + 1 - 1/\Theta} - 1\right\}$.

After the flame–shock collision, the solution of Eq. (7.1.8) for the perturbations at the flame front consists of the growing mode (label g) and the decaying mode (label d)

$$f' = f_0'(1-A)\exp(\sigma_g t) + A f_0' \exp(-\sigma_d t), \tag{7.1.10}$$

where the relative amplitude of the decaying mode is

$$A = \frac{\Theta - 1}{\Theta + 1}\frac{\Delta U_1 k}{\sigma_g + \sigma_d}, \tag{7.1.11}$$

and the decrement of this mode is

$$\sigma_d = \sigma_g + \frac{2\Theta}{\Theta + 1}\left(1 + \frac{\Theta \ln \Theta}{\Theta - 1} k L_f\right) U_f k. \tag{7.1.12}$$

It should be noted that the result of a flame–shock collision given by Eq. (7.1.10) differs essentially from the well-known Richtmyer–Meshkov (RM) instability [9, 10], which concerns the instability of the contact surface between two gases accelerated by a shock wave. In case of the RM instability, perturbations of the contact surface grow linearly in time, while the result of a flame–shock interaction is a superposition of two exponentially growing and decaying modes. The exponential growth rate is the result of the DL instability inherent to a flame.

According to Eq. (7.1.10), the amplitude of the decaying mode goes to zero after a time interval of about $1/\sigma_d$, and therefore, the outcome of the flame–shock collision depends on the amplitude of the growing mode $f_0'(1-A)$. This amplitude depends on the intensity of the shockwave colliding with the flame, on the flame expansion coefficient Θ, and on the perturbation wavenumber k

$$A = \frac{\Delta U_0}{U_f}\sqrt{\Theta}\left(\sqrt{\Theta} - 1\right)\left[(\Theta + 1)\Gamma\left(1 - \frac{k\lambda_c}{2\pi}\right) + \Theta + \frac{\Theta^2 \ln \Theta}{\Theta - 1} k L_f\right]^{-1}. \tag{7.1.13}$$

If $A = 1$, then the amplitude of the growing mode vanishes, and the flame–shock collision leads to complete suppression of initial perturbations of the flame front. According to the linear theory, the complete suppression of initial perturbations corresponds to the intensity of the primary shock

$$\Delta U_0 = \Delta U_c = \frac{U_f}{\sqrt{\Theta}\left(\sqrt{\Theta} - 1\right)}\left[(\Theta + 1)\Gamma\left(1 - \frac{k\lambda_c}{2\pi}\right) + \Theta + \frac{\Theta^2 \ln \Theta}{\Theta - 1} k L_f\right]. \tag{7.1.14}$$

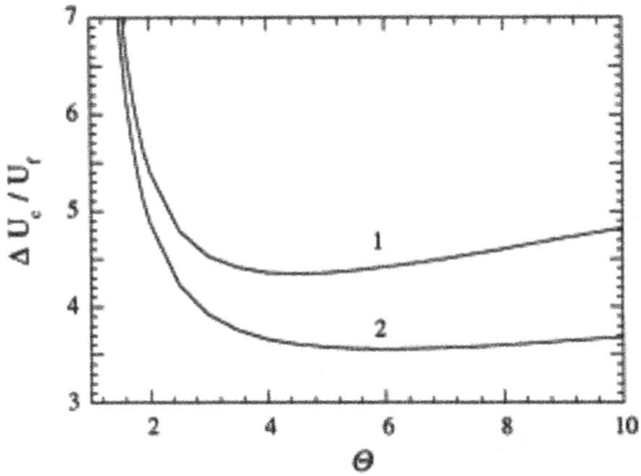

Fig. 7.2 The scaled critical intensity of the shock wave stabilizing the flame perturbations vs the flame expansion coefficient for the long-wavelength perturbations $k\lambda_c \ll 1$ (curve 1) and for the wavelength of perturbation $k\lambda_c = \pi$ (curve 2); from [7]

The dependence of the critical intensity of the shock wave ΔU_c on the flame expansion coefficient Θ obtained in the linear theory is shown in Fig. 7.2 for the long wavelength perturbation $k\lambda_c \ll 1$, and for the perturbations of a relatively short wavelength $k\lambda_c = \pi$. That condition $k\lambda_c = \pi$, corresponds to the critical width D_c of the tube with ideal slip walls such that for the narrower tube $D < D_c = \lambda_c$ all permitted perturbations belong to the stable part of the dispersion Eq. (7.2.43). If the shock intensity is larger than the critical value $\Delta U_0 > \Delta U_c$, then the relative amplitude of the growing mode is negative $(1 - A) < 0$, and this implies that the flame–shock collision leads to the inversion of the flame shape, when the convex parts of the flame front becoming concave and vice versa. It is seen in Fig. 7.2 that realistic $\Theta = 5 - 10$ shock waves of a relatively small intensity can stabilize the DL instability for the case of narrower tubes, and the critical shock intensity depends weakly on the tube width and the perturbation wavenumber. If the shock intensity is smaller than the critical one $\Delta U_0 < \Delta U_c$, then the flame–shock collision leads to a temporary stabilization of the DL instability.

As one can see from Eq. (7.1.10), in this case, the perturbation amplitude may decrease during the time

$$t = \frac{1}{\sigma_g + \sigma_d} \ln\left(\frac{A\sigma_d}{(1 - A)\sigma_g}\right), \tag{7.1.15}$$

but after this time the growth of instability recovers. Using Eqs. (7.1.10) and (7.1.15), we can calculate the efficiency of the shock stabilization for the shock intensities $\Delta U_0 < \Delta U_c$, how much the amplitude of the flame front perturbations was decreased during the temporary stabilization

$$f'/f'_0 = \frac{\sigma_d(1 - A)}{\sigma_g + \sigma_d} \left(\frac{A\sigma_d}{(1 - A)\sigma_g}\right)^{\sigma_g/(\sigma_g+\sigma_d)}. \tag{7.1.16}$$

If the shock intensity is larger than the critical value $\Delta U_0 > \Delta U_c$, then the relative amplitude of the growing mode is negative $(1 - A) < 0$, and this implies that the flame–shock collision leads to the inversion of the flame shape, when the convex parts of the flame front becoming concave and vice versa.

For the case $\Delta U_0 = \Delta U_c$, the amplitude of the growing mode after the flame–shock collision vanishes, and after the time $t \propto 1/\sigma_d$, when the decaying mode vanished, the initial perturbations at the flame front are completely suppressed. Figure 7.3 shows the evolution of the flame perturbations depending on the intensity of colliding shock wave. Temporal evolution of the flame front perturbation amplitudes for various scenarios in Fig. 7.3 after the flame–shock collision is shown in Fig. 7.4.

If a flame propagates in a closed tube, the accelerated flame front produces pressure waves which can steepen into a primary shock wave. The intensity of the primary shock is determined by the flame velocity and the expansion coefficient of the flame: $\Delta P = (\Theta - 1)\rho_f c_f U_f$. It is seen that the effect of the flame–shock collision depends mostly on the expansion coefficient. When the shock reaches the opposite end of the tube, it will be reflected from the closed end of the tube and moves toward the flame with the same intensity. The reflected shock wave slows down or even stops the flow of the unburned mixture that flows toward the far end of the tube and raises the pressure in the flow ahead of the flame up to $P = P_f + 2(\Theta - 1)\rho_f c_f U_f$.

Figure 7.5 shows the relative amplitude of the growing mode after the flame–shock collision versus the flame expansion coefficient for the case of the flame–shock interaction in relatively narrow tubes and for the case of wide tubes obtained in the two-dimensional simulations [7]. In agreement with the linear theory, the numerical simulations [7] show that the perturbed flame front collision with weak shocks generated by the flame in a closed tube leads to inversion of the flame shape. For flames with sufficiently large expansion coefficients $\Theta = 8 - 10$, the inversion may be accompanied by additional destabilization of the flame front in the case of tubes of moderate widths $R = 2R_c$. If the shock intensity generated by the flame is close to the critical value ΔU_c, the linear theory predicts a strong reduction of the DL instability by the weak shocks. Indeed, the numerical simulations [7] show that the first Fourier harmonic perturbations with $k = \pi/2R$ imposed on the flame front were suppressed because of the flame–shock collisions. Instead of the first harmonic, the second Fourier harmonic with the wave number $k = \pi/R$ became dominating at the flame front, though the amplitude of this harmonic also experienced strong pulsations in the flame–shock collisions.

The linear theory predicts that the flame–shock collisions in closed tubes lead to inversion of the flame shape, and the inversion may be accompanied by additional destabilization of the flame in the tubes of moderate widths $R = 2R_c$. The interaction of a flame with weak shock results in stabilization of the DL instability if the absolute value of the growing mode after the flame–shock collision is smaller than it was before the collision $|1 - A| < 1$, while in the opposite case $|1 - A| > 1$ the shock destabilizes the flame front. If the shock intensity is sufficiently large, $\Delta U > 2\Delta U_c$, the

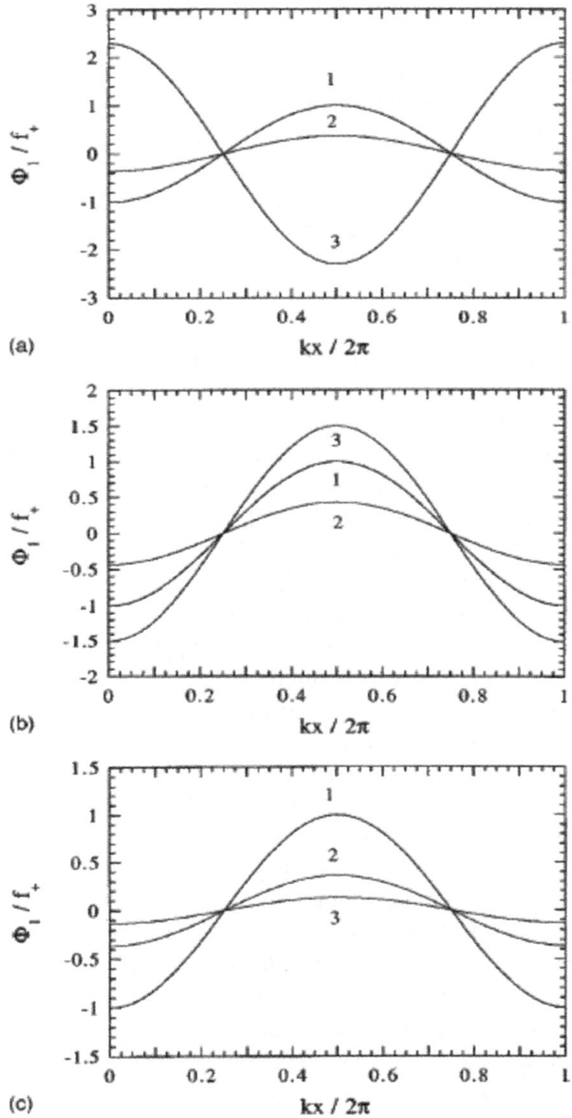

Fig. 7.3 Evolution of the flame front perturbation with wave number $k\lambda_c = \pi$ after the flame collision the shocks of intensity: **a** $\Delta U_0/U_f = 9$, curves 1, 2, 3 correspond instants $U_f t/L_f = 0,\ 0.5,\ 4$; **b** $\Delta U_0/U_f = 3$, curves 1, 2, 3 correspond to $U_f t/L_f = 0,\ 4,\ 15$; **c** $\Delta U_0/U_f = 3.68$, curves 1, 2, 3 correspond to $U_f t/L_f = 0,\ 2,\ 4$

effect of the shock–flame collision is destabilizing, the shock enhances the DL instability. For the smaller shock intensities $\Delta U < 2\Delta U_c$, a temporary flame stabilization is possible.

For the flames with large expansion coefficients ($\Theta \sim 10$), the shock intensity generated by a flame under confinement exceeds the critical value ΔU_c, the amplitude of the growing mode in Eq. (7.1.10) becomes negative, which may lead to the flame shape inversion. The flame shape inversion may be related to the formation of the "tulip" flame; a phenomenon that has been studied in numerous experimental and

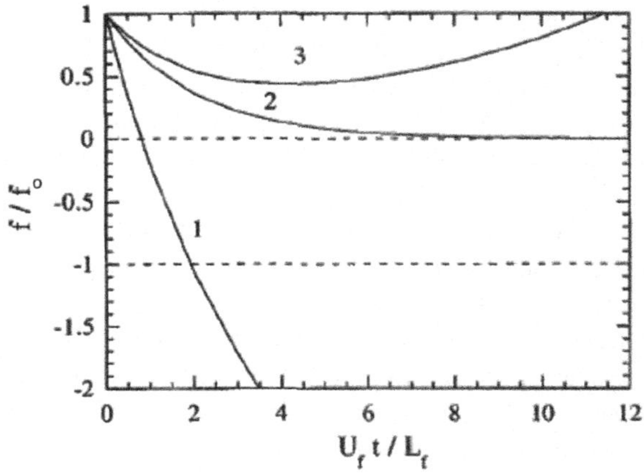

Fig. 7.4 Temporal evolution of the flame front perturbation after the flame–shock collision. Curves 1, 2, 3 correspond to cases (**a**), (**c**), and (**b**) in Fig. 7.3; from [6]

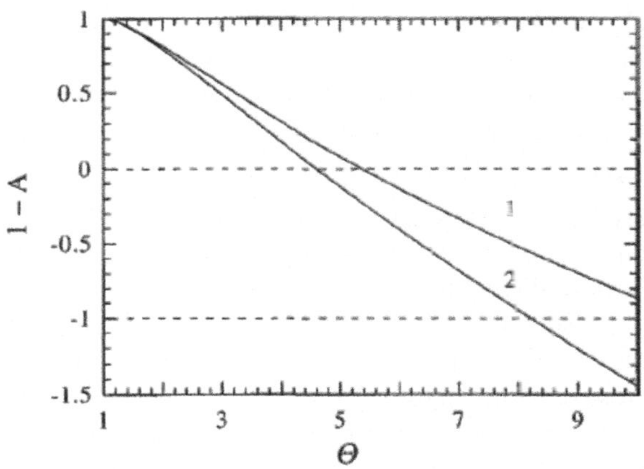

Fig. 7.5 The relative amplitude of the growing mode $(1 - A)$ after the flame–shock collision in a closed tube versus the flame expansion coefficient Θ. Curve 1 corresponds to wide tubes $R \gg R_c$, curve 2 is for $R = 2R_c$; from [6]

numerical studies [12–18], although there is still no consensus on the mechanism of the "tulip" flame formation.

7.2 Nonlinear Effects of Flame–Shock Interactions

Interaction of curved flames with weak shocks of controlled intensity in open tubes has been investigated using 2D numerical simulations [7]. Compared to the configuration of a flame propagating in a closed tube, this setup of the problem eliminates additional complications such as the monotonous pre-compression of the unburned mixture ahead of the flame front, variation of the planar flame velocity, and limited time of the burning process. For the flame propagating in a tube with ideal walls the tube width determines half of the maximal possible wavelength of perturbations $\lambda = 2R/j$, and the corresponding wave numbers $k = \pi j/R$ with $j = 1, 2, 3...$ The fastest growth of small perturbations on the linear stage of the DL instability is for the perturbation wavelength $\lambda = 2\lambda_c$ which corresponds to the tube width $D = 2R = \lambda_c$. Since the linear theory shows that the result of a flame–shock collision depends only slightly on the tube width, the tube width in simulations [7] was taken $R = 2R_c$. In this case, the main parameters influencing the effect of a head-on curved flame collision with a shock are the shock intensity ΔU and the flame expansion coefficient Θ.

The dynamics of the flame and the flame–shock interaction was studied in [7] by solving the two-dimensional, time-dependent, reactive Navier–Stokes equations for compressible flow including the effects of viscosity, thermal conduction, molecular diffusion, for the ideal gas equation of state and a one-step chemical reaction. When a planar stationary flame is established, 2D perturbations are imposed at the flame front, $u_z \to u_z + u_z'(z, x)$ with the initial perturbation of the flame velocity $u_z'(z, x) \propto A_0 u_z(z)$, and the initial dimensionless perturbation amplitude $A_0 = 10^{-3}$. The small perturbations grow because of the DL instability until the flame acquires a curved stationary shape and moves in the direction of negative z with the constant velocity $U_w > U_f$. A typical configuration of the flow and flame colliding with a weak shock is shown in Fig. 7.1. A step-like distribution of the velocity and pressure in the fuel mixture ahead of the flame front is imposed to generate a weak shock. The velocity jump ΔU determining the change in pressure and density of the shock at a surface of discontinuity $\rho_f c_f \Delta U = \Delta P / \sqrt{1 + \frac{\gamma+1}{2\gamma}\frac{\Delta P}{P_f}} \approx \Delta P$, $\Delta P/P \ll 1$ is used as a characteristic of the shock intensity. After the flame–shock collision, the flame front, the flow of the fresh mixture, and the combustion products move in the positive direction z due to the flow created by the shock, as it is seen in Fig. 7.1b, after the flame–shock collision, the flame front, the flow of the fresh mixture, and the combustion products move in the positive direction along z-axis due to the flow created by the shock.

To study the flame evolution resulting from the flame–shock collision, the flame shape is expanded into the series of Fourier harmonics $F(x, t) = \sum F_j(t)\cos(j\pi x/R)$, and then we consider the time history of the amplitudes $F_j(t)$ normalized by the respective amplitude of the first harmonic for the curved stationary flame. Figure 7.6 shows the time history of the first three Fourier harmonic amplitudes

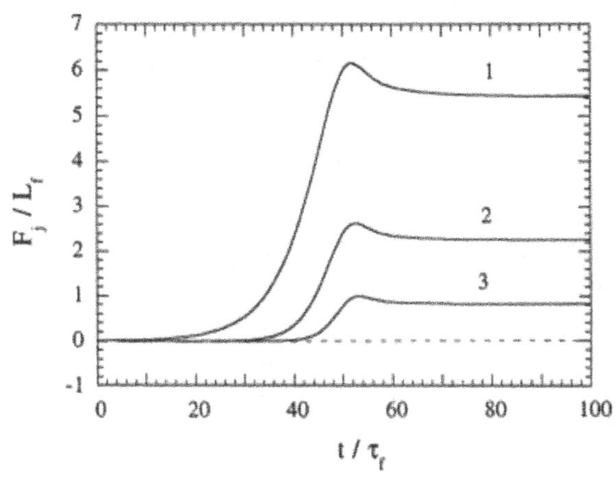

Fig. 7.6 Time history of the scaled amplitudes of the first three Fourier harmonics (curves 1, 2, 3 correspond $j = 1, 2, 3$) with the first harmonic being imposed on an initially planar flame front with the expansion coefficient $\Theta = 8$

$j = 1, 2, 3$ with the first harmonic being imposed on the initially planar flame front with the expansion coefficient $\Theta = 8$. At the beginning, the first Fourier harmonic with $j = 1$ grows exponentially at the flame front because of the DL instability. When the amplitude of the harmonic has reached a considerable value, it induces growth of higher harmonics because of the nonlinear effects at the flame front. Higher harmonics with $j = 2, 3, \ldots$, etc. are stable with respect to the DL instability, therefore, the development of these harmonics slows down the growth of the first unstable harmonic until the instability becomes completely stabilized in the nonlinear stage and the flame front acquires a curved stationary shape.

We denote the amplitude of the first Fourier harmonic corresponding to the curved stationary flame as F_s and the characteristic time associated with a flame front by $\tau_f = L_f/U_f$. An interesting feature of the nonlinear stabilization of the DL instability observed in simulations is that the amplitudes of the Fourier harmonics overshoot temporarily the final values corresponding to the curved stationary flame. The stationary flame shape corresponding to the scenario in Fig. 7.6 is shown in Fig. 7.7, where the flame front propagates to the left.

The stationary shape in Fig. 7.7a can be described as a smooth hump directed toward the fresh mixture and a cusp pointing to the combustion products. To make

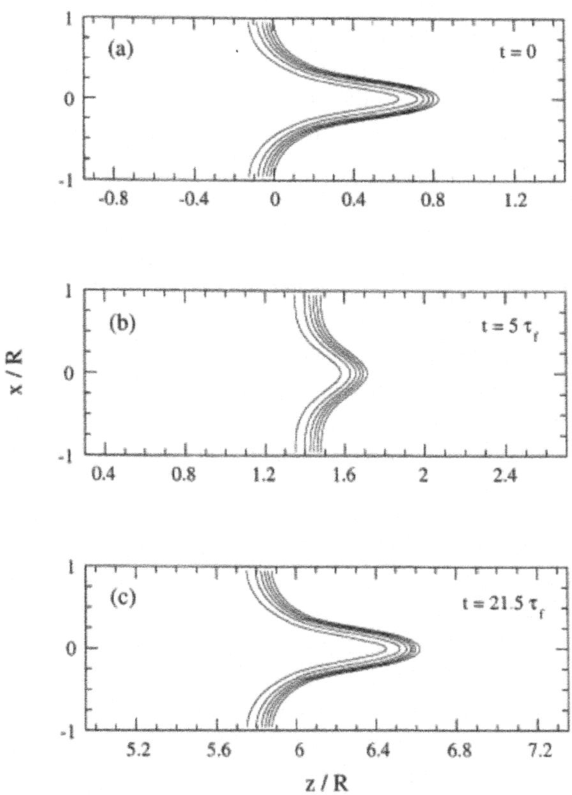

Fig. 7.7 The flame shape after collision with a weak shock of intensity $\Delta U = 3U_f$ at the time instants shown in the slides. The isotherms correspond to the temperatures from 600 to 2100 K with the step 300 K

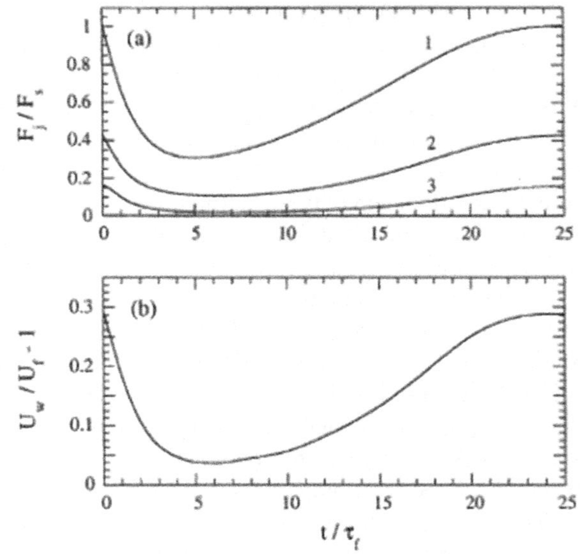

Fig. 7.8 The time history of the amplitude of the Fourier harmonics (**a**), and the scaled flame velocity (**b**) after the flame collision with a shock of intensity $\Delta U = 3U_f$

the cusp clearly seen, it is shown not only the flame front in the domain of calculations $0 < x/D < 1$, but also the flame front in the domain $-1 < x/D < 0$, which was obtained from the numerical results by the use of the symmetry transformation. After the collision of the flame front with a weak shock, the flame shape changes and the flame curvature may increase or decrease depending on the flame and shock parameters. Since the curved flame shape is the result of the DL instability, we will call the decrease of the flame curvature as the flame stabilization and an increase in the flame curvature as the flame destabilization.

The flame shape evolution is shown in Fig. 7.7 for the shock intensity $\Delta U = 3U_f$. The flame curvature is reduced after the shock–flame collision, as seen in Fig. 7.7b. The temporary stabilization of the flame is also seen in Fig. 7.8a, which shows the time history of the amplitudes of the first three Fourier harmonics normalized on the amplitude of the first harmonic of the curved stationary flame F_s. It is seen that the amplitude decreases down to the minimal value $F_1 \approx 0.3F_s$. Figure 7.8b shows the change of the flame velocity caused by the change of the flame shape. Since the velocity amplification of the curved flame depends on the flame curvature, which defines the increase of the flame front "surface", the flame velocity at the instant of minimal curvature is only slightly larger than the velocity of the planar flame. Still, the flame stabilization by the shock is temporary, and after some time, the amplitude of the first harmonic starts to grow again until the flame acquires the initial curved stationary shape as it was before the flame–shock collision. The time needed for the flame to recover its shape is about $25\tau_f$, which is about 3–4 times larger than the inverse instability increment $1/\sigma_g = 7.9\tau_f$. It is interesting that the flame curvature decreases after the collision much faster than the flame recovers its shape: the corresponding time intervals are about $5\tau_f$ and $20\tau_f$. This difference is understandable because according to the linear theory, the time scale of the amplitude decrease is determined by the inverse decrement of the decaying mode $1/\sigma_d$, while the time necessary for the flame shape recovering is characterized by the inverse growth rate of instability $1/\sigma_g$.

For the parameters used in calculations of Fig. 7.7, the ratio of the increment and the decrement is $\sigma_d/\sigma_g = 3.7$, which agrees with the ratio of the time intervals needed for the flame shape recovering and for the decrease of the flame curvature.

For a shock wave of larger intensity, the response of the flame front to the flame–shock collision is different. While the simulation of the flame–shock collisions for the flame with expansion coefficient $\Theta = 8$ and the shock of intensity $\Delta U = 3U_f$ causes temporary flame stabilization, the result of the same flame collision with the shock of larger intensity $\Delta U = 4.2U_f$ leads to the inverted flame shape shown in Fig. 7.10. Similar to the case $\Delta U = 3U_f$, at the beginning, just after the flame–shock collision, the flame curvature temporary decreases and the flame front becomes almost planar, but the following instability development leads to the formation of the inverted stationary flame shape with the negative amplitude of the Fourier harmonics, as shown in Fig. 7.9a. At the same time the absolute value of the amplitudes after the inversion equals to their absolute values before the flame–shock collision, and as it is seen in Fig. 7.9b, the curved stationary flame with inverted shape propagates with the same velocity as the initial stationary flame.

Though there is no explicitly defined boundary separating the period of the flame curvature decrease from the following stage of the flame front inversion, but similar to the previously considered case one can distinguish in Fig. 7.10 the time interval of the fast decrease of the Fourier harmonic amplitudes and the interval of slow increase

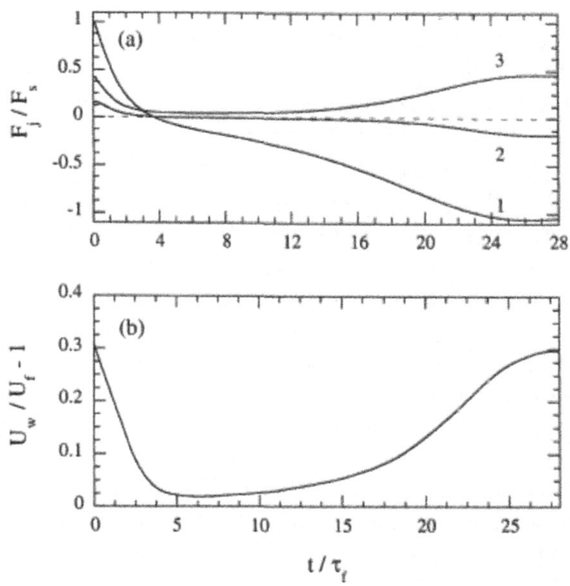

Fig. 7.9 **a** the time history of the scaled amplitude of the first three Fourier harmonics curves 1, 2, 3; **b** the scaled flame velocity after the flame collision with a weak shock of intensity $\Delta U = 4.2U_f$

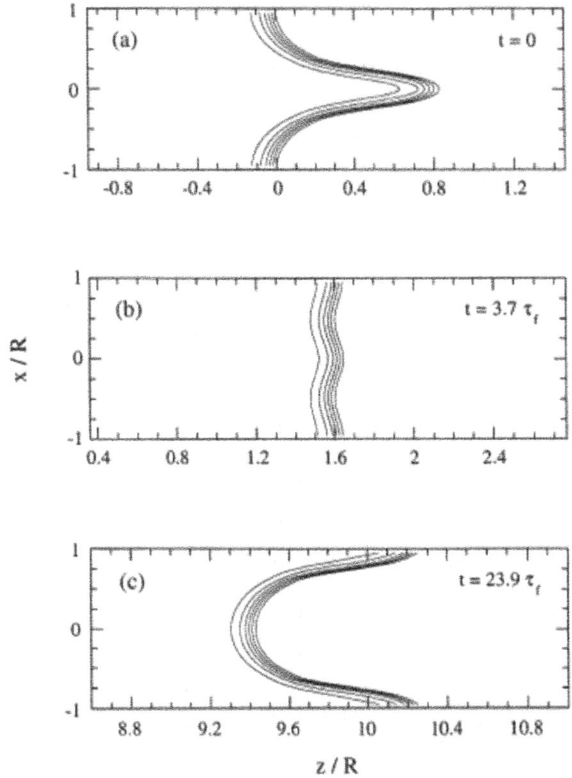

Fig. 7.10 The flame shape after the collision with a shock of intensity $\Delta U = 4.2U_f$. The isotherms correspond to the temperature isolines with the step 300 K

of the absolute value of their amplitudes to the values corresponding to the inverted shape of the stationary flame. An approximate ratio of these time intervals is about 4, which agrees with the ratio $\sigma_d/\sigma_g = 3.7$, predicted by the linear theory.

Figure 7.11 shows the evolution of the flame front shape as a result of a flame with expansion coefficient $\Theta = 8$ collision with the shock of intensity $\Delta U = 9U_f > 2\Delta U_c$. The corresponding time history for the amplitudes of the first three Fourier harmonics is shown in Fig. 7.12. In this case, the sharp decrease of the amplitude cannot be associated with flame stabilization.

The critical shock intensity needed to invert the flame shape is expected to be in the range $3U_f < \Delta U_c < 4.2U_f$ for the flame with an expansion coefficient $\Theta = 8$. Studies of the flame–shock collisions for intermediate shock intensities have shown that, for the expansion coefficient $\Theta = 8$, the critical shock intensity is $\Delta U_c \approx 3.7U_f$, which is surprisingly well consistent with the critical intensity of the shock $\Delta U_c \approx 3.6U_f$ obtained from the linear theory. According to the linear theory, the intensity of a weak shock which may destabilize a flame should be $\Delta U > 2\Delta U_c$. Instead, the flame shape becomes inverted after a relatively short time interval $\approx 6\tau_f$, and for some time the amplitude of the first Fourier harmonic exceeds the respective amplitude for the stationary configuration by the absolute value $F_1 = 1.7F_s$, and

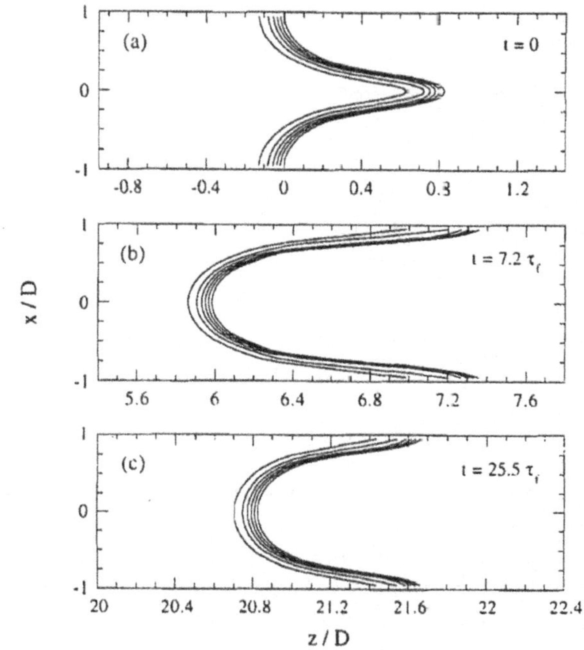

Fig. 7.11 The flame shape after the collision with a weak shock of intensity $\Delta U = 9U_f$ at the time instants $t/\tau_f = 0,\ 7.2,\ 25.5$ The isotherms correspond to the temperatures from 600 to 2100 K with the step 300 K, $\Theta = 8$

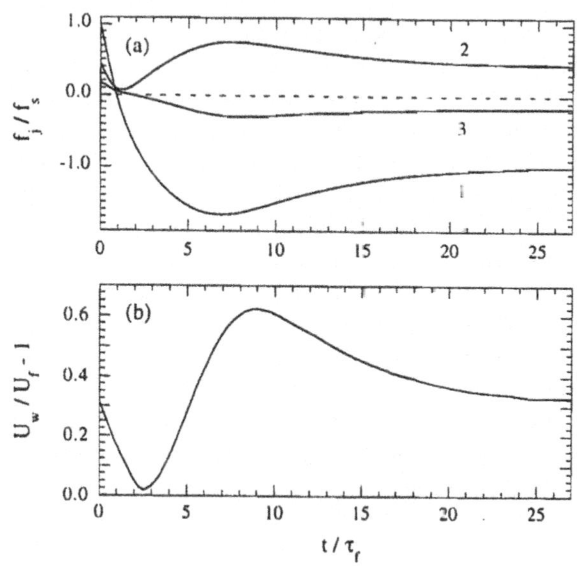

Fig. 7.12 **a** The time history of the scaled amplitude of the first three Fourier harmonics (curves 1, 2, 3); **b** the scaled flame velocity after the flame–shock collision for the conditions of Fig. 7.11

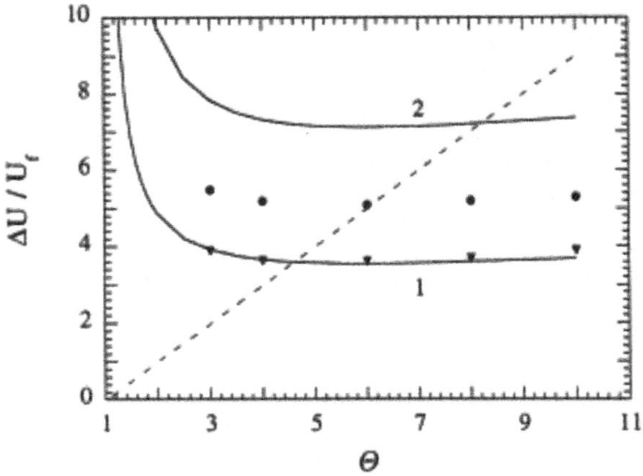

Fig. 7.13 The critical shock intensities (curve 1) and the shock intensities destabilizing flame (curve 2) were obtained in the linear theory. The triangles and the circles show respective shock intensities found numerically. The dashed line is the scaled intensity of the shock generated by a flame in a closed tube, $\Delta U = (\Theta - 1)U_f$; from [7]

flame velocity increases up to $U_w = 1.68U_f$. However, the achieved state is not stationary, and therefore, the sharp increase in the flame curvature is followed by a long period of relaxation to the inverted stationary shape with a reduction of the absolute value of the first harmonic and the flame velocity.

The numerical simulations [7] show that the nonlinear theory allows three possible outcomes of the flame–shock collision: (1) a curved flame may be temporarily stabilized and flattened by the flame-shock collision, which implies stabilization of the DL instability; (2) the stabilization can be accompanied by inversion of the flame shape; (3) the flame front can be inverted and the flame is destabilized by a passing shock. The calculated critical shock intensities ΔU_c needed to invert the shape of the curved stationary flames with the expansion coefficients $\Theta = 3 - 10$ are shown in Fig. 7.13 by triangles; circles are for the shock intensities destabilizing flame; the solid lines show the respective results obtained from the linear theory. It is seen that the critical shock intensities obtained from the linear theory are qualitatively consistent with the numerical simulations of the nonlinear problem.

Particularly, both the linear theory and the numerical simulations show that the ratio of the critical shock intensity and the flame velocity remains almost constant, $\Delta U_c / U_f = 3.6 - 3.9$, for flames with expansion coefficients $\Theta = 5 - 10$. For smaller expansion coefficients, which are not realistic, this ratio increases. However, unlike the predictions of the linear theory, the destabilization of the flame front by a weak shock was observed in simulations with smaller than the critical value $2\Delta U_c$ of shock intensities $\Delta U \geq 1.4\Delta U_c$.

In the previous section, it was shown that a flame propagating in a closed tube produces a weak shock of intensity $\Delta U = (\Theta - 1)U_f$ which is shown in Fig. 7.13 by a dashed line. One can see that in closed tubes, a weak shock stabilizes the flames with expansion coefficients $\Theta < 4.5$ without flame inversion. If a flame in a closed tube has the expansion coefficient in the range $4.5 < \Theta < 6$, then a collision with the flame-generated shock inverts the flame shape and stabilizes it temporarily. Flames with the expansion coefficients $\Theta = 6 - 10$ propagating in closed tubes are not only inverted, but also destabilized by a weak shock. The strength of the stabilization

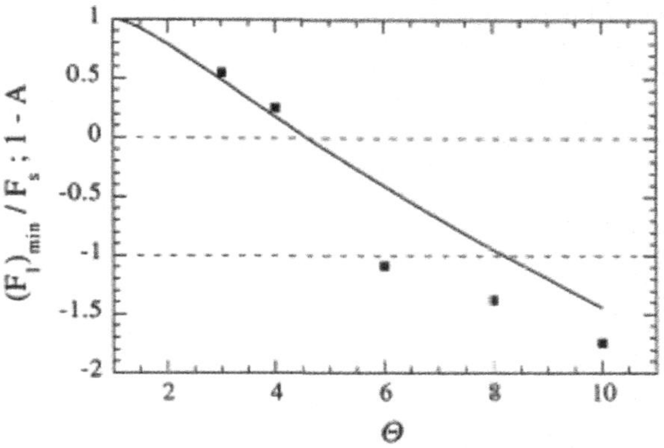

Fig. 7.14 The extreme (minimal or maximal) amplitude of the first Fourier harmonic of the flame front after the flame collision with the shock of the intensity $\Delta U = (\Theta - 1)U_f$. The solid line shows the relative amplitude of the growing mode $(1 - A)$ just after the flame–shock collision calculated using Eq. (7.1.13) of the linear theory; from [6]

(destabilization) of a flame by a weak shock can be characterized by the relative minimal (maximal) amplitude $(F_1)_{min}/F_s$ of the first Fourier harmonic of the flame isotherms. For the flame propagating in a closed tube, these extreme amplitudes are shown in Fig. 7.14 versus the expansion coefficient Θ.

Since the decrement σ_d of the decaying mode in Eq. (7.1.12) is usually considerably larger than the increment σ_g of the growing mode, then in a short time about $1/\sigma_d$ after the collision, the decaying mode almost vanishes, while the amplitude of the growing mode changes only slightly. As one can see in Fig. 7.14, the value of the extreme amplitudes found in numerical simulations is in good agreement with the values $(1 - A)$ of the growing mode just after the collision calculated from Eq. (7.1.13). However, one should remember that the linear theory provides only qualitative results for the nonlinear problem. The agreement is much better for the flame stabilization by a shock, since in this case, the similarity between the linear and the nonlinear problems is stronger. The difference between the linear and the nonlinear problems is more pronounced in the case of flame destabilization, still, some qualitative agreement remains. Both the linear theory and the numerical simulation of the nonlinear problems show that the strongest destabilization of a flame in a closed tube by the shock wave is expected for flames with larger expansion coefficients. Particularly, for the expansion coefficient $\Theta = 10$, the maximal amplitude of the first Fourier harmonic may exceed the stationary amplitude by the factor 1.8. The velocity of the curved flame at this instant becomes as large as $U_w = 1.7U_f$.

If we consider a flame–shock interaction in closed tubes, the secondary shock generated during the first flame–shock collision propagates through the burnt mixture, reflects from the closed end of the tube, and comes to the flame from the side of the burned hot combustion products. Collision of the reflected secondary shock with the curved flame affects the flame shape and velocity. The intensity of the reflected shock is defined by Eq. (7.1.7) and for the expansion coefficient $\Theta = 8$, it equals $\Delta U_1 = 10.3U_f$. The result of the curved flame collision with the secondary reflected shock is shown in Fig. 7.15, and the time corresponding time history of the amplitudes of the first three Fourier harmonics is shown in Fig. 7.16. It is seen that, in this case, the flame–shock collision results in a significant increase in the flame curvature and

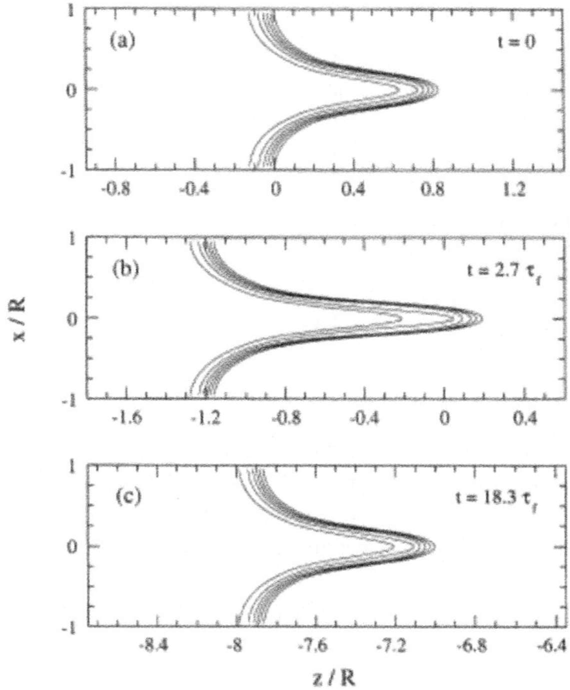

Fig. 7.15 The change of the flame shape after the collision with a weak shock of intensity $\Delta U_1 = 10.3 U_f$. The expansion coefficient of the flame is $\Theta = 8$, from [7]

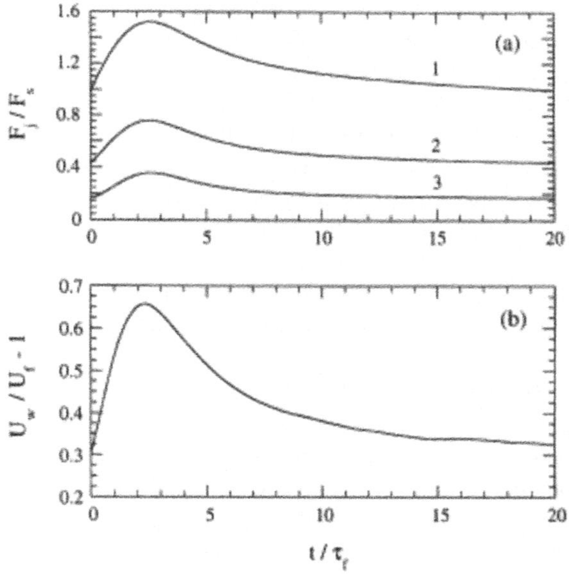

Fig. 7.16 **a** The time history of the scaled amplitude of the first three Fourier harmonics (curves 1, 2, 3); **b** the scaled flame velocity after the flame–shock collision for the conditions in Fig. 7.15, from [7]

velocity up to the value $U_w = 1.7U_f$. The flame response to a shock propagating in the same as the flame direction may be qualitatively understood on the basis of the linear theory. In such a collision, the pulse acceleration produced by the passing shock is negative, the amplitude of the decaying mode is negative $A < 0$ (see Eq. (7.1.13)), and after the collision, the amplitude of the growing mode exceeds the initial amplitude $1 - A > 1$. Therefore, any such collision of the flame with a weak shock coming from the burnt mixture should result in the flame destabilization independently of the shock intensity. For a flame with the expansion coefficient in the range $\Theta = 8 - 10$, the flame destabilization is shown in Fig. 7.15. A single collision of a flame and a weak shock is unlikely to produce an important effect on the flame dynamics in closed chambers, since after some time, a flame relaxes to the stationary shape. The flame–shock interaction in a closed tube may become important if the relaxation time about $4/\sigma_g$ is comparable to the time needed for a weak shock to traverse back and forth the tube. In this case, the flame–shock interaction may result in continuous acceleration of the flame. A single flame–shock collision may be crucial for flame dynamics only if the shock intensity is sufficiently large. Sufficiently strong shock waves of an external origin may strongly distort the flame shape [1] changing the internal flame structure and possibly quenching the reaction [2].

Finally, it is interesting to compare the results of numerical simulations [7] with the experiments on flame interaction with shocks of controlled intensity described by Markstein [1]. The experiments were performed with slow butane-air flames, while relatively large intensity shocks were generated with the pressure increase $\Delta P/P = 0.1; 0.3$. From 2D numerical simulations [7], the order of magnitude of the relative pressure jump corresponding to the critical shock intensity is $\Delta P/P = 10^{-2}$. The intensity of shocks used in the experiments [1] is much larger than the critical shock intensity, so that collisions of these shocks with curved flames should lead to strong destabilization of the flame accompanied by inversion of the flame shape and the development of mushroom shape of the flame front, similar to that observed in numerical simulations of the pressure pulse induced instabilities in premixed flames [19].

7.3 Hydrodynamic Instability of Planar Flame in Closed Chambers

Flame dynamics in closed tubes is one of the most important issues of combustion science since it is directly related to the modeling of combustion in car engines and many industrial processes. Particularly, the flame propagation in a closed chamber is a key issue for the understanding of the "knocking" combustion in car engines [20–25]. One of the most important problems is the interaction of a flame with pressure waves under confinement. It is a general belief that the flame dynamics in a closed chamber are considerably defined by the intrinsic hydrodynamic instabilities of a flame. The DL instability is especially important since it results in the flame front distortion, augmentation of the flame front surface, following by the increase of the flame velocity. A flame accelerates when it acquires a curved shape, and therefore, its surface increases. The accelerating flame, acting as a semitransparent piston, produces the pressure and acoustic waves, which after reflections from the walls, interact with the flame. Even in the case of ideal walls, there are many specific factors that influence the development of the DL instability in closed chambers. Flames propagating in open tubes consume fuel in a nearly isobaric regime since the expansion of the combustion products is not restricted by the walls. The situation is different for a flame propagating in a closed tube: expansion of the hot gas of combustion products causes adiabatic compression of the unburned fuel ahead of the flame. Compression of the unburned gaseous fuel can be accompanied by a significant build-up of pressure P and temperature T, compared to their initial values P_0 and T_0

$$P = P_0\left(1 + \frac{\gamma Q}{C_P T_0}\right) = (\gamma\Theta_0 + 1 - \gamma)P_0, \tag{7.3.1}$$

$$T = (\gamma\Theta_0 + 1 - \gamma)^{1-1/\gamma}T_0. \tag{7.3.2}$$

The increase of pressure and temperature in the unburned mixture alters all parameters characterizing the flame. Typically, the pressure and temperature changes in the unburned mixture are slow compared to the time scale $\tau_f = L_f/U_f$. Therefore, the internal structure of the flame front has enough time to adjust to the change of external conditions and at any time instant. Thus, at each instant, the flame can be considered as a freely propagating flame in a mixture with slowly varying temperature and pressure. The normal velocity of a planar laminar flame in ZFK theory depends on the fuel parameters. Let U_0 is the flame velocity in the mixture with temperature and pressure (T_0, P_0). Using (7.3.1), (7.3.2), and (4.4.17), the flame velocity at the time when the temperature and pressure are equal (T, P) will be

$$\begin{aligned} U_f(T,P) = U_0\left(\frac{P}{P_0}\right)^{\frac{n-2}{2}}\frac{T}{T_0}\left(\frac{T+Q/c_p}{T_0+Q/c_p}\right)^{\frac{2n+1}{2}} \times \\ \exp\left(\frac{E(T-T_0)}{2(T_0+Q/c_p)(T+Q/c_p)}\right), \end{aligned} \tag{7.3.3}$$

where n is the reaction order. It is seen that the velocity of laminar flame with a first-order reaction decreases with the pressure increase, while for the third-order reaction the flame velocity increases with the pressure. The flame thickness decreases considerably to the end of burning for flames of both the third-order reaction and the first-order reaction, though in the former case, the decrease is much stronger. For a flame with the reaction order $n = 1$, the decrease in the flame thickness is mostly due to the increase in the fuel density ($L_f = \kappa/\rho_f C_P U_f$). On the contrary, in the case of a third-order reaction, $n = 3$, both extensions of the fuel density and the flame velocity contribute to the decrease in the flame thickness. The scaled flame velocity

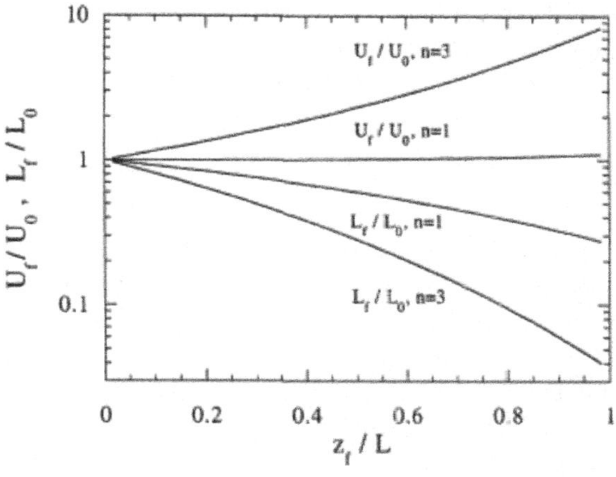

Fig. 7.17 The scaled velocity U_f/U_o of a planar flame and the scaled flame thickness L_f/L_0 versus the flame position for the reaction of the first and the third order, from [26]

and thickness are shown in Fig. 7.17 versus the flame front position in a tube for flame with the first, $n = 1$ and the third $n = 3$ order reactions [26].

The eventual decrease of the flame thickness in a closed tube makes the competition of perturbation modes on the nonlinear stage of the DL instability more complicated. This conclusion follows already from the observation that the decrease of the flame thickness leads to the decrease of the cutoff wavelength λ_c, so that a perturbation model that has been stable at the beginning of combustion, may later become unstable. As a result, the nonlinear stage of the DL instability in closed tubes may be characterized by the transitions from one quasi-stationary nonlinear structure to another one with a different wavelength.

Flame dynamics under confinement may also be affected by the flame acceleration or deceleration that can reach significant values in short tubes leading to strong suppression or enhancement of flame instabilities. The velocity of the flame in the laboratory reference frame U_l is quite different from the normal flame velocity [26]. In the coordinate system of the flame, the fuel mixture flows into the flame front with velocity U_f and the combustion products flow out with the velocity ΘU_f. At the beginning, the burnt products are at rest, the fuel mixture is pushed forward and the flame velocity in the laboratory reference frame is $U_l = \Theta_0 U_0$. Close to the end of combustion, when the flame comes close to the opposite wall of the tube, the fuel mixture is at rest, and the burnt products are pushed backward and the flame velocity relative to the walls equals the normal flame velocity $U_l = U_1$. For example, in 2D simulations [13] of the "tulip flames" phenomenon for the flame with $U_f = 4.5$ m/s, $\Theta_0 = 7.67$, $\gamma = 1.4$, $E = 4.36\Theta_0 T_0$, and the first-order reaction, the flame propagated with the velocity of about 40 m/s just after ignition. However, the final flame velocity at the end of combustion was about 5 m/s in agreement with the above estimates. For $\Theta_0 = 5$, $\gamma = 1.4$, $E = 5\Theta_0 T_0$ adopted in 2D simulations of flames in closed tubes [11] the flame velocity in the laboratory reference frame drops from $5U_0$ at the beginning to $1.1U_0$ for the flame with the first reaction order (the flame decelerates); while for the flame with the third reaction order the flame velocity in the laboratory reference frame increases from $5U_0$ to about $10U_0$ (the flame accelerates). The acceleration (deceleration) of the flame front with the initial flame thickness L_0 in a tube of length L can be estimated as

$$g = \frac{U_1^2 - \Theta_0^2 U_0^2}{2L}. \tag{7.3.4}$$

For the flame parameters used in [11], the scaled acceleration equals to $gL_0/U_0^2 = 1/Fr = -12L_0/L$ for the first-order reaction, and $gL_0/U_0^2 = 36L_0/L$ for the third-order reaction. Therefore, for the third-order reaction flames under confinement, the DL instability is enhanced by the RT instability of the flame front in the effective gravitational field. On the contrary, for the first-order reaction flames, the deceleration is a stabilizing factor. The complete suppression of the DL instability by a negative acceleration (in this chapter) is expected for $g = \pi\Theta_0 U_f^2/2\lambda_c$. For flames considered in [11], the complete suppression of the DL instability is expected for tubes shorter than $L = 40L_0$, so that in order to observe the DL instability and "tulip flames" in numerical simulations, one has to study flame dynamics in rather longer tubes, about $L \geq 100L_0$.

The effect of acceleration is not the only restriction on the tube length needed to observe the DL instability. Since a flame traverses a short tube in a short time, it may happen that the instability has no time to develop. The DL instability is a slow process compared to the time scale of a flame front L_f/U_f. From Eq. (7.2.4), we can estimate the maximum instability growth rate for a flame with the expansion coefficient $\Theta = 5$ and $\mathrm{Le} = 1$ as $\sigma_m = 0.07U_f/L_f$. At the beginning of combustion, the flame propagates with the velocity ΘU_f with respect to the tube walls. Therefore, the amplitude of perturbations depends on the distance z_f passed by the flame as $\exp(0.07z_f/\Theta L_f)$. To reduce the influence of the "arbitrary" initial conditions on the instability development, we can consider the initial perturbation amplitude at least 10 times smaller than the saturation amplitude F_s achieved on the nonlinear stage for a curved stationary flame. Since the perturbation amplitude is proportional to $\exp(0.014z_f/\Theta L_f)$, then it may increase by a factor of 10 only in a tube with a length of about $160L_0$, if there are no other stabilizing effects. Thus, to observe a strongly curved "tulip flame" resulting from the DL instability in a closed tube, we have to consider tubes of lengths of $200L_0$ and longer. The influence of the aspect ratio of a closed tube on the development of "tulip flames" has been pointed out in many experimental studies and 2D simulations [12—18], but the restriction on the tube length proposed in [11] is more severe than the aspect ratio of a tube. It implies that not all "tulip flames" observed in experiments and numerical simulations up to now resulted from intrinsic flame properties. Any "tulip flame" observed in a shorter closed tube can be probably just an outcome of imposed initial conditions and may be eliminated by another choice of ignition geometry.

When a flame is in the vicinity of the walls at the very beginning and at the very end of the combustion process, then the DL instability growth rate may be directly affected by the boundary conditions at the rigid walls. This effect is stabilizing similar to the influence of a wall on the RT instability in an inert gas [28]. Taking the characteristic wavenumber of the RT instability as determined by the tube width $k = \pi/R$, its relative role can be estimated as $\exp(-2kz_f)$. Therefore, when the flame front has passed the distance of about the tube width $z_f = R$, the direct influence of the boundary conditions at the walls on the DL instability becomes negligible.

A common theoretical approach to the flame dynamics under confinement implies that a sound speed is infinite and the pressure is balanced at both ends of the tube instantly [26, 27]. Obviously, this assumption neglects non-stationary effects related to acoustic and weak shock waves and it is valid only if the instability time scale is much longer than the time needed for the acoustic wave to travel along the tube forth and back $2L\sigma_m/c_s \ll 1$. Figure 7.17 shows the pressure distribution in a closed tube of the length $L = 500L_0$ calculated in [11] for the time instant when a flame with parameters $\Theta_0 = 5$, $\gamma = 1.4$, $E = 5\Theta_0 T_0$, $n = 3$ and the initial Mach numbers $M_0 = 10^{-2}$ or $M_0 = 10^{-3}$ reaches the middle of the tube, $z = 250L_C$. It is seen that in the case of a smaller flame velocity ($M_0 = 10^{-3}$), the pressure is balanced at the tube ends, while for the flame with a larger velocity ($M_0 = 10^{-2}$) the main acoustic mode with the wavelength $\lambda_s = 2L$ has a noticeably large amplitude. The estimate of the pressure perturbations in the acoustic wave is given by the amplitude of the weak shock generated by a flame that moves from a closed end of a tube $\Delta P = (\Theta - 1)\rho c_s U_f$ (Fig. 7.18).

Still even in short tubes with a small period of the main acoustic mode, pressure waves generated by the flame can significantly affect the dynamics of the flame. It was shown in Chap. 6, that the pressure waves take a form of acoustic oscillations if the frequency of the main acoustic mode is larger than the maximum growth rate of the DL instability, which implies the condition $2L\sigma_m/c_s < 1$. In the opposite case $2L\sigma_m/c_s > 1$ the pressure waves can be considered as weak shocks. For the flame parameters used in the 2D simulations [11], the characteristic tube length that separates the domain of applicability of acoustic waves and shock waves is about $L = 500L_0$: for shorter tubes, we can consider sound waves, while for much longer tubes pressure waves manifest themselves more like shock waves. In Sect. 6.3, we considered the flame-acoustic interaction and found that the interaction of acoustic

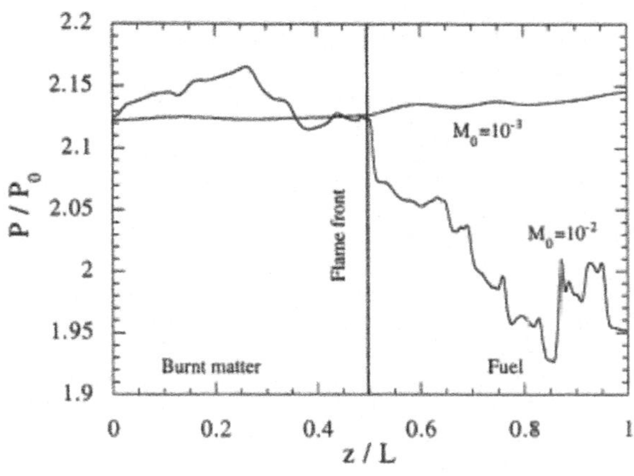

Fig. 7.18 Pressure distribution in a closed tube of length $L = 500L_0$ at a time instant when the flame front is at $z = 250L_0$. The initial Mach numbers are $M = 10^{-2}$ and $M = 10^{-3}$, the reaction order is $n = 3$; from [11]

waves with the flame with $\Theta_0 = 5$ and $M_0 = 10^{-2}$ in a short tube $L < 500L_0$ results in considerable stabilization of the flame front. According to Sect. 7.1, a head-on collision of the flame with $\Theta_0 = 5$ and the flame-generated shock also substantially suppresses the DL instability, though in this case, this is only a temporary suppression. However, these conclusions about the flame–shock and the flame–acoustic interactions are applicable only as long as the flame is far enough from the end of the tube. When a flame comes to the tube end, a strong nonlinear coupling between the flame and the pressure waves may develop for sufficiently fast flames, similar to that observed by Searby in experiments [27].

Finally, the increase of the fuel temperature due to the pre-compression may cause spontaneous ignition of the reaction ahead of the flame front and even the following development of a detonation is possible.

References

1. G.H. Markstein, *Nonsteady Flame Propagation* (Pergamon, Oxford, 1964)
2. G.A. Batley, A.C. McIntosh, J. Brindley, The time evolution of interactions between Ultra Short length scale pressure disturbances and premixed flames. Combust. Sci. Technol. **92**, 367–388 (1993)
3. G.A. Batley, A.C. McIntosh, J. Brindley, Baroclinic distortion of laminar flames. Proc. Roy. Soc. London **A452**, 199–221 (1996)
4. M.A. Liberman, V.V. Bychkov, S.M. Golberg, L.E. Eriksson, Numerical study of curved flames under confinement. Combust. Sci. Technol. **136**, 221–251 (1998)
5. V.V. Bychkov, Stabilization of the hydrodynamic flame instability by a weak shock. Phys. Fluids **10**, 2669–2675 (1998)
6. V.V. Bychkov, M.A. Liberman, Dynamics and stability of premixed flames. Phys. Reports **325**, 115–237 (2000)
7. O.Y. Travnikov, V.V. Bychkov, M.A. Liberman, Interaction of curved flames and weak shocks. Combust. Sci. Technol. **142**, 1–28 (1999)
8. L.D. Landau, E.M. Lifshitz, *Course of Theoretical Physics*, vol. 6 (Fluid Mechanics. Pergamon Press, Oxford, 2011)
9. R.D. Richtmyer, Taylor instability in shock acceleration of compressible fluids. Commun. Pure Appl. Math. **13**, 297–319 (1960)
10. E.E. Meshkov, Instability of the interface of two gases accelerated by a shock wave. Izv. Acad. Sci. USSR Fluid Dyn. **4**, 101–104 (1969)
11. M.A. Liberman, V.V. Bychkov, S.M. Golberg, L.E. Eriksson, Numerical study of curved flames under confinement. Combust. Sci. Tech. **136**, 221–251 (1998)
12. D. Dunn-Rankin, R.F. Sawyer, Tulip flames: changes in shapes of premixed flames propagating in closed tubes. Exp. Fluids **24**, 130–140 (1998)
13. M. Gonzales, R. Borghi, A. Saouab, Interaction of a flame front with its self-generated flow in an enclosure: the "tulip flame" phenomenon. Combust. Flame **88**, 201–220 (1992)
14. J.L. McGreevy, M. Matalon, The initial development of the tulip flame. Symp. (Int.) Combust. **25**, 1407–1413 (1994)
15. B. N'Konga, G. Fernandez, H. Guillard, B. Larrouturou, Numerical investigations of the tulip flame instability—comparisons with experimental results. Combust. Sci. Tech. **87**(1993), 69–89 (1993)
16. C. Clanet, G. Searby, On the 'tulip flame' phenomenon. Combust. Flame **105**, 225–238 (1996)
17. D. Dunn-Rankin, Tulip flames—the shape of deflagrations in closed tubes, in *Combustion Phenomena: Selected Mechanisms of Flame Formation, Propagation and Extinction* (CRC Press, Boca Raton, FL. 2009)

18. B. Ponizy, A. Claverie, B. Veyssière, Tulip flame—the mechanism of flame front inversion. Combust. Flame **161**, 3003–3266 (2014)
19. N.R. Edwards, A.C. McIntosh, J. Brindley, The development of pressure induced instabilities in premixed flames. Combust. Sci. Technol. **99**, 179–199 (1994)
20. G.L. Borman, K.W. Ragland, *Combustion Engineering* (McGraw-Hill, New York, 1998)
21. M.A. Liberman, L.-E. Eriksson, M.F. Ivanov, O. Peil, Numerical modeling of the propagating flame and knock occurrence in spark-ignition engines. Comb. Sci. Tech. **177**, 151–182 (2005)
22. M.A. Liberman, L.-E. Eriksson, M.F. Ivanov, Hot spot formation by the propagating flame and the influence of EGR on knock occurrence in SI-engines. Comb. Sci. Tech. **178**(2006), 1613–1647 (2006)
23. Z. Wang, H. Liu, R.D. Reitz, Knocking combustion in spark-ignition engines. Prog. Energy Combust. Sci. **61**, 78–112 (2017)
24. G. Kalghatgi, Knock onset, knock intensity, superknock and preignition in spark ignition engines. Int. J. Engine Res. **19**, 7–20 (2018)
25. D. Bradley, C. Morley, X.J. Gu, D.R. Emerson, Amplified pressure waves during autoignition: relevance to CAI engines, SAE Technical Paper, 2002–01–2868 (2002)
26. V.V. Bychkov, M.A. Liberman, On the stability of a flame in a closed chamber. Phys. Rev. Letters **78**, 1371–1734 (1997)
27. G. Searby, Acoustic instability in premixed flames. Combust. Sci. Technol. **81**, 221–231 (1992)
28. S. Chandrasekhar, *Hydrodynamic and Hydromagnetic Stability* (Clarendon press, Oxford, 1961)

Chapter 8
Dynamics of Curved Flames Propagating in Tubes

8.1 Nonlinear Stage of Instability; Cellular Flame's Structure

A flame ignited at some location can propagate from the location where it was ignited either being confined by the walls of the vessel, or as a freely propagating flame. Both of these configurations correspond to typical experimental situations of a flame propagating in closed or open tubes, or in the form of a spherical expanding "fireball". The dynamics of a propagating flame is one of the most fundamental problems in combustion theory, which is a rather complex mathematical problem. Flame propagation is described by the system of the reactive Navier–Stokes equations for the fuel mixture and combustion products, together with the energy release in chemical reactions.

It is well established both theoretically and experimentally that a planar front of a flame propagating in a gaseous combustible mixture is intrinsically unstable against small perturbations bending the flame front. One of the main reasons for the loss of its planar configuration by a flame is the DL instability, which is inherent to all flames in gas mixtures, since it is associated exclusively with the gas expansion in exothermic reactions. At the linear stage, small perturbations of the planar flame front grow exponentially with the growth rate of the instability, which depends on the wave number of the perturbation. Later, at the nonlinear stage, saturation of the growth of perturbations due to nonlinear effects leads to the formation of curved stationary flames or to the flame with a cellular structure. A wrinkled or curved flame consumes fresh fuel over a larger surface area per unit time, resulting in a noticeable increase in the rate of heat release per unit time and to a higher effective burning rate.

The physical mechanism of nonlinear stabilization, which is schematically shown in Fig. 8.1, was proposed by Ya. Zel'dovich [1] and Markstein [2] using a model of an infinitely thin flame front. At all points, the flame propagates along the normal to the flame surface. According to the Huygens principle, the convex parts (humps) of the flame front diverge, while the concave parts (cusps pointing toward the burnt gas) converge until angle points are formed. As it is seen in Fig. 8.1, the angular points

M. A. Liberman, *Combustion Physics*,
https://doi.org/10.1007/978-3-030-85139-2_8

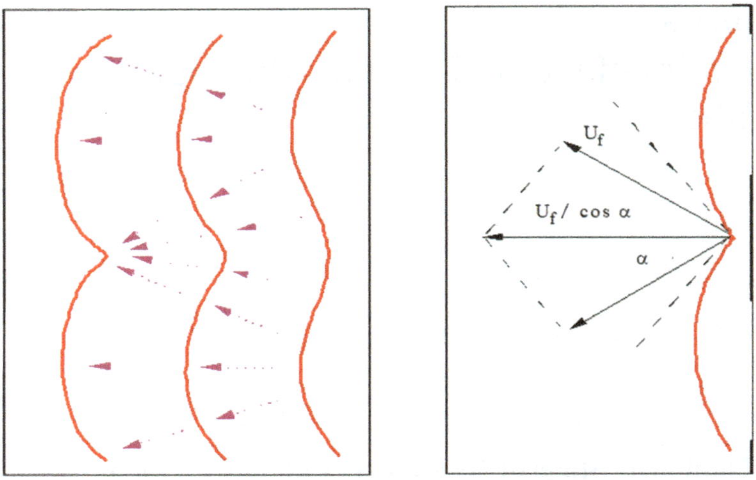

Fig. 8.1 Sketch illustrating the nonlinear stabilization of the DL instability

propagate faster than the smooth parts of the flame, the velocity increase at the corners balances the growth of the humps and the flame acquires a curved stationary shape. In reality, a flame front has finite thickness, and the cusps are smoothed out by thermal conduction.

The development of perturbations at the nonlinear stage was an important topic of many publications. The general opinion, confirmed by many experiments, is that the DL instability leads to the formation of a curved flame and, under certain circumstances, can lead to turbulent flame, although often other reasons, such as rough walls, obstacles, etc., can be the main source of turbulence. An important effect of the DL instability is the amplification of a curved flame velocity in comparison with the planar flame. The velocity increase of a flame when it acquires a curved shape was well known from the very first studies of flames propagation in tubes. Almost 150 years ago Mallard and Le Chatelier [3] found that the velocity of a methane-air flame propagating in a tube increases progressively with an increase in the tube diameter. Later, Coward and Hartwell [4–6] conducted detailed experimental studies of the flame shape and the flame velocity depending on the tube diameter, shown in Fig. 8.2, where blue lines are numerical simulations [7].

In these experiments, the stationary velocity of the methane-air flame was observed in tubes of different diameters from 2.5 cm till 24 cm. Smoothly curved stationary flames resulting from the development of the DL instability were later observed in many experimental and numerical studies where the formation of cells occurs due to the growth of larger cells at the expense of neighboring cells of the smaller size. It was generally accepted that the saturation of the DL instability during nonlinear stage leads to the formation a stationary monocellular structure with characteristic sizes much larger than the most unstable wavelength of the order $2\lambda_c$ predicted by the linear theory. All the same, Markstein [2] observed that for flames propagating

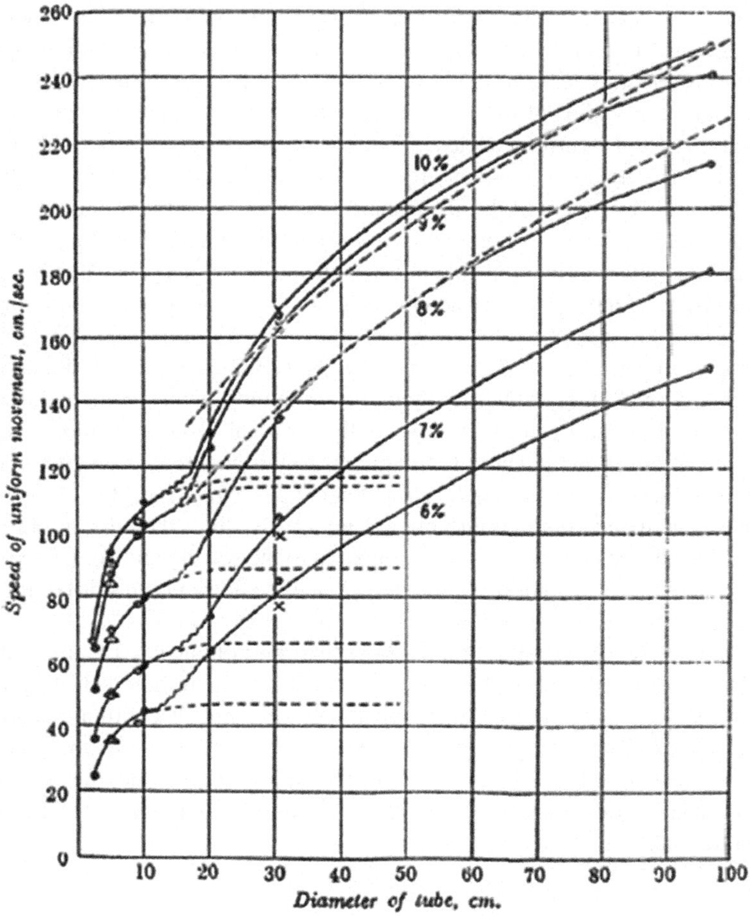

Fig. 8.2 Flame speed versus tube diameter for the methane–air mixture. Solid lines are experiments by Coward and Hartwell [3]. Dashed lines are theory and simulations from Liberman et al. [7]

in sufficiently large vessels the process of cells merging and splitting never leads to the stationary monocellular structure observed in small-scale experiments. On the contrary, in large vessels, the merging of the cells is permanently accompanied by their splitting, when they reached an excessive size, and the creation of new smaller cells. An illustrative example is the formation of a fractal structure of the freely expanding spherical flame [8] (see Chap. 11).

The possibility of analytical solution of the nonlinear problem of the DL instability is very attractive, since it allows a deeper understanding of the problem for a wide range of problem parameters, which is difficult to obtain using numerical simulations. The main reason for the complexity of the problem is that the evolution of the flame front cannot be prescribed in advance. Instead, it should be determined in the course of joint analysis of the flow dynamics outside the flame front and the heat conduction and species diffusion processes inside the flame front. In general, the nonlinear interaction of different perturbation modes under the smoothing influence of thermal conduction leads to the formation of a steady curved flame configuration with the curvature radius of the order $20L_f$ estimated from the linear theory of the DL instability for the cutoff wavelength. However, as indicated earlier, the situation is different for a flame in a sufficiently large vessel. Strictly speaking, analysis of the nonlinear development of the DL instability cannot be carried out perturbatively for the arbitrary coefficient of gas expansion. From the mathematical point of view, perturbation analysis is inapplicable because the assumptions of weak nonlinearity and stationarity of the flame contradict each other. Their simultaneous use turns out to be inconsistent, except for the case $(\Theta - 1) \to 0$, while for real flames the gas expansion coefficient is $\Theta = \rho_u/\rho_b \sim (6-10) \gg 1$. Finite vorticity production in the DL instability of a flame implies that the flow dynamics downstream, and therefore the flame front dynamics itself, is essentially nonlocal. The latter means that the nonlocality of equations governing a flame propagation is more complex than that used in the linear theory. Only in the case of small gas expansion, $(\Theta - 1) \to 0$, can the problem be treated both perturbatively and locally, since then the amplitudes of perturbations can be considered small compared to the wavelengths of perturbations at all stages of the DL instability development, and the flow at both sides of the flame can be considered as potential in the lowest order in $(\Theta - 1) \ll 1$.

Sivashinsky [9–11] was the first to show that the nonlinear problem of the DL instability can be reduced to the nonlinear equation for the flame front position, $z = f(x, t) - U_f t$. Considering the flame as a gasdynamic discontinuity, and using the asymptotic expansion of the complete system of equations in the limit of a small parameter $\varepsilon = (\Theta - 1)/\Theta \to 0$, the nonlinear equation at first order in $\varepsilon \ll 1$ for the flame propagating in z-direction with normal velocity U_f with respect to an initially uniform fuel takes the form

$$\frac{\partial f}{\partial t} + \frac{U_f}{2}(f')^2 = \frac{4\sigma_M}{k_c}\left[\frac{1}{k_c} f'' + \hat{\Phi}(f, x)\right]. \tag{8.1.1}$$

Here, σ_M is the growth rate of the fastest unstable mode, k_c is a cut-off wavenumber, $f' \equiv \partial f/\partial x$, and the operator $\hat{\Phi}$ is the Landau-Darrieus operator defined as

$$(\hat{\Phi} f)(x) = \int_{-\infty}^{\infty} dk|\mathbf{k}| f_k \exp(ikx), \tag{8.1.2}$$

with f_k being the Fourier transform of f.

Already the first numerical experiments on this equation by Michelson and Sivashinsky [10] have shown that a disturbed initially plane flame evolves into the essentially unsteady cellular flame with the cell size about $\lambda_c \approx 2\pi/k_c$. Surprisingly, the results of the numerical solution of Eq. (8.1.1), which is formally a small expansion model, qualitatively rather well reproduced the chaotic structure of the cellular propane-air flame observed experimentally in [12, 13] despite a crude assumption of a negligibly small expansion, which is used to derive the evolution equation.

In the limit $\varepsilon \to 0$ ($\Theta \to 1$), Eq. (8.1.1) is an exact solution. The phenomenological extension of Eq. (8.1.1) for realistic values of Θ was considered by Joulin and Cambray [14]. Moreover, later it was shown that despite new terms correcting the mean front displacement, the modified equations keep almost the same functional structure up to the second order [14, 15]. Further important progress has been made in recent studies by Almarcha et al. [16] and Radisson et al. [17]. Using a quasi-two dimensional Hele-Shaw burner for experimental study the interaction between cusps developing at the flame front, they have shown that the small expansion model [9, 10] provides a good description of the flame evolution. They showed that analytical solutions of Eq. (8.1.1) in the form of pole decomposition [18] make the flame dynamics analytically tractable and can be interpreted as the dynamics of cusps and their interaction on the flame surfaces. All the same extrapolating Θ to realistic values in Eq. (8.1.1), e.g. $\Theta \sim (6-10)$, results in essentially lower velocities than those observed in experiments. Much better results corresponds to use $(\Theta - 1)$ as an expansion parameter[1] instead of $\varepsilon = (\Theta - 1)/\Theta$, used originally in [9].

Let us consider a 2D steady flame of zero thickness propagating in the negative direction of the z-axis with the normal velocity U_f with respect to an initially uniform fuel at rest. In the reference frame of the stationary flame front, the flow equations upstream and downstream, together with the conservation laws at the flame front, imply the following relation between the flame front position $z = f(x)$ and z-component of the fuel velocity $\mathbf{v} = \mathbf{v}(x, z)$ normalized on U_f. Just ahead of the front

$$u_- = v_z(x, f(x) - 0) = V + \frac{(\Theta - 1)}{2}\hat{\Phi} f, \tag{8.1.3}$$

where V is the absolute value of the flame velocity. Combining Eq. (8.1.3) with the condition that the normal flame velocity with respect to the fuel to the leading order in the front slope f' is $u_- = \partial f/\partial t + (f')^2/2$, for a steady flame propagation, $\partial u_-/\partial t = 0$ and $\partial f/\partial t = -W$, we obtain

$$-W + \frac{1}{2}(f')^2 = \frac{(\Theta - 1)}{2}\hat{\Phi} f, \tag{8.1.4}$$

where $W \sim (f')^2$ is the flame velocity increase due to the flame front curvature. Since the right-hand side of Eq. (8.1.4) is linear in the slope, while the left-hand side is only quadratic, nonlinearity can be considered small only if $Le = 1$ is small, in which case $f' = O(\varepsilon)$ and $W = O(\varepsilon^2)$. Thus, for arbitrary Θ, the weak nonlinearity approach to the stationary flames turns out to be self-contradictory [19]. A consistent perturbative treatment of the stationary flames is possible only in the framework of the small the small ε expansion.

[1] G. Sivashinsky, private communication.

8.2 Nonlinear Equation for Curved Stationary Flames

The stabilizing role of nonlinear effects is best illustrated in the case of stationary flame. We will consider a two-dimensional flame propagating in an initially homogeneous pre-mixed gaseous fuel-oxygen mixture. Since the main problem is the effect of the finite expansion coefficient on the flame structure, we will further assume the equation of state of an ideal gas, a constant coefficient of thermal conductivity, and a constant specific heat capacity. The ratio of the heat and mass diffusivities, the Lewis number, is taken $Le = 1$. We consider a flame propagating in the 2D tube with ideally slip adiabatic walls. The development of the DL instability ends up with the formation of a steady flame configuration. We consider first a zero-thickness flame. The Cartesian coordinates (x, z) are chosen so that z-axis is parallel to the tube walls. It is convenient to introduce the following dimensionless variables

$$(\eta, \xi) = (x/R; z/R), \ (w; u) = (v_x/U_f; v_z/U_f), \ \Pi = \frac{P - P_f}{\rho_- U_f^2}, \tag{8.2.1}$$

where U_f is the velocity of a plane flame, P_f is the initial pressure in the fuel mixture far ahead of the flame, and R is some characteristic length of the problem, e.g., the cutoff wavelength. The boundary conditions at the tube walls are $f(\eta = 0, R) = 0$, $w(\eta = 0, R) = 0$.

Then, the velocity and pressure fields obey the following equations

$$\frac{\partial u}{\partial \xi} + \frac{\partial w}{\partial \eta} = 0, \tag{8.2.2}$$

$$u\frac{\partial u}{\partial \xi} + w\frac{\partial u}{\partial \eta} = -\frac{1}{\rho}\frac{\partial \Pi}{\partial \xi}, \tag{8.2.3}$$

$$u\frac{\partial w}{\partial \xi} + w\frac{\partial w}{\partial \eta} = -\frac{1}{\rho}\frac{\partial \Pi}{\partial \eta}, \tag{8.2.4}$$

where ρ is the fluid density scaled on the fuel mixture density in the upstream flow ρ_-.

The flow equations are complemented by the conservation laws to be satisfied across the flame front:

$$u_+ - \Theta u_- - f'(w_+ - \Theta w_-) = 0, \tag{8.2.5}$$

$$w_+ - w_- + f'(u_+ - u_-) = 0, \tag{8.2.6}$$

$$\Pi_+ - \Pi_- + (\Theta - 1) = 0, \tag{8.2.7}$$

where the flame front position in dimensionless variables is $\xi = f(\eta)$, the subscripts "+" and "−" mean that the corresponding quantity should be evaluated at $\xi = f(\eta) + 0$ and $\xi = f(\eta) - 0$, respectively.

The normal velocity of the fuel at the flame front is unity, which is expressed in the form of the evolution equation

$$u_- - f' w_- = N. \tag{8.2.8}$$

where $N = \sqrt{1 + (f')^2}$ and $f' \equiv df/d\eta$.

The upstream flow is assumed to be potential at $\xi = -\infty$ with the flame velocity in the laboratory frame of reference, $u = V$, and $w = 0$. In an ideal fluid vorticity is conserved along the stream lines according to the Kelvin-Thomson theorem [20] and in the case of the two-dimensional incompressible flow, also the value of vorticity transferred by the fluid is conserved. This means that the upstream flow is potential everywhere for $\xi < f(\eta)$. Since the flow is potential upstream, the general solution of Eqs. (8.2.2)–(8.2.4) can be written as

$$u \equiv V + \tilde{u} = V + \int_{-\infty}^{\infty} \tilde{u}_k \exp(|k|\xi + ik\eta) dk, \tag{8.2.9}$$

$$w = \hat{H}\tilde{u}, \tag{8.2.10}$$

$$\Pi + \frac{1}{2}\left(u^2 + w^2\right) = const_1, \tag{8.2.11}$$

where $\widehat{H} = \hat{\Phi}^{-1}\partial/\partial x$ is the Hilbert operator

$$(\hat{H} f)(\eta) = \frac{1}{\pi} p.v. \int_{-\infty}^{\infty} d\zeta \frac{f(\zeta)}{\zeta - \eta}, \tag{8.2.12}$$

where "p.v." denotes the principal value. Although the relation (8.2.10) between the velocity components upstream is nonlocal, it is expressed in terms of the transverse coordinate η only.

To derive the equation for the flame front position $f(\eta)$ we must solve the bulk Eqs. (8.2.2)–(8.2.4) for the fuel and combustion products with appropriate boundary conditions on the walls, subject to the conservation laws (8.2.5)–(8.2.7) which are the boundary conditions at the flame front. The requirement that the obtained solution satisfies Eq. (8.2.8) then gives an equation for the function $f(\eta)$.

Because of the curved flame generates vorticity, the downstream flow is not potential, and therefore the general solution of the flow equations for the burnt matter cannot be written in a closed form, since the value of the vorticity downstream is not known in advance. The presence of vorticity makes the problem essentially nonlocal, so that there is no relation for the variables u, w, Π, which can be expressed in terms of the coordinate η alone, since the ξ-dependence of these variables is unknown. As in [21] it is assumed that a potential mode $v_p = (w_p, u_p)$ can be extracted from the downstream velocity field $v = (w, u)$, such that the Bernoulli-type relation holds between v_p and Π:

$$\Pi + \frac{1}{2\Theta}\left(u_p^2 + w_p^2\right) = const_2. \tag{8.2.13}$$

Then, Eqs. (8.2.2)–(8.2.4) can be written as

$$\frac{\partial u}{\partial \xi} + \frac{\partial w}{\partial \eta} = 0, \quad \frac{\partial u_p}{\partial \xi} + \frac{\partial w_p}{\partial \eta} = 0, \tag{8.2.14}$$

$$u\frac{\partial u}{\partial \xi} + w\frac{\partial u}{\partial \eta} - u_p\frac{\partial u_p}{\partial \xi} - w_p\frac{\partial w_p}{\partial \eta} = 0, \tag{8.2.15}$$

$$u\frac{\partial w}{\partial \xi} + w\frac{\partial w}{\partial \eta} - u_p\frac{\partial w_p}{\partial \xi} - w_p\frac{\partial w_p}{\partial \eta} = 0. \tag{8.2.16}$$

The solution for the potential mode can be written similar to Eqs. (8.2.9) and (8.2.10)

$$u_p \equiv \Theta V + \tilde{u}_p = \Theta V + \int_{-\infty}^{\infty} \tilde{u}_k \exp(-|k|\xi + ik\eta)dk, \tag{8.2.17}$$

$$w_p = -\hat{H}\tilde{u}_p. \tag{8.2.18}$$

The Eq. (8.2.13) does not uniquely define the potential mode, since a constant term in the ξ-component of the velocity can be assigned either to the potential mode or to the vorticity mode. The value of "$const_2$" in Eq. (8.2.13) depends on the choice of "$const_1$" in Eq. (8.2.11). The choice of the constant term ΘV in (8.2.17) is fixed up to the second order by the requirement that the mass flow at $\xi \to +\infty$ where $u = u_p = const$ and $w_p = 0$, equals to the mass flow at $\xi \to -\infty$.

Using the continuity Eq. (8.2.14) we can write (8.2.15) in the form in which ξ-dependence is implicit

$$u\frac{\partial w}{\partial \eta} - w\frac{\partial u}{\partial \eta} - u_p\frac{\partial w_p}{\partial \eta} + w_p\frac{\partial u_p}{\partial \eta} = 0. \tag{8.2.19}$$

The Eqs. (8.2.5)–(8.2.8), (8.2.10)–(8.2.11), (8.2.13), (8.2.18) and (8.2.19) is written now in the form in which the ξ dependence of the flow variables is purely parametric. The equations for the potential mode upstream and downstream, together with Eq. (8.2.19) and the conservation laws at the flame front constitute the system of equations sufficient to derive the equation for $f(\eta)$ in a closed form. However, in its general form such strongly nonlinear integrodifferential equation is very complicated and can be solved only numerically. We can say that to derive the equation for $f(\eta)$ it is not necessary to know the explicit dependence of the variables describing the flow on the coordinate ξ because the equation for the flame front shape can always be reduced to $\xi - f(\eta) = 0$, which means that it is sufficient to know only the dependence of velocities and pressure on η. This fact allows one to reduce the above set of equations to the single equation for $f(\eta)$. Details of rather cumbersome calculations can be found in [22]. The third order expansion in $(\Theta - 1) \ll 1$ we obtain

$$(\Theta + 1)W - \Theta(f')^2 = (\Theta - 1)\hat{H} f'. \tag{8.2.20}$$

Since $\widehat{H} = \Phi^{-1}\partial/\partial x$, Eq. (8.2.20) coincides with equation derived in [15] for the flame of zero thickness

$$(f')^2 - (\Theta + 1)W = (\Theta - 1)\hat{\Phi} f' \tag{8.2.20a}$$

Equation (8.2.20) can be generalized for the flame front of small thickness L_f. Using the general form of the evolution equation and the conservation laws derived in Chap. 5 we first obtain expression for the jump $[\sigma]_-^+$ of vorticity $\sigma = (\partial u/\partial \eta - \partial w/\partial \xi)$ across the flame front, which in the 2D stationary case can be written as

$$[\sigma]_-^+ = -\frac{\Theta - 1}{\Theta N}\left(\hat{D}w_- + f'\hat{D}u_- + \frac{1}{N}\hat{D}f'\right), \tag{8.2.21}$$

where $\hat{D} = \left(w_- + \frac{f'}{N}\right)\frac{\partial}{\partial \eta}$.

The conservation laws at the flame front in the first order approximation with respect to the flame front thickness can be written as

$$u_+ - u_- - f'(w_+ - w_-) = (\Theta - 1)N, \tag{8.2.22}$$

$$w_+ - w_- + f'(u_+ - u_-) = \varepsilon ln\Theta\left(\hat{D}w_- + f'\hat{D}u_- + \frac{1}{N}\hat{D}f'\right), \tag{8.2.23}$$

$$\begin{aligned}\Pi_+ - \Pi_- &= -(\Theta - 1) + \varepsilon(\Theta - 1)\left(\frac{f'}{N}\right)\\ &+ \frac{\varepsilon \ln \Theta}{N}\left(w_-^2 f'' + 2\hat{D}N - \frac{f'N'}{N}\right),\end{aligned} \tag{8.2.24}$$

and the evolution equation

$$u_{-} - f'w_{-} = N - \varepsilon \frac{\Theta \ln \Theta}{\Theta - 1} \frac{d}{d\eta}(Nw_{-} + f'). \tag{8.2.25}$$

where $\varepsilon = L_f/R$ is the small dimensionless parameter, the ratio of the flame thickness to the characteristic length of the problem, the width of the tube, R.

The set of the above equations written in the form, in which the ξ dependence of the flow variables is only parametric, can be reduced to a single equation for $f(\eta)$ and after a simple but lengthy algebra [22], for the flame of a small but finite thickness, we obtain the equation for the flame front position in the third order expansion

$$\Theta(f')^2 - (\Theta + 1)W = (\Theta - 1)\left(-\hat{H}f' + \frac{\lambda^{(1)}}{2\pi}f''\right), \tag{8.2.26}$$

where $\lambda^{(1)} = \frac{4\pi\varepsilon}{\Theta-1}\left(1 + \frac{3(\Theta-1)}{2}\right)$ is the first order approximation to the exact value of the cutoff wavelength obtained in the linear theory of the DL-instability (Chap. 5)

$$\lambda_c = \frac{2\pi\varepsilon}{\Theta - 1}\left(\Theta - 1 + \Theta \ln \Theta \frac{\Theta + 1}{\Theta - 1}\right).$$

The linear terms in Eq. (8.2.26) describe the dispersion relation of the DL instability in the limit $\Theta - 1 \ll 1$ with the cut-off wavelength $\lambda_c = \frac{4\pi L_f}{\Theta-1}$.

A nonlinear, nonstationary equation describing evolution of weakly curved premixed flames with arbitrary gas expansion was derived in [23]. Further progress in obtaining a nonlinear equation for an arbitrary gas expansion is based on the fact that, using the Kelvin-Thomson theorem, one can obtain an explicit expression for the vorticity mode in the downstream flow [24, 25]. Then, using the dispersion relation for the potential mode at the flame front, the system of bulk equations and jump conditions for the flow variables can be reduced to a set of integrodifferential equations for the flame front position. For a flame of zero thickness a single equation for the flame front position can be obtained in the third and fourth orders expansion in $(\Theta - 1) \ll 1$, as well as in the asymptotic expansion with respect to the small flame front thickness. In particular, it was pointed in [23] that the constant term $V\Theta$ in the potential mode decomposition in Eq. (8.2.17) is of the third order accuracy in $(\Theta - 1) \ll 1$. The corrected in [24] equation of the fourth order accuracy in $(\Theta - 1)$ is

$$\frac{4\Theta^2(f')^2}{(\Theta + 1)^2} - 2\Theta W - W^2 = (\Theta - 1)\left(-\hat{H}f' + \frac{\lambda_c}{2\pi}f''\right) \tag{8.2.27}$$

At the same time, one should keep in mind that we cannot expect that analytical theory can provide a quantitative agreement with either real or numerical experiments for such complex phenomena. An analytical approach is inevitably based on simplifying assumptions that do not necessary fully adequate to real physical processes.

8.3 Velocity of 2D Curved Flame

The solution to Eq. (8.2.26) allows to find the velocity of a 2D curved stationary flame propagating in a tube of width R with ideally slip adiabatic walls. The configuration of the flame in a tube with ideal walls is equivalent also to the flame with a periodic structure, for which half of the period is determined by the tube width. It was first shown by Thual et al. [18] that nonlinear equations of the type considered in Sect. 8.2, admit analytical solutions in the form of pole decomposition with periodic boundary conditions. More details on application of the method of pole decomposition can be found in [26–28]. According to this method we consider a flame propagating in a tube of width R with ideal walls at $\eta = 0$ and $\eta = R$, and look for the $2R$ periodic solution with the boundary conditions at the walls

$$f'(\eta = 0) = f'(\eta = 1) = 0. \tag{8.3.1}$$

It should be noted that for the adiabatic boundary conditions at the ideal walls, the flame front touches the walls at the right angle, which is a direct consequence of the effect of thermal conduction accounted, for example, in (8.2.26) by the term proportional to a finite value of L_f. If one considers a wide tube neglecting by the thermal conduction, then the flame front would touch one of the walls at a sharp angle leading to formation of a stagnation zone in the flow close to the wall behind the flame. The sharp angle between the flame is smoothed out due to the finite thickness of the flame front, and the stagnation zone near the wall is reduced to a region of large velocity gradient.

We look for the 2R-periodic solution satisfying the boundary conditions (8.3.1) in the form

$$f(\eta) = A \sum_{i=1}^{2M} \ln\left[\sin\left(\frac{\pi}{2R}(\eta - \eta_i)\right)\right], \tag{8.3.2}$$

where the amplitude A and the complex poles η_i, with $i = 1, \ldots 2M$ will be determined substituting (8.3.2) into Eq. (8.2.26). Since the function $f(\eta)$ is real for real η, the poles appear as M conjugate pairs. The condition of the 2R-periodic solutions to be symmetric with respect to the reflection $\eta \to -\eta$, leads to periodic as well as non-periodic solutions of Eq. (8.2.26) with the boundary conditions $f'(\eta = 0) = f'(\eta = 1) = 0$. Using the formulas [18]

$$\cot a \cot b = -1 + \cot(a - b)(\cot b - \cot a), \quad \mathrm{sign}(x) \equiv x/|x|,$$

$$\hat{H} f' = -\frac{\pi A}{2R} \sum_{k=1}^{2M} \left\{ 1 + i \cdot \text{sign}[\text{Im}\eta_k] \cot\left(\pi \frac{\eta - \eta_k}{2R}\right) \right\}, \quad \text{sign}(x) \equiv x/|x|,$$

it can be verified that $f(\eta)$ in the form (8.3.2) is a solution to Eq. (8.2.26) for

$$A = -\frac{\lambda_c(\Theta - 1)}{2\pi\Theta}, \quad W = \frac{(\Theta - 1)^2}{2\Theta^2} \frac{M\lambda_c}{2R} \left(1 - \frac{M\lambda_c}{2R}\right), \tag{8.3.3}$$

and the poles η_k, $k = 1, \ldots., 2M$ satisfy the following set of equations

$$i \cdot \text{sign}[\text{Im}\eta_k] + \frac{\lambda_c}{2R} \sum_{i=1 (i \neq k)}^{2M} \cot\left(\frac{\pi}{2R}(\eta_k - \eta_i)\right) = 0, \quad k = 1, \ldots., 2M. \tag{8.3.4}$$

It is seen that the obtained solution is not unique: there is a number of solutions corresponding to different numbers M of poles. The problem arises how to treat these solutions since such parameter, as the number of poles does not have clear physical interpretation. To identify the physical solution, a stability analysis of the solutions is required which, cannot be carried out in the framework of the stationary theory. It was shown in [29–31] that for the stationary Sivashinsky equation there is only one stable solution corresponding to the number of poles for the maximum flame velocity for a particular tube width. Since the structure of Eq. (8.2.26) is similar to the stationary Sivashinsky equation, it is reasonable to assume that the nonstationary versions of these equations [23] and the stability analysis are also similar. Using Eq. (8.3.3) for W we conclude that the maximum of the flame velocity corresponds to the maximum number of the pole pairs, $M_{\max} = \text{Int}[\frac{R}{\lambda^{(1)}} + \frac{1}{2}]$, where Int$[x]$ denotes the integer part of x. Then, for the flame velocity increase we obtain

$$W = U_w - U_f = 4W_{\text{m}} \frac{M_{\max}\lambda^{(1)}}{2R} \left(1 - \frac{M_{\max}\lambda^{(1)}}{2R}\right). \tag{8.3.5}$$

The maximum velocity increase depending on the expansion coefficient is

$$W_s = \frac{U_m}{U_f} = \frac{(\Theta - 1)^2}{4\Theta(\Theta + 1)}. \tag{8.3.6}$$

Figure 8.3 shows the dependence of the flame velocity increase on the inverse tube width (solid line) obtained from Eq. (8.3.5) in comparison with the numerical simulations [32] shown by stars. In the same way the velocity of a stationary flame can be calculated in the fourth order approximation (8.2.27). Looking for a periodic solution of Eq. (8.2.27) in the form (8.3.2), we find that Eq. (8.3.3) for the pole positions remain the same, and therefore for values of A and W we have

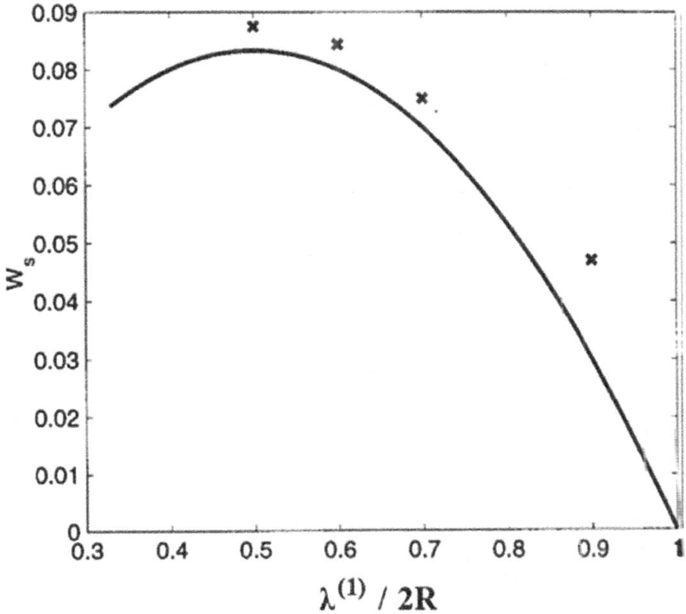

Fig. 8.3 Dependence of the flame velocity increase on the inverse tube width, given by Eq. (8.3.5) for the case $\Theta = 3$ is shown by solid line, and the results of numerical experiments [32] (stars)

$$A = -\frac{\lambda_c(\Theta^2-1)}{2\pi\Theta^2}(\Theta+1), \quad W^2+\Theta W = \frac{(\Theta-1)^2}{4\Theta^2}\frac{M\lambda_c}{2R}\left(1-\frac{M\lambda_c}{2R}\right). \tag{8.3.7}$$

Since the stable solution corresponds to the largest flame velocity for a given tube width and a given value of Θ, for the maximum flame velocity increase we obtain

$$W_s = \frac{U_{\max}}{U_f} = \frac{\left(\Theta^2-1\right)^2}{32\Theta^3}. \tag{8.3.8}$$

The maximal increase of the flame velocity versus the gas expansion coefficient Θ given by the Sivashinsky equation (dashed line), by Eq. (8.3.6) (dashed dotted line) and by the Eq. (8.3.8) (solid line), and the results of numerical simulations [32] are shown in Fig. 8.4. It should be remembered that formulas (8.3.6) and (8.3.8) were derived only as the asymptotic expansion for sufficiently small values of the expansion coefficient $(\Theta-1) \ll 1$. However, it is interesting that the derived nonlinear Eqs. (8.2.26) and (8.2.27) are valid not only for the case of ideal nonviscous fluids with a constant coefficient of thermal conduction, but they are also applicable to viscous mixtures, with arbitrary dependence of the transport coefficients on temperature and an arbitrary equation of state [33].

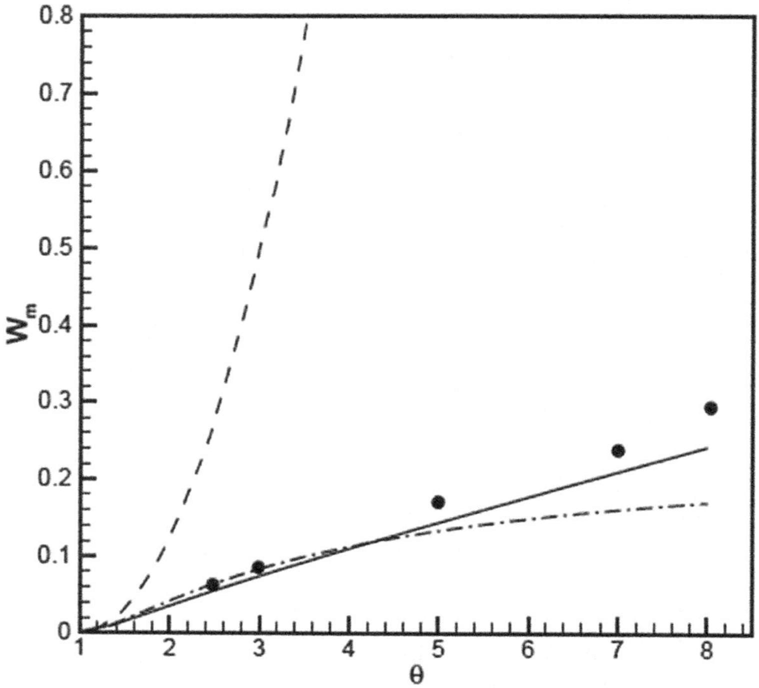

Fig. 8.4 Dependence of the maximum flame velocity increase on the expansion coefficient: dashed line—the Sivashinsky equation, dashed-dotted line—Eq. (8.3.6), solid line—Eq. (8.3.8), dots—numerical experiments [32]

The corresponding corrections can be made for the expression of the cut-off wavelength that follows from the linear dispersion relation (Chap. 5, Eq. (5.3.17)) while the mathematical structure of Eqs. (8.2.26) and (8.2.28) remains unchanged.

8.4 Velocity and Shape of 2D Flame

Dynamics of the DL instability development leading to the formation of curved flames was studied using 2D numerical simulations [7, 32–34] for different widths of the tube and different values of the expansion coefficient. In the reference frame comoving with the flame, the fuel mixture enters the tube as a uniform flow with a velocity of the planar flame. Because of the ideal boundary conditions at the walls the tube width R determines a half of the largest possible wavelength of permitted perturbations, i.e. the wavelength of the perturbation may take the values $\lambda = 2R/j$, $j = 1, 2, 3$, etc. The development of the DL instability for perturbations of a wavelength shorter than the cut-off wavelength is suppressed by the thermal conduction. This means that if a tube is sufficiently narrow, such that $R < R_c = \lambda_c/2$, then all permitted perturbations belong to the stable part of the dispersion Eq. (7.2.43).

According to Eq. (8.3.5), which is convenient to rewrite in dimensionless form

$$(U_w - U_f)/U_{\max} = 4\frac{M\lambda_c}{2R}\left(1 - \frac{M\lambda_c}{2R}\right), \tag{8.4.1}$$

the dependence of the curved flame velocity on the inverse tube width should look like a combination of parabola pieces with maximums at the points $R = M\lambda_c$, $M = 1, 2, 3, \ldots$ shown in Fig. 8.5 obtained using 2D numerical simulations [32, 34]. In agreement with the theory, the numerical simulations of a flame propagating in the tubes with ideal walls [32, 34] demonstrate that flame evolution in narrow tubes $R < R_c$ always leads to a planar flame. Even in the case when the initially planar flame front was disturbed by the perturbations of a relatively large amplitude, the flame front restored its initial planar configuration.

In wider tubes $R > R_c$ the hydrodynamic instability results in a stationary curved shape of a flame front [7, 32]. As the amplitude of the first perturbation harmonic with the wavelength $\lambda = 2R$ grows, it induces the growth of higher harmonics with smaller wavelengths $\lambda = R, 2R/3$, etc. due to the nonlinear interaction. Since harmonics with higher wavenumbers are stable, the rate of the perturbation growth slows down, saturated and stops when the final amplitudes corresponding to the

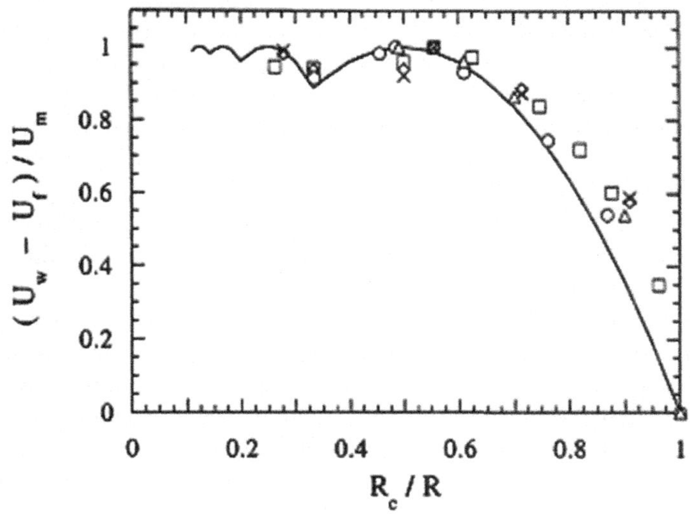

Fig. 8.5 Scaled velocity of a 2D curved stationary flame in the tubes with ideal walls versus the inverse tube width. The solid line is the flame velocity given by Eq. (8.3.9); triangles, circles, squares, diamonds, crosses show results of numerical simulations for expansion coefficients $\Theta = 3, 5, 6, 8, 10$, respectively

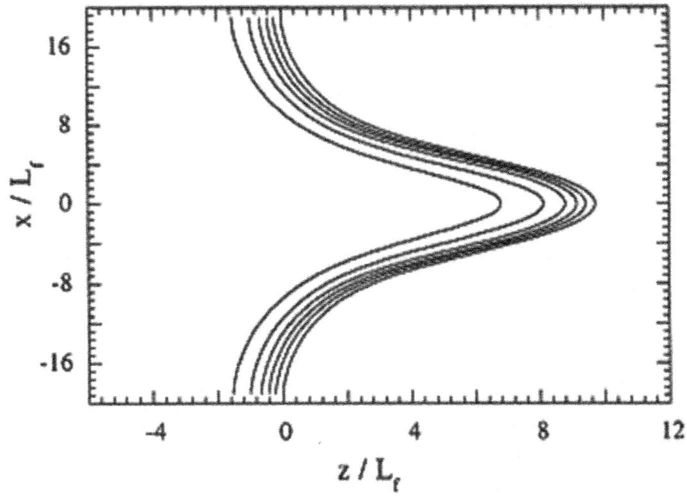

Fig. 8.6 The shape of a curved steady flame with the expansion coefficient $\Theta = 6$ in the tube of width $R = 2R_c$ shown by isotherms corresponding to the temperatures from 600 to 2100 K with the step 300 K, [32]

stationary curved flame are achieved. The resulting shape of the stationary curved flame with the expansion coefficient $\Theta = 5$ is shown in Fig. 8.6.

In [7, 32] the increase of the flame velocity caused by the DL instabilities of the flame front was studied by means of numerical simulations of hydrodynamics equations with chemical kinetics of a one-step irreversible reaction, and with a constant coefficient of thermal conduction, constant Prandtl number and unit Lewis number. An infinite length of the tube was assumed, which was achieved in simulations by an appropriate choice of the computational intervals. The boundary condition at the ideally adiabatic and slip walls of the tube of width R are $u_x = 0,\ u_z \neq 0,\ \partial T/\partial x = 0\ at\ x = 0, R$, the z-axis z is directed along the wall and the x-axis in the transverse direction. Then the main parameters of the simulations are the tube width R and the expansion coefficient Θ.

When a planar stationary flame was established, 2D perturbations were imposed at the flame front, $u_z = u_z + A_0 u_z(z)\varphi(x)\exp\left[-\pi^2(z - z_f)^2/R^2\right]$ with $\varphi(x) = \cos(\pi x/L)$ and the dimensionless perturbation amplitude $A_0 = 10^{-3} - 10^{-4}$. Small perturbations grow due to the DL instability until the flame acquires a curved shape and the flame velocity increases. To check the dependence on the initial conditions, several calculations were performed with different kinds of initial perturbations decomposed into a set of the Fourier harmonics $\varphi(x) = \sum_{i=1}^{N} \tilde{a}_i \cos(i\pi x/R)$. In tubes of a moderate width $1 < R/R_c < 3$ the DL instability leads to a smooth curved stationary flame shown in Fig. 8.6. The flame front may be described as a hump directed towards the fresh fuel mixture and a cusp pointing to the combustion products. It is seen in Fig. 8.5 that the velocity of the curved stationary flame equals to the planar flame velocity U_f for $R < R_c$ in agreement with Eq. (8.3.5). An increase in the width of the tube leads to the increase in the flame velocity until it reaches the first local maximum. According to (8.3.5) the first local maximum of the flame velocity is achieved for the tube width equal to the doubled first critical value $R = 2R_c$ in a good agreement with the first local maximum in Fig. 8.5. The point of the first local maximum of a curved stationary flame also corresponds to the largest growth rate of instability for the tube of width $R = 2R_c$ in the linear theory of the DL instability. The maximum velocity amplification for curved stationary flames depends on the expansion coefficient: the larger the expansion coefficient, the stronger the DL instability and the greater the velocity amplification. If the tube width increases, the radius of curvature of the stationary flame also increases, and the stabilizing effect of the curved flame shape weakens: the flame front resembles locally a planar flame, and the DL instability should occur on a new scale. This secondary DL instability can lead to additional wrinkling of the flame front and to the additional increase of the flame velocity.

The result of the secondary DL instability is the abrupt change in the dependence of the flame velocity on the tube width near the point corresponding to $R \approx 1.8\lambda_c$ in Fig. 8.5. The structure of the stable solution for a wide tube is similar to the stationary flame shape shown in Fig. 8.7 for $R = 8\lambda_c$, $\Theta = 8$. The wider the tube, the stronger the flame velocity increases and the flame front inclines more towards the tube walls. The evolution of the flame shape propagating in the tube of width

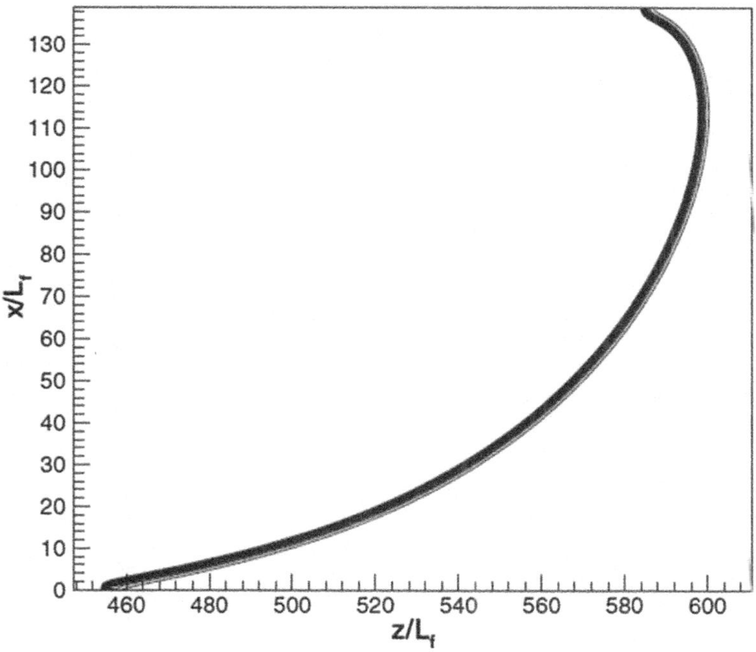

Fig. 8.7 The stationary flame front shape in a tube of width $R = 8\lambda_c$ and $\Theta = 8$

$R = 4\lambda_c$, for $\Theta = 8$ is shown in Fig. 8.8, where the flame front isotherms are shown at dimensionless time $\tau = U_f t/R$. The last plot at $\tau = 8.49$ shows the stationary flame. The "two-hump" shape of the flame develops during the earlier time but turns out to be unstable and converts to the stable "single-hump" flame front with the flame velocity $U_w > U_m$. A 'multi-hump' shape of the flames in wide tubes possible only during the nonstationary stage of the flame propagation. In the stationary regime only a 'single-hump' flame inclined to the tube walls is possible. The tilted shape of flames propagating in wide tubes is similar to that observed in experiments [4–6], see Fig. 8.2. The transition from the intermediate nonstationary 'multi-hump' configuration to a convex one looks like sweeping out of one hump by another one.

The flame shape resulting from the growth of the most unstable harmonic is quasi-stationary. The flame front at this stage consists of two identical humps and its velocity can be estimated with a high accuracy [34]. The velocity evolution is similar to the velocity evolution in tubes of moderate width. But later, harmonics with initially

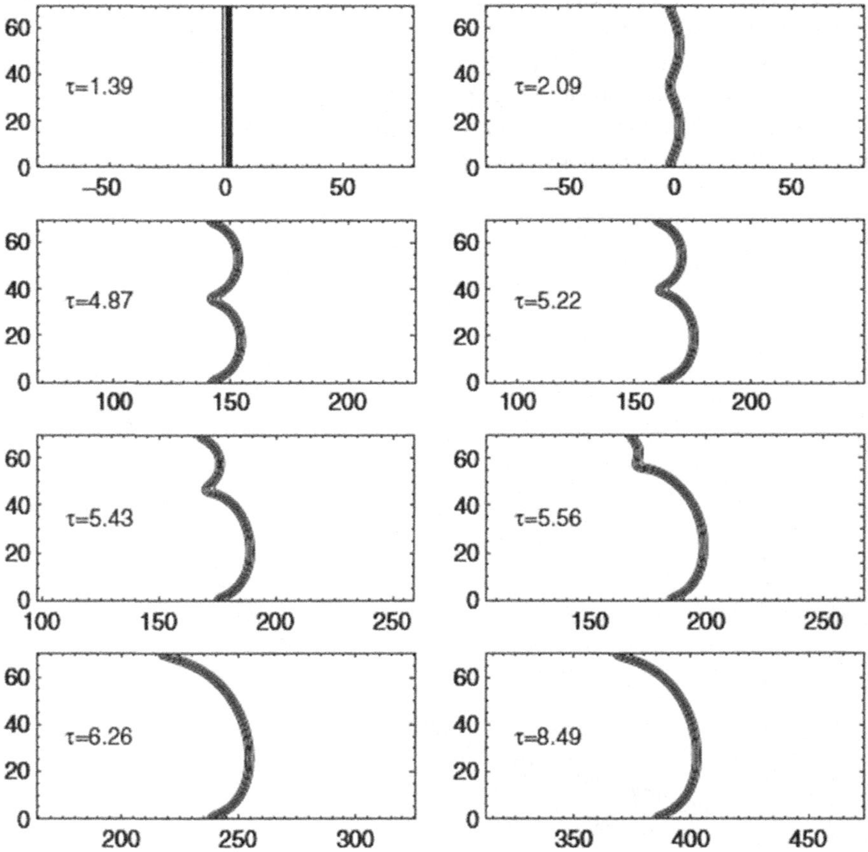

Fig. 8.8 Evolution of the flame front in a tube of width $R = 4\lambda_c$, $\Theta = 8$. The isotherms are shown at different time instants with the last frame showing a stationary flame [7]

lower amplitudes begin to grow, and the quasi-stationary flame structure decays. Pulsations of the flame velocity refers to a change in the depth of the cusps at the flame front. The secondary DL instability results in a considerable growth of the Fourier harmonic, while the first harmonic losses its dominant role. As a rule, the new extra cusp developing at the flame front has noticeably larger depth than the first.

In Fig. 8.9 it is seen that, contrary to the velocity dependence from the tube width of a curved stationary flame predicted by the solution of nonlinear equation and shown in Fig. 8.5, the flame velocity considerably increases when the tube width exceeds $R > 3.4\lambda_c$, while the flame shape remains smooth and inclined in the tube without forming a multi-hump structure. This velocity increase is similar to the experimentally observed dependence of the flame speed shown in Fig. 8.2. The increase of the curved flame velocity U_w with the tube width larger than $R \geq D_{c2} \approx 4\lambda_c$ can be obtained from the closed form of nonlinear equation, which can be solved only numerically. Finally, it is worth noting that the simulation results shown in Figs. 8.6, 8.7, 8.8, 8.9 were obtained for a slow flame with the Mach number $M_f = U_f/a_s \approx 0.001$ to avoid the influence of compressibility.

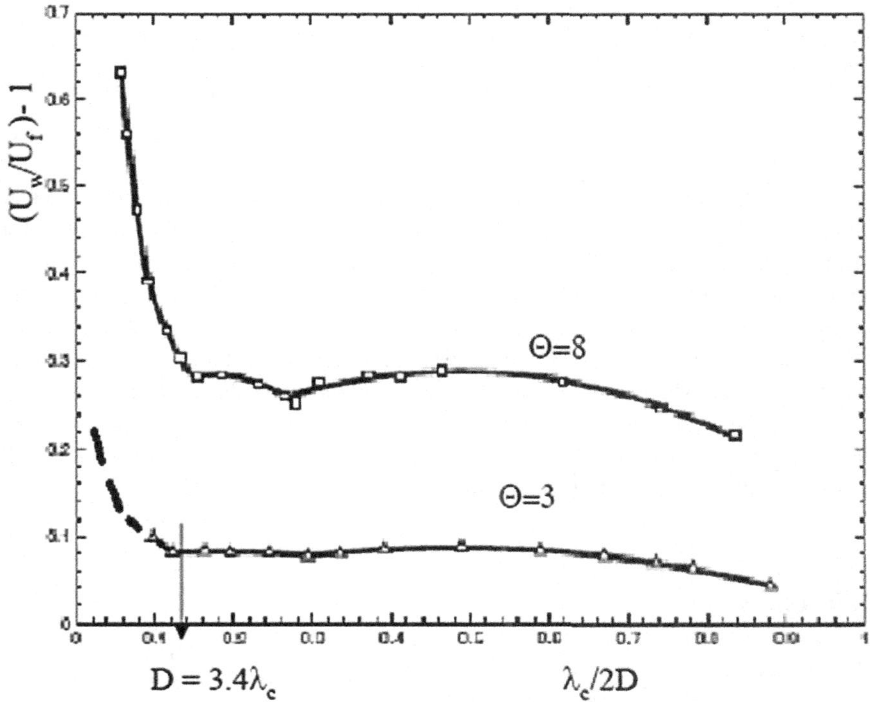

Fig. 8.9 Dependence of the scaled flame velocity increase on the inverse tube width ($R \equiv D$) for wide tubes. Squares and triangles correspond to solutions for $\Theta = 8$ and $\Theta = 3$, respectively [7]

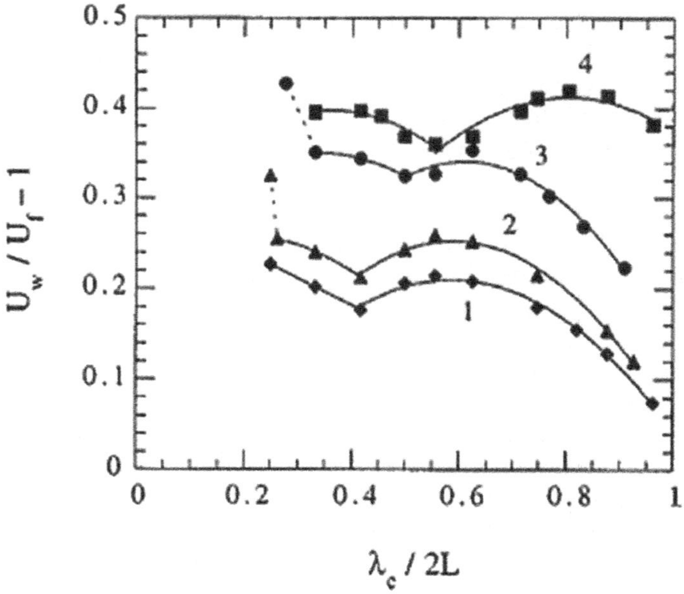

Fig. 8.10 The scaled velocity of a curved stationary flame with $\Theta = 6$ as a function of the inverse tube width for different Mach numbers of the initial planar flame: (1) $M_f = 0.01$; (2) $M_f = 0.1$; (3) $M_f = 0.15$; (4) $M_f = 0.17$. The markers show the results of 2D simulations, the solid lines present the best parabolic approximation

Figure 8.10 shows the dependence of the reduced velocity of a curved stationary flame with $\Theta = 6$ on the inverse tube width scaled by the cutoff wavelength for different Mach numbers of the initial planar flame [35]. The dependence of the curved flame velocities on the inverse tube width in compressible flows qualitatively resembles the results given by solutions of nonlinear equation for isobaric curved flames. In agreement Sect. 5.5.4 the growth rate of the DL instability increases with an increase in the Mach number, and also the speed increases due to the curved shape of the flame becomes stronger. This means that even simple solutions of the nonlinear equation obtained in a closed form in isobaric approximation for an arbitrary Θ will provide reliable results only for very slow flames.

References

1. Y.B. Zeldovich, An effect stabilizing the curved front of a laminar flame. Zh. Prikl. Mekh. Tekh. Fiz. **1**, 102–104 (1966). ((in Russian))
2. G.H. Markstein, Experimental and theoretical studies of flame front stability. J. Aero. Sci. **18**, 199–209 (1951)
3. E. Mallard, H. Le Chatelier, Recherches Experimentales et Theoriques sur la Combustion des Melanges Gaseoux Explosifs. Ann. Mines **8**, 274–568 (1883)

4. H.F. Coward, H.P. Greenwald, Propagation of flame in mixtures of natural gas and air. U.S. Bureau of Mines, Tech. Paper **427**, 28 (1928)
5. H.F. Coward, F.J. Hartwell, Studies in the mechanism of flame movement. Part I. The uniform movement of flame in mixtures of methane and air, in relation to tube diameter. J. Chem. Soc. 1996–2004 (1932)
6. H.F. Coward, F.J. Hartwell, Studies in the mechanism of flame movement. Part II. The fundamental speed of flame in mixtures of methane and air. J. Chem. Soc. 2676–2684 (1932)
7. M.A. Liberman, M.F. Ivanov, O. Peil, D. Valiev, L.-E. Ericsson, Numerical modeling of flame propagation in wide tubes. Comb. Theory Modeling **7**, 653–676 (2003)
8. Y.A. Gostintsev, A.G. Istratov, Y.V. Shulenin, Self-similar propagation of a free turbulent flame in mixed gas mixtures. Combust. Explos. Shock Waves **24**, 563–569 (1989)
9. G.I. Sivashinsky, Nonlinear analysis of hydrodynamic instability in laminar flames-I Derivation of basic equations. Acta Astronaut. **4**, 1177–1206 (1977)
10. D.M. Michelson, G.I. Sivashinsky, Nonlinear analysis of hydrodynamic instability in laminar flames-II. Num. Exper. Acta Astronaut. **4**, 1207–1221 (1977)
11. G.I. Sivashinsky, Instabilities, pattern formation, and turbulence in flames. Ann. Rev. Fluid Mech. **15**, 179–199 (1983)
12. F. Sabathier, L. Boyer, P. Clavin, Experimental study of a weak turbulent premixed flame. Prog. Aeronaut. Astronaut. **76**, 246–258 (1981)
13. J. Quinard, G. Searby, L. Boyer, Cellular structures on premixed flames in a uniform laminar flow. In "Cellular structures in instabilities", Lecture Notes in Physics, J.E. Weisfreid and S. Zaleski Ed., Springer Verlag, vol. 210, pp. 331–341 (1984)
14. G. Joulin, P. Cambray, On a tentative, approximate evolution equation for markedly wrinkled premixed flames. Combust. Sci. Technol. **81**, 243–256 (1992)
15. G.I. Sivashinsky, P. Clavin, On the nonlinear theory of hydrodynamic instability in flames. J. Phys. (Paris) **48**, 193–198 (1987)
16. C. Almarcha, B. Radisson, E. Al Sarraf, E. Villermaux., B. Denet, J. Quinard, Interface dynamics, pole trajectories, and cell size statistics. Phys. Rev. E. **98**(1–5), 030202 (2018)
17. B. Radisson, B. Denet, C. Almarcha, Nonlinear dynamics of premixed flames: from deterministic stages to stochastic influence. J. Fluid Mech. 903(1–31), 903 A17 (2020)
18. O. Thual, U. Frish, M. Henon, Application of pole decomposition to an equation governing the dynamics of wrinkled flames. J. Phys. France **46**, 1485–1494 (1985)
19. K.A. Kazakov, M.A. Liberman, Effect of vorticity production on the structure and velocity of curved stationary flames. Phys. Rev. Letters. **88**, 064502 (2002)
20. L.D. Landau, E.M. Lifshitz, *Fluid Mechanics* (Pergamon, Oxford, 1987)
21. S.K. Zhdanov, B.A. Trubnikov, Nonlinear theory of instability of a flame front. J. Exp. Theor. Phys. **68**, 65–78 (1989)
22. K.A. Kazakov, M.A. Liberman, Nonlinear equation for curved stationary flames. Phys. Fluids **14**, 1166–1180 (2002)
23. K.A. Kazakov, M.A. Liberman, Non-stationary nonlinear equation for a curved flame. Comb. Sci. Tech. **174**, 157–179 (2002)
24. K.A. Kazakov, On-shell description of stationary flames. Phys. Fluids. **17**(1–15), 032107 (2005)
25. G. Joulin, H. El-Rabii, K.A. Kazakov, On-shell description of unsteady flames. J. Fluid Mech. **608**, 217–242 (2008)
26. Y.C. Lee, H.H. Chen, Nonlinear dynamical models of plasma turbulence. Phys. Scripta. **2**, 41–47 (1982)
27. M. Renardy, A model equation in combustion theory exhibiting an infinite number of secondary bifurcations. Physica. D. **28**, 155–167 (1987)
28. G. Joulin, On the Zhdanov-Trubnikov equation for premixed flame instability. J. Exp. Theor. Phys. (Sov. Phys. JETP) **73**, 234–236 (1991)
29. M. Rahibe, N. Aubry, G.I. Sivashinsky, Stability of pole solutions for planar propagating flames. Phys. Rev. E. **54**, 4958–4972 (1996)
30. M. Rahibe, N. Aubry, G.I. Sivashinsky, Instability of pole solutions for planar propagating flames in sufficiently large domains. Combust. Theory Modell. **2**, 19–41 (1998)

31. D. Vaynblat, M. Matalon, Stability of pole solutions for planar propagating flames: I. Exact eigenvalues and eigenfunctions, SIAM J. Appl. Math. **60**, 679–702 (2000). Stability of pole solutions for planar propagating flames: II. Properties of eigenvalues and eigenfunctions with implication to flame stability," ibid. **60**, 703–728 (2000)
32. V.V. Bychkov, S.M. Golberg, M.A. Liberman, L.E. Eriksson, Propagation of curved stationary flames in tubes. Phys. Rev. E. **54**, 3713–3724 (1996)
33. V.V. Bychkov, M.A. Liberman, On the dynamics of a curved deflagration front. J. Exp. Theor. Phys. **84**, 281–288 (1997)
34. O. Yu. Travnikov, V.V. Bychkov, M.A. Liberman, Numerical studies of flames in wide tubes: stability limits of curved stationary flames. Phys. Rev. E. **61**, 468–474 (2000)
35. OYu. Travnikov, V.V. Bychkov, M.A. Liberman, Influence of compressibility on propagation of curved flames. Phys. Fluids **11**, 2657–2666 (1999)

Chapter 9
Dynamics of Flames Under Confinement

9.1 Hydrodynamic Instability of Planar Flame in Closed Chambers

Flame dynamics in closed tubes is one of the most fundamental problems in combustion science, as it is directly related to the modeling of combustion in car engines and many other industrial processes. In particular, flame propagation in a closed chamber is a key issue for understanding "knocking" combustion in car engines [1–5]. One of the most important problems is the interaction of the flame with pressure waves under confinement conditions. It is generally accepted that the flame dynamics in a closed chamber can be considerably affected by the intrinsic hydrodynamic instabilities of the flame. The DL instability is especially important because it can result in the flame-front distortion, augmentation of the flame-front surface, followed by the increase of the flame velocity. A flame accelerates when it acquires a curved shape and, therefore, its surface increases. The accelerating flame, acting as a semitransparent piston, produces the pressure and acoustic waves, which after reflections from the walls interact with the flame. When a flame front acquires a curved shape, the interaction of acoustic waves with the flame front becomes more effective, which makes easier the spontaneous reaction in the fresh fuel ahead of the flame. Even in the case of ideal walls, there are many specific factors that influence the development of the DL instability in closed chambers. Flames propagating in open tubes consume fuel in a nearly isobaric regime since the expansion of the combustion products is not restricted by the walls. The situation is different for a flame propagating in a closed tube: expansion of the hot gas of combustion products causes adiabatic compression of the unburned fuel ahead of the flame. Compression of the unburned gaseous fuel can be accompanied by a significant build-up of pressure P and temperature T, compared to their initial values P_0 and T_0, so that close to the tube end the pressure and the temperature of the unburned fuel are amplificated:

$$P = P_0\left(1 + \frac{\gamma Q}{C_P T_0}\right) = (\gamma \Theta_0 + 1 - \gamma) P_0, \qquad (9.1.1)$$

M. A. Liberman, *Combustion Physics*,
https://doi.org/10.1007/978-3-030-85139-2_9

$$T = (\gamma\Theta_0 + 1 - \gamma)^{1-1/\gamma} T_0, \tag{9.1.2}$$

where the initial fuel parameters in a closed tube are denoted by the label 0. For example, for $\Theta_0 = 5$ and $\gamma = 1.4$, the final pressure is $P/P_0 = 10.8$, and for $\Theta_0 = 8$, it is $P_1 \approx 11\ P_0$ exceeding by order of magnitude the initial pressure. The increase in pressure and temperature of the unburned mixture alters all parameters characterizing the flame, such as the flame velocity and the thickness of the flame front. Typically, the change in pressure and temperature in the unburned mixture is slow compared to the characteristic time scale $\tau_f = L_f/U_f$. Consequently, the internal structure of the flame front has enough time to adapt to changes in external conditions. Therefore, the internal structure of the flame front has enough time to adjust to the changing external conditions and at any time instant, a planar flame front in the enclosure has the properties of a freely propagating quasi-isobaric flame with some new values of the fuel pressure and temperature. The normal velocity of a planar laminar flame in ZFK theory depends on the fuel parameters as given by Eq. (6.4.17). Denoting by U_{f0} the normal velocity of a laminar flame in the mixture with temperature and pressure (T_0, P_0) and using (9.1.1) and (9.1.2), the flame velocity at the moment when the temperature and pressure became (T, P) can be written as

$$U_f(T, P) = U_{f0}\left(\frac{P}{P_0}\right)^{\frac{n-2}{2}} \frac{T}{T_0}\left(\frac{T + Q/c_p}{T_0 + Q/c_p}\right)^{\frac{2n+1}{2}} \times \exp\left(\frac{E(T - T_0)}{2(T_0 + Q/c_p)(T + Q/c_p)}\right), \tag{9.1.3}$$

where n is the reaction order. It is seen that the velocity of a laminar flame of the first-order reaction remains almost the same and only slightly increases at the end of burning with the pressure increase, while for the third-order reaction, the flame velocity considerably by an order of magnitude increases. The flame thickness decreases considerably to the end of burning for flames of both the third-order reaction and the first-order reaction, though in the former case, the decrease is much stronger. For a flame with the first-order reaction, the decrease of the flame thickness is mostly due to the increase of the fuel density ($L_f = \kappa/\rho_f C_P U_f$). On the contrary, in the case of a third-order reaction, both amplifications of the fuel density and the normal flame velocity contribute to the decrease of the flame thickness.

The scaled flame velocity and thickness are shown in Fig. 9.1 versus the flame-front position in a tube for flames with the first- and the third-order reactions [7]. Possible reduction of the flame thickness in a closed tube complicates the competition of perturbation modes at the nonlinear stage of DL instability. This already follows from the observation that a decrease in the thickness of the flame leads to a decrease in the cutoff wavelength λ_c, so that the perturbation mode, which was stable at the beginning of combustion, can subsequently become unstable. As a result, the nonlinear stage of the DL instability in closed tubes may be characterized by the

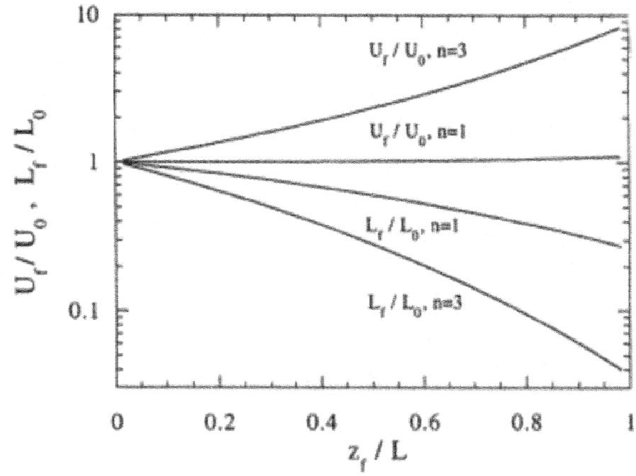

Fig. 9.1 The scaled velocity U_f/U_o of a planar flame and the scaled flame thickness L_f/L_0 versus the flame position for the reaction of the first and the third order [7]

transition from one quasi-stationary nonlinear structure to another one with a different wavelength.

Flame dynamics under confinement may be also affected by the flame acceleration or deceleration that can reach significant values in short tubes leading to strong suppression or enhancement of flame instabilities. The velocity of the flame in the laboratory reference frame U_l is quite different from the normal flame velocity. In the reference frame of a planar flame front, the fuel mixture flows into the flame front with velocity U_f and the combustion products flow out with the velocity ΘU_f. In the beginning, the burned products are at rest, the fuel mixture is pushed forward, and the flame velocity in the laboratory reference frame is $U_{fl} = \Theta_0 U_{f0}$. Close to the end of combustion, when the flame comes close to the opposite closed end of the tube, the fuel mixture is at rest, and the burned products are pushed backward and the flame velocity relative to the walls equals the normal flame velocity $U_{fl} = U_{f0}$. For example, in 2D simulations of "tulip flames" phenomenon [8] for the flame with $U_f = 4.5$ m/s, $\Theta_0 = 7.67$, $\gamma = 1.4$, $E = 4.36\Theta_0 T_0$ and the first-order reaction, the flame propagated with the velocity of about 40 m/s just after ignition. However, the final flame velocity at the end of combustion was about 5 m/s in agreement with the above estimates. For $\Theta_0 = 5$, $\gamma = 1.4$, $E = 5\Theta_0 T_0$ adopted in 2D simulations of flames in closed tubes [9], the flame velocity in the laboratory reference frame drops from $5U_0$ at the beginning to $1.1U_0$ at the end for the first-order reaction (the flame decelerates); while for the third-order reaction, the flame velocity in the laboratory reference frame increases from $5U_0$ to about $10U_0$ (the flame accelerates). The acceleration (deceleration) of the flame front with the initial flame thickness L_0 in a tube of length L can be estimated as

$$g = \frac{U_1^2 - \Theta_0^2 U_0^2}{2L}. \tag{9.1.4}$$

For the flame parameters used in [9], the scaled acceleration equals $gL_0/U_0^2 = 1/Fr = -12L_0/L$ for the first-order reaction, and $gL_0/U_0^2 = 36L_0/L$ for the third-order reaction. Therefore, for the third-order reaction flames under confinement, the DL instability is enhanced by the RT instability of the flame front in the effective gravitational field. On the contrary, for the first-order reaction flames, the deceleration is a stabilizing factor. The complete suppression of the DL instability by a negative acceleration is expected for $g = \pi\Theta_0 U_f^2/2\lambda_c$. For flames considered in [9], the complete suppression of the DL instability is expected for tubes shorter than $L = 40L_0$, so that in order to observe the DL instability and "tulip flames" in numerical simulations, one has to study flame dynamics in longer tubes, about $L \geq 100L_0$.

The effect of acceleration is not the only restriction on the tube length needed to observe the DL instability. Since a flame traverses a short tube in a short time and it may happen that the instability has no time to develop. The DL instability is a slow process compared to the time scale of a flame front L_f/U_f. From Eq. (7.2.4), we can estimate the maximum instability growth rate for a flame with the expansion coefficient $\Theta = 5$ and Le = 1 as $\sigma_m = 0.07U_f/L_f$. At the beginning of the combustion process, the flame propagates with velocity ΘU_f with respect to the tube walls. Therefore, the amplitude of perturbations depends on the distance z_f passed by the flame as $\exp\left(0.07z_f/\Theta L_f\right)$. To reduce the influence of the "arbitrary" initial conditions on the instability development, we can consider the initial perturbation amplitude at least 10 times smaller than the saturation amplitude F_s achieved on the nonlinear stage for a curved stationary flame. Since the perturbation amplitude is proportional to $\exp\left(0.014z_f/\Theta L_f\right)$, then it may increase by a factor of 10 only in a tube with a length of about $160L_0$, if there are no other stabilizing effects. Thus, to observe a strongly curved "tulip flame" resulting from the DL instability in a closed tube, we have to consider tubes of lengths of $200L_0$ and longer. The influence of the aspect ratio of a closed tube on the development of "tulip flames" has been pointed out already in the early experimental studies and 2D simulations [8, 10, 11], but the restriction on the tube length proposed in [9] is more severe than the aspect ratio of a tube. It implies that not all "tulip flames" observed in experiments and numerical simulations up to now resulted from intrinsic flame properties. The "tulip flame" observed in a shorter closed tube can probably be simply the result of the chosen initial conditions and may be eliminated by another choice of ignition geometry. We will consider the mechanism of tulip flame formation in the following sections. The mechanism of tulip flame formation will be discussed in Sect. 9.4.

When a flame is in the vicinity of the walls at the very beginning and at the very end of the combustion process, then the DL instability growth rate may be directly affected by the boundary conditions at the rigid walls. This effect is stabilizing similar to the influence of a wall on the RT instability in an inert gas [12] and its relative role may be estimated as $\exp\left(-2kz_f\right)$, with the characteristic wave number of the RT instability determined by the tube width $k = \pi/R$. Therefore, when the flame front has passed the distance of about the tube width $z_f = R$, the direct influence of the boundary conditions at the walls on the DL instability becomes negligible.

A common theoretical approach to the flame dynamics under confinement implies that a sound speed is infinite and the pressure is balanced at the both ends of the tube instantly. Obviously, this assumption neglects nonstationary effects related to acoustic and weak shock waves and it is valid only if the instability time scale is much longer than the time needed for the acoustic wave to travel along the tube forth and back $2L\sigma_m/c_s \ll 1$. In such a case, one should expect rather smooth development of the LD instability under confinement. On the contrary, in the case when the time scales of the LD instability and the acoustic waves are close, $L\sigma_m/c_s \sim 1$, the interaction of the LD instability with the main acoustic mode of the wavelength $\lambda_s = 2L$ may result in strong oscillations of the perturbation amplitude. For example, for the parameters $\Theta_0 = 5, \gamma = 1.4, E = 5\Theta_0 T_0$ used in 2D simulations of flames in closed tubes [9], strong oscillations of the 2D flame structure are expected for tubes of a length larger than $\approx 500L_0$ for flames with the initial Mach number $M_0 = 10^{-2}$. The characteristic length scale of these oscillations is about $2LU_f/c_s$, which is the distance passed by the flame in a period of the acoustic oscillations. Presumably, the moving flame front will also generate weaker sound waves with such a wavelength. Figure 9.2 shows the pressure distribution in a closed tube of the length $L = 500L_0$ calculated in [9] for the time instant when a flame with parameters $\Theta_0 = 5, \gamma = 1.4$, $E = 5\Theta_0 T_0, n = 3$ and the initial Mach numbers $\mathrm{M}_0 = 10^{-2}$ or $\mathrm{M}_0 = 10^{-3}$ reaches the middle of the tube, $z = 250L_0$. It is seen that in the case of a lower flame velocity ($\mathrm{M}_0 = 10^{-3}$), the pressure at the tube ends is balanced, while for the flame with larger velocity ($\mathrm{M}_0 = 10^{-2}$), the main acoustic mode with the wavelength $\lambda_s = 2L$ has a noticeably large amplitude.

An estimate of the pressure perturbations in the acoustic wave can be obtained similarly to the amplitude of the weak shock generated by a flame moving from the closed end of the tube $\Delta P = (\Theta - 1)\rho c_s U_f$.

Still, even in short tubes with a small period of the main acoustic mode, pressure waves generated by the flame can significantly affect the flame and the flame

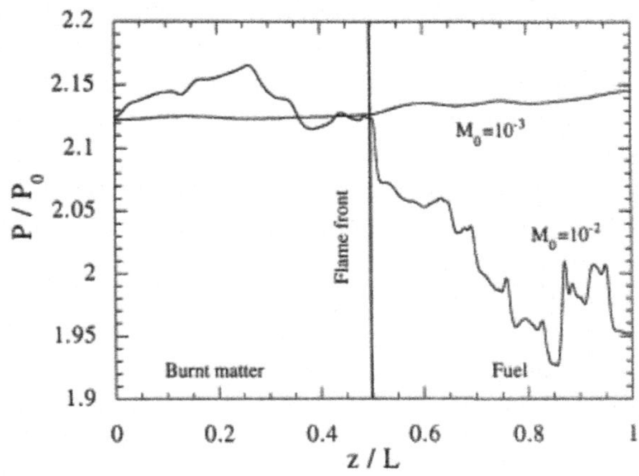

Fig. 9.2 Pressure distribution in a closed tube of length $L = 500L_0$ at a time instant when the flame front is at $z = 250L_0$. The initial Mach numbers are $M = 10^{-2}$ and $M = 10^{-3}$, the reaction order is $n = 3$, from [9]

dynamics. In Chap. 6, it was shown that pressure waves take the form of acoustic oscillations if the frequency of the main acoustic mode is greater than the maximum growth rate of the DL instability, which implies the fulfillment of the condition $2L\sigma_m/c_s < 1$. In the opposite case, $2L\sigma_m/c_s > 1$, the pressure waves can be considered as weak shocks. For the flame parameters used in the 2D simulations [11], the characteristic tube length that separates the domain of applicability of acoustic waves and shock waves is about $L = 500L_0$: for shorter tubes, we can consider sound waves, while for much longer tubes, pressure waves manifest themselves more like shock waves. In Sect. 6.3, we considered the flame–acoustic interaction and found that the interaction of acoustic waves with the flame with $\Theta_0 = 5$ and $\mathrm{M}_0 = 10^{-2}$ in a short tube $L < 500L_0$ results in considerable stabilization of the flame front. According to Sect. 7.1, a head-on collision of the flame with $\Theta_0 = 5$ and the flame-generated shock also substantially suppresses the DL instability, though in this case, this is only a temporary suppression. However, these conclusions about the flame–shock and the flame–acoustic interactions are applicable only as long as the flame is far enough from the end of the tube. When a flame comes to the tube end, a strong nonlinear coupling between the flame and the pressure waves may develop for sufficiently fast flames, similar to that was observed experimentally by Searby [13]. Finally, an increase in fuel temperature due to compression of the fresh mixture can cause spontaneous ignition of the reaction ahead of the flame and even subsequent transition to detonation.

9.2 Numerical Simulations of the DL Instability in Closed Tubes

Simulations of the linear stage of DL instability for an infinitely thin-flame front in a closed tube have been performed in [14]. In these simulations, the approximation of an incompressible fluid and isobaric flame was used, and an infinitely large speed of the sound in the combustible mixture was assumed, which means an infinitely high speed of sound in a combustible mixture. The influence of acoustic waves and shocks on the flame dynamics is not considered in such a model. Therefore, the acceleration or deceleration of the flame and the finite time of the combustion process are the main factors influencing the development of DL instability and flame dynamics. The authors [14] considered the flame with a one-step, first-order chemical reaction and the mixture/flame parameters $\Theta_0 = 7$, $\gamma = 1.4$, $E/\Theta_0 T_0 = 2.87$. As it was discussed in Sect. 9.1, in this case, the flame decelerates, according to Eqs. (9.1.1)–(9.1.3) and the value of the negative acceleration is determined by Eq. (9.1.4). The simulations [13] have shown reduction of the DL instability growth rate due to the negative acceleration. Two-dimensional numerical simulations of the "tulip flame" phenomenon have been reported in [8, 11]. These studies were mainly aimed at elucidating the mechanisms responsible for the formation of the tulip flame.[1] In simulations [8], a "tulip flame" was observed in a closed tube of length $L = 9\,\text{cm}$ filled with a mixture at an initial temperature, pressure, and Prandtl number, $T = 300\,\text{K}$, $P = 10^5\,\text{Pa}$, $\text{Pr} = 1$, $\Theta_0 = 7.7$; viscosity coefficient $\mu = 3 \cdot 10^{-3}\,\text{kg/m s}$; and adiabatic exponent $\gamma = 1.4$. A one-step irreversible reaction of the first order was assumed with the activation energy $E/\Theta_0 T_0 = 4.35$ and the pre-exponential factor $3.3 \cdot 10^7\,\text{s}^{-1}$. The laminar flame velocity and thickness for these parameters can be estimated according to the Zel'dovich–Frank-Kamenetskii theory as $U_f = 4.4\,\text{m/s}$ and $L_f = 0.6\,\text{mm}$. They adopted in [8] that the ratio of the tube length to the initial flame thickness was $L/L_{f0} = 150$. Consequently, the formation time of the tulip flame is mainly associated with the initial conditions and features of the flow near the flame, but not with the DL instability of the flame. Besides, stabilization of "tulip flames" observed in [8] is coupled to the high-frequency velocity pulsations typical for acoustic waves. In [11], the combustible mixture parameters have been chosen for simulations in such a way that the velocity and thickness of a planar flame were more or less similar to the methane–air flame velocity and thickness ($U_f = 38\,\text{cm/s}$, $L_f = 0.45\,\text{mm}$, $\Theta = 10.6$). At the same time, the tube length in [11] was taken $L < 150\,\text{mm}$, so that the "tulip flames" observed in [11] are likely to be the outcome of the ignition conditions.

In [9], the DL instability has been studied in tubes of different lengths from a short $L = 120L_{f0}$ to a relatively long tube $L = 2000L_0$, for the flame parameters $\Theta_0 = 5$.

In [9], the DL instability has been studied in tubes of different lengths from a short $L = 120L_{f0}$ to a relatively long tube $L = 2000L_0$, for the flame parameters $\Theta_0 = 5$, $\gamma = 1.4$, $E/\Theta_0 T_0 = 5$, for the one-step reactions of the first and the third orders, and for the initial Mach number $\text{M}_0 = 10^{-2}$. After initiation of a planar flame near the one end of the tube, when it passed a distance of about $3L_{f0}$ and a planar stationary flame front was developed, the flame front was disturbed by imposing the velocity perturbations in the form

$$u_z(z) = u_z(z)\left[1 + A_0 \cos(\pi x/D) \exp\left(-\frac{\pi^2}{D^2}(z - z_f)^2\right)\right], \tag{9.2.1}$$

where $A_0 \ll 1$. The development of the DL instability was studied considering the time history of the Fourier harmonic amplitudes $F_j(t)$ of the flame isotherms $F(x,t) = \sum F_j(t)\cos(j\pi x/R)$. The instant velocity of curved nonstationary flames in closed tubes is difficult to measure with good accuracy, unlike amplitudes of the perturbation harmonics.

However, the instant velocity of a curved flame is related to the total area of the flame surface and, consequently, to the amplitude of the first harmonic of the flame isotherms. This relation calculated in [9] for the flame propagating in a tube of infinite length and the width $R = 2R_c$ is shown in Fig. 9.3.

[1] The mechanism of inversion of the flame front and the formation of a tulip flame is discussed in Sect. 11.4.

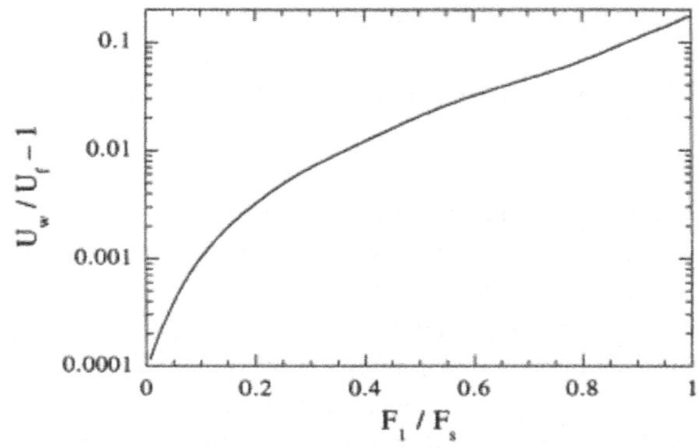

Fig. 9.3 The scaled velocity of a curved flame in an open tube versus the scaled amplitude of the first Fourier harmonic of the flame-front isotherms, from [9]

9.2.1 *Flames of the First-Order Reaction*

The theoretical estimates [7] predict that in short tubes, $L \leq 40L_0$, the DL instability of a flame with the first-order reaction is completely suppressed due to the flame deceleration. Indeed, flame instabilities were not observed in 2D simulations [9] in such short tubes. Moreover, initially imposed perturbations attenuated and vanished rather quickly. In longer tubes with the lengths $L/L_0 = 100{-}500$, the perturbation amplitudes at the flame front are growing at the beginning of the flame propagation; the reduction of the DL instability due to the flame deceleration does not complete until the end of flame propagation.

The perturbation amplitudes for the flames with the first-order reaction versus the flame-front position in tubes are shown in Fig. 9.4 for the tube lengths $L/L_0 = 120,\ 200,\ 500$. It is seen that in spite of some initial growth, the perturbation amplitude starts to decrease long before it reached the saturation limit F_s on the nonlinear stage. The DL instability becomes completely suppressed in the second half of the tubes for the tube lengths $L/L_0 = 120,\ 200,\ 500$. The pulsations of the perturbation

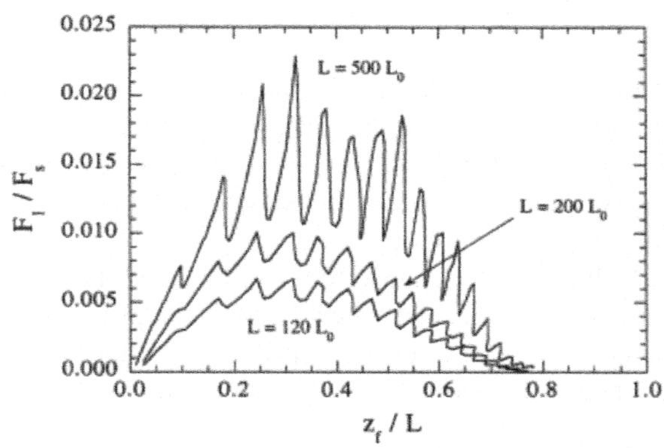

Fig. 9.4 The scaled amplitude of the Fourier harmonics of the flame-front isotherms F_1/F_s versus the scaled flame position for the tube lengths $L/L_0 = 120,\ 200,\ 500$, and the flame reaction order $n = 1$. Adopted from [9]

amplitude indicate that the suppression of the instability occurs due to the flame–acoustic interaction, as was discussed in Sect. 6.5. The distance traveled by the flame front during the period of the main acoustic mode $\approx 2LU_l/c_s$ determines the characteristic length scale of the amplitude oscillations. An increase of the characteristic perturbation amplitude at the flame front with the increase in the tube length can also be explained due to the flame–acoustic interaction. From the analogy with the pendulum stabilization by an oscillating effective gravitational field [15], it follows that the smaller the frequency of the gravity oscillations, the weaker is the stabilizing effect. Similarly, in the case of flames propagating in closed tubes, the longer the tube, the lower the frequency of acoustic oscillations, $\omega = \pi c_s/L$, and the weaker the stabilizing role of the acoustic waves. For this reason, the typical perturbation amplitudes at the flame front achieved in the tube of the length $L = 500L_0$ are considerably larger than the perturbation amplitudes achieved in the shorter tubes with lengths $L/L_0 = 120, 200$.

These tendencies are more pronounced for the case of a very long tube $L = 2000L_0$, for which the acoustic oscillation time is considerably smaller than the typical time of the DL instability. For such a long tube, it is convenient, similar to Chap. 7, to treat the pressure waves as weak shocks with rarefaction waves. When the flame starts to propagate from the closed end of the tube, the accelerating flame acts as an accelerating semitransparent piston producing a pressure wave in the upstream flow, which steepens into a weak shock of intensity $\Delta P = (\Theta_0 - 1)\rho_0 c_0 U_0$. The reflection of this shock from the walls causes the secondary weak shock and rarefaction waves that influence the perturbation amplitudes. The perturbation amplitudes of different harmonics for the flame propagating in a long tube, $L = 2000L_0$, are shown in Fig. 11.5 as a function of the distance passed by the flame [9]. As it is seen in Fig. 9.5, the first collision of the flame with a weak shock happens at the flame position $z_f \approx 200L_0$ and somewhat retards the perturbation growth. As it was shown in Chap. 7, the collision might cause stronger stabilization, but the secondary shock reflected from the tube end at $z = 0$ destabilizes the flame front again. Still, the perturbations keep growing and by the time of the second collision,

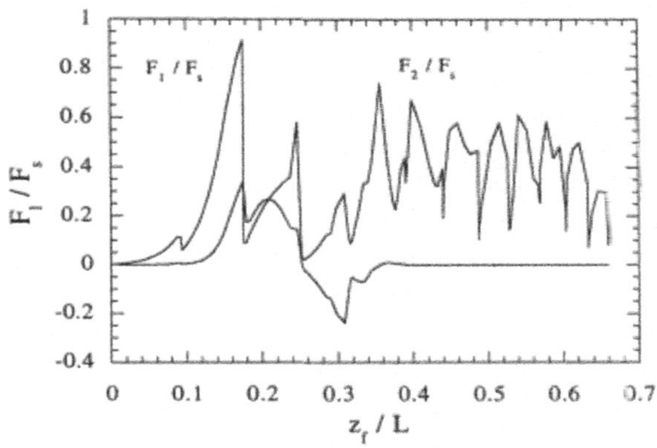

Fig. 9.5 The scaled amplitude of the Fourier harmonics of the flame-front isotherms F_1/F_s versus the scaled flame position for the tube length $L = 2000L_0$, and the flame reaction order $n = 1$, from [9]

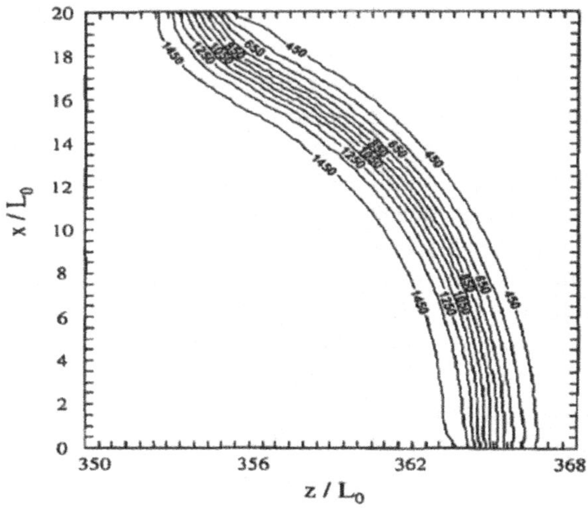

Fig. 9.6 Isotherms of the curved flame front at $z_f \approx 370L_0$ for the tube length $L = 2000L_0$. The reaction order $n = 1$, the initial Mach number $M_0 = 10^{-2}$, from [9]

the perturbation amplitude at the flame front has reached already an appreciable value. The second head-on collision happens at the flame position $z_f \approx 370L_0$ with the perturbation amplitudes at this moment being close to the values corresponding to the isotherm of the curved stationary flame F_s. The shape of the curved flame at this time instant shown in Fig. 9.6 is similar to the flame shape in an open tube. The nonlinear stationary structure at this time has been almost formed, but it was destroyed by the coming weak shock. The collisions of the flame front with weak shocks play a crucial role in the nonlinear competition of the perturbation modes. It is seen in Fig. 9.5 that when the weak shock has passed, the second perturbation mode F_2 with the wavelength $\lambda = R$ dominates at the flame front instead of the principal mode F_1 with the wavelength $\lambda = 2R$.

At this time, both the first and the second perturbation modes belong to the unstable domain of the dispersion Eq. (5.2.43) due to a significant decrease in the flame-front thickness L_f. Subsequently, the second perturbation mode remains the principal one in the nonlinear structure of the flame front in spite of all other weak shocks colliding with the flame front. When recovered after collision with another shock, the perturbation amplitude may reach considerable values $F_2 \approx 0.7F_s$, though it does not come close to the amplitude corresponding to the stationary curved flame anymore. However, when the flame comes closer to the end of the tube, the perturbation amplitude decays and the nonlinear structure vanishes similar to the case of short tubes $L/L_0 = 120, 200, 500$. Thus, for the flames with the first-order reaction considered in [9], the influence of the confinement and the acoustic waves on the flame instabilities is only stabilizing in agreement with the theoretical predictions. At the same time, one should remember that for flames with larger expansion coefficients $\Theta = 7 - 10$, the analytical theory (Sect. 9.1) predicts the destabilizing influence of the weak shocks [16], while the flame–acoustic interaction may also result in parametric instability for these flames.

9.2.2 Flames of the Third-Order Reaction

According to theoretical predictions [7], if the flame with a third-order chemical reaction accelerates under confinement, the DL instability of the flame front in short tubes $L/L_0 = 120,\ 200$ is enhanced by additional action of the RT instability. Still, despite this amplification of the flame instability, the time available for perturbations to grow in a short tube is not sufficient to form a quasi-stationary nonlinear structure. The development of the perturbation amplitudes of the flame front in short tubes of the length $L/L_0 = 120,\ 200$ is shown in Fig. 9.7 versus the flame-front position [9]. Unlike the flame with the first-order reaction, perturbations at the flame front with the third-order reaction are growing almost monotonically with the instability growth rate exceeding the respective growth rate of the freely propagating flame instabilities.

Even more, for the tube length $L = 200L_0$, the perturbations might achieve the saturation amplitude F_s. The spontaneous reaction ahead of the flame front at the end of burning was artificially suppressed in [9] in order to avoid an explosion that could be ignited ahead of the flame. For this purpose, the chemical reaction rate was used in the form

$$\frac{d}{dt}(\rho Y) = \begin{cases} -(\rho Y)^3 \exp(-E/RT)/\rho_0^2\tau_R, & Y < 0.99 \\ 0, & Y > 0.99 \end{cases}.$$

Such artificial form of the fuel mass fraction Y does not distort the flame dynamics and corresponds to the reaction propagation controlled by the thermal conduction and the fuel diffusion.

For flames in tubes of a moderate length $L = 500L_0$, the interaction of the flame instability and the acoustic waves becomes noticeable. Figure 9.8a shows the amplitudes of the Fourier harmonics of the flame front versus the distance passed by the flame for the fast flames $M_0 = 10^{-2}$. Similar dynamics for the slow flames $M_0 = 10^{-3}$ is shown in Fig. 9.8b. It is seen that in both cases, the tendencies of the perturbation development are similar.

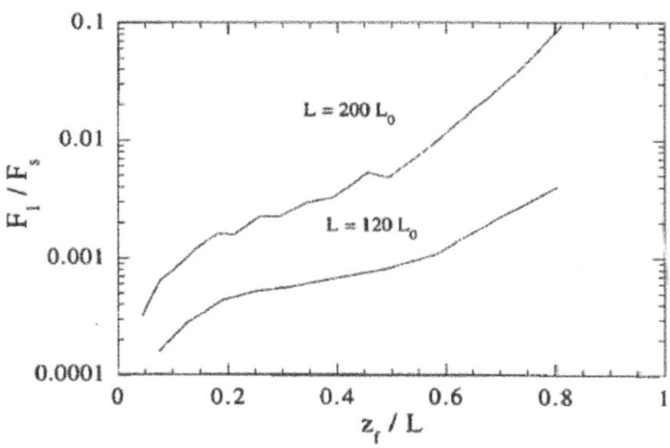

Fig. 9.7 The scaled amplitude of the Fourier harmonics of the flame-front isotherms F_1/F_s versus the scaled flame position for the tube length $L/L_0 = 120,\ 200$. The reaction order is $n = 3$, the initial Mach number is $M_0 = 10^{-2}$ [9]

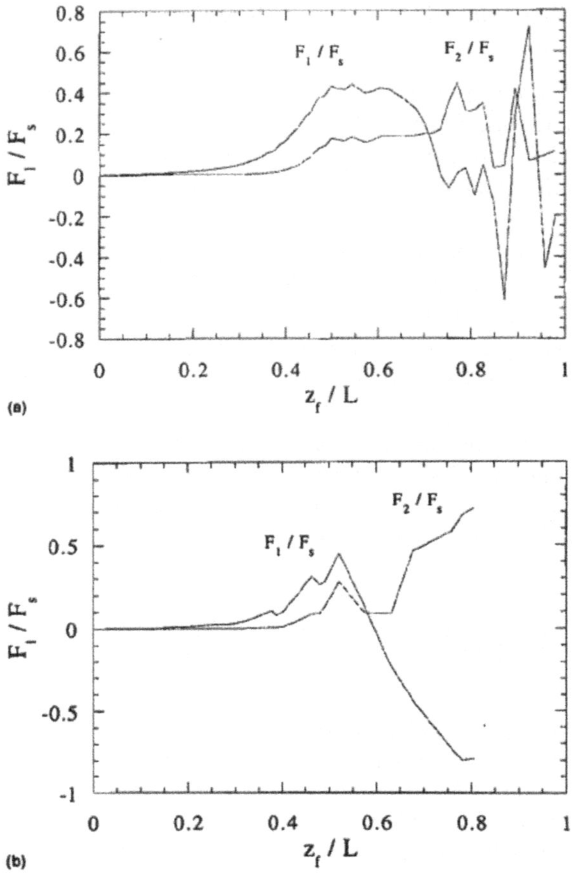

Fig. 9.8 (a, b) The scaled amplitude of the Fourier harmonics of the flame isotherms F_1/F_s versus the scaled flame position z_f/L for the tube length $L/L_0 = 500$. The reaction order is $n = 3$, the initial Mach number $M_0 = 10^{-2}$, from [9]. The same as in Fig. 9.8a, but for a slow flame $M_0 = 10^{-3}$

It is seen that the influence of the acoustic waves on the flame-front perturbations in the case of the third-order reaction is different depending on the flame position. Close to the tube center, the acoustic waves reduce the DL instability similar to the case of the flames with the first-order reactions.

As a result, the perturbation growth is saturated at the amplitudes $F_1 = (0.4 - 0.5)F_s$, which is noticeably smaller than the saturation amplitude for freely propagating flames. However, at the end of burning, the interaction of the DL instability with the acoustic waves is different for flames with the first-order and the third-order reactions. While for the first-order reaction, the DL instability remains suppressed till the very end of burning, in the case of the third-order reaction, there is again a strong growth of the flame instability induced by the acoustic oscillations in

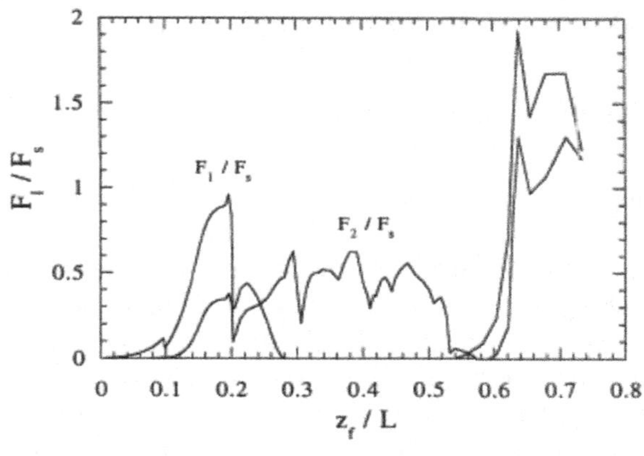

Fig. 9.9 The scaled amplitude of the Fourier harmonics of the flame-front isotherms F_1/F_s versus the scaled flame position z_f/L for the tube length $L/L_0 = 2000$. The reaction order is $n = 3$

the second part of the tube. Similar behavior has been observed by Searby in experiments [13] for the flame velocity exceeding 25 cm/s when the DL instability did not recover after being suppressed by the acoustic oscillations close to the tube center. In this sense, the difference in the interaction of the DL instability with the acoustic waves observed in [9] for the flames with the first- and the third-order reactions may be related to the difference in their velocities at the end of burning. The reason is that at the end of burning, the flame velocity in the case of the third-order reaction is almost 10 times larger than in the case of the first-order reaction.

For a flame with the third-order reaction propagating in a long tube $L = 2000L_0$, we can clearly distinguish the period of the free development of the DL instability during the time needed for the primary weak shock to pass the tube twice (Fig. 9.9). As in the case of the first-order reaction, the first collision with the shock does not stop the instability development. But at the time of the second collision, occurring at $z_f \approx 370L_0$, the nonlinear stationary structure with the wavelength $\lambda = 2R$ is almost formed at the flame front.

Like for the first-order reaction, the destruction of the nonlinear structure by the shock finally leads to the development of the structure with another wavelength $\lambda = R$ and a smaller amplitude $F_1 = (0.3 - 0.7)F_s$ oscillating because of the interaction with the pressure waves. However, in the second part of the tube, there is considerable difference between the flames with the first-order and the third-order reactions. For the third-order reaction at the end of burning, the DL instability perturbations start growing once more due to the interaction with the acoustic waves, and at the end of burning, the perturbation amplitude exceeds the saturation value almost two times $F_1 = (1.7 - 1.9)F_s$.

9.3 Flammability Limits and Flame Quenching

9.3.1 Flammability Limits

Flammability limits are of considerable interest not only for fundamental combustion research but also for fire safety control. Accurate knowledge of the flammability limits is needed to estimate safety conditions in temperature and pressure, for example, for storage of flammable gases. In view of its fundamental importance and its importance for practical applications, many theoretical, numerical, and experimental studies of flammability limits have been carried out. It is well known that depending on the thermodynamics conditions, the planar flame propagation is impossible in the too rich and too lean combustible mixture. The existence of flammability limits is determined by the chemical kinetics and depends on the thermodynamics conditions. The well-known example is a hydrogen–oxygen mixture, where such limits are related to the competition between the chain-branching reactions controlled by reaction producing active radicals (H and O) and the chain termination reactions (see Sect. 3.4.1). The flammability limits can be defined as the mixture critical compositions beyond which planar flame propagation becomes impossible. Earlier studies of flammability limits date back to the famous paper by Mallard and Le Chatelier [17]. The theoretical understanding of the flammability limits is based on the works by Daniell [18], Lewis and Elbe [19], and Zel'dovich [20]. Coward and Jones [21] published the first extensive data on flammability limits in different combustible mixtures, which they defined as "a borderline composition: a slight change in one direction produces a flammable mixture, in the other direction a non-flammable mixture".

Sustainable combustion waves are possible not for all mixtures of fuels and oxidizers. Addition of diluent gases in sufficient quantity may limit the flammability of the mixture. These diluents may consist of excess of either fuel or excess of oxidants or they can be inert gases. Being ignited, a stoichiometric combustible mixture can reach a sufficiently large temperature, moves into the explosive regime, and can support the flame propagation. However, being diluted to the sufficiently lean or rich concentration, the heat of the reaction is spent on raising the temperature of the diluted components, and the mixture is not able to reach a sufficiently high temperature for a self-sustaining reaction and is not able to support sustaining combustion waves. With the increased dilution of the combustible mixture, when the mixture ratio falls into the lean or rich ends of the concentration spectrum, the burning temperature decreases and as a consequence also decreases the combustion wave speed. The ability of the given fuel–oxidizer stoichiometry mixture to support a propagating reaction is related to the chemical rate of a given mixture ratio and the rate of heat generated in the reaction. The flammability limit can be defined either by a competition between the rate of heat generated in the reaction and the external rate of heat losses. On the other hand, the flammability limits beyond which the self-sustained ignition is impossible are defined by the competition between the chain-branching and chain termination reactions, which leads to the crossover temperature T^*. The nature of this phenomenon is similar to the explosion limits considered in Chap. 2. The flammability limit is often defined as such composition of the combustible mixture for which the temperature achieved due to the energy released in the chemical reaction is equal to the crossover temperature T^*. Regarding the fuel–oxidizer mixture ratio, the flammability limit is defined by the minimum concentration of a combustible substance that is capable to support self-sustained propagation of the reaction wave. In real conditions, the flammability limit is determined by the competition between the rate of heat generated by reaction and the external rate of heat loss by the flame in addition to the heat of the reaction spent on raising the temperature of the diluted components. As a result of the external heat losses (heat losses to the walls, radiation, etc.) the flame temperature becomes lower than the adiabatic flame temperature, and correspondingly the flame speed is smaller than the normal flame speed corresponding to a stoichiometric mixture. The external heat losses in turn depend on the velocity of the flame propagation. The leanest and richest concentrations, which are just able to support flame, are called the lean and the rich flammability limits.

In such phenomena as stretching of the flame front and the flame instabilities, the wall catalytic effect of capturing radicals complicate the experimental and theoretical study of the flammability limits [22, 23]. The theoretical study of flammability limits requires knowledge of the detailed chemical mechanism for a particular combustible mixture. One of the first such studies was carried out by He and Clavin [24], who investigated numerically the flammability limits and the structure of adiabatic 1D planar flames in premixed hydrogen–oxygen gases using the chemical kinetic mechanism responsible for the explosion limits of hydrogen–oxygen mixtures, which defines the crossover flame temperature. They numerically investigated the 1D structure of a laminar flame and the burning velocity near the explosion limits, where the flame velocity decreases to negligible values ~mm/s. Since then, significant progress has been made in understanding flammability limits, in particular thanks to improved experimental techniques, powerful computers available, and a database of detailed chemical mechanisms for various fuels. A theoretical definition of flammability limit was proposed by Williams [25], which involves the failure of the propagation of the ideal 1D planar, laminar adiabatic flame. Egolfopoulos et al. [26] proposed a "chain-thermal" theory, based on the assumption that near the flammability limits, the competition between branching and termination reactions determines the flame velocity, when the flame propagation speed slows down, and the heat losses become important. Considerable contributions to understanding the impact of flame-front stretching on the burning velocity and flammability limits have been due to Law [27, 28] and Dixon-Lewis [29] with many other subsequent experimental and numerical studies [22]. Currently, there is a large number of computational studies of the flammability limits, motivated mainly by the evaluation of safety operating conditions and the risks of explosion in various combustible mixtures [30–34]. These investigations are greatly benefited by the existing huge database of detailed chemical mechanisms for various fuels [35, 36] and corresponding software [37]. The minimum concentration of a particular combustible gas necessary to support its combustion in air is defined as the lower explosive limit (LEL).

Table 9.1 Lower and upper explosive limits

Gas	The lower limit	The upper limit
Acetylene	2.5	100.0
Ethanol	3.3	19.0
Ethylene	2.7	36.0
Gasoline	1.2	7.1
Hydrogen	4.0	75.0
Methane	5.0	15.0
Propane	2.1	9.5
Methyl Chloride	7.0	17.4
Carbon Monoxide	12.5	74.0

Below this level, the mixture is too "lean" to burn. The maximum concentration of a combustible gas that will burn in the air is defined as the upper explosive limit (UEL). Above this level, the mixture is too "rich" to burn. These limitations are understandable: too much heat is consumed to heat excess air (LEL) or fuel (UEL), which leads to a decrease in temperature below which combustion becomes impossible. The lower and upper explosive limits for several combustible gases are shown in Table 9.1.

9.3.2 *Heat Losses and Flame Quenching*

A flame propagating in tubes can be extinguished if mechanisms of species diffusion and heat transfer, which are responsible for flame propagation, are affected by the heat and/or loss of radicals to the tube wall. Heat and momentum losses to the walls can reduce the reaction rate to the point when it cannot sustain the propagating reaction. If a flame propagates in a wide tube, one can neglect the heat losses to the cold walls. In narrow tubes, the heat losses increase due to the increase of the transverse temperature gradient, and for some critical tube diameter (quenching distance), the self-supported flame propagation becomes impossible. We will assume that the reaction propagates in the deflagration regime in the form of a planar flame in a tube of some given diameter. If the heat losses to the tube walls, which are at the room temperature, will be the only mechanism responsible for the flame quenching, then we can estimate a critical tube diameter. For a simple case of the 2D problem, this is the minimal distance between two parallel plates when the reaction can propagate. The quenching distance corresponds to the distance between the plates, such that the rate of heat generated in the reaction equals the rate of heat losses to the walls. If the heat of the reaction per mole for the given stoichiometric mixture is Q_R, and τ_R is the reaction rate ($mole/t \cdot L^3$), then the rate of heat generating in the flame is

$$\dot{q} \propto Q_R \tau_R (dS \cdot d), \tag{9.3.1}$$

where dS is a surface element at a wall, and d is the width between walls. The rate of heat losses to the wall is

$$q_T = \lambda \frac{dT}{dx} dS. \tag{9.3.2}$$

Let T_q be the lowest temperature at which the flame still can propagate, and T_0 is the temperature of the wall. Approximating the rate of the heat losses to the wall as

$$q_T = \lambda \frac{dT}{dx} dS \cong 2\lambda \frac{T_q - T_0}{(d_q/2)} dS, \tag{9.3.3}$$

we obtain the equation for the minimal distance between the wall when flame still can propagate (quenching distance):

$$4\lambda \frac{T_q - T_0}{d_q} dS \propto \tau_R (d_q dS) Q_R. \tag{9.3.4}$$

Taking $Q_R = C_P(T_f - T_0)$, we obtain the quenching distance in the form

$$d_q^2 \propto \frac{4\lambda}{C_P \tau_R} \frac{T_q - T_0}{T_f - T_0}. \tag{9.3.5}$$

Using the estimate for the laminar flame speed in the form $U_f \propto \sqrt{\lambda_f / \rho_f C_p \tau_R}$, we obtain that $d_q \propto U_f$. Thus, the higher the speed of the laminar flame, the greater the extinguishing distance between the walls for a particular mixture. This can be formulated in another way: the shorter the time of the heat losses to the wall, the greater and larger the quenching distance. As the initial temperature of the unburned mixture increases, the flame is able to pass through a narrower channel. The quenching distance also depends on the pressure as $d_q \propto 1/P$.

It should be noted that for the same combustible mixture, the minimum ignition energy (see Chaps. 4 and 15) is proportional to the third degree of the quenching distance. This is clear because in order to ignite a flame, it is necessary to deliver heat in the volume exceeding d_q^3.

Flame propagation in channels is of great interest to study the fundamental features of the process in practical combustion devices. Since the pioneering work of Ya. B. Zel'dovich [38] who was the first to predict the limiting conditions, an extensive study of flame propagation in pipes and flame extinguishing due to heat loss to cold walls was conducted by many researchers. The flame behavior near the walls may play an important role affecting the burning rate, velocity of the flame, the flame quenching, and emission of chemical pollutants. As a result of heat losses, the flame moves away from the wall with the quenching region about the thickness of the thermal boundary layer, which depends on the Prandtl number [39]. The effect of heat loss on a wall depends on the surface-to-volume ratio of the particular system. In a wide tube, heat losses to the wall make up an insignificant part of the heat released during combustion in the volume, and the adiabatic boundary conditions can be used with good accuracy. On the contrary, in the case of a flame propagating in a narrow channel, a non-negligible part of the heat is transferred by the thermal conduction to the solid walls. This may lead to total flame extinction, but the boundary between total extinction and partial extinction in the vicinity of the walls is not well determined. Due to various technological applications, in particular in the lighting industry, new burner technologies, micro-power generators, etc., there is currently a significant amount of literature on experimental studies and numerical simulations on the impact of heat losses to the wall on the flame dynamics in narrow channels; see, e.g. review [40] and comprehensive work [41] and references within. For the Prandtl number $\mathrm{Pr} = \nu/\chi$ of the order of unity, the thermal boundary layer δ_T is of the order of the boundary layer width δ_l. Hence, the heat flux to the wall can be estimated as $\kappa_g(T_g - T_w)/\delta_l$. Usually, a simplified boundary condition is used at the inner surface of the wall

$$\kappa_g \frac{\partial T}{\partial n}\bigg|_{z=0} = \kappa_w (T_{z=0} - T_{ext})/d_w, \tag{9.3.6}$$

where κ_g and κ_w are the coefficients of thermal conduction of the gas mixture and the wall material, respectively, d_w is the thickness of the channel wall, $T_{z=0}$ is the temperature of the internal surfaces of the channel walls, T_{ext} is the temperature of the external surface. The limiting cases of adiabatic and cold isothermal walls correspond to $\kappa_w = 0$ and $\kappa_w = \infty$, respectively.

Figure 9.10a, b shows the pressure (a) and temperature profiles (b) on the channel axis obtained by high-resolution simulations of the hydrogen/air flame propagation in a channel with boundary conditions for adiabatic walls and for heat losses to the wall. The effect of heat losses to the walls on flame acceleration and the transition from deflagration to detonation (DDT) in narrow channels has been studied by several authors [42–47]. At the same time, it should be noted that most of these works

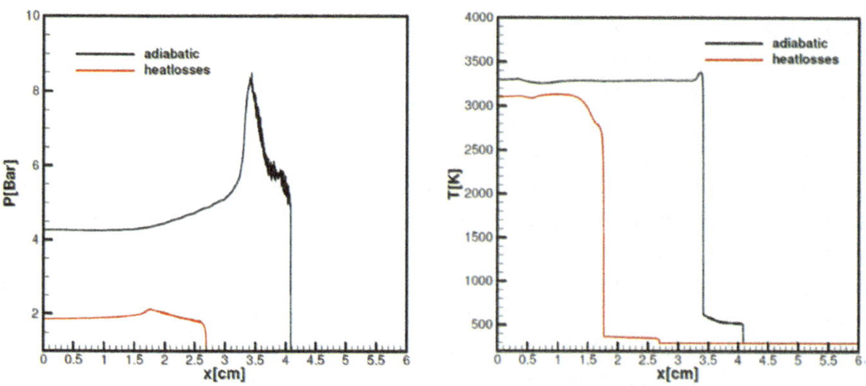

Fig. 9.10 Pressure (**a**) and temperature (**b**) profiles at the channel axis for H_2/air flame propagating in a channel with adiabatic walls and for the case of heat losses to the walls

used simplified transport and chemical models to reduce computational costs. As is known, to obtain reliable simulation results, it is necessary to use a detailed model of chemical reactions. On the other hand, Bongers and De Goey [48] have shown that for an accurate prediction of the burning velocity and other related quantities, a detailed multicomponent transport model is necessary. For some types of combustion mixture, neglecting the influence of the Soret and Dufour effects may lead to large errors in the prediction of the burning velocity.

9.4 Tulip Flames. Effect of the Boundary Layer

9.4.1 Experimental, Theoretical, and Numerical Studies

The dynamics of premixed flames propagating in tubes is important for understanding the behavior and characteristics of combustion processes in confined regions, including many applications, from explosions and safety issues to combustion regimes in gas turbines and internal combustion engines. From the beginning, a flame that is ignited near the center of the closed end of a tube and propagates to the opposite closed or open end acquires a convex shape toward the unburned gas due to interaction with the boundary layer. The thermal expansion of combustion products causes the flow in the unburned gas ahead of the flame. The velocity of the unburned gas is parallel to the tube walls, uniform in the bulk, and drops to zero in the boundary layer. The flame surface increases due to its stretching along the boundary layer near side walls, and the flame velocity increases proportionally to the increase in the flame surface. Since the rate of the flame surface increase caused by the flame surface stretching is much less than the transit time L_f/U_f of a fluid particle across the flame, the internal structure of the flame can be considered quasi-stationary. When the flame skirt touches the side walls of the tube, it slows down owing to the quenching of the rear part of the flame skirt and the decrease of the flame surface. The reduced expansion of the burned gases and deceleration of the flame cause rarefaction waves, which are spread in the unburned and the burned gases, and can be reflected toward the flame front. During this stage, the flame front begins to invert, forming a cusp that points toward the burned gas, with the leading edges of the flame front near the walls and the center part of the flame near the axis lagging behind. In sufficiently long tubes, the inversion of the flame front can reverse again from a concave to a convex shape. This behavior of a flame has been observed in many experimental works, and different theoretical explanations have been proposed by various researchers. The development of a tulip-shaped hydrogen–oxygen flame propagating in a stainless-steel tube with a rectangular cross section of 5×5 cm is shown in the shadow photographs in Fig. 9.11.

The first experimental observation and photographic study of the "tulip" flame formation were published by Ellis and his collaborators [50–52] almost a century

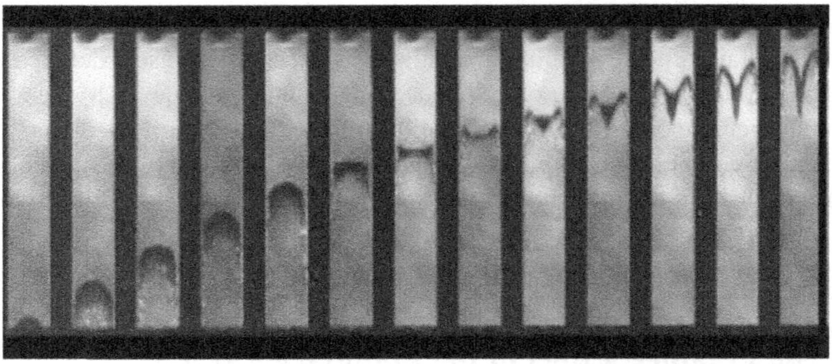

Fig. 9.11 A sequence of shadow photographs [49] shows the evolution of the H_2/O_2 flame front and the development of a tulip flame. Photographs were taken from 0.5 ms to 1.75 ms with 0.1 ms between the frames; initial pressure is $P_0 = 0.75$ bar

ago using a newly invented at that time rotating shutter camera. Later, this inward-pointing cusp-shaped flame was called a "tulip flame" by Salamandra et al. [53], and it is usually applied in studies of the deflagration-to-detonation transition (Chap. 15). In any case, a salient feature of flame propagation is a change in the shape of the flame surface associated with deceleration and subsequent re-acceleration if the flame propagates in a sufficiently long tube.

The formation of a "tulip" flame was observed both in many experiments and numerical simulations of a flame propagating in tubes with both closed ends or from the ignition (closed) end to the open end. A typical example of a tulip flame formation obtained in 3D numerical simulations [50] is shown in Fig. 15.14a in Chap. 15. When the flame propagates from a local ignition source at the closed end of the tube, the shape of the flame front is characterized by the inversion of the flame curvature from convex to concave toward the unburned mixture. In long closed tubes with an aspect ratio greater than 20, the shape of a flame can at some point be inverted and become convex again toward the unburned mixture, and later it can turn back into a tulip flame [10]. If a large ignition source is used, the flame will reach the walls more rapidly, so that the maximum flame surface area will be smaller than in the case of a point source. The most pronounced flame-inversion behavior and the tulip flame formation occur when the deflagration is ignited at one end of a tube that is closed at both ends and has a length to width ratio $L/D \approx 10 \div 20$. Tulip flame formation begins after a rapid decrease in flame surface area as a result of the extinguishing of the rear part of the flame skirt on the side walls of the tube.

After the earliest publications by Ellis et al. [50–52], the process of the "tulip" flame formation has been experimentally studied by many researchers. The first attempt to explain the mechanism of tulip flame formation was undertaken by Markstein [10, 54], who showed that the inversion of the flame shape can be the result of the interaction of a curved flame front with a planar shock wave. Starke and Roth [55] hypothesized that the onset of the tulip flame in Markstein's experiments is associated with the Rayleigh–Taylor instabilities. However, the relatively high intensity of shock waves, as in Markstein experiments, do not appear in the early stage of flame propagation, when the tulip shape is formed.

Detailed information on the flow, pressure field, and the flame evolution was obtained by Dunn-Rankin et al. in the series of experimental studies [56–59]. From early experiments [57], Dunn-Rankin concluded that the flame-front inversion is an initial distortion caused by the Landau–Darrieus instability, which grows, developing in the form of a cusp directed backward, thus forming a tulip shape. However, in later publications [58, 59], Dunn-Rankin interpreted the flame-front inversion as associated with the recirculation of burned gases observed in schlieren images. It was hypothesized that the expansion of burned gases along the normal to the convex flame surface deflects the flow behind the flame front toward the tube axis. Kerampran et al. [60] studied a flame in different combustible mixtures propagating from the closed to the open end of a tube and found that it is always accompanied with pressure oscillations. Hariharan and Wichman [61] carried out a detailed experimental study and 2D numerical simulations of the flame propagating in a rectangular channel with both closed ends with an aspect ratio $L/D = 6$. They concluded that the transition to the tulip shape is influenced by the "stagnation point"—when the flame front is almost planar and the flame speed is minimal. They emphasized that the further inversion of the flame front and the development of the tulip shape are dictated by the presence of vortices in the flow.

A thorough experimental study was conducted with a half-open tube by Clanet and Searby [62]. Their experiments showed that the formation time of a tulip flame mainly depends on the aspect ratio of the tube, L/D, which can be considered as an indication that the process is influenced by the structure of the boundary layer in the unburned gas ahead of the flame, and, probably, by the time of vortices formation in the boundary layer in the burned combustion products behind the flame. At the same time, the authors concluded that neither acoustic effects nor the boundary layer are dominant in the formation of the tulip flame. While this conclusion may be true for acoustic effects, the conclusion about the boundary layer is questionable because many experiments show a close relationship between the formation of a tulip flame and the upstream flow ahead of the flame. In addition, it should be emphasized that a tulip flame has never been observed in narrow tubes; since the Poiseuille flow with a parabolic velocity distribution is established very quickly ahead of the flame in a narrow channel, there is only an acceleration phase of the flame and there is no deceleration phase.

The quantitative data obtained by Dunn-Rankin [57, 58]; Starke and Roth [55]; and Jeung, Cho, and Jeong [63] from velocity measurements during the tulip formation have stimulated a series of numerical simulations of the tulip phenomenon [11, 64, 65]. A number of explanations for the phenomenon of tulip formation have been proposed in the literature. Some authors explain it by the cooling of the burned gases, which would cause expansion of the fresh gases, others believe it is caused by a compression wave generated by the flame, which is reflected by the downstream end of the tube and returns to strike the flame front. One of the first hypotheses [10, 24, 26, 54] were based on the similarity of the tulip phenomenon and the flame-front inversion as a result of the interaction of a curved flame with a shock wave. In both cases, a slowdown of the flame is recorded before the inversion, which led to the idea that the tulip flame formation is the result of the Rayleigh–Taylor instability. However, this hypothesis was not confirmed by later experiments and theoretical analysis. Various other hypotheses have been proposed, such as the effect of the Darrieus–Landau and Taylor instabilities [7, 9, 11, 14, 25, 64, 65] and the effect of a flame interaction with acoustic waves [25, 66]. It was believed [7, 64] that DL instability is always present in the flame-front inversion, but this is incompatible with the fact that the characteristic time scale of the DL instability, $\tau \sim 1/kU_f$, is much longer than the characteristic time of the flame-front inversion observed in experiments and simulations. At the same time, it should be stressed that even earlier simulations have shown the presence of a reverse flow in the burned gases behind the flame and a transverse velocity gradient along the flattened flame front near the side walls, as well as the slowdown of the central part of the flame. However, the simulation resolutions at that time were too low to distinguish causes from consequences and did not confirm or exclude the participation of DL instability in the inversion of the flame front. It should be noticed also that the onset of flame-front inversion cannot be explained simply by linear stability analysis of the DL instability. Dold and Joulin [67] hypothesized that the tulip flame has little or nothing to do with the deceleration in gas velocity caused by the reduction of the flame area. By means of numerical simulation of the modified Michelson–Sivashinsky equation, they demonstrated that the combined influence of the flame-front curvature, geometric nonlinearity, and the Darrieus–Landau instability is sufficient to produce the inversion of the flame front. Also, experiments and direct numerical simulations show that the formation of a tulip flame is accompanied by vortices formed in the burned gas, while the Michelson–Sivashinsky equation assumes that both flows ahead of the flame and behind it are potential.

Another theoretical explanation of the tulip flame formation was proposed by Matalon and Metzener [68, 69]. They derived a nonlinear evolution equation, valid in the limit of weak thermal expansion, $q = QY_0/c_pT_o \ll 1$, where Q is the total heat of combustion and Y_0 is the initial mass fraction of the reactants. This equation describes the motion and the instantaneous shape of the 2D flame front as a function of its mean position and contains a destabilizing term that results from the gas motion induced by thermal expansion and a memory term associated with vorticity generation. Numerical solutions to this equation demonstrated the development of convex and tulip flames. In the burned gas, the flow is no longer potential because vortices can be produced by the curved flame front. In [69], the authors emphasized that although the evolution equation contains a destabilizing term arising from the DL instability, it is the vortex motion that can lead to the formation of a tulip flame, depending on the initial conditions. The approach developed by Matalon and Metzener [68, 69] is the only reasonable analytical model developed in an attempt to explain the mechanism of tulip flame formation. However, since the process essentially depends on many parameters, such as the aspect ratio, the laminar flame velocity, the intensity of the ignition source (the size of the ignition spot), and the interaction of rarefaction waves in the unburned and burned gases, an analytical solution to the problem is apparently impossible.

Despite a large number of experimental, theoretical, and numerical studies, there is still no unambiguous explanation of the physical mechanism responsible for the flame-front inversion. Recent experimental studies of tulip flame formation by Ponizy, Claverie, and Veyssière [70] were focused primarily on the flame-front inversion mechanism. The experiments were carried out with a stoichiometric propane–air mixture at an initial atmospheric temperature and pressure using cylindrical tubes of different aspect ratios closed at both ends, as well as for a tube closed at the ignition end with an open opposite end. A detailed study of gas dynamics in the upstream and downstream flows was carried out using a high-speed time-resolved particle image velocimetry (PIV) system based on a dual cavity laser Nd:YLF. The authors concluded that the inversion of the flame front is a pure hydrodynamic process and that the intrinsic flame-front instabilities (Darrieus–Landau, Rayleigh–Taylor, and Richtmyer–Meshkov) are not involved in the process. The authors stressed that "the convex flame front has no influence on the phenomenon, but only the lateral parts of the flame front with radial expansion of burned gases are necessary". They found that the expansion of burned products, especially gases burned near the wall, are the main factors in the formation of a tulip flame. Inversion occurs by the action of two opposed flows: the reverse flow of burned gases and the forward directed flow of unburned, previously accelerated gases.

9.4.2 *Mechanisms of the Tulip Flame Formation*

In a typical experiment, a flame is ignited at the tube axis near the closed end of the tube. Different stages of flame propagation can be distinguished before the development of the tulip flame shape. After the flame was ignited near the closed end of the tube, the flame front rapidly acquires a near hemispherical shape. Clanet and Searby [62] have shown that the first stage is the formation of the finger shape flame, with the axial coordinate of the flame tip X_{tip} depending on time as

$$X_{tip} \propto \exp(4\Theta U_f t/D), \tag{9.4.1}$$

where D is the tube diameter (width in the 2D case), $\Theta = \rho_u/\rho_b$, and U_f is the velocity of a laminar flame. This stage takes place in a short time $t \approx D/2\Theta U_f$ after ignition and ends when the rear edge of the flame skirt touches the side wall. The expansion of the high-temperature burned products between the flame and the closed end of the tube pushes the unburned gas toward the opposite end of the tube, thus creating an upstream flow ahead of the flame. A flame initiated near the closed end of the tube controls the upstream flow that forms ahead of the flame front, which leads to the flame acceleration compatible with the boundary conditions. From the conservation of the mass flux, it follows that the velocity at the upstream flow is equal to $u_u = (\Theta - 1)U_f$, while the flame propagates with the velocity $U_{Lf} = \Theta U_f$ relative to the tube wall and with the normal velocity of the laminar flame U_f relative to the upstream flow. In the beginning, during the accelerating stage, the flow of the burned gas, $u_b(x, y, t)$, is directed toward the flame tip, and it vanishes at the tube walls, $y = y_w = \pm D/2$, so that $u_b(x, y_w, t) = 0$. Because of the wall friction, the velocity in the upstream flow is maximal near the tube axis and drops to zero at the tube walls. In the laboratory reference frame, the local velocity of the flame front at the point (x, y) is

$$U_{fl}(y) = U_f + u_u(x, y), \tag{9.4.2}$$

where U_f is the normal velocity of a laminar flame. Every point at the flame front moves relative to the unreacted mixture with normal velocity U_f and simultaneously it is dragged by the upstream flow with its local velocity $u_u^+(x, y)$ just ahead of this point of the flame front. The velocity distribution in the upstream flow (ahead of the flame) depends on the tube width. In a wide tube, the velocity is almost uniform in the bulk and drops to zero in a thin boundary layer of thickness $\delta_l \ll D$, while in a narrow tube, the Poiseuille flow with a parabolic velocity profile in the upstream flow develops during a short time $\approx D^2/100\nu$. The shape of the flame front is determined by the speed of its various parts. The flame sheet to some extent "repeats" the shape of the velocity profile in the upstream flow, remaining almost flat in the bulk (actually retaining the convex shape acquired earlier after ignition) with the trailing edges of the flame skirt extending backward in the boundary layer. The stretched flame consumes fresh fuel over a larger surface area, which results in an increase in the rate of heat release per unit projected flame area. This in turn results in a higher volumetric burning rate and a higher effective burning velocity based on the average heat release rate per frontal area of the stretched flame sheet. A higher burning velocity leads to an increase of the flow velocity ahead of the flame, which in turn leads to an increase in the stretching of the flame surface and a further increase in the burning rate. Thus, a positive feedback coupling is established between the upstream flow and the burning velocity, a process known as flame self-acceleration. Within the model of a thin-flame front, the increase in the burning rate is proportional to the relative increase in the flame surface (length in the 2D case). With an accuracy $\delta_l/D \ll 1$, the flame surface grows linearly in time. Note that in a relatively wide channel, the area within the boundary layer is the main contributor to the increase of the flame surface, which leads to an exponential increase of the flame velocity [49], similar to formula (9.4.1)

$$U_{fL} \propto \Theta U_f \exp\bigl(\alpha U_f t/D\bigr), \tag{9.4.3}$$

where α is a dimensionless coefficient of the order of unity. Interesting is that in reality, when combustion is ignited by a spark, the flame front is not flat in the bulk, but rather finger-like, but the velocity profile in the upstream flow is flat in the bulk (see Figs. 15.45 and 15.46 in Chap. 15). The formula $U_{fL} \propto \Theta U_f \exp\bigl(\alpha U_f t/D\bigr)$ is in good agreement with the results of numerical simulations [49].

After the rear edges of the flame skirt touched the side walls and the flame skirt in the boundary layer has elongated along the side walls, the angle between the flame skirt and the wall decreases and the rear part of the flame skirt becomes nearly parallel to the side walls. Shortly thereafter, the rear parts of the flame skirt touch the side walls and extinguish, which leads to a sharp decrease in the flame surface area and, as a result, to a sharp decrease in the average speed of the combustion wave.[2] It should be noticed that, unlike a stationary flame, the flow with the accelerating flame is not isobaric. In the case of an accelerating flame, the pressure increases at about the same rate as the flame velocity, producing pressure waves and pressure gradient that support the increase of the flow velocity ahead of the flame. During the acceleration phase, the maximum pressure is near the flame surface and it decreases toward the opposite end of the tube.

The decrease of the flame surface area due to quenching of the rear parts of the flame skirt at the side walls leads to a sharp decrease in the average speed of the combustion wave and to the reduced expansion of the burned gases, in which velocity can be roughly estimated in a 1D model as $\langle u_b \rangle = \frac{2L}{D}\Theta U_f$, where L is the distance from the flame tip to the rear side of the flame skirt [62]. The decelerating flame causes rarefaction waves that are generated in the upstream flow of the unburned mixture. They are reflected from the closed end andspread to the flame, where they are partially transferred to the burned gas and partially reflected back in the upstream. It should be noted fruitful hypothesis proposed by Guénoche [6] that during the phase of the flame deceleration owing to the quenching the part of the flame skirt, the flame produces rarefaction waves. Guénoche considered the interaction of rarefaction waves with the flame as a key process in the formation of a tulip flame. The rarefaction waves may cause a reversed flow of the unburned gas directed toward the flame in the case of a tube with both closed ends, and the revered velocity field behind the flame caused by the rarefaction and pressure waves explains the formation of a tulip flame in long tubes or in tubes with the open end.

A few comments should be made about a tulip flame formation depending on a channel width. The thickness of a laminar boundary layer in the upstream flow is $\delta_l \approx 5X/\sqrt{\text{Re}}$, where $\text{Re} \approx \Theta U_f X/\nu$ is the Reynolds number and X is the coordinate along the tube. The mass of the unburned mixture passing through the tube cross section is the same. Since the thickness of the boundary layer increases with X, the inner part of the flow, where the velocity is uniform, accelerates with the increase of δ_l. In general, this leads to the development of a Poiseuille flow with a parabolic velocity profile in the upstream flow. The Poiseuille flow will be formed when the boundary layer thickness becomes equal to half the width of the tube. This leads to an estimate of the distance that the flame travels before establishing a Poiseuille flow in the upstream flow: $X_P \sim \Theta U_f D^2/100\nu$. The corresponding time is of the order of $t_P \sim D^2/100\nu$. A more detailed analysis [71, 72] leads to the same functional dependence with a slightly different numerical coefficient: $X_P \approx \Theta U_f D^2/104\nu$, and $t_P \approx D^2/104\nu$. Using the relation $U_f L_f \sim \nu$, we can rewrite these formulas in the form: $X_P \approx \Theta D^2/100\, L_f$ and $t_P \approx D^2/100\nu$. Consider the numerical values of X_P and t_P for the channel of width $D = 1$ cm. The Poiseuille flow in a hydrogen/air mixture ($L_f = 0.035$ cm, $\Theta = 7.8$) will be established after a

[2] Depending of the size of the ignition kernel and the tube width, the flame skirt may contact the side walls earlier, during "finger shape" phase.

rather long time $t_P \approx 50$ ms compared to the characteristic times of the flame-front inversion $t_{tulip} \approx 1.0$ ms $t_{DDT} \approx 2.0$ ms or time of transition to detonation. On the contrary, in narrow channels, $D \leq 1$ mm, the Poiseuille flow is established in a short time $t_P \sim 0.1$ ms. The velocity profiles ahead of the hydrogen/oxygen flame obtained in numerical simulations [73, 74] in a wide channel, $D = 1$ cm, and in a narrow channel, $D = 0.5$ mm, are shown in Figs. 15.5 and 15.6 (Chap. 15).

Note that in narrow tubes, $D \leq 1$ mm, where the Poiseuille flow is established at the earliest stage of flame propagation, the flame accelerates monotonically without the deceleration stage, and the formation of tulip flame in narrow tubes has never been observed experimentally. This confirms the assumption that the stretching of the flame skirt along the boundary layer plays a key role in the mechanism of the tulip flame formation. Since the Rayleigh-Taylor (RT) instability is still discussed as a possible mechanism leading to the formation of a tulip or distorted tulip flame (DTF) (see e.g. 77), it is useful to verify whether RT instability might be the mechanism of the tulip flame formation. The flame front is distorted by RT instability if fla me acceleration is directed from the unburned fuel to the combustion products. This occurs when the side parts of the flame front skirt approach and touch the tube walls and extinguish, resulting in a decrease in flame velocity. The duration of the negative flame acceleration caused by extinguishing the side parts of the flame skirt on the tube walls, can be estimated as $\Delta t \approx \delta_1/U_f$, where δ_1 is a thickness of the boundary layer and U_f is a normal flame speed. The flame acceleration can be estimated as $g \sim U_f^2/\delta_1$. In order the effect of RT instability to be noticeable, the duration of the stage of flame propagation with negative acceleration should be much longer than the characteristic time of development of RT instability $1/\sigma = \sqrt{gk}$ should be much longer than the characteristic time of development of RT instability. This leads to the limitation on the wavelength of perturbations $\lambda \ll \delta_1$. Thus, only perturbations with a wavelength less than or of the order of the boundary layer thickness c an develop. But such short-w avelength perturbations are suppressed by viscosity, while perturbations of the wavelength of the order of the half-width of the tube, which are assumed to be responsible for the formation of a tulip flame, do not have enough time to develop.

9.4.3 Numerical Simulations and Mechanism of Tulip Flame Formation

The mechanism of the tulip flame formation was studied in detail by Qian, Wang and Liberman [75]. The dynamics of a laminar stoichiometric hydrogen–air flame ignited near the left closed end of the 2D channel of length $L = 8.4$ cm and width $D = 0.6$ cm was investigated using high-resolution numerical simulations. The computations solved the multidimensional, time-dependent, reactive compressible Navier–Stokes equations including molecular diffusion, thermal conduction, viscosity, and energy release in chemical reactions.

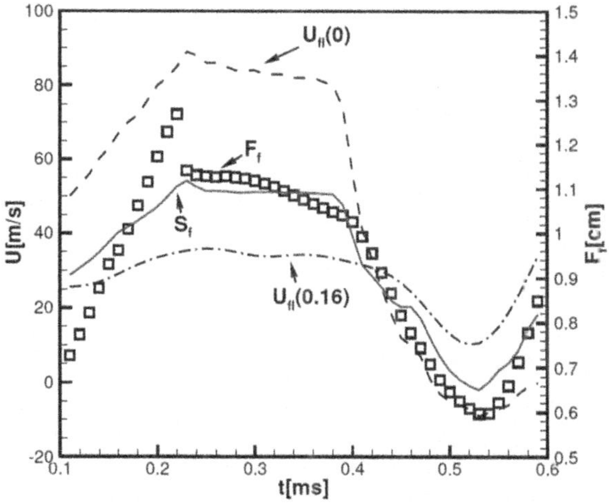

Fig. 9.12 Time evolution of combustion wave velocity S_f (solid line), local x-component of velocities of the flame front $U_{fl}(x, y)$ at the points $y = 0$ and $y = 0.16$ cm, and the flame surface area (length) shown by squares

It is well known that a one-step Arrhenius chemical model is computationally much easier for multidimensional simulations if it can satisfactorily reproduce flame dynamics. Since, according to experiments [70], the formation of a tulip flame is expected to be a pure hydrodynamic phenomenon, a one-step Arrhenius chemical model was used in simulations. Therefore, the parameters of the one-step chemical model in simulations [75] of a premixed hydrogen/air flame were taken the same as in [76] for the stoichiometric hydrogen–air flame with the Lewis number equal to unity. There is a quantitative difference in the laminar flame velocities obtained for a one-step model, $U_f = 3.1$ m/s, while for a detailed model, the laminar flame velocity is $U_f = 2.36$.

Figure 9.12 shows the time evolution of the local speeds of the flame front at the channel axis ($y = 0$): $U_{fL}(0)$, near the side wall at $y = 0.16$ cm: $U_{fL}(0.16)$, the combustion wave velocity defined as $S_f = \frac{1}{D}\int_0^D U_{fL}\cdot\sqrt{1+(x')^2}dy$ (the burning rate), and the flame surface area (length) F_f during the tulip shape formation, where $U_{fl}(x, y)$ is the x-component of the local speed of the flame front at the point (x, y) on the flame-front surface in the laboratory reference frame. In Fig. 9.12, the initial exponential increase in the combustion wave velocity and the velocity of the flame tip ($y = 0$), followed by a subsequent sharp decrease of the flame surface area (length) at $t \approx 0.22$ ms, and the transition to the deceleration stage, during which the flame surface area and local velocities of different parts of the flame surface continue to diminish, can be seen. The calculated evolution of the velocities of different parts of the flame surface is shown in Fig. 9.13, from $t = 0.2$ ms, when the rear part of the flame skirt was extinguished on the side wall, to the inversion of the flame front and the formation of a tulip flame. It is seen that while the local velocities of the flame

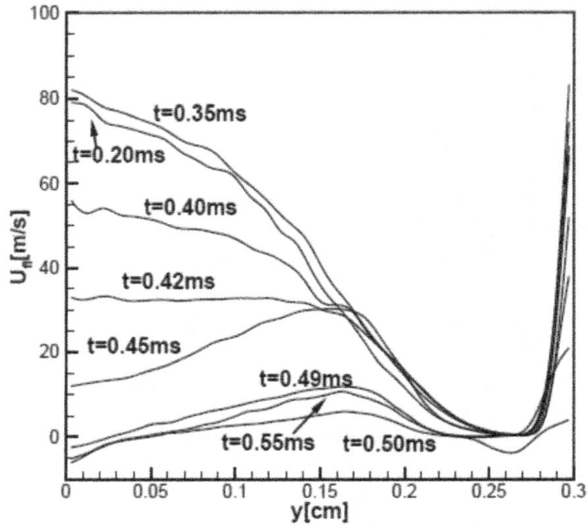

Fig. 9.13 Time evolution of local velocities of the flame front $U_{fl}(x, y)$

front near the tube axis ($y = 0$) decrease significantly, the decrease in the velocities of the flame front near the side walls is considerably less.

This behavior of the local velocities of the flame front, leading to the formation of a tulip-shaped flame front, is associated with rarefaction waves created in the unburned gas mixture by the flame during the deceleration stage. These rarefaction waves travel to the right closed end, reflect from the closed end, propagate back toward the flame, and so on. The result is the development of a reverse flow of the unburned gas mixture directed toward the flame. This can be seen in Fig. 9.14, where the flow velocity of the unburned gas in the close vicinity ahead of the flame front, $U_+ = u(X_{f+}, Y_f)$, is shown.

It is seen that rarefaction waves propagate in the positive direction of the x-axis, from the flame to the opposite closed end of the tube. After reflection from the closed end, the rarefaction wave returns to the flame front, partly reinforces the effect of the original rarefaction wave, and partly transmits into the burned gases as a rarefaction wave. The rarefaction waves reduce the flow velocity of unburned gas ahead of the flame stronger near the axis of the tube, $y = 0$, and finally, they cause the reverse flow of the unburned gas. Because of the no-slip walls, the largest negative velocity in the flow of the unburned gas is near the axis, $y = 0$. Accordingly, the local velocity of the flame front $U_{fl}(y) = U_f + u_u(x, y)$ decreases more strongly at $y = 0$ and decreases less at the side wall, where the tulip petal is formed at $y \approx 0.22$ cm.

The overall process of the tulip flame formation is shown in Fig. 9.15 with more details of the flow ahead and behind the flame shown in Fig. 9.16.

It is seen that after the first stage of the flame deceleration, $0.2\,\text{ms} < t < 0.4\,\text{ms}$, the pressure wave is coming from the right closed end approaches the flame (vertical line at frame $t = 0.39$ ms), and after reflection from the flame and side walls, two

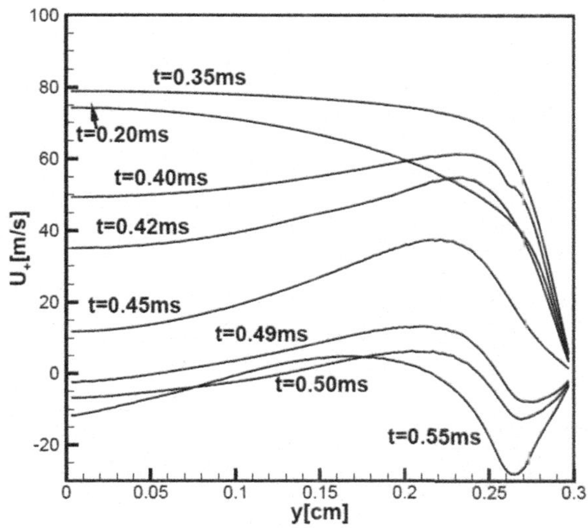

Fig. 9.14 Velocities of the unburned gas in the close vicinity ahead of the flame front

vortices in the boundary layer ahead of the flame are created at $t \geq 0.41$ ms. At the same time, two vortices are generated in the flow behind the flame near the rear end of the tube. In both cases, these vortices are not caused by the flame instability, but by the interaction of opposite flows of the burned gas between the flame and the rear end of the tube and the transverse flow of the burned gas originated after quenching the rear part of the flame skirt at $t = 0.22$ ms. At later times, vortices behind the flame disappear. Unlike the Matalon and Metzener theory [68, 69], which attributed the tulip flame formation to vortices behind the flame that drive and invert the flame front into a tulip shape, vortices do not appear to play an important role, if any, in the formation of the tulip shape. Also, vortices do not appear due to the intrinsic instabilities of the flame and, apparently, are not key factors in the inversion of the flame front, as suggested in previous explanations of the tulip flame formation.

The scenario of the formation of a tulip flame in a long tube (large aspect ratio L/D) or in a tube closed at the ignition end and open at the other end, as well as a tulip shape formation for slow flames in a slow velocity flame in the tube with both closed ends, is different from the case of a fast flame in the tube with both closed ends, considered above. As an example, Fig. 9.17 shows the evolution of combustion wave velocity, local flame-front velocities at $y = 0$ and at $y = 0.16$ cm, and the flame surface area for a hydrogen/air flame propagating in the tube $D = 0.8$ cm from the closed ignition end to the other open end. It is seen that when the propagating flame has reached the side walls of the tube, it slows down owing to the decrease of the flame surface area. The reduced surface of the flame leads to flame deceleration and causes the reduced expansion of the burned products, which generates rarefaction waves in the unburned mixture ahead of the flame.

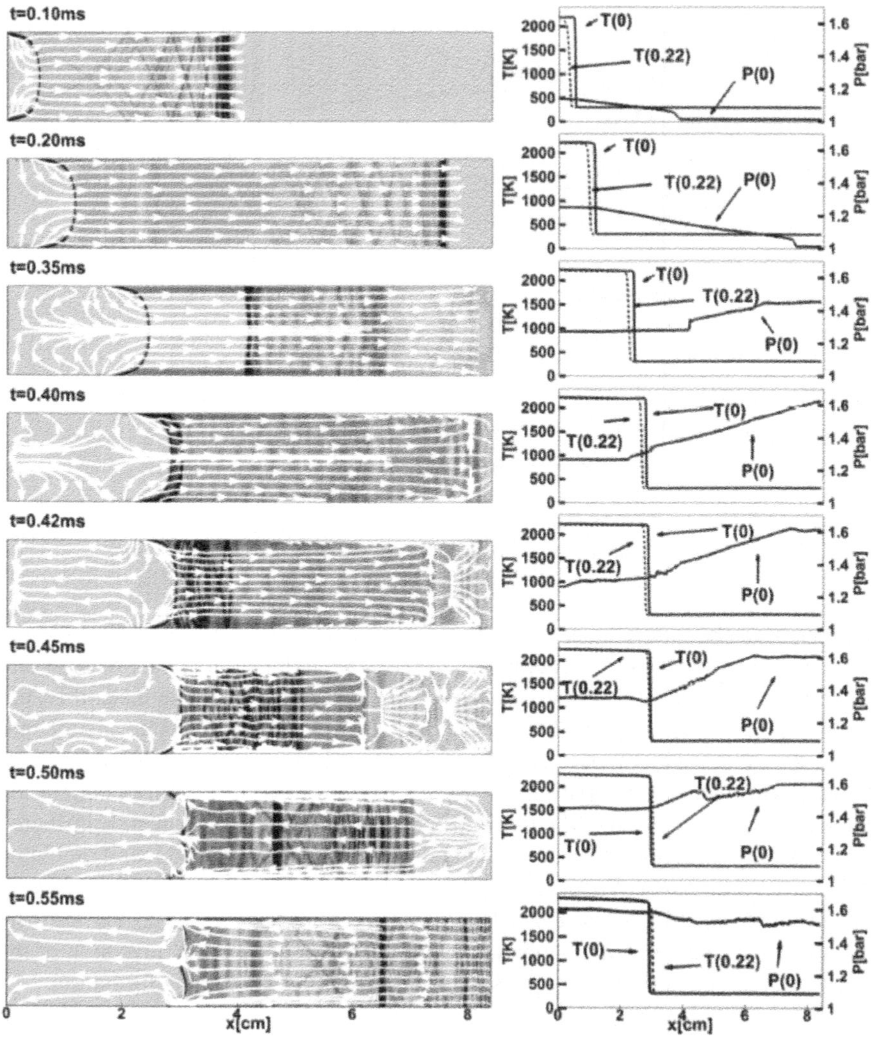

Fig. 9.15 Time sequences of "schlieren images" (calculated density gradients) and streamlines (left column) and evolution of temperature and pressure profiles at $y = 0$ (solid lines) and $y = 0.22$ cm (dashed lines) in the flow ahead of the flame (right column)

The rarefaction waves generated by the deceleration flame reduce the velocities in the flow of the unburned mixture as it is shown in Fig. 9.18. Accordingly, the velocities of different parts of the flame front shown in Fig. 9.19 also change but in a more complex manner due to the interaction of the flame with pressure and rarefaction waves in the downstream flow.

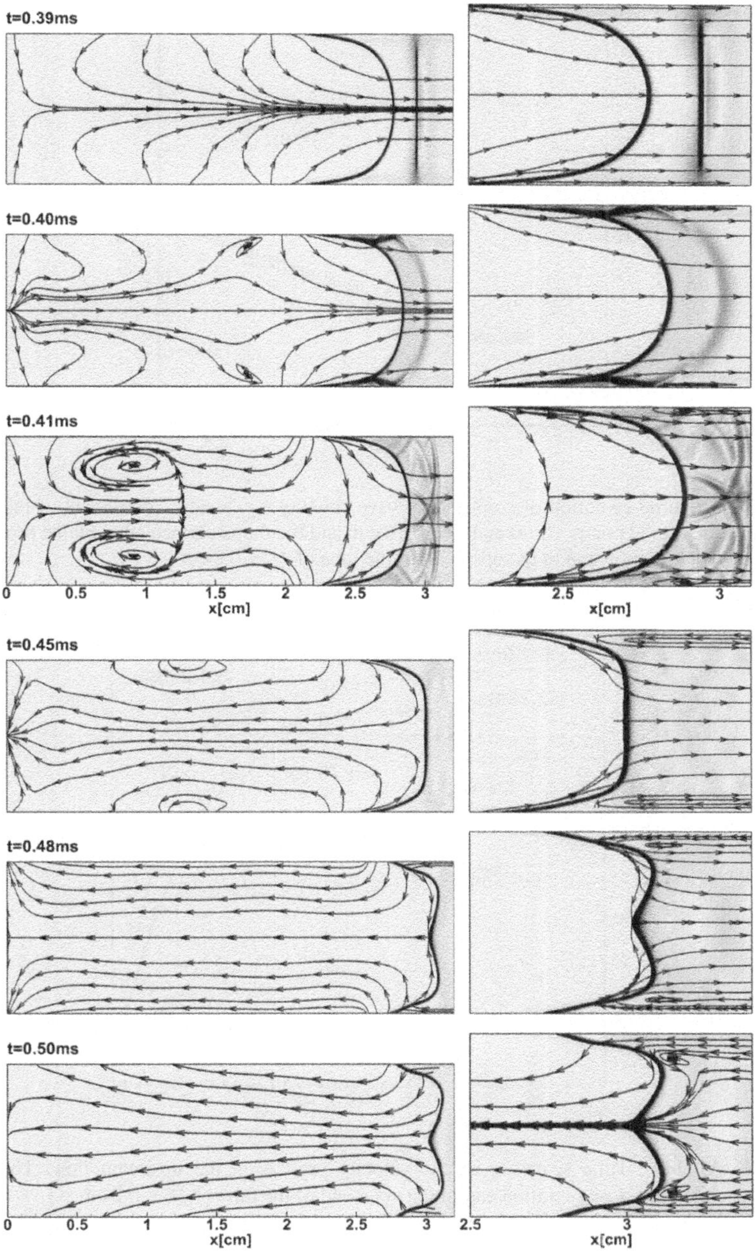

Fig. 9.16 Density gradient and streamlines behind and in the vicinity of the flame front during the tulip-shaped flame formation

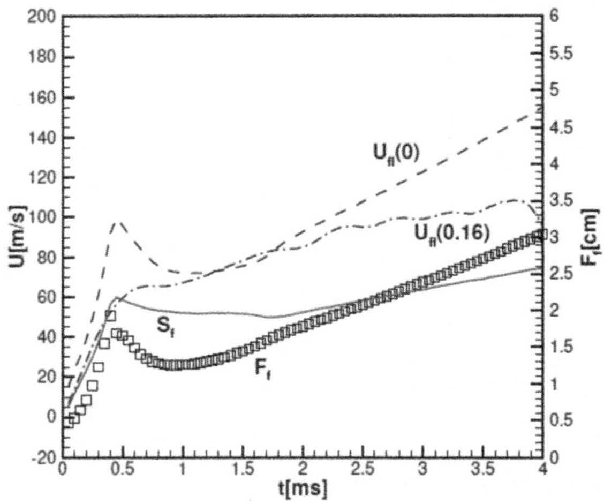

Fig. 9.17 The time evolution of combustion wave velocity S_f, local x-component of the flame-front velocities at the points $y = 0$ and $y = 0.16$ cm, and the flame surface area. Flame propagates from the closed ignition end to the open end of the tube of $D = 0.8$ cm

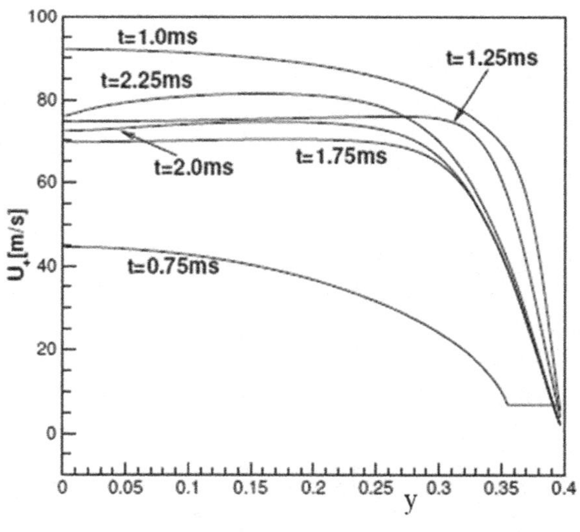

Fig. 9.18 A. (left). Time sequence of the velocity profiles in the upstream flow. The flame propagates from the closed ignition end to the open end of the tube of $D = 0.8$ cm

The interaction of the rarefaction wave with the flame leads to the same effect, whether a flame propagates in a sufficiently long tube with both closed ends or toward the open end, and also in the case of a low-velocity flame (e.g. methane/air) in a tube with both closed ends and not too small aspect ratio L/D.

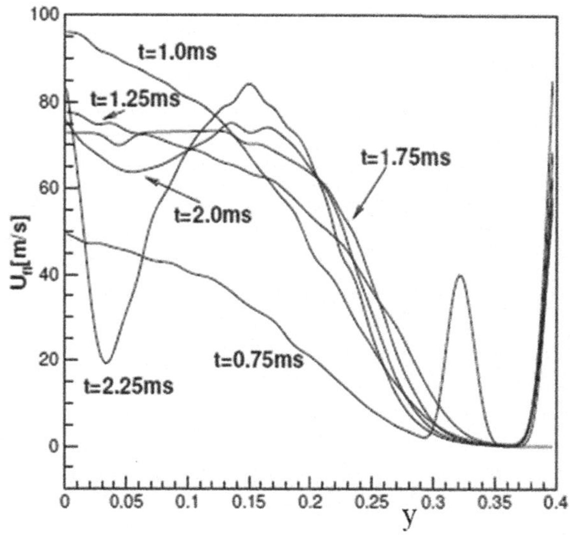

Fig. 9.19 A (right). Time sequence of x-components of the velocities of different parts of the flame front during the formation of a tulip shape in a half-open tube

The calculated dynamics of the tulip flame formation in the hydrogen/air flame propagating to the open end of the tube of width $D = 0.8$ cm is shown in Fig. 9.20. In this case, a reversed flow does not form ahead of the flame; however, the compression and rarefaction waves change the direction of the flow behind decelerating flame toward the closed end. As in the case of the tube with both closed ends shown in Figs. 9.15 and 9.16, two vortices appear, which are seen in the $t = 1.25$ ms frame. However, these vortices arise not due to the flame instability, but as a result of the interaction of the flows of burned gases leaving the flame front and the flows reflected from the side walls and the closed end of the tube.

Comparison of simulations using a one-step chemical model [76] with simulations that used a detailed chemical model showed that the difference in the obtained results is mainly related to the large difference in the velocity of laminar flame for the one-step model, $U_f = 3.1$ m/s, and for the detailed chemical model, $U_f = 2.35$ m/s. For example, the scenario of the tulip flame formation shown in Fig. 9.5 was obtained for $U_f = 3.1$ m/s and for the tube width $D = 0.6$ cm and length $L = 8.4$ cm. However, for the flame velocity $U_f = 2.1$ m/s, such tube will be "long", and the scenario of the tulip formation is similar to that shown in Fig. 9.20. A scenario similar to Fig. 9.15 for the flame velocity $U_f = 2.1$ m/s requires more than two times shorter tube.

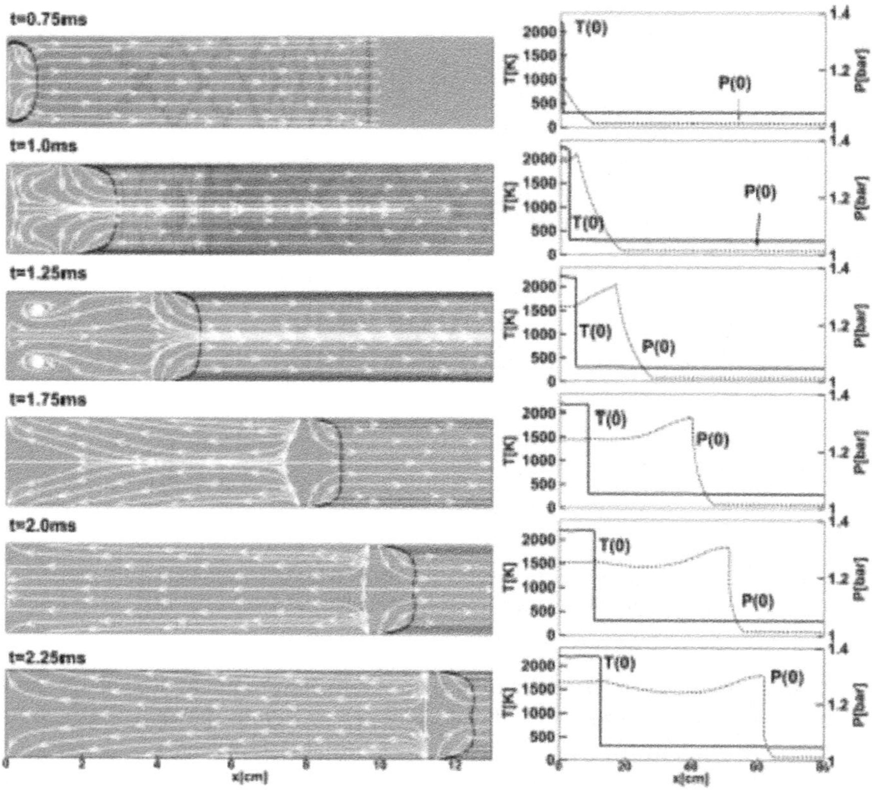

Fig. 9.20 Flame dynamics during the formation of a tulip shape of the hydrogen/air flame propagating from the closed to open end of the tube of width $D = 0.8$ cm. Left column: the sequences of density gradient and streamlines; right column: evolution of temperature and pressure (dashed lines) at $y = 0$

References

1. G.L. Borman, K.W. Ragland, *Combustion Engineering* (McGraw-Hill, New York, 1998)
2. M.A. Liberman, L.-E. Eriksson, M.F. Ivanov, O. Peil, Numerical modeling of the propagating flame and knock occurrence in spark-ignition engines. Comb. Sci. Tech. **177**, 151–182 (2005)
3. M.A. Liberman, L.-E. Eriksson, M.F. Ivanov, Hot spot formation by the propagating flame and the influence of EGR on knock occurrence in SI-engines. Comb. Sci. Tech. **178**(2006), 1613–1647 (2006)
4. Z. Wang, H. Liu, R.D. Reitz, Knocking combustion in spark-ignition engines. Prog. Energy Combust. Sci. **61**, 78–112 (2017)
5. J.I. Ramos, *Internal Combustion Engine Modelling* (Hemisphere Publishing Corporation, New York, 1989)
6. V.N. Kurdyumov, M. Matalon, Self-accelerating flames in long narrow open channels. Proc. Combust. Inst. **35**, 921–928 (2015)
7. V.V. Bychkov, M.A. Liberman, On the stability of a flame in a closed chamber. Phys. Rev. Letters **78**, 1371–1734 (1997)

8. M. Gonzales, R. Borghi, A. Saouab, Interaction of a flame front with its self-generated flow in an enclosure: the "tulip flame" phenomenon. Combust. Flame **88**, 201–220 (1992)
9. M.A. Liberman, V.V. Bychkov, S.M. Golberg, L.E. Eriksson, Numerical study of curved flames under confinement. Combust. Sci. Technol. **136**, 221–251 (1998)
10. H. Guénoche, Flame propagation in tubes and in closed vessels, in *Non-Steady Flame Propagation*, ed. By G.H. Markstein (Pergamon, London, 1964), pp. 107–181
11. B. N'Konga, G. Fernandez, H. Guillard, B. Larrouturou, Numerical investigations of the tulip flame instability—comparisons with experimental results. Combust. Sci. Tech. **87**(1993), 69–89 (1993)
12. S. Chandrasekhar, *Hydrodynamic and Hydromagnetic Stability* (Clarendon press, Oxford, 1961)
13. G. Searby, Acoustic instability in premixed flames. Combust. Sci. Technol. **81**, 221–231 (1992)
14. M. Matalon, J.L. McGreevy, The initial development of the tulip flame. Symp. (Int.) Combust. **25**, 1407–1413 (1994)
15. L.D. Landau, E.M. Lifshitz, *Course of Theoretical Physics. Mechanics*, vol. 1 (Pergamon Press, Oxford, 2011)
16. OYu., Travnikov, V.V. Bychkov, M.A. Liberman, Interaction of curved flames and weak shocks. Combust. Sci. Technol. **142**, 1–28 (1999)
17. E. Mallard, H.L. Le Chatelier, On the propagation velocity of burning in gaseous explosive mixture. Compt. Rend. **93**, 145–148 (1881)
18. P. Daniell, The theory of flame motion. Proc. R. Soc. **A126**, 393–405 (1930)
19. B. Lewis, G. Elbe, On the theory of flame propagation. J. Chem. Phys. **2**, 537–546 (1934)
20. Ya.B. Zel'dovich, Theory of combustion and gas detonation. Akad. Nauk SSSR, Moscow (1944)
21. H.F. Coward, G.W. Jones, Limits of flammability of gases and vapors. Bureau of Mines Report **503**, 1–124 (1952)
22. J. Jarosinski, Flammability limits: history and mechanism of flame extinction, in *Combustion Phenomena*, ed. by J. Jarosinski, B. Veyssiere (CRC Press, Taylor & Francis Group, 2009), pp. 15–25
23. I. Wierzba, Q. Wang, The flammability limits of H_2–CO–CH_4 mixtures in air at elevated temperatures. Int. J. Hydrogen Energy **31**, 485–489 (2006)
24. L. He, P. Clavin, Premixed hydrogen-oxygen flames. Part I&II. Combust. Flame **93**, 391–420 (1993)
25. F.A. Williams, *Combustion Theory* (Benjamin/Cummings, Menlo, CA/Reading, MA, 1985)
26. F.N. Egolfopoulos, A.T. Holley, C.K. Law, An assessment of the lean flammability limits of CH_4/air and C_3H_8/air mixtures at engine-like conditions. Proc. Combust. Inst. **31**, 3015–3022 (2007)
27. C.K. Law, Dynamics of stretched flames. Proc. Combust. Inst. **22**, 1381–1402 (1989)
28. C.K. Law, F.N. Egolfopoulos, A unified chain-thermal theory of fundamental flammability limits. Proc. Combust. Inst. **24**, 137–144 (1992)
29. G. Dixon-Lewis, Laminar premixed flame extinction limits. I combined effects of stretch and upstream heat loss in the twin-flame unburnt-to-unburnt opposed flow configuration. Proc. R. Soc. Lond. A **452**, 1857–1884 (1996)
30. Y.N. Shebeko, W. Fan, I.A. Bolodian, V. Navzenya, An analytical evaluation of flammability limits of gaseous mixtures of combustible-oxidizer-diluent. Fire Safety J. **37**, 549–568 (2002)
31. A.D. Benedetto, The thermal/thermodynamic theory of flammability: the adiabatic flammability limits. Chem. Eng. Sci. **99**, 265–273 (2013)
32. M. Grabarczyk, A. Teodorczyk, V.D. Sarli, A. Benedetto, Effect of initial temperature on the explosion pressure of various liquid fuels and their blends. J. Loss Prevention Process Ind. **44**, 775–779 (2016)
33. H. Tian, M. Wu, G. Shu, Y. Liu, X. Wang, Experimental and theoretical study of flammability limits of hydrocarbon CO_2 mixture. Int. J. Hydrogen Energy **42**, 29597–29605 (2017)
34. A. Bertolino, A. Stagni, A. Cuoci, T. Faravelli, A. Parente, A. Frassoldati, Prediction of flammable range for pure fuels and mixtures using detailed kinetics. Combust. Flame **207**, 120–133 (2019)

35. R.J. Kee, F.M. Rupley, J.A. Miller, The Chemkin thermodynamic data base. Sandia National Laboratories Report SAND87-8215B (1990)
36. G.P. Smith, D.M. Golden, M. Frenklach, N.W. Moriarty, B. Eiteneer, M. Goldenberg, C.T. Bowman, R.K. Hanson, S. Song, W.C. Gardiner, Jr., V.V. Lissianski, Z. Qin, GRI-Mech 3.0. http://www.me.berkeley.edu/gri_mech/
37. B.J. McBride, S. Gordon, FORTRAN IV program for calculation of thermodynamic data. NASA Report NASA TN-4097 (1967)
38. Ya. B. Zel'dovich, Zh. Eksp. Theor. Fiz. **11**, 159–168 (1941). English translation: Selected Works of Ya. B. Zel'dovich, Princeton University Press, **1**, 271–287 (1992)
39. C.R. Ferguson, J.C. Keck, On laminar flame quenching and its application to spark ignition engines. Combust. Flame **28**, 197–205 (1977)
40. J. Jarosinski, A survey of recent studies on flame extinction. Progr. Energy Combust. Science **12**, 81–116 (1986)
41. K. Bioche, L. Vervisch, G. Ribert, Premixed flame–wall interaction in a narrow channel: impact of wall thermal conductivity and heat losses. J. Fluid Mech. **856**, 5–35 (2018)
42. L. Kagan, G. Sivashinsky, The transition from deflagration to detonation in thin channels. Combust. Flame **120**, 222–232 (2003)
43. K.V. Dobrego, I.M. Kozlov, V.V. Vasiliev, Flame dynamics in thin semi-closed tubes at different heat loss conditions. Int. J. Heat Mass Transfer **49**, 198–208 (2006)
44. L. Kagan, D. Valiev, M. Liberman, G. Sivashinsky, Effect of hydraulic resistance and heat losses on the deflagration-to-detonation transition, in *Proceedings of Pulsed and Continuous Detonations*, ed. by G. Roy, S. Frolov, J. Sinibaldy (Torus Press Ltd., 2006), pp. 119–123
45. L. Kagan, G. Sivashinsky, On the transition from deflagration to detonation in narrow tubes. Flow Turbul. Combust. **84**, 423–437 (2010)
46. M. Fukuda, E. Dzieminska, A.K. Hayashi, E. Yamada, N. Tsuboi, Effect of wall conditions on DDT in hydrogen–oxygen mixtures. Shock Waves **23**(2013), 191–200 (2013)
47. C. Wang, Y. Zhao, W. Han, Effect of heat-loss boundary on flame acceleration and deflagration-to-detonation transition in narrow channels. Combust. Sci. Tech. **189**, 1605–1623 (2017)
48. H. Bongers, L.P.H. De Goey, The effect of simplified transport modelling on the burning velocity of laminar premixed flames. Combust. Sci. andTech. **175**, 1915–1928 (2003)
49. M.A. Liberman, M.F. Ivanov, A.D. Kiverin, M.S. Kuznetsov, A.A. Chukalovsky, T.V. Rakhimova, Deflagration-to-detonation transition in highly reactive combustible mixtures. Acta Astronaut. **67**, 688–701 (2010)
50. O.C. Ellis, H.A. Robinson, New method of flame analysis. J. Chem. Soc. **127**, 760–767 (1925)
51. O.C. Ellis, R.V. Wheeler, The movement of flame in closed vessels. J. Chem. Soc. **127**, 764–772 (1925)
52. O. Ellis, Flame movement in gaseous explosive mixtures. Fuel Sci. **7**, 502–508 (1928)
53. G.D. Salamandra, T.V. Bazhenova, I.M. Naboko, Formation of a detonation wave in gas combustion in tubes. Symp. (Int.) Combust. **7**, 851–855 (1959)
54. G.H. Markstein, *Nonsteady Flame Propagation* (Pergamon, Oxford, 1964)
55. R. Starke, P. Roth, An experimental investigation of flame behavior during cylindrical vessel explosions. Combust. Flame **66**, 249–259 (1986)
56. D. Dunn-Rankin, The interaction between a laminar flame and its self-generated flow. Ph.D. Dissertation (University of California, Berkeley, 1985)
57. D. Dunn-Rankin, P.K. Barr, R.F. Sawyer, Numerical and experimental study of "tulip" flame formation in a closed vessel. Symp. (Int.) Combust. **21**, 1291–1301 (1986)
58. D. Dunn-Rankin, R.F. Sawyer, Tulip flames: changes in shapes of premixed flames propagating in closed tubes. Exp. Fluids **24**, 130–140 (1998)
59. D. Dunn-Rankin, Tulip flames—the shape of deflagrations in closed tubes, in *Combustion Phenomena: Selected Mechanisms of Flame Formation, Propagation and Extinction* (CRC Press, Boca Raton, FL, 2009)
60. S. Kerampran, D. Desbordes, B. Veyssere, Study of the mechanisms of flame acceleration in a tube of constant cross section. Combust. Sci. Technol. **158**, 71–91 (2000)

61. A. Hariharan, I.S. Wichman, Premixed flame propagation and morphology in a constant volume combustion chamber. Combust. Sci. Technol. **186**, 1025–1040 (2014)
62. C. Clanet, G. Searby, On the 'tulip flame' phenomenon. Combust. Flame **105**, 225–238 (1996)
63. I. Jeung, K. Cho, K. Jeong, Role of flame generated flow in the formation of tulip flame, in *Paper AIAA 89–0492, 27th AIAA Aerospace Sciences Meeting*, Reno, Nevada, January 9–12, 1989
64. M. Gonzalez, R. Borghi, A. Saouab, Interaction of a flame front with its self-generated flow in an enclosure: the "tulip flame" phenomenon. Combust. Flame **88**, 201–220 (1992)
65. F.S. Marra, G. Continillo, Numerical study of premixed laminar flame propagation in a closed tube with a full Navier-Stokes approach. Twenty-Sixth Symp. (Int.) Combust. **26**, 907–913 (1996)
66. M. Gonzalez, Acoustic instability of a premixed flame propagating in a tube. Combust. Flame **107**, 245–259 (1996)
67. J.W. Dold, G. Joulin, An evolution equation modeling inversion of tulip flames. Combust. Flame **100**, 450–456 (1995)
68. M. Matalon, P. Metzener, The propagation of premixed flames in closed tubes. J. Fluid Mech. **336**, 331–350 (1997)
69. P. Metzener, M. Matalon, Premixed flames in closed cylindrical tubes. Combust. Theor. Model. **5**, 463–483 (2001)
70. B. Ponizy, A. Claverie, B. Veyssière, Tulip flame: the mechanism of flame front inversion. Combust. Flame **161**, 3051–3062 (2014)
71. H. Schlichting, K. Gersten, *Boundary-Layer Theory*, 9th edn. (Springer-Verlag, Berlin Heidelberg, 2017)
72. Ya. B. Zeldovich, On the theory of establishing a detonation in gases (in Russian Zh. Tekhn. Fiz.) J. Tech. Phys. **17**, 3–26 (1947)
73. M.F. Ivanov, A.D. Kiverin, M.A. Liberman, Flame acceleration and DDT of hydrogen-oxygen gaseous mixtures in channels with no-slip walls. Intern. J. Hydrogen Energy **36**, 7714–7728 (2011)
74. M.F. Ivanov, A.D. Kiverin, M.A. Liberman, Hydrogen-oxygen flame acceleration and transition to detonation in channels with no-slip walls for a detailed chemical reaction model. Phys. Rev. E **83**, 056313 (1–16) (2011)
75. C. Qian, C. Wang and M. Liberman, On the mechanism of tulip flames formation. Preprint NORDITA 2021-073, arXiv:2108.09970v1 [physics.flu-dyn]. Submitted to Combust. Flame (2021)
76. V.N. Gamezo, T. Ogawa, E.S. Oran, Numerical simulations of flame propagation and DDT in obstructed channels filled with hydrogen–air mixture. Proc. Combust. Inst. **31**, 2463–2471 (2007)
77. X. Huahua, R.W. Houim, E.S. Oran, Formation and evolution of distorted tulip flames. Combust. Flame **162**(11), 4084–4101 (2005). https://doi.org/10.1016/j.combustflame.2015.08.020

Chapter 10
Flame in a Gravitational Field

10.1 The Rayleigh–Taylor Instability

The Rayleigh-Taylor (RT) instability develops at the interface between a heavy and a light fluid when the fluid (gas) with a higher density is supported by the fluid with smaller density in a gravitational field. The RT instability is of great interest for many problems in physics, astrophysics, and technology. RT instability is the main factor preventing to achieve a high degree of spherically symmetric compression of laser-irradiated targets [1, 2] or strong compression of the Z-pinch plasma [3] in inertial confinement fusion.

The linear stage of RT instability was considered in Sec. 5.3. It was shown that in the case of an upward propagating flame, small perturbations of the infinitely thin flame-front grow exponentially in time with the instability growth rate

$$\frac{\sigma_0}{kU_f} = \frac{\Theta}{\Theta + 1}\left(\sqrt{\Theta + 1 - \frac{1}{\Theta} + \frac{\Theta^2 - 1}{\Theta^2 k L_f Fr}} - 1\right), \tag{10.1.1}$$

where the Froude number $Fr = U_f^2/gL_f$. The dispersion relation (10.1.1) coincides with the growth rate of DL instability if the gravitational acceleration tends to zero, $Fr \to \infty$. In the opposite case of large gravitation, $Fr \to 0$, the growth rate goes over to the expression for the classical RT instability $\sigma_{RT} = \sqrt{A_t gk}$, where $A = (\rho_2 - \rho_1)/(\rho_2 + \rho_1)$ is the Atwood number. The dispersion relation of RT instability for a flame of finite thickness is given by Eq. (5.3.9) and the growth rate is given by Eq. (5.3.10).

During the early nonlinear stage, strong interface corrugations appear, which eventually develop in upward-moving mushroom-like structures. The next "late stage" of the RT instability begins with formation of spikes and bubbles of the lighter gas floating upward and transforming potential energy of the gravitational field into the kinetic energy of the upward flow. Despite more than 20 years of hard efforts and impressive work (see reviews [2, 4, 5]), there is no complete analytical theory for the

M. A. Liberman, *Combustion Physics*,
https://doi.org/10.1007/978-3-030-85139-2_10

nonlinear RT instability even for the simplified problem of ideal incompressible and inviscid fluid. However, many advanced numerical simulations (e.g. [6–9]) confirm our qualitative understanding of the RT instability development at the nonlinear stage. In the final stage of the nonlinear RT instability, the bubbles of the light gas are developed and rise upward with a constant velocity through the heavy gas. The velocity of a bubble rising in the incompressible fluid in tubes can be calculated using simple analytical methods developed in [10–12] for 2D and 3D problems. Numerical simulations [13, 14] have shown that the bubble velocity specifies the asymptotic behavior of the flame velocity propagating in a tube with a relatively large radius in a gravitational field. Experimental studies [15–17] have shown that a flame in the mixture 10% CH_4 and air propagates in a horizontal tube of diameter 90 cm with the velocity 245 cm/s exceeding considerably the normal velocity of the flame, which is 43 cm/s. Such a strong increase in the flame velocity cannot be explained only by the DL instability, but it can be associated with the stretching of the flame-front in the inhomogeneous flow created ahead of the flame due to wall friction or due to the RT instability of the flame in the gravitational field.

The linear stage of the RT instability of an upward propagating flame was considered in Chap. 5. The nonlinear stage of the instability is a more complex problem, although it can be qualitatively described as a bubble rising. A flame propagating in the cold combustible mixture converts it into hot combustion products of lower density, which tend to move upward in the gravitational field. It can be assumed that the rising light combustion products form a bubble, the surface of which is the flame-front. Let us evaluate how the velocity of the rising bubble depends on the gravitational acceleration g. There are only two values characterizing the problem: the gravitational acceleration g and the bubble size R, from which a single value of the velocity dimension can be formed, $U_b \propto \sqrt{gR}$. This should be the velocity of the rising bubble with the accuracy of a numerical factor of the order of unity, which can be found from numerical simulations. The bubble velocity depends also on the geometry of a bubble, which may be characterized as an open bubble or as a closed bubble shown in Fig. 10.1. The result is expected to be different also for 2D or 3D cases.

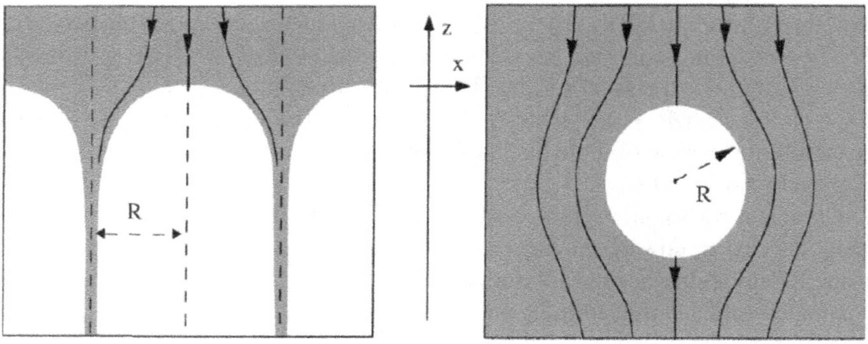

Fig. 10.1 Configurations of open (left) and closed rising bubbles

10.2 Velocities of Rising Bubbles

Let us calculate the velocity U_b of an open 3D bubble of size R formed by a light gas rising upward in an ideal tube of width D. We assume that the light gas density is negligible compared to the density of the heavy gas (fluid) and consider a stationary flow caused by a semi-infinite bubble propagating upward in a channel of width D, as shown schematically in Fig. 10.1 (left). In the reference frame of the bubble with coordinates $x = 0, z = 0$, which is the top of the bubble, the fluid at infinity flows toward the bubble with a constant velocity $u_z = U_b$ and then flows around the bubble and forms a thin jet. At the surface of a massless bubble, the Bernoulli equation is

$$\frac{1}{2}u^2 + gz = 0. \tag{10.2.1}$$

At rigid walls, $x = (0, D)$, the x-component of velocity is equal to zero, $u_x = 0$. Since there are no external sources of vortices, the flow must remain potential everywhere. In order to calculate the bubble velocity, we need to solve the Laplace equation for the velocity potential $\nabla^2\varphi = 0$ in the whole flow with the boundary conditions at the walls and at the bubble surface. This is rather difficult mathematical problem, which was not solved yet. We consider an approximation, which can be used both for closed [10] and for open bubbles [11]. The velocity potential and the stream function for the incoming flow are taken as

$$\varphi = U_b z + U_b\frac{R}{\pi}\exp\left(-\frac{\pi z}{R}\right)\cos\left(\frac{\pi x}{R}\right), \tag{10.2.2}$$

$$\psi = -U_b x + U_b\frac{R}{\pi}\exp\left(-\frac{\pi z}{R}\right)\sin\left(\frac{\pi x}{R}\right). \tag{10.2.3}$$

The corresponding velocity components are

$$u_x = -U_b\exp\left(-\frac{\pi z}{R}\right)\sin\left(\frac{\pi x}{R}\right), \tag{10.2.4}$$

$$u_z = U_b\left[1 - \exp\left(-\frac{\pi z}{R}\right)\cos\left(\frac{\pi x}{R}\right)\right]. \tag{10.2.5}$$

From the equation $\psi = 0$ for the streamline that passes through the stagnation point at the bubble top steams that the bubble surface near the top is determined by the formula $z = -\pi x^2/6R$. Substituting (10.2.4) and (10.2.5) into the Bernoulli equation and taking into account the expression for the bubble shape close to the top, we find for the velocity of the bubble

$$U_{2b} = \sqrt{gR/3\pi} = 0.33\sqrt{gR}. \tag{10.2.6}$$

The velocity of 3D bubbles is larger [11]: for the open bubbles, $U_{3b} = 0.51\sqrt{gR}$, and for the closed bubbles, $U_{3b} = 0.67\sqrt{gR}$. For the non-zero ratio $\rho_b/\rho_f = 1/\Theta$ of the burned gas density ρ_b and the fresh fuel density, ρ_f, assuming that the gas inside the bubble is at rest, we find for the pressure inside the bubble: $P = \rho_b g z$. From the Bernoulli equation (10.2.1) we have for the pressure balance at the bubble surface

$$\frac{1}{2}\rho_f u^2 + \rho_f g z = \rho_b g z. \tag{10.2.7}$$

Then, the term with acceleration g in the formula for the bubble velocity (10.2.6) should be replaced by $g(\Theta - 1)/\Theta$ so that Eq. (10.2.6) becomes

$$U_b = 0.33\sqrt{(\Theta - 1)gR/\Theta}. \tag{10.2.8}$$

A more detailed analysis, e.g. [8, 18, 19], predicts the velocity of 2D bubbles $U_{2D} = \sqrt{A_t gR/3\pi(A_t + 1)}$, and for 3D geometry, $U_{3D} = \sqrt{A_t gR/\pi(A_t + 1)}$. Here, the bubble radius is related to the wave number $k = 2\pi/R$ so that 3D bubbles propagate in $\sqrt{3} = 1.73$ times faster than 2D bubbles. It is interesting that, taking into account the nonzero velocity of the light gas in the bubble leads to nonzero vorticity in the downstream flow and increases the bubble speed so that instead of expression (10.2.8) for the velocity of 2D bubbles, we obtain Eq. (10.2.6).

It should be noticed that, in the case of non-zero density of the gas inside a bubble the stationary flow is unstable also against the Kelvin–Helmholtz instability of the heavy matter sliding along the surface of the light matter. As a consequence, the nonlinear stage of the RT instability is often accompanied by the development of mushroom shaped structures with the jets of the falling heavy matter and by mixing of the heavy and light gases.

10.3 Curved Flames in Vertical Tubes

Studying the propagation of a flame in vertical tubes in a gravitational field, we are interested mostly in the configuration of open bubbles. Since bubble velocity increases with the tube radius according to Eq. (10.2.8), then for a sufficiently wide tube this velocity may be much larger than the normal velocity of a planar flame-front. In this case, the velocity of flame propagation is mainly determined by the velocity of the light bubble of burnt matter. Although the problem of bubble motion does not include the process of burning and fuel consumption, it provides a fairly good approximation of the flame velocity for the case of a large acceleration or large tube radius, $gR/U_f^2 >> 1$. If we consider the solution of the complete system of equations of the flame dynamics in a strong gravitational field, then the bubble velocity represents the principal term in the expansion of the flame velocity in a power series of the small parameter of the Froude number $U_f^2/gR << 1$ [20, 21].

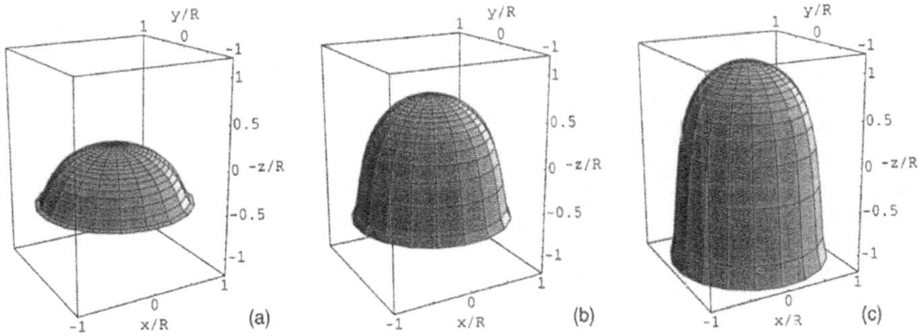

Fig. 10.2 The shapes of the convex flames with an expansion coefficient $\Theta = 5$ propagating upward in a cylindrical tube with a diameter $R = 2R_c$. Dimensionless accelerations $gL_f/U_f^2 = 0, 1, 3$ correspond to (a), (b), (c) respectively [13].

The opposite case of small acceleration and small expansion coefficient of the fuel $\Theta - 1 << 1$ was considered in [22], but the case of small acceleration and small expansion coefficient is far from a real flame. The velocities and shape of the upward propagating flames in vertical tubes have been investigated in numerical experiments [13, 14]. Figure 10.2 shows typical shapes of axisymmetric curved flames with the expansion coefficient $\Theta = 5$ in a cylindrical tube with the diameter $R = 2R_c$ obtained in [14] for different values of the dimensionless acceleration $gL_f/U_f^2 = 0, 1, 3$.

It can be seen that the flame-front becomes more and more curved with increasing dimensionless acceleration so that for a large value of $gL_f/U_f^2 = 3$ corresponding to $gR/U_f^2 = 75$, the shape of the flame looks like a rising semi-infinite bubble. The only important difference of the flow caused by the rising flame from the bubble flow is the absence of infinitely long jets of the falling heavy matter. The heavy fuel of the jets is consumed by the flame at some distance from the bubble top due to the "fire-polishing" effect, and the larger the dimensionless gravity acceleration, the longer the fuel jets. Similar behavior has been observed in numerical simulations of 2D curved flames propagating upward [13].

As in the case of zero gravity, two types of stationary solutions were observed for an upward propagating flame in cylindrical tubes with larger and smaller velocities corresponding to a convex and a concave flame. The shape of a convex flame (Fig. 10.2) is similar to the shape of a curved flames observed experimentally in vertical tubes [23, 24]. Concave flames propagating upward have never been reported in experimental studies. Presumably, the wall friction, the thermal losses to the walls, etc. make initiation of a concave flame not favorable for the upward propagating flame in a gravitational field. Nevertheless, in the numerical simulations of the flame dynamics in tubes with ideal walls, both convex and concave flames can propagate stationary without any change in shape or velocity, which implies the stability of both solutions.

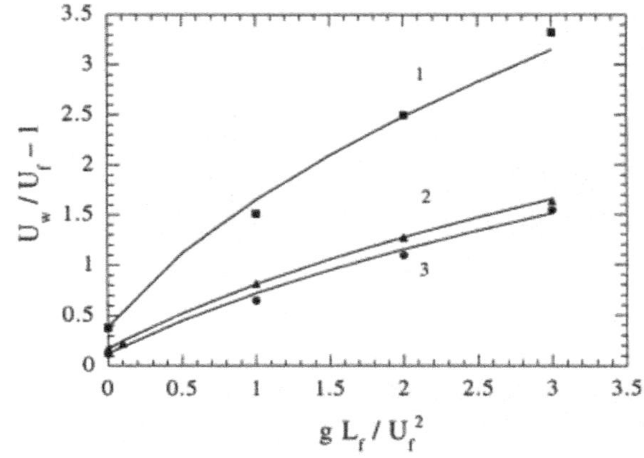

Fig. 10.3 The scaled velocity of an upward propagating flame ($\Theta = 5$, $R = 2R_c$) vs the dimensionless acceleration. Squares, circles, and triangles correspond to simulations of convex, concave, and 2D curved flames. Curves 1, 2, and 3 are the approximation by Eq. (10.3.1)

Figure 10.3 (from [14]) shows the velocities of a curved flame with the expansion coefficient $\Theta = 5$ as a function of the dimensionless acceleration gL_f/U_f^2, propagating upward in a cylindrical tube of radius $R = 2R_c$. As expected, the velocity of the convex flame rising in the cylindrical tube is considerably larger than the velocity of the 2D flame, while the velocity of the concave flame is quite close to the 2D case. The velocities of both the convex and concave flames increase with the increase in the gravity acceleration, but not as fast as the velocity of an inert bubble $\propto \sqrt{g}$. This can be explained by the influence of the specific properties of the flame and the DL instability. With account of both the flame properties and the effect of bubble motion, the velocity of the upward propagating flame in a cylindrical tube can be approximated as

$$U_w = \sqrt{U_{DL}^2 + U_b^2} = \sqrt{U_{DL}^2 + \beta^2 g R}, \tag{10.3.1}$$

where U_{DL} is the velocity of the curved flame for the case of zero gravity and the numerical factor β is estimated as $\beta_{3D} = 0.51$ for the convex flames, $\beta_{3D} = 0.29$ for the concave flames, and $\beta_{2D} = 0.35$ for the 2D flames. The values of the numerical factors obtained in simulations of convex flames and 2D flames are in good agreement with the estimates found in the theory of inert rising bubbles $\beta_{3D} = 0.51$ and $\beta_{2D} = 0.33$. In addition, Eq. (10.3.1) is in good agreement with the experimental results, predicting the same velocity $U_w = 25$ cm/s for an upward propagating propane flame in the tube with diameter 5 cm as the velocity measured experimentally in [24].

The influence of bubble motion on the flame dynamics in vertical tubes is shown in Fig. 10.4, which shows the dependence of the convex flame velocity on the inverse tube radius for the cases of zero gravity, $gL_f/U_f^2 = 0$ and for $gL_f/U_f^2 = 1$. In agreement with the linear theory of the RT instability of a flame-front, a curved stationary flame is possible in much narrower tubes for the case of non-zero gravity compared to the case of zero gravity. For wide tubes the flame velocity increases much faster for the case of an upward propagating flame, since the effect of bubble rising

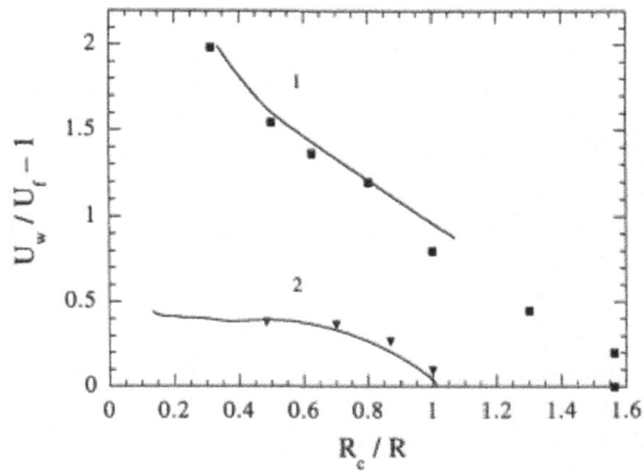

Fig. 10.4 The scaled velocity of an upward propagating convex fame, $\Theta = 5$, versus the inverse tube radius R_c/R for $gL_f/U_f^2 = 0$ (triangles) and $gL_f/U_f^2 = 1$ (squares). Curve 1 shows the analytical approximation (10.3.1); curved 2 is for the case $g = 0$

becomes more and more important as the tube diameter becomes larger. Figure 10.4 demonstrates a good accuracy of Eq. (10.3.1) for sufficiently wide tubes, when the growth of the RT instability is not influenced by thermal conduction. As it follows from Eq. (10.3.1), even if the gravity effects are small in tubes of moderate widths, the bubble motion may become a dominant effect for wider tubes. The same behavior of the flame velocity is obtained for the case of a concave flame propagating upward and for 2D flames.

From Eq. (10.3.1) it can be concluded that the influence of the earth's gravity on the flame propagating upward is much weaker than it is usually assumed. It should be emphasized that even for rather large values of the scaled acceleration gR/U_f^2 the relative increase of the flame velocity is quite moderate. For example, for $gR/U_f^2 = 75$ and the tube radius $R = 2R_c$, the scaled velocity increase is only $U_w/U_f - 1 = 3.5$. The surprisingly weak effect of gravity may be attributed to the small numerical factor β, which appears in the theory of bubble motion.

As a result, for example, for the experimental conditions used in [24] with the tube radius $R = 2.5$ cm, the gravity is much more important than the DL instability only for very slow flames with velocities $U_f < 20$ cm/s, such as propane flames used in [24]. On the contrary, the effect of gravity is expected to be small even for flames of moderate velocities like the flame in the mixture 7.7% C_2H_4O + Air, which velocity is $U_f = 70$ cm/s. Finally, for the case of fast hydrogen–oxygen flames with the velocities up to 10 m/s the influence of gravity on the flame velocity is negligible. In contrast to fast flames, the photograph images of lean methane-air flames propagating upward in 50 mm and 24 mm diameter tubes obtained in the experimental study of the effect of flame stretching on the flammability limits of lean methane-air flames [25–27], where the normal flame speed is lowest, are very similar to the shapes of convex flames shown in Fig. 10.2b, c.

Another effect that complicates the nonlinear stage of the RT instability at a flame-front is the secondary Kelvin–Helmholtz instability of the jet of heavy matter sliding down along the surface of a light matter. Development of the secondary instability results in non-stationary mushroom shaped structures at the ends of the jets observed frequently at the nonlinear stage of the RT instability [4]. Obviously, in the case of a flame in a very strong gravitational field the fire polishing effect fails and the mushroom shaped structures develop. Development of such structures has been observed in many numerical simulations, probably the first one in [28], where an effective gravitational acceleration has been modeled by a constant pressure gradient. For the flame parameters considered in [28] the pressure gradient for which the mushroom shaped structures appeared corresponds to the scaled accelerations of about $gR/U_f^2 = 10^5 - 10^6$. Still even in this case it is not certain that the mushroom structure development can be considered as turbulization of a flame, since the non-stationary motion takes place only locally at the ends of the cusps, while most of the flow remains stationary similar to that in Fig. 10.2. The complete turbulization of a flame due to the RT instability has been reported by Khokhlov [29] in the simulation of thermonuclear flames in Supernovae. Although the scaling of the turbulent flame velocities obtained in [29] is $U_w \approx 0.5\sqrt{gR}$, the observed flame shapes contradict the observations of most experimental studies and numerical simulations of the upward propagating flames. This is different for recently published simulations [30, 31] where a very strong turbulization was observed even for terrestrial flames.

10.4 Flame in Horizontal Tubes

Similar to flames propagating upward in a vertical tube, flames propagating in horizontal tubes of large diameters is also influenced by the gravity. In this case, the hot burned gas of smaller density tends to occupy the upper part of the tube, while the heavy fuel extends along the lower part. As a result, the flame-front acquires a curved shape and propagates stationary with a velocity considerably exceeding the velocity of a planar flame, as it is shown schematically in Fig. 10.5.

One of the first theoretical models of flame propagation in a horizontal channel [32, 33] was based on a modified Sivashinsky equation, which assumes a small difference in the density of the unburned mixture compared to the density of combustion products, $(\Theta - 1) << 1$, moderate gravitational acceleration, and moderate tube

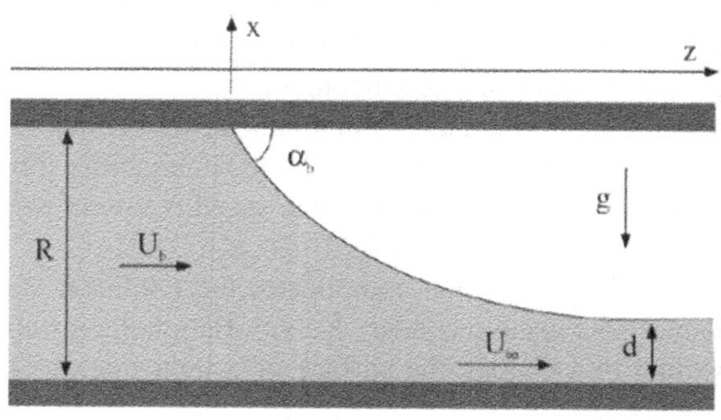

Fig. 10.5 Schematic configuration of an open bubble in 2D horizontal channel

width. As a result of these assumptions a flame-front remains almost planar and the effects of gravity result in only small corrections to the flame velocity. Obviously, the model of small expansion of the burning matter yields a poor extrapolation to the high acceleration limit. When gravitational effects prevail, a better approximation of the flame velocity can be obtained in the model of an inert massless open bubble moving in a horizontal channel.

Let us consider a stationary flow caused by a 2D open bubble propagating in a horizontal channel of width R as shown schematically in Fig. 10.5. A characteristic feature of the flame in horizontal tubes is the sharp angle $\alpha_b = \pi/3$ at the top between the flame-front and the upper wall of the channel, which is clearly seen in the snapshot photographs of a flame in horizontal tubes. The sharp angle can be clearly distinguished on the bright part of the flame-front, where effects of thermal conduction and losses to the walls do not influence the flame structure. Dimensional consideration implies that the bubble velocity is $U_b = \beta\sqrt{gR}$. The numerical coefficient β can be found form the solution of the problem in a way similar to that used in Sec.10.2. Let $x = 0$ and $y = 0$ be coordinates of the top of the bubble. In the reference frame of the bubble the fluid at infinity flows toward the bubble with the velocity U_b, rounds the bubble, and forms a thin jet of width d. The velocity of the fluid in the jet is $U_\infty = U_b/a$, where $a = d/R$. At the surface of a massless bubble the Bernoulli equation takes the form

$$\frac{1}{2}\mathbf{v}^2 + gx = 0, \tag{10.4.1}$$

so that the velocity of the fluid in the jet can be calculated as

$$U_\infty = \sqrt{2g(R-d)} = U_b\frac{\sqrt{2}}{\beta}\sqrt{1-a}. \tag{10.4.2}$$

Equation (10.4.2) couples the dimensionless coefficient β to the ratio of the jet width and the tube width:

$$\frac{1}{2}\beta^2 = a^2(1-a). \tag{10.4.3}$$

Since there are no sources of vorticity, the flow is potential everywhere. Close to the top of the bubble the velocity potential and the stream function can be written in polar coordinates in the following form:

$$\varphi = Nr^n\cos[n(\alpha+\alpha_b)]\, and\, \psi = Nr^n\sin[n(\alpha+\alpha_b)], \tag{10.4.4}$$

where $n = \pi/(\pi - \alpha_b)$, and the velocity components are

$$v_r = nNr^{n-1}\cos[n(\alpha+\alpha_b)]\, and\, v_\alpha = -nNr^{n-1}\sin[n(\alpha+\alpha_b)]. \tag{10.4.5}$$

Substituting the velocities into the Bernoulli equation we obtain the equation for the coefficient N and for the power exponent n

$$N^2 = 4g\sqrt{3}/9, \ \ n = 3/2, \ \ \alpha_b = \pi/3. \tag{10.4.6}$$

Therefore, for the scaled complex potential $\Phi = \frac{\varphi + i\psi}{U_b R}$ close to the bubble top we obtain

$$\Phi = \frac{2\sqrt[4]{3}}{3\beta}\zeta^{3/2}\exp(i\pi/2), \tag{10.4.7}$$

where $\zeta = \frac{x+iy}{R}$ is the scaled complex coordinate. An interesting consequence of Eq. (10.4.6) should be noted: a horizontally propagating bubble touches the wall at a sharp angle $\alpha_b = \pi/3$, while the bubbles rising upward always have a flat top.

In order to obtain the bubble velocity, we have to find the solution of the Laplace equation $\Delta\Phi = 0$ for the whole flow, with the boundary conditions at the walls and the bubble surface. Instead, similar to the case of upward propagating flames we use an approximate solution for the flow around the bubble and match this solution with the solution at the bubble top. In terms of the scaled complex velocity $u = u_x + iu_y = (v_x + iv_y)/U_b$ we have to construct a flow, which moves from infinity with the dimensionless velocity $u = 1$, rounds the bubble top of an angle $\alpha_b = \pi/3$, and forms a jet with the dimensionless velocity $u = 1/a$. It can be verified that the flow with such properties close to the bubble top is the flow described by the complex potential and the corresponding velocity in the form

$$\Phi = \frac{2}{3}\sqrt{\frac{2\pi}{3}}\frac{\zeta^{3/2}}{(1-a)^{3/2}}\exp(i\pi/2), \tag{10.4.8}$$

$$u = \frac{1}{(1-a)^{3/2}}\sqrt{\frac{2\pi}{3}}\zeta\exp(i\pi/2). \tag{10.4.9}$$

Then, using Eqs. (10.4.7) and (10.4.8) we obtain relation between the dimensionless parameters β and a

$$\beta^2 = \frac{3\sqrt{3}}{2\pi}(1-a)^3. \tag{10.4.10}$$

Finally, from Eqs. (10.4.3) and (10.4.10) we obtain $a = 0.39$ and $\beta = 0.43$. Thus, a massless bubble in a horizontal 2D channel of width R propagates with the velocity $U_b = 0.43\sqrt{gR}$. In the case of the non-zero bubble density the velocity of the bubble is $U_b = 0.43\sqrt{\frac{\Theta-1}{\Theta}gR}$, where the correction coefficient $\frac{\Theta-1}{\Theta}$ can be obtained in the way similar to Eq. (10.2.8) of Sec. 10.2. The same velocity is expected for a flame in a horizontal tube in the case of sufficiently wide tubes $R >> 5.4U_f^2/g$.

The obtained velocity of a bubble moving horizontally in a tube is close to the velocity of open bubbles moving upward in ideal 2D vertical tubes $\beta\sqrt{gR}$, with $\beta = 0.33$ for the 2D case. Thus, similar to the case of vertical tubes the dimensionless coefficient β is rather small which implies weaker influence of gravity on the flame dynamics than it is generally believed. Effect of the buoyancy is more pronounced for slower flames and becomes noticeable for the tube widths exceeding $\mathrm{R} \approx 5.4\mathrm{U}_f^2/\mathrm{g}$. For example, for slow flames such as 6% CH_4/air flame with the normal velocity $U_f = 5$ cm/s, when the Froude number $U_f^2/gR << 1$ is sufficiently small and/or the tube is sufficiently wide. For such a flame the condition $R >> 5.4U_f^2/g \approx 0.13$ cm is satisfied for a commonly used laboratory experiments and the velocity of flame propagation is mainly determined by the effects of bubble motion. If a flame has larger normal velocity, then the gravity effects become important for wider tubes. For example, the flame in the mixture $10\%\mathrm{CH_4}/air$ propagates normally with the velocity $U_f = 43$ cm/s, the bubble velocity is equal to the normal flame velocity for the tube width $R = 5.4U_f^2/g = 10$ cm, and the flame velocity would be defined by the gravity only for much wider tubes, $R \sim 30-50\,cm$. For example, the experiments [15–17] show that in a wide horizontal tube of diameter from 24 cm up to 90 cm the curved flame in the mixture 10%CH_4 propagates with the velocity $U_u = 245$ cm/s considerably exceeding the normal laminar velocity of the flame. The theoretical analysis based on the bubble motion predicts the bubble velocity $U_b = 130$ cm/s for the 2D tube of the width $R = 90\,cm$. The discrepancy of the theoretical and the experimental results is presumably caused by a well pronounced 3D structure of the experimental flow. If we suppose that a 3D bubble moves approximately $\sqrt{3} = 1.73$ times faster than a 2D bubble as it is for bubbles in vertical tubes (Sec. 10.2), then we obtain the flame velocity $U_w = 225$ cm/s, which is only slightly smaller than the experimentally measured flame velocity. It is quite natural that the bubble velocity somewhat underestimates the velocity of gravity influenced curved flame. The joint effect of the bubble motion and fuel consumption by the flame should give a slightly larger velocity than the velocity of an inert bubble. An additional increase of the flame velocity may be caused by the DL hydrodynamic instability of the flame-front inherent to all gaseous flames and may be caused by stretching of the flame-front due to nonslip boundary conditions which is not taken into consideration in the present analysis.

References

1. S. Atzeni, J. Meyer-ter-Vehn, *The Physics of Inertial Fusion: Beam Plasma Interaction Hydrodynamics* (Oxford University Press, Oxford, Hot Dense Matter, 2004)
2. S.I. Abarzhi, Review of theoretical modelling approaches of Rayleigh-Taylor instabilities and turbulent mixing. Phil. Trans. R. Soc. **368**, 1809–1828 (2010)
3. M.A. Liberman, J. DeGroot, A. Toor, R. Spielman, *Physics of High-Density Z-Pinch Plasmas* (Springer-Verlag, New York, 1999)
4. H. Kull, Theory of the Rayleigh-Taylor instability. Phys Rep. **206**, 197–325 (1991)

5. N. Inogamov, The role of Rayleigh-Taylor and Richtmyer-Meshkov instabilities in astrophysics: an introduction. Astrophys. Space Phys. **10**, 1–335 (1999)
6. R. Betti, J. Sanz, Bubble acceleration in the ablative Rayleigh-Taylor instability. Phys Rev. Lett. **97**, 205002 (2006)
7. L.F. Wang, W.H. Ye, W.Y. Zhang, X.T. He, Numerical investigation of nonlinear ablative single-mode Rayleigh–Taylor instability in the presence of preheating. Phys. Scr. 014018 (2013)
8. W.H. Liu, L.F. Wang, W.H. Ye, X.T. He, Temporal evolution of bubble tip velocity in classical Rayleigh-Taylor instability at arbitrary Atwood numbers. Phys. Plasma. **20**(1–11), 062101 (2013)
9. H.G. Lee, J. Kim, Numerical simulation of the three-dimensional Rayleigh-Taylor instability. Computers and Math. Applications. **66**, 1466–1474 (2013)
10. R. Davies and G. Taylor, The mechanism of rising bubbles. Proc. Roy. Soc. A200, 375–390 (1950)
11. D. Layzer, On the Instability of Superposed Fluids in a Gravitational Field. Astrophys. J. **122**, 1–12 (1955)
12. V.N. Goncharov, Analytical Model of Nonlinear, Single-Mode, Classical Rayleigh-Taylor Instability at Arbitrary Atwood Numbers. Phys. Rev. Lett. **88**(1–4), 134502 (2002)
13. V. Bychkov, S. Golberg, M. Liberman, L.E. Eriksson, Propagation of Curved Stationary Flames in Tubes. Phys. Rev. E **54**, 3713–3724 (1996)
14. V.V. Bychkov, S.M. Golberg, M.A. Liberman, A.I. Kleev, L.E. Eriksson, Numerical simulation of curved flames in cylindrical tubes Combust. Sci. Technol. **129**, 217–242 (1997)
15. H.F. Coward,H.P. Greenwald, Propagation of flame in mixtures of natural gas and air. U.S. Bureau of Mines, Tech. Paper. **427**,88 (1928)
16. H.F. Coward, F.J. Hartwell, Studies in the mechanism of flame movement. Part I. The uniform movement of flame in mixtures of methane and air, in relation to tube diameter, J. Chem. Soc. 1996–2004 (1932)
17. H.F. Coward, F.J. Hartwell, Studies in the mechanism of flame movement. Part II. The fundamental speed of flame in mixtures of methane and air, J. Chem. Soc. 2676–2684 (1932)
18. K.O. Mikaelian, Analytic Approach to Nonlinear Rayleigh-Taylor and Richtmyer-Meshkov Instabilities. Phys. Rev. Lett. **80**, 508–512 (1998)
19. K.O. Mikaelian, Explicit expressions for the evolution of single-mode Rayleigh-Taylor and Richtmyer-Meshkov instabilities at arbitrary Atwood numbers. Phys. Rev. E 67 (2003) pp. 026319.
20. P. Pelce, Effect of gravity on the propagation of flames in tubes. J. Physique **46**, 503–510 (1985)
21. Y.M. Shtemler, G.I. Sivashinsky On Upward Propagating Flames. Combust. Sci. Technol. **102**, 81–93 (1994)
22. Z. Rakib, G. Sivashinsky, Instabilities of upward propagating flames. Combust. Sci. Technol. **54**, 69–84 (1987)
23. A. Levy, An optical study of flammability limits. Proc. R. Soc. London A **283**, 134–145 (1965)
24. C. Pelcé-Savornin, J. Quinard, G. Searby, The flow field of a curved flame propagating freely upwards. Combust. Sci. Tech. **58**, 337–346 (1988)
25. Y.J. Shoshin, L. Tecce, J. Jarosinski, Experimental and computational study of lean limit methane-air flame propagating upward in a 24 mm diameter tube. Combust. Sci. Technol. **180**, 1812–1828 (2008)
26. Y.J. Shoshin, J.A. van Oijen, A.V. Sepman, L.P.H. de Goey, Experimental and computational study of the transition to the flame ball regime at normal gravity. Proc. Combust. Inst. **33**, 1211–1218 (2011)
27. V. Muntean, F.J. Higuera, Lean methane-air flames propagating upward in vertical tubes. Combust. Sci. Technol. **188**, 818–830 (2016)
28. N.R. Edwards, A.C. McIntosh, J. Brindley, The Development of Pressure Induced Instabilities in Premixed Flames. Combust. Sci. Technol. **99**, 179–199 (1994)
29. A.M. Khokhlov, Propagation of Turbulent Flames in Supernovae. Astrophys. J. **449**, 695–713 (1995)

30. A.Y. Poludnenko, J. Chambers, K. Ahmed, V.N. Gamezo, B.D. Taylor, Unified Theory of Detonation Initiation in Type Ia Supernovae and Terrestrial chemical systems. Proc. Combust. Inst. **37**, 3495–3502 (2019)
31. A.Y. Poludnenko, J. Chambers, K. Ahmed, V.N. Gamezo, B.D. Taylor, A unified mechanism for unconfined deflagration-to-detonation transition in terrestrial chemical systems and type Ia supernovae. Science. **366**, 7365 (2019)
32. Z. Rakib, G. Sivashinsky, Instabilities in Upward Propagating Flames. Combust. Sci. Technol. **54**, 69–84 (1987)
33. Z. Rakib, G. Sivashinsky, Flame propagation in a horizontal channel. Combust Sci. Technol. **59**, 247–260 (1988)

Chapter 11
Stability Limits; Spherically Expanding Flames

11.1 Flame Instabilities in Wide Tubes

Propagating laminar flames are subjected to the intrinsic flame instabilities, of which the most common modes are hydrodynamic—the Darrieus-Landau instability and the thermal diffusion instability. The DL instability arises due to a sharp change in density across the flame front and occurs for the flame thickness being much smaller than the characteristic spatial scales of the problem. The thermal diffusive instability is related to the discrepancy between molecular and thermal diffusivity defined by the Lewis number $Le = \chi/D$. A generally accepted classification of mixtures according to Lewis numbers is as following: if the oxidizer is air, then it has been established that rich mixtures of hydrogen, methane, and the higher hydrocarbons belong to $Le > 1$ mixtures, while the respective lean mixtures are $Le < 1$ mixtures. As was discussed in Chap. 5, while the DL instability is inherent feature of any flames, the occurrence of the thermal diffusive instability depends on the mixture composition. In contrast to the DL instability, which is characterized by a relatively large scale and can be suppressed in small-scale systems, the thermal diffusive instability causes a cellular structure on the flame-front. The presence of these cells increases the surface area of the flame and can amplify the flame propagation speed compared to the speed of a planar flame. If diffusion is the dominant process, that is if $Le = \chi/D < 1$, then diffusion of a fresh fuel is greater in the convex parts, increasing the reaction rate, and enhancing the bending of the flame front. On the contrary, if $Le > 1$, then thermal conductivity is the dominating process. In the latter case, when the flame front bends, the convex parts of the flame lose more heat than the concave parts, so the speed of the convex parts will be reduced and the concave parts are accelerated, thus smoothing out the curvature of the flame surface.

In very wide tubes of the width much greater than the cut-off wavelength of the DL instability $R/\lambda_c \gg 1$, a stationary curved flame similar to that shown in Fig. 10.9 is not observed in real life. The radius of curvature near the hump of these flames tends to infinity with increasing tube width, and such flames are locally similar to an infinitely thin planar flame and, therefore, they should be unstable against the DL

M. A. Liberman, *Combustion Physics*,
https://doi.org/10.1007/978-3-030-85139-2_11

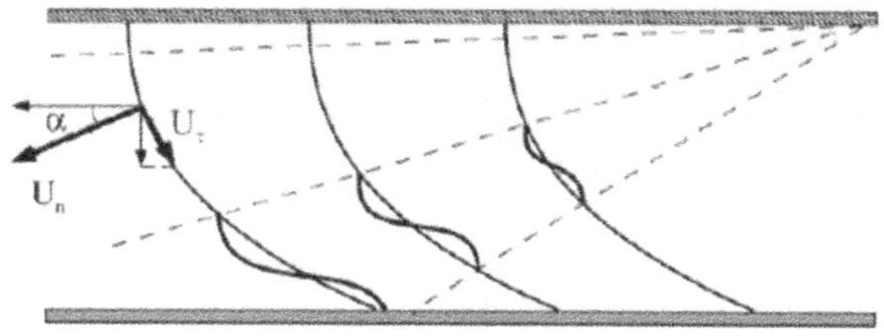

Fig. 11.1 Stretching of a curved stationary flame front in a wide tube

instability. The development of DL instability of an infinitely thin curved flame in very wide tubes was analyzed by Ya. B. Zeldovich et al. [1]. A qualitative scenario of the development of flame front instabilities in a wide tube is shown in Fig. 11.1.

It is seen that due to the curved shape of the flame, the flam front stretches, its surface increases, and the distance between different points of the front also increases with time, similar to an expanding spherical flame. We can say that any element of the flame front effectively moves from the hump to the cusp. As a result, perturbations at the flame front are also stretched and their wavelengths increase, which leads to the decrease in the growth rate of instability according to Eq. (5.1.22). In addition, the perturbations have finite time to develop since any localized perturbation is suppressed as soon as it "falls" into the main cusp and reaches the side wall. The time available for perturbations of any finite wavelength λ to grow can be estimated as the time needed for a point at the flame front to move from the initial location $x \approx \lambda$ close to the hump to the "final" location at the wall $x = R$. The tangential velocity U_τ characterizing motion along the flame front can be evaluated as the tangential fuel velocity just ahead of the flame front. The value of U_τ can be estimated as $U_\tau = U_f \tan\alpha = U_f \frac{dF}{dx}$, where α is the angle between the tube axis and the normal to the flame front (see Fig. 11.1). Since locally in time and space the perturbations of the flame surface $\tilde{F}$ grow exponentially

$$\frac{d\tilde{F}}{dt} = \sigma\tilde{F}, \tag{11.1.1}$$

the maximum perturbation amplitude $\tilde{F}_w$ in the Wentzel-Kramers-Brillouin approximation [2] achieved at the wall is related to the initial perturbation amplitude $\tilde{F}_0$ as

$$\ln(\tilde{F}_w/\tilde{F}_0) = \int_{l=\lambda_0}^{l=R} (\sigma/U_\tau)dl, \tag{11.1.2}$$

where $dl = U_\tau dt$ and l is the coordinate alone the curved flame front, and λ_0 is the initial perturbation wavelength. In Eq. (11.1.2) the instability growth rate $\sigma = \sigma(\lambda)$ and the perturbation wavelength λ are functions of l. The stretch of the perturbation wavelength can be expressed through the stretch velocity U_τ as

$$d\lambda = \lambda \frac{dU_\tau}{dl} dt = \lambda \frac{dU_\tau}{U_\tau}. \tag{11.1.3}$$

The main part of the integral (11.1.2) for an infinitely thin flame comes from the regions close to the flame hump. The stretch increases with the decrease of the tube width, therefore there must be a critical tube width R_w for which all perturbations of the curved flame are suppressed. Numerical evaluation of the integral (11.1.2) for curved infinitely thin flames performed in [1] shows that for the expansion coefficient $\Theta = 8.3$ and for Le $= 1$ the critical tube width is approximately $R_w = 50L_f$. The last estimate can be also expressed through the other critical value of the tube width R_c for which a planar flame becomes unstable $R_w = 5R_c$. Thus, according to [1] curved stationary flames with expansion coefficients $\Theta = 5 - 10$ remain stable for tubes of width $R < 5R_c$, while for wider tubes $R > 5R_c$ the secondary DL instability develops at a curved flame front.

For very wide tubes far away from the stability limits $R \gg 5R_c$ the secondary DL instability presumably leads to a fine time-dependent structure of small humps and cusps imposed at the main hump, as shown in Fig. 11.2 from Ref. 6 in Chap. 7.

Similar structures have been observed in numerical simulations [3] of the nonlinear Sivashinsky equation [4] with periodical and adiabatic boundary conditions for the effective tube width up to $R = 163R_c$. The observed small humps were in the state of permanent motion moving from the hump to the cusp of the

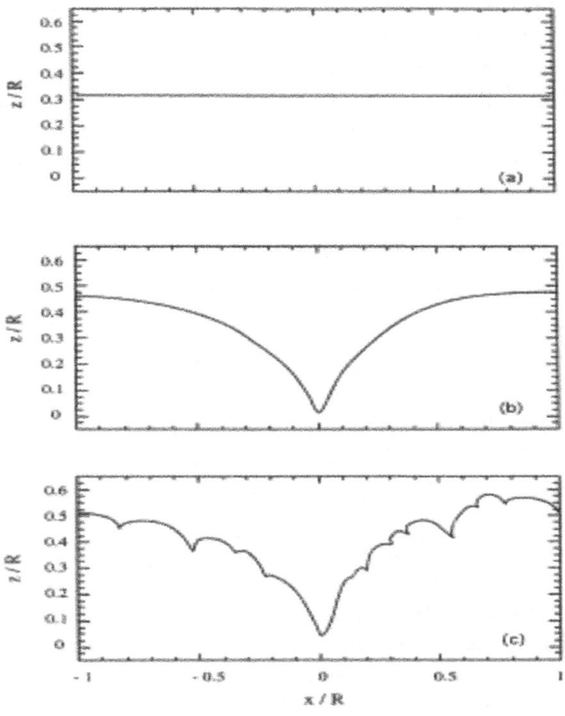

Fig. 11.2 Characteristic shapes of the flame propagating in tubes of width **a** $R < R_c$; **b** $R_c < R < R_w$; **c** $R_w < R$

principal large-scale structure. Development of small humps leads to increase of the total surface area of the flame, which results in additional amplification of the flame propagation velocity. Particularly, for the effective tube width of $R = 163R_c$ the velocity amplification observed in [3] was 2.8 times larger than the maximal velocity amplification for a curved stationary flame with similar parameters.

Close to the stability limits $R = R_w$ one should expect the secondary DL instability to develop in a much smoother way. Presumably, in this case the instability takes the form of an extra cusp emerging at the hump of the initial stationary curved flame. Similar evolution of the flame shape has been observed in simulations [5] for a flame with the expansion coefficient $\Theta = 6.5$ propagating in the tube of width $R = 4R_c$ (see Figs. 8.8 and 8.9). The value $R = 4R_c$ is quite close to the evaluation $R_w = 5R_c$ of the stability limits of curved stationary flames with expansion coefficients $\Theta = 5-10$ obtained in [1]. Another important point is that the observed transition in the flame shape was accompanied by a noticeable increase of the flame velocity above the maximal velocity of a curved stationary flame.

It should be emphasized that the analytical solution of the 2D version of the nonlinear Eq. (8.1.1) for flames with small expansion coefficients $\Theta - 1 \ll 1$ demonstrated that curved stationary flames are linearly stable, no matter how wide is the tube. This surprising result contradicts the above physical ideas [1] on the stability of a curved flame as well as the results of numerical simulations [3] of the nonlinear Eq. (8.1.1) and of the complete system of hydrodynamic flame equations [5]. Later studies of flame stability [6–8] using Eq. (8.1.1) did not change the basic conclusions about linear stability of curved stationary flames at all scales. In particular, in [7] instability of curved stationary flames in wide tubes was observed, but the unstable stationary solutions investigated in [7] were not the main stationary solutions with minimal possible wavenumber and maximal possible number of poles providing the maximal velocity amplification for a given tube width.

To avoid the contradiction between the analytical solution of the nonlinear Eq. (8.1.1) and the physical understanding of the flame stability Joulin [6] proposed that curved stationary flames in wide tubes are linearly stable, but nonlinearly unstable against perturbations of a finite amplitude. In addition, it was shown in [6] that the larger the tube width and the local curvature of the flame front, the smaller the perturbation amplitude needed to induce the nonlinear instability. Still the analysis made in [6] did not provide an estimate for the critical tube width R_w, for which a curved stationary flame becomes unstable nonlinearly. The idea of the nonlinear secondary instability has been proposed in [7] where it was shown that the threshold perturbation amplitude needed to induce the instability for wide tubes changes with the tube width as $\Delta F \propto R(R_c/R)^3$. Numerical simulations of the nonlinear Eq. (8.1.1) performed in [7] for very wide tubes $R/R_c > 10^3$ show also an increase of the flame velocity in comparison with the maximal velocity of a curved stationary flame. On the basis of the numerical simulations it was claimed that the wrinkled flame velocity increases with the tube width as $U_w \propto R^{0.42}$ without external noise and even faster in presence of a sufficiently strong noise. It was shown that the wider the tube is, the more sensitive flame dynamics to the noise level with the critical level separating the regimes of "weak" and "strong" noise being inversely proportional to the

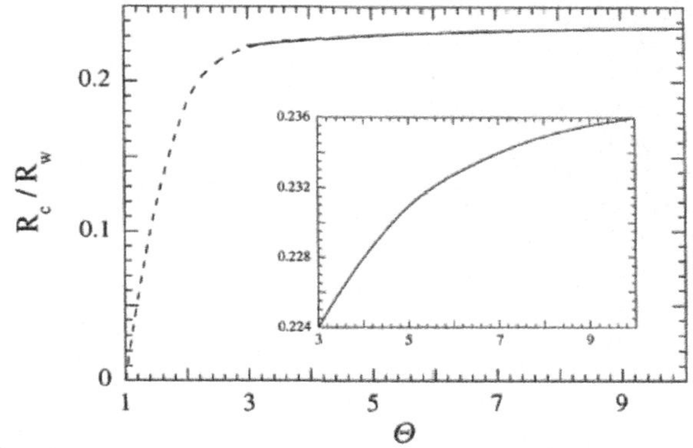

Fig. 11.3 Stability limits of the curved stationary flame versus the expansion coefficient. The dashed line shows extrapolation of the numerical results into the domain of small expansion coefficients

tube width. Theoretical evaluations [8] of the dependence of the flame velocity on the tube width gave $U_w \propto R^{0.37}$, though these evaluations were based mostly on the results of numerical simulations of the flame dynamics in the presence of noise. Since the nonlinear Eq. (8.1.1) is valid only for flames with small expansion coefficients $\Theta - 1 \ll 1$, the study [6–8] on stability limits of curved flames are also restricted to the same peculiar limiting case.

The problem of stability limits can be formulated as an eigenvalue problem of the linearized nonlinear Eq. (8.2.27) with the inverse critical tube width R_c/R_w playing the role of an eigenvalue and perturbation of the flame front playing the role of the eigenfunction. Strictly speaking the problem should be considered using a closed form of the nonlinear equation valid for an arbitrary expansion coefficient [9–13]. The critical tube width, for which the principal solution loses stability is shown in Fig. 11.3 versus the expansion coefficient of the flame.

As one can see, the theoretical value of the critical tube width R_w/R_c depends only slightly for flames with expansion coefficients in the range $\Theta = 5-10$ being about $R_w/R_c = 4.1-4.2$. The critical tube width is almost constant for realistic expansion coefficients, $\Theta = 5-10$, and only slightly increases with the increase of Θ, which implies that the smaller the expansion coefficient, the larger the stability limits of curved stationary flames. Taking into account that for very small expansion coefficients $\Theta - 1 \ll 1$ curved stationary flames are stable independent of the tube width one should expect that the critical tube width goes to infinity $R_w \to \infty$ at $(\Theta - 1) \to 0$. However, such expansion coefficients does not correspond to a real flames and have only academic interest. The obtained stability limits are quite close to the stability limits evaluated for the secondary DL instability in [1, 5]. In particular, the curved flames with an extra cusp similar to that shown in Fig. 8.9 were observed in the 2D numerical simulations [5] for the tube width $R = 4R_c$ and the flame expansion coefficient $\Theta = 6.5$. According to these results the critical tube width for which the stationary curved flame front with $\Theta = 6.5$ loses stability is $R_u = 4.29R_c$, which is in a good agreement with the value $R_w = 4.29R_c$ in Fig. 11.3 for $\Theta = 6.5$.

The perturbation of the flame front $\tilde{F}$ at the stability limit $R = R_w$ for a flame with the expansion coefficient $\Theta = 5$ is shown in Fig. 11.4. As one can see, the

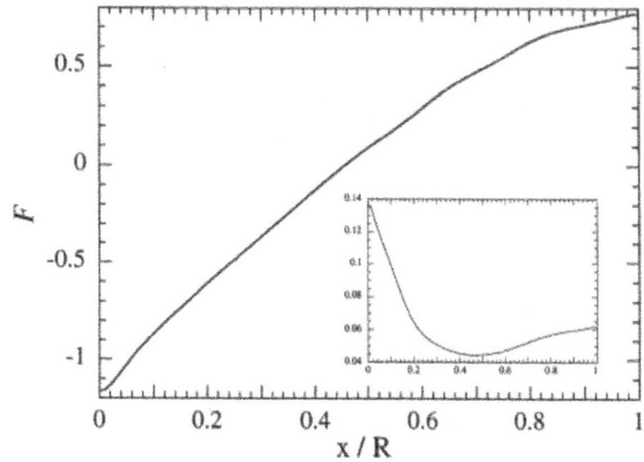

Fig. 11.4 The perturbation of the flame front $\tilde{F}$ at the stability limit $R = R_w$ for the expansion coefficient of the flame $\Theta = 5$. The inset shows possible flame shape resulting from the instability on the nonlinear stage

perturbation function is rather smooth without well-pronounced wrinkles, but even such smooth perturbations may lead to an extra cusp at a flame front at the nonlinear stage of the instability. Possible development of the instability at the nonlinear stage is illustrated in the insert of Fig. 11.4, where the perturbations multiplied by a small factor were added to the stationary solution. The flame front in the insert propagates downwards. The resulting flame shape looks like the flame shape at $\tau = 6.26$ and $\tau = 8.49$ in Fig. 8.9 which is an intermediate step in the development of the secondary DL instability.

Thus, a good agreement of the stability limits obtained using the nonlinear equation and the direct numerical simulations [5] with a semi-qualitative analysis [1] confirms the concept of the secondary DL instability. Nevertheless, the final conclusion about the resulting instability is possible only on the basis of numerical simulation of the nonlinear time dependent equation, which has not yet been carried out. Although the critical tube widths found in simulations, $R_w/R_c = 3.7-3.95$ for $\Theta = 6-10$, are slightly lower than the theoretical values, they are in good agreement with theoretical predictions. Beyond the stability limits the simulations demonstrate considerable increase of the flame velocity in comparison with the maximal velocity amplification for stationary flames. For example, a flame with $\Theta = 8$ in a tube of width $R_w = 4R_c$ propagates with velocity $U_w = 1.8U_f$ noticeably larger than the respective value of maximal velocity for stationary flames $U_w = 1.3U_f$. One of the possible regimes of flame dynamics beyond the stability limits observed in [5] is the pulsating regime, when velocity of flame propagation oscillates in time.

Interesting is that numerical solution of the Sivashinsky equation for a two-dimensional planar flame revealed that the hydrodynamically unstable flame leads to the fractal flame, whereas the diffusive-thermally unstable flame does not. Numerical simulations [14], which studied the DL instability of a flame in wide tubes by solving the Navier–Stokes equations with approximation of a one-step Arrhenius reaction, have shown that the increase of the flame propagation speed with the increase of the tube width can be fitted by a power law indicating the mean fractal properties of the flame front structure.

11.2 Morphology of Unconfined Spherically Expanding Flames

When the combustible mixture is centrally ignited, a spherical flame expanding outward is created. From practical point of view the flame propagating from the ignition point is of great interest in many applications, such as e.g. engines, explosions, safety etc. From a theoretical point of view, the development of cells in a spherically expanding flame illustrates the difference between thermal-diffusive and the Darrieus-Landau instabilities.

The morphology of spherically expanding wrinkled flames is essentially different from that occurring in planar flames. In the latter case short-wavelength corrugations merge forming a single cusp, whose scale is controlled by the overall size of the system. In the case of an expanding spherical flame, the tendency to merge is compensated by the expansion of the spherical flame. The evolution of a spherically expanding flame can be considered as consisting of three stages: ignition, wrinkling of the flame front, and a stage of self-acceleration, followed by the transition to a fractal structure. First, the energy deposition leads to the formation of a small flame ball with radius comparable to the width of the flame front. From the beginning, when radius of the spherically expanding flame front is small, the DL instabilities are stabilized due to the stretching of the flame surface, which has a positive curvature towards the fresh mixture, so that the flame front is initially smooth. The flame front evolution depends on the Lewis number. If the Lewis number is smaller than unity, $Le = \chi/D < 1$, the additional effect of the thermal-diffusive instability leads to more complicated behaviour of the flame front, when small-scale cells over the flame surface can be produced. Typically, the thermal diffusive instability leads to the formation of a cellular flame surface and the corresponding acceleration of the flame expansion. Cells cannot grow indefinitely, since the stabilizing effect of diffusive processes becomes stronger at small length scales and the cell amplitude becomes saturated. The formation of cells over the flame front increases its surface area and, as a consequence, increases the flame speed.

Small-scale thermal diffusion instabilities prevail at the initial stage of flame propagation, when the flame radius is small. Later, when the flame expands and the flame radius increases so that the thickness of the flame front becomes small compared to the radius of the flame ball, the Darrieus-Landau hydrodynamic instabilities become dominant regardless of the value of the Lewis number. The transition to this stage can be characterized by the critical Péclet number $Pe_c = R/L_f \gg 1$. On this stage the DL instability dominates the formation of cells on the nonlinear stage, which leads to further increase of the flame surface and the flame speed. Since new cells continually appear on the surface of the flame as the radius of the flame increases, the speed of the flame also continually increases. Cells of different scales are continuously born and grow on the flame surface independently. The growth rate of the cell amplitude can be estimated similar to the growth rate of perturbations on the linear stage of the instability development [15]. At this stage, the range of possible length scales available on the surface of the flame expands as its radius increases and the total surface area increases, and the dynamics of the expanding flame is determined by the formation of new length scales caused by the growth and splitting of cells. The onset of instability and the dynamics of a spherically expanding flame depends on the competition between the growth rate of the flame radius (and the corresponding decrease in the local curvature of the flame front) and the growth rate of instabilities. As the flame expanded, the stretching effect progressively decreased, the flame wrinkled with the increasing influence of the cellular instabilities, and consequently, the flame speed increases. The corresponding non-dimensional parameter of the stretch rate for a spherical flame can be defined as the Karlovitz number

$$Ka = \frac{2}{R}\frac{dR}{dt}\frac{L_f}{U_f}. \tag{11.2.1}$$

For the growth of instabilities and the formation of cells, the rate of their growth must be greater than the rate of growth of the flame radius. Since during the initial phase when the flame radius is small, the stretch of the spherically expanding flame is strongest, the tendency for cells to develop increases as the flame radius increases. The non-linear evolution starts in the late stage of the flame acceleration, when the small-scale thermal diffusional structures develop on the large-scale DL structures.

The onset of instability in a spherically expanding flame was investigated analytically and numerically. One of the first studies was carried out by Istratov and Librovich [16, 17] who considered the DL instability model with Markstein number correction [18]. The critical conditions for the onset of instability on spherically expanding flames, growth rates of the instabilities for both linear and non-linear stages, cut-off limits, and the subsequent cell development were analysed in [19–25].

Based on the analysis of many experimental studies, Gostintsev et al. [26] proposed that the dynamics of a spherically expanding flame can be described by a power-law dependence of the average flame radius versus time in the form

$$\langle R \rangle = R_0 + At^{\alpha}, \tag{11.2.2}$$

where $\langle R(t) \rangle$ is the average radius of the wrinkled flame ball, R_0, A are fitting constants. The analysis of experimental data [26] has shown that $\alpha \approx 3/2$, and the critical flame radius R_c associated with the onset of the self-similar regime of a stoichiometric hydrogen-air flame propagation in large atmospheric explosions was found $R_c \approx 1.2\,\text{m}$. Moreover, in [27, 28] was shown that the flame dynamics is self-similar and that the wrinkled flame surface can be considered as a fractal surface if the Peclet number, $Pe = R/L_f$ exceeds some critical value, which depends on the particular mixture, and if α is a constant, with the fractal excess, $d = (\alpha - 1)/a$. The fractal structure of a flame develops when the flame radius exceeds considerably the critical flame radius R_c. The structure and similarity with turbulence of the spherically expanding "turbulent" flame were discussed in [29–31]. Below the critical Peclet number, which is estimated from different experimental data as $Pe_{cr} \sim 10^3 - 10^4$, the measured value of the exponent α in Eq. (11.2.2) can be less than 3/2. It should be stressed that the theoretical predictions obtained in [27, 32] for a particular value of $\alpha = 3/2$ corresponds to the fractal dimension of a three-dimensional spherically expanding flame, while for cylindrically expanding 2D flames the predicted value of the exponent is $\alpha = 1.25$.

Considerable experimental and computational efforts have been made to study spherically expanding flames because of their fundamental and practical importance. In particular, studies of the expanding flames have been mostly focused on the self-acceleration nature of the expanding flame, especially on the possibility that the temporal exponent describing the time variation of the flame radius could asymptotically attain a constant value. The experiments were conducted using large-scale explosions for various mixtures at atmospheric and elevated pressures [33–41], as well as small-scale confined flames in closed chambers at elevated initial pressures up to 60 atm. The purpose of these experiments was to reduce the width of the flame from the very beginning and, therefore, an earlier transition to the self-similar regime [25, 42–56]. The self-similar propagation of expanding spherical flames in lean hydrogen-air mixtures at elevated pressures $P = 0.1-0.5$ Mpa and the transition to a self-similar regime depending on the equivalence ratio $\varphi = 0.5-1.0$ was investigated in experimental studies [40]. The experimentally measured values of α corresponding to the fractal dimension in lean hydrogen-air mixtures at elevated pressure was found within a wide range of $r/R_c = 2 \div 15$. The manifestation of the thermal-diffusional instability leads to the increase in α with a decrease in the equivalence ratio and an increase in initial pressure. The value of α increases with the increase in r/R_c and saturated at $\alpha = 1.4-1.5$ for $r/R_c > 10$.

Interesting phenomenon of global pulsation of the speed of spherically expanding flames, presumably caused by the nonlinear averaging of the local variation in geometry of the cellular flame fronts, have been discovered in laboratory experiments [44–48, 53, 54] and has been reported in large-scale of a few meters in dimension experiments at ambient pressure for propane and hydrogen flames [41, 57].

Majority of experimental studies aimed to verify the theoretically predicted value $\alpha = 3/2$ and hence possible self-turbulization. Specifically, in [29, 30, 59] it was shown theoretically and experimentally that for fully-developed Kolmogorov turbulence the fractal dimension, defined as $2+d$, takes a value of 7/3, which corresponds to $\alpha = 3/2$. Although in all these studies a self-similar acceleration and fractalization of the flame front were reported, the measured values of exponent α in Eq. (11.2.2) were reported to be smaller than $\alpha = 3/2$. For example, in experimental studies of the stoichiometric hydrogen/air expanding flame at elevated pressures [47], the maximum value was found of about $\alpha = 1.37$. In experiments [49] the evolution and self-acceleration of constant-pressure, spherically expanding $H_2/O_2/N_2$ flames were studied at pressures 1, 2.5, 5.0, and 10.0 atm. The change in the morphology and propagation mode of the expanding flame was defined by two critical Peclet numbers: Pe_c which designates the minimum flame radius when the cellular structure develops and below which the flame-front is smooth, and the saturation Peclet number Pe_s, which designates the flame radius when the cellular structure saturates the flame-front, leading to a distinct slowdown in acceleration from that in the transition stage. The acceleration exponent for the fully developed saturated instability was found to be around $\alpha = 1.2-1.4$, which is smaller than $\alpha = 1.5$, the suggested value for self-turbulization. Similar results were obtained in experimental studies [56]. Some uncertainty of the experimental values of the exponent α, can be caused by not too accurate interpretation of the experimental, or because the asymptotic regime was not achieved.

11.3 Fractal Structure and Self-similar Regime

11.3.1 Fractal Dimension of 2D and 3D Expanding Flames

A spherical flame expanding out from the ignition point is one of the most fundamental manifestations of combustion. As discussed earlier, when the regime of hydrodynamic instability is reached, the growth of perturbations is limited due to nonlinear effects, and the outcome of the DL instability at the nonlinear stage can be wrinkling of the initially smooth flame front. For the case of the freely propagating flames this wrinkling may lead to cellular structure of a flame. In the self-similar regime, the radius and velocity of the flame front changes with time as:

$$R(t) = At^{\alpha},\ U_w = \dot{R} = \alpha A^{1/\alpha} R^{(\alpha-1)/\alpha}, \tag{11.3.1}$$

where A is a coefficient that can be a function of the parameters characterizing the flame. In particular, for comparison with experimental measurements usually used the following expressions

$$R = R_0 + At^{\alpha},\ U_w = \frac{dR}{dt} = \alpha A^{1/\alpha}(R - R_{cr})^{(\alpha-1)/\alpha}, \tag{11.3.1a}$$

where $A = c_g \Theta^2 U_f^2 / \chi^{1/2}$, $c_g \approx 0.002$, and $\chi = \kappa/\rho C_p$ is the thermal diffusivity.

The approximate value of the exponent according to [26–29] is $\alpha \approx 3/2$. The self-similar regime is associated with the development of a fractal structure on the flame front with total surface of the flame,

$$S = 4\pi R^2 \frac{\alpha A^{1/\alpha}}{\Theta U_f} R^{(\alpha-1)/\alpha}, \tag{11.3.2}$$

where U_f is the normal velocity of laminar flame and Θ is the ratio of the fuel density ρ_u to the density of the combustion products, ρ_b. According to Eq. (11.3.2) the fractal dimension of the unstable flame front is $2 + (\alpha - 1)/\alpha$, so that the fractal excess is $d = (\alpha - 1)/\alpha$.

The fractal structure of a flame front can be described as cascading cells: cells of smaller scale are imposed on cells of large scale and so on. Let every step of the cascade decreases the size of triangles by a factor b: $L_{k+1} = L_k/b$, and increases the total length of the curve by a factor β, $S_{k+1} = \beta S_k$. The cascade process is limited from below by the smallest triangle size λ_{min}, since the cell size cannot be less than the cutoff wavelength λ_c, for which the instability is suppressed by thermal conduction. The cell size is limited from above too (the largest triangle size λ_{max}) since perturbations with a length scale larger than $2\pi R/n_c$ are stable due to the spherical geometry, where n_c is the critical number of a spherical function [6, 16, 19]. Assuming that the whole cascade involves a very large number of steps $N \gg 1$ we have

$$\lambda_{max} = b^N \lambda_{min}. \tag{11.3.3}$$

The general idea of the process [58] can be understood from the Koch curved constructed as a cascade of triangles as shown in Fig. 11.5. The total number of steps can be estimated as $N = \ln(\lambda_{max}/\lambda_{min})/\ln(b)$. Then the total length of the curve is

$$S_N = S_0\beta^N = S_0(\lambda_{max}/\lambda_{min})^d, \tag{11.3.4}$$

where the fractal excess is $d = (\ln\beta)/(\ln b)$ and the fractal dimension is $d + 1$ for the case of 2D geometry and it is $d + 2$ in the case of 3-D geometry.

Note that the total length of the fractal structure increases as $S_N \propto \lambda_{max}^{d+1}$. When considering fractal flames we are interested in the increase of flame velocity rather than in flame surface area, though these values are closely related. In scope of a simple model of an infinitely thin flame front amplification of the flame velocity is equal to the amplification of the total flame surface area. It can be seen from the results of numerical simulation (Fig. 11.8) that cascading cusps and humps on the front of freely propagating cylindrical (2D) flame resemble step-by-step the Koch curve: cusps of smaller scale develop on cusps of larger scales, resulting in a self-similar picture. At the certain moment small disturbances begin to form on the background of disturbance with critical wavelength, and later secondary disturbances

Fig. 11.5 Three steps of the Koch curve

grow until their wavelengths become critical. For spherically expanding flame, its fractal structure implies a large number of cascades $N = \ln(2\pi R/\lambda_c n_c)/\ln(b) \gg 1$, and the total flame surface is

$$S_f = S_N = 4\pi R^2 \beta^N = 4\pi R^2 \left(\frac{2\pi R}{n_c \lambda_c}\right)^d, \qquad (11.3.5)$$

In Eq. (11.3.5) the cutoff wavelength $\lambda_c(\Theta)$ is known from the DL linear theory of a planar flame instability, while b and β should be determined from the nonlinear theory of a curved stationary flame. To obtain the critical number n_c one must consider stability of a spherical flame in the self-similar regime of the flame propagation, Eq. (11.3.1).

To evaluate the increase of the total flame surface area on one step of the cascade we can use the maximal velocity amplification for a curved initially planar stationary flame. The increase of the length scale of the fractal generator corresponds to the stability limit of the curved stationary flames: $b = D_w/D_c$, where D_w is the tube width corresponding to the stability limit of a curved stationary flame in tube, D_c is the critical tube width. The corresponding maximal velocity amplification is $\beta = U_w/U_f$, where U_w is the velocity of the curved stationary flame, and U_f is the normal velocity of a planar laminar flame. For example, for flames with $\Theta = 8$ propagating in 2D channel numerical simulations give $b = D_w/D_c \approx 3.4$ and the corresponding value $\beta_{2D} = U_w/U_f \approx 1.28$. For these values, the fractal excess is $d_{2D} = (\ln \beta_{2D})/\ln b \approx 0.2$. Consequently, the fractal dimension in the 2D case is $d + 1 = 1.2$, $\alpha = 1/(1-d) = 1.25$. This means that in the 2D case the radially expanding cylindrical flame accelerates as $R \propto t^{1.25}$. The fractal dimension for the case of expanding spherical flames can be estimated, considering that the fractal excess in 3D geometry should be about two times larger compared to 2D geometry [28]. Thus, the approximate value of fractal excess for 3D case is $d_{3D} \approx 0.4$, and the evaluated fractal dimension of the spherical expanding flame is equal to 2.4, which is slightly larger, but still consistent with the value 2.33 suggested from the analysis of experimental data [26, 29].

11.3.2 Stability of Self-similar Spherically Expanding Flames

The acceleration of the flame in the self-similar regime determined by Eq. (11.3.1) changes stability properties of the flame, which can affect both the smallest and the largest length scales of the fractal structure. Since locally the expanding spherical flame of a large radius is quasi-planar, then the smallest length scale of such a flame is given by the cut-off wavelength of a planar flame front in an effective destabilizing gravitational field resulting from the self-similar propagation regime. However, for the self-similarity exponent $\alpha = 3/2$ the flame acceleration decreases in time as $d^2R/dt^2 \propto \alpha(\alpha - 1)t^{\alpha-2}$ and its influence becomes negligible in the asymptotic and, therefore, the cut-off wavelength λ_c is determined entirely by the DL instability.

Unlike the smallest length scale of the fractal flame structure, the largest length scale does experience influence of the flame self-acceleration. The largest length scale λ_{max} is determined by the flame radius and by the critical number of the spherical harmonic n_c characterizing the wavelength of unstable perturbations $\lambda_{max} = 2\pi R/n_c$. For $n > n_c$ linear perturbations are unstable, while perturbations of a smaller spherical number $n < n_c$ and a larger length scale are damped by the spherical shape and by the expansion of the flame front. To find the critical number n_c we consider, following [27], the stability of an infinitely thin flame propagating in the self-similar regime.

For the flame velocity in Eq. (11.3.1) less than the sound speed, we can consider the incompressible flow, so that the hydrodynamic equations for the radial u and the angular $\mathbf{w}$ velocity components can be written in the form

$$\frac{1}{r^2}\frac{\partial}{\partial r}\left(r^2 u_r\right) + \nabla_{\perp}\mathbf{w} = 0, \tag{11.3.6}$$

$$\frac{\partial \mathbf{w}}{\partial t} + u_r\frac{\partial \mathbf{w}}{\partial r} + (\mathbf{w}\nabla_{\perp})u_r = -\frac{1}{\rho}\nabla_{\perp}P, \tag{11.3.7}$$

$$\frac{\partial u_r}{\partial t} + u_r\frac{\partial u_r}{\partial r} + (\mathbf{w}\nabla_{\perp})u_r = -\frac{1}{\rho}\frac{\partial P}{\partial r}, \tag{11.3.8}$$

where $\nabla_{\perp}$ is the angular part of the operator ∇. For the unperturbed spherical flame $\mathbf{w} = 0$ and the flow ahead of the flame front is given by

$$u_r = B(t)/r^2,\ (P - P_{\infty})/\rho_1 = \dot{B}(t)/r - B^2(t)/2r^4. \tag{11.3.9}$$

Behind the flame front the combustion products are at rest: $u_r = 0$ for $r < R$. The boundary conditions at the spherical flame front are

$$\rho_1(u_{r1} - \dot{R}) = -\rho_2\dot{R}, \tag{11.3.10}$$

$$P_1 + \rho_1(u_{r1} - \dot{R})^2 = P_2 + \rho_2\dot{R}^2. \tag{11.3.11}$$

From (11.3.10) and (11.3.11) we obtain $B(t) = R^2\dot{R}(\Theta - 1)/\Theta$, and for the unperturbed flow ahead of the flame front, for $r > R$ we obtain

$$u_r = \frac{\Theta - 1}{\Theta}\left(\frac{R}{r}\right)^2\dot{R}, \tag{11.3.12}$$

$$\frac{P}{\rho_1}(\dot{R})^2 = \frac{\Theta - 1}{\Theta}\frac{3\alpha - 1}{\alpha}\frac{R}{r} - \frac{(\Theta - 1)^2}{2\Theta^2}\left(\frac{R}{r}\right)^4. \tag{11.3.13}$$

From Eqs. (11.3.11) to (11.3.13) we obtain for pressure behind the flame front

$$\frac{P}{\rho_1}(\dot{R})^2 = \frac{\Theta - 1}{\Theta}\frac{3\alpha - 1}{\alpha} + \frac{\Theta^2 - 1}{2\Theta^2}\left(\frac{R}{r}\right)^4. \tag{11.3.14}$$

The linearized equations for small perturbations of the spherical self-similar flame are

$$\frac{1}{r}\frac{\partial}{\partial r}(r^2\tilde{u}_r) + \hat{l}\tilde{\mathbf{w}} = 0, \tag{11.3.15}$$

$$\frac{\partial}{\partial t}\hat{l}\tilde{\mathbf{w}} + u_r\frac{\partial}{\partial r}\hat{l}\tilde{\mathbf{w}} = -\frac{1}{\rho r}\hat{l}^2\tilde{P}, \tag{11.3.16}$$

$$\frac{\partial \tilde{u}_r}{\partial t} + \frac{\partial}{\partial r}(u_r\tilde{u}_r) = -\frac{1}{\rho}\frac{\partial \tilde{P}}{\partial r}, \tag{11.3.17}$$

where perturbations are denoted by a tilde and the operator $\hat{l} = r\nabla_\perp$ is introduced. The perturbed boundary conditions at the flame front are

$$\tilde{u}_{r1} - \frac{\partial \tilde{R}}{\partial t} + \left(\frac{\partial u_r}{\partial r}\right)_1 \tilde{R} = \frac{1}{\Theta}\left(\tilde{u}_{r2} - \frac{\partial \tilde{R}}{\partial t}\right), \tag{11.3.18}$$

$$\tilde{P}_1 + \left(\frac{\partial P}{\partial r}\right)_1 \tilde{R} = \tilde{P}_2, \tag{11.3.19}$$

$$\hat{l}\tilde{\mathbf{w}}_1 + \frac{u_{r1}}{R}\hat{l}^2\tilde{R} = \hat{l}\tilde{\mathbf{w}}_2, \tag{11.3.20}$$

where labels 1 and 2 refer to the fuel and to the burnt matter respectively. As an additional boundary condition, we take the DL condition that the velocity of flame propagation is not changed by the small perturbations, i.e.

$$\tilde{u}_{r1} - \frac{\partial \tilde{R}}{\partial t} + \left(\frac{\partial u_r}{\partial r}\right)_1 \tilde{R} = 0. \tag{11.3.21}$$

It should be emphasized that the condition (11.3.21) applied to the self-similar regime (11.3.12) does not mean a constant velocity of the flame, as it was in the original statement in the derivation of the DL instability. Instead it means that the instantaneous normal velocity $\dot{R}$ remains unchanged by perturbations of an infinitely thin front in the linear regime.

The flow ahead of the flame front is potential, $\tilde{u}_{r1} = \partial\tilde{\varphi}/\partial r$, $\tilde{\mathbf{w}}_1 = \nabla_\perp\tilde{\varphi}$, $\Delta\tilde{\varphi} = 0$. Considering a spherical symmetry of the unperturbed flow, we obtain perturbations ahead of the flame front in the form

$$\tilde{\varphi} = \dot{R}R\Phi_1(t)Y_{n,m}(R/r)^{n+1}, \tag{11.3.22}$$

$$\tilde{u}_{r1} = -(n+1)\dot{R}\Phi_1(t)Y_{n,m}(R/r)^{n+2}, \tag{11.3.23}$$

$$\hat{l}\tilde{\mathbf{w}} = -n(n+1)\dot{R}\Phi_1(t)Y_{n,m}(R/r)^{n+2}, \tag{11.3.24}$$

where $Y_{n,m}$ are the spherical harmonics, $\hat{l}^2 Y_{n,m} = -n(n+1)Y_{n,m}$. The perturbation of pressure ahead of the flame front follows from Eqs. (11.3.17) and (11.3.23)

$$\frac{P}{\rho_1(\dot{R})^2} = Y_{n,m}(R/r)^{n+1}\Phi_1 \times$$
$$\left[-\frac{t}{\alpha\Phi_1}\frac{d\Phi_1}{dt} - n - 2 - \frac{\alpha-1}{\alpha} + (n+1)\frac{\Theta-1}{\Theta}\left(\frac{R}{r}\right)^3 \right] \tag{11.3.25}$$

It follows from Eqs. (11.3.16) and (11.3.17) that the perturbation of pressure behind the flame front satisfies the Laplace equation

$$\Delta\tilde{P} = 0. \tag{11.3.26}$$

Then solution of Eqs. (11.3.15)–(11.317) behind the flame front is

$$\frac{\tilde{P}}{\rho_1(\dot{R})^2} = Y_{n,m}(r/R)^n\Phi_2\left[-\frac{t}{\alpha\Phi_2}\frac{d\Phi_2}{dt} - n + 1 + \frac{\alpha-1}{\alpha} \right], \tag{11.3.27}$$

$$\tilde{u}_{r2} = -n\Theta\dot{R}Y_{n,m}(r/R)^{n-1}\Phi_2 + \Theta Y_{n,m}\Phi_3(r), \tag{11.3.28}$$

$$\hat{l}\tilde{\mathbf{w}}_2 = n(n+1)\Theta\dot{R}Y_{n,m}(r/R)^{n-1}\Phi_2 - \Theta Y_{n,m}\left(r\frac{d\Phi_3}{dr} + 2\Phi_3 \right). \tag{11.3.29}$$

Taking the perturbations of the flame front in the form

$$\tilde{R} = RY_{n,m}\Phi_4, \tag{11.3.30}$$

and substituting Eqs. (11.3.23)–(11.3.25), (11.3.27)–(11.3.30) into the boundary conditions (11.3.18)–(11.3.21), we obtain the system of ordinary equations for the unknown functions Φ_1, Φ_2, Φ_3, Φ_4. Looking for the solution of these equations in the form $\Phi_1, \Phi_2, \Phi_4 \propto t^\sigma$, $\Phi_3 \propto t^{\sigma+\alpha-1}$ we come to the system of uniform algebraic equations. The condition of existing a unique solution leads to the dispersion relation for the instability growth rate

$$\left(\frac{\sigma}{\alpha}\right)^2(1+n+\Theta n) + \frac{\sigma}{\alpha}[3(n+1) + d(1+n+\Theta n) + 2n(2\Theta+n)] -$$
$$\frac{\Theta-1}{\Theta}n^2(n+\Theta d) - 2n^2 - (3\Theta+2\Theta d+3)n + \frac{n}{\Theta} - 2 - d = 0. \tag{11.3.31}$$

where $d = (\alpha - 1)/\alpha$ is the fractal excess of the self-accelerating flame.

For the case $\alpha = 1$ the dispersion relation (11.3.31) coincides with that obtained in [17], and in this case the fractal excess is $d = 0$. The critical spherical number n_c for which $\sigma = 0$ is determined by the equation

$$n_c^2(n_c + \Theta d)(\Theta - 1) - 2\Theta n_c^2 - (3\Theta + 2\Theta d + 3)\Theta n_c + n_c - 2\Theta - \Theta d = 0. \tag{11.3.32}$$

The critical number n_c as a function of the expansion coefficient Θ is shown in Fig. 11.6 for a flame with constant velocity ($\alpha = 1$) and for the self-accelerating flames ($\alpha = 3/2$).

The stability limits are almost the same for flames with expansion coefficient $\Theta < 2$, $n_c \approx \frac{2}{\Theta-1}$, but it does not correspond to real flames. The self-similar regime, $\alpha = 3/2$, becomes essentially more unstable for flames with realistic expansion coefficients $\Theta = 6-10$. The instability growth rate is shown in Fig. 11.7 versus

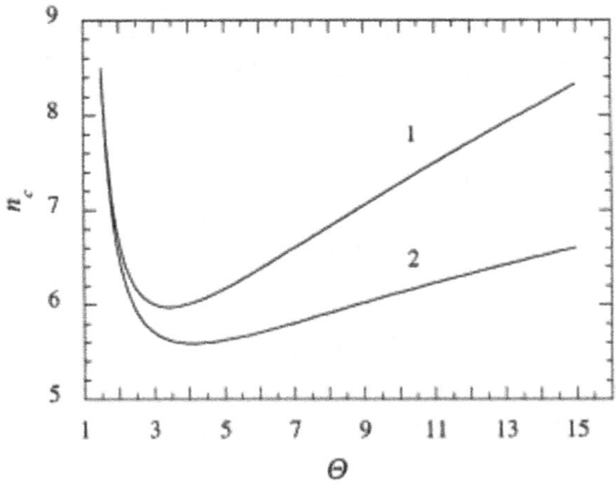

Fig. 11.6 Stability limits for a spherical flame (1) with a constant velocity; (2) propagating in the self-similar regime with $\alpha = 3/2$

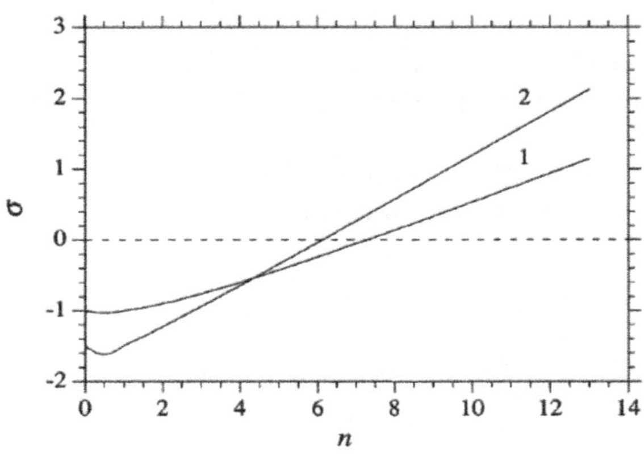

Fig. 11.7 The growth rate σ versus the number of spherical harmonics n for a flame: (1) with a constant velocity; (2) in the self-similar regime, $\alpha = 3/2$. For $\Theta = 10$, from [27]

the number of the spherical harmonic n for the flame with the expansion coefficient $\Theta = 10$. The instability growth rate for the self-similar regime a spherical flame with the fractal structure differs noticeably from the increment for the flame with a constant velocity even for the case when the stability limits change slightly.

For high order harmonics $n \to \infty$ the instability growth rate is proportional to the exponent of the self-similar regime

$$\sigma \to \alpha n \sigma_{DL}/\Theta, \tag{11.3.33}$$

where

$$\sigma_{DL} = \frac{\Theta}{\Theta - 1}\left(\sqrt{\Theta + 1 - 1/\Theta} - 1\right), \tag{11.3.34}$$

is the coefficient obtained in the Darrieus–Landau theory of flame stability.

The obtained results give an explicit analytical expression for the factor A in the self-similar law, Eq. (11.3.1). From Eqs. (11.3.2) and (11.3.5) we obtain

$$A = \left(\frac{2\pi\Theta u_f}{\alpha n_c \lambda_c}\right)^{\alpha} \frac{n_c \lambda_c}{2\pi}. \tag{11.3.35}$$

Thus, the velocity of a spherical flame with the fractal structure is

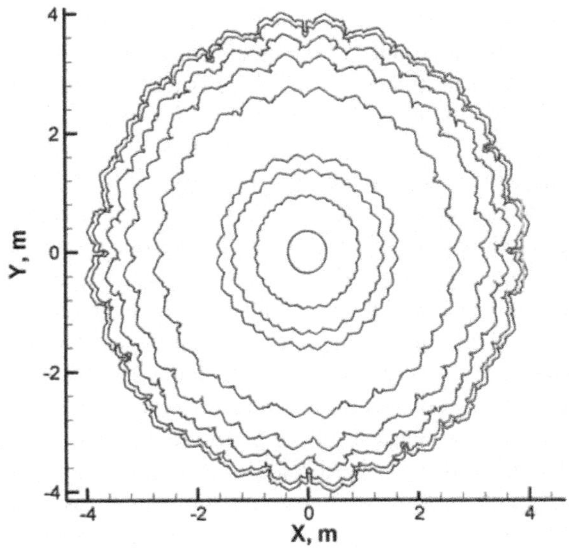

Fig. 11.8 Flame front evolution for the grid cell size $10L_f$, $\Theta = 8$, is shown at selected time instants: from $t = 0$ up to the fully developed fractal structure (from [32])

$$U_w = \Theta U_f \left(\frac{2\pi R}{n_c \lambda_c} \right)^d, \tag{11.3.36}$$

where the critical number of spherical harmonics $n_c = n_c(\Theta)$ is determined by Eq. (11.3.33), and the cut-off wavenumber λ_c follows from the stability theory of planar stationary flames of finite thickness. Of course, the obtained results, Eqs. (11.3.35) and (11.3.36), are valid for flames unstable only against the Darrieus-Landau instability for which the self-similar regime was suggested in [26]. Additional influence of the thermal-diffusion instability makes behavior of the flame surface even more complicated.

11.4 Self-acceleration and Fractal Structure of Expanding Flames

The fractal dimension can be determined from the slope of the double-logarithmic plot of the measured experimentally or found from numerical simulations flame radius versus time. According to Mandelbrot [58] the fully-developed Kolmogorov turbulence has a fractal character, and the turbulent velocity depends on the length scale as $u_t \propto \ell^{1/3}$, which is similar to the fractal dependence of the velocity of spherically expanding flames given by Eq. (11.3.1) for $\alpha = 3/2$: $dR/dt \propto R^{1/3}$. Consequently, if the development of fractal flames due to cellular instability corresponds to Eq. (11.3.1) with $\alpha = 3/2$, it can be treated as spontaneous flame self-turbulization. In view of fundamental importance of the above consideration there was numerous experimental and numerical studies, aiming to verify if α tends to approach a constant value, providing a self-similar regime, and if this value is $\alpha = 3/2$, and the fractal excess is $d = (\alpha - 1)/\alpha = 1/3$. Experimental studies [25, 37–39, 53, 54] reported slightly smaller than the critical value $\alpha = 1.5$ suggested for self-turbulization, or from 1.23 to 1.36 depending on mixture stoichiometry and pressure (1–15 atm) [47, 49]. It should be noted that the approach to the asymptotic regime with $\alpha = 3/2$ as well as manifestation of pulsations depends on the particular combustible mixture used in experimental studies and the possibility to achieve rather high values of the Peclet number. Also, experimental measurements of the fractal excess are complicated by the influence of gravity, confinement, flame acoustic interaction, etc. A qualitative agreement was obtained between theoretical predictions [8, 60–63] and experimental observations. Because of the unaffordable high computational cost, numerical simulations have been performed for the 2D cylindrical instead of the 3D spherical configuration. Most studies used the model of Sivashinsky or Frankel equations [28, 61–67] or the G-equation [68, 69] to model the flame front as a surface of discontinuity. Two-dimensional Lagrangian simulations of flame propagation performed in [69] demonstrated the radial growth with $\alpha = 3/2$. But it should be stressed that some if not all these studies have erroneously asserted the $\alpha = 3/2$

value, obtained from studies of 2D cylindrical flames, while this value has to do with 3D spherical flames.

The first 2D direct numerical simulation of the compressible Navier–Stokes equations, with one-step chemical model was performed in [32], where also was pointed to the difference in fractal dimension between 2D cylindrical and 3D spherical radially expanding flames, leading to $\alpha = 5/4$ in 2D case and $\alpha = 3/2$ in 3D case. Initial flame in simulations [32] was set as an unperturbed circular flame of radius $R_0 = 100L_f$ with zero angular (φ) velocity component u_φ of the unperturbed circular flame. Initial perturbation with spectral properties similar to a white noise were taken in the form:

$$f(\varphi) = \sum_{k=1}^{N} \tilde{A}_0 \cos(k\varphi), \qquad (11.3.37)$$

where amplitudes $\tilde{A}_0 \sim 0.001$, N is a maximal number of harmonics. It was verified that numerical noises appearing during simulation also contribute to the perturbation of the flame front. These noises can affect the shape of the flame at the initial stage of evolution, but it was found that development of perturbations on convex cylindrical front implies that perturbations of such small amplitudes cannot affect the developed fractal flame structure.

Figure 11.8 shows a representative plot of the flame front evolution obtained in simulations [32] with large grid cells: wrinkling of the flame front and development of fractal structure is clearly seen at the consecutive instants of time. It is seen that as a result of the DL instability the initially smooth flame front becomes wrinkled and the front surface area increases during expansion, exhibiting cusps that multiply and grow in amplitude.

It is seen from Fig. 11.8 that formation of the wrinkling cells is limited from below. In all simulations the smallest hump size was of the order of $\lambda_c \approx 20L_f$, which is the cut-off wavelength of DL instability. The perturbation size is also limited from above since perturbations with a length scale larger than $2\pi R/n_c$ are suppressed due to the finite curvature of the flame front, where n_c is the critical number of a spherical harmonic. It should be noted that if perturbations grow slower compared to the rate of the radial flame expansion, which corresponds to linear temporal dependence $R(t)$ for $\alpha = 1$, then the flame front does not acquire fractal structure and remains smooth.

A typical plot of the stream lines, flow field, and vortexes generated by wrinkled flame front is shown in Fig. 11.9. The fractal excess and dimension were calculated by best fitting the obtained dependence $R(t)$ to power function $at + c$, where a, b, c are fitting parameters. To extrapolate the obtained results from the 2D to 3D geometry of spherically expanding flames we should take into account that sum of two solutions of the nonlinear equation for the flame front is also solution to the equation.

We conclude that velocity amplification for spherical flames should be twice of the velocity amplification for cylindrical flames, which means that for spherically expanding flame $\beta_{2D} = U_w/U_f \cong 1.56$, and for the spherically expanding flame $d = \ln \beta_{2D}/\ln b \cong 0.36$. This gives fractal dimension 2.36, which is close to fractal

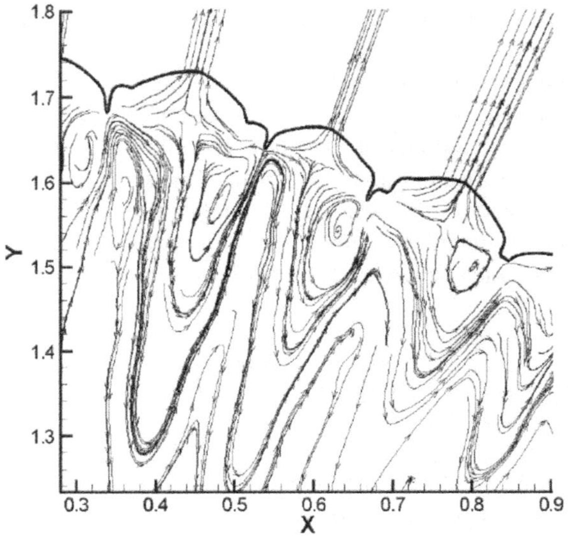

Fig. 11.9 Structure of the flow field and vortexes generated by wrinkled outward propagating 2D cylindrical flame obtained in simulations with fine grid cells size equal to $1L_f$; $\Theta = 8$, $t = 0.7$ s (from [32])

dimension 2.33 found in the experimental studies for $R(t) \propto t^{3/2}$. Obtained results were compared with the analytical estimates Eq. (11.3.1), $d_{2D} = (\ln \beta_{2D})/\ln b \approx 0.2$, $R(t) \propto t^{1.25}$. The calculated growth of the flame surface obtained in simulations with small ($\Delta = 1L_f$) and large ($\Delta = 10L_f$) grid cells are shown in Fig. 11.10, which shows dependence of the average normalized flame radius relative to its initial value versus normalized time. The dashed-dotted line of constant slope in Fig. 11.10 corresponds to the regime of outwardly propagating 2D cylindrical flame, $R(t) \propto t^{1.25}$. For comparison, the temporal dependence of the radius of outwardly propagating spherical flame with the corresponding power-law $R(t) \propto t^{1.5}$ is shown by the dashed line of constant slope.

Comparison of two calculations for flame fronts of "zero" thickness (large grid cells) and of finite thickness (small grid cells) shows that in the case of large grid cells the humps developed on the flame front are sharper, while calculations with small grid cells resolving the front structure show more smooth humps. These features result in different velocities of the flames in both cases and in different rates of approaching the self-similar regime corresponding to $R(t) \propto t^{1.25}$.

Since the transition to the self-similar regime is smooth, the value of the critical radius R_{cr} corresponding to the transition to the self-similar regime can be obtained only approximately. Critical radius can be evaluated as $R_{cr} \approx 2.0$ m, which corresponds to the critical Reynolds number $\mathrm{Re}_{cr} \approx 0.33 \cdot 10^5$. This value of the critical Reynolds number is consistent with that experimental values of the critical Reynolds numbers $\mathrm{Re}_{\exp} \sim 10^4{-}10^5$. Subsequent high-resolution numerical simulations of self-accelerating cylindrical hydrogen–air flames with detailed chemistry

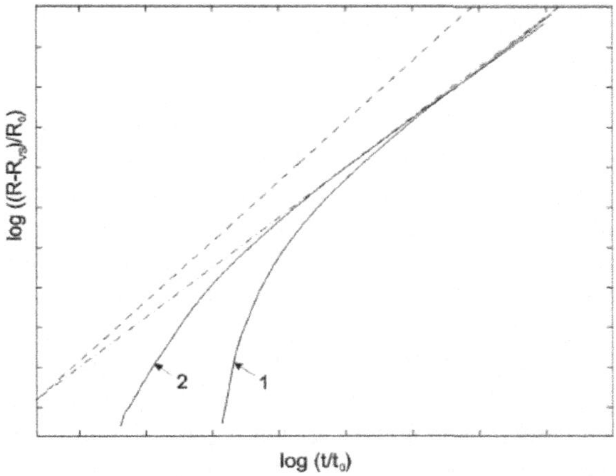

Fig. 11.10 Normalized flame radius plotted versus normalized time: curve (1) is simulations with large grid cells ($\Delta = 10L_f$); curve (2) is simulations with small grid cells ($\Delta = 1L_f$). R_{vs} is the distance to the virtual source. The dotted line of constant slope corresponds to the regime of expanding cylindrical flame with 1.25 power-law. Dashed line of constant slope corresponds to the regime of spherical expanding flame with the 3/2 power-law [32]

[70] have shown almost the same acceleration exponent $\alpha = 1.22$ as was obtained in [32]. The corresponding acceleration exponent for the spherical expanding flame is 1.564, which exceeds the self turbulization threshold of 1.5. It is unclear whether the distinction from 1.5 is due to the inaccuracy of extrapolating the result obtained for the two-dimensional case to the three-dimensional fractal excess, or the value of the exponent obtained for the two-dimensional case is inaccurate.

In [71] the transition to cellular structure and the dynamics of lean premixed hydrogen/air flames propagating outwards in 2D circular domains under the combined influence of the hydrodynamic and thermo-diffusive instabilities was investigated numerically using detailed chemistry and transport. It was found that the rate at which the flame radius grows is non-monotonic, reflecting the transitions in the dynamic behavior of the expanding front through the creation and annihilation of secondary cellular structures. The average speed of the wrinkled fronts reaches a mean value around $7U_f$, that is independent of the cellular patterns formed along the propagating front. Also, the temporal evolution of the mean flame radius varies almost linearly in time, at least during the time interval of the simulations run. This contradicts the available experimental data, which show unlimited flame acceleration.

In an attempt to answer the long-standing question of whether a self-accelerating, cellular-unstable, spherically expanding laminar flame can achieve "turbulent" behavior known as self-turbulization, experimental measurements of the flow characteristics of the flame-generated "turbulence" have been recently conducted [72]. From a theoretical point of view, the possibility of establishing self-turbulization is associated with whether the wrinkled flame front accelerating as $R_f \propto t^{\alpha}$ can reach the fractal excess $d = (\alpha - 1)/\alpha = 1/3$, which corresponds to the value of Kolmogorov turbulence, or fractal dimension $2 + d = 7/3$ and the acceleration exponent $\alpha = 3/2$, which inherently suggests that self-turbulization should lead to the Kolmogorov turbulence-like behavior. The experimental studies [72] used a centrally ignited, outwardly propagating spherical premixed flame in diffusively neutral stoichiometric hydrogen/oxygen/nitrogen mixtures, $Le \approx 1$. The experiments were designed using dual-chamber combustion vessel, which makes possible to study outwardly propagating spherical flames in a constant pressure environment for different pressures 1, 3, 5, 7 atm. Similar to the previous studies of the same group [49–54], three different stages were observed: (1) free from instabilities initial sooth expansion stage at $Pe < Pe_1 \approx 1000$; (2) a transition stage, when the flame front is fully covered with cells at $Pe_1 < Pe < Pe_2 \approx 1900$; the saturated stage of self-similar acceleration, when the flame front is completely covered with fissile cells at $Pe > Pe_2$. It was found that the far field flow is dominated by a radial flow, owing to thermal expansion with the rather weak RMS velocity. Only in the saturated stage and rather close to the flame front the flow shows turbulence-like behavior for the tangential velocity caused by the wrinkling of the cellular flame front with the power-law dependence in the energy spectra directly related to the fractal dimension of the wrinkled flame.

References

1. Y.B. Zeldovich, A.G. Istratov, N.I. Kidin, V.B. Librovich, Flame propagation in tubes: hydrodynamics and stability. Combust. Sci. Technol. **24**, 1–13 (1980)
2. L.D. Landau, E.M. Lifshitz, *Quantum Mechanics* (Pergamon Press, Oxford, 1976)
3. S. Gutman, G.I. Sivashinsky, The cellular nature of hydrodynamic instability. Physica D **43**, 129–139 (1990)
4. G.I. Sivashinsky, Nonlinear analysis of hydrodynamic instability in laminar flames-I. Derivation of basic equations. Acta Astronaut. **4**, 1177–1206 (1977)
5. OYu. Travnikov, V.V. Bychkov, M.A. Liberman, Influence of compressibility on propagation of curved flames. Phys. Fluids **11**, 2657–2666 (1999)
6. G. Joulin, On the hydrodynamic stability of curved premixed flames. J. Phys. **50**, 1069–1082 (1989)
7. M. Rahibe, N. Aubry, G.I. Sivashinsky, Stability of pole solutions for planar propagating flames. Phys. Rev. E **54**, 4958–4972 (1996)
8. Z. Olami, B. Galanti, O. Kupervasser, I. Procaccia, Random noise and pole-dynamics in unstable front propagation. Phys. Rev. E **55**, 2649–2663 (1997)
9. K.A. Kazakov, On-shell description of stationary flames. Phys. Fluids, **17**, 032107 (1–15) (2005)
10. G. Joulin, H. El-Rabii, K.A. Kazakov, On-shell description of unsteady flames. J. Fluid Mech. **608**, 217 (2008)
11. K.A. Kazakov, M.A. Liberman, Nonlinear equation for curved stationary flames. Phys. Fluids **14**, 1166–1180 (2002)
12. K.A. Kazakov, M.A. Liberman, Effect of vorticity production on the structure and velocity of curved stationary flames. Phys. Rev. Lett. **88**, 064502 (1–4) (2002)
13. K.A. Kazakov, M.A. Liberman, Non-stationary nonlinear equation for a curved flame. Comb. Sci. Tech. **174**, 157–179 (2002)
14. R. Yu, X.S. Bai, V. Bychkov, Fractal flame structure due to the hydrodynamic Darrieus-Landau instability. Phys. Rev. E **92**, 063028 (1–6) (2015).

15. Y.B. Zeldovich, G.I. Barenblatt, V.B. Librovich, G.M. Makhviladze, *Mathematical Theory of Combustion and Explosion* (Consultants Bureau, New York, 1985)
16. A.G. Istratov, V.B. Librovich, On the stability of propagation of spherical flames. J. Appl. Mech. Tech. Phys. **1**, 67–78 (1966)
17. A.G. Istratov, V.B. Librovich, On the stability of gasdynamic discontinuities associated with chemical reactions. The case of spherical flame. Acta Astronaut. **14**, 453–467 (1969).
18. G.H. Markstein, *Nonsteady Flame Propagation* (Pergamon Press, Oxford, 1964)
19. M. Matalon, B.J. Matkowsky, On the stability of plane and curved flames. Siam J. Appl. Math. **44**, 327–343 (1984)
20. J.K. Bechtold, M. Matalon, Hydrodynamic and diffusion effects on the stability of spherically expanding flames. Combust. Flame **67**, 77–90 (1987)
21. R. Addabbo, J.K. Bechtold, M. Matalon, Wrinkling of spherically expanding flames. Proc. Combust. Inst. **29**, 1527–1535 (2002)
22. J. Bechtold, C. Cui, M. Matalon, The role of radiative losses in self-extinguishing and self-wrinkling flames. Proc. Combust. Inst. **30**(1), 177–184 (2005)
23. M. Matalon, Intrinsic flame instabilities in premixed and nonpremixed combustion. Annu. Rev. Fluid Mech. **39**, 163–191 (2007)
24. M. Matalon, Flame dynamics. Proc. Combust. Inst. **32**, 57–63 (2009)
25. J. Beeckmann, R. Hesse, S. Kruse, A. Berens, N. Peters, H. Pitsch, M. Matalon, Propagation speed and stability of spherically expanding hydrogen/air flames: experimental study and asymptotics. Proc. Combust. Inst. **36**, 1531–1538 (2017)
26. Yu. Gostintsev, A. Istratov, Yu. Shulenin, Self-similar propagation of a free turbulent flame in mixed gas mixtures. Combust. Expl. Shock Waves **24**, 563–569 (1988)
27. V.V. Bychkov, M.A. Liberman, Stability and the fractal structure of a spherical flame in a self-similar regime. Phys. Rev. Lett. **76**, 2814–2817 (1996)
28. S.I. Blinnikov, P.V. Sasorov, Landau-Darrieus instability and the fractal dimension of flame fronts. Phys. Rev. E **53**, 4827–4841 (1996)
29. Yu.A. Gostintsev, A.G. Istratov, V.E. Fortov, On fractal structure of turbulent spherical flame. Dokl. Akad. Nauk USSR **353**, 55–56 (1997)
30. Yu.A. Gostintsev, A.G. Istratov, N.I. Kidin, V.E. Fortov, Structure of spherical turbulent flames. Teplofiz. Vysok. Temp. (High Temp. Phys.) **37**, 306–312 (1999).
31. Y.A. Gostintsev, V.E. Fortov, Y.V. Shatskikh, Self-similar propagation law and fractal structure of the surface of a free expanding turbulent spherical flame. Doklady Phys. Chem. **397**, 141–144 (2004)
32. M.A. Liberman, M.F. Ivanov, O.E. Peil, D.M. Valiev, L.-E. Eriksson, Self-acceleration and fractal structure of outward freely propagating flames. Phys. Fluids **16**, 2476–2482 (2004)
33. D. Bradley, C.M. Harper, The development of instabilities in laminar explosion flames. Combust. Flame **99**, 562–572 (1994)
34. D. Bradley, R.A. Hicks, M. Lawes, C.G.W. Sheppard, R. Woolley, The measurement of laminar burning velocities and Markstein numbers for iso-octane–air and iso-octane–n-heptane–air mixtures at elevated temperatures and pressures in an explosion bomb. Combust. Flame **115**, 126–144 (1998)
35. D. Bradley, Instabilities and flame speeds in large-scale premixed gaseous explosions. Philos. Trans. R. Soc. London Ser. A **357**, 3567–3581 (1999)
36. D. Bradley, T. Cresswell, J. Puttock, Flame acceleration due to flame-induced instabilities in large-scale explosions. Combust. Flame **124**, 551–559 (2001)
37. O. Kwon, G. Rozenchan, C.K. Law, Cellular instabilities and self-acceleration of outwardly propagating spherical flames. Proc. Combust. Inst. **29**, 1775–1783 (2002)
38. W.K. Kim, T. Mogi, K. Kuwana, R. Dobashi, Self-similar propagation of expanding spherical flames in large scale gas explosions. Proc. Combust. Inst. **35**, 2051–2058 (2015)
39. W. Kim, T. Imamura, T. Mogi, R. Dobashi, Experimental investigation on the onset of cellular instabilities and acceleration of expanding spherical flames. Int. J. Hydrogen Energy **42**, 14821–14828 (2017)

40. W. Kim, T. Namba, T. Johzaki, T. Endo, Self-similar propagation of spherically expanding flames in lean hydrogen-air mixtures Int. J. Hydrogen Energy **45**, 25608–25614 (2020)
41. C.R. Bauwens, J.M. Bergthorson, S.B. Dorofeev, Experimental study of spherical-flame acceleration mechanisms in large-scale propane–air flames. Proc. Combust. Inst. **35**, 2059–2066 (2015)
42. C.R. Bauwens, J.M. Bergthorson, S.B. Dorofeev, Experimental investigation of spherical-flame acceleration in lean hydrogen-air mixtures. Int. J. Hydrogen Energy **42**, 7691–7697 (2016)
43. C.K. Law, G. Jomaas, J.K. Bechtold, Cellular instabilities of expanding hydrogen/propane spherical flames at elevated pressures: theory and experiment. Proc. Combust. Inst. **30**, 159–167 (2005)
44. G. Jomaas, J.K. Bechtold, C.K. Law, Spiral waves in expanding hydrogen–air flames: experiment and theory. Proc. Combust. Inst. **31**, 1039–1046 (2007)
45. G. Jomaas, C.K. Law, J.K. Bechtold, On transition to cellularity in expanding spherical flames. J. Fluid Mech. **583**, 1–26 (2007)
46. G. Jomaas, C.K. Law, Observation and regime classification of pulsation patterns in expanding spherical flames. Phys. Fluids **22**, 124102 (1–7) (2010).
47. F. Wu, G. Jomaas, C.K. Law, An experimental investigation on self-acceleration of cellular spherical flames. Proc. Combust. Inst. **34**, 937–945 (2013)
48. J. Wang, Y. Xie, X. Cai, Y. Nie, C. Peng, Z. Huang, Effect of H2O addition on the flame front evolution of syngas spherical propagation flames. Combust. Sci. Technol. **188**, 1054–1072 (2016)
49. S. Yang, A. Saha, F. Wu, C.K. Law, Morphology and self-acceleration of expanding laminar flames with flame-front cellular instabilities. Combust. Flame **171**, 112–118 (2016)
50. S.D. Tse, D.L. Zhu, C.K. Law, Morphology and burning rates of expanding spherical flames in H_2/O_2/inert mixtures up to 60 atmospheres. Proc. Combust. Inst. **36**, 1549–1557 (2017)
51. Y. Xie, J. Wang, X. Cai, Z. Huang, Self-acceleration of cellular flames and laminar flame speed of syngas/air mixtures at elevated pressures. Int. J. Hydrogen Energy **41**(48), 18250–18258 (2016)
52. S. Yang, A. Saha, Z. Liu, C.K. Law, Role of Darrieus-Landau instability in propagation of expanding turbulent flames. J. Fluid Mech. **850**, 784–802 (2018)
53. J. Huo, A. Saha, Z. Ren, C.K. Law, Self-acceleration and global pulsation in hydrodynamically unstable expanding laminar flames. Combust. Flame **194**, 419–425 (2018)
54. J. Huo, A. Saha, T. Shu, Z. Ren, C.K. Law, Self-acceleration and global pulsation in expanding laminar H_2-O_2-N_2 flames. Phys. Rev. Lett. **4**, 043201 (1–4) (2019).
55. X. Cai, J. Wang, H. Zhao, M. Zhang, Z. Huang, Flame morphology and self-acceleration of syngas spherically expanding flames. Int. J. Hydrogen Energy **43**, 17531–17541 (2018)
56. X. Cai, J. Wang, Z. Bian, H. Zhao, H. Dai, Z. Huang, On transition to self-similar acceleration of spherically expanding flames with cellular instabilities. Combust. Flame **215**, 364–375 (2020)
57. C.R.L. Bauwens, J.M. Bergthorson, S.B. Dorofeev, Experimental investigation of spherical-flame acceleration in lean hydrogen-air mixtures. Int. J. Hydrog. Energy **42**, 7691–7697 (2017)
58. B.B. Mandelbrot, *The Fractal Geometry of Nature* (Freeman, San Francisco, 1983).
59. K.R. Sreenivasan, R. Ramshanker, C. Meneveau, Mixing, entrainment, and fractal dimension of interfaces in turbulent flows. Proc. R. Soc. **A421**, 79–108 (1989)
60. Z. Olami, I. Procaccia, R. Zeitak, Interface roughening in systems with quenched disorder. Phys. Rev. E **52**, 3402–3414 (1995)
61. B. Galanti, O. Kupervasser, Z. Olami, I. Procaccia, Dynamics and wrinkling of radially propagating fronts inferred from scaling laws in channel geometries. Phys. Rev. Lett. **80**, 2477–2480 (1998)
62. O. Kupervasser, Z. Olami, I. Procaccia, Stability analysis of flame front: dynamical systems approach in complex plane. Phys. Rev. E **59**, 2587–2593 (1999)
63. L. Filyand, G. Sivashinsky, M. Frankel, On self-acceleration of outward propagating wrinkled flames. Physica D **72**, 110–118 (1994)
64. M. Rahibe, N. Aubry, G.I. Sivashinsky, R. Lima, Formation of wrinkles in outwardly propagating flames. Phys. Rev. E **52**, 3675–3686 (1995)

65. V. Karlin, G. Sivashinsky, The rate of expansion of spherical flames. Combust. Theory Modell. **10**, 625–637 (2006).
66. V. Karlin, G. Sivashinsky, Asymptotic modelling of self-acceleration of spherical flames. Proc. Combust. Inst. **31**, 1023–1030 (2007)
67. R. Aldredge, B. Zuo, Flame acceleration associated with the Darrieus-Landau instability. Combust. Flame **127**, 2091–2101 (2001)
68. K.L. Pan, R. Fursenko, Characteristics of cylindrical flame acceleration in outward expansion. Phys. Fluids **20**, 094107 (2008).
69. W.T. Ashurst, Darrieus-Landau instability, growing cycloids and expanding flame acceleration. Combust. Theory Modelling **1**, 405–428 (1997)
70. Y.X. Xin, C.S. Yoo, J.H. Chen, C.K. Law, A DNS study of self-accelerating cylindrical hydrogen–air flames with detailed chemistry. Proc. Combust. Inst. **35**, 753–776 (2015)
71. C. Altantzis, C. Frouzakis, A. Tomboulides, S. Kerkemeier, K. Boulcuchos, Direct numerical simulation of circular expanding premixed flames in a lean quiescent hydrogen-air mixture: Phenomenology and detailed flame front analysis. Combust. Flame **162**, 331–344 (2015)
72. Z. Liu, V.R. Unni, S. Chaudhuri, R. Sui, C.K. Law, A. Saha, Self-turbulization in cellularly unstable laminar flames. J. Fluid Mech. **917**, A53 (1–12) (2021).

Chapter 12
Detonation Waves

12.1 Evolutionarity Condition and Possible Combustion Modes

A general condition of stability of the solution, which is considered as a discontinuous surface, is given by the evolutionarity condition [1]. The evolutionarity condition requires that a small perturbation of the initial state should cause only an infinitely small change of the flow field at least during some small but finite time interval. This condition is a necessary but not sufficient condition. Evolution of small perturbations of the discontinuity surface is defined by the set of linearized boundary conditions, which must be satisfied at the discontinuity surface. The stability condition is satisfied if the number of these equations is the same as the number of independent parameters in these equations. A small perturbation of the discontinuity surface consisting of an infinitesimal displacement in the direction of the normal to the discontinuity surface can be considered as superposition of small perturbations of gas velocity, pressure and temperature on both sides of the discontinuity surface. The perturbation originated at the discontinuity surface then propagate away from both sides of the discontinuity surface. Thus, a perturbation can be considered as a superposition of sound waves propagating with the local sound speed (pressure, density and velocity) and perturbation of entropy, which is carried by the flow. The change of quantities, which are carried by sound waves, are not independent but they are related by the relations arising from the equations of motion. In general, the necessary condition for the existence and uniqueness of the solution in the form of a discontinuity surface, the evolutionarity condition requires that the number of possible outgoing simple waves from both side of the discontinuity surface $s_1 + s_2$ to be less by one than the number r of boundary conditions:

$$s_1 + s_2 = r - 1. \tag{12.1.1}$$

For example, considering a shock wave we have three boundary conditions: continuous of mass flux, momentum flux and energy flux. Two waves are always

M. A. Liberman, *Combustion Physics*,
https://doi.org/10.1007/978-3-030-85139-2_12

propagating in the downstream flow away from the front. These are: acoustic waves, propagating with velocity, $a_s + u_2$, and the entropy wave, propagating together with the downstream flow, u_2. In this case, the evolutionary condition requires the absence of outgoing waves in the upstream flow ahead of the shock front. The relation (12.1.1) will be satisfied only if there are no outgoing simple waves ahead of the shock wave. This means that ahead of the shock must be: $u_0 - a_{s0} > 0$, $M_0 = u_0/a_{s0} > 1$, and in the downstream flow behind the shock, $u_2 - a_{s2} < 0$, $M_2 = u_2/a_{s2} < 1$. If the condition (12.1.1) is not satisfied ($r - 1 < s_1 + s_2$ or $r - 1 > s_1 + s_2$), then the initial discontinuity surface will split into some other discontinuity surfaces.

An exothermic reaction in a gas mixture can propagate in the form of a slow combustion wave (deflagration) propagating with velocity smaller than the speed of sound. Another, supersonic regime of combustion wave—*detonation* consists of a leading shock wave followed by the reaction zone. Since the width of the combustion zone for both combustion modes deflagration and detonation is usually much smaller than the scales characterizing the environment, we can neglect by the width of the reaction zone, considering both deflagration and detonation as a surface of zero thickness separating the unburned gas from combustion products. This discontinuity surface propagates relative to the unburned gas in the direction normal to the front with some velocity u_0. The conditions of continuity of the momentum and energy fluxes at the discontinuity surface written in dimensionless variables are

$$u - 1 + \frac{1}{\gamma M_0^2}\left(\frac{T}{u} - 1\right) = 0, \tag{12.1.2}$$

$$u^2 - 1 + \frac{2}{(\gamma - 1)M_0^2}[T - 1] + \alpha\frac{q}{M_0^2} = 0. \tag{12.1.3}$$

Here $q = Q/c_0^2$ is nondimensional reaction heat, and for the sake of simplicity the adiabatic constant γ is taken the same for the unburned gas and for the combustion products. We will also assume that the heat of chemical reaction is high enough, q > > 1. In Eq. (12.1.3) ahead of the reaction front $\alpha = 1$ and in the combustion products $\alpha = 0$. Eliminating the final temperature from Eqs. (12.1.2) and (12.1.3) we obtain the equation for the velocity of combustion products

$$(1 - u_k)(u_k - u_{GD}) = \frac{2(\gamma - 1)}{(\gamma + 1)}\frac{q}{M_0^2}. \tag{12.1.4}$$

Here index $k = 1,\ 2$ denotes two possible final states, and $u_{GD} = \frac{2+(\gamma-1)M_0^2}{(\gamma+1)M_0^2}$ is the dimensionless velocity in the flow behind the discontinuity.

Equation (12.1.4) has two real roots. One root (u_1) corresponds to a supersonic mode (detonation) $M_0 \geq M_{CJ}^{(1)} = \sqrt{2q(\gamma^2 - 1)}$, another root ($u_2$) corresponds to subsonic mode (deflagration) $M_0 \leq M_{CJ}^{(2)} = 1/\sqrt{2q(\gamma^2 - 1)}$. Signs of equality in these formulas correspond to the particular cases: Chapman-Jouguet (CJ) detonation and Chapman-Jouguet deflagration, which will be considered in the subsequent sections of this chapter.

$$u_{CJ}^{(1)} = \sqrt{2q(\gamma^2 - 1)}. \tag{12.1.5}$$

The temperature of products in the detonation wave is

$$C_v T_{CJ}^{(1)} = q\frac{2\gamma}{\gamma + 1}. \tag{12.1.6}$$

The velocity and temperature of CJ deflagration are

$$u_{CJ}^{(2)} = 1/\sqrt{2q(\gamma^2 - 1)}. \tag{12.1.7}$$

$$C_P T_{CJ}^{(2)} = 2q/(\gamma + 1). \tag{12.1.8}$$

Remember, that to simplify the final formulas, we used the same value of γ for the fresh gas and combustion products and assumed $q \gg C_P T_2$. The velocity of products behind the CJ detonation and behind the CJ deflagration equals to the local sound speed.

Form (12.1.6) and (12.1.8) we find the ratio of temperatures of the products of CJ detonation and CJ deflagration

$$T_{CJ}^{(1)}(\mathrm{det})/T_{CJ}^{(2)}(def) = \gamma^2. \tag{12.1.9}$$

For comparison, the temperature of products in deflagration is $C_V T^{(2)}(def) = q/\gamma$, and the ratio of the products temperature of CJ deflagration and "normal" deflagration is

$$T_{CJ}^{(2)}(def)/T^{(2)}(def) = 2\gamma/(\gamma + 1). \tag{12.1.10}$$

Since $\gamma > 1$ this ratio is always larger than unity.

12.2 Structure of Detonation Waves; Detonation Adiabatic

In the previous chapters we have considered deflagrations (flames) which are slow combustion waves that propagate with velocities much smaller than the sound speed. In a deflagration the chemical reaction propagates through the gaseous mixture of fuel and oxidants due to diffusion processes (thermal conduction or molecular diffusion), which result in the transfer of heat or radicals from the burning gas to the fresh unburned mixture. The diffusive character of the mechanism of a deflagration wave propagation makes the deflagration velocity smaller than the sound speed in the fresh (unburned) mixture. The entirely different mechanism of the reaction wave propagation is also possible due to shock waves. The shock wave heats and compresses the gas as it passes through the shock front so that the temperature and density of the combustible mixture behind the shock increase. If a shock wave is sufficiently strong, then an increase in temperature may be enough to initiate combustion (the Neumann state). The unburned mixture is ignited when it moves through the shock wave; that is the reaction wave will be propagated with the velocity of the shock wave, which exceeds the sound speed and much faster than the velocity of a deflagration. This mechanism of combustion wave propagation is called *detonation*. The theory of detonation has been developed by Ya. B. Zel'dovich [2] and independently by J. von Neumann [3] and W. Döring [4].

When the shock wave passes some point in the gas, the reaction begins at this point (the Neumann point) and continues until all the fuel will be burned during some time τ_r, which characterizes the reaction rate of this particular fuel-oxidants mixture. The value of τ_r dependents on the temperature and pressure, and therefore on the strength of the shock. It decreases for a stronger shock, on account of the increase in the reaction rate with rising temperature. It is therefore clear that a layer of chemical reaction zone moving with the shock in which combustion is occurring will follow the shock wave and the width of this layer can be estimated as the speed of the shock wave multiplied by the characteristic reaction time τ_r.

Typical structure of a detonation, known as the Zel'dovich- Neumann-Döring model (ZND), consisting of a leading shock wave followed by the reaction zone, is shown in Fig. 12.1.

Since typically the characteristic dimensions of the problem are much larger than the width of detonation wave (together with reaction zone), we can consider the shock wave together with the reaction zone following it, as a single surface of discontinuity, which separates the burned and unburned gases. Across a detonation wave the fluxes of mass density, energy and momentum must be continuous. Since it is always possible to choose a coordinate system in which the velocity is directed along the normal to the shock front, we can consider a one-dimensional problem. Then, similar to an ordinary shock adiabatic we can write relations, which follow from continuity conditions alone

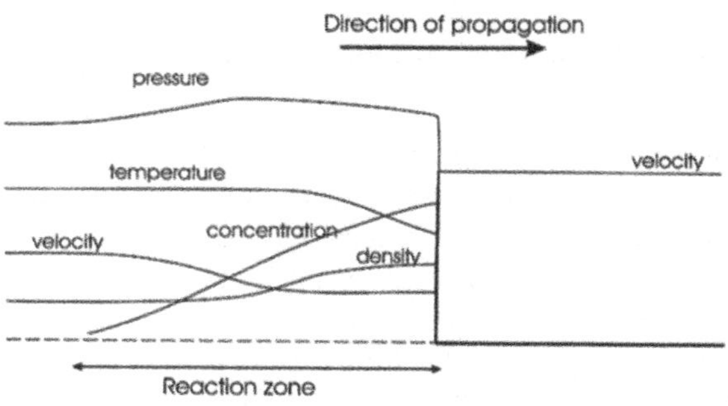

Fig. 12.1 Schematic structure of the Zel'dovich-von Neumann-Döring detonation

$$j \equiv \rho_1 u_1 = \rho_2 u_2 \tag{12.2.1}$$

$$P_1 + \rho_1 u_1^2 = P_2 + \rho_2 u_2^2, \tag{12.2.2}$$

$$H_1 + \frac{u_1^2}{2} = H_2 + \frac{u_2^2}{2}. \tag{12.2.3}$$

Here we introduced specific volume as $V = 1/\rho$, and using the suffix 1 that always pertains to the unburned gas in the upstream flow ahead the shock front, and the suffix 2 pertains to the combustion products in the downstream flow behind detonation front. From Eq. (12.2.1) and Eq. (12.2.2) we obtain tangent of the angle of the cord connecting two points (P_1, V_1) and (P_2, V_2)

$$j^2 = \frac{P_2 - P_1}{V_1 - V_2}, \tag{12.2.4}$$

In particular, from Eqs. (12.2.1)–(12.2.3) we obtain

$$H_1 - H_2 + \frac{1}{2}(V_1 - V_2)(P_2 + P_1) = 0 \tag{12.1.5}$$

The curve of P_2 as a function of V_2 given by the Eq. (12.2.5) is called the *detonation adiabatic*. Unlike the shock adiabatic, this curve does not pass through the given initial point (P_1, V_1). The fact that the shock adiabatic passes through this point was due to the fact that the enthalpy H_1 and H_2 were the same functions of P_1, V_1 and P_2, V_2 respectively, whereas this does not now hold in detonation because of the chemical difference between the unburned and burned gases. In Fig. 12.2 the continuous line shows the detonation adiabatic and the ordinary shock adiabatic for

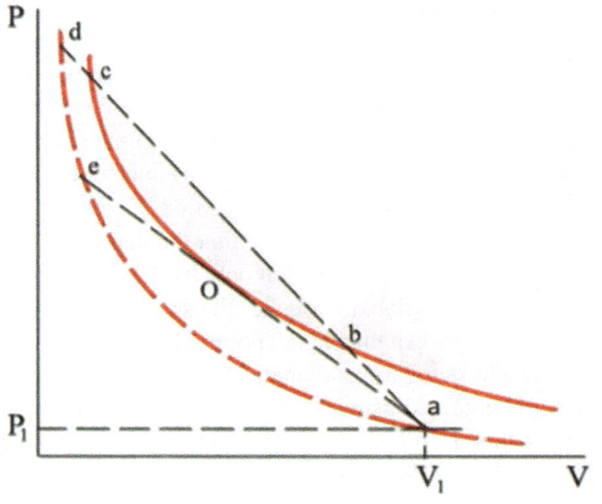

Fig. 12.2 Detonation adiabatic

the unburned gas mixture is shown by dashed line for comparison purpose. The last one passes through the point P_1, V_1 as it must be for the shock adiabatic for the unburned gas mixture. The detonation adiabatic always lies above the shock adiabatic, because a high temperature is reached in combustion, and the gas pressure is therefore greater than it would be in the unburned gas for the same specific volume.

It is obvious that Eq. (12.2.4), which is a consequence of the equations of continuity for density and momentum fluxes only, is valid for both shock wave and detonation. In the Fig. 12.2 for detonation adiabatic j^2 is the slope of the chord passing from the point P_1, V_1 to any other point P_2, V_2. Such chord in Fig. 12.2 is shown by the dashed line a-b-c-d. The immediate consequence, which is seen from Fig. 12.2 is that j^2 cannot be less than the slope of the tangent $a - O - e$ to a curve of the detonation adiabatic. It is clear that the flux j^2 is the mass of the gas mixture, which is ignited per unit time per unit area on the surface of the detonation wave. We can conclude that, in a detonation, this quantity cannot be less than a certain limiting value $j_{\min}$, which depends on the initial state of the unburned gas mixture. The Eq. (12.1.4) is valid also not only for the final state of the combustion products for a given initial state, but for all intermediate states, in which only part of the reaction has been evolved. In other words, the pressure P and specific volume V of the gas in any state obey the linear relation

$$P = P_1 + j^2(V_1 - V), \tag{12.2.6}$$

which corresponds to the chord (a-b-c-d) in Fig. 12.2.

The variation of the gas mixture states through the layer of finite width representing a detonation wave has been studied by Ya. B. Zel'dovich. The leading front of the detonation wave is a true shock wave in the unburned gas (with index 1). In this region, the gas is compressed and heated to a state represented by the point d in Fig. 12.2 on the shock adiabatic of gas 1 this is end point at the shock adiabatic. The chemical reaction begins in the compressed gas (the Neumann point), and as the reaction proceed the state of the gas is represented by a point which moves down the chord (d-a), from the point d to point c, since the heat is evolved, the gas expands, and its pressure decreases. This continues until combustion is completed and the whole heat of the reaction has been evolved. The corresponding point is c, which lies on the detonation adiabatic representing the final state of the combustion products. The lower point b at which the chord (a-d) intersects the detonation adiabatic cannot be reached for a gas in which combustion is caused by compression and heating in a shock wave. Thus, we conclude that the detonation is represented, not by the whole of the detonation adiabatic, but only by the upper part, lying above the point O where the detonation adiabatic touches the straight-line a-O drawn from the initial point a.

It can be shown that, at the point where $d\,j^2/dP_2 = 0$, i.e. where the chord a-O-e touches the detonation adiabatic, the velocity u_2 is equal to the sound speed a_{s2}. This result is obtained using only the conservation laws for the surface of discontinuity, and is therefore entirely applicable to the detonation wave also. On the ordinary shock adiabatic for a nonreacting gas there are no points where $dj^2/dP_2 = 0$. On the detonation adiabatic, however, there is such a point, namely the point O. At the same time, at such a point $u_2 = a_{s2}$, and in addition at this point takes place the inequality $d(u_2/a_{s2})/dP_2 < 0$, which means that $u_2 < a_{s2}$ above point O, at higher values of P_2. Since detonation corresponds to the upper part only of the adiabatic, above point O, we conclude that behind the detonation

$$u_2 \leq a_{s2}, \tag{12.2.7}$$

which means that a detonation wave moves relative to the gas just behind it with a velocity equal to or less than that of sound velocity. The equality $u_2 = a_{s2}$ holds for a detonation, for which the final state corresponds to the point O, which is called the *Chapman-Jouguet* point, and this particular detonation is called the *Chapman-Jouguet detonation* [5–7]. The velocity of the detonation wave relative to unburned gas 1 is always supersonic, including for the point O:

$$u_1 > a_{s1}. \tag{12.2.8}$$

Validity of inequality (12.2.8) is seen directly from Fig. 12.2. The velocity of sound a_{s1} in the figure is given by the slope of the tangent to the shock adiabatic for gas 1 at the point a. On the other hand, the slope of the chord a-c gives velocity u_1, because $u_1 = jV_1$. Since all the chords going from point a are steeper than the tangent, we conclude that $u_1 > a_{s1}$. Moving with supersonic velocity, the detonation wave, like a shock wave, does not affect the state of the gas in front of it. The velocity u_1 with which the detonation wave moves relative to the unburned gas at rest is the velocity of the detonation propagation in the fresh unburned gas. From this consideration we also conclude that the Chapman-Jouguet detonation propagates with the minimal velocity compared with other possible detonation waves, which corresponds to steeper cord going from the initial state of the point a.

Since $u_1/V_1 = u_2/V_2 \equiv j$, and $V_1 > V_2$ (recall that $V_1 = 1/\rho_1$, and $V_2 = 1/\rho_2$), it follows that $u_1 > u_2$. The difference $u_1 - u_2$ is the velocity of the combustion products relative to the unburned gas. This difference is positive, i.e. the combustion products move in the direction of the detonation wave propagation. It should be also mentioned the following important property of the Chapman-Jouguet detonation. Since $ds_2/dj^2 > 0$, where s is entropy, then at the point where j^2 has a minimum, the entropy s_2 therefore also has a minimum. It is obvious that, this point is point O, and this point corresponds to the minimal value of the entropy s_2 on the detonation adiabatic. The entropy s_2 also has a maximum at point O if we consider the change in the state along the line a-O-e, since the slopes of the curve and the tangent at O are the same. A displacement from point e to point O corresponds to the change of state as the combustion reaction occurs in the compressed gas, and this is accompanied by the production of heat in the exothermic reaction and the increase in entropy. A passage from point O to point a, however, would correspond to the endothermic conversion of the combustion products into the original gases, which is a process of a decrease in entropy. If the detonation is caused by a shock wave, which is produced by some external source then any point on the upper part of the detonation adiabatic (above point O) may correspond to the detonation. A detonation, which is due to the combustion process itself, is the most important case. Later we will show that detonation propagating freely from the closed end of a tube or from the open end of a tube is always the Chapman-Jouguet detonation, so that the velocity of the detonation wave relative to the combustion products behind it is equal to the sound velocity, while the velocity u_1 relative to the unburned gas has its minimal possible value. In a sense we can say that Chapman-Jouguet detonation is the weakest possible detonation.

12.3 Velocity of a Detonation Wave

Let us now consider an ideal polytropic gas and derive the relationships between various quantities in a detonation wave. Substituting in Eq. (12.2.5) the heat function in the form

$$H_{1,2} = H_{01,2} + c_p T_{1,2} = H_{01,2} + \frac{\gamma P_{1,2} V_{1,2}}{\gamma - 1}, \tag{12.3.1}$$

(for an ideal gas $P = (\gamma - 1) c_V \rho T$, and adiabatic constant is $\gamma = c_p / c_V$) we obtain

$$\frac{\gamma_2 + 1}{\gamma_2 - 1} P_2 V_2 - \frac{\gamma_1 + 1}{\gamma_1 - 1} P_1 V_1 - P_2 V_1 + P_1 V_2 = 2Q, \tag{12.3.2}$$

where $Q = H_{01} - H_{02}$ is the heat of the reaction, reduced to the absolute zero of temperature.

The curve $P_2 = P_2(V_2)$ given by the Eq. (12.3.2) is a rectangular hyperbola that is detonation adiabatic. For a strong detonation, $P_2/P_1 \gg 1$, the ratio of densities tends to a finite limit

$$\frac{\rho_2}{\rho_1} = \frac{V_1}{V_2} = \frac{\gamma_2 + 1}{\gamma_2 - 1}. \tag{12.3.3}$$

This is the highest compression that can be achieved in a detonation wave similar to that achieved in a shock wave.

All formulas are simplified in the case of strong detonation waves, which correspond to the case when the heat released in the reaction is large compared to the internal thermal energy of the source gas, i.e. when $Q \gg c_V T_1$. In this case we can neglect the terms containing P_1 in Eq. (12.3.2) and we obtain

$$P_2 \left(\frac{\gamma_2 + 1}{\gamma_2 - 1} V_2 - V_1 \right) = 2Q. \tag{12.3.4}$$

12.4 The Chapman-Jouguet Detonation

As it was shown, at the Chapman-Jouguet point

$$j^2 = \frac{P_2 - P_1}{V_1 - V_2} = \frac{a_{s2}^2}{V_2^2}. \tag{12.4.1}$$

From this equation together with Eq. (12.2.4) we can express P_2 and V_2 in the form

$$P_2 = \frac{P_1 + j^2 V_1}{\gamma_2 + 1}, \quad and \quad V_2 = \frac{\gamma_2 (P_1 + j^2 V_1)}{j^2(\gamma_2 + 1)} \tag{12.4.2}$$

Substituting these expressions in Eq. (12.3.2), replacing the flux j by $u_1 = jV_1$, and using again the expression $PV = (\gamma - 1)c_V T$, we obtain the equation for the velocity u_1:

$$u_1^4 - 2u_1^2[(\gamma_2^2 - 1)Q + (\gamma_2^2 - \gamma_1)c_{V1}T_1] + \gamma_2^2(\gamma_1 - 1)^2 c_{V1}^2 T_1^2 = 0. \tag{12.4.3}$$

This equation gives the velocity of detonation as a function of parameters characterizing physical and chemical properties of the unburned mixture (temperature and the heat release in chemical reaction). From four roots of this biquadratic equation, we are interested in two positive roots only:

$$u_1 = \left\{\frac{\gamma_2 - 1}{2}\left[(\gamma_2 + 1)Q + (\gamma_1 + \gamma_2)c_{v1}T_1\right]\right\}^{1/2} + \left\{\frac{\gamma_2 + 1}{2}\left[(\gamma_2 - 1)Q + (\gamma_2 - \gamma_1)c_{v1}T_1\right]\right\}^{1/2}. \tag{12.4.4}$$

From these two roots that differ by sign plus and minus in front of the second root square in Eq. (12.4.4), the root with sign plus corresponds to the tangent that passes upwards from the point a in Fig. 12.2. The two signs in this case correspond to the fact that two tangents can be drawn from the point a to the detonation adiabatic: one upwards, as shown in Fig. 12.2, and the other downwards. The upward tangent, in which we are interested, has the steeper slope and corresponds to detonation. Another root corresponds to the Chapman-Jouguet deflagration and will be discussed later. Formula (12.4.4) determines the detonation velocity in terms of the temperature T_1 of the unburned gas mixture. We can use Eqs. (12.4.1) and (12.4.2) to express P_2 and V_2 in terms of u_1 and T_1. For this purpose, we rewrite these equations in the form

$$\frac{P_2}{P_1} = \frac{u_1^2 + (\gamma_1 - 1)c_{v1}T_1}{(\gamma_2 + 1)(\gamma_1 - 1)c_{v1}T_1}, \tag{12.4.5}$$

$$\frac{V_2}{V_1} = \frac{\gamma_2[u_1^2 + (\gamma_1 - 1)c_{v1}T_1]}{(\gamma_2 + 1)u_1^2}. \tag{12.4.6}$$

Equations (12.4.5) and (12.4.6) together with Eq. (12.4.4) determine the pressure and density ratios between combustion products and unburned gas for a detonation wave propagating in a gas mixture with a given particular pressure, density and temperature.

Using formulas (12.4.4), (12.4.5) and (12.4.6) and, taking into account that velocity of the flow behind the detonation is $u_2 = u_1 V_2/V_1$, we can calculate u_2. The result is

$$u_2 = \left\{\frac{\gamma_2 - 1}{2}[(\gamma_2 + 1)Q + (\gamma_1 + \gamma_2)c_{v1}T_1]\right\}^{1/2} + \frac{\gamma_2 - 1}{\gamma_2 + 1}\left\{\frac{\gamma_2 + 1}{2}[(\gamma_2 - 1)Q + (\gamma_2 - \gamma_1)c_{v1}T_1]\right\}^{1/2} \tag{12.4.7}$$

The velocity of the combustion products relative to the unburned gas mixture, i.e. the difference $u_1 - u_2$ can be written as

$$u_1 - u_2 = \left\{\frac{2}{\gamma_2 + 1}[(\gamma_2 - 1)Q + (\gamma_2 - \gamma_1)c_{v1}T_1]\right\}^{1/2}. \tag{12.4.8}$$

Finally, for the temperature of the combustion products we obtain

$$T_2 = \frac{u_2^2}{\gamma_2(\gamma_2 - 1)c_{V2}}, \tag{12.4.9}$$

where for the Chapman-Jouguet detonation $u_2 = a_2$.

All these formulas become simpler for strong detonation waves, when t $Q \gg c_V T_1$. In this case the velocities are

$$u_1 = \sqrt{2(\gamma_2^2 - 1)Q}, \tag{12.4.10}$$

$$u_1 - u_2 = \frac{1}{(\gamma_2 + 1)}u_1. \tag{12.4.11}$$

For density, pressure and temperature of the combustion products we obtain

$$\frac{V_2}{V_1} = \frac{\gamma_2}{\gamma_2 + 1}, \tag{12.4.12}$$

$$\frac{P_2}{P_1} = \frac{2(\gamma_2 - 1)}{(\gamma_1 - 1)}\frac{Q}{c_{V1}T_1}, \tag{12.4.13}$$

$$T_2 = \frac{2\gamma_2}{(\gamma_2 + 1)} \frac{Q}{c_{V2}}. \tag{12.4.14}$$

We also write the relation between the temperature of the combustion products for detonation and for deflagration, for the case $Q \gg c_V T_1 \sim c_{P1} T_1$

$$\frac{T_2^{\text{det}}}{T_2^{\text{flame}}} = \frac{2\gamma_2^2}{\gamma_2 + 1}. \tag{12.4.15}$$

Since $\gamma_2 > 1$, the right-hand side of (12.4.15) is always larger than one. For example, $2\gamma_2^2/(\gamma_2+1) = 1.309$ for $\gamma_2 = 1.2$, and $2\gamma_2^2/(\gamma_2+1) = 1.633$ for $\gamma_2 = 1.4$, which means that the temperature of the combustion products is always considerably higher behind detonation wave compared to the temperature of combustion products behind a deflagration.

We will show that for the most important practical cases, the regime of a detonation wave propagating in a gas mixture is the Chapman-Jouguet detonation. Let us consider first the case of a detonation propagating from the closed end of the tube. The boundary conditions in this case are that the gas velocity is zero in front of the detonation wave, since a supersonic detonation does not affect the state of the gas in front of it. The boundary conditions also require that the gas velocity must be zero at the closed end of the tube. Since the gas acquires a non-zero velocity when the detonation wave passes, the velocity must diminish in the region between the detonation wave and the closed end of the tube. In order to determine the resulting flow pattern, we notice that in this case there is no parameters of the length scale, which might characterize the conditions of flow along the tube. Therefore, the gas velocity can change either in a shock wave separating two regions where the velocity is constant or in a rarefaction wave.

If we assume that a detonation is not the Chapman-Jouguet detonation, then the detonation velocity relative to the gas behind the detonation is $u_2 < a_{s2}$. However, in this case it is impossible to satisfy the boundary conditions, because in this case, neither a shock wave nor a weak discontinuity, which is the forward front of a rarefaction wave, can follow the detonation wave. Indeed, the shock wave would propagate relative to the gas in front of it with a velocity exceeding a_{s2} and therefore overtake the detonation wave. A weak discontinuity of a rarefaction wave, the front of which propagates with a velocity equal to the local sound speed a_{s2} also would overtake the detonation wave. In both cases, the velocity of the gas behind the detonation wave cannot decrease, i.e. the boundary condition at the closed end of the tube, at $x = 0$, cannot be satisfied. The only way to satisfy the boundary condition at the closed end of the tube for a detonation wave corresponds to the Chapman-Jouguet detonation. In the latter case, $u_2 = a_{s2}$, and a rarefaction wave follows the detonation wave. The rarefaction wave is formed at $x = 0$ when the detonation begins, and its forward front coincides with the "back side" of the detonation wave. Thus, we arrived to the conclusion that the detonation wave ignited at the closed end of the tube must be the Chapman-Jouguet detonation. It moves relative to the gas behind it with a velocity equal to the local sound speed. The detonation wave is adjoined by a rarefaction wave a rarefaction wave, in which the gas velocity (relative to the tube) decreases monotonically to zero. The point where the velocity becomes zero is a weak discontinuity. Behind this discontinuity the gas is at rest.

Let us consider the detonation wave propagating from the open end of a tube. The pressure of the gas in front of the detonation wave must be equal to the original atmospheric pressure as it is outside of the tube, because any perturbations cannot run ahead of the detonation propagating with supersound speed. The velocity behind the detonation must decrease behind the detonation wave and the velocity must change the sign. If the gas velocity would be constant between the detonation wave and the open end of the tube, then the gas was being sucked into the open end of the tube from outside atmosphere. However, this is impossible, since the pressure behind the detonation wave increases and it is greater than the external pressure. For the same reasons as in the previous case, the detonation wave must be the Chapman-Jouguet detonation. The resulting flow behind the detonation should be as following. Immediately behind the detonation wave there is a rarefaction wave, in which the velocity decreases monotonically towards the open end of the tube and it changes sign at some point. This means that, in the end section of the tube, the gas moves towards the open end and it flows out of the open end. The speed with which the gas leaves the tube is equal to the local speed of sound, and the pressure in the flow behind the detonation is everywhere higher than the external pressure.

12.5 Possible Modes of Exothermal Reactions Propagation

Let us consider possible combustion modes and the number of boundary conditions required to satisfy the evolutionarity condition.[1] There are four possible combinations of speeds ahead and behind of the discontinuity surface. For supersonic waves – detonation:

$$u_0 > a_{s0},\ \ u_2 \le a_{s2}, \tag{12.5.1}$$

$$u_0 > a_{s0},\ \ u_2 > a_{s2}. \tag{12.5.2}$$

For subsonic wave - deflagration:

$$u_0 < a_{s0},\ \ u_2 < a_{s2}, \tag{12.5.3}$$

$$u_0 < a_{s0},\ \ u_2 > a_{s2}. \tag{12.5.4}$$

---

[1] The structure of detonation and deflagration waves, the evolutionarity conditions and realization of possible combustion regimes has been studied by M.A. Liberman and A.L. Velikovich [1], chaps. 1.7, 1.8, 1.9.

Three boundary conditions follow from continuous flows of mass, momentum and energy. The case (12.5.1) is similar to a shock wave and for the detonation mode (in fact this is "strong detonation") the number of boundary conditions and the conditions required for evolutionarity remain the same as for an ordinary shock wave. The case (12.5.2) corresponds to supersonic mode, but the evolutionarity condition in this case requires one additional "boundary" condition, since the velocity in the flow behind the wave is supersonic, which means that the propagation velocity of this mode is artificially kept higher than it would be for a freely propagating leading shock in a detonation wave. The subsonic mode $u_0 < a_{s0},\ u_2 < a_{s2}$ corresponds to deflagration. The evolutionarity condition in this case requires one additional boundary condition, which is the value of the propagation velocity of the deflagration wave. The evolutionarity of the subsonic mode $u_0 < a_{s0},\ u_2 > a_{s2}$ would require two additional conditions, and therefore such regime does not exist.

Writing the mass flux density (12.2.1) at the discontinuity surface as

$$j = \rho_0 u_0 = \rho_2 u_2, \tag{12.5.5}$$

we will present it using Eq. (12.2.2) through the pressures and densities of the gas on both sides of the discontinuity surface (here we will use formulas (12.2.1)–(12.2.3) with subscripts 0 instead of 1)

$$j^2 = (P_2 - P_0)/(1/\rho_0 - 1/\rho_2). \tag{12.5.6}$$

Vertical and horizontal lines coming from the initial state (P_0, V_0) intersect the adiabatic hyperbola in Fig. 12.3 at points O_1 and O_2. In this part of the adiabatic curve $P_2 - P_0 > 0$ and $(1/\rho_0 - 1/\rho_2) < 0$, so that the relation (12.5.) corresponds to imaginary values for j and u_0, therefore, in this part of the Hugoniot curve there are no solutions that have a physical sense.

The value of u_2 depends on the slope of the chord drawn from the initial point (P_0, V_0) to the adiabatic hyperbola. The flux j and velocities u_0, u_2 of the "weak detonation" increase from the CJ detonation point and tends to infinity at the point O_1. Modes of the "weak detonation" (between the CJ and O_1 points) require an additional "boundary" condition, which is a prescribed value of the weak detonation velocity. While the detonation velocity above CJ-detonation point is determined by the shock wave velocity, the propagation velocity of the "weak detonation" is determined, for example, by artificially forced ignition of a chemical reaction and depends on the thermodynamic characteristics and chemical kinetics of the gas mixture.

The part of the adiabatic curve below O_2 corresponds to a slow combustion - deflagration wave, where $u_0 < a_{s0}$ and $u_2 < a_{s2}$. In this case there are three simple waves outgoing from the discontinuity surface: sound waves from the both side of the discontinuity surface and entropy wave carried with the outgoing mass flux behind the wave front. One additional boundary condition is required for evolutionarity of this mode, that is the prescribed value of the flame velocity, which is defined by chemical kinetics and transport coefficients.

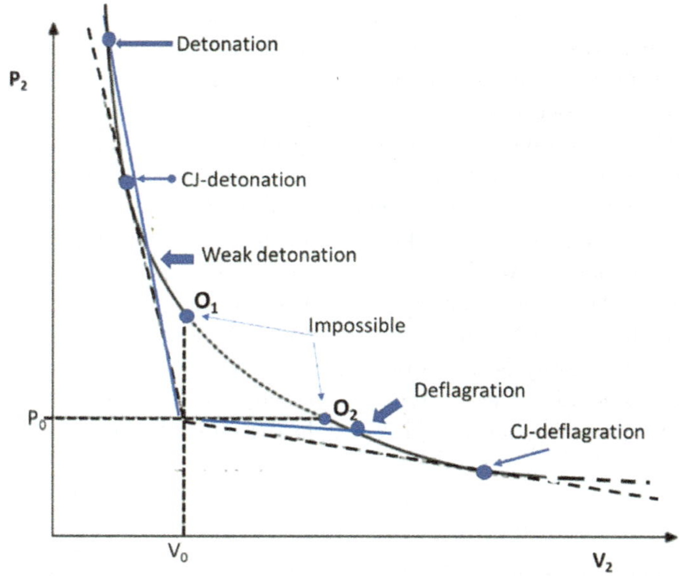

Fig. 12.3 Hugoniot adiabatic and possible combustion modes

At the point O_2 we have $P_0 = P_2$ and $j = 0$. The mass flux j increases towards the point CJ-deflagration ($u_2 = a_{s2}$), which corresponds to the maximum possible velocity of the deflagration mode. Below point O_2 inequalities, $u_0 > u_2$, $P_2 > P_0$, $P_1 > P_0$, $\rho_2 > \rho_0$ for the compression waves of supersonic discontinuities are reversed to: $u_0 < u_2$, $P_2 < P_0$, and $\rho_2 < \rho_0$. Below the CJ-deflagration point $u_0 < a_{s0}$, $u_2 > a_{s2}$, the evolutionarity conditions cannot be satisfied, which means that the solution does not exist (it is absolutely unstable).

It is difficult or even unlikely to achieve the CJ-deflagration for a freely propagating flame. On the contrary, a flame propagating in a tube with rough walls or obstacles is accelerated due to stretching of the flame front, turbulence produced by the rough walls, etc. and can reach the velocity of the CJ deflagration. However, since the CJ-deflagration is the maximum velocity of the deflagration mode, further acceleration of the flame would lead to an absolutely unstable combustion mode. As a result, the so-called choking regime will develop with the flame velocity oscillating near the CJ-deflagration speed. Depending on the particular condition, the next stage can be either a quasi-stationary fast deflagration or a transition from deflagration to a detonation regime.

12.6 CJ-Deflagration and CJ-Detonation

The problem of the transition from deflagration to detonation (DDT), which is important for safety and energy related issues, has been studied for many years using theoretical analyses, experiments, and extensive numerical simulations. Currently, there is a considerable amount of literature devoted to the topic of DDT. The question of how detonation can form and what is the physical mechanism of the transition from deflagration to detonation remains open until the mechanism of the formation of a strong shock wave inherent in detonation is understood. The problem of the deflagration-to-detonation transition (DDT), including experimental studies, numerical simulations, possible transition mechanisms, and influence of different physical processes on DDT, will be considered in Chap. 15. Here, we restrict ourselves by considering the dynamics of a deflagration wave propagating in a tube and those features that can affect the transition from a subsonic to a supersonic combustion mode.

For a flame propagating from the closed end of the tube with cross section S, the increase in volume occupied by the combustion products compared to the volume of fuel mixture burned by the flame is

$$\Theta S U_f - S U_f = S U_f (\Theta - 1), \tag{12.6.1}$$

where the expansion coefficient is $\Theta = T_b/T_u = \rho_u/\rho_b \approx 8 - 10$. The increase in the volume of combustion products produces a flow ahead of the flame with the velocity $(\Theta - 1)U_f$. In the laboratory system flame propagates with velocity ΘU_f, and the flame velocity relative to the flow of unburned fuel ahead of the flame is U_f. The fresh gas mixture far ahead of the flame front is at rest. All these conditions can be satisfied only if ahead of the flame there is a shock wave which involves gas in motion, as it is shown schematically in Fig. 12.4.

This is true as long as the flame velocity U_f is much less than the sound speed and the pressure is almost constant across the flame front. The formation of flow ahead of the flame and the flame acceleration is accompanied by a precursor shock wave ahead of the flame. We consider the development of CJ deflagration according to the schematic scenario shown in Fig. 12.4. The velocity, density and pressure behind the shock is defined by the conservation of the mass, momentum and energy fluxes

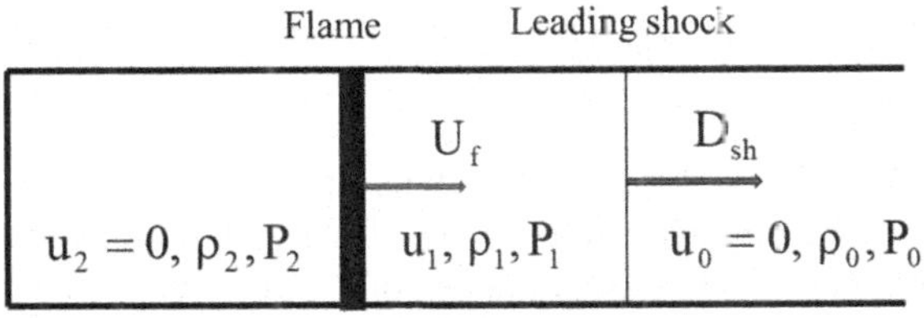

Fig. 12.4 Schematic of one-dimensional deflagration wave propagating from the closed end of a tube

$$\rho_0 D_{sh} = \rho_1 (D_{sh} - u_1), \tag{12.6.2}$$

$$P_0 + \rho_0 D_{sh}^2 = P_1 + \rho_1 (D_{sh} - u_1)^2, \tag{12.6.3}$$

$$\varepsilon_0 + P_0/\rho_0 + D_{sh}^2/2 = \varepsilon_1 + P_1/\rho_1 + (D_{sh} - u_1)^2/2. \tag{12.6.4}$$

Parameters of the flow ahead and behind the flame front are defined by similar relations

$$\rho_1 (U_f - u_1) = \rho_2 U_f, \tag{12.6.5}$$

$$P_1 + \rho_1 (U_f - u_1)^2 = P_2 + \rho_2 U_2^2, \tag{12.6.6}$$

$$\varepsilon_1 + P_1/\rho_1 + (U_f - u_1)^2/2 + Q = \varepsilon_2 + P_2/\rho_2 + U_f^2/2. \tag{12.6.7}$$

The designations of the quantities in Eqs. (12.6.2)–(12.6.6) are shown in Fig. 12.4. Enthalpy in Eqs. (12.6.2)–(12.6.7) was expressed via the internal energy $h = \varepsilon + P/\rho$ and Q is the heat of chemical reaction. Such a one-dimensional "double discontinuities" model was considered in [8–10] and it admits simple solutions.

For simplicity, we assume a perfect gas with the same adiabatic constant γ in all three regions 0, 1, 2, and a large chemical energy release, $Q \gg C_v T_1$, or in dimensionless units $q/a_{s1}^2 \gg 1$. Solutions for different values of adiabatic constant in regions 0, 1, 2 can be found in [9, 10], however, the qualitative results are not affected by accounting of different values of γ in regions 0, 1, 2. Since the flame propagates from the closed end, the boundary condition at the end of the tube is $u_2 = 0$, and obviously $u_2 = 0$ everywhere in the region 2 between the closed end and the flame front. Parameters of the flow in the region 1 can be calculated using corresponding formulas for a shock wave

$$u_1 = D_{sh}(1 - \rho_0/\rho_1), \tag{12.6.8}$$

$$\rho_1 = \rho_0 \frac{(\gamma + 1) M_0^2}{2 + (\gamma - 1) M_0^2}, \tag{12.6.9}$$

$$T_1 = T_0 \frac{[2 + (\gamma - 1) M_0^2][2\gamma M_0^2 - (\gamma - 1)]}{(\gamma + 1)^2 M_0^2}, \tag{12.6.10}$$

where $M_0 = D_{sh}/a_0$ is Mach number, and a_0 is the sound speed in the region 0.

Experimental studies of DDT show that the precursor shock is relatively weak, $M_0 \le 3$. Then, the parameters of the flow between the flame front and the precursor shock, in region 1, estimated using Eqs. (12.6.8)–(12.6.9) for $M_0 \sim 2 \div 3$ are:

$$u_1 = (2.1 - 2.22)a_{s0}, \quad \rho_1 = (3.33 - 3.86)\rho_0, \quad T_1 = (2.16 - 2.68)T_0. \quad (12.6.11)$$

The burning rate increases since the mixture entering the flame from the region 1 is compressed and heated by the precursor shock. This leads to the flame acceleration and the corresponding increase of the deflagration velocity U_f. Accelerating flame acts as a piston producing pressure waves which steepen into weak shocks propagating ahead of the flame and amplifying the leading precursor shock. However, even with account of this process, the amplification of the leading shock wave remains more or less insignificant, so that its intensity remains moderate, $M_0 \leq 3$.

In fact, we must keep in mind that the real problem is three-dimensional. At the initial stage, the expanding hot combustion products push the gas away from the end wall. This leads to establishing the flow of fresh mixture ahead of the flame along the channel walls involving the flame into the motion. Mechanisms of the flame acceleration, which propagates from the closed end in a tube with no-slip walls include: stretching of the flame front [11] within the boundary layer,[2] turbulence in case of rough walls or obstructed channels in addition to the increased temperature and the mixture density in the post-shocked flow.

It was previously shown that the velocity of CJ-deflagration is the maximum possible velocity of deflagration. This means that as soon as the deflagration velocity approaches the velocity of CJ deflagration, the deflagration wave either decays or becomes a chocking deflagration because solutions with $U_f > a_{s2}$ does not satisfy the evolutionarity condition and become absolutely unstable. This should lead to extinguishing and reignition of deflagration propagating at the velocity close to the value of the CJ deflagration speed, its subsequent quenching, reignition, etc. Such a periodic process of deflagration quenching and re-initiation in the vicinity of the CJ deflagration known as "chocking flame" has been observed in many experiments [12–14] as a double wave structure consisting of the leading shock wave and the reaction front lagging behind the leading shock. Earlier it was said that the boundary conditions of vanishing velocity at the closed end of the tube requires vanishing velocity everywhere behind the flame front in the region 2. For the CJ deflagration, which in general case is formed at rather large distance from the closed end, velocity drops to zero in the rarefaction wave propagating with velocity a_{s2} behind the flame. The experimental observations suggest that the formation of high-speed deflagrations in the choking regime depend mainly on the combustible mixture and the detailed of the flow structure. Typical experimental conditions for observing the chocking flame mode is a flame re-initiation after the detonation decays during its transmission [13, 14].

To understand the mechanism of the deflagration to detonation transition, it is necessary to find out how a strong shock wave is formed in the process of flame propagation. The formation of the chocking flame with near-sonic speed during the flame evolution can influence the final regime of combustion. The flame velocity

---

[2] Stretching of the flame front within the boundary layer as the main mechanism of flame acceleration has been pointed out for the first time by Ya. B. Zel'dovich in [11]. Unfortunately, this paper was published in Russian and remained unnoticed by researchers in western countries.

in the laboratory coordinate system, U_{Lf} equals to sum of the flame velocity, U_f relative to the flow in region 1, and the flow velocity, u_1 carrying the flame. When the accelerating flame approaches the CJ deflagration and the chocking flame mode is established, the flame velocity in the laboratory coordinate system (relative to the tube walls) is

$$U_{Lf} = u_1 + U_f \approx u_1 + a_{s2}. \qquad (12.6.12)$$

Assuming that the intensity of the leading shock, $M_0 \approx 3$, and taking into account that $a_{s2} \approx a_{s0}\sqrt{T_2/T_1} \approx 1.79a_{s0}$, we obtain from Eq. (12.6.12) $U_{Lf} \approx 3.9c_0$. Such a fast flame can catch the leading shock and its coupling with the shock wave can lead to the mutual shock-flame amplification similar to SWACER mechanism [15, 16]. Such a scenario of the shock-flame coupling and mutual amplification leading to the establishment of DDT was considered in [17]. The shock-flame amplification leading to DDT can occur before the transition to a chocking flame mode is achieved if the flame propagates at near-sonic speed and a shock wave is formed close to the flame front. This requires that the flame speed at some instant starts to increase as $U_f \propto t^n$ with $1 > n > 0$ [17]. This scenario of the deflagration-to-detonation transition was found to be in a good agreement with experimental studies of DDT in highly reactive hydrogen–oxygen and ethylene–oxygen mixtures [17].

The importance of understanding DDT mechanisms for industrial safety and for various engineering applications, e.g. micro-scale propulsion and power devices, has stimulated numerous numerical studies of DDT. One of the main difficulties in simulating multidimensional gas dynamic combustion problems is the need to use detailed models of chemical reactions, involving a large number of chemical reactions for many species. For this reason, many researchers have used simplified chemical models, either with reduced kinetics of chemical reactions, or with a simplified model in which the exothermic reaction proceeds through a one-step Arrhenius reaction. Numerous simulations of DDT with a one-step chemical models, for which the induction time inevitably by orders of magnitude less than the experimental value, led to the conclusion that the mechanism of DDT is always the formation of "hot spots" involving gradients of reactivity in nearby unreacted mixture [18, 19], which then produce a detonation through the Zel'dovich gradient mechanism [20, 21]. This was the major trend in DDT studies until it was shown experimentally [17, 22, 23] that for a stoichiometric hydrogen/oxygen and ethylene/air mixtures the temperature ahead the flame front remains too low to ignite exothermic reaction all the time prior to DDT. 2D and 3D numerical simulations of DDT in hydrogen/oxygen with detailed chemical models [17, 24–26] have shown that the mechanism of DDT is the coupling and mutual amplification of a weak shock formed close ahead to the flame front with the flame reaction zone, which is basically similar to the SWACER mechanism. The coupling of the flame with the shock wave becomes possible when during the flame acceleration the deflagration approaches the maximum speed of the CJ deflagration. In simulations on DDT in the obstructed channel filled with hydrogen/air mixture [27], the authors compared results of simulations obtained using a one-step chemical model with the simulations using a detailed chemical kinetics. It was found that a one-step chemical model can lead to detonation onset from a hot spot for the conditions of temperature much lower than the ignition threshold obtained from the experimental data. On the other hand, in simulations with a detailed chemical scheme the transition to detonation was always triggered inside the flame brush due to the flame-shock coupling according to the mechanism of DDT described above.

It was shown in [28–31] that the minimum scales of temperature gradients (hot spot sizes) that are sufficient to initiate detonation through the Zel'dovich gradient mechanism [21] obtained in calculations with simplified (one-step, two-step or four-step) chemical models are in orders of magnitude smaller than those calculated using detailed chemical models. This, apparently, is one of the reasons why simulations of flame propagation and DDT using a one-step chemical model cannot correctly describe the experiment. Another reason is the higher values for temperature and pressure obtained in simulations using a one-step model. For a one-step model an exothermic chemical reaction has no temperature threshold and starts at an arbitrarily low temperature, while in reality the exothermic reactions occur only at the temperature, that is higher than a certain threshold value - the crossover temperature.

12.7 Explosions and Spherically Expanding Detonation

A spherically expanding detonation from the point where it was ignited is an important model of an unconfined explosion. which for the first time was considered by Ya. B. Zel'dovich [32]. Direct initiation of detonation refers to the formation of a detonation in the decay of the strong blast wave produced by a concentration of a large amount of energy in a small volume. Similar to the problem of the flame and shock wave propagating in a tube, considered in the previous section, there is no characteristic values having dimension of length (we consider the detonation wave of zero thickness). This means that the problem is self-similar and all variables are functions of combination (r/t) only. For a spherically symmetric motion, velocity have only radial component $\mathbf{u} = (u_r(r, t), 0, 0)$, and the Euler equations together with conservation of entropy (S) equation in spherical coordinates are

$$\frac{\partial \rho}{\partial t} + \frac{\partial (u_r \rho)}{\partial t} + \frac{2u_r \rho}{r} = 0, \tag{12.7.1}$$

$$\frac{\partial u_r}{\partial t} + u_r \frac{\partial u_r}{\partial r} + \frac{1}{\rho}\frac{\partial P}{\partial r} = 0, \tag{12.7.2}$$

$$\frac{\partial S}{\partial t} + u_r \frac{\partial S}{\partial t} = 0. \tag{12.7.3}$$

Using the self-similar coordinate $\xi = r/t$, we can write Eqs. (12.7.1)–(12.7.3) in the form

$$(\xi - u_r)\frac{1}{\rho}\frac{d\rho}{d\xi} = \frac{du_r}{d\xi} + \frac{2u_r}{\xi}, \tag{12.7.4}$$

$$(\xi - u_r)\frac{du_r}{d\xi} = -\frac{1}{\rho}\frac{dP}{d\xi}, \tag{12.7.5}$$

$$(\xi - u_r)\frac{dS}{d\xi} = 0. \tag{12.7.6}$$

From the last equation steams that the flow is isentropic, e.g. $S = const$. Another possibility is $(\xi - u_r) = 0$, but this contradicts to Eqs. (12.7.4)–(12.7.5). For isentropic flow $P' = a_s^2\rho'$, where we used the prime to denote differentiation with respect to ξ. The Eqs. (12.7.4) and (12.7.5) with the help of relation $P' = a_s^2\rho'$ are reduced to

$$\left[\frac{(\xi - u_r)^2}{a_s^2} - 1\right]\frac{du_r}{d\xi} = \frac{2u_r}{\xi}, \tag{12.7.7}$$

$$(\xi - u_r)\frac{du_r}{d\xi} = \frac{a_s^2}{\rho}\frac{d\rho}{d\xi}. \tag{12.7.8}$$

These equations have no analytical solution, but it is easy to understand the qualitative property of the solution. The solution must consist of two spherical surfaces with the center at $r = 0$. The outer sphere is the front of the expanding detonation wave, and the inner sphere must be the surface of a weak discontinuity where velocity tends to zero, because the boundary condition at $r = 0$. From Eq. (12.7.7) steams that if $u_r = 0$, then $\xi = a_s$. This means that around the center $r = 0$ there is a region inside the inner sphere $\xi = a_s$ where $u_r = 0$. Moreover, this means that similar to the detonation wave propagating in a tube, the spherically expanding detonation must be also the Chapman-Jouguet detonation.

We will not consider here the dynamics of detonations, which include interesting phenomena of galloping detonations, the multi-dimensional and one-dimensional instabilities, as well as cellular structures of detonations. There are many excellent books and articles [33–38] that deal with the nonlinear dynamics of detonation fronts, pulsating and/or cellular structure of detonations, experimental, theoretical and numerical studies of detonation waves.

12.8 Strong Explosion in Homogenous Atmosphere

When the energy delivered in a small volume is large, compression wave became a shock wave after it left the location of the energy release. In this case thermal conductivity of the gas does not important and the problem is pure gasdynamic one. The idealized problem of strong explosion in a homogeneous atmosphere is a typical example of this kind of problem.

Let us consider an ideal gas of density ρ_0 with constant specific heat, in which a large amount of energy E was released in a small volume during a short time interval. As a result, a shock wave will be formed around location where energy was released, and it propagates in form of a strong spherically expanding shock wave from the location where the energy was released. We shall consider the process at the stage when the shock wave, while still remaining very strong, moved away from the explosion site at a distance much larger than the size of the region in which the energy was initially released. This means that we can neglect by the initial pressure of the gas in comparison with the pressure behind the shock wave, which is equivalent to neglecting the initial internal energy of the gas, which was involved in motion in comparison to the explosion energy. For such formulated problem, the energy can be considered to be instantaneously released at the point. Assumption of a very strong shock wave

$$\frac{P_2}{P_0} \gg \frac{\gamma+1}{\gamma-1}, \tag{12.8.1}$$

means that ratio of the densities in the shock wave is equal to its limiting value $\rho_2/\rho_0 = (\gamma+1)/(\gamma-1)$. For the problem formulated in this way, the gas motion is determined by only two parameters: the initial density of the surrounding atmosphere ρ_0 and the amount of energy E released during the explosion. From these two parameters and two independent variables, the time t and the radial coordinate r can be constructed only one dimensionless combination

$$\xi = r(\rho_0/Et^2)^{1/5}. \tag{12.8.2}$$

Therefore, such a motion is *self-similar* and we can claim that the shock wave at every moment of time must correspond to certain constant value of the self-similar co-ordinate ξ, so that the shock wave location R = R(t) is defined as

$$R(t) = \xi_0 \left(\frac{Et^2}{\rho_0}\right)^{1/5}, \tag{12.8.3}$$

where ξ_0 is some numerical constant, which can be found by solving equation of motion.

The propagation velocity of the shock wave in the laboratory coordinate system is

$$D = \frac{dR}{dt} = \frac{2R}{5t} = \xi_0 \frac{2}{5}\left(\frac{E}{\rho_0 t^3}\right)^{1/5} = \frac{2}{5}\xi_0^{5/2}\left(\frac{E}{\rho_0}\right)^{1/2} R^{-3/2}. \tag{12.8.4}$$

The density, pressure and velocity of the gas behind the shock wave can be expressed using formulas for a strong shock

$$\rho_2 = \rho_0 \frac{\gamma+1}{\gamma-1}, \quad P_2 = \frac{2}{\gamma+1}\rho_0 D^2, \quad u_2 = \frac{2}{\gamma+1} D. \tag{12.8.5}$$

As it is seen from (12.8.5) the pressure decreases with time as

$$P_2 \propto \rho_0 D^2 \propto \rho_0 \left(\frac{E}{\rho_0}\right)^{2/5} \propto \frac{E}{R^3}. \tag{12.8.6}$$

The fact that we neglected by the initial pressure of the gas means that we consider initial energy of the gas being negligibly small compared to the energy of the explosion. It is therefore obvious that in this approximation the total energy in the region bounded inside the sphere of shock wave remains constant and equal to the energy released in the explosion

$$E = \int_0^R 4\pi r^2 \rho(\varepsilon + u^2/2) dr = \int_0^R 4\pi r^2 dr \rho\left(\frac{a_s^2}{\gamma(\gamma-1)} + \frac{u^2}{2}\right) dr. \tag{12.8.7}$$

Equation (12.8.7) is the condition of the energy conservation, which can be used to determine unknown constant ξ_0 for exact solution.

To solve the problem, we need to solve equations for spherically symmetric adiabatic flow, similar to Eqs. (12.7.1)–(12.7.3) with boundary conditions at the surface of a strong shock wave

$$u_{r2} = \frac{2}{\gamma+1} D, \quad \rho_2 = \frac{\gamma+1}{\gamma-1}\rho_0, \quad P_2 = \frac{2}{\gamma+1}\rho_0 D^2. \tag{12.8.8}$$

Instead of exact solution, which can be obtained only numerically, we consider a simple model, which gives the solution with a good accuracy. Let us assume that the entire mass of the gas involved in the motion behind the shock wave is concentrated in a thin layer behind the shock front, and that the density of the layer is constant and equal to $\rho_2/\rho_0 = (\gamma+1)/(\gamma-1)$. The thickness of the layer is determined from conservation of mass

$$4\pi R^2 \Delta r \rho_2 = \frac{4\pi R^3}{3}\rho_0, \quad or \quad \Delta r = \frac{R}{3}\frac{\rho_0}{\rho_2} = \frac{R}{3}\frac{\gamma-1}{\gamma+1}. \tag{12.8.9}$$

For example, $\Delta r/R = 0.055$ for $\gamma = 1.4$, and $\Delta r/R = 0.0435$ for $\gamma = 1.3$.

We denote the pressure at the inner side of the layer P_C and assume that this is the fraction β of the pressure P_2 behind the shock wave, i.e. $P_c = \beta P_2$. We shall consider the layer as being infinitely thin with all the mass concentrated within the layer. Then, the equation of motion is the Newton law

$$\frac{d}{dt}(M u_2) = 4\pi R^2 P_c = 4\pi R^2 \beta P_2, \tag{12.8.10}$$

where $M = \frac{4}{3}\pi R^3 \rho_0$ is the total mass in the layer, which is some function of time.

The only one external force acts on the layer from inside, $4\pi R^2 P_c$, with P_c being the force per unit are, since the external pressure is assumed negligible. Expressing u_2 and P_2 in terms of the shock velocity $D = dR/dt$, and using (12.8.5), we obtain

$$\frac{1}{3}\frac{d}{dt}(R^3 D) = \beta D^2 R^2. \tag{12.8.11}$$

Using new independent variable, by noting that

$$\frac{d}{dt} = \frac{dR}{dt}\frac{d}{dR} = D\frac{d}{dR},$$

we obtain the equation

$$\frac{1}{3}D\frac{d}{dR}(R^3 D) = \beta D^2 R^2, \tag{12.8.12}$$

which being integrated, gives

$$D = C\, R^{-3(1-\beta)}. \tag{12.8.13}$$

The constants C and β are determined from the condition of energy conservation. Note that the pressure inside the sphere of radius R is equal to $P_c = \beta P_2$, which means that although we assumed that almost all the mass is concentrated in the layer, the space inside the sphere contains some small amount of the gas with the internal energy $E_{\text{int}} = \frac{1}{\gamma-1}\frac{4}{3}\pi R^3 P_c$. The kinetic energy of the gas is $E_{kin} = M u_2^2/2$. For the total energy, using $P_c = \alpha P_2$, expressing u_2 in terms of D, and substituting for D its expression (12.8.13) we obtain

$$E = \frac{4\pi}{3}\rho_0 C^2\left[\frac{2\beta}{\gamma^2-1} + \frac{2}{(\gamma+1)^2}\right] \cdot R^{3-6(1-\beta)}. \tag{12.8.14}$$

Since the energy released in the explosion is constant, the exponent of the variable R in (12.8.14) must be zero; that gives $\beta = 1/2$. Then, the Eq. (12.8.14) defines the constant C:

$$C = \left[\frac{3}{4\pi}\frac{(\gamma-1)(\gamma+1)^2}{3\gamma-1}\frac{E}{\rho_0}\right]^{1/2}. \tag{12.8.15}$$

Thus, with account of Eq. (12.8.13) we have the following relations

$$D \propto R^{-3/2},\ \ P_2 \propto R^{-3},\ \ u_2 \propto R^{-3/2},\ \ R \propto t^{2/5}. \tag{12.8.16}$$

The coefficients in these relations can be expressed using (12.8.15). For example

$$R = \left[\frac{75}{16\pi} \frac{(\gamma - 1)(\gamma + 1)^2}{3\gamma - 1} \frac{E}{\rho_0} \right]^{1/5} t^{2/5} = \xi_0 \left(\frac{E}{\rho_0} \right)^{1/5} t^{2/5}. \quad (12.8.17)$$

The approximation we used here gives the results, which are very close to the exact solution of the problem. For example, the numerical values of ξ_0 for the shock wave propagation are 1.033 and 0.89 for $\gamma = 1.4$ and $\gamma = 1.2$, for exact solution, and they are 1.014 and 0.89 for the approximate solution, correspondingly.

References

1. M. Liberman, A. Velikovich, *Physics of Shock Waves in Gases and Plasmas* (Springer-Verlag, Berlin—New York, 1985)
2. Y.B. Zel'dovich, On the theory of propagation of detonation in gaseous systems. Zh. Eksp. Teor. Fiz.10 (1940) pp. 542–568. (English translation: NACA TM 1261, 1950)
3. J. von Neumann, Progress report on "Theory of detonation waves". OSRD. Rept. No. 549, National Defense Research Committee Div. B.1942. Reprinted in The Collected Works of John von Neumann, A. J. Taub (ed.), Pergamon (1963)
4. W. Döring, On detonation processes in gases. Ann. Phys. **43**, 421–436 (1943)
5. D.L. Chapman, On the rate of explosion in gases. The London, Edinburgh, and Dublin Philosophical Magazine J. Sci. **47**, 90–104 (1899)
6. E. Jouguet, Sur la propagation des réactions chimiques dans les gaz [On the propagation of chemical reactions in gases]. J. de mathematiques Pures et Appliquees **1**, 347–425 (1905). ((in French))
7. E. Jouguet, Sur la propagation des réactions chimiques dans les gaz [On the propagation of chemical reactions in gases]. Journal de Mathématiques Pures et Appliquées, series 6 (in French) **2**, 5–85 (1906)
8. A.K. Oppenheim, Gasdynamic analysis of the development of gaseous detonation and its hydraulic analogy. 4th Symp. Int. Combust. Symp. Proc. Combust Inst. **4**, 471–480 (1953)
9. K.I. Shchelkin, K. Ya. Troshin, Gasdynamics of combustion. Izd. Akad. Nauk, Moscow, 1964 (English translation: National Aeronautics and Space Administration, Washington, DC, 1964)
10. R.S. Chue, J.F. Clarke, J.H. Lee, Chapman-Jouguet deflagrations. Proc. Roy. Soc. Lond. A **441**, 607–623 (1993)
11. B. Ya. Zel'dovich, On the theory of transition to detonation in gases. Zh. Tekh. Fiz. **17**, 3–26 (1947) (In Russian)
12. V.I. Alekseev, M.S. Kuznetsov, Yu.G. Yankin, S.B. Dorofeev, Experimental study of flame acceleration and the deflagration-to-detonation transition under conditions of transverse venting. J. Loss Prev. Process Ind. **14**, 591–596 (2001)
13. S. Khomik, B. Veyssiere, S. Medvedev, V. Montassier, H. Olivier, Limits and mechanism of detonation re-initiation behind a multi-orifice plate. Shock Waves **22**, 199–205 (2012)
14. M. Kellenberger, G. Ciccarelli, Propagation mechanisms of supersonic combustion waves. Proc. Combust. Inst. **35**, 2109–2116 (2015)
15. J.H.S. Lee, R. Knustautas, N. Yoshikawa, Photochemical initiation of gaseous detonations. Acta Astron. **5**, 971–982 (1978)
16. J.H. Lee, I.O. Moen, The mechanism of transition from deflagration to detonation in vapor cloud explosions. Prog. Energy Combust. Sci. **6**, 359–389 (1980)
17. M.A. Liberman, M.F. Ivanov, A.D. Kiverin, M.S. Kuznetsov, A.A. Chukalovsky, T.V. Rakhimova, Deflagration-to-detonation transition in highly reactive combustible mixtures. Acta Astronaut. **67**, 688–701 (2010)
18. E.S. Oran, V.N. Gamezo, Origins of the deflagration-to-detonation transition in gas-phase combustion. Combust. Flame **148**, 4–47 (2007)

19. R.W. Houim, A. Ozgen, E.S. Oran, The role of spontaneous waves in the deflagration-to-detonation transition in submillimetre channels. Combust. Theory Model. **20**, 1068–1087 (2016)
20. B. Ya. Zel'dovich, V.B. Librovich, G.M. Makhviladze, G.I. Sivashinsky, On the de- velopment of detonation in a non-uniformly preheated gas. Acta Astron. **15**, 313–321 (1970)
21. B. Ya. Zel'dovich, Regime classification of an exothermic reaction with nonuniform initial conditions, Combust. Flame **39**, 211–214 (1980)
22. M. Kuznetsov, M. Liberman, I. Matsukov, Experimental study of the preheat zone formation and deflagration to detonation transition. Combust. Sci. Technol. **182**, 1628–1644 (2010)
23. M. Kuznetsov, V. Alekseev, I. Matsukov, S. Dorofeev, DDT in a smooth tube filled with a hydrogen-oxygen mixture. Shock Waves **14**, 205–215 (2005)
24. M. Ivanov, A. Kiverin, M. Liberman, Flame acceleration and DDT of hydro- gen-oxygen gaseous mixtures in channels with no-slip walls. Int. J. Hydrogen Energy **36**, 7714–7727 (2011)
25. M.F. Ivanov, A.D. Kiverin, M.A. Liberman, Hydrogen-oxygen flame acceleration and transition to detonation in channels with no-slip walls for a detailed chemical reaction model, Phys. Rev. E **83**, 056313 (2011)
26. M.F. Ivanov, A.D. Kiverin, I.S. Yakovenko, M.A. Liberman, Hydrogen-oxygen flame acceleration and deflagration-to-detonation transition in three-dimensional rectangular channels with no-slip walls. Int. J. Hydrogen Energy **38**, 16427–16440 (2013)
27. O. Dounia, O. Vermorel, A. Misdariis, T. Poinsot, Influence of kinetics on DDT simulations. Combust. Flame **200**, 1–14 (2019)
28. M. Liberman, A. Kiverin, M. Ivanov, On detonation initiation by a temperature gradient for a detailed chemical reaction models. Phys. Lett. A **375**, 1803–1808 (2011)
29. M.A. Liberman, A.D. Kiverin, M.F. Ivanov, Regimes of chemical reaction waves initiated by nonuniform initial conditions for detailed chemical reaction models. Phys. Rev. E **85**, 056312 (2012)
30. M. Liberman, C. Wang, C. Qian, J. Liu, Influence of chemical kinetics on spontaneous waves and detonation initiation in highly reactive and low reactive mixtures. Combust. Theor. Model. **23**, 467–495 (2019)
31. C. Wang, C. Qian, J. Liu, M.A. Liberman, Influence of chemical kinetics on detonation initiating by temperature gradients in methane/air. Combust. Flame **197**, 400–415 (2018)
32. B. Ya. Zel'dovich, On the distribution of pressure and velocity in the products of a detonation explosion, particularly with a spherical propagation of the detonation wave, Zh. Éksp. Teor. Fiz. **12**(9), 389–406 (in Russian) (1942)
33. P. Clavin, Theory of gaseous detonations. Chaos **14**, 825–838 (2004)
34. L. He, J.H.S. Lee, The dynamical limit of one-dimensional detonations. Phys. Fluids **7**, 1151–1158 (1995)
35. L. He, Theoretical determination of the critical conditions for the direct initiation of gaseous detonations in hydrogen-oxygen mixtures. Combust. Flame **104**, 401–418 (1996)
36. H.D. Ng, M.I. Radulescu, A.J. Higgins, N. Nikiforakis, J.H.S. Lee, Numerical investigation of the instability for one dimensional Chapman-Jouguet detonations with chain-branching kinetics. Combust. Theor. Model. **9**, 385–401 (2005)
37. Y.B. Zel'dovich, A.S. Kompaneets, *Theory of Detonation* (Academic Press, New York, 1960)
38. John H. S. Lee, The Detonation Phenomenon (Cambridge University Press, 2010). https://doi.org/10.1017/CBO9780511754708

Chapter 13
Ignition

Ignition processes are important both in the technology of optimal design of thermal devices and in safety issues and prevention of accidental explosions. Usually, two different scenarios are distinguished: spontaneous or autoignition and forced ignition. Autoignition occurs through the self-heating of the reactants, whereas forced ignition occurs with the assistance of an ignition source. For example, in diesel and homogeneous charge compression ignition (HCCI) engines, fuel is injected into a hot mixture and autoignition occurs during the compression stroke when the temperature reaches a certain value. In spark-ignition (SI) engines, the forced ignition is initiated by an electric spark producing a hot spot of the initial flame kernel. On the contrary, the unwanted knock in SI engines is due to the autoignition of the unburned air–fuel mixture.

If the energy of an electric spark released in a finite volume of reactive gas (hot spot) is sufficient to create an initial flame core that can grow, then combustion will propagate as a self-sustaining flame. The minimum spark energy that can lead to successful ignition is defined as the minimum ignition energy. There have been many studies of spark ignition; the dependence of the minimum ignition energy on the rate of transient thermal energy release, hot spot size, fuel mixture composition, and initial pressures. Earlier experimental studies and the relationship between minimum ignition energy and extinguishing distance are presented in the book by Lewis and von Elbe "Combustion, Flame and Explosion of Gases" [1].

13.1 Spark Ignition and Minimum Ignition Energy

The ignition source, which is usually a heat source, raises the temperature and initiates the reaction. A heat source such as an electric spark can also produce free radicals that can act as chain carriers in the chemical reaction. The heat flux, as well as the flux of chain carriers from the ignition source, initiates a chemical reaction in an adjacent layer of the combustible mixture, so that this layer itself becomes a

M. A. Liberman, *Combustion Physics*,
https://doi.org/10.1007/978-3-030-85139-2_13

source of heat and/or chain carriers capable of initiating chemical reaction in the next layer, and so on. Despite the qualitative agreement of the predicted minimum ignition energy calculated using simplified chemical schemes, there was a large discrepancy between such calculations and experimentally measured values. Sloane and Ronney [2] showed that the ignition energy for methane–air, computed using a one-step model, differs by two orders of magnitude from the experimentally measured value, but when a detailed chemical mechanism is used, the predicted minimum ignition energies are close to the experimental values. Ignition processes in the hydrogen–oxygen system were studied by Warnatz and Maas [3] using a 1D numerical model with a detailed reaction mechanism and a multispecies transport model. Spark ignition of hydrogen–air mixtures for SI engines was investigated in [4] using a 2D model with detailed chemistry and molecular transport in the gas phase as well as heat conduction to electrodes.

Let us consider uniform initial conditions and the transient external source of energy localized on the scale of "hot spot" $0 \leq x \leq L$, where the amount of energy Q_{ig} is added during the time Δt_Q. Gas dynamics of the explosion initiated by the localized energy deposition is characterized by the interrelationship between the rate of energy transfer with the time characterizing energy deposition Δt_Q in the hot spot of size L and the characteristic times of the problem. For simplicity, we assume that the rate of energy addition is linear in time, so that the total energy deposition into the hot spot is equal to $Q_{ig} = W \Delta t_Q$, where W is the power of the external source of energy, and Δt_Q is the transient heating time of the hot spot. The characteristic acoustic time $t_a = L/a_s(T)$ determines the setting of the accompanying motion in the gas mixture, where $a_s(T)$ is the speed of sound. If $\Delta t_Q \ll t_a$, local heat addition occurs as a constant-volume process and the temperature elevation within the hot spot is accompanied by a concomitant increase in pressure. Subsequent expansion of the hot spot driven by the large pressure gradient between the hot spot and the ambient gas causes compression and shock waves in the surrounding ambient gas. When $\Delta t_Q \gg t_a$, the acoustic waves have enough time to equalize the pressure and the local heat addition occurs as an adiabatic process at nearly constant pressure. The possibility for ignition to occur depends on the competition between the ability to initiate a chemical reaction, which can be measured in terms of the homogeneous ignition (induction) time $\tau_r(T)$, and the rate of heat losses resulting in a decrease in temperature and induction time. Depending on the dominant mechanism, two ignition regimes can be distinguished: a slow diffusive regime and a fast non-diffusive one. Typically, the Damköhler number $Da \gg 1$ since diffusion is slow and the reaction can be considered as an instantaneous process. The diffusive regimes were studied in [5] and the non-diffusive regimes were studied in [6]. In both studies [5] and [6], the chemical reaction was modeled by a one-step Arrhenius reaction with large activation energy. It was shown that there is a critical value of Damköhler numbers for thermal runaway.

There are four characteristic time scales: the induction time $\tau_{\text{ind}}(T)$ characterizing the time scale of the reaction development, the acoustic time $t_a = L/a_s$, the time characterizing energy deposition into a hot spot, Δt_Q, and the time $t_T \sim L^2/\chi \sim L^2/\ell a_s$ characterizing heat losses from the hot spot. First, we notice that the ratio

t_a/t_T is the Knudsen number $Kn = \ell/L \ll 1$, which is the ratio of the mean free path ℓ to the hot spot size L. The condition for the applicability of the equations of hydrodynamics is $Kn = \ell/L \ll 1$, and in what follows, we will assume that this condition is satisfied, so that $t_a \ll t_T$.

The local heating of the gas mixture at the hot spot extends to the surrounding gas. For the 1D problem, the temperature distribution is given by the well-known solution of the thermal conduction equation

$$T(x,t) \propto \frac{1}{2\sqrt{\pi\chi t}} \exp(-x^2/4\chi t) \tag{13.1.1}$$

where $\chi = \kappa(T)/\rho C_P$ is the thermal diffusivity coefficient, which is assumed constant. From Eq. (13.1.1), it follows that heat propagates over a distance $x \sim \sqrt{4\chi t}$ and the temperature in the surrounding gas increases noticeably due to the propagation of a thermal wave during time

$$t_T \sim x^2/\chi. \tag{13.1.2}$$

At the very beginning, the thermal wave propagates with the velocity $dx/dt \sim \sqrt{\chi/t}$ overtaking a shock wave or pressure wave, which were created by the energy deposition and propagate with the velocity of about sound speed. However, very soon the shock wave overtakes the thermal wave, so that the characteristic time of thermal wave time is effectively much longer than the acoustic time. For example, in the hydrogen–oxygen mixture, at initial pressure $P_0 = 1$ atm, for the hot spot of size $\delta x \simeq 1$ mm, we have $t_T \sim (\delta x)^2/\chi \approx 0.1$ sec and $t_a \sim \delta x/a_s \approx 2\,\mu$s. Whether a chemical reaction will start and what processes determine the regime of combustion wave depend on the interrelationship between Δt_Q, t_a, t_T, and the induction time $\tau_{\text{ind}}(T)$ at the given temperature and pressure. The induction time is the time scale for the stage of endothermic chain initiation and branching reactions (in the case of a global one-step reaction, this is the time scale determined by the achievement of the maximum reaction rate).

A simple estimate for the ignition of a deflagration can be obtained using the ZFK theory [7, 8]. We consider the spark as a point heat source and assume that Δt_Q is the smallest time scale. The spherically symmetric distribution of the heat is obtained from the equation of thermal conduction

$$\frac{\partial T}{\partial t} = \chi \Delta T, \tag{13.1.3}$$

with the boundary conditions

$$\frac{\partial T}{\partial r} = 0 \text{ at, } r = 0, \; T(r = \infty) = T_0, \tag{13.1.4}$$

and the conservation for the deposited amount of energy Q is

$$Q = 4\pi\rho C_p \int_0^\infty (T - T_0)r^2 dr. \tag{13.1.5}$$

The solution of Eq. (13.1.3) is

$$(T - T_0) = \frac{Q}{\left(C_p\rho 4\pi\chi t^{3/2}\right)^{3/2}} \exp\left(-r^2/4\chi t\right). \tag{13.1.6}$$

Obviously, the maximum temperature is at $r = 0$. For a one-step Arrhenius reaction with activation energy $E \gg RT$, the Frank-Kamenetskii temperature $\Delta T = RT_{\max}^2/E$ determines the order of magnitude of the temperature change, which leads to the reaction time changing by a factor of order unity. Evaluating the characteristic cooling time as the time of changing $T_{\max}$ by a small value of the Frank-Kamenetskii temperature $RT_{\max}^2/E$ and taking the amount of deposited energy needed to heat uniformly a spherical volume of radius r_f from T_0 to T_f as

$$Q = \frac{4}{3}\pi r_f^3 \rho C_p (T_f - T_0), \tag{13.1.7}$$

we obtain an estimate for the cooling time t_T

$$\tau_T \sim \frac{\Delta T}{6\pi\chi(T_f - T_0)}\left(\frac{Q}{\rho C_p(T_f - T_0)}\right)^{2/3}. \tag{13.1.8}$$

For ignition to occur, the cooling time t_T must be longer than the characteristic time of the chemical reaction in a laminar flame, which in the ZFK theory can be estimated as

$$\tau_r \sim \frac{2\Delta T}{(T_f - T_0)}(\chi/U_f)^2. \tag{13.1.9}$$

From Eqs. (13.1.8) and (13.1.9), the condition for the successful ignition, $\tau_r > t_T$, becomes [7, 8]

$$r_f > 3.7L_f, \tag{13.1.10}$$

where $L_f = \chi/U_f$ is the width and U_f is the normal velocity of a laminar flame. This result means that for ignition to occur, a hot spot several times larger than the width of the laminar flame must be heated to the adiabatic temperature of the flame. This criterion is closely related to the condition of a flame quenching. Qualitatively, quenching of the flame occurs if heat losses to the surrounding walls exceed the heat flux in the direction normal to the flame front, which is necessary for flame propagation. For example, a flame can propagate in a tube only if the tube width is several times wider than the thickness of the flame front.

13.2 Zel'dovich's Gradient Mechanism; Spontaneous Reaction Wave

The various modes of the self-sustaining reaction wave discussed in previous chapters propagate either as subsonic (deflagration) or supersonic (detonation) waves with a well-defined speed, which is determined either by the diffusion of heat or reagents or by an increase in the reaction rate due to strong compression and heating of the mixture in the shock wave. These combustion modes represent a thin layer of the reaction wave propagating in a medium that is assumed to be uniform compared to the large gradients within the wave front. This approach is based on a significant difference between the time scale of chemical reactions and characteristic hydrodynamic time scales. Let the reaction occur in a vessel of characteristic size L. If the reaction wave propagates in deflagration regime, then the characteristic reaction time is $\tau_R = L/U_f$, and the characteristic hydrodynamic time scale is $\tau_a = L/a_s$. Since the flame velocity is much less than the sound speed, also holds the inequality $\tau_a \ll \tau_R$. In this case, sound waves equalize the pressure over the whole volume, and the process is isobaric with accuracy $\Delta P/P \propto M^2 \ll 1$. Accordingly, the movement of matter and the associated hydrodynamic effects are insignificant. The opposite limiting case corresponds to the self-ignition of a homogeneous reacting pocket, which is a thermal explosion. The maximum pressure jump in a thermal explosion is of the order of $P_2/P_0 \approx T_b/T_0 \sim 10$, so that the shock waves generated by the thermal explosion are relatively weak to initiate detonation ($P_2/P_0 \sim 10$ corresponds to the shock wave intensity $M_{sh} \sim 2$).

In most practical cases, ignition occurs in a small area of the combustible mixture that is locally heated by an electric spark, hot wire, focused laser light, etc. Such a local release of energy leads to the formation of an initial nonuniform distribution of temperature (or concentration of reagents), which determines the further development of the reaction depending on the reactivity of the mixture and the initial pressure. Already from the previous discussion, it can be concluded that a much larger amount of thermal energy is required to ignite a detonation wave. Since the width of the reaction zone in the detonation wave is of the order of $\tau_R a_s$, and it is about a_s/U_f times larger than the width of the deflagration reaction zone, it can be expected that the size of the hot spot required for ignition a detonation is approximately a_s/U_f times greater than the size of the hot spot required to ignite a deflagration.

The question of what combustion modes a hot spot can ignite remained open until Zel'dovich et al. [9] demonstrated using numerical simulations that a hot spot with a sufficiently shallow temperature gradient can initiate a detonation. The concept of the spontaneous reaction wave propagating through the reactive mixture along a spatial gradient of reactivity, which Zel'dovich introduced in [10], is of great fundamental and practical importance since it opened an avenue to study ignition of different regimes of the reaction wave initiated by the initial nonuniformity in temperature or reactivity caused by the local energy release.

For simplicity, we consider an Arrhenius type with the reaction time depending on the initial fuel temperature as $\tau_R \equiv \tau_{\text{ind}} = \tau_0 \cdot \exp(E/T)$, where we use energy units and assume that activation energy is large, $E/T \gg 1$, and τ_0 is a constant in the dimension of time. When the reaction starts, the concentration of the fuel changes slightly, and the energy release is small. The reaction is completed as an abrupt explosion within a short time interval of about $\tau_{\text{exp}} \simeq (T/E)\,\tau_{\text{ind}}$. In a region with a nonuniform temperature distribution, the reaction begins at the point of the minimum ignition delay time, where the temperature is maximal. The adjacent portion of the mixture ignites at its prescribed (longer) induction time, so that the ignition propagates in the direction of the diminishing temperature gradient by spontaneous autoignition in adjacent locations where the autoignition delay time $\tau_{\text{ind}}(T)$ is longer. The basic idea of the Zel'dovich concept is that the front of spontaneous autoignition propagates relative to the unburned gas at the velocity, which is inversely proportional to the gradient of the induction (autoignition) time [10]:

$$U_{sp} = |(d\tau_{\text{ind}}/dx)|^{-1} = \left|(\partial\tau_{\text{ind}}/\partial T)^{-1}(\partial T/\partial x)^{-1}\right|. \tag{13.2.1}$$

Since there is no causal link between successive autoignitions, there is no restriction on the value of U_{sp}, which depends only on the steepness of temperature gradient $\partial T/\partial x$ and the value of $\partial\tau_{\text{ind}}/\partial T$. It is obvious that a very steep gradient (hot wall) ignites a deflagration mode (flame), while a zero gradient corresponds to the ignition of the uniform thermal explosion, which occurs in the induction time. It was recognized by Zel'dovich [10] that there is a critical value of the temperature gradient, $(\partial T/\partial x)_{cr}$, such that the autoignition front moves into the unburned mixture at the velocity, which is about the local sound speed, a_s. In this case, the pressure wave generated by the heat release in chemical reactions can couple with the autoignition reaction front, with mutual amplification of both fronts resulting in the development of a detonation. The critical slope of the temperature gradient of a hot spot for such a resonant mutual amplification yields from Eq. (13.2.1)

$$\left(\frac{\partial T}{\partial x}\right)_{cr} = 1/[a_s(\partial\tau_{\text{ind}}/\partial T)]. \tag{13.2.2}$$

The reaction region can be considered as a wave with a well-defined front if the thickness of the reaction zone $\Delta_{sp} = u_{sp}\tau_R$ is small compared to the scale of the temperature change $L_T \approx T/|\nabla T|$. This condition always holds for the large activation energy, except the region of the temperature maximum, or if $\nabla T = 0$. Since

$$\Delta_{sp} = U_{sp}\tau_{\text{ind}} = \frac{T}{E}\frac{T}{|\nabla T|} \approx \frac{T}{E}L_T,\ \Delta_{sp} \approx (T/E)L_T \ll L_T. \tag{13.2.3}$$

For a nearly uniform temperature distribution corresponding to a very small temperature gradient, when the width of the spontaneous reaction wave is very large and the velocity is much larger than the local acoustic velocity, a nearly constant-volume thermal explosion takes place. In the opposite limiting case, when the velocity of the spontaneous reaction is much smaller than the local acoustic velocity, which corresponds to a large temperature gradient, the explosion is almost isobaric, which corresponds to a deflagration regime of reaction wave propagation. For an intermediate temperature gradient, a detonation wave can be initiated due to the initial temperature gradient if the generated shock wave is synchronized with the energy released by the self-ignition, which occurs at different times at each place.

Zel'dovich [10] also identified five modes of reaction initiated by the initial gradient of the induction time depending on the interrelation of U_{sp}, the laminar flame speed U_f, the sound speed a_s, and the Chapman–Jouguet detonation speed D_{CJ}:

i. $\partial T/\partial x = 0$, the constant-volume thermal explosion;
ii. $U_{sp} > D_{CJ}$, the spontaneous wave transition in a weak detonation.
iii. $D_{CJ} \geq U_{sp} > a_s$, the reaction wave couples with a shock wave with a transition to the CJ detonation;
iv. $D_{CJ} > a_s > U_{sp} > U_f$, shock wave followed by a fast deflagration;
v. $U_{sp} < U_f$, the spontaneous wave transition in a deflagration.

Obviously, formula (13.2.2) can be considered only as an approximate condition of the critical slope of the temperature gradient. During the induction period, both the initial temperature gradient and the corresponding distribution of $\tau_{\text{ind}}(T)$ are changing due to the thermal conduction and diffusion of species.

The temperature-gradient model [9, 10] received considerable attention not only as a fundamental concept in the combustion theory, but also as a possible mechanism explaining such processes as, for example, the deflagration-to-detonation transition or the mechanism explaining the knock occurrence in SI engines. Although the first numerical study in [9] used a low resolution, presumably the highest available that time, nevertheless, the authors have shown that for sufficiently shallow temperature gradients, a fast-reactive wave decelerated and coupled with the shock wave to form a detonation, but for larger temperature gradients, the reaction wave and the shock failed to couple together.

Numerous computational and analytical studies have been carried out over several decades, mainly focusing on the conditions leading to detonation initiated by a hot spot with a reactivity gradient [11–33]. The approach used in these studies was based on a simple mathematical model of an ideal gas undergoing a one-step exothermic Arrhenius reaction in a planar configuration with the reactivity gradient being modeled by the linear temperature gradient in the initial hot spot. The activation energy was assumed large, in view of the observed strong temperature sensitivity of the reaction rate. Subsequent studies provided more detailed information on the initiation of detonation by a temperature gradient with the main purpose of finding the critical conditions for the steepness of the temperature gradient (hot spot size) and the survival of detonation after it has left the hot spot and propagated into the

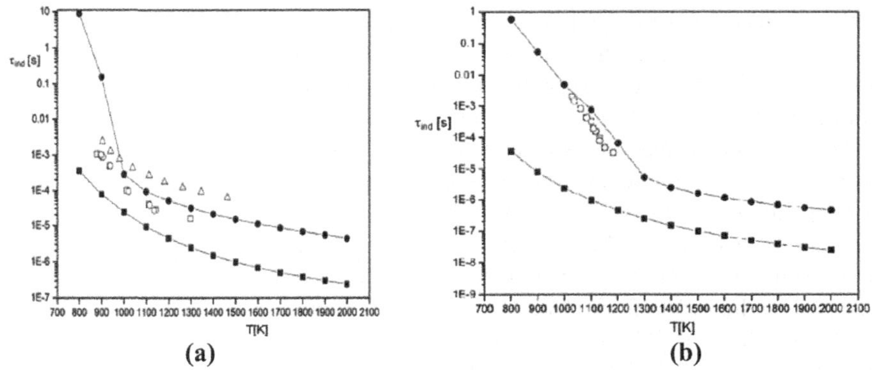

Fig. 13.1 **a** Induction times for stoichiometric hydrogen–air mixture at 1 atm: (■)—one-step model; (•)—detailed model; symbols are experimentally measured induction time. **b** Induction times for stoichiometric hydrogen–air mixture at 10 atm: (■)—one-step model; (•)—detailed model; symbols are experimentally measured induction time

cold ambient gas. In particular, He and Clavin [17–19] examined the phenomenon of spontaneous quenching of the developing detonation depending on the critical size of the hot spot and the temperature of the ambient fresh mixture outside the hot spot. Later it became obvious that the one-step chemical models cannot adequately describe combustible mixtures governed by chain-branching reaction [20], and some authors [21–23] used more sophisticated two-step and three-step models, which, to some extent, are believed to imitate the chain-branching kinetics with simplified reaction schemes between a set of pseudo-species. In particular, the role of chain-branching crossover temperature in the shock-induced ignition was studied using numerical simulations in [23] and it was found that the results obtained using two-step and three-step models are qualitatively different from that obtained in simulations with a one-step model. However, the quantitative difference between one-step and detailed chemical models, important for practice and industrial application, remained unclear. One of the first studies of the critical conditions for the development of detonation initiated by a hot spot and the classification of possible combustion modes was carried out in [31] using detailed chemical kinetic schemes for the autoignition of stoichiometric H_2–CO–air and H_2–air mixtures under high pressures. Since the tendency of knock and super-knock is the main limitation for the development of internal combustion engines with higher thermal efficiency and lower fuel consumption, there is extensive research on the autoignition mode due to its close relationship with knocking combustion [32–43]. Bradley and coworkers [32, 33] proposed an operational peninsula, which consists of two dimensionless parameters: the normalized temperature gradient and the ratio of the acoustic time to the excitation time of the chemical energy release for the knock and super-knock occurrence due to the end gas autoignition and the development of detonation.

Figure 13.1a, b shows the induction times for the hydrogen–air at pressures 1 atm and 10 atm as a function of temperature, calculated using one-step and detailed chemical models [44]. A single-step chemistry approach suggests that the complex set of reactions can be modeled by a one-step Arrhenius reaction. The induction time for a one-step model in Fig. was calculated for the one-step model widely used for 2D simulations of the hydrogen/air flame acceleration and DDT in smooth and obstructed channels [45]. The detailed mechanism chosen to model the hydrogen–air chemistry is the mechanism developed by Kéromnès et al. [44], which consists of 19 reactions and 9 species. An advantage of this mechanism is that it was extensively validated over a large number of experimental conditions, and it shows an excellent agreement between the modeling and experimental measurements of the flame velocity–pressure dependence over a wide range of pressures and equivalence ratios.

An advantage of this mechanism is that it has been extensively validated for various experimental conditions, and has shown an excellent agreement between numerical simulations and experimental measurements of the flame velocity–pressure dependence over a wide range of pressures and equivalence ratios.

According to Eq. (13.2.1), the velocity of spontaneous reaction wave depends on $(\partial \tau_{\text{ind}}/\partial T)$ rather than on τ_{ind}. Figure 13.2 shows the value of $(\partial \tau_{\text{ind}}/\partial T)$ as a function of temperature calculated using one-step models and detailed chemical models for hydrogen–air. It is seen that the induction times and the temperature derivations of the induction time calculated using a detailed chemical model differ from the corresponding values calculated using a single-step model by two–four orders of magnitude [46].

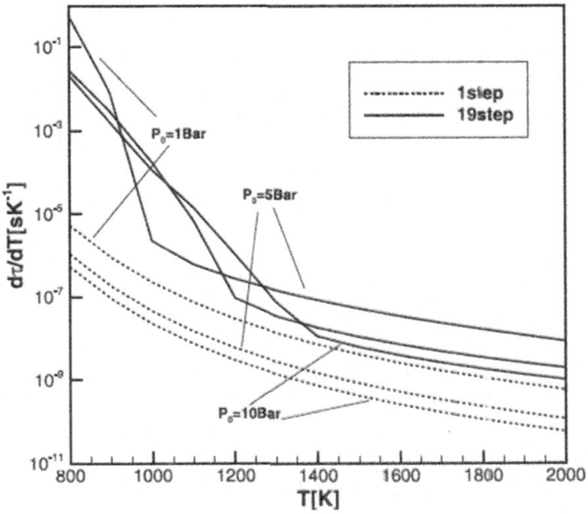

Fig. 13.2 $(\partial \tau_{\text{ind}}/\partial T)$ for H_2/air at initial pressure 1, 5, and 10 atm calculated for detailed (solid lines) and one-step (dashed lines) models

13.3 Combustion Modes Initiated by Hot Spots

We consider uniform initial conditions apart from a linear temperature gradient within a hot spot. The model of the linear temperature gradient is convenient for analysis and it has been widely used in many studies. The initial conditions at $t = 0$, prior to ignition, are constant pressure and zero velocity of the unburned mixture. At the left boundary at $x = 0$, the conditions are for a solid reflecting wall, where $u(0, t) = 0$ and the initial temperature $T = T*$ exceeds the ignition threshold value. Thus, the initial conditions are quiescent and uniform, except for a linear gradient in temperature (and hence density):

$$T(x, 0) = T * -(T * -T_0)(x/L),\ 0 \leq x \leq L, \tag{13.3.1}$$

$$P(x, 0) = P_0,\ u(x, 0) = 0. \tag{13.3.2}$$

The initial temperature gradient is characterized by the temperature $T(0, 0) = T*$ at the left end, by the mixture temperature outside the gradient, $T(x \geq L, 0) = T_0$, and by the gradient steepness, $(T * -T_0)/L$. The "length" L, which characterizes the gradient steepness can be viewed as the hot spot size, where the initial temperature gradient was formed.

The 1D direct numerical simulations are performed to solve the set of the 1D time-dependent, fully compressible reactive Navier–Stokes equations and chemical kinetics.

$$\frac{\partial \rho}{\partial t} + \frac{\partial(\rho u)}{\partial x} = 0, \tag{13.3.3}$$

$$\frac{\partial Y_i}{\partial t} + u\frac{\partial Y_i}{\partial x} = \frac{1}{\rho}\frac{\partial}{\partial x}\left(\rho D_i \frac{\partial Y_i}{\partial x}\right) + \left(\frac{\partial Y_i}{\partial t}\right)_{ch}, \tag{13.3.4}$$

$$\rho\left(\frac{\partial u}{\partial t} + u\frac{\partial u}{\partial x}\right) = -\frac{\partial P}{\partial x} + \frac{\partial \sigma_{xx}}{\partial x}, \tag{13.3.5}$$

$$\begin{aligned}\rho\left(\frac{\partial E}{\partial t} + u\frac{\partial E}{\partial x}\right) &= -\frac{\partial (Pu)}{\partial x} + \frac{\partial}{\partial x}(\sigma_{xx}u) + \frac{\partial}{\partial x}\left(\kappa(T)\frac{\partial T}{\partial x}\right) + \\ &+ \sum_k \frac{h_k}{m_k}\left(\frac{\partial}{\partial x}\left(\rho D_k(T)\frac{\partial Y_k}{\partial x}\right)\right),\end{aligned} \tag{13.3.6}$$

$$P = R_B T n = \left(\sum_i \frac{R_B}{m_i} Y_i\right)\rho T = \rho T \sum_i R_i Y_i, \tag{13.3.7}$$

$$\varepsilon = c_v T + \sum_k \frac{h_k \rho_k}{\rho} = c_v T + \sum_k h_k Y_k,\ \sigma_{xx} = \frac{4}{3}\mu\left(\frac{\partial u}{\partial x}\right). \tag{13.3.8}$$

In Eqs. (13.3.3)–(13.3.8), the usual notations were used: P, ρ, u, are pressure, mass density, and flow velocity, $Y_i = \rho_i/\rho$—the mass fractions of the species, $E = \varepsilon + u^2/2$—the total energy density, ε—the inner energy density, R_B—the universal gas constant, m_i—the molar mass of i-species, $R_i = R_B/m_i$, n—the molar density, σ_{ij}—the viscous stress tensor, $c_v = \sum_i c_{vi} Y_i$—the constant-volume specific heat, c_{vi}—the constant-volume specific heat of i-species, h_i—the enthalpy of formation of i-species, $\kappa(T)$ and $\mu(T)$ are the coefficients of thermal conductivity and viscosity, $D_i(T)$—the diffusion coefficients of i-species, and $(\partial Y_i/\partial t)_{\text{ch}}$—the variation of i-species concentration (mass fraction) in chemical reactions.

The equations of state for the reactive mixture and for the combustion products were taken with the temperature dependence of the specific heats and enthalpies of each species borrowed from the JANAF tables (Joint Army Navy NASA Air Force Thermochemical Tables) [47] and interpolated by the fifth-order polynomials [48]. The ideal gas equation of state was used in the case of a single-step model. The viscosity and thermal conductivity coefficients of the mixture are calculated from the gas kinetic theory using the Lennard–Jones potential [49]. Coefficients of the heat conduction of ith species $\kappa_i = \mu_i c_{pi}/\text{Pr}$ are expressed via the viscosity μ_i and the Prandtl number, $\text{Pr} = 0.75$.

When using a one-step model, it is necessary to select a method for calibrating the model parameters. For a stationary problem, for example, for a stationary propagating plane flame, one can choose the model parameters corresponding to the normal flame velocity and flame thickness, obtaining at least a qualitatively correct description of the flame dynamics. The situation is different for non-stationary processes, such as ignition, transient processes, and the like. In a trivial way, taking, for example, an unrealistically high activation energy in order to more or less accurately match the induction time in accordance with the experimental value, the model parameters can be chosen so as to reproduce the induction time of constant volume or a homogeneous explosion. However, this choice of a high activation energy results in completely incorrect ignition energy, and, in addition, the main reaction zone becomes exponentially thin which considerably affects the velocity, thickness, dynamics, and stability of the flame. For example, when calibrating the one-step model used in [45], a set of parameters is selected that most accurately reproduces the properties of the laminar flame (flame velocity and thickness) and the size of the detonation cell. However, it was pointed out in [46] that the disadvantage of such calibrated one-step model is the incorrect dependence of the flame velocity on pressure. Another disadvantage of the one-step model is that the reaction is exothermic at all temperatures for such a model, whereas in reality, the reactions begin with a neutral or endothermic induction stage, representing chain initiation and branching. Therefore, the gas dynamics is effectively "switched-off" during the induction stage. This means that the early stage of ignition, determined by the evolution of spontaneous wave, is not affected by the gas dynamics. On the contrary in the case of a one-step chemical model, gas dynamics from the very beginning has a significant effect on the propagation of a spontaneous wave. Because of this, the velocity of the spontaneous wave produced by the same temperature gradient on the early stage of the development of adiabatic explosion, when there is no gas-dynamic evolution, is considerably different and smaller for the chain-branching reaction compared with that for a one-step model. These differences between the one-step models and the detailed models of chemical chain-branching reactions significantly affect the evolution of the spontaneous combustion wave, especially below the "crossover" temperature. Because of this, all combustion modes initiated by the temperature gradient correspond to much shallower temperature gradients than those predicted by the studies with a one-step model.

Figure 13.3(a, b) from [50] shows typical trajectories of the spontaneous and pressure waves initiated in hydrogen–oxygen stoichiometric mixture by the temperature gradient for a single-step model (Fig. 13.1 (a, b)a $T^* = 1500\,\mathrm{K}$, $T_0 = 300\,\mathrm{K}$, $L = 0.1\,\mathrm{cm}$) and for a detailed model (Fig. 13.3b): $T^* = 1500\,\mathrm{K}$, $T_0 = 300\,\mathrm{K}$, $L = 0.5\,\mathrm{cm}$). For the steep (small L) temperature gradient, the pressure wave for a one-step model arises not behind the spontaneous wave, as it is for the case of the detailed model, but from the very beginning, it is formed ahead of the spontaneous wave. This feature of a one-step model remains similar for the shallower gradients. Therefore, the gas dynamics with a one-step model is always becoming the determining factor. As a result, the steepness of the temperature gradient initiating all possible combustion modes will be much steeper (L will be much smaller) for a one-step model compared to the gradient steepness calculated with detailed chemical kinetics.

To examine possible regimes of chemical reaction wave initiated by the temperature gradient in stoichiometric hydrogen–oxygen and hydrogen–air mixtures, we consider temperature gradients of different steepness (different size L of the hot spots) at the initial pressure 1 atm with $T^* = 1500\,\mathrm{K}$ and $T_0 = 300\,\mathrm{K}$. The calculated evolution of the reaction wave velocity (solid line) for the gradient of the scale $L = 8\,\mathrm{cm}$ in the H_2/O_2 gaseous mixture is shown in Fig. 13.4 along with the pressure-wave velocity (dash-dotted line). The velocity of the spontaneous wave was determined

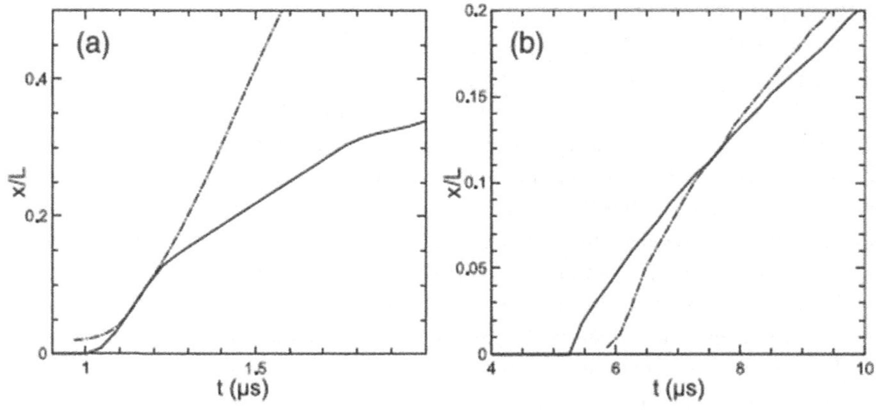

Fig. 13.3 Trajectories of the reaction wave (solid lines) and pressure wave (dash-dotted lines) for small values of U_{sp}: **a** calculated for a one-step model, $T^* = 1500\,\mathrm{K}$, $L = 0.1\,\mathrm{cm}$; **b** a detailed model, $T^* = 1500\,\mathrm{K}$, $L = 0.5\,\mathrm{cm}$ (from [50])

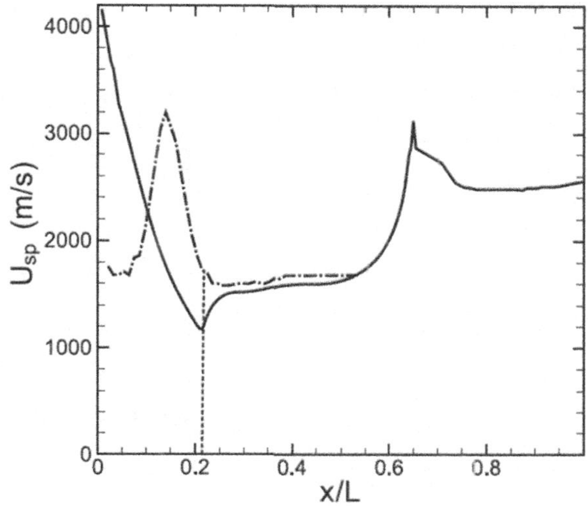

Fig. 13.4 Velocities of the spontaneous reaction wave (solid line) and pressure wave (dash-dotted lines) computed for the temperature gradient $L = 8\,\text{cm}$, $T^* = 1500\,\text{K}$ and $T_0 = 300\,\text{K}$ in H_2/O_2 gaseous mixture at the initial pressure $P_0 = 1\,\text{atm}$. The vertical dashed line shows the location of the maximum slowdown of the spontaneous wave, after which the pressure wave steepens into a shock wave [50]

from the trajectory of the reaction front position (position of the maximum H-radical iso-line). The velocity of the pressure wave was determined from the trajectory of the maximum pressure point in the pressure-wave profile. The velocity of spontaneous wave initiated by the initial temperature gradient decreases while the wave propagates along the gradient, and reaches the minimum value at the point close to the crossover temperature where it is caught up with the pressure wave which was generated behind the spontaneous wave front.

After the spontaneous wave front and the pressure wave intersected, the spontaneous wave transforms into a combustion wave and the pressure wave steepens into the shock wave. For a shallower gradient, such that the minimum speed of the spontaneous wave is of the order of the sound speed $a_* = a(T^*)$ at the top of the gradient, the intensity of the shock wave formed ahead of the reaction wave is sufficient to accelerate the reaction in the flow formed behind the shock. As a result, a pressure peak is formed at the reaction front, which grows at the expense of energy released in the reaction.

After the pressure peak becomes large enough, it steepens into a shock wave, forming an overdriven detonation wave (the peak of the maximum velocity U_{sp} at the point $x/L = 0.65$ in Fig. 13.4). The evolution of the temperature and the pressure profiles corresponding to Fig. 13.4 is shown in Fig. 13.5.

If the initial temperature gradient is steeper (e.g. the gradient scale $L = 7\text{cm}$), then the velocity of spontaneous wave in the minimum point, where the pressure wave overtakes the reaction wave, is not sufficient to sustain synchronous amplification of the pressure pulse in the flow behind the shock. As a result, the pressure wave runs ahead of the reaction wave as shown in Fig. 13.6, and the velocity of the reaction wave decreases. Possible modes of the propagating combustion wave inspired by the spontaneous wave initiated by the temperature gradient depend on the gradient

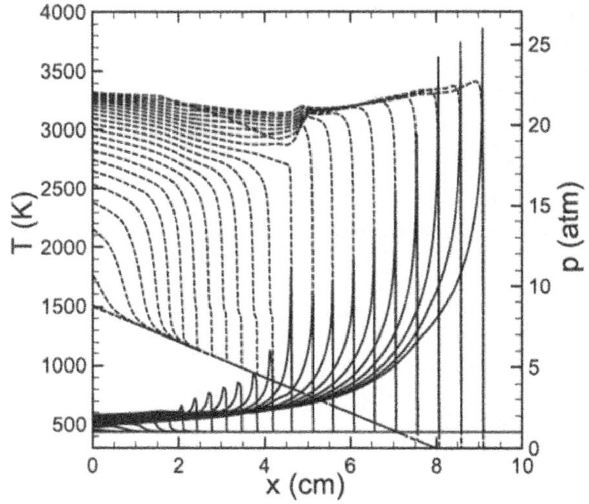

Fig. 13.5 Evolution of the temperature (dashed lines) and pressure (solid lines) profiles during the formation of the detonation shown at intervals 2 μs

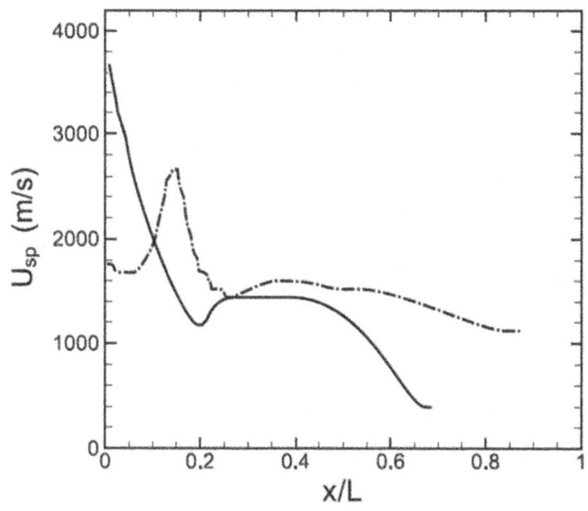

Fig. 13.6 Evolution of the spontaneous wave (solid curve) and pressure-wave (dash-dotted line) velocities calculated for temperature gradient, $L = 7\,\text{cm}$, $T^* = 1500\,\text{K}$ in H_2/O_2 mixture, and $P_0 = 1$ atm [50]

steepness. The pressure waves generated during the exothermic stage of the reaction can couple and evolve into a self-sustained detonation wave or produce a flame and a decoupled shock depending on the gradient steepness. The outcome depends on the gradient steepness and the relation between the speed of the spontaneous wave at the point where its velocity reaches a minimum (minimum point) and the characteristic velocities of the problem: the laminar flame speed U_f; sound speeds $a_0 = a(T_0)$, $a* = a(T^*)$, and a_N, a_{CJ} at the Newman point and at the Chapman–Jouguet point; and the velocity of the Chapman–Jouguet detonation U_{CJ}. Possible modes of the combustion initiated by the temperature gradients of different steepness in hydrogen–oxygen mixture at the initial pressure $P_0 = 1$ atm obtained from

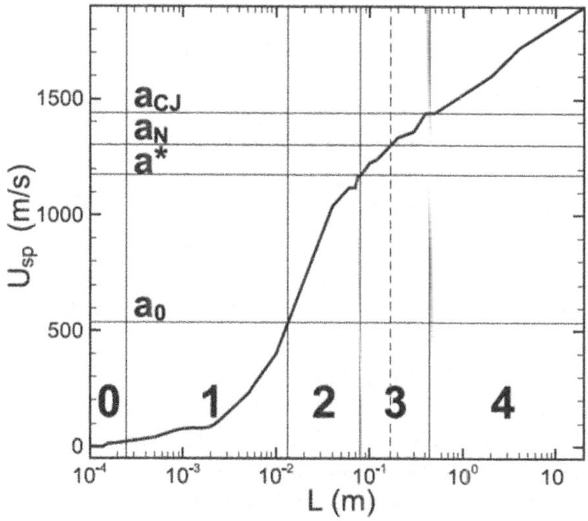

Fig. 13.7 Modes of the combustion wave initiated by temperature gradients of different steepness in hydrogen–oxygen mixture, $P_0 = 1$ atm. From [50]

the numerical studies with detailed chemical kinetics [50] are shown in Fig. 13.7. The figure represents a diagram of the spontaneous wave speed U_{sp} relative to the characteristic velocities of the problem versus the inverse gradient steepness $L = T/(dT/dx)$.

In summary, there are the following modes of reaction front propagation. For $U_{sp} < U_f \ll a_s$ (domain 0), the rate of the heat transfer by thermal conduction is greater than the spontaneous wave velocity, and the resulting regime is a deflagration wave propagating due to thermal conduction with the normal flame velocity $U_f = 10\,\mathrm{m/s}$. Domain (0) in Fig. 13.7 corresponds to the regime (5) in the Zel'dovich classification. Ignition of the deflagration is bounded from below by the minimum size of the hot region, for which the rate of heat removal from the "hot wall" is less than the normal velocity of a deflagration. If the spontaneous wave velocity in the minimum point is greater than the normal flame speed but less than the sound speed in the unperturbed media $U_f < U_{sp} < a_0$ (domain 1), then the "fast" deflagration is formed. The pressure wave overtakes the deflagration wave, and the fast deflagration wave propagates at nearly constant pressure. If $a_0 < U_{sp} < a^*$, then pressure wave overtakes the reaction wave to form a weak shock wave that compresses and heats the gas further speeding up the deflagration wave (domain 2). There are two different scenarios for domain 3, where $a^* < \min\{U_{sp}\} < a_{CJ}$. If $a^* < \min\{U_{sp}\} < a_N < a_{CJ}$, the reaction wave accelerates behind the shock and the transition to detonation occurs due to the formation and amplification of the pressure peak at the front of the reaction wave. If $a_N < \min\{U_{sp}\} < a_{CJ}$, then a quasi-stationary structure consisting of a shock wave followed by the reaction zone transforms into a detonation propagating down the temperature gradient. In both cases, the spontaneous reaction wave is accelerated in the flow behind the shock wave and transits to a detonation wave. Both of these regimes correspond to more detailed classification of the regime 3 in the Zel'dovich's classification [10]. If $a_{CJ} < \min\{U_{sp}\}$ corresponding to the domain 4, then the intersection of the pressure wave and the spontaneous reaction wave creates a classical structure of a detonation wave with the leading shock wave initiating the reaction. Finally, the limiting case of very shallow temperature gradient $\nabla T \to 0$, $U_{sp} \to \infty$ corresponds to the adiabatic explosion. The described regimes of the combustion wave initiated by the initial temperature gradient provide more detailed classification of possible propagating modes of combustion disclosed first by Zel'dovich [10] for a one-step chemical model.

As can be seen in Fig. 13.4, the spontaneous reaction wave begins at the upper point of the gradient, where the temperature is maximum and the ignition time is minimal, and then its speed decreases and reaches a minimum value. The speed of a spontaneous wave that starts at the top ($T = T^*$) of the temperature gradient decreases along the gradient ($\partial\tau_{\mathrm{ind}}/\partial T$ increases with decreasing temperature) and reaches its minimum value at the point close to the crossover temperature (see Eq. (13.2.1) and Figs. 13.2, 13.4, and 13.6). Therefore, the necessary condition for initiating detonation by the spontaneous reaction wave is that the spontaneous wave can be caught up and coupled with the pressure wave, which was generated behind the high-speed spontaneous wave front. Since the exothermic stage of the reaction begins and produces pressure pulse at the temperature slightly above the crossover temperature, the necessary condition for triggering a detonation can be written in the form

$$U_{sp}(T^*_{cr}) = \left(\frac{\partial\tau}{\partial T}(T^*_{cr})\right)^{-1}\left(\frac{\partial T}{\partial x}(T^*_{cr})\right)^{-1} = \left(\frac{\partial\tau}{\partial T}(T^*_{cr})\right)^{-1}\frac{L}{T^* - T_0} \geq a_s(T^*_{cr}), \tag{13.3.9}$$

where the temperature T^*_{cr} is slightly above the crossover temperature T_{cr}, and $a_s(T^*_{cr})$ is the sound speed at the point corresponding to $\min\{U_{sp}\}$. Indeed, Eq. (13.3.9) gives the minimal steepness of the temperature gradient (minimal value of the hot spot L) with quite good accuracy for a wide range of initial pressures [45]. Figure 13.8 shows velocities of the spontaneous wave at the minimum point as the function of the hot spot size L calculated for the detailed chemical models at initial pressures $P_0 = 1$, 5, and 10 atm; $T^* = 1500\,\mathrm{K}$, $T_0 = 300\,\mathrm{K}$. The crossover temperature corresponding to the transition from the induction to termination stage, known as the extended second explosion limit [49], shifts to higher temperatures at higher pressures. Since at higher pressures the induction time decreases (more precisely, $\partial\tau_{\mathrm{ind}}/\partial T$ decreases), the minimal steepness of the temperature gradients necessary for detonation initiation increases.

13.4 Ignition by Transient Energy Deposition

When considering ignition, we want to know how combustion starts and how the transient energy deposition influences the regime of the reaction wave which propagates out from the ignition zone—the hot spot. What type of combustion wave is

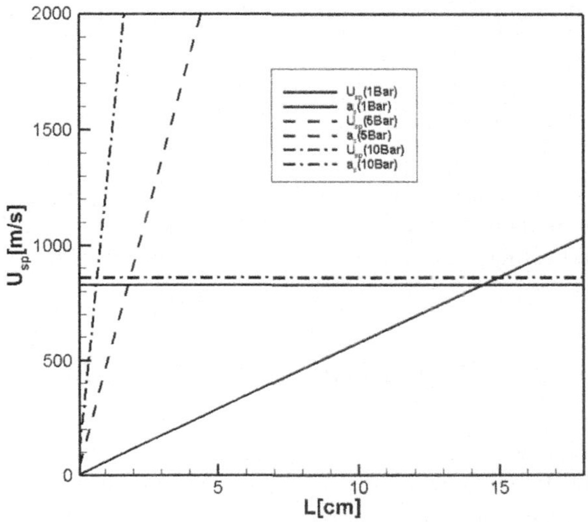

Fig. 13.8 The minimum hot spot size $L = L_{cr}$ producing detonation in hydrogen–air at $P_0 = 1, 5, 10$ atm calculated for the detailed chemical kinetics model; $T^*_{cr}(1\,\text{atm}) = 1300\,\text{K}$, $T^*_{cr}(5\,\text{atm}) = 1400\,\text{K}$, and $T^*_{cr}(10\,\text{atm}) = 1410\,\text{K}$. See Ref. [57] of Chap. 2

formed depending on: the amount of energy actually added to a finite volume of reactive gas on a specific time scale, the power deposition, and the ignition conditions, e.g. size of the hot spot, initial pressure, etc.?

Let us consider combustion regimes initiated by the energy deposition at the initial pressure and temperature $P_0 = 1$ atm and $T_0 = 300$ K in the hydrogen–oxygen stoichiometric mixture [51]. We consider uniform initial conditions and transient external source of energy localized on the scale of "hot spot" $0 \le x \le L$ where energy Q_{ig} is added during the time Δt_Q. For a 1D formulation, the hot spot represents a region $0 \le x \le L$, where energy is deposited with the rate $W(t) = dQ(t)/dt$. The specific inner energy of the mixture is changed as $Q(t) = Q(t - \Delta t) + \Delta Q$ within the hot spot during the time interval Δt. The total energy at the end of the energy deposition is

$$Q_{\text{total}} = \sum_{t=0}^{\Delta t_Q} \sum_{x=0}^{L} \Delta Q \cdot \rho(x, t) \cdot \Delta x. \tag{13.4.1}$$

The total energy transfer till the formation of a 1D stationary combustion wave far from the hot spot (at $x \gg L$) can be viewed as the ignition energy required for initiation of the particular combustion regime.

The governing equations are the 1D, time-dependent, multispecies reactive Navier–Stokes equations including the effects of compressibility, molecular diffusion, thermal conduction, viscosity, and chemical kinetics. They are the same as Eqs. (13.3.3–13.3.8), except the energy equation (13.3.6) has an additional term associated with the deposition of external energy

$$\rho\left(\frac{\partial E}{\partial t}+u\frac{\partial E}{\partial x}\right)=-\frac{\partial(Pu)}{\partial x}+\frac{\partial}{\partial x}(\sigma_{xx}u)+\frac{\partial}{\partial x}\left(\kappa(T)\frac{\partial T}{\partial x}\right)+ \\ \sum_k\frac{h_k}{m_k}\left(\frac{\partial}{\partial x}\left(\rho D_k(T)\frac{\partial Y_k}{\partial x}\right)\right)+W(t). \tag{13.4.2}$$

The initial conditions at $t = 0$ are constant pressure and zero velocity of the unburned mixture. At the left boundary at $x = 0$, the conditions are for a solid reflecting wall, where $u(0, t) = 0$ and the initial temperature $T(t = 0) = T_0$.

13.4.1 Rapid Energy Deposition—Microsecond Time Scale

First, we consider cases when the time scale of the energy deposition is comparable to or shorter than the acoustic time scale and less than the induction time at the ignition temperatures. The acoustic time is $t_a \approx 20\,\mu s$ for the hot spot of size $L = 1$ cm, and $t_a \approx 2\,\mu s$ for $L = 1$ mm [$a_0(T = 300\,K) = 539\,m/s$]. The rapid addition of energy to the hotspot at time scales much shorter than acoustic time causes an almost uniformly rapid increase in pressure and temperature, resulting in a volumetric explosion.

Figure 13.9 shows the calculated temporal evolution of temperature at the center of the hot spot ($x = 0$) for different values of the transmitted energy when the energy addition time is very short: $\Delta t_Q \sim 0.1\,\mu s < t_{ign} \ll t_a$. The temperature and pressure of the mixture in the hot spot depend on the energy transmitted to the hot spot. After the end of the energy deposition process, the induction period reaction starts. After $10\,\mu s$, the stationary combustion regimes are established.

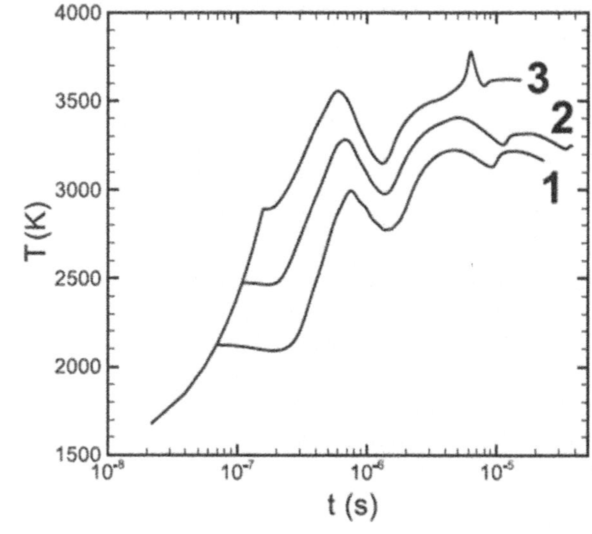

Fig. 13.9 Temperature evolution in the center of the hot spot for rapid sub-microsecond energy deposition $\Delta t_Q = (0.1 \div 0.2)\,\mu s$; 1: deflagration ($Q = 1.9\,kJ/m^2$), 2: deflagration ($Q = 2.4\,kJ/m^2$), and 3: detonation ($Q = 3.0\,kJ/m^2$) [51]

It should be noted that for each value of deposited energy, there is some definite temperature and pressure at which the reaction starts and the combustion regime is produced by the volumetric explosion at these conditions.

For a slower process of energy deposition ($\Delta t_Q = 5 \mu s < t_{ign} < t_a$), a large pressure jump is formed at the boundary of the hot spot. If the power is large enough, subsequent events represent the decay of the initial discontinuity consisting of a compression wave propagating in the direction $x > 0$, which steepens into the shock wave and the rarefaction wave propagating to the left from the boundary of the hot spot (in the direction $x = 0$) with the velocity equal to the local sound speed. Such a scenario, which is similar to a strong point explosion [52, 53], results in the direct triggering of a detonation wave if a concomitant shock wave at the right boundary is strong enough.

A more interesting scenario emerges in the case of a weaker shock. Figure 13.10a, b shows the transient evolution of the temperature (Fig. 13.10a) and pressure (Fig. 13.10b) profiles inside the hot spot ($L = 1$ cm) during the energy deposition.

The rarefaction wave propagating to the left creates shallow temperature and pressure gradients on the scale of about the size of the hot spot. At an initial pressure of $P_0 = 1$ atm, the temperature gradient with the temperature difference and length scale $L \sim 1$ cm as in Fig. 13.10a, b cannot trigger detonation. However, since the pressure of the heated mixture increased during the heating up to ≈ 4 atm, this temperature gradient is sufficient to produce a detonation through the Zel'dovich gradient mechanism. Time evolution of the temperature and pressure profiles shown in Figs. 13.10 and 13.11 demonstrates also the emergence of the spontaneous reaction

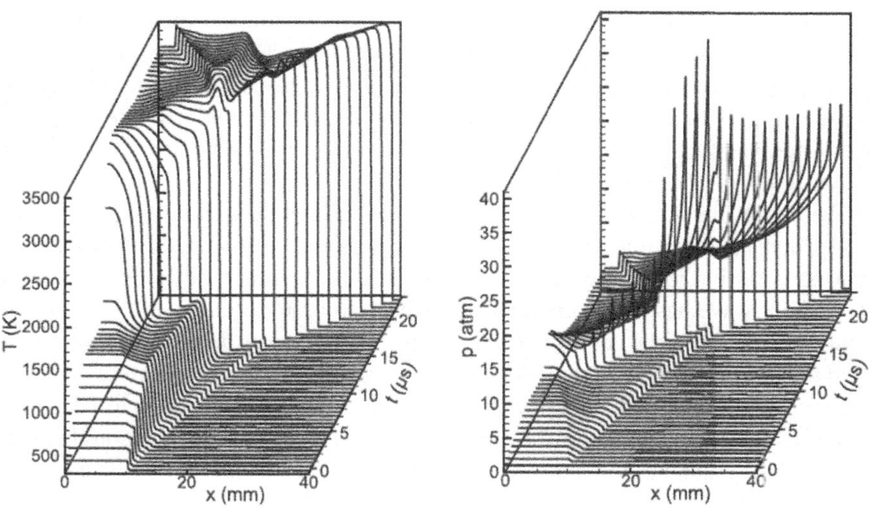

Fig. 13.10 **a** Evolution of temperature (left) and **b** pressure (right) profiles during detonation formation in the energy release region ($L = 1$ cm, $\Delta t_Q = 5\,\mu s$). Time interval $\Delta t = 0.5\,\mu s$ [51]

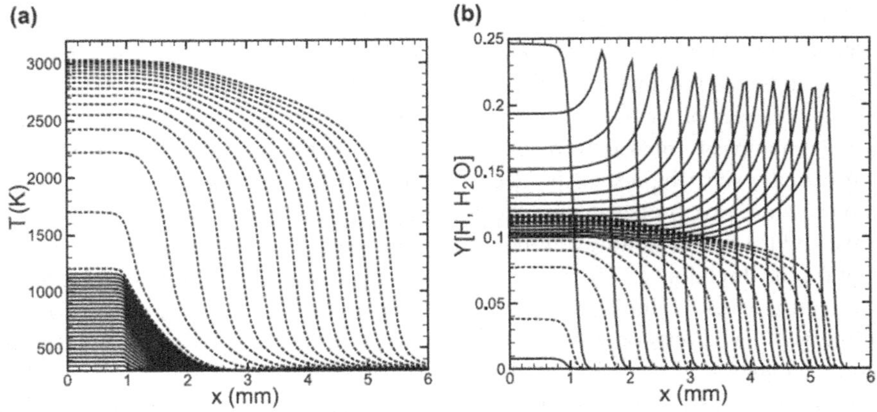

Fig. 13.11 **a** Evolution of temperature profiles in the hot spot during energy deposition (solid lines) and during the combustion wave formation (dashed lines); $L = 1$ mm, $\Delta t_Q = 1000\,\mu$s, and $\Delta t = 5\,\mu$s. **b** Evolution of H_2O concentration (dashed lines) and H-radical concentration (solid lines) profiles corresponding to the combustion wave formation in (**a**)

wave and its coupling with the pressure wave leading to the detonation initiation through the Zel'dovich mechanism.

In the opposite case when the ignition time is greater than acoustic time and the energy addition time is less than the acoustic time $\Delta t_Q < t_a < t_{\text{ign}}$, the gradient induced by the rarefaction wave forms on the stage following the end of energy addition. During the relatively long induction phase, the acoustic perturbations equalize pressure in the energy deposition zone and further ignition processes evolve at constant pressure from the steady temperature gradient. The temperature in the top of the gradient remains nearly constant till the ignition takes place. The combustion regime forming in this case depends on the environmental conditions: at $P_0 = 1$ atm, such a temperature gradient causes deflagration wave formation; at $P_0 = 10$ atm, it causes a detonation wave.

An interesting scenario takes place when the energy deposition time is slightly longer than acoustic time so that $t_a < \Delta t_Q < t_{\text{ign}}$. In the beginning, the gas dynamics process is similar to the previous case. However, the heating time is longer, and the rarefaction wave has time to reach the left wall, to be reflected from the wall, and return to the edge of the hot spot during the time of the energy deposition. As a result, even before the reaction started, the hot spot expands with the temperature profile in the mixture consisting of a shallow temperature gradient in the direction to $x = 0$ and a steep temperature gradient in the direction $x > 0$. By the time the reactions begin at the top of the gradients, the right-side gradient is too steep to trigger a detonation, and instead, a deflagration wave propagating to the right is ignited. At the same time due to the elevated pressure, the left shallow gradient can facilitate triggering detonation through the Zel'dovich gradient mechanism [51].

The rapid energy deposition such that the heating time, Δt_Q, is comparable to the characteristic acoustic time scale always results in the shock waves propagating away from the hot spot. A particular scenario of the resulting combustion regime depends on the size of the hot spot, though the basic physics appears to be similar to that described above. In the case of the smaller size of the hot spot ($L \sim 1$ mm), the scenario of the combustion regime ignition may differ because the scale of the temperature gradient created by the rarefaction wave may not be compatible with the detonation formation in real mixtures. There are many different scenarios that include direct detonation formation by a strong enough shock wave in the context of a thermal explosion or the shock waves propagating away from the hot spot producing ignition of the fast deflagration propagating behind the shock waves (regime 3 according to the classification in Sect. 13.3).

13.4.2 Millisecond Time Scale of Energy Deposition

Sufficiently rapid and large amount of thermal energy deposition into a reactive gas can trigger either direct initiation of detonation through a constant-volume explosion or through the Zel'dovich gradient mechanism due to the shallow gradient formed by the rarefaction wave at the increased pressures in the hot spot region. In the case of the rapid but relatively small energy deposition, the resulting regime is a fast deflagration wave. The scenario for low-power thermal energy addition over a longer period of time is different. If the acoustic time is much less than the energy deposition time, $t_a \ll \Delta t_Q \leq t_{\text{ign}}$, then there is enough time for pressure to be spatially homogenized by acoustic waves. In this case, there are no strong compression waves emitted from the hot spot, and the combustion regimes initiated by the energy deposition into the hot spot depend essentially on the steepness of the temperature gradient, which is formed by the thermal wave and gas expansion in the vicinity of the hot spot.

During the time of energy deposition Δt_Q, the thermal wave propagates away from the hot spot at the distance $x_T/\text{mm} = (\chi \Delta t_Q)^{1/2} \approx 0.9(\Delta t_Q/\text{ms})^{1/2}$. Some of the mass in the volume of the hot spot heated by the added energy, therefore, of higher temperature flows away as the temperature increases and the density falls. The expelled mass together with the thermal wave gives rise to the temperature gradient in the surrounding mixture behind the boundary of the hot spot. The temperature profile is almost linear because of the weak temperature dependence of the coefficient of thermal conduction ($\kappa \propto T^{0.75}$). Figure 13.11a shows the temperature gradient formed at the end of the energy deposition for $\Delta t_Q = 1000\,\mu\text{s}$, $L = 1$ mm (solid lines). A greater distance compared to that created by the thermal wave alone is due to the hot spot expansion, the decreased density, and increased temperature in the hot spot during the process of energy deposition. However, this linear temperature gradient is too steep to initiate detonation through the Zel'dovich gradient mechanism, and as a result, a deflagration wave is initiated. The thermal wave and the gas expansion are too slow to expand temperature and to form a temperature gradient compatible with the detonation formation in real mixtures at atmospheric or lower pressures. Long before the thermal wave moves away at a sufficiently long distance, the temperature of the mixture rises to ignite the reaction, so that either

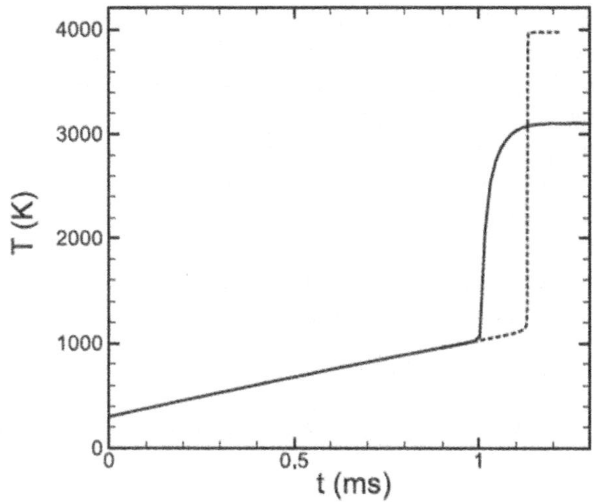

Fig. 13.12 Temperature evolution in the hot spot center for slow energy deposition ($t_a \ll \Delta t_Q$); $L = 1$ mm, $P_0 = 1$ atm (solid line), $P_0 = 10$ atm (dashed line);$\Delta t_Q = 1\,ms$, [51]

a deflagration wave or a fast deflagration wave is initiated. In Fig. 13.11a, dashed lines show the deflagration wave formation out from the formed temperature gradient. Figure 13.11b shows the evolution of H-radical and H_2O concentration profiles while the combustion wave is forming.

While the sound speed does not depend on pressure, the induction time $t_{\text{ind}}(T)$ at the temperature range (1100–1200) K is considerably longer at larger pressures. This leads to a significant delay of ignition for the same energy deposition regime at higher pressures.

As an example, the growth of temperature at the center of the hot spot $L = 1$ mm at the instants when the reaction starts is shown in Fig. 13.12 for $\Delta t_Q = 1\,ms$ and for initial pressures $P_0 = 1$ atm and $P_0 = 10$ atm.

Since the coefficient of thermal conductivity does not depend on pressure, and the steepness of the temperature gradient for direct detonation initiation through the Zel'dovich gradient mechanism decreases considerably with the increase in pressure, the temperature gradient created by the thermal wave can trigger detonation at high enough initial pressure. To elucidate the process, we consider a relatively small hot spot of size $L = 1$ mm at initial pressure $P_0 = 10$ atm. For a sufficiently long energy deposition time, the thermal wave creates a gradient appropriate for the detonation initiation through the Zel'dovich mechanism at $P_0 = 10$ atm (dashed line in Fig. 13.12).

Figure 13.13a shows the calculated temporal evolution of temperature and pressure profiles illustrating the formation of the temperature gradient outside of the hot spot, $L = 1$ mm, the development of the spontaneous wave along the gradient, and transition to detonation for the energy deposition time $\Delta t_Q = 1ms$. Figure 13.13b shows the corresponding evolution of the concentration profiles for H-radicals and the combustion products (H_2O).

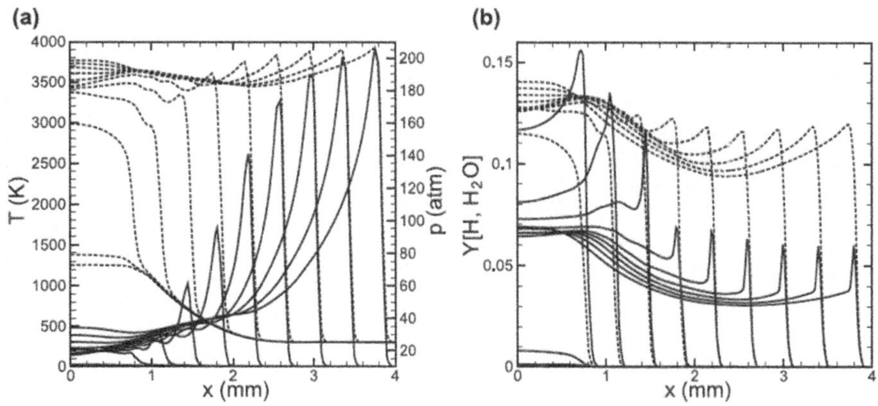

Fig. 13.13 (a, b) **a** Evolution of temperature (dashed lines) and pressure (solid lines) profiles illustrating detonation formation on the gradient formed in the energy release region. $L = 1$ mm, $\Delta t_Q = 1100\,\mu$s, $P_0 = 10$ atm, and $\Delta t = 0.15\,\mu$s. **b** Evolution of H_2O concentration (dashed lines) and H-radical concentration (solid lines) profiles for the conditions of (**a**) [51]

13.4.3 Energy of Ignition

The results obtained in the previous paragraph allow us to estimate the energy required for the ignition of a particular combustion regime. At the same time, it should be noted that the amount of the ignition energy obtained by extrapolating results of the 1D problem most likely will not match the actual value of the ignition energy for the 3D problem where the process is associated with a 3D expansion and converging rarefaction wave.[1] This difference is particularly important for the initiation of detonation. In this case, the 3D expansion additionally enhances the rarefaction, leading to less suitable conditions for a detonation initiation. A large number of 3D simulations of ignition due to the energy addition to the spherical hot spot were used to verify realization of different combustion regimes, to compare the ignition energy obtained from 3D spherically symmetric model to that extrapolated using the 1D model, and to assess the effect of spherical expansion on the ignition process.

Figure 13.14 shows the temperature evolution in the center of the hot spot for rapid energy deposition ($t_a \sim \Delta t_Q$) in a 1D case where $L = 1$ mm and in the 3D spherical hot spot of radius $R = 1$ mm. In both cases, the final temperature in the center of the hot spot tends to the same value. Temperature oscillations in the spherical hot spot are caused by convergence of the rarefaction wave to the center and reflection, which is enhanced by the expansion of the gaseous spherical hot spot, where the density decreases approximately as $\rho \propto \left(1 - (r/R)^2\right)^{1/(\gamma-1)}$. Since the rarefaction is much stronger for the spherical expansion, the final temperature of the spherical

[1] Clavin and Denet [54] have recently analytically studied, using an asymptotic analysis, the direct initiation of gaseous detonations in a spherical geometry for the case of a small heat release.

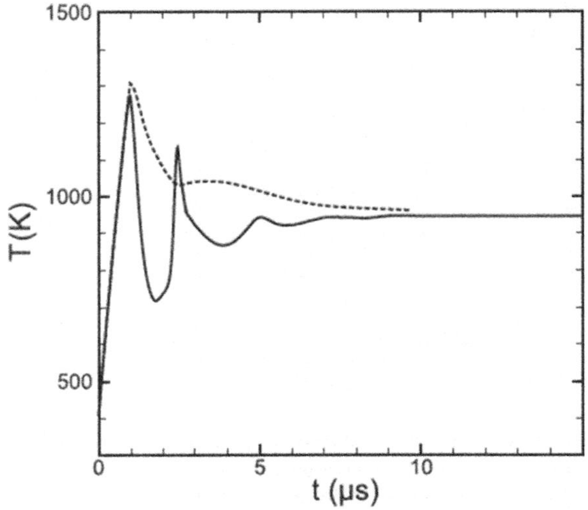

Fig. 13.14 Temperature evolution in the center of the hot spot for rapid energy deposition ($t_a \sim \Delta t_Q$): 1D case, $L = 1$ mm (dashed line) and 3D case: spherical hot spot of radius $R = 1$ mm (solid line) [51]

hot spot is considerably lower than in the planar case at the same parameters of energy deposition if the rarefaction reflects before the energy is deposited ($t_a < \Delta t_Q$). For the same reason, the actual energy required to initiate a detonation is larger for the spherical target than that in the planar case. Even if the temperature of the hot spot rises high enough for initiating detonation in the planar case, it is reduced by the converging and reflecting rarefaction wave and detonation cannot be ignited. An example of such a scenario is shown in Fig. 13.15, where for the same conditions, the detonation is initiated in the plane geometry and it is not for a spherical target.

The larger the acoustic time compared to the time of energy deposition, the less the influence of the rarefaction wave on the detonation initiation. This means, for

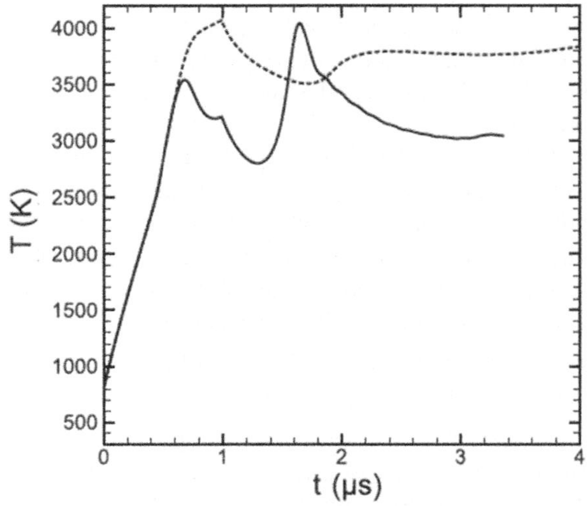

Fig. 13.15 Temperature evolution in the hot spot center for the case of rapid energy deposition ($t_a > \Delta t_Q$). For 1D case $L = 1$ mm (dashed line); 3D case $R = 1$ mm (solid line)

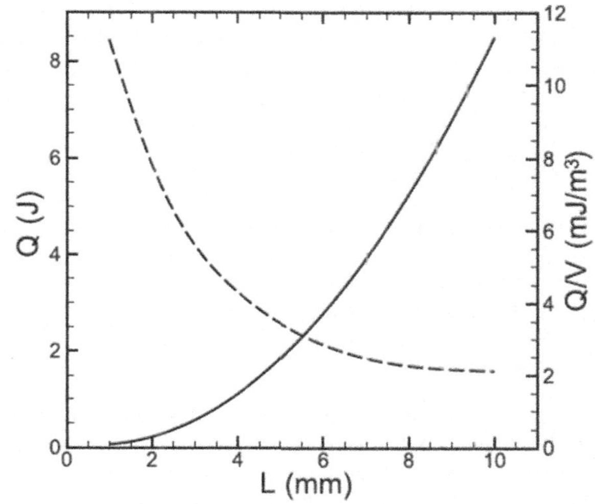

Fig. 13.16 The minimal energy required for detonation initiation in H_2/O_2 depending on the size (radius) of the hot spot for $\Delta t_Q = 1\,\mu$s. The solid line is total deposited energy; dashed line is the specific energy (Q/V [J/m^3]) [51]

example, that with the increase of the hot spot size ($t_a \propto L$), the initiation of detonation requires less energy deposition into the specific volume of the hot spot for a given time of power deposition Δt_Q, although the total deposited energy will be larger than in the case of detonation initiation in a smaller hot spot using a higher level of power deposition.

This tendency is illustrated in Fig. 13.16. For a shorter period of power deposition, the energy amount per unit volume of a hot spot capable of detonation initiation decreases. The lower limit for hydrogen detonation initiation can be obtained using a short sub-microsecond laser pulse focused in a sub-millimeter area and it is estimated in [1] as $\sim 10^{-2}$ mJ, which agrees with the extrapolation of the dependence shown in Fig. 13.16. In the 3D case, spherical expansion of the hot spot weakens the generated shock wave in favor of an intensified rarefaction wave. It results in sufficient drop in temperature and pressure in the hot spot on the time scales of the order of acoustic time. For the same conditions as in the 1D case, a less intensive combustion regime arises. However, to ignite detonation, one should increase sufficiently the power of the energy source.

References

1. D. Lewis, G. von Elbe, *Combustion, Flames and Explosion of Gases*, 2nd ed., Part 1 (Academic Press, New York, 1961)
2. T.M. Sloane, P.D. Ronney, A comparison of ignition phenomena modelled with detailed and simplified kinetics. Combust. Sci. Technol. **88**, 1–13 (1993)
3. J. Warnatz, U. Maas, Ignition processes in hydrogen-oxygen mixtures. Combust. Flame **74**, 53–69 (1988)
4. M. Thiele, J. Warnatz, A. Dreizler, S. Lindenmaier, R. Schießl, U. Maas, A. Grant, P. Ewart, Spark ignited hydrogen/air mixtures: two dimensional detailed modeling and laser based

diagnostics. Combust. Flame. **128**, 74–87 (2002)
5. D.R. Kassoy, A.K. Kapila, D.S. Stewart, A unified formulation for diffusive and nondiffusive thermal explosion theory. Combust. Sci. Technol. **63**, 33–43 (1989)
6. C. Vázquez-Espí, A. Liñán, Fast, non-diffusive ignition of a gaseous reacting mixture subject to a point energy source. Combust. Theory Modell. **5**, 485–498 (2001)
7. I. Glassman, R.A. Yetter, *Combustion*, 5th edn. (Elsevier, 2014)
8. Y.B. Zel'dovich, Theory of a limit of slow flame propagation. Zh. Exp. Theor. Fiz. 1, 159–169 (1941) (in Russian)
9. Y.B. Zel'dovich, V.B. Librovich, G.M. Makhviladze, G.I. Sivashinsky, On the development of detonation in a non-uniformly preheated gas. Acta Astronaut. **15**, 313–321 (1970)
10. Y.B. Zel'dovich, Regime classification of an exothermic reaction with nonuniform initial conditions. Combust. Flame **39**, 211–214 (1980)
11. B.E. Gelfand, A.N. Polenov, S.M. Frolov, S.A. Tsyganov, On detonation onset in non-uniformly preheated mixture. Combust. Expl. Shock Waves **21**, 488–492 (1985)
12. B.E. Gelfand, A.N. Polenov, S.M. Frolov, S.A. Tsyganov, Occurrence of detonation in a nonuniformly heated gas mixture. Combust. Expl. Shock Waves **22**, 118–123 (1986)
13. B.E. Gelfand, S.M. Frolov, A.N. Polenov, S.A. Tsyganov, Detonation initiation in systems with non-uniform temperature and concentration. Sov. Chem. Phys. **9**, 1277–1284 (1986)
14. Y.B. Zeldovich, B.E. Gelfand, S.A. Tsyganov, S.M. Frolov, A.N. Polenov, Concentration and temperature nonuniformities (CTN) of combustible mixtures as a reason of pressure waves generation. Prog. Astronaut. Aeronaut. **14**, 99–123 (1988)
15. G. Singh, J.F. Clarke, Transient phenomena in the initiation of a detonation. Proc. Roy. Soc. Lond. Ser. A **438**, 23–46 (1992)
16. G.M. Makhviladze, D.I. Rogatykh, Nonuniformities in initial temperature and concentration as a cause of explosive chemical reactions in combustible gases. Combust. Flame **87**, 347–356 (1991)
17. L. He, P. Clavin, Critical conditions for detonation initiation in cold gaseous mixtures by nonuniform hot pockets of reactive gases, in *Proceedings 24th Symposium (International) on Combustion* (1992), pp. 1861–1867
18. L. He, P. Clavin, Theoretical and numerical analysis of the photochemical initiation of detonations in hydrogen–oxygen mixtures, in *Proceedings 25th Symposium (International) on Combustion* (1994), pp. 45–51
19. P. Clavin, L. He, Direct initiation of gaseous detonations. J. Phys. IV Coll. **5**, 431–440 (1995). C4
20. L. He, Theoretical determination of the critical conditions for the direct initiation of detonations in hydrogen-oxygen mixtures. Combust. Flame **104**, 401–418 (1996)
21. J.W. Dold, A.K. Kapila, Comparison between shock initiations of detonation using thermally-sensitive and chain-branching chemical models. Combust. Flame **85**, 185–194 (1991)
22. G.J. Sharpe, M. Short, Detonation ignition from a temperature gradient for a two-step chain-branching kinetics model. J. Fluid Mech. **476**, 267–292 (2003)
23. G.J. Sharpe, N. Maflahi, Homogeneous explosion and shock initiation for a three-step chain-branching reaction mode. J. Fluid Mech. **566**, 163–194 (2006)
24. M. Short, The initiation of detonation from general nonuniformly distributed initial conditions. Phil. Trans. Roy. Soc. Lond. A **353**, 173–203 (1995)
25. A.M. Khokhlov, E.S. Oran, J.C. Wheeler, A theory of deflagration-to-detonation transition in unconfined flames. Combust. Flame **108**, 503–517 (1997)
26. N. Nikiforakis, J.F. Clarke, Numerical studies of the evolution of detonations. Math. Comput. Modell. **24**, 149–164 (1996)
27. M. Short, On the critical conditions for the initiation of a detonation in a nonuniformly perturbed reactive fluid. SIAM J. Appl. Math. **57**, 1242–1280 (1997)
28. A.M. Bartenev, B.E. Gelfand, Spontaneous initiation of detonations. Prog. Energy Combust. Sci. **26**, 29–55 (2000)
29. A.K. Kapila, D.W. Schwendeman, J.J. Quirk, T. Hawa, Mechanisms of detonation formation due to a temperature gradient. Combust. Theory Modell. **6**, 553–594 (2002)

30. T.M. Sloane, P.D. Ronney, A comparison of ignition phenomena modelled with detailed and simplified kinetics. Combust. Sci. Technol. **88**, 1–13 (1992)
31. X.J. Gu, D.R. Emerson, D. Bradley, Modes of reaction front propagation from hot spots. Combust. Flame **133**, 63–74 (2003)
32. D. Bradley, G.T. Kalghatgi, Influence of auto-ignition delay time characteristics of different fuels on pressure waves and knock in reciprocating engines. Combust. Flame **156**, 2307–2318 (2009)
33. G.T. Kalghatgi, D. Bradley, Pre-ignition and "super-knock" in turbo-charged spark-ignition engines. Int. J. Engine Res. **13**, 399–414 (2012)
34. Z. Wang, Y. Qi, X. He, J. Wang, S. Shuai, C.K. Law, Analysis of pre-ignition to super-knock: hotspot-induced deflagration to detonation. Fuel **144**, 222–227 (2015)
35. L. Bates, D. Bradley, G. Paczko, N. Peters, Engine hot spots: Modes of auto-ignition and reaction propagation. Combust. Flame **166**, 80–85 (2016)
36. A. Robert, S. Richard, O. Colin, T. Poinsot, LES study of deflagration to detonation mechanisms in a downsized spark ignition engine. Combust. Flame **162**, 2788–2807 (2015)
37. P. Dai, Z. Chen, S. Chen, Y. Ju, Numerical experiments on reaction front propagation in n-heptane/air mixture with temperature gradient. Proc. Combust. Inst. **35**, 3045–3052 (2015)
38. M.A. Liberman, M.F. Ivanov, D.M. Valiev, Hot spot formation by the propagating flame and the influence of EGR on knock occurrence in SI engines. Combust. Sci. Technol. **178**, 1613–1647 (2006)
39. J. Pan, H. Wei, G. Shu, R. Chen, Effect of pressure wave disturbance on auto-ignition mode transition and knocking intensity under enclosed conditions. Combust. Flame **185**, 63–74 (2017)
40. K. Tanoue, T. Jimoto, T. Kimura, M. Yamamoto, J. Hashimoto, Effect of initial temperature and fuel properties on knock characteristics in a rapid compression and expansion machine. Proc. Combust. Inst. **36**, 3523–3531 (2017)
41. J. Rudloff, J.M. Zaccardi, S. Richard, Analysis of pre-ignition in highly charged SI engines: emphasis on the auto-ignition mode. Proc. Combust. Inst. **34**, 2959–2967 (2013)
42. W.Q. Sun, S.H. Won, X.L. Gou, Y. Ju, Multi-scale modeling of dynamics and ignition to flame transitions of high pressure stratified n-heptane/toluene mixtures. Proc. Combust. Inst. **35**, 1049–1056 (2015)
43. P. Dai, Z. Chen, X. Gan, M.A. Liberman, Autoignition and detonation development from a hot spot inside a closed chamber: effects of end wall reflection. Proc. Combust. Inst. (2020). https://doi.org/10.1016/j.proci.2020.09.025
44. A. Kéromnès, W.K. Metcalfe, K.A. Heufer, N. Donohoe, A.K. Das, C.-J. Sung, J. Herzler, C. Naumann, P. Griebel, O. Mathieu, M.C. Krejci, E.L. Petersen, W.J. Pitz, H.J. Curran, An experimental and detailed chemical kinetic modeling study of hydrogen and syngas mixture oxidation at elevated pressures. Combust. Flame **160**, 995–1011 (2013)
45. E.S. Oran, V.N. Gamezo, Origins of the deflagration-to-detonation transition in gas-phase combustion. Combust. Flame **148**, 4–47 (2007)
46. M. Liberman, C. Wang, C. Qian, J. Liu, Influence of chemical kinetics on spontaneous waves and detonation initiation in highly reactive and low reactive mixtures. Combust. Theor. Modell. **23**, 467–495 (2019)
47. The JANAF tables are available online: http://kinetikcs.nist.gov/janaf/
48. J. Warnatz, U. Maas, R.W. Dibble, *Combustion. Physical and chemical fundamentals, modeling and simulations, experiments, pollutant formation* (Springer, New York, USA, 2001)
49. J.O. Hirschfelder, C.F. Gurtiss, R.B. Bird, *Molecular Theory of Gases and Liquids* (Wiley, New York, 1964)
50. M.A. Liberman, A.D. Kiverin, M.F. Ivanov, Regimes of chemical reaction waves initiated by nonuniform initial conditions for detailed chemical reaction models. Phys. Rev. E **85**, 056312(1–11) (2012)
51. A.D. Kiverin, D.R. Kassoy, M.F. Ivanov, M.A. Liberman, Mechanisms of ignition by transient energy deposition: regimes of combustion waves propagation. Phys. Rev. E **87**, 033015(1–10) (2013)

52. Y.B. Zeldovich, Yu.P. Raizer, *Physics of Shock Waves and High-Temperature Hydrodynamic Phenonema* (Academic Press, New York, 1966)
53. P. Reinicke, J. Meyer-ter-Vehn, The point explosion with heat conduction. Phys. Fluids A **3**, 1807–1818 (1991)
54. P. Clavin, B. Denet, Analytical study of the direct initiation of gaseous detonations for small heat release. J. Fluid Mech. **897**, A30(1–50) (2020)

Chapter 14
Regimes of Premixed Flames

A premixed combustion can occur uniformly throughout the volume, as an explosion. However, a distinctive feature of the premixed combustion is its ability to form a self-sustained reaction wave initiated at some location and then propagating either sub-sonically (deflagration) or supersonically (detonation) at a well-defined speed. Once fuel and oxidizer have been homogeneously mixed and the heat source is supplied, it becomes possible for a reaction wave to propagate through the combustible mixture. There are many important applications of premixed combustion, the principal being spark-ignition engines, lean-burn gas turbines, household burners, recently renewed interest in pulsed detonation propulsion, and many others. Apart from their technological relevance, premixed combustion constitutes a fascinating dynamical system, displaying an amazingly rich variety of phenomena such as non-uniqueness of possible propagation regimes, their birth (ignition) and destruction (extinction), chaotic self-motion and fractal-like growth, and various transitions.

14.1 Regimes of Premixed Turbulent Combustion

Although turbulent combustion is very common in technical applications, the theory of turbulent flames is much less developed because of its complexity. One of the reasons is that after more than a hundred years of intensive research by such outstanding scientists as Osborn Reynolds, Ludwig Prandtl, Theodore von Karman, Werner Heisenberg, Lev Landau, Andrey Kolmogorov, and many others, so far there is no complete theory of even turbulence in inert gases, and turbulence remains one of the most difficult unsolved problems of classical physics. The current state of the art of turbulent combustion is presented in a number of specialized books and review articles. We will restrict ourselves to presenting only the basic concepts and principles that give an idea of the physics of turbulent combustion.

In a sense, the theory of laminar flames should be the natural basis of the turbulent flame theory in order to avoid unjustified phenomenological assumptions about

M. A. Liberman, *Combustion Physics*,
https://doi.org/10.1007/978-3-030-85139-2_14

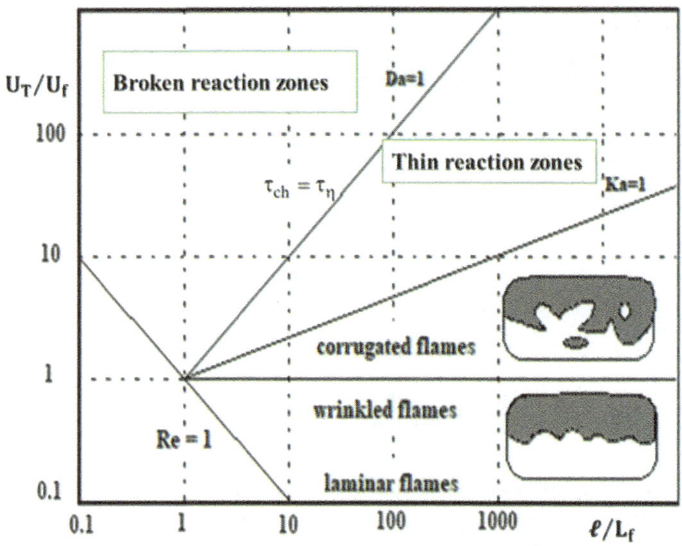

Fig. 14.1 Borghi diagram

the properties of turbulent flame. As a matter of fact, the words "turbulent flames" encompass a wide variety of different combustion regimes, from near-laminar flames to well-stirred reactors, as shown in Fig. 14.1. Turbulence can significantly affect the propagation of the combustion wave. While in the case of a freely propagating laminar flame, the flow conditions do not alter the rate of energy release in reactions or chemical mechanisms, the situation may be different in the presence of turbulence. In contrast, fluctuating components of velocity, temperature, and concentration can affect the chemical reaction and heat release rate. Turbulent combustion can be characterized by velocities scaled by the laminar flame velocity and length scales scaled by the laminar flame thickness. If the velocity of a laminar flame, U_f, is a physicochemical quantity inherent in a particular combustible mixture, the velocity of turbulent flame, U_T, may depend mainly on the properties of the external turbulent flow field in which it exists and is defined as the mass flow rate per unit area divided by the density of the unburned mixture. The turbulent flame structure depends upon the relation of characteristic time associated with the chemical reaction ($\tau_{Ch} = L_f/U_f$) and the characteristic time associated with turbulent velocity fluctuating components ($\tau_T = \lambda/u'$). If $\tau_{Ch} \ll \tau_T$, then the components of the velocity fluctuations do not have a significant effect on chemistry, in the opposite case, $\tau_{Ch} \gg \tau_T$, turbulence can affect the rate of a chemical reaction, the energy release, and the flame structure. The concept of turbulent reactive flow may have many different meanings and depends on the interaction range, which is governed by the general nature of the turbulent flow. For example, in the case of very strong turbulent pulsations, the fluctuating temperature and concentrations may affect the chemical reactions and heat release, and then the combustion products may be rapidly mixed with reactants within a time interval shorter than the chemical reaction time. The case of a strong influence of turbulence on the chemical reactions and heat release, which considerably changes the inner structure of the flame, is called *stirred reactor* regime. In this case, the conception of the flame structure is meaningless. If the external turbulent flow is relatively weak, which means that large-scale small-intensity turbulence prevails, then the result of interaction with turbulence will be a wrinkled flame front. Turbulence may either enhance the hydrodynamic instability inherent to a flame or independently create wrinkles in the flame surface. In this case, the flame remains locally laminar in the sense that the inner structure of the flame is not changed by turbulence. As the intensity of turbulence increases, the local laminar structure of the flame does not change as long as the scale of the smallest turbulent eddies remains greater than the thickness of the laminar flame. This quasi-laminar flame regime in the presence of turbulence is called a *flamelet* regime. Most of the typical turbulent combustion in engines, gas turbines, etc. are composed of wrinkled flamelets.

To understand the effect of turbulence, we consider the quantities that determine the characteristics of a turbulent flow. For simplicity, it is convenient to assume equal diffusivities for all reactive components, assuming the Schmidt number, $\mathrm{Sc} = \nu/D = 1$, the Prandtl number, $\mathrm{Pr} = \nu/\chi = 1$, and the Lewis number $\mathrm{Le} = \chi/D = 1$. One set of quantities is the characteristics of a laminar flame: its thickness L_f, the laminar flame speed U_f, and the chemical time $\tau_\mathrm{Ch} = L_\mathrm{f}/U_\mathrm{f}$. A turbulent flow is characterized by integral quantities of velocity, U, and length scale, L, that correspond to the size of the device in which the flow occurs, density ρ, and kinematic viscosity $\nu = \mu/\rho$. These quantities can be combined into a dimensionless quantity, the Reynolds number, $\mathrm{Re} = UL/\nu$. The flow becomes turbulent if the Reynolds number is sufficiently large.

It will be appropriate here to briefly recall the concept of turbulence and what length scales characterize turbulence (see, for example, [1, 2]). We assume that turbulence consists of eddies of different sizes. The largest eddies size is characterized by the length scale ℓ_0 which is comparable to the characteristic dimension of the flow, L, e.g. the tube diameter, and is called the integral length scale. The Reynolds number for the large eddies $\mathrm{Re} = U_0\ell_0/\nu$ is assumed to be large. The turbulent kinetic energy is defined as

$$k = \frac{1}{2}\langle u_i u_i\rangle = \frac{1}{2}\left(\overline{u_x'^2} + \overline{u_y'^2} + \overline{u_z'^2}\right). \tag{14.1.1}$$

For isotropic turbulence, $u_x'^2 = u_y'^2 = u_z'^2$, and the characteristic velocity is on the order of the root-mean-square (rms) velocity fluctuation $U_\mathrm{rms} \equiv u'$

$$u' = \sqrt{2k/3} \sim k^{1/2}. \tag{14.1.2}$$

Since a large Reynolds number is equivalent to a small kinematic viscosity, the effects of viscous dissipation on large scales are negligible. According to the concept of an energy cascade proposed by Richardson [3], there is a continuous flux of energy from larger to smaller eddies, so that kinetic energy is ultimately dissipated in the smallest eddies, where kinetic energy is converted into heat. Using the concept of an energy cascade, Kolmogorov [4–8] suggested that in the limit of very high Reynolds numbers, turbulent kinetic energy is locally isotropic; the motions of small-scale eddies are universal and independent on the mean flow field and boundary conditions. Kolmogorov's theory postulates universal, homogeneous, and isotropic distributions for small-scale velocity fluctuations and a self-similar cascade of kinetic energy from large spatial scales to small scales. The Kolmogorov cascade concept is local in scale, and all statistical information related to large scales is lost, with the exception of the rate of average energy cascade ε. This means that, although the dissipation is caused by viscosity, the rate of energy dissipation, the rate of energy flux from larger eddies to smaller eddies, should not depend on kinematic viscosity but should be determined only by the quantities characterizing large eddies. The only quantity having the dimension of the energy dissipation rate per unit mass, namely [erg/g s] $=$ [cm^2/s^3], that can be constructed from the fluid density ρ, velocity fluctuations u', and dimension ℓ_0 is

$$\varepsilon \sim \frac{(u')^3}{\ell_0}. \tag{14.1.3}$$

In a similar way, the only length scale that can be constructed from parameters ε and ν is Kolmogorov's length scale:

$$\ell_\eta = \left(\frac{\nu^3}{\varepsilon}\right)^{1/4} = \left(\frac{\nu^3 \ell_0}{(u')^3}\right)^{1/4} = \left(\frac{\nu^3 \ell_0^4}{(u')^3 \ell_0^3}\right)^{1/4} = \ell_0/\mathrm{Re}^{3/4}. \tag{14.1.4}$$

The corresponding velocity and time are

$$u_\eta \sim (\varepsilon\nu)^{1/4} = u'/\mathrm{Re}^{1/4}, \quad \tau_\eta \sim \ell_\eta/u_\eta = (\nu/\varepsilon)^{1/2} = (\ell_0/u')/\mathrm{Re}^{1/2}. \tag{14.1.5}$$

Note that ℓ_η and u_η are the length and velocity scales of the smallest eddies, on which energy is dissipated due to viscosity. Indeed, from (14.1.4) and (14.1.5), we find $u_\eta = \nu/\ell_\eta$ and $\mathrm{Re}_\eta = \ell_\eta u_\eta/\nu = 1$. Kolmogorov's scales are related to integral scales as

$$\ell_\eta = \ell_0/\mathrm{Re}^{3/4}, \quad u_\eta = u'/\mathrm{Re}^{1/4}, \quad \tau_\eta = \tau_0/\mathrm{Re}^{1/2}. \tag{14.1.6}$$

Since Kolmogorov's theory is valid for very large Reynolds numbers, so that $\ell_\eta \ll \ell_0$, this means that there is a wide range of intermediate scales ℓ that are small compared to ℓ_0 and large compared to ℓ_η. This is the inertial range, in which the velocity and time scales decrease as the length scale ℓ decreases.

An additional length intermediate in size between ℓ_η and ℓ_0, which often arises in formulating equations for mean quantities in turbulent flows, is the Taylor length λ, which represents the dimension over which strain occurs in a particular viscous medium. The Taylor length is followed from the conservation equation for turbulent kinetic energy

$$\varepsilon = 2\nu\overline{(\sigma_{ik}\sigma_{ik})}, \quad \sigma_{ik} = \frac{1}{2}\left(\frac{\partial u_i'}{\partial x_k} + \frac{\partial u_k'}{\partial x_i}\right). \tag{14.1.7}$$

The Taylor microscale is defined by estimating

$$\left(\overline{\frac{\partial u_i'}{\partial x_i}}\right) \sim \frac{\overline{u'^2}}{\overline{\lambda^2}} = \frac{u'^2}{\lambda^2}. \tag{14.1.8}$$

For isotropic turbulence, Eq. (14.1.7) gives

$$\varepsilon = 15\nu\overline{(\partial u_1/\partial x_1)^2}. \tag{14.1.9}$$

Then, for the Taylor length scale, we obtain

$$\varepsilon = 15\nu u'^2/\lambda^2. \tag{14.1.10}$$

Using Eqs. (14.1.2) and (14.1.3), we can express λ in Eq. (14.10) as $\lambda \approx (10\nu k/\varepsilon)^{1/2}$. Calculating the relations between λ, ℓ_0, ℓ_η and neglecting the numerical factor $\sqrt{10}$, which is not important in comparison with Re $\gg 1$, we find

$$\lambda \sim \ell_0/\mathrm{Re}^{1/2} = \ell_\eta \mathrm{Re}^{1/4}. \tag{14.1.11}$$

The ratio of the characteristic time of the turbulent flow and the chemical time $\tau_{\mathrm{ch}} = L_{\mathrm{f}}/U_{\mathrm{f}}$ is a dimensionless parameter—the Damköhler number [9]

$$\mathrm{Da} = \frac{\text{turbulent time}}{\text{chemical time}} = \frac{\tau_{\mathrm{T}}}{\tau_{\mathrm{ch}}} = \frac{\ell}{u'}\frac{U_{\mathrm{f}}}{L_{\mathrm{f}}}. \tag{14.1.12}$$

In the limit of small Damköhler numbers, Da $\ll 1$, chemical reactions are slow, and turbulent fluid pulsations change the chemical composition over a time interval less than the time of the chemical reaction on the scale of the flame thickness. In this regime, the turbulent eddies are nearly all embedded in the reaction zone, so that the meaning of the "flame front" has no sense anymore; reactants and products are mixed by the turbulent flow before reactions occur. This regime is a well-stirred reactor. On the contrary, in the limit of large Damköhler numbers, the chemistry is fast (i.e. the reaction time τ_{Ch} is short); the internal structure of the flame is not strongly affected by turbulence and can be considered locally as an element of a laminar flame called a *flamelet*. In this limit, the turbulent flow wrinkles and strains the flame surface. The flamelet concept is applicable to turbulent combustion when the chemistry is fast, so that burning occurs in an asymptotically thin layer embedded within the external turbulent flow field. In the flamelet regime, the flame front is strongly distorted on large scales compared to the thickness of the flame, while the internal flame structure remains locally similar to that of a laminar flame.

To make a diagram of combustion regimes, where different modes are presented depending on dimensionless parameters, we introduce the turbulent Karlovitz number, Ka, which is the ratio of the chemical time scale $t_{\mathrm{ch}} \approx L_{\mathrm{f}}/U_{\mathrm{f}} \sim \nu/U_{\mathrm{f}}^2$ to the Kolmogorov time scale $\tau_\eta = \ell_\eta/u_\eta = \ell_\eta^2/\nu$ (recall: $\mathrm{Re}_\eta = \ell_\eta u_\eta/\nu = 1$, and $L_{\mathrm{f}} U_{\mathrm{f}} \sim \nu$):

$$\mathrm{Ka} = \frac{\tau_{\mathrm{ch}}}{\tau_\eta} = \frac{L_{\mathrm{f}}^2}{\ell_\eta^2} = \frac{u_\eta^2}{U_{\mathrm{f}}^2}. \tag{14.1.13}$$

Equation (14.1.13) shows that the turbulent Karlovitz number quantifies the ratio of the laminar flame scale to the Kolmogorov scale ℓ_η and, therefore, the flame stretching. It is also appropriate to define another Karlovitz number as

$$Ka_\ell = \frac{L_f^2}{\ell^2} = \frac{u\prime^2}{U_f^2}. \tag{14.1.14}$$

Different combustion regimes are shown in Fig. 14.1, where the dimensionless ratio u'/U_{f} is plotted along the vertical axis and the dimensionless ratio ℓ/L_{f} is plotted along the horizontal axis. Similar diagram defining turbulent combustion regimes in terms of the scale ratio of velocity and length was first proposed by Borghi [10] and then analyzed and modified by many authors [11–13]. The region of laminar and wrinkled flames is located below the line $\mathrm{Re} = 1$ and $u'/U_{\mathrm{f}} = 1$. The line $\mathrm{Re} = 1$ separates all turbulent flame regimes characterized by $\mathrm{Re} > 1$ from the regime of laminar flames, which is situated in the lower-left corner of the diagram. For the wrinkled flamelet regime, the turn-over velocity u' of the large eddies is not large enough to compete with the flame front propagating at a laminar velocity U_{f}. The line corresponding to $\mathrm{Da} = 1$ separates the region of the well-stirred reactor regime. The intermediate regime between the lines $\mathrm{Ka} = 1$ and $\mathrm{Da} = 1$ corresponds to thick flames when the intensity of turbulence is high, so that turbulence changes the transport properties of the gas, although the flame still represents a configuration with burned gas being separated from the unburned mixture.

The flamelet concept was introduced and elaborated by various authors independently [11–18], and the flamelet regimes are of great practical interest. The region of corrugated flames is characterized by inequalities $\mathrm{Re} > 1$ and $\mathrm{Ka} < 1$. According to Eq. (14.1.13), the inequality $\mathrm{Ka} < 1$ means that $L_{\mathrm{f}} < \ell_\eta$ and the laminar structure of the flame does not change by turbulent eddies, but the entire flame structure is embedded within turbulent eddies, where the flame structure remains quasi-laminar. The boundary between the corrugated flamelets regime and the thin-flame regime is the line $\mathrm{Ka} = 1$. According to Eq. (14.1.13), above the line $\mathrm{Ka} = 1$, the flame thickness is larger than the Kolmogorov length scale $L_{\mathrm{f}} > \ell_\eta$. Also, for the thin reaction zone regime, the chemical time is larger than the Kolmogorov time, $\tau_{\mathrm{f}} > \tau_\eta$. The condition $\mathrm{Ka} = 1$ is also known as the Klimov–Williams criterion [18–20].

Klimov [19] and Williams [20] also analyzed laminar flames in a shear flow using the Karlovitz flame stretch factor defined as

$$\mathrm{Ka}_2 = \frac{L_\mathrm{f}}{U_\mathrm{f}} \frac{1}{\Lambda} \frac{\partial \Lambda}{\partial t}, \tag{14.1.15}$$

where Λ is the area of the flame surface, and $\partial \Lambda / \partial t$ is the rate of the increase in the surface area. In general, for $\mathrm{Ka}_2 > 0$, stretching decreases the flame thickness and increases the reactant consumption per unit area of the flame. On the other hand, compression (for $\mathrm{Ka}_2 < 0$) increases the flame thickness and reduces reactant consumption per unit incoming reactant area. It was shown in [19, 20] that a conventional propagating flame can exist only if the stretch factor Ka_2 is less than a critical value of unity. In the broken reaction region, both $\mathrm{Da} > 1$ and $\mathrm{Re} > 1$, which means that at least the smallest eddies of size ℓ_η can penetrate into the reactive–diffusive flame structure, changing the laminar transport.

14.2 G-Equation

The flamelet regime of premixed turbulent combustion can be described using a phenomenological scalar G-equation. Consider the surface defined by the scalar function $G(\mathbf{r}, t) = G_0$, such that the region of the burned gas corresponds to $G(\mathbf{r}, t) > G_0$, and the region of the unburned gas corresponds to $G(\mathbf{r}, t) < G_0$, where $\mathbf{r} = (x, y, z)$. Let $\mathbf{n} = -\nabla G / |\nabla G|$ is the unit normal vector to the front surface in the direction of the unburned gas. The time derivative of the function $G_0 = G(\mathbf{r}, t)$ is

$$\frac{dG_0}{dt} = \frac{\partial G}{\partial t} + \nabla G \cdot \frac{d\mathbf{r}_f}{dt} = 0, \tag{14.2.1}$$

where $\mathbf{r}_\mathrm{f}$ is the vector coordinate at the surface. Obviously, the propagation velocity of the flame front $d\mathbf{r}_\mathrm{f}/dt$ is equal to the sum of the flow velocity $\mathbf{u}$ and the component of the local burning velocity normal to the flame front

$$\frac{d\mathbf{r}_\mathrm{f}}{dt} = \mathbf{u} + \mathbf{n} \cdot U_\mathrm{f} = \mathbf{u} - \frac{\nabla G}{|\nabla G|} U_\mathrm{f}, \tag{14.2.2}$$

where the normal vector is $\mathbf{n} = -\nabla G / |\nabla G|$. Substituting (14.2.2) into (14.2.1), we obtain the equation for G in the form [18, 21]

$$\frac{\partial G}{\partial t} + \mathbf{u} \cdot \nabla G = U_f |\nabla G|. \tag{14.2.3}$$

When deriving Eq. (14.2.3), we used the simplifying assumptions. It was implicitly assumed that the density is constant and that the flame is much thinner than the hydrodynamic length scales. The first assumption of constant density is equivalent to the assumption of negligible heat release in the flame and, obviously, contradicts the meaning of a conventional flame itself. Nevertheless, the use of the G-equation often greatly simplifies the description of flame dynamics.

As an example of the application of G-equation, let's calculate the angle of the flame cone in the Bunsen burner shown in Fig. 14.2. We consider a 2D problem with the angle 2α of the flame cone above the burner tube, the axis x being the symmetry axis directed along the central line of the burner channel, and the burner rim is located at $x = 0$, $y = \pm d$. For a steady flow with the vertical flow velocity of the unburned mixture u, G-Eq. (14.2.3) takes the form

$$u\frac{\partial G}{\partial x} = U_{\mathrm{f}}\sqrt{\left(\frac{\partial G}{\partial x}\right)^2 + \left(\frac{\partial G}{\partial y}\right)^2}. \tag{14.2.4}$$

For our purpose, it is enough to consider only the inner part of the flame cone, i.e. the region of the unburned mixture. We will seek a solution to Eq. (14.2.4) in the form

$$G = G_0 + x + F(y), \tag{14.2.5}$$

where G_0 is some constant. Substituting (14.2.5) into Eq. (14.2.4), we obtain

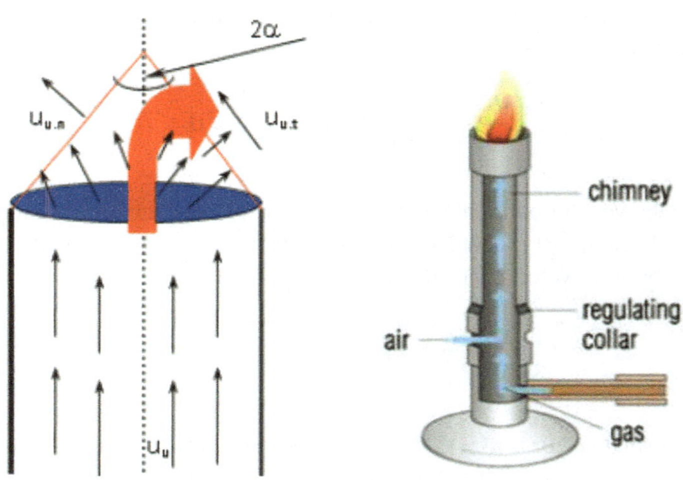

Fig. 14.2 Schematic flow in Bunsen burner

$$u = U_{\mathrm{f}}\sqrt{1 + \left(\frac{\partial F}{\partial y}\right)^2}. \tag{14.2.6}$$

Integrating Eq. (14.2.6), we obtain

$$F(y) = \frac{\sqrt{u^2 - U_{\mathrm{f}}^2}}{U_{\mathrm{f}}}(|y| - \mathrm{const}). \tag{14.2.7}$$

Taking into account the boundary conditions at the burner rim: $F(x = 0, y = \pm d) = 0$, and substituting this into (14.2.6), we will find the solution of the G-equation

$$G = G_0 + x + \frac{\sqrt{u^2 - U_{\mathrm{f}}^2}}{U_{\mathrm{f}}}\left(|y| - \frac{d}{2}\right). \tag{14.2.8}$$

According to the derivation of G-equation, the flame is the surface $r = r(x_{\mathrm{f}}, y_{\mathrm{f}})$, where $G(x, y) = G_0$; therefore, for the flame surface, we have

$$x_{\mathrm{f}} = \frac{\sqrt{u^2 - U_{\mathrm{f}}^2}}{U_{\mathrm{f}}}\left(\frac{d}{2} - |y_{\mathrm{f}}|\right). \tag{14.2.9}$$

It is clear from this formula that a flame can exist only if the flow velocity in the tube is greater than the normal flame velocity, and for the flame tip, which is located at $y = y_{\mathrm{f}} = 0$, we obtain

$$x_{\mathrm{f}0} = \frac{d}{2}\frac{\sqrt{u^2 - U_{\mathrm{f}}^2}}{U_{\mathrm{f}}}. \tag{14.2.10}$$

From geometrical considerations, it is clear that the angle at the flame tip is

$$\tan\alpha = \frac{d}{2x_{\mathrm{f}0}} = \frac{U_{\mathrm{f}}}{\sqrt{u^2 - U_{\mathrm{f}}^2}}. \tag{14.2.11}$$

It should be noted that this expression for the cone angle is valid independently of either a flame is laminar or it is turbulent.

We now use the G-equation to discuss flame stretching as suggested by Peters [18]. If we do know the flow velocity at the flame surface, the flame speed, and the flame geometry, then the G-equation describes the dynamics of the flame. Flame stretching occurs due to the flame curvature and due to the flow divergence. The flame curvature is defined as

$$\Bbbk = \nabla \cdot n = -\nabla \cdot \left(\frac{\nabla G}{|\nabla G|} \right). \tag{14.2.12}$$

This formula for $\Bbbk$ can be transformed as

$$\Bbbk = -\frac{\nabla^2 G + n \cdot \nabla (n \cdot \nabla G)}{|\nabla G|}. \tag{14.2.13}$$

A stretch rate at any point on the flame surface given by $G = 0$ is defined as the time derivative of the logarithm of an infinitesimal element A of the surface [14]

$$K = \frac{1}{A} \frac{\mathrm{d}A}{\mathrm{d}t}. \tag{14.2.14}$$

Let the flame surface is $\mathbf{z}(p, q, t)$, and $\mathrm{d}\mathbf{z}(p, q) = \mathbf{e}_p \mathrm{d}p + \mathbf{e}_q \mathrm{d}q$, where $\mathbf{e}_p$ and $\mathbf{e}_q$ are the unit vectors in the directions of the curvilinear coordinates p and q on the surface. An elemental area of the surface at time t is $\mathbf{A}(p, q, t) = (\mathbf{e}_p dp) \times (\mathbf{e}_q dq) = \mathbf{n}(dpdq)$, and the instantaneous velocity of the surface is $\mathbf{u}_f(p, q, t) = \partial \mathbf{z}(p, q, t) \partial t$. Here, $\mathbf{n} = \mathbf{e}_p \times \mathbf{e}_q$ is the unit vector pointed in the direction of normal to the elemental surface, in which the surface is propagating. After a simple algebra, we obtain for the change of an elementary surface during time δt:

$$\mathbf{A}(p, q, t + \delta t) = \left(\mathbf{e}_p + \frac{\partial \mathbf{u}_f}{\partial p} \delta t \right) \times \left(\mathbf{e}_q + \frac{\partial \mathbf{u}_f}{\partial q} \delta t \right) \mathrm{d}p \mathrm{d}q. \tag{14.2.15}$$

Taking $A = \mathbf{A} \cdot \mathbf{n}$, and using Eq. (14.2.15), we can write formula (14.2.14) for the stretch rate as

$$K = \frac{1}{A(t)} \lim_{\delta t \to 0} \frac{A(t + \delta t) - A(t)}{\delta t} = \left(\mathbf{e}_p \frac{\partial \mathbf{u}_f}{\partial p} + \mathbf{e}_q \frac{\partial \mathbf{u}_f}{\partial q} \right) \tag{14.2.16}$$

Next, we use $\partial \mathbf{e}_p / \partial p + \partial \mathbf{e}_q / \partial q = -(\nabla \cdot \mathbf{n})\mathbf{n}$, introduce the tangential component of the gradient operator ∇_t, such that $\nabla = \nabla_t + \nabla_n$, and decompose $\mathbf{u}_f$ into its tangential, $\mathbf{u}_{f,t}$, and normal components $\mathbf{u}_f = \mathbf{u}_{f,t} + (\mathbf{u}_f \cdot \mathbf{n})\mathbf{n}$. In addition, we assume that $\mathbf{u}_{f,t}$ is equal to the tangential component of the flow velocity $\mathbf{u}_{f,t} = \mathbf{V}_{s,t}$ at the flame surface, which means that we choose the coordinate system in which the flame propagates along the normal direction to its surface. Then Eq. (14.2.14) becomes

$$K = (\nabla_t \cdot \mathbf{V}_{s,t}) + (\mathbf{u}_f \cdot \mathbf{n})(\nabla \cdot \mathbf{n}). \tag{14.2.17}$$

Equation (14.2.17) shows that there are two sources of stretch: the influence of flow nonuniformity along the flame surface is represented by the first term; the second term represents a stretch due to the flame curvature, otherwise, for a plane flame, the second term consisting $(\nabla \cdot \mathbf{n})$ vanishes. We note that because $\mathbf{V}_{s,t} = n \times (\mathbf{V}_s \times n)$, the effect due to a flow nonuniformity exists only if $\mathbf{V}_s \times n \neq 0$, e.g. if the flow is oblique to the flame surface. Since $\nabla_n \cdot \mathbf{V}_{s,t} = 0$, we can rewrite Eq. (14.2.17) as

$$K = (\nabla \cdot \mathbf{V}_{s,t}) + (\mathbf{u}_f \cdot \mathbf{n})(\nabla \cdot \mathbf{n}), \tag{14.2.18}$$

or using $\mathbf{V}_{s,t} = n \times (\mathbf{V}_s \times n)$, it can be written as

$$K = -n \cdot \nabla \times (\mathbf{V}_s \times n)) + (\mathbf{u}_f \cdot \mathbf{n})(\nabla \cdot \mathbf{n}). \tag{14.2.19}$$

Let's consider an example of the stretch rate for a spherically expanding flame. For a spherical flame propagating either outward (+) or inward (−) along the normal to its surface, the instantaneous radius of the flame is $R_f(t)$. We have: $\mathbf{V}_{s,t} = 0$, $(\mathbf{u}_f \cdot \mathbf{n}) = U_f = \mathrm{d}R_f(t)/\mathrm{d}t$, and $(\nabla \cdot \mathbf{n}) = \pm 2/R_f$. Therefore, from Eq. (14.2.19), we obtain

$$K = \pm \frac{2}{R_f} \frac{dR_f}{dt}. \tag{14.2.20}$$

Note that the stretching effect weakens for outward expanding flames and it increases for inwardly propagating flames.

Returning to the G-equation, consider the effect of curvature on the speed of a spherical flame. Since the flow velocity is zero in the burned gas, it is convenient to formulate the G-equation with respect to the burned gas. Then for the radial flame position $R_f(t)$, we can write

$$\frac{\mathrm{d}R_f}{\mathrm{d}t} = U_{f,b} \tag{14.2.21}$$

As it was shown according to expression (14.2.17), the flame stretch is due to the flame curvature and due to the flow divergence. It can be shown [22–24] that for a one-step chemical model with a large activation energy and under the condition $L\mathbb{k} \ll 1$, these two effects modify the burning velocity $U_{f,b}$ as

$$U_f = U_{f,0} - L\mathbb{k}U_{f,0} + L(\mathbf{n} \cdot \nabla)\mathbf{V}_s \cdot \mathbf{n}, \tag{14.2.22}$$

where $U_{f,0}$ is the velocity of the unstretched flame and L is the Markstein length. In particular, considering a spherical flame, we can write Eq. (14.2.3) in spherical coordinates as

$$\frac{\partial G}{\partial t} = U_{f,0}\left(\left| \frac{\partial G}{\partial r} \right| + \frac{2L}{r}\frac{\partial G}{\partial r} \right). \tag{14.2.23}$$

Taking in Eq. (14.2.23) $G = R_f(t) - r$, we obtain for $R_f(t)$

$$\frac{\mathrm{d}R_f}{\mathrm{d}t} = U_{f,b}\left(1 - \frac{2L}{R_f}\right). \tag{14.2.24}$$

Integrating this equation with the initial condition $R_f(0) = R_{f,0}$, we obtain

$$U_{\mathrm{f,b}}t = R_{\mathrm{f}} - R_{\mathrm{f,0}} + 2L\ln\left(\frac{R_{\mathrm{f}} - 2L}{R_{\mathrm{f,0}} - 2L}\right). \tag{14.2.25}$$

It is seen that the solution exists only if $R_{\mathrm{as0}} > 2L$ and together with the condition $L\Bbbk \ll 1$, this means that the curvature corrections to the flame velocity are important only during the early time of the spherical flame expansion.

14.3 Velocity of Turbulent Flames

It should be noted that, contrary to a laminar flame, the turbulent flame velocity measured experimentally is not a local, but an integral quantity characterizing the entire flame front. Below we will use the definition of rms velocity fluctuations $U_{\mathrm{rms}} = \sqrt{\overline{u'(t)^2}}$. Damköhler [9] analyzed the experimental data that existed at that time and for the first time identified two different regimes of premixed turbulent combustion, which he called large-scale and small-scale turbulent regimes. He proposed that the main result of the large-scale, small-intensity turbulence is the wrinkling of a thin reaction layer, which retains the local characteristics, including transport properties of a laminar flame, so that the laminar flame structure would not be affected. In this case, the velocity of a turbulent flame can be defined as the mass consumption rate per unit area divided by the density of the unburned gas mixture. This means that the fuel mixture is locally consumed at a rate $\rho_{\mathrm{u}}U_{\mathrm{f}}$ over a large surface area S_f of the wrinkled flamelet.

From the equality of the mass flux $\rho_{\mathrm{u}}S_{\mathrm{f}}U_{\mathrm{f}}$ through the instantaneous wrinkled area of the flame surface S_f, propagating with the local speed of the laminar flame, and the mass flux $\rho_{\mathrm{u}}S_0U_{\mathrm{T}}$ through the time-averaged surface area of the turbulent flame S_0 (in Fig. 14.3, it is a planar cross section of the tube), we obtain the turbulent flame velocity

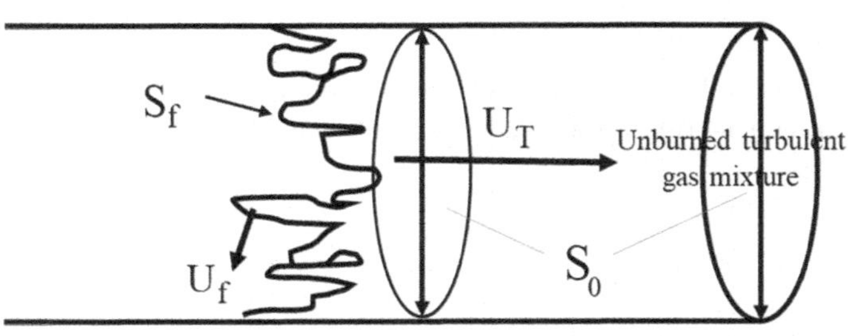

Fig. 14.3 Schematic of the wrinkled surface area of the turbulent flame brush—the flamelet surface area (S_{f}) and the envelope turbulent flame surface area (S_0)

$$U_T = U_f(S_f/S_0). \tag{14.3.1}$$

Damköhler also assumed that the interaction between the wrinkled flame front and the turbulent flow field can be approximated as

$$U_T/U_f \sim 1 + U_{rms}/U_f. \tag{14.3.2}$$

Then, combining Eqs. (14.3.2) and (14.3.1), we obtain

$$U_T = U_f + U_{rms}. \tag{14.3.3}$$

From Eq. (14.3.3), it follows that in the limit of strong turbulence, $U_{rms}/U_f \gg 1$, the turbulent flame speed is $U_T \sim U_{rms}$, but the linear growth at high turbulent intensities contradicts experimental observations.

In the flamelet regime, the laminar flame front is wrinkled by turbulence, its surface area increases, and every local part of the flame front propagates along the normal to its surface with the velocity of the laminar flame. This approach is called the "*thin-flame*" model. The flamelet regime itself consists of the regime of wrinkled flamelets and corrugated flamelets (Fig. 14.1). In the regime of wrinkled flames, $U_{rms} < U_f$, and U_{rms} can be interpreted as the circumferential velocity of the large eddies ($u'(\ell_0) \equiv U_{rms}$), which are bending the flame front but which are too large to convolute the flame front so that to form islands or multiply reaction sheets. The interaction of weak turbulence with the flame front in the regime of wrinkled flames can be described using asymptotic methods for high activation energies. In the case of high activation energies, the flame velocity is very sensitive to temperature changes in the thin reaction layer caused by flame stretching, which is caused by interaction with turbulent vortices, since it changes the relative diffusion of heat and reactant. This means that positive stretching increases the enthalpy and hence the temperature in the reaction layer, and this leads to the flame acceleration if Le < 1 or if the flame stretching is negative and Le > 1. Correspondingly, the flame decelerates if the flame stretch is negative and Le < 1 or if the flame stretching is positive and Le > 1. In the flamelet regime, a flame front is strongly distorted on large scales in comparison with the flame thickness L_f, while the inner flamelet structure remains similar to the structure of laminar flames.

As the turbulent intensity increases, the speed of the turbulent flame reaches its maximum value. The interaction between the flame and large-scale turbulent eddies leads to the increase of the flame surface and, consequently, to the increase in the combustion rate. Nevertheless, it has long been noted that for each combustible mixture there is a certain level of turbulence intensity at which the flame velocity reaches its maximum, and a further increase in the turbulence intensity leads to saturation of the velocity growth rate, followed by flame decomposition and even extinction. The flamelet model suggests a plausible explanation for this phenomenon. The positive stretch caused by a high-intensity turbulence reduces the reaction rate. The sign of the stretch changes along the flame front, and turbulent eddies carry the reactants to the flame, which leads to a steeper gradient at the flame front, which in turn causes larger diffusion losses. Finally, with increasing turbulence intensity, the finite rate of the chemical reaction is unable to produce products as fast as reactants are delivered. As a consequence, the flame temperature drops, which further lowers the reaction rate. Finally, this can lead to the extinguishing of the flame.

Damköhler [9] also argued that the small-scale turbulent eddies modify the transport and, therefore, the structure of the flame front. For a high-intensity turbulence, Damköhler proposed that similar to the laminar flame velocity scaling, $U_\mathrm{f} \sim \sqrt{D/t_\mathrm{ch}}$, where $D \sim L_\mathrm{f}U_\mathrm{f}$ is the molecular diffusivity (or thermal conductivity χ, for the unity Lewis number), the turbulent burning velocity should be given by the expression with the turbulent diffusivity D_T instead of the molecular diffusivity D

$$U_\mathrm{T} \sim \sqrt{D_\mathrm{T}/t_\mathrm{ch}}. \tag{14.3.4}$$

Then, the ratio of the turbulent flame speed to the laminar flame velocity will be

$$U_\mathrm{T}/U_\mathrm{f} \sim \sqrt{D_\mathrm{T}/D}. \tag{14.3.5}$$

Estimating the turbulent diffusivity as $D_\mathrm{T} \sim u_\mathrm{rms}\ell$, and the molecular diffusivity $D \sim L_\mathrm{f}U_\mathrm{f}$, Eq. (14.3.5) can be written as

$$U_\mathrm{T}/U_\mathrm{f} \sim \sqrt{u_\mathrm{rms}\ell/U_\mathrm{f}L_\mathrm{f}}. \tag{14.3.6}$$

Note that this formula as well as Eq. (14.3.4) are valid only until all turbulent scales are larger than the laminar flame thickness, $\ell > L_\mathrm{f}$, and the smallest Kolmogorov eddies do not penetrate into the reaction zone. This implies that there is an upper limit for the regime of corrugated flames, which according to definition (14.1.14) is given by expression $\mathrm{Ka}_\ell = (L_\mathrm{f}^2/\ell^2) < 1$.

Shchelkin [25] assumed that flame surface is distorted by the large-scale, small-intensity turbulence with their bases proportional to the square of the average turbulent eddies' diameter, and from the geometry consideration, Shchelkin proposed

$$U_\mathrm{T} = U_\mathrm{f}[1 + (U_\mathrm{rms}/U_\mathrm{f})^2]^{1/2}. \tag{14.3.7}$$

Shchelkin argued that Damköhler's formula is valid only at high-intensity turbulence, $U_\mathrm{rms}/U_\mathrm{f} \gg 1$, where it coincides with the Damköhler formula, $U_\mathrm{T} \sim U_\mathrm{rms}$, but differs for low-intensity turbulence, $U_\mathrm{rms}/U_\mathrm{f} \ll 1$

$$U_\mathrm{T}/U_\mathrm{f} = 1 + \frac{1}{2}(U_\mathrm{rms}/U_f)^2. \tag{14.3.8}$$

Clavin and Williams [24] used a multiscale perturbative method and received a similar result but with a different numerical factor at $(U_\mathrm{rms}/U_\mathrm{f})^2$

$$U_\mathrm{T}/U_\mathrm{f} = 1 + (U_\mathrm{rms}/U_\mathrm{f})^2. \tag{14.3.9}$$

However, later it was shown in [26–28] that for low-intensity turbulence, the quadratic dependence on turbulence intensity is only transient and was proposed an alternative dependence $U_T/U_f - 1 \propto (U_{rms}/U_f)^{4/3}$. In the limit of large rms velocity, we have $U_T \sim U_{rms}$. This result was generalized by Pocheau [29], who used a renormalization procedure to show that only an expression such as $U_T/U_f = \left[1 + C(U_{rms}/U_f)^n\right]^{1/n}$ satisfies scale invariance in the corrugated flamelet regime.

14.3.1 Modeling of the Premixed Turbulent Flame Speed

The fundamental work by Damköhler and Shchelkin laid the foundation for extensive theoretical [11–24, 26–54], experimental [55–76], and computational studies [13, 76–101] devoted to the quantitative assessment of the turbulent flame speed depending on turbulent intensity. However, despite great efforts and a large number of publications using various approaches and techniques, the problem has not yet been resolved and still, there are disagreements between the proposed theories and experiments. One of the main reasons is that the problem is greatly complicated by the large scatter of experimental results even for the same experimental setup, which has been discovered in the very first detailed analysis of experimental data [55]. A long history of theoretical, computational, and experimental efforts to solve the problem is summarized in many review articles and books [18, 102–122].

Damköhler's hypothesis that the speed of turbulent flame is mainly determined by the flame surface area, and the idea that the speed of a turbulent flame can be obtained in the form of some universal formula representing simple functional dependence such as

$$U_T/U_f = f(u'/U_f), \tag{14.3.10}$$

inspired many attempts to improve the earliest scaling proposed by Damköhler and to derive expressions that would reproduce the existing experimental data on turbulent burning velocities. The search for a universal scaling law for the turbulent speed of a premixed flame is still under discussion, though a wide scatter of experimental results, which depend on various parameters characterizing the experimental conditions and the particular type of device used (Bunsen combustor, V-flames, combustion bomb, etc.) [55–76] make questionable the very possibility of such a universal formula. The scaled turbulent flame velocity U_T/U_f measured by different experimental groups may differ by order of magnitude for the same scaled turbulent intensity u'/U_f.

Since the development of a rigorous theory of turbulent combustion is an extremely difficult task, some simplified assumptions and models were used. From the beginning, many theoretical approaches considered a model of turbulent flame using an artificial assumption of zero thermal expansion of the burning matter and zero thickness of the flame front [24, 26, 29, 32–39]. This model implies that the expansion coefficient, which is the density ratio of unburned to burned gas mixtures, is unity,

$\Theta = \rho_u/\rho_b = 1$. In this case, the problem of flame propagation is considerably simplified, since the flame does not affect the external turbulent flow, which can be considered as being prescribed in advance.

Flame dynamics in a turbulent flow was studied in [21, 32–37, 104] using G-Eq. (14.2.3). To describe a turbulent flame propagating in the z-direction with the flame position given by $G(z, x, t) = z - f(x, t) = 0$, a simplified version of the G-equation was used [21, 26], which describes a 2D problem of flame bending in a 2D turbulent flow

$$\frac{\partial f}{\partial t} + u_x \frac{\partial f}{\partial x} = U_f \sqrt{1 + (\partial f/\partial x)^2} + u_z + \nu \frac{\partial^2 f}{\partial x^2}. \tag{14.3.11}$$

The last diffusion term with the factor ν proportional to the flame thickness has been added [46, 50] in order to stabilize numerical simulations. The turbulent velocity was modeled as ($u_x = 0$ in the case of 2D model)

$$u_z = \sum U_i \cos(k_i z + U_T k_i t + \Omega_i t + \varphi_{iz}) \cos(k_i x + \varphi_{ix}) \cos(k_i y + \varphi_{iy}). \tag{14.3.12}$$

It was assumed that turbulence is well developed with Kolmogorov's energy spectrum, so that amplitudes of the turbulent modes U_i depend on the wave number as $U_i \propto k_i^{-5/6}$, while $\varphi_{ix}, \varphi_{iy}, \varphi_{iz}$ are random phases of the corresponding modes. The rms turbulent velocity ($U_{rms} \equiv u'$) in one direction is related to the amplitudes of turbulent harmonics $U_{rms}^2 = \sum U_i^2/4$. This approach leads to the Clavin–Williams [24] formula (14.3.9).

Usually, the theoretical studies of the turbulent flame speed adopted the Taylor hypothesis, which states that the turbulent flow is stationary in the laboratory reference frame and allows to neglect the explicit time dependence of the turbulent flow. Assumption of the Taylor hypothesis means that in this case $U_T k_i \gg \Omega_i$ which considerably simplifies calculations [104], and the solution can be reduced to the Clavin–Williams formula [24]

$$U_T/U_f - 1 = \sum \frac{k_i^2 U_i^2}{8\omega_i^2} = \sum \frac{U_i^2}{8U_f^2} = \frac{U_{rms}^2}{U_f^2}. \tag{14.3.13}$$

The Clavin–Williams formula was obtained for a weak turbulence, $U_T/U_f - 1 \ll 1$, assuming zero flame thickness and zero thermal expansion, $\Theta = 1$. It can be extrapolated to strong turbulence and strongly corrugated flames assuming self-similar properties of the corrugated flame according to [36, 37]. To avoid misunderstanding, it should be pointed out that in the Clavin–Williams formula, $U_T/U_f - 1 = U_{rms}^2/U_f^2$, the expression $U_{rms} = u'(\ell_0)$, where ℓ_0 is the integral length scale of the turbulent flow, is used. Various authors [18, 21, 32–39] used scaling arguments and renormalization techniques to derive expressions of the form (14.3.10) for the turbulent flame speed.

For strongly corrugated flamelets an effective method was proposed by Yakhot [32]. Assuming isotropic well-developed turbulence with the Kolmogorov turbulent spectrum density $\varepsilon_T(k) \propto k^{-5/3}$, we can express U_{rms}^2 via spectral density $\varepsilon_T(k)$ as

$$U_{rms}^2 = \int_{k_T}^{k_\eta} \varepsilon_T(k)dk, \tag{14.3.14}$$

where $k_T = 2\pi/\ell_0$ and $k_\eta = 2\pi/\ell_\eta$ are the wave numbers corresponding to the integral, ℓ_0, and the Kolmogorov ℓ_η length scales, respectively. Using Eq. (14.3.14), the Clavin–Williams formula (14.3.19) can be written as

$$\frac{U_T}{U_f} - 1 = \frac{1}{U_f^2} \int_{k_T}^{k_\eta} \varepsilon_T(k)dk. \tag{14.3.15}$$

Following [32], the entire spectrum of flame wrinkles can be decomposed into narrow bands with different wave numbers, assuming that each band provides an infinitely small increase in the flame velocity. Since every band in the spectrum of flame wrinkles causes an infinitesimal increase in the flame velocity, successive velocity amplifications can be written as $dU_T = U_T d \sum \frac{k_i^2 U_f^2}{8\omega_i^2}$, which leads to the solution obtained by Yakhot [32]

$$ln(U_T/U_f) = U_{rms}^2/U_T^2. \tag{14.3.16}$$

Pocheau [37] considered turbulent flame propagation in a scale-invariant regime, extrapolating Eq. (14.3.13) to the case of strong turbulence and assuming self-similar properties of the corrugated front. The method used in [37] consists in decomposing the turbulent energy density spectrum into infinitely narrow bands, so that different components of the flame wrinkles correspond to different wave numbers of the corresponding narrow bands in the spectrum. Assuming that each of them provides a similar small increase of the flame velocity, we can write Eq. (14.3.14) in the form

$$\frac{dU}{U} = -\varepsilon_T(k)\frac{dk}{U^2}. \tag{14.3.17}$$

Integrating Eq. (14.3.17) over the whole turbulent spectrum, we obtain for the turbulent velocity of strongly corrugated flames

$$U_T^2 = U_f^2 + 2U_{rms}^2. \tag{14.3.18}$$

At the same time, it should be noted that one of the problems that complicate theoretical calculations of the turbulent flame velocity is the small difference between the integral, ℓ_0, and Kolmogorov ℓ_η length scales under real experimental conditions.

This means that the total spectral density in terms of the wave numbers $k_T = 2\pi/\ell_0$ and $k_\eta = 2\pi/\ell_\eta$ is too narrow, and the separation between small and large scales is too small for a certain universality expected for small-scale turbulent combustion.

14.3.2 Effects of Darrieus–Landau Instability and Thermal Expansion

As it was said, most of the earlier theoretical models were limited by the conditions of zero thermal expansion of the burning matter, $\Theta = \rho_u/\rho_b = 1$ (constant density approximation), and zero flame thickness $L_f = 0$. With these restrictions, the flame itself does not influence the turbulent flow but propagates in a given external turbulent flow with the local velocity of a laminar flame. These models do not account for variations of the laminar flame velocity caused by the flame-front stretch, and at zero thermal expansion $\Theta = \rho_u/\rho_b = 1$, the hydrodynamic flame instabilities do not develop. Taking into account the realistic thermal expansion, $\Theta \approx 6-10$ in turbulent models is inevitably accompanied by an increase in the intrinsic hydrodynamic instability of the Darrieus–Landau (DL) flame instability. The effect of the flame-front wrinkling due to DL instabilities can be comparable with turbulent wrinkling at least for weak turbulence.

The effects of a finite thermal expansion were considered by Searby and Clavin [31] who investigated the linear response of a weakly wrinkled flame front to weak external turbulence. Aldredge and Williams [40] used linear theory to study the flame velocity increase caused by external turbulence. In both studies [31, 40], the authors considered a flame propagating downwards, so that the hydrodynamics instability was stabilized due to the effects of thermal diffusion and gravity. At the same time, already the very first numerical simulations [49–54] have demonstrated a noticeable increase of the flamelet velocity because of the Darrieus–Landau instability. The influence of the DL instability on the velocity of turbulent flames was recognized in a number of experimental studies [58, 59, 64, 123–135] in cases of weak turbulence intensities. The DL instability may be significant if the integral turbulence length scale significantly exceeds the cutoff wavelength λ_c. The DL instability does not contribute to the increase of the turbulent flame speed if the size of the experimental device is less than the cutoff wavelength, as it was observed in [57] for a methane–air flame propagating in a turbulent flow created by two coaxial cylinders with a 1 cm gap between the cylinders.

Attempts to extend the theoretical analysis to take into account the finite values of the expansion coefficient and the flame thickness were based on the model nonlinear equation [136–138] obtained in the approximation of weak nonlinearity [54, 139]. These studies used the Pelce–Clavin equation for the growth of flame perturbations in time (Ref. [11] of Ch.5) and the model nonlinear equation [136] for the velocity and shape of a stationary curved flame to calculate the linear response of a weakly wrinkled flame front to weak turbulence using the Searby–Clavin equation [30].

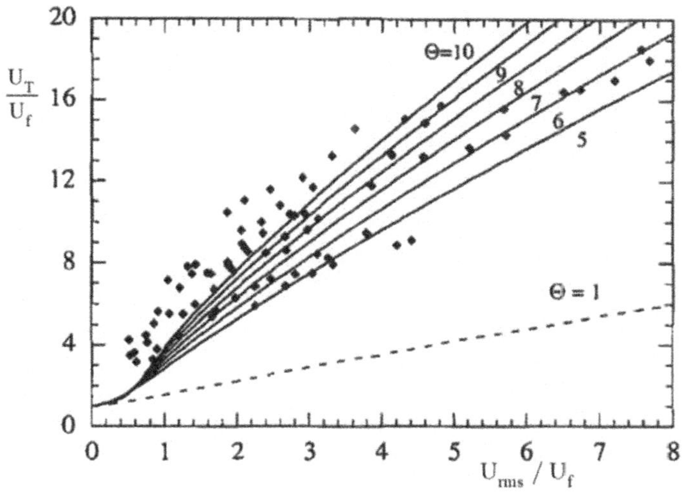

Fig. 14.4 Scaled velocity U_T/U_f of a turbulent flame of finite thickness versus the scaled velocity of external turbulence for different expansion coefficients Θ. The markers show the experimental results from [57]

Strictly speaking, this approach has one significant disadvantage: the model nonlinear equation is applicable only in the limit $\Theta \to 1$. From the mathematical point of view, the assumption of weak nonlinearity and stationarity of the flame made in the derivation of the model equation is valid only for the constant density, $\Theta = \rho_u/\rho_b \to 1$. From the physical point of view, this means that a weakly curved flame with finite gas expansion cannot be stationary (Ch. 8, Refs [12, 13]).

Thus, the results of [136–138] applied for the case $\Theta > 1$ should be considered as a qualitative assessment of the tendency of the dependence of the turbulent velocity on the intensity of turbulence. As an illustration, Figs. 14.4 and 14.5 show the normalized turbulent velocity versus the normalized turbulent intensity calculated in [137] for different values of Θ in comparison with the experimentally measured turbulent flame velocities in a methane–air mixture for almost isotropic turbulence [57] in Fig. 14.4 and with the experimental results collected in [55] in Fig. 14.5.

Experimental studies [58, 64, 124] examined the effect of pressure variations on turbulent Bunsen flames and proposed a scaling law in the form

$$U_T/U_f \propto (U_{rms}/U_f)^{0.38}. \tag{14.3.19}$$

Numerical simulations [90–95] have revealed the existence of well-defined scaling laws for turbulent velocity and the enhancement of the turbulent flame speed due to variations of the laminar flame velocity caused by the flame-front stretch. Depending on the interplay of the characteristic length scales: the size of the device, L, and the DL cutoff wavelength λ_c, the planar laminar flame is unstable if the wavelengths

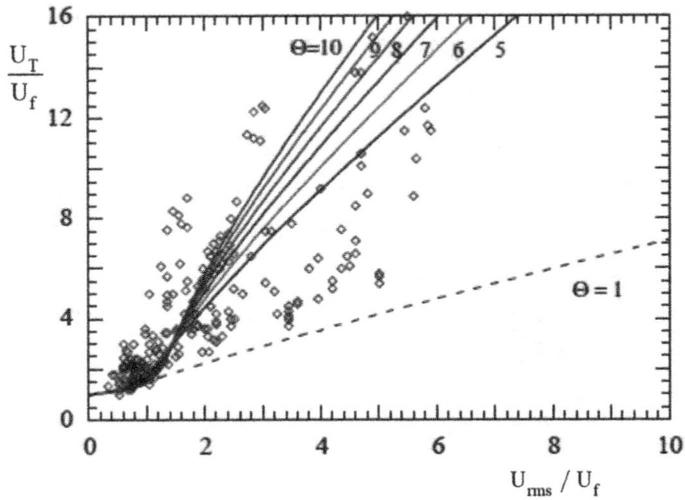

Fig. 14.5 Scaled velocity of turbulent flame versus the scaled velocity of external turbulence for different fuel expansion Θ. The markers show the experimental results from [55]

of perturbations λ are within the interval $\lambda_c < \lambda < L$, where the cutoff wavelength, in general, depends on the Markstein length. The interplay between the DL instability-induced corrugation and turbulence-induced wrinkling of the flame front was reviewed in [114]. In [90–95], two different turbulent combustion modes characterized by distinct scaling laws of turbulent speed were revealed. A *supercritical* turbulent mode is a mode in which the DL instability is dominant, and therefore, in the presence of external turbulence, it exhibits a stronger corrugation. A *subcritical* turbulent mode is a mode in which the originally laminar planar flame remains statistically planar. It was found [90–95] that these two turbulent modes can be characterized by distinct scaling laws for the turbulent speed in the form

$$U_T/U_f \propto \left(U_{rms}/U_f\right)^n. \tag{14.3.20}$$

In particular, for these two turbulent modes, $(U_T/U_f)_{sup} > (U_T/U_f)_{sub}$, where subscripts "sup" and "sub" indicate correspondingly supercritical and subcritical modes. As the intensity of turbulence increases, the DL mechanism gradually weakens, resulting in a common behavior for both subcritical and supercritical flames. Experimental studies of the expanding turbulent spherical flames [129, 130] also indicated a limited range of dominance of the DL instability. However, if DL instabilities can develop in weaker turbulence, then two distinct scaling laws for the turbulence speed, derived through spectral closure techniques and self-similarity arguments, can emerge.

A similar scaling was proposed in [131, 140] for the outwardly expanding spherical flame using the fact that DL instability leads to a self-similar fractal-like structure of a spherically expanding flame. From experimental measurements of constant-pressure expanding turbulent flames, it was concluded that considering the Markstein diffusivity as the leading dissipation mechanism for the large wave number flame surface fluctuations and replacing the thermal diffusivity with the Markstein diffusivity in the definition of the turbulence Reynolds number (Re_T), the normalized turbulent flame speeds are scaled as

$$\frac{U_\mathrm{T}}{U_\mathrm{f}} \propto \left(\frac{U_\mathrm{rms}}{U_\mathrm{f}}\frac{\overline{R}}{L_\mathrm{f}}\right)^{1/2} \sim \left(\frac{U_\mathrm{rms}\overline{R}}{\nu}\right)^{1/2} \sim \sqrt{\mathrm{Re}_T}. \tag{14.3.21}$$

An experimental investigation of the Darrieus–Landau instability effects on turbulent premixed flame propagation was performed using propane–air Bunsen flames at atmospheric pressure in [133], and for near-isotropic turbulence in a dual-chamber and fan-stirred vessel in a number of premixed flame mixtures in [131, 134].

Attempts to obtain a universal dependence of the turbulent flame speed on the intensity of turbulence still continue, but the problem has not yet been resolved. Some of the existing turbulent flame speed formulas are based either on numerical modeling or semi-empirical scaling methods, including dependence on additional parameters such as the mean stretch rate [133], non-local behavior resulting from the advection of wrinkles, the brush thickness, and stretch factor [66, 73, 74, 113–117], but the problem is still elusive. Despite substantial progress in understanding the discussed issues, further theoretical and experimental researches are needed.

In general, the scaled turbulent flame velocity $U_\mathrm{T}/U_\mathrm{f}$ depends more or less linearly on the scaled turbulent intensity u'/U_f for weak turbulence, and the value of $U_\mathrm{T}/U_\mathrm{f}$ is saturated at higher turbulent intensities. However, it is seen in Figs. 14.4 and 14.5 that the experimental data are greatly scattered and depend heavily on the flow configurations, so that the values of $U_\mathrm{T}/U_\mathrm{f}$ can differ by order(s) of magnitude for the same value of u'/U_f. It should be noted that the assumption of some universal formula with functional dependence similar to Eq. (14.3.10) means also that the turbulent flame speed should be independent of the specific geometry of the experimental setup. In this case, the experimental results for $U_\mathrm{T}/U_\mathrm{f}$ can be directly compared for different experiments, regardless of whether measurements were carried out in a combustion bomb [55], for Bunsen flames [56, 57], for flames propagating in tubes [58, 59], or in other configurations. But in reality, the experimental data are highly scattered and depend heavily on the flow configurations. Already the first review of experimental works [55] demonstrated considerable scatter of $U_\mathrm{T}/U_\mathrm{f}$ values measured by different groups for the same value of u'/U_f, and subsequent experiments [57, 65, 66, 128] cast doubt on the very possibility of a simple functional dependence in the form of Eq. (14.3.10). Many authors try to modify Eq. (14.3.10) by including the Reynolds and/or Karlovitz number in the dependence of $U_\mathrm{T}/U_\mathrm{f}$ from u'/U_f, but later experiments [56–59, 64] demonstrated that the turbulent flame speed in tubes depends strongly on the boundary conditions and on the Darrieus–Landau instability at low turbulent intensity. A major obstacle to progress in the field of premixed turbulent combustion has been the expectation that segments of real flames should have the properties of an ideal "geometry-independent" turbulent flame. However, experiments [66] indicate that the wrinkling process is "geometry-dependent" and has a "memory" of whatever wrinkling occurred at upstream locations [115]. It is well established that experimental results on the turbulent flame velocity are quite different for different experimental setups (flames propagating in a tube, Bunsen flames, or spherically expanding flames), and even for different flow conditions within the same experimental setup. Since the experimental data are highly scattered (the difference in the values of U_T / U_f for the same value of u' / U_f is about an order of magnitude) and strongly depend on the chosen flow configurations, the comparison is made primarily to the various correlations suggested in the literature based on the collected data. Turbulence not only wrinkles the flame surface, which leads to an increase of its surface area and the velocity of flame propagation, but can also affect the inner structure of the flame, making it thinner or wider and changing the local heat release rate within it. The effect has been considered in a number of experimental, computational, and theoretical studies, but the problem of whether variations in the local flame thickness are controlled by the local stretch rate or by the penetration of small-scale turbulent eddies into the flame still remains intricate and unresolved [113–115, 141].

14.3.3 Numerical Simulations of Turbulent Combustion

In this section, we briefly describe the main methods used for numerical simulations of turbulent combustion. Readers interested in a more detailed discussion of premixed and non-premixed turbulent flame modeling are referred to the review articles [18, 106–114, 116–121]. In practice, the majority of combustion devices operate in the thin reaction zones regime, the advantage of which is a higher volumetric heat release and a shorter combustion time. In the classification of combustion regimes, this is consistent with the assumption that small turbulent eddies do not alter the chemistry and transport within the flame, which retains its laminar structure with transport and chemistry accounted for through the Markstein length. This is the large-scale or the flamelet regime of turbulent combustion in the classification of combustion regimes. Due to the complexity of turbulent combustion simulations, special computational methods are required.

A central challenge in modeling is to ensure that the computations resolve all of the most important hydrodynamic and chemical time and space scales, which requires adequate gridding and time step. Since in direct numerical simulations (DNS) all scales of turbulent motion are resolved, it is an ideal tool to investigate the flame–turbulence interaction. The advantage of DNS is that it does not require any models or empirical constants to study the interaction of turbulence with flames. In various DNS studies, the interaction of the flame front with turbulence has been investigated in order to clarify the effects of flame stretch and curvature on burning rates, to assess the accuracy of reduced chemical models, and to develop models for closing the unresolved scales in RANS and LES [142]. However, the scales in combustion processes may span about six orders in space, from the smallest turbulent scale, ℓ_η, and the chemical reaction width of a laminar flame to an integral turbulent scale (about the size of the combustion chamber), and even larger range of time scales. Turbulent scales are adequately resolved if the largest and smallest eddies are captured by the mesh. The ratio of the largest and smallest turbulent scales is $Re^{3/4}$, which means that DNS of the flow characterized by a high turbulent Reynolds number is a complex computational problem. Even for the case of incompressible nonreacting flow, the size of numerical mesh required for performing 3D DNS of a turbulent flow scales as Re^3. This means that for an overall length scale of the system of about 10 cm, the total number of mesh points should be $10^5 \times 10^5 = 10^{10}$ for 2D and 10^{15} for 3D problems, respectively. Therefore, the high computational cost of DNS limits its application to simple model problems within relatively small domains and short time intervals, and most DNS studies have used a one-step chemical mechanism or performed simulated 2D geometry, often using both simplifications. In the case of combustion, the main challenge lies not only in a significant increase in the number of transport equations to be solved if a detailed chemical scheme is used, but also in the expansion of the range of spatial scales to be resolved. Because DNS of turbulent combustion is too expensive, in practice, other techniques such as the RANS or LES are used for most applications.

The traditional strategy used to model combustion processes deals with the mean flame characteristics, which are determined by averaging them over time. However, this technique, known as the Reynolds averaging Navier–Stokes (RANS) method, is not always sufficient for complex flows. The flamelet-structure approach includes two RANS models: the coherent flamelet model (CFM) [143–145], which solves the flame surface density balance equation, or another RANS approach [43, 44], which solves the G-equation.

Large eddy simulation (LES) [81, 111] has emerged as an alternative mathematical framework for the modeling of complex multiscale problems and has already been applied to a variety of combustion problems of technical interest [111, 146]. In the LES approach, the largest flow structures are computed explicitly, as in the DNS, while the effects of small-scale structures are modeled. LES simulations are more expensive than RANS calculations, but faster than corresponding DNS. A spatial filtering operation applied to the instantaneous turbulent fields removes turbulent motions of length scales smaller than the filter size Δ. The large-scale turbulence, which contains most of the turbulent kinetic energy and controls the dynamics of turbulence, is resolved and computed directly. Therefore, knowledge of the large-scale dynamics and the assumption that an applied model should be valid independently of the filter size lead to the formulation of the so-called dynamic models. All the same, the LES depends heavily on the adopted closure assumptions or sub-grid models, making it difficult to assess the accuracy of the obtained results. The LES numerical technique offers an opportunity to straightforwardly resolve eddies of large and medium scales and requires modeling of small-scale processes, which are usually considered to be universal and less sensitive to boundary conditions. Nevertheless, it is worth remembering that since most of the heat is released in the thin reaction zone, the most important effects associated with the interaction of flame with turbulence are small-scale effects that are not resolved and require separate modeling [113, 115]. If the combustion effects on small and large scales can be separated, then under the limiting assumption of isotropy and a wide turbulent spectrum, the renormalization method can be applied for small scales which makes it possible to obtain analytical solutions for turbulent combustion. The small-scale turbulent combustion can be described separately using an equation similar to Eq. (14.3.10) for the local average turbulent flame velocity.

An efficient computational method of a *thickened flame,* developed in [79, 82, 147], was applied in large eddy simulations. The main idea of the thickened flame method is that adopting the renormalization idea and applying the concept of turbulent flame velocity in the form similar to Eq. (14.3.10) for the local average turbulent flame velocity, one can expect that the numerical results will differ from the experimental data by a numerical factor caused by large-scale effects. In the thickened flame approach, the chemical-transport interaction is artificially modified to obtain a thickened flame that can be resolved using not too small cells of the computational grid. To achieve this, the turbulent transport coefficients are multiplied by a constant factor, appropriately increasing the viscosity value, thermal conductivity, diffusion, and flame thickness. The chemical source term is divided by the same factor to obtain the same flame velocity as in the unmodified case and replace the laminar flame velocity with a turbulent one. Although this method resolves the artificial flame problem, the flame–turbulence interaction must be changed from a transport-controlled to a chemistry-controlled combustion regime, for an adequate description of the heat release, which is achieved using the so-called efficiency function, which is determined from the DNS of the flame–vortex interaction. In addition, another important simplification, which is widely used to model a premixed turbulent combustion, consists in describing the state of the reacting mixture using a single scalar variable, which significantly reduces the number of reaction species and transport equations to be solved. In order to justify such simplification, combustion chemistry should be reduced to a single irreversible reaction.

Another numerical method that is used to overcome the problem of flame resolution when simulating the flamelet model is the G-equation model [104] for premixed turbulent combustion. In the G-equation model, the flame-front position is represented with a constant value G_0 of the level set function $G(\mathbf{r})$. The value of G is chosen so that $G = 0$ at the flame front, $G < 0$ in the unburned mixture, and $G > 0$ in the burned gases. The surface represented by the level set function can be chosen as a surface of constant temperature, or as a reaction progress variable, or other similar quantity. The assumption of an infinitely thin flame is not required. As for the reaction progress variable equation, closure for the G-equation describing the mean evolution of the flame front is still required, which was done by Peters [18, 44] for Reynolds averaging. The unfiltered G-equation is valid only at the instantaneous flame-front location. Therefore, a filtering procedure is applied to derive the appropriate LES equation. The filter kernel is used to derive the G-equation for the filtered flame-front location. This equation has two unclosed terms, including the flame-front conditional averaged flow velocity and the filtered propagation term. A model for the conditional velocity was derived, expressing this quantity in terms of the Favre-filtered flow velocity [111].

14.4 Influence of Chemical Reactions on Turbulent Transport

Due to the complexity of turbulent combustion, many researchers use turbulent combustion models in which the chemistry is modeled by a one-step reaction between oxidizer and fuel. Obviously, such simplified combustion models cannot be used, for example, for predicting pollutant emissions or for engineering optimization purposes for internal combustion engines (IC) or gas turbines. It is now widely accepted that a correct description of physical processes in both laminar and turbulent combustion can be achieved only by including detailed chemical kinetics in the model [109, 148, 149].

The effect of chemical reactions on the turbulent diffusion of chemically reacting gaseous mixtures in developed turbulence has been studied analytically using the path-integral approach for a δ-correlated-in-time random velocity field [150]. It also was studied applying the spectral τ-approach, which is valid for large Reynolds and Péclet numbers [151]. It was shown that the turbulent diffusion of reacting species can be strongly suppressed with the increase of the Damköhler number.

In the case of subsonic velocities, the flow can be considered as incompressible. Then, the equation for the scalar field in the incompressible chemically reacting flow reads

$$\frac{\partial c}{\partial t} + (\mathbf{v} \cdot \nabla)c = W(c) + D\Delta c, \tag{14.4.1}$$

where $c(t, \mathbf{x})$ is a scalar field, $\mathbf{v}(t, \mathbf{x})$ is the instantaneous velocity field, D is the constant diffusion coefficient based on molecular Fick's law, and W(c) is the source (or sink) term. To illustrate the influence of chemical reactions on turbulent transport, consider a simplified model of a one-step reaction with a function W(c) chosen as the Arrhenius law, which is often used in DNS of turbulent combustion. The velocity of the gas mixture is determined by the Navier–Stokes equation:

$$\frac{\partial \mathbf{v}}{\partial t} + (\mathbf{v} \cdot \nabla)\mathbf{v} = -\frac{1}{\rho}\nabla p + \nu\Delta\mathbf{v} + f, \tag{14.4.2}$$

where $\mathbf{f}$ is the external force to support turbulence, $\nu$is the kinematic viscosity, p and ρ are the fluid pressure and density, respectively, and the gas density is constant for an incompressible flow.

14.4.1 Turbulent Flux

To determine turbulent transport coefficients, Eq. (14.4.1) is averaged over an ensemble of turbulent velocity fields. In the framework of a mean-field approach, the scalar field $c(t, \mathbf{x})$ is decomposed into the mean field $\langle c\rangle$ and fluctuations c', where $\langle c'\rangle = 0$ and angular brackets designate ensemble averaging. This can be considered, for example, as the averaging over an ensemble of independent snapshots taken in numerical simulations with the time interval that is larger than the turbulent correlation time. In particular, for comparison with the DNS results, we employ a spatial averaging over a plane perpendicular to the direction of the flame-front propagation. The velocity field is decomposed in a similar fashion, $\mathbf{v} = \langle \mathbf{U}\rangle + \mathbf{u}$, assuming, for simplicity, a vanishing mean fluid velocity $\langle \mathbf{U}\rangle$, where $\mathbf{u}$ are the gas velocity fluctuations.

Using the equation for fluctuations $c' = c - \langle c\rangle$ of the scalar field and the Navier–Stokes equation for fluctuations $\mathbf{u}$ of the velocity field written in $\mathbf{k}$ space, we derive an equation for the second-order moment $\langle c'\mathbf{u}\rangle_{\mathbf{k}} \equiv \langle c'(\mathbf{k})u_i(-\mathbf{k})\rangle$:

$$\frac{\partial \langle c' u_i\rangle_{\mathbf{k}}}{\partial t} = -\left[\tau_c^{-1} + (\nu + D)k^2\right]\langle c' u_i\rangle_{\mathbf{k}} + \widehat{N}\langle c' u_i\rangle - \langle u_i u_j\rangle_{\mathbf{k}} \nabla_j \langle c\rangle, \quad (14.4.3)$$

where $\langle u_i u_j\rangle_{\mathbf{k}} \equiv \langle u_i(\mathbf{k}) u_j(-\mathbf{k})\rangle$, $\tau_c = \langle c\rangle/\langle W\rangle$, is the chemical time, $\langle W\rangle$ is the mean source function, and $\widehat{N}\langle c' u_i\rangle$ includes the third-order moments caused by the nonlinear terms:

$$N\langle c' u_i\rangle = -\langle [\nabla \cdot (c'\mathbf{u})] u_i\rangle_{\mathbf{k}} - \langle c'[(\mathbf{u}\cdot\nabla) u_i]\rangle_{\mathbf{k}} - \langle c'[\rho^{-1}\nabla_i p']\rangle_{\mathbf{k}}. \quad (14.4.4)$$

The procedure of derivation of the turbulent fluxes is described in detail in [151], with an account of a large yet finite Reynolds number. In particular, we use a multi-scale approach, that is, we separate fast and slow variables, where fast small-scale variables correspond to fluctuations and slow large-scale variables correspond to mean fields. Since the ratio of the spatial density of species is assumed to be much smaller than the density of the surrounding gas (i.e. small mass-loading parameter), there is only one-way coupling, i.e. no effect of species on the fluid flow. For the same reason, the energy release caused by chemical reactions is much smaller than the internal energy of the surrounding gas mixture. We also assume that the deviations of the source term W from its mean value $\langle W\rangle$ are not large. Although this assumption does not hold in a typical premixed turbulent flame, we will show that the theory predicts well the effect of a chemical reaction on turbulent transport, even if the difference between W(c) and $\langle W\rangle(\langle c\rangle)$ is significant, as it usually occurs in the case of premixed combustion.

The equation for the second-order moment (14.4.3) includes the first-order spatial differential operators applied to the third-order moments $\widehat{N}\langle c' u_i\rangle$. To close the system of equations, it is necessary to express the third-order terms $\widehat{N}\langle c' u_i\rangle$ through the lower order. This can be done using the spectral-τ approximation [153] that postulates that the deviations of the third-order moments $\widehat{N}\langle c' u_i\rangle$ from the contributions to these terms afforded by the background turbulence $\widehat{N}\langle c' u_i\rangle^{(0)}$ can be expressed through the similar deviations of the second-order moments $\langle c' u_i\rangle_{\mathbf{k}} - \langle c' u_i\rangle_{\mathbf{k}}^{(0)}$,

$$\widehat{N}\langle c' u_i\rangle - \widehat{N}\langle c' u_i\rangle^{(0)} = -\frac{1}{\tau_r(k)}[\langle c' u_i\rangle_k - \langle c' u_i\rangle_k^{(0)}], \tag{14.4.5}$$

where $\tau_r(k)$ is the scale-dependent relaxation time, which can be identified with the correlation $\tau(k)$ of the turbulent velocity field at large Reynolds and Peclet numbers. The functions with the superscript (0) correspond to the background turbulence with zero gradients of the mean scalar field. The validation of the τ approximation for different situations has been verified in various numerical simulations and analytical studies. When the gradients of the mean scalar field are zero, the turbulent flux vanishes, and the contributions of the corresponding fluctuations [the terms with the superscript (0)] vanish as well. Consequently, Eq. (14.4.5) reduces to $\widehat{N}\langle c' u_i\rangle_{\mathbf{k}} = -\langle c'(\mathbf{k}) u_i(-\mathbf{k})\rangle/\tau(k)$.

We also assume that the characteristic times of variation of the second-order moments are substantially larger than the correlation time $\tau(k)$ for all turbulence scales. This allows us to consider the steady-state solution of Eq. (14.4.5), which yields the following expression for the turbulent flux $\langle c' u_i\rangle_{\mathbf{k}} = \langle c'(\mathbf{k}) u_i(-\mathbf{k})\rangle$ in $\mathbf{k}$ space:

$$\langle c' u_i\rangle_{\mathbf{k}} = -\tau_{\text{eff}}(k)\,\langle u_i u_j\rangle_{\mathbf{k}}^{(0)} \nabla_j \langle c\rangle, \tag{14.4.6}$$

where $\tau_{\text{eff}}(k) = \left[\tau_c^{-1} + (\nu + D)k^2 + \tau^{-1}(k)\right]^{-1}$ is the effective time.

We assume also that background turbulence is isotropic and homogeneous, so that $\langle u_i u_j\rangle_{\mathbf{k}}^{(0)} \equiv \langle u_i(\mathbf{k}) u_j(-\mathbf{k})\rangle$, and

$$\langle u_i(\mathbf{k}) u_j(-\mathbf{k})\rangle = \frac{E_T(k)}{8\pi k^2}\left[\delta_{ij} - \frac{k_i k_j}{k^2} + \frac{i}{2k^2}(k_i \nabla_j - k_j \nabla_i)\right], \tag{14.4.7}$$

where

$$E_T(k) = \frac{2}{3k_0}\left(1 - \text{Re}^{-1/2}\right)^{-1}\left(\frac{k}{k_0}\right)^{-5/3} \tag{14.4.8}$$

is the energy spectrum function for $k_0 \le k \le k_0 \text{Re}^{3/4}$, $\tau(k) = 2\tau_0(k/k_0)^{-2/3}$ is the turbulent correlation time, $k_0 = \ell_0^{-1}$, and $\text{Re} = u_0 \ell_0/\nu \gg 1$ is the Reynolds number, and $u_0 = \sqrt{\langle \mathbf{u}^2\rangle}$ is the characteristic turbulent velocity in the integral scale, ℓ_0, of turbulence. The last two terms in Eq. (14.4.7) determine contributions from inhomogeneous turbulence. To compare the theory with DNS, we do not neglect the small term $\text{Re}^{-1/2}$ in Eq. (14.4.8).

The turbulent flux $\langle c'\mathbf{u}\rangle$ is obtained by integration in $\mathbf{k}$ space,

$$\langle c'\mathbf{u}\rangle = \int \langle c'u_i\rangle_{\mathbf{k}} d\mathbf{k} = -D_T \nabla \langle c\rangle, \tag{14.4.9}$$

where the coefficient of turbulent diffusion D_{T} of the scalar field is

$$D_T = \frac{D_0^T}{Da}\left[1 - \frac{\Phi(Da,\,\mathrm{Re},\,\mathrm{Pr})}{1-\mathrm{Re}^{-1/2}}\right], \tag{14.4.10}$$

$D_0^T = \tau_0 u_0^2/3$ is the characteristic value of the turbulent diffusion coefficient without chemical reactions, $\tau_0 = \ell_0/u_0$ is the characteristic turbulent time, and the function $\Phi(\mathrm{Da},\,\mathrm{Re},\,\mathrm{Pr})$ is

$$\Phi(\mathrm{Da},\,\mathrm{Re},\,\mathrm{Pr}) = \int_{\mathrm{Re}^{-1/2}}^{1} \frac{X^2 + a(\mathrm{Re},\,\mathrm{Pr})}{2\mathrm{Da}X^3 + X^2 + a} dX, \tag{14.4.11}$$

where $\mathrm{Pr} = \nu/D$ is the Prandtl number and $a(\mathrm{Re},\mathrm{Pr}) = 2\left(1+\mathrm{Pr}^{-1}\right)/\mathrm{Re}$.

The integral in Eq. (14.4.11) can be evaluated by expanding the integrand over a small parameter $a(\mathrm{Re},\,\mathrm{Pr})$ for large yet finite Reynolds numbers. Then, we obtain the following dependence of turbulent diffusion coefficient on Damköhler and Reynolds numbers:

$$\begin{aligned} D_T = \frac{D_0^T}{\mathrm{Da}}&\left[1 - \frac{1}{2\mathrm{Da}\left[1-\mathrm{Re}^{-1/2}\right]} \ln \frac{1+2\mathrm{Da}}{1+2\mathrm{Da}\mathrm{Re}^{-1/2}}\right] \\ &-2D_0^T\left(1+\mathrm{Pr}^{-1}\right)\frac{\ln \mathrm{Re}}{\mathrm{Re}}. \end{aligned} \tag{14.4.12}$$

In the limit of very large Reynolds numbers, we obtain

$$D_T = \frac{D_0^T}{Da}\left[1 - \frac{\ln(1+2Da)}{2Da}\right]. \tag{14.4.13}$$

Equation (14.4.12) implies that for small Damköhler numbers, $Da \ll 1$, the function $D_T(\mathrm{Da})$ is given by

$$D_T = D_0^T[1 - \frac{4\mathrm{Da}}{3} - 2D_0^T\left(1+\mathrm{Pr}^{-1}\right)\frac{\ln \mathrm{Re}}{\mathrm{Re}} + \mathrm{Re}^{-1/2}], \tag{14.4.14}$$

while for large Damköhler numbers, $\sqrt{\mathrm{Re}} \gg \mathrm{Da} \gg 1$, it is

$$D_T = \frac{D_0^T}{\mathrm{Da}}\left[1 - \frac{\ln 2\mathrm{Da}}{2\mathrm{Da}} - 2\left(1+\mathrm{Pr}^{-1}\right)\frac{\mathrm{Da}\ln \mathrm{Re}}{\mathrm{Re}}\right], \tag{14.4.15}$$

and for very large Damköhler numbers, $1 \ll \sqrt{\text{Re}} \ll \text{Da}$, it is

$$D_{\text{T}} = \frac{D_0^T}{\text{Da}} \left[1 - 2\left(1 + \text{Pr}^{-1}\right) \frac{\text{Da} \ln \text{Re}}{\text{Re}} \right]. \tag{14.4.16}$$

Equations (14.4.15) and (14.4.16) show that turbulent diffusion for a large Damköhler number, $\text{Da} \gg 1$, is considerably reduced, i.e. $D_{\text{T}} = D_0^{\text{T}}/\text{Da} = \tau_c u_0^2/3$. This implies that turbulent diffusion for a large turbulent Damköhler number is determined by the chemical time. The underlying physics of the strong reduction of turbulent diffusion becomes obvious if we consider as an example a first-order chemical reaction $\text{A} \to \text{B}$. The species A of the reactive mixture are consumed and their concentration decreases much faster during the chemical reaction, so that the usual turbulent diffusion based on the turbulent time $\tau_0 \gg \tau_c$ does not contribute to the mass flux of a reagent A.

14.4.2 Comparison with Numerical Simulations

The results of the analytical theory of the influence of chemistry on turbulent transport were compared with DNS of a reaction wave of finite thickness, modeled using Eq. (14.4.1) for a single progress variable c ($c = 0$ in reactants and $c = 1$ in products) propagating in forced, homogeneous, isotropic, and incompressible turbulence for chemical reactions of the first order. To mimic a highly nonlinear dependence of the reaction rate W on the scalar field c in a typical premixed flame characterized by significant variations in the density ρ and temperature T, the following expression for the reaction rate was used

$$W = \frac{1-c}{\tau_R(1+\tau)} \exp\left[-\frac{Ze(1+\tau)^2}{\tau(1+c\tau)} \right], \tag{14.4.17}$$

where τ_{R} is a reaction time scale, while parameters $\text{Ze} = 6.0$ and $\tau = 6.0$ are the Zel'dovich number $\text{Ze} = E_{\text{a}}(T_{\text{b}} - T_{\text{u}})/\text{RT}_{\text{b}}^2$ and the heat-release factor $(\rho_{\text{u}} - \rho_{\text{b}})/\rho_{\text{b}}$, respectively. Here, subscripts u and b designate unburned and burned mixtures. Substitution of $c = (T - T_{\text{u}})/(T_{\text{b}} - T_{\text{u}})$ in the exponent in Eq. (14.4.17) leads to the usual expression for the Arrhenius law

$$W = \frac{1-c}{\tau_{\text{R}}(1+\tau)} \exp\left(-\frac{E_{\text{a}}}{\text{RT}} \right). \tag{14.4.18}$$

The speed U_{L} of the reaction wave propagation in a laminar flow, the wave thickness $L_{\text{F}} = D/U_{\text{L}}$, and the wave time scale $\tau_{\text{F}} = L_{\text{F}}/U_{\text{L}}$ were varied by changing the diffusion coefficient D and the reaction time scale τ_{R}, which were constant input

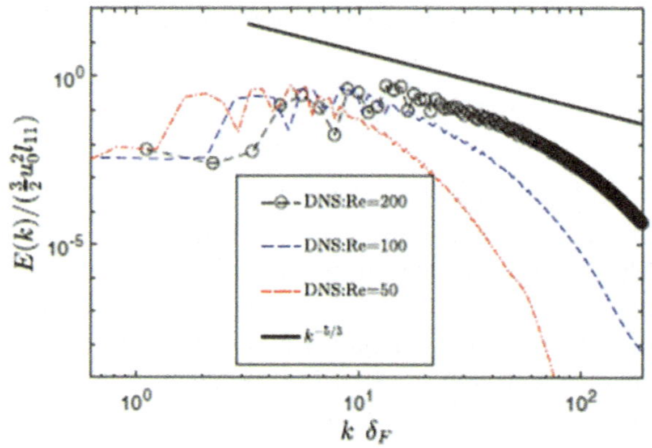

Fig. 14.6 Spectrum of turbulent kinetic energy

parameters for each DNS run. The speed U_L was determined by solving numerically 1D Eq. (14.4.1) with $\mathbf{v} = 0$.

The initial turbulent field was generated by synthesizing the prescribed Fourier waves with the initial rms velocity u_0 and the forcing scale $\ell_f = L/4$, with $L = L_y = L_z = L_x/4$ being the width of the computational domain. Subsequently, a forcing function was invoked to maintain statically stationary turbulence. It was verified that: (i) the rms velocity u_0 was maintained as the initial value, (ii) the normalized dissipation rate $\langle\varepsilon\rangle\ell_f/u_0^3$averaged over the computational domain fluctuated slightly above 3/2 after a short period ($t < \tau_f = \ell_f/u_0$) of rapid transition from the initial artificially synthesized flow to the fully developed turbulence, (iii) the forced turbulence achieved good statistical homogeneity and isotropy over the entire domain, and (iv) the energy spectrum showed a range of the Kolmogorov scaling (−5/3) at the Reynolds number $\mathrm{Re} \equiv u_0\ell_f/\nu = 200$ based on the scale ℓ_f as shown in Fig. 14.6.

In order to study a fully developed reaction wave, a planar wave $c(\mathbf{x}, t = 0) = c_L(\xi)$ was initially ($t = 0$) released at $x^0 = L_x/2$ such that $\int_{-\infty}^{0} c_L(\xi)\mathrm{d}\xi = \int_0^{\infty} [1 - c_L(\xi)]\mathrm{d}\xi$, where $\xi = x - x^0$ and $c_L(\xi)$ is the precomputed laminar-wave profile. Subsequently, the evolution of this field $c(\mathbf{x}, t)$ was simulated by solving Eq. (14.4.1). To enable periodic propagation of the c-field along the x-direction, the field is extrapolated outside the axial boundaries of the computational domain at each time step t^n as $c(x', y, z, t^n) = c(x, z, t^n)$, where $x' = x + I\mathrm{L}_x$ and I is an arbitrary (positive or negative) integer number. Consequently, Eq. (14.4.1) is solved in the interval $x' \in [\zeta(t^n) - \Delta, \zeta(t^n) + \Delta]$, where $\zeta(t^n)$ is the mean coordinate of a reaction wave on the x'-axis and $\Delta = 0.45\mathrm{L}_x$ in order to avoid numerical artifacts in the vicinity of $x' = \zeta(t^n) \pm 0.5\mathrm{L}_x$.

Three turbulent fields were generated by specifying three different initial turbulent Reynolds numbers Re = 50, 100, and 200, which were increased by increasing the domain size L. The increase in L resulted in the increase of the longitudinal integral length scale ℓ_{11}, the Taylor length scale $\lambda = \sqrt{15\nu u_0^2/\langle\varepsilon\rangle}$, the Taylor scale Reynolds

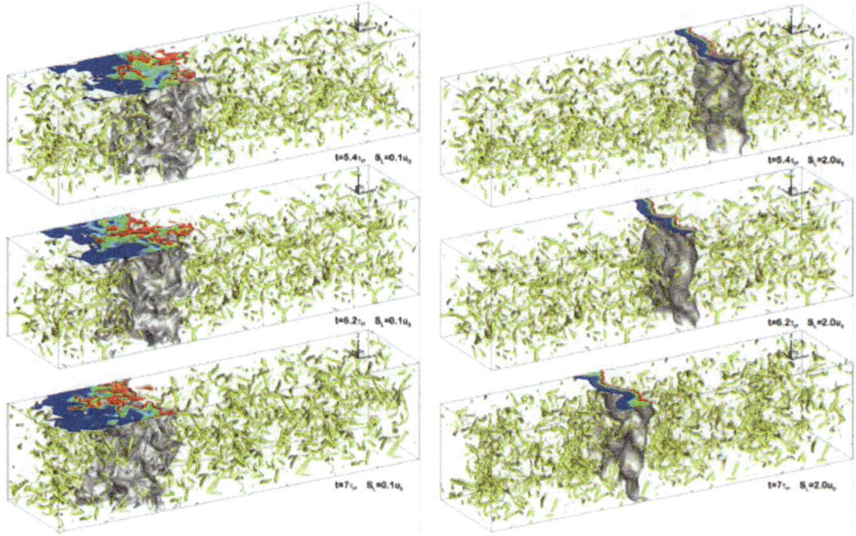

Fig. 14.7 Instantaneous c field obtained in the DNS at $t/\tau_f = 5.4$ (top row), 6.2 (middle row), and 7 (bottom row) for Re = 200, with $U_L/u_0 = 0.1$ (left column) and $U_L/u_0 = 2.0$ (right column). Gray iso-surfaces are associated with the peak reaction rate, i.e. $c(x, y, z, t) = c^*$, where $W(c^*)$ is the maximum value. Top-sliced planes are colored from blue to red in order to represent an increase in the local c from zero to unity. Yellow vortex tubes visualize turbulence field by showing regions characterized by large values $Q > Q^* = 2u_0^2/\tau_f\nu$ of the Q criterion [152]

number $\mathrm{Re}_\lambda = u_0\lambda/\nu$, the turbulent time scale $\tau_{11} = \ell_{11}/u_0$, and, therefore, the Damköhler number $\mathrm{Da}_{\mathrm{DNS}} = \tau_{11}/\tau_F$.

Figure 14.7 shows the evolution of c fields for Re = 200 obtained in simulations [152]. The background turbulent flow structures are visualized by showing regions characterized by large values $Q > Q^* = 2u_0^2/\tau_f\nu$ of the Q criterion. Such a flow structure is typical for homogenous isotropic turbulence. A comparison of these figures shows that a lower normalized laminar-wave speed U_L/u_0 is associated with more wrinkled surfaces of the reaction zone and a thicker mean turbulent wave brush that propagates at a lower speed.

Dependences of the normalized turbulent scalar flux $\langle u'c'\rangle/u_0$ on the mean reaction progress variable are shown in Fig. 14.8: (**a**) Re = 50, (**b**) Re = 100, (**c**) Re = 200. In the unburned or burned mixtures, the instantaneous progress variable is constant, $c = 0$ or $c = 1$, respectively. This implies that in the two regions, turbulent flux $\langle c'\mathbf{u}'\rangle = 0$. Inside the mean reaction wave, the mean progress variable $\langle c\rangle$ varies between 0 and 1. In this region, the gradient $\nabla\langle c\rangle$ does not vanish. Since $\nabla\langle c\rangle$ is positive in this region, the turbulent flux $\langle c'\mathbf{u}'\rangle = -D_T\nabla\langle c\rangle$ is negative inside the mean reaction wave. The absolute value of the turbulent flux $|\langle c'\mathbf{u}'\rangle|$ reaches a maximum at the point where the gradient $\nabla\langle c\rangle$ is maximum. If the probability of deviation of the reaction wave from its mean position is described by the Gaussian

Fig. 14.8 Dependences of the normalized turbulent scalar flux $\langle u'c'\rangle/u_0$ on the mean reaction progress variable ⟨c⟩, computed at different ratios of U_L/u_0 specified in the legend for (**a**) Re = 50, (**b**) Re = 100, (**c**) Re = 200 [152]

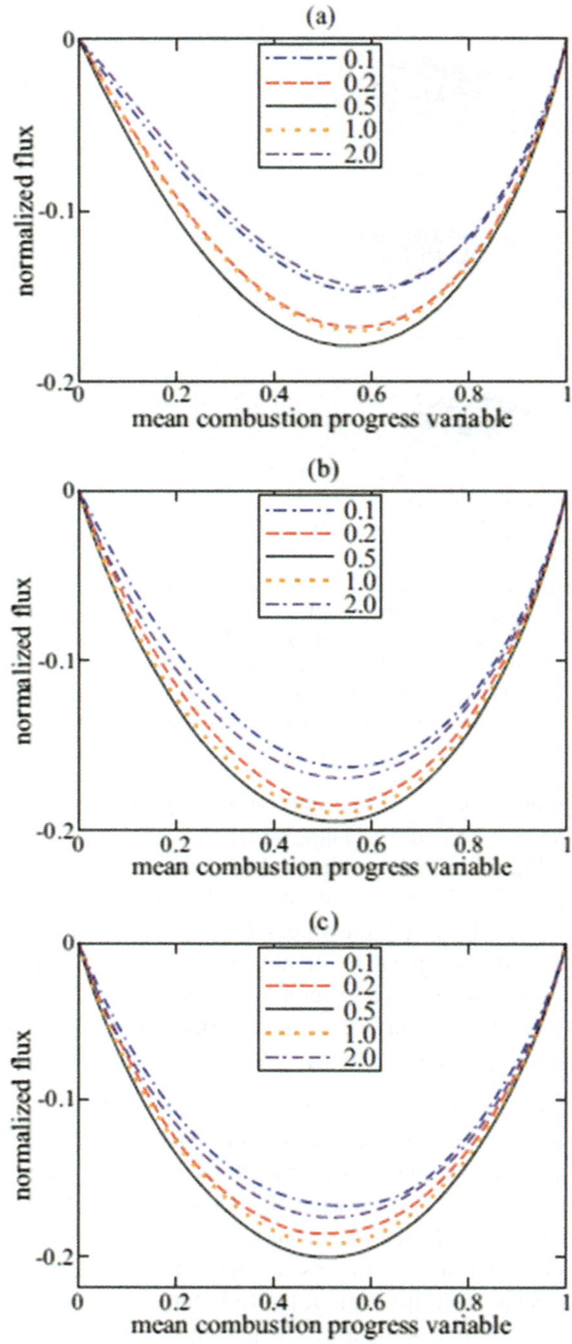

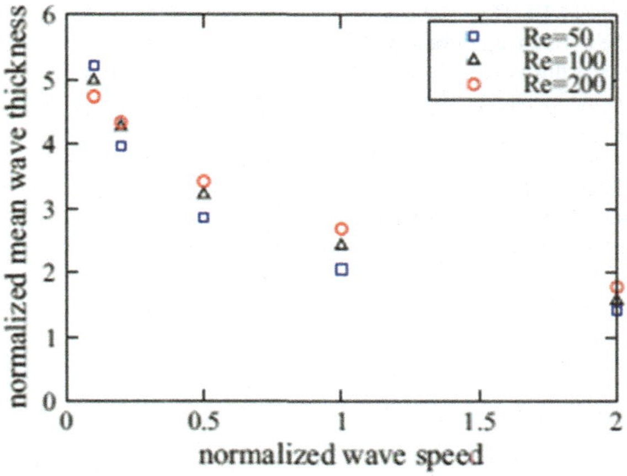

Fig. 14.9 Dependences of the normalized mean turbulent wave thickness δ_t/ℓ_{11} on the normalized wave speed U_L/u_0 computed at three different turbulent Reynolds numbers specified in legend [152]

distribution, the gradient $\nabla\langle c\rangle$ is maximum at $\langle c\rangle = 0.5$. For instance, in various premixed turbulent flames, $\nabla\langle c\rangle$ does peak at $\langle c\rangle = 0.5$ according to [121]. While the flux magnitude depends on U_L/u_0 and hence on Da_{DNS}, such variations in the flux magnitude are sufficiently weak and nonmonotonic, with the peak magnitude being obtained at a medium $U_L/u_0 = 0.5$.

On the contrary, the mean turbulent wave thickness δ_t defined using the maximum gradient method,

$$\delta_t = \frac{1}{\max\{\nabla_x\langle c\rangle\}}, \tag{14.4.19}$$

decreases rapidly with the increase of the normalized wave speed U_L/u_0 and hence Da_{DNS} (shown in Fig. 14.9). This numerical result is fully consistent with the theory, which predicts a decrease in D_T with the increasing Damköhler number. Under the DNS conditions, an increase in U_L/u_0 results in increasing Da and, therefore, decreasing D_T. Consequently, $\delta_t \propto [D_T(\mathrm{Da})\,\tau_c]^{1/2}$ decreases with increasing U_L/u_0.

Accordingly, the gradient of the mean reaction progress variable is increased by U_L/u_0 (or Da_{DNS}), whereas turbulent diffusivity evaluated as

$$D_T(\langle c\rangle) = -\frac{\langle u'c'\rangle}{\nabla_x\langle c\rangle} \tag{14.4.20}$$

decreases with increasing U_L/u_0 and Da_{DNS} as shown in Fig. 14.10.

The decrease of the turbulent diffusion coefficient D_T with the increase of the Damköhler number observed in DNS agrees well with the theory. Moreover, Fig. 14.10 indicates that D_T evaluated using Eq. (14.4.20) depends weakly on $\langle c\rangle$, thus implying that the influence of the reaction on the turbulent diffusion coefficient may be characterized by a single mean turbulent diffusivity defined as

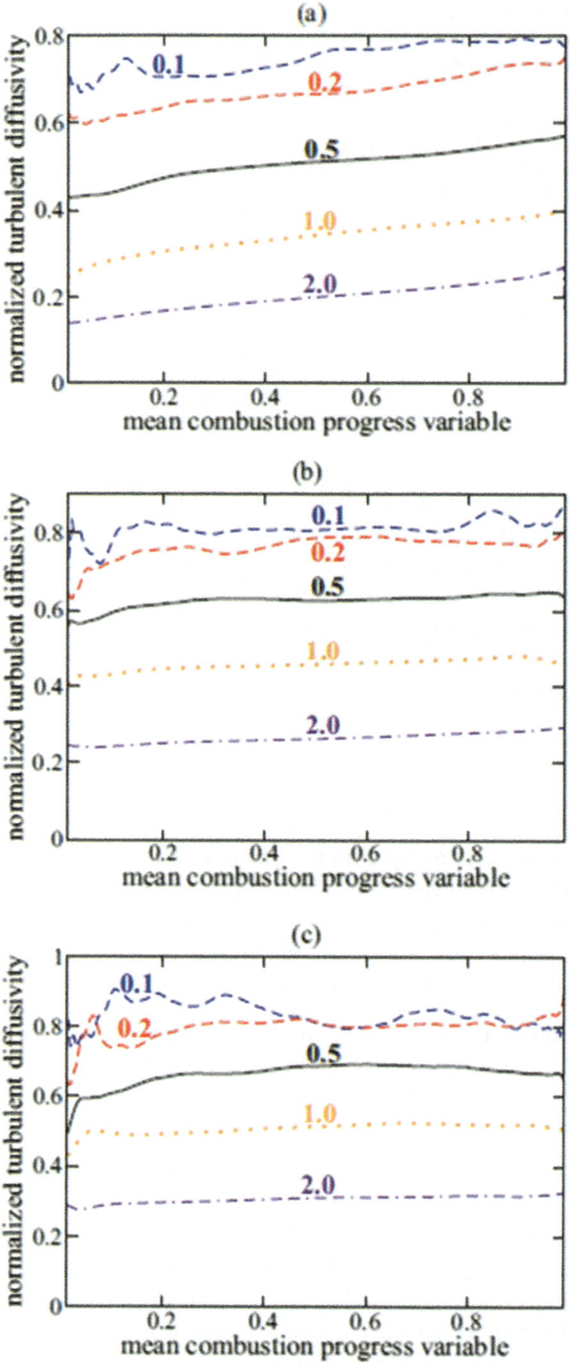

Fig. 14.10 Dependences of the normalized turbulent scalar diffusivity $D_T/(u_0\ell_{11})$ on the mean reaction progress variable $\langle c \rangle$, computed for (**a**) Re = 50, (**b**) Re = 100, and (c) Re = 200. Values of U_L/u_0 are specified near the curves

$$\hat{D}_{\mathrm{T}} = \int_0^1 D_{\mathrm{T}}(\xi)d\xi. \qquad (14.4.21)$$

To compare values of the mean turbulent diffusion coefficient, $\hat{D}_{\mathrm{T}}$, obtained in the simulations with the theoretical predictions for $D_{\mathrm{T}}(\mathrm{Da})$, we must take into account that the Damköhler number, Da_{DNS}, used in DNS is different from the Damköhler number, Da, used in the theoretical analysis in Sect. 14.4.1. In the theoretical analysis, Da involves the time scale τ_c that characterizes the peak mean rate $\langle W\rangle$. In the DNS, due to strong fluctuations in the scalar field $c(t, x)$ and especially $W[c(t, x)]$, the peak mean rate $\max\{\langle W\rangle(\langle c\rangle)]\}$ is much less than the peak local rate $\max\{W[c(t, x)]\}$, as is well known in combustion theory [117, 120, 121]. Accordingly, the peak mean reaction rate is characterized by a significantly larger chemical time scale $\langle\tau_c\rangle$ when compared to the time scale τ_f associated with the laminar $W(c)$. A ratio of these time scales can be estimated as $\theta = \tau_f/\langle\tau_c\rangle = \mathrm{Da}/\mathrm{Da}_{\mathrm{DNS}}$. The reaction-rate ratio θ versus U_{L}/u_0 is shown in Fig. 14.11 for different Reynolds numbers. Using the values of θ obtained in the DNS and plotted in Fig. 14.11, we relate the Damköhler number Da used in the theory with the $\mathrm{Da}_{\mathrm{DNS}}$ used in DNS: $\mathrm{Da} = \theta \mathrm{Da}_{\mathrm{DNS}}$. The mean turbulent diffusion coefficient $\hat{D}_{\mathrm{T}}$ versus $\mathrm{Da} = \theta\mathrm{Da}_{\mathrm{DNS}}$ obtained in the simulations (symbols) is compared with the theoretical predictions for D_{T} given by Eq. (14.4.12) in Fig. 14.12.

It can be seen that chemistry significantly affects the turbulent transport if the Damköhler number is large, $\mathrm{Da} > 1$, as expected. However, the influence is quite noticeable even for small Damköhler number, $\mathrm{Da} < 1$. Although for small Damköhler numbers, the hydrodynamic description of the flame assumes that turbulent eddies do not penetrate the flame and do not change the chemistry and transport inside the flame, which retains its laminar structure, however, as it is seen in Fig. 14.12, even for small Damköhler numbers, the chemistry itself can noticeably modify the transport.

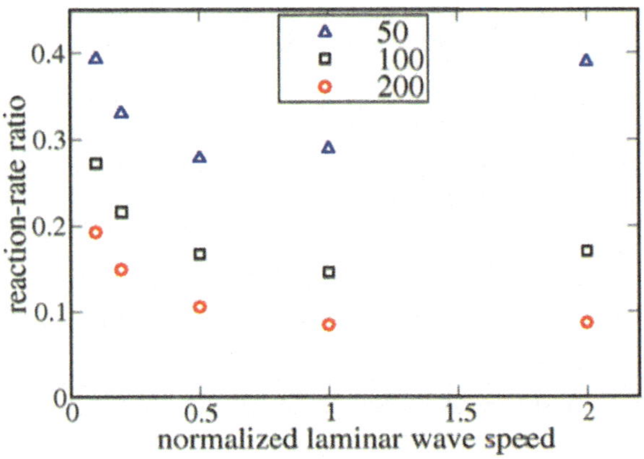

Fig. 14.11 Reaction-rate ratio $\theta = \max\{\langle W\rangle(\langle c\rangle)]\}/\max\{W(c)\}$ versus U_{L}/u_0. Symbols show DNS data, with the Reynolds numbers Re being specified in the legend [152]

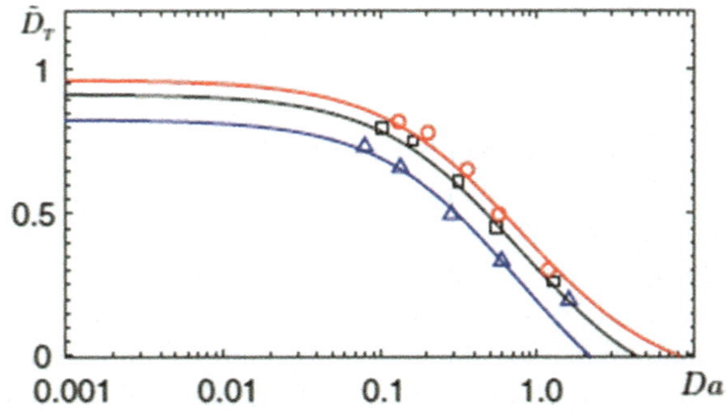

Fig. 14.12 Theoretical dependence $\hat{D}_{\mathrm{T}} \equiv D_{\mathrm{T}}/D_0^{\mathrm{T}}$ on the Damköhler number Da determined by Eq. (14.1.12) for different values of the Reynolds number: Re = 50 (blue), Re = 100 (black), and Re = 200 (red) at Pr = 1. The DNS data for $\hat{D}_{\mathrm{T}}$ normalized using the turbulent scalar diffusivity $D_{\mathrm{T}}/(u_0\ell_{11})$ are shown in blue triangles, black squares, and red circles

References

1. G.K. Batchelor, *An Introduction to Fluid Dynamics* (Cambridge University Press, Cambridge, United Kingdom, 1967)
2. S.B. Pope, Turbulent Flows. Cambridge University Press 2000.
3. L.F. Richardson, *Weather Prediction by Numerical Process* (Cambridge University Press, Cambridge, United Kingdom, 1922)
4. A.N. Kolmogorov, Energy dissipation in locally isotropic turbulence. Dokl. Akad. Nauk SSSR **32**, 19–21 (1941)
5. A.N. Kolmogorov, Local structure of turbulence in an incompressible fluid at very high Reynolds numbers. Dokl. Akad. Nauk SSSR **30**, 299–303 (1941)
6. A.N. Kolmogorov, Equations of turbulent motion of an incompressible fluid. Izvestia Acad. Sci. USSR, Phys. 6 (1942) pp. 56–58.
7. For English translation of Refs. [4–6] see V.M. Tikhomirov (ed.) Selected Works of A. N. Kolmogorov, vol. 1, p. 242– 270. Kluwer, Dordrecht (1991).
8. L. Prandtl, Nachr. Akad. Wiss. Götingen Math- Phys. **K1**, 6–19 (1945)
9. G. Damköhler, Der Einfluss der Turbulenz auf die Flammengeschwindigkeit in Gasgemischen (The influence of turbulence on the flame speed in gas mixtures). Z Elektrochem. **46**, 601–652 (1940)
10. R.W. Borghi, On the structure and morphology of turbulent premixed flames. In: Recent Advances in the Aerospace Science (Ed. C. Bruno and C. Casci) pp. 117–138. Plenum, New York, 1985.
11. N. Peters, Laminar flamelet concepts in turbulent combustion. Twenty-First Symposium (International) on Combustion, The Combustion Institute, Pittsburgh, pages 1231–1250, 1986.
12. R.G. Abdel-Gayed, D. Bradley, F.K.-K. Lung, Combustion regimes and the straining of turbulent premixed flames. Combust. Flame **76**, 213–218 (1989)
13. T. Poinsot, D. Veynante, S. Candel, Quenching process and premixed turbulent combustion diagrams. J. Fluid Mech. **228**, 561–606 (1991)
14. F.A. Williams, Recent advances in theoretical descriptions of turbulent diffusion flames, in *Turbulent mixing in nonreactive and reactive flows*. ed. by S.N.B. Murthy (Plenum Press, New York, 1975), pp. 189–208
15. N. Peters, Local quenching of diffusion flamelets and non-premixed turbulent combustion. Western States Section of the Combustion Institute, paper WSS 80–4, Spring Meeting, Irvine, 1980.

16. S.B. Pope and M.S. Anand, Flamelet and distributed combustion in premixed turbulent flames. Twentieth Symposium (International) on Combustion, The Combustion Institute, Pittsburgh, 25 (1984) pp. 403–410.
17. K. N. C. Bray, N. Peters, Laminar flamelets in turbulent flames. In P. A. Libby and F. A. Williams, editors, Turbulent Reacting Flows, pages 63–113. Academic Press, 1994.
18. N. Peters, Turbulent Combustion. Cambridge University Press, 2000.
19. A.M. Klimov, Laminar flame in a turbulent flow. Zh. Prikl. Mekh. Tekh. Fiz. **3**, 49–58 (1963)
20. F.A. Williams, Criteria for existence of wrinkled laminar flame structure of turbulent premixed flames. Combust. Flame **26**, 269–270 (1976)
21. A.R. Kerstein, W.T. Ashurst, F.A. Williams, Field equation for interface propagation in an unsteady homogeneous flow field. Phys. Rev. A **37**, 2728–2731 (1988)
22. M. Matalon, On flame stretch. Combust. Sci. Technol. **31**, 169–181 (1983)
23. M. Matalon, B.J. Matkowsky, Flames as gasdynamics discontinuities. J. Fluid Mech. **124**, 239–259 (1982)
24. P. Clavin, F.A. Williams, Theory of premixed flame propagation in large-scale turbulence. J. Fluid Mech. **90**, 589–604 (1979)
25. K.I. Shchelkin, On combustion in a turbulent flow. Russian J. Technical Physics, 13 (1943) pp. 520–530 (in Russian). English translation: K.I. Schelkin, On combustion in a turbulent flow, NACA Tech. Memorandum No. 1110, 1947].
26. A.R. Kerstein, W.T. Ashurst, Propagation rate of growing interfaces in stirred fluids Phys. Rev. Lett. **68**, 934–938 (1992)
27. J.R. Mayo, A.R. Kerstein, Scaling of Huygens-front speedup in weakly random media. Phys. Lett. A **372**, 5–11 (2007)
28. J.R. Mayo, A.R. Kerstein, Fronts in randomly advected and heterogeneous media and nonuniversality of Burgers turbulence: Theory and numerics. Phys. Rev. E 78 (2008) pp. 056307 (1–5).
29. A. Pocheau, Front propagation in a turbulent medium. Europhys. Lett. **20**, 401–406 (1992)
30. A.M. Klimov, Premixed turbulent flames – interplay of hydrodynamic and chemical phenomena. Prog. Astronaut. Aeronaut. **8**, 133–146 (1983)
31. G. Searby, P. Clavin, Weakly Turbulent, Wrinkled Flames in Premixed Gases. Combust. Sci. Technol. **46**, 167–193 (1986)
32. V. Yakhot, Propagation velocity of premixed turbulent flames. Combust. Sci. Technol. **60**, 191–214 (1988)
33. G.I. Sivashinsky, Cascade renormalization theory of turbulent flame speed. Combust. Sci. Technol. **62**, 77–96 (1988)
34. A.R. Kerstein, Simple derivation of Yakhot's turbulent premixed flame speed formula. Combust. Sci. Technol. **60**, 163–165 (1988)
35. K.N.C. Bray, Studies of the turbulent burning velocity. Proc. R. Soc. Lond. **A431**, 315–335 (1990)
36. R. Ronney, V. Yakhot, Flame broadening effects on premixed turbulent flame speed. Combust. Sci. Tech. **86**, 31–43 (1992)
37. A. Pocheau, Scale invariance in turbulent front propagation. Phys. Rev. E **49**, 1109–1122 (1994)
38. J.W. Dold, O.S. Kerr, I.P. Nikolova, Flame propagation through periodic vortices. Combust. Flame **100**, 359–366 (1995)
39. L. Kagan, G. Sivashinsky, Flame propagation and extinction in largescale vortical flows. Combust. Flame **120**, 222–232 (2000)
40. R.C. Aldredge, F.A. Williams, Influence of wrinkled premixed-flame dynamics on large-scale, low intensity turbulent flow. J. Fluid Mechanics **228**, 487–511 (1991)
41. R.C. Aldredge, The Propagation of Wrinkled Premixed Flames in Spatially Periodic Shear Flow. Combust. Flame **90**, 121–133 (1992)
42. K.N.C. Bray, R.S. Cant, Some applications of Kolmogorov's turbulence research in the field of combustion. Proc. Roy. Soc. London A **434**, 217–240 (1991)

43. N. Peters, A spectral closure for premixed turbulent combustion in the flamelet regime. J. Fluid Mech. **242**, 611–629 (1992)
44. N. Peters, The turbulent burning velocity for large-scale and small-scale turbulence. J. Fluid Mech. **384**, 107–132 (1999)
45. N. Peters, H. Wenzel, F.A. Williams, Modification of the turbulent burning velocity by gas expansion. Proc. Combust. Inst. **28**, 235–243 (2000)
46. W.T. Ashurst, Flow-frequency effect upon Huygens front propagation. Combust. Theory Modell. **4**, 99–105 (2000)
47. L. Kagan, P. Ronney, G. Sivashinsky, Activation energy effect on flame propagation in large-scale vortical flows. Combust. Theory Modell. **6**, 479–488 (2002)
48. A.R. Kerstein, Turbulence in combustion processes. Comput. Phys. Communic. **148**, 1–14 (2002)
49. P. Cambray, G. Joulin, On moderately-forced premixed flames. Proc. Combust. Inst. **24**, 61–67 (1992)
50. A. Trouvé, T. Poinsot, The evolution equation for the flame surface density in turbulent premixed combustion. J. Fluid Mech. **278**, 1–31 (1994)
51. R.N. Paul, K.N.C. Bray, *Study of Premixed Turbulent Combustion Including Landau-Darrieus Instability Effects, Twenty-Sixth Symposium (International) on Combustion* (The Combustion Institute, Pittsburgh, PA, 1996), pp. 259–266
52. H. Boughanem, A. Trouvé, The domain of influence of flame instabilities in turbulent premixed combustion. Proc. Combust. Inst. **27**, 971–978 (1998)
53. B.T. Helenbrook, C.K. Law, The Role of Landau-Darrieus Instability in Large Scale Flows. Combust. Flame **117**, 155–169 (1999)
54. B. Denet, Frankel equation for turbulent flames in the presence of a hydrodynamic instability. Phys. Rev. E **55**, 6911–6916 (1997)
55. R.G. Abdel-Gayed, D. Bradley, M. Lawes, Turbulent burning velocities: a general correlation in terms of straining rates. Proc. R. Soc. Lond. A **414**, 389–413 (1987)
56. R.K. Cheng, I.G. Shepherd, The Influence of Burner Geometry on Premixed Turbulent Flame Propagation. Combust. Flame **85**, 7–26 (1991)
57. R.C. Aldredge, V. Vaezi, P.D. Ronney, Premixed flame propagation in turbulent Taylor-Couette flow. Combust. Flame **115**, 395–405 (1998)
58. H. Kobayashi, Y. Kawabata, K. Maruta, Experimental study of general correlation of turbulent burning velocity at high pressure. Proc. Combust. Inst. **27**, 941–948 (1998)
59. H. Kobayashi, T. Kawahata, K. Seyama, T. Fujimari, J.S. Kim, Relationship between the smallest scale of flame wrinkles and turbulent characteristics of high-pressure turbulent premixed flames. Proc. Combust. Inst. **29**, 1793–1800 (2002)
60. H. Kido, M. Nakahara, K. Nakashima, J. Hashimoto, Influence of local flame displacement velocity on turbulent burning velocity. Proc. Combust. Inst. **29**, 1855–1861 (2002)
61. D. Bradley, M. Haq, R. Hicks, T. Kitagawa, M. Lawes, C. Sheppard, R. Woolley, Turbulent burning velocity, burned gas distribution, and associated flame surface definition. Combust Flame **133**, 415–430 (2003)
62. Y.-C. Chen, R.W. Bilger, Experimental investigation of three-dimensional flame front structure in premixed turbulent combustion-I: hydrocarbon air Bunsen flames. Combust Flame **131**, 400–435 (2002)
63. Y.-C. Chen, R.W. Bilger, Experimental investigation of three-dimensional flame front structure in premixed turbulent combustion II: lean hydrogen air Bunsen flames. Combust Flame **138**, 155–174 (2004)
64. H. Kobayashi, K. Seyama, H. Hagiwara, Y. Ogami, Burning velocity correlation of methane/air turbulent premixed flames at high pressure and high temperature. Proc. Combust. Inst. **30**, 827–834 (2005)
65. T.W. Lee, S.J. Lee, Direct comparison of turbulent burning velocity and flame surface properties in turbulent premixed flames. Combust. Flame **132**, 492–502 (2003)
66. S.A. Filatyev, J.F. Driscoll, C.D. Carter, J.M. Donbar, Measured properties of turbulent premixed flames for model assessment, including burning velocities, stretch rates, and surface densities. Combust. Flame **141**, 1–21 (2005)

67. F.T.C. Yuen, Ö.L. Gülder, Dynamics of lean turbulent premixed flames at high turbulence intensities. Combust. Sci. Technol. **182**, 544–558 (2010)
68. D. Bradley, M. Lawes, M. Mansour, Correlation of turbulent burning velocities of ethanol–air, measured in a fan-stirred bomb up to 1.2 MPa, Combust. Flame 158 (2011) pp. 123–138.
69. P. Venkateswaran, A. Marshall, D.H. Shin, D. Noble, J. Seitzman, T. Lieuwen, Measurements and analysis of turbulent consumption speeds of H2/CO mixtures. Combust. Flame **158**, 1602–1614 (2011)
70. P. Venkateswaran, A.D. Marshall, J.M. Seitzman, T.C. Lieuwen, Turbulent consumption speeds of high hydrogen content fuels from 1–20 atm. Trans. ASME J. Engng. Gas Turbines Power 136 (2014) pp. 011504.
71. F.T.C. Yuen, Ö.L. Gülder, Turbulent premixed flame front dynamics and implications for limits of flamelet hypothesis. Proc. Combust. Inst. **34**, 1393–1400 (2013)
72. P. Tamadonfar, Ö.L. Gülder, Experimental investigation of the inner structure of premixed turbulent methane/air flames in the thin reaction zones regime. Combust. Flame **162**, 115–128 (2015)
73. T.M. Wabel, A.W. Skiba, J.F. Driscoll, Turbulent burning velocity measurements - extended to extreme levels of turbulence. Proc. Combust. Inst. **36**, 1801–1808 (2017)
74. T.M. Wabel, A.W. Skiba, J.E. Temme, J.F. Driscoll, Measurements to determine the regimes of premixed flames in extreme turbulence. Proc. Combust. Inst. **36**, 1809–1816 (2017)
75. D. Fries, B.A. Ochs, A. Saha, D. Ranjan, S. Menon, Flame speed characteristics of turbulent expanding flames in a rectangular channel. Combust. Flame **199**, 1–13 (2019)
76. A. Tyagi, I. Boxx, S. Peluso, J. O'Connor, Pocket formation and behavior in turbulent premixed flames. Combust. Flame **211**, 312–324 (2020)
77. T. Poinsot, Using direct numerical simulations to understand premixed turbulent combustion. Proc. Combust. Inst. **26**, 219–232 (1996)
78. R.C. Aldredge, Premixed flame propagation in a high-intensity, large scale vortical flow. Combust. Flame **106**, 29–40 (1996)
79. O. Colin, F. Ducros, D. Veynante, T. Poinsot, A thickened flame model for large eddy simulations of turbulent premixed combustion. Phys. Fluids **12**, 1843–1863 (2000)
80. B. Denet, Pockets in turbulent premixed flames. Combust. Theory Modell. **5**, 85–95 (2001)
81. H. Pitsch, L. Duchamp De Lageneste, Large-Eddy Simulation of Premixed Turbulent Combustion Using a Level-Set Approach. Proc. Combust. Inst. **29**, 2001–2008 (2002)
82. L. Selle, G. Lartigue, T. Poinsot, R. Koch, K.U. Schildmacher, W. Krebs, B. Prade, P. Kaufmann, D. Veynante, Compressible large eddy simulation of turbulent combustion in complex geometry on unstructured meshes. Combust. Flame **137**, 489–505 (2004)
83. E.R Hawkes, R. Sankaran, J.C Sutherland, J.H Chen, Direct numerical simulation of turbulent combustion: fundamental insights towards predictive models. J. Phys.: Conf. Ser. 16 (2005) pp. 65–79.
84. J.B. Bell, M.S. Day, I.G. Shepherd, M.R. Johnson, R.K. Cheng, J.F. Grcar, V.E. Beckner, M.J. Lijewski, Numerical simulation of a laboratory-scale turbulent V-flame. Proc. Natl. Acad. Sci. USA **102**, 10006–10011 (2005)
85. R. Sankaran, E.R. Hawkes, J.H. Chen, T. Lu, C.K. Law, Structure of a spatially developing turbulent lean methane–air Bunsen flame. Proc. Combust. Inst. **31**, 1291–1298 (2006)
86. J.B. Bell, M.S. Day, J.F. Grcar, M.J. Lijewski, J.F. Driscoll, S.A. Filatyev, Numerical simulation of a laboratory-scale turbulent slot flame. Proc. Combust. Inst. **31**, 1299–1307 (2007)
87. V.B. Akkerman, V.V. Bychkov, L.E. Eriksson, Numerical study of turbulent flame velocity. Combust. Flame **151**, 452–471 (2007)
88. V.B. Akkerman, V.V. Bychkov, R.J.M. Bastiaans, L.P.H. de Goey, J.A. van Oijen, L.E. Eriksson, Flow-flame interaction in a closed chamber. Phys. Fluids. 20 (2008) pp. 055107 (1–21).
89. Y. Shim, S. Tanaka, M. Tanahashi, T. Miyauchi, Local structure and fractal characteristics of H2–air turbulent premixed flame. Proc. Combust. Inst. **33**, 1455–1462 (2010)

90. F. Creta, M. Matalon, Propagation of wrinkled turbulent flames in the context of hydrodynamic theory. J. Fluid Mech. **680**, 225–264 (2011)
91. F. Creta, N. Fogla, M. Matalon, Turbulent propagation of premixed flames in the presence of Darrieus-Landau instability. Combust. Theory Model. **15**, 267–298 (2011)
92. M. Matalon, F. Creta, The "turbulent flame speed" of wrinkled premixed flame. C.R. Mec. **340**, 845–858 (2012)
93. N. Fogla, F. Creta, M. Matalon, Influence of the Darrieus-Landau instability on the propagation of planar turbulent flames. Proc. Combust. Inst. **34**, 1509–1517 (2013)
94. N. Fogla, F. Creta, M. Matalon, Effect of folds and pockets on the topology and propagation of premixed turbulent flames. Combust. Flame **162**, 2758–2777 (2015)
95. N. Fogla, F. Creta, M. Matalon, The turbulent flame speed for low-to-moderate turbulence intensities: Hydrodynamic theory vs. experiments. Combust. Flame 175 (2017) pp. 155–169.
96. A.Y. Poludnenko, E.S. Oran, The interaction of high-speed turbulence with flames: global properties and internal flame structure. Combust. Flame **157**, 995–1011 (2010)
97. A.Y. Poludnenko, E.S. Oran, The interaction of high-speed turbulence with flames: turbulent flame speed. Combust. Flame **158**, 301–326 (2011)
98. J.H. Chen, Petascale direct numerical simulation of turbulent combustion - fundamental insights, towards predictive models. Proc. Combust. Inst. **33**, 99–123 (2011)
99. T. Poinsot, Prediction and control of combustion instabilities in real engines. Proc. Combust. Inst. **36**, 1–28 (2017)
100. W. Han, H. Wang, G. Kuenne, E.R. Hawkes, J.H. Chen, J. Janicka, C. Hasse, Large eddy simulation/dynamic thickened flame modeling of a high Karlovitz number turbulent premixed jet flame. Proc. Comb. Inst. **37**, 2555–2563 (2019)
101. A.J. Aspden, M.S. Day, J.B. Bell, Towards the distributed burning regime in turbulent premixed flames. J. Fluid Mech. **871**, 1–21 (2019)
102. P. Clavin, Dynamic behavior of premixed flame fronts in laminar and turbulent flows. Progr. Energy Combust. **11**, 1–59 (1985)
103. R. Borghi, Turbulent combustion modelling. Prog. Energy Combust. Sci. **14**, 245–292 (1988)
104. F.A. Williams, *Combustion Theory*, 2nd edn. (The Benjamin Cummings Publishing Co., Menlo Park, 1985)
105. R. Borghi, M. Destriau, *Combustion and Flames, Chemical and Physical Principles* (Editions Technip, Paris, 1998)
106. T. Poinsot, S. Candel, A. Trouve, Applications of direct numerical simulations to premixed turbulent combustion. Prog. Energy Combust. Sci. **21**, 531–576 (1995)
107. S.B. Pope, Turbulent flows. Cambridge University Press; 2000.
108. S. Ishizuka, Flame propagation along a vortex axis. Prog. Energy Combust. Sci. **28**, 477–542 (2002)
109. R. Hilbert, F. Tap, H. El-Rabii, D. Thèvenin, Impact of detailed chemistry and transport models on turbulent combustion simulations. Prog. Energy Combust. Sci. **30**, 61–117 (2004)
110. T. Poinsot, D. Veynante, *Theoretical and Numerical Combustion*, 2nd edn. (PA, Edwards, Philadelphia, 2005)
111. H. Pitsch, Large eddy simulations of turbulent combustion. Ann. Rev. Fluid Mech. **38**, 453–482 (2006)
112. S. Kadowaki, T. Hasegawa, Numerical simulation of dynamics of premixed flames: flame instability and vortex-flame interaction. Prog. Energy Combust. Sci. **31**, 193–241 (2005)
113. A.N. Lipatnikov, J. Chomiak, Turbulent flame speed and thickness: phenomenology, evaluation, and application in multidimensional simulations. Prog Energy Combust. Sci. **28**, 1–74 (2002)
114. A.N. Lipatnikov, J. Chomiak, Molecular transport effects on turbulent flame propagation and structure. Prog Energy Combust. Sci. **31**, 1–73 (2005)
115. J.F. Driscoll, Turbulent premixed combustion: flamelet structure and its effect on turbulent burning velocities. Prog. Energy Combust. Sci. **34**, 91–134 (2008)
116. T. Echekki, E. Mastorakos, *Turbulent Combustion Modeling* (Fluid mechanics and its applications. Springer Verlag, Dordrecht, New York, 2011)

117. A.N. Lipatnikov, J. Chomiak, Effects of premixed flames on turbulence and turbulent scalar transport. Prog. Energy Combust. Sci. **36**, 1–102 (2010)
118. J.A. van Oijen, A. Donini, R.J.M. Bastiaans, J.H.M. ten Thije Boonkkamp, L.P.H. de Goey, State-of-the-art in premixed combustion modeling using flamelet generated manifolds. Prog. Energy Combust. Sci. 57 (2016) pp. 30–74.
119. E.D. Gonzalez-Juez, A.R. Kerstein, R. Ranjan, S. Menon, Advances and challenges in modeling high-speed turbulent combustion in propulsion systems. Prog. Energy Combust. Sci. **60**, 26–67 (2017)
120. Turbulent Premixed Flames, *Edited by N* (Cambridge University Press, Swaminathan and K. N. C. Bray, 2011)
121. A. Lipatnikov, Fundamentals of premixed turbulent combustion, Taylor Francis, 2013.
122. J.F. Driscoll, J.H. Chen, A.W. Skiba, C.D. Carter, E.R. Hawkes, H. Wang, Premixed flames subjected to extreme turbulence: Some questions and recent answers. Prog. Energy Combust. Sci. 76 (2020) pp. 100802 (1–36).
123. R.N. Paul, K.N.C. Bray, *Study of Premixed Turbulent Combustion Including Landua-Darrieus Instability Effects, Twenty-Sixth Symposium (International) on Combustion* (The Combustion Institute, Pittsburgh, PA, 1996), pp. 259–266
124. H. Kobayashi, T. Tamura, K. Maruta, T. Niioka, F. Williams, Burning velocity of turbulent premixed flames in a high-pressure environment. Proc. Combust. Inst. **26**, 389–396 (1996)
125. D. Bradley, Instabilities and flame speeds in large-scale premixed gaseous explosions, Phil. Trans. Roy. Soc. Lond. 357 (1999) pp. 3567–3581. (experiments)
126. D. Bradley, T.M. Cresswell, J.S. Puttock, Flame acceleration due to flame-induced instabilities in large-scale explosion. Combust. Flame **124**, 5551–5559 (2001)
127. H. Kobayashi, H. Kawazoe, Flame instability effects on the smallest wrinkling scale and burning velocity of high-pressure turbulent premixed flames. Proc. Combust. Inst. **28**, 375–381 (2000)
128. V. Savarianandam, C. Lawn, Burning velocity of premixed turbulent flames in the weakly wrinkled regime. Combust. Flame **146**, 1–18 (2006)
129. A.Al. Shahrany, D. Bradley, M. Lawes, K. Liu, R. Woolley, Darrieus-Landau and thermo-acoustic instabilities in closed vessel explosions. Combust. Sci. Technol. 178 (2006) pp. 1771–1802.
130. D. Bradley, M. Lawes, K. Liu, M. Mansour, Measurements and correlations of turbulent burning velocities over wide ranges of fuels and elevated pressures Proc. Combust. Inst. **34**, 1519–1526 (2013)
131. S. Chaudhuri, F. Wu, C.K. Law, Scaling of turbulent flame speed for expanding flames with Markstein diffusion considerations. Phys. Rev. E 88 (2013) pp. 033005 (1–12).
132. J. Quinard, G. Searby, B. Denet, J. Graña-Otero, Self-Turbulent Flame Speeds. Flow, Turbulence Combust. **89**, 231–247 (2012)
133. G. Troiani, F. Creta, M. Matalon, Experimental investigation of Darrieus-Landau instability effects on turbulent premixed flames. Proc. Combust. Inst. **35**, 1451–1459 (2015)
134. S. Yang, A. Saha, Z. Liu, C.K. Law, Role of Darrieus-Landau instability in propagation of expanding turbulent flames. J. Fluid Mech. **850**, 784–802 (2018)
135. C.R. Bauwens, J.M. Bergthorson, S.B. Dorofeev, On the interaction of the Darrieus-Landau instability with weak initial turbulence. Proc. Combust. Inst. **36**, 2815–2822 (2017)
136. V. Bychkov, Velocity of Turbulent Flamelets with Realistic Fuel Expansion. Phys. Rev. Lett. **84**, 6122–6125 (2000)
137. V. Bychkov, M.A. Liberman, R. Reinmann, Velocity of turbulent flamelets of finite thickness. Comb. Sci. Tech. **168**, 113–129 (2001)
138. V. Bychkov, Importance of the Darrieus-Landau instability for strongly corrugated turbulent flames. Phys. Rev. E 68 (2003) pp. 066304 (1–12).
139. B. Denet, V. Bychkov, Low-vorticity and small gas expansion in premixed flames. Comb. Sci. Tech. **177**, 1543–1566 (2005)
140. S. Chaudhuri, V. Akkerman, C. Law, Spectral formulation of turbulent flame speed with consideration of hydrodynamic instability. Phys. Rev. E 84 (2011) pp. 026322 (1–14).

141. R. Yu, T. Nillson, X.-S. Bai, A.N. Lipatnikov, Evolution of averaged local premixed flame thickness in a turbulent flow. Combust. Flame **207**, 232–249 (2019)
142. D. Veynante, J.M. Duclos, J. Piana, Experimental analysis of flamelet models for premixed turbulent combustion. Proc. Combust. Inst. **25**, 1249–1256 (1994)
143. P. Trisjono, H. Pitsch, Systematic analysis strategies for the development of combustion models from DNS: a review. Flow Turbul. Combust. **95**, 231–259 (2015)
144. C. Meneveau, T. Poinsot, Stretching and quenching of flamelets in premixed turbulent combustion. Combust. Flame **86**, 311–332 (1991)
145. J.M. Duclos, D. Veynante, T. Poinsot, A comparison of flamelet models for premixed turbulent combustion. Combust. Flame, 95)1993) pp. 1–16.
146. L.Y.M. Gicquel, G. Staffelbach, T. Poinsot, Large eddy simulations of gaseous flames in gas turbine combustion chambers. Prog. Energy Combust. Sci. **38**, 782–815 (2012)
147. F. Proch, P. Domingo, L. Vervisch, A.M. Kempf, Flame resolved simulation of a turbulent premixed bluff-body burner experiment. Part II: A-priori and a-posteriori investigation of sub-grid scale wrinkling closures in the context of artificially thickened flame modeling. Combust. Flame, 180 (2017) pp. 340–350.
148. B. Fiorina, D. Veynante, S. Candel, Modeling combustion chemistry in large eddy simulation of turbulent flames. Flow Turb. Combust. **94**, 3–42 (2015)
149. B. Rochette, F. Collin-Bastiani, L. Gicquel, O. Vermorel, D. Veynante, T. Poinsot, Influence of chemical schemes, numerical method and dynamic turbulent combustion modeling on LES of premixed turbulent flames. Combust. Flame **191**, 417–430 (2018)
150. T. Elperin, N. Kleeorin, I. Rogachevskii, Effect of Chemical Reactions and Phase Transitions on Turbulent Transport of Particles and Gases. Phys. Rev. Lett. **80**, 69–72 (1998)
151. T. Elperin, N. Kleeorin, M.A. Liberman, I. Rogachevskii, Turbulent diffusion of chemically reacting gaseous admixtures. Phys. Rev. E 90 (2014) pp. 053001 (1–9).
152. T. Elperin, N. Kleeorin, M. Liberman, A.N. Lipatnikov, I. Rogachevskii, R. Yu, Turbulent diffusion of chemically reacting flows: Theory and numerical simulations. Phys. Rev. E **96**, 053111–053118 (2017)
153. A.S. Monin, A.M. Yaglom, *Statistical Fluid Mechanics*, vol. 2 (MIT Press, Cambridge, 1975)

Chapter 15
Flame Acceleration and Deflagration-To-Detonation Transition

15.1 Explosions and Detonation

Although the term "explosion" is commonly used, its meaning is vague. An explosion can be defined as a scenario in which, due to the rapid release of energy (over time t_E), the pressure increases much faster than it can be balanced by pressure waves propagating at the speed of sound, which means $t_E \ll L/a_s$, where L and a_s are the system size and the speed of sound. This means that the explosions should be accompanied by the formation of large pressure gradients and shock waves of moderate strength.

A self-sustained wave of the exothermic chemical reaction propagating through a homogeneous combustible mixture is known to occur either as a subsonic deflagration (premixed flame) or supersonic detonation. In unconfined obstacle-free space, the concrete realization of the specific propagation mode is, as a rule, controlled by the ignition conditions. The deflagration wave is a subsonic combustion mode and therefore proceeds at an almost constant pressure. Typical velocities of deflagration modes are from a few cm/s to 10 m/s, and therefore $\Delta P/P \sim M^2 \ll 1$. The detonation is a supersonic combustion mode, in which chemical reactions and energy release occur as a result of the mixture heating in the strong shock wave and therefore the shock wave and the trailing reaction zone are coupled and propagate at the same velocity. For stoichiometric fuel–air mixtures, the pressure jump across the detonation wave is about $P_2/P_0 \sim 20 \div 30$. Typical intensity (the Mach number) of the leading shock in a detonation wave is $M \sim 5-7$. Many useful detonation data including initiation energy, minimum and critical tube diameter, and detonation cell size can be found in the Detonation Database [1].

Normally, a deflagration mode is initiated by a mild energy release, e.g. by an electrical spark, while detonations are triggered by strong shock waves through a localized explosion initiated by the deposition of a large amount of energy in a small volume, for example, using a solid explosive (detonator). As it was discussed in Chapter 12, both deflagration and detonation represent stationary solutions of the system of hydrodynamic equations for a compressible reactive flow, and therefore

M. A. Liberman, *Combustion Physics*,
https://doi.org/10.1007/978-3-030-85139-2_15

appear to be stable attractors each being linked to its own base of initial data. Possible combustion modes of deflagration and detonation were analyzed in Sect. 12.5.

Although a detonation wave can be initiated directly by a strong shock wave, it has been known a long time that a premixed flame ignited near the closed end of a tube and propagating to the open or closed end can accelerate until a sudden abrupt increase of the flame velocity, which ends up by the transition to detonation. This phenomenon is known as the transition from slow combustion (deflagration) to detonation or deflagration-to-detonation transition (DDT).

The first experimental observations of the deflagration-to-detonation transition date back to the studies of the late nineteenth century by Mallard and Le Chatelier [2–4] and Bertelot and Vieille [5, 6], who tried to understand the origin of catastrophic explosions in coal mines in France. Almost simultaneously, the Scottish physicist William Rankine [7] and the French scientist Pierre Hugoniot [8, 9] developed the theory of shock waves, and the English physicist David Chapman [10] and the French engineer Émile Jouguet [11, 12] created the basis for detonation waves. A more detailed detonation model, known as the ZND detonation model, was independently developed in the 1940s by Yakov Zel'dovich [13], John von Neumann [14], and Werner Döring [15].

From the beginning, investigations of the origin of detonation and DDT were motivated by catastrophic explosions in coal mines. Military applications also required a better understanding of the origins of detonation and explosion. Since the first observations of the transition to detonation more than 100 years ago, a huge amount of experimental, theoretical, and numerical research has been carried out in an attempt to understand what causes the transition to detonation. Nevertheless, even now, after decades of experimental, theoretical and numerical researches, the physical processes involved in the mechanism of the transition from deflagration to detonation are still not fully understood and remain one of the main unsolved problems in the science of combustion. DDT can be found in a wide variety of systems, from internal combustion engines and random explosions in various industrial facilities to thermonuclear explosions of white dwarfs, in which Type Ia supernovae are born. DDTs can be different in their particular physical content, but have much in common with respect to the mathematical structure of the governing equations and the basic physical mechanism. Apart from their technological relevance, DDT constitutes fascinating dynamical system displaying variety of nonlinear phenomena. Over the past decades, the deflagration-to-detonation transition has been intensively studied, since detonation represents the most severe form of explosive hazard in industrial accidents, and understanding how DDT occurs is important for industrial safety [16, 17], and because potential applications of DDT in microscale propulsion and power devices [18]. It is also of fundamental importance for predicting critical conditions for the hydrogen safety of nuclear power plants and the use of hydrogen as an energy carrier [17]. In astrophysics, this is one of the main unanswered questions about the life and death of thermonuclear supernovae [19, 20].

DDT is a stochastic process, meaning that there is some uncertainty regarding the time and location of the transition to detonation. Exactly how DDT occurs can vary depending on specific experimental conditions. Since the time of the classical works of K.I. Shchelkin [21–23], it is known that the presence of obstacles along the channel walls leads to a considerable increase in the flame acceleration and reduction of the run-up distance. Experimental studies of DDT demonstrate that the flame accelerates more rapidly if it passes through an array of turbulence-generating baffles. This was probably one of the reasons why the first attempts to explain DDT were heavily based on the assumption that strong acceleration of flame and DDT can occur only in the case of turbulent flames, and there was a general opinion that a sufficiently strong flame acceleration for DDT is impossible without external turbulent flow. Oppenheim and co-workers found that in DDT experiments in long channels containing hydrogen–oxygen mixtures [24–26], which was ignited near the closed end, the flame accelerated, could become turbulent, and eventually DDT occurred, which manifested itself as a sudden explosion in the immediate vicinity of the flame brush, which the authors called "explosion in the explosion". Numerous experimental studies have shown the following stages of the DDT process. Ignition of a combustible mixture near the closed end of the tube results in the initially laminar flame, which accelerates toward the opposite end of the tube (open or closed) and can become turbulent. The accelerating flame acts as a piston, producing pressure waves which steepen into the weak shocks far ahead of the flame front. These shocks eventually merge and overlap in the flow ahead of the flame and become stronger. Experimental Schlieren photographs showed that the place of transition to detonation always occurs somewhere within the shock–flame complex, either near the wall in the boundary layer, or in the bulk upstream flow.

Since the early 80s, when more realistic gas dynamics computations became available, much effort has been made to understand the mechanism of DDT using 2D and sometimes 3D modeling. Many authors have used analytical approaches trying to account for various factors that can influence the flame acceleration and DDT, including boundary conditions in smooth or obstructed channel walls, turbulence, hydraulic resistance, viscous stress heating, and chemical kinetics. Nevertheless, many issues are still poorly understood, and the identification of the mechanisms controlling the transition from deflagration to detonation remains one of the main problems discussed in numerous reviews [26–39]. To understand the mechanism of the transition from deflagration to detonation, it is necessary to answer two basic questions: what are the processes that cause flame acceleration; how is formed a shock wave, which is strong enough to become the leading shock in detonation?

15.2 Experimental Study of DDT in Highly Reactive Mixtures

Extensive experimental study of DDT in highly reactive stoichiometric hydrogen–oxygen and ethylene–oxygen mixtures [40–44] were focused on the key characteristics of DDT: (1) flame acceleration, which determine the strength and location of shock waves generated by the accelerating flame, (2) the formation of a heated and compressed unreacted mixture ahead of the flame, (3) the mechanism of transition to detonation. The experimental facility consisted of a stainless-steel detonation tube with a rectangular cross section of 5 × 5 cm and a length 3.4 and 6.065 m. The optical Schlieren system with stroboscopic pulse generator and high-speed camera was used to record the flame evolution. Ignition was initiated near the closed end of the tube by means of a low power electrical spark. Length of the tube was chosen long enough to avoid possible interaction of the flame with the shock waves reflected from the opposite end of the tube. The thermodynamic properties of the hydrogen–oxygen and ethylene–oxygen mixtures used in the experiments were calculated using STANJAN code [45] and Cantera code based on the oxidation mechanism [46] are shown in Tables 15.1 and 15.2.

Three different stages of the combustion wave velocity evolution are clearly seen in the experimentally measured velocity–time dependence for the hydrogen–oxygen flame shown in Fig. 15.1 for different initial pressures and different roughness of the walls. During the first stage, the flame velocity grows exponentially and the flame produces weak shocks in the flow far ahead of the flame front. On the second stage, the rate of the flame acceleration decreases and the compression waves generated by the flame steepen into shocks close ahead of the flame front, creating a pocket of unreacted compressed material adjacent to the flame front. The third stage is an abrupt increase in the flame velocity and the actual transition to overdriven detonation, which subsequently turns into the Chapman–Jouguet detonation. The sound velocities in H_2-O_2 are 530 m/s in the unreacted gas at $T = 298K$ and 1350 m/s in the combustion products. The overall reaction order for the stoichiometric hydrogen–oxygen is $n = 2.73$. In the laboratory reference frame, the flame speed increases from 55 to 74 m/s.

During the first stage the flame velocity increases almost linearly (in fact exponentially). At the end of the first stage, the visible flame velocities in H_2/O_2 mixture are 270 m/s for $P_0 = 0.2bar$, 375 m/s for $P_0 = 0.6bar$, and 420 m/s for $P_0 = 0.75bar$. At the end of the second stage, just before the transition to detonation, the visible flame velocity is 850-900 m/s for $P_0 = 0.6bar$, and 1100 m/s for $P_0 = 0.2bar$.

Figure 15.2 shows the velocity of H_2/O_2 flame obtained using the high-speed photo processing, for initial pressures $P_0 = 0.2,\ 0.6,\ 0.75bar$. The flame velocity as function of the distance from the closed end grows almost linearly in Fig. 15.2 (left), but in fact it grows exponentially with time as it is seen in Fig. 15.2 (right). Figure 15.2 shows also that the flame velocities decrease after 0.6, 0.7, and 1 ms for initial pressures $P_0 = 0.2,\ 0.6,\ 0.75bar$, correspondingly. The decrease of the flame tip velocity is associated with the development of the tulip flame (Sect. 9.4). Though the flame tip velocity decreases during the formation of the tulip flame, the burning rate, which is defined as the speed of the combustion wave based on the average heat release rate per frontal area of the wrinkled or stretched flame sheet, can continue to increase due to the increase of the tulip petals and therefore the total flame surface, or it also can slightly decrease.

The increased rate of heat release gives rise to a higher volumetric burning rate and a higher velocity of combustion wave, which controls the upstream flow ahead of the advancing flame.

Table 15.1 Stoichiometric hydrogen–oxygen mixture; $T_0 = 293$ K

Initial pressure	Density	Sound speed in reactants	Sound speed in products	Adiabatic combustion temperature	Adiabatic combustion pressure	Expansion ratio	Laminar flame speed	Laminar flame thickness	Overall reaction order
P	ρ(kg/m^3)	a_{su} (m/s)	a_{sb} (m/s)	T_b, K (K)	P_{icc} (bar)	Θ	U_{f0} (m/s)	L_f (mm)	n
0.2	0.10	530	1347	2877	1.88	8.10	6.84	1.56	2.74
0.5	0.25	530	1368	2989	4.84	8.33	8.25	0.54	2.74
0.6	0.30	530	1372	3012	5.85	8.37	8.54	0.44	2.73
0.75	0.37	530	1378	3041	7.37	8.43	8.88	0.34	2.73

Table 15.2 Stoichiometric ethylene–oxygen mixture; $T_0 = 293$ K

Initial pressure	Density	Sound speed in reactants	Sound speed in products	Adiabatic combustion temperature	Adiabatic combustion pressure	Expansion ratio	Laminar flame speed	Laminar flame thickness	Overall reaction order
P_0	$\rho(kg/m^3)$	a_{su} (m/s)	a_{sb} (m/s)	T_b, K (K)	P_{icc} (bar)	Θ	U_{f0} (m/s)	L_f (mm)	n
0.05	0.064	325	1173	3053	0.77	14.1	5.28	1.540	1.99
0.075	0.095	325	1170	3014	1.17	14.0	5.29	1.025	1.98
0.1	0.127	325	1166	2986	1.58	13.9	5.28	0.762	1.98
0.12	0.153	325	1164	2975	1.90	13.9	5.20	0.615	1.97
0.15	0.191	325	1162	2968	2.40	13.8	5.20	0.506	1.96
0.2	0.255	325	1165	2977	3.24	13.9	5.23	0.379	1.96
0.25	0.318	325	1166	2993	4.08	13.9	5.15	0.304	1.95
0.3	0.382	325	1167	3011	4.93	14.0	5.12	0.255	1.94

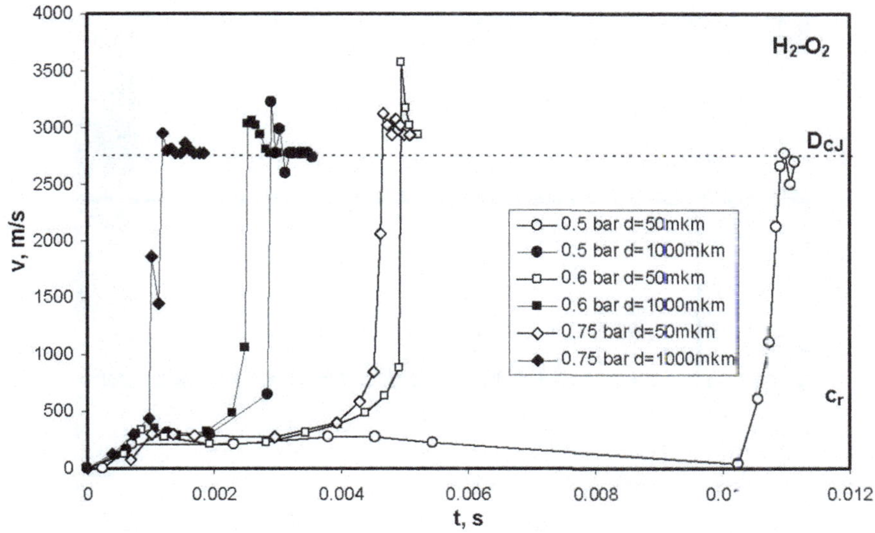

Fig. 15.1 Velocity–time dependence during the detonation initiation in hydrogen–oxygen mixtures for different initial pressures P_0 and wall roughness (d shown in legend), from [44]

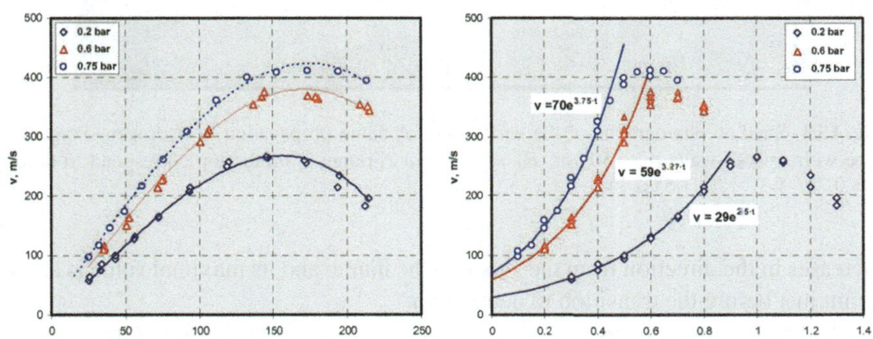

Fig. 15.2 Flame velocity along the tube (left) and against time (right) during the first stage of the flame acceleration [44]

Figure 15.3 shows the sequence of shadow photographs of the flame-generated shock waves (the top image) and pressure waves during the first stage for the flame acceleration. Coordinates and velocities of flame and shock waves are $X_f = 229, 371, 410, 452\,mm$, $U_f = 353, 375, 395, 424\,m/s$, $X_{sh} = 420, 826, 903, 980mm$. The intensity of the shock waves during the first stage is $M_{sh} = 1.4 \div 1.55$, which corresponds to the pressure and temperature increase across the shock: $P_2/P_0 = 2.1 - 2.6$ and $T_2/T_{u0} = 1.24 - 1.32$. A thin boundary layer, which is seen as a white strip near the lower wall, appears at the distance 150 mm from the ignition point. The thickness of the boundary layer gradually

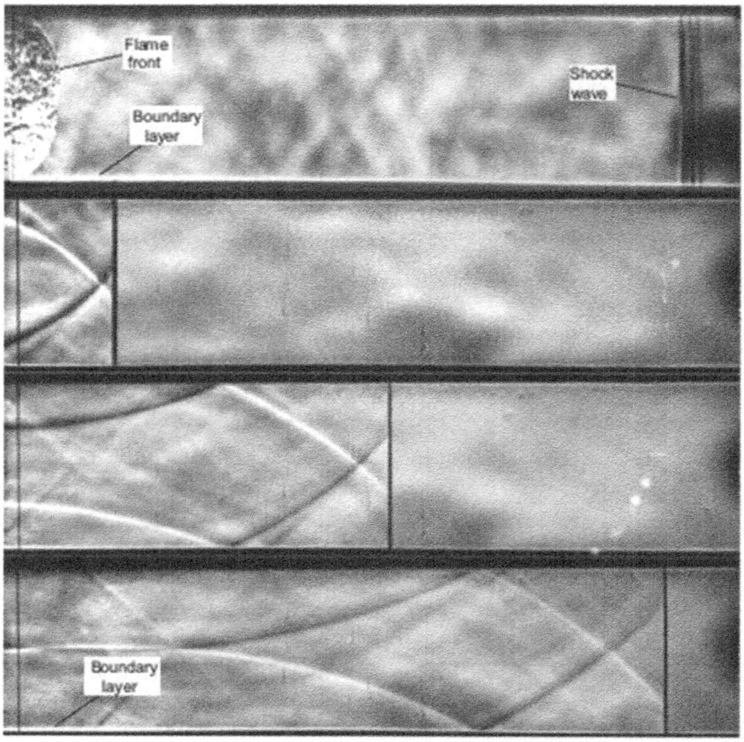

Fig. 15.3 Shadow photographs of the early stage of flame propagation in hydrogen–oxygen in tube with smooth walls $d \approx 50\,\mu m$, $P_0 = 0.75\,bar$. Frames from the top correspond to times: $t = 0.75,\ 1.3,\ 1.4,\ 1.5\,ms$ [44]

increases in the direction from the shock to the flame, and its maximal value is about 3 mm just before the transition to detonation.

Shadow photographs in Fig. 15.4 shows the evolution of the flame front during the first and at the beginning of the second stage for $P_0 = 0.2 bar$. The first image from the left is taken at 0.5 ms, and subsequent images are taken with time interval

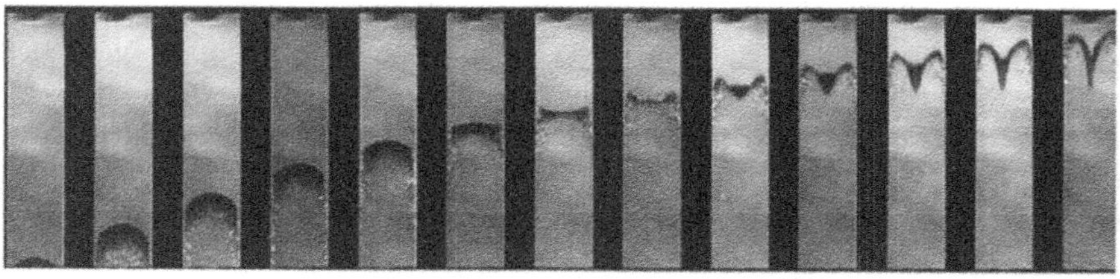

Fig. 15.4 Sequence of shadow photographs of the evolution of the hydrogen–oxygen flame during transition from the first to the second stage, from 0.75 to 1.45 ms with 0.1 ms between the frames; initial pressure is $P_0 = 0.2bar$, [44]

0.1 ms. The flame velocity measured at the tube axis is 284 m/s at $t = 0.6\,ms$; it increases to 298 m/s at $t = 0.7\,ms$, and later it decreases till $t = 2ms$ during the formation of the tulip flame.

A high space resolution of the shadow photos makes possible to resolve fine details of the flow field ahead of the flame. The flow remains laminar in the bulk with a thin boundary layer throughout the entire process.

The flame ignited at the closed end produces a weak shock which runs away building up the upstream flow ahead of the flame. Initially, the flow ahead of the flame is almost uniform, the flow velocitiesare parallel to the tube wall and drop to zero in a thin boundary layer. In the reference frame co-moving with the shock wave, the gas "enters" the tube in the direction to the flame front. Consequently, the boundary layer thickness increases along the tube from the leading shock to the flame, and a parabolic Poiseuille velocity profile could develop asymptotically. The measured and calculated maximum boundary layer thickness in the flow close ahead of the flame front for different pressures and wall roughness are shown in Fig. 15.5. Turbulence within the boundary layer is clearly seen in shadow and schlieren photographs, but the bulk flow ahead of the flame remains laminar all the time until the transition to detonation.

It should be noted that the flame velocity increases in 1.3 ÷ 1.4 times due to the pre-compression of the fresh mixture in the shock wave. Elevated temperature of the fresh mixture, 360–390 K, behind the shock with $M_{sh} = 1.4 \div 1.55$ also promotes the increase of the flame velocity. X-t diagram in Fig. 15.6 shows the flame acceleration for the case: $P_0 = 0.6\,bar$, wall roughness $d \approx 1000\mu m$. The shock trajectory was calculated, all other lines were obtained experimentally: FF-flame front; DW-detonation wave; RW-retonation wave; SW-precursor shock wave, blue line is the

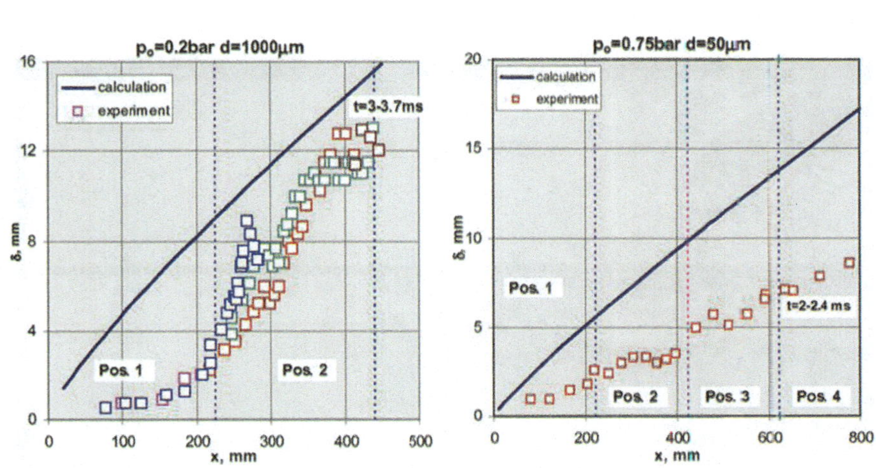

Fig. 15.5 The calculated and measured thickness of the boundary layer in H_2/O_2 experiments. Experimental points correspond to maximum values for different window positions along the tube. $P_0 = 0.2\,bar$ (left); $P_0 = 0.75\,bar$ (right)

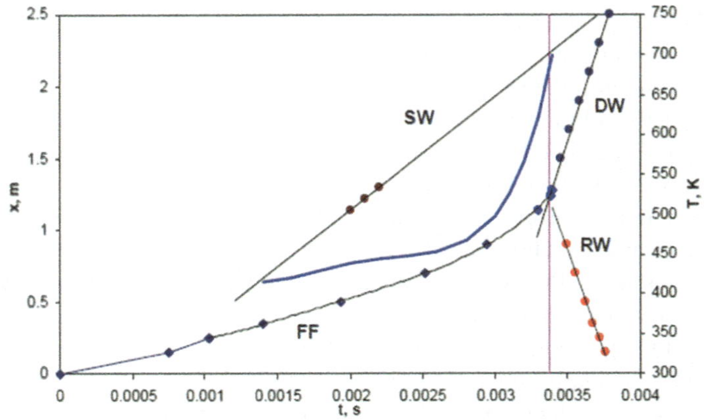

Fig. 15.6 X-t diagram of the H_2/O_2 flame acceleration and DDT for the conditions: $P_0 = 0.6\,bar$, wall roughness $d = 1000\,\mu m$; blue line is maximal temperature behind the shock and ahead of the flame front

maximum temperature ahead of the flame front. It is seen that the temperature ahead of the flame begins to rise shortly before the DDT point.

Shadow photos in Fig. 15.7 show the end of the second stage, just before the transition to detonation in H_2–O_2 mixture for $P_0 = 0.6 bar, d = 1\,mm$. The 17 mm wide preheat zone between the flame and the leading shock of intensity $M_{sh} = 2.3$ is seen in the top image taken at 3.3 ms. The maximum temperature, pressure and the shock wave intensity ahead of the flame front are $T \approx 600\,K$, $P \approx 3.6\,bar$, $M_{sh} = 2.3$. These temperature and pressure are too low to ignite the reaction ahead

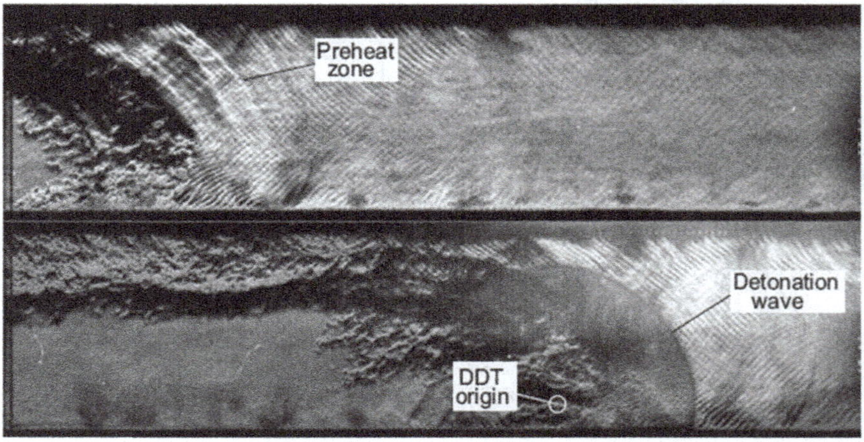

Fig. 15.7 Shadow photographs of DDT in hydrogen–oxygen: $P_0 = 0.6\,bar$. The top image is taken at 3.3 ms, the width of the preheat zone between the flame and shock is 17 mm. The second image is taken at 3.4 ms, [44]

of the flame front. In the second image taken at 3.4 ms, the transition to detonation has already occurred, which takes less than 0.1 ms.

The dynamics of flame acceleration and transition to detonation in a stoichiometric ethylene–oxygen mixture is similar to DDT in a hydrogen–oxygen.

Because the experimental studies were focused on the mechanism of DDT, vertical position of an optical knife and a slit was installed in a schlieren system to register horizontal density gradients. Typical X-t diagram of the DDT process based on photodiodes records and data on arrival time obtained using high-speed schlieren photos for stoichiometric ethylene–oxygen mixture is similar to that shown in Fig. 15.6 for hydrogen–oxygen. The three stages of the flame velocity evolution measured in experiments with C_2H_4/O_2 for different initial pressures are shown in Fig. 15.8. In this case, the sound speed is 325 m/s in the reactants and 1170 m/s in the combustion products. Since the overall reaction order for $C_2H_4 - O_2$ is $n \simeq 2$, the laminar flame velocity depends only slightly on the initial pressure. At the end of the first stage, the visible flame velocity is 265 m/s for $P_0 = 0.05bar$, 330 m/s for $P_0 = 0.12bar$ and 375 m/s for $P_0 = 0.3bar$. At the end of the second stage, shortly before the transition to detonation, the visible flame velocity is 1020 m/s for $P_0 = 0.05bar$, 1060 m/s for $P_0 = 0.12bar$, and 860 m/s for $P_0 = 0.3bar$, which are close to the CJ deflagration speed. This evolution of the flame velocity, consisting of three different stages, is a common feature of the flame velocity evolution for DDT experiments in wide tubes. This is illustrated in Fig. 15.9, which shows the dependence of the dimensionless velocity $M = U_f/a_{s0}$ on the dimensionless coordinate normalized on the laminar flame width x/L_f. It is seen that the dynamics of combustion waves is nearly the same and depends only slightly on the initial pressures.

The sequence of shadow photos in Fig. 15.10, taken with interval 0.1 ms, shows the flow evolution during the end of the second stage in ethylene–oxygen mixture for the initial pressure $P_0 = 0.12bar$. In this case, the optical knife position was chosen such

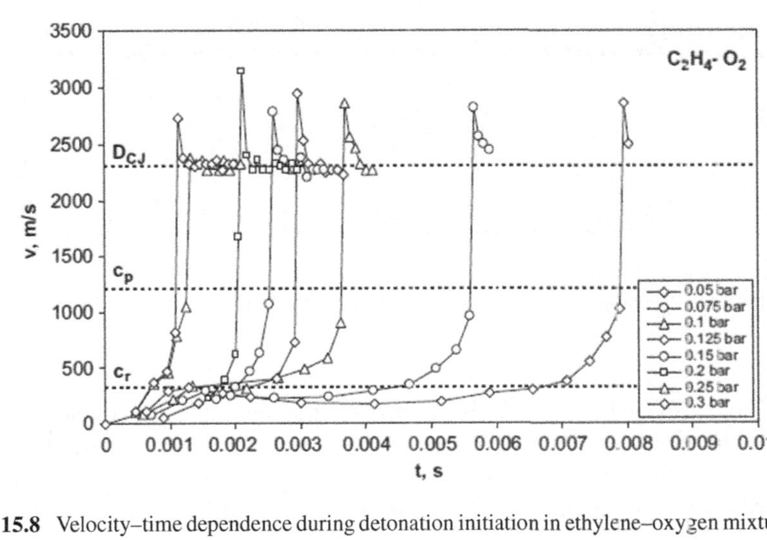

Fig. 15.8 Velocity–time dependence during detonation initiation in ethylene–oxygen mixture, [44]

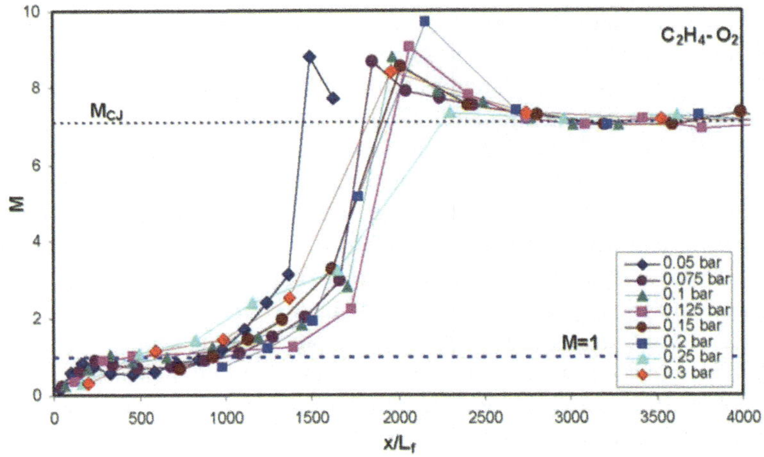

Fig. 15.9 Dimensionless combustion wave velocity $M = U_f/a_{s0}$ versus dimensionless coordinates, x/L_f

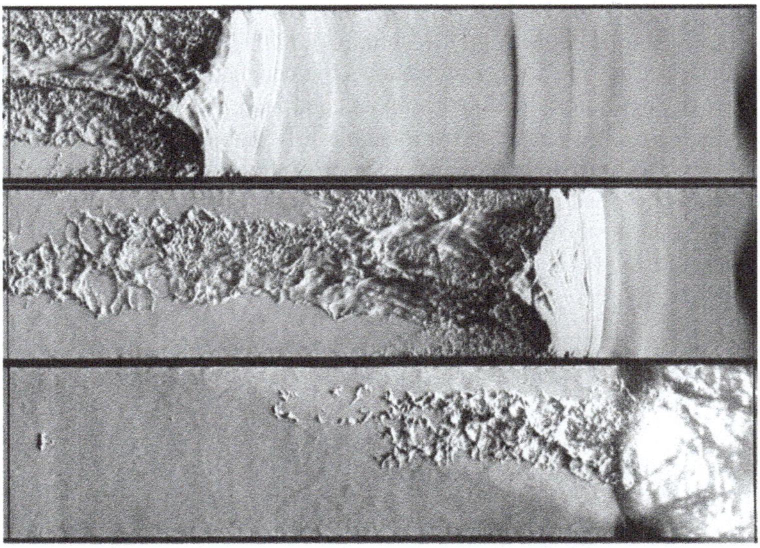

Fig. 15.10 Sequence of shadow photographs during the second stage and the transition to detonation in ethylene–oxygen mixture, $P_0 = 0.12 bar$. Time from the top is $t = 3.75,\ 3.85,\ 3.95\,ms$, [44]

that it would provide better resolution of density gradients in the direction along the tube axis. The preheat zone of the compressed unreacted mixture between the shock wave (white strips) and the flame front is well seen. In the top image taken at 3.75 ms the coordinates of the flame and shock are $X_f = 1081\,mm$ and $X_{sh} = 1094\,mm$, correspondingly, the visible velocity of the flame and shock wave are $U_f = 589\, m/s$ and $M_{sh} = 2.13$.

One can see that unlike the first stage, the compression waves steepen into shocks very close, almost at the flame front. In the second image, at 3.85 ms, the shocks are coalesced creating a pocket of compressed unreacted gas which appears close ahead to the flame. The width of the preheat zone is 16 mm ($X_f = 1177\, mm$, $X_{sh} = 1193\, mm$), the shock intensity is $M_{sh} = 3.03$, and the maximal pressure and temperature ahead of the flame front are: $P_u \approx 10 bar$, $T_u \approx 800 K$. In the last frame, taken at 3.95 ms, the transition to detonation has already occurred and only the retonation wave is visible. Similar scenario can be seen in the Schlieren photographs of earlier experiments [24–26]. However, this feature of the flow between the accelerating flame and the leading shock, which is important for understanding the mechanism of DDT was overlooked in previous studies.

15.3 Flame Acceleration in Channels with No-Slip Walls

Experimental studies [42–44] and simulations [44, 70–73] show three distinct stages of the process: (1) the flame accelerates exponentially, producing shock waves far ahead of the flame, (2) flame acceleration decreases and shocks are formed in the immediate vicinity ahead of the flame, (3) the final third stage is the actual transition to detonation due to coupling and mutual amplification of the shock wave and the reaction wave (flame).

The first explanation of the flame acceleration in tubes with no-slip walls was proposed by Shchelkin [22]. He argued that the thermal expansion of the burned products caused a flow ahead of the flame that became turbulent. Shchelkin considered turbulence as the main reason of the flame acceleration. All the same, DDT occurs in smooth-walled channels [42–44] and in even in thin capillary tubes [47–50], where the flow obviously remains laminar until the transition to detonation. The first qualitative study of flame acceleration caused by the interaction of a flame with the upstream flow in a channel with no-slip walls was carried out by Zel'dovich [51]. Analyzing Shchelkin's experiments Zel'dovich emphasized the fruitfulness of the Shchelkin idea about the role of a flame interaction with the upstream flow, and questioned Shchelkin's idea about turbulence as the main reason of the flame acceleration. Zel'dovich considered the stretching of the flame front due to the interaction with a nonuniform velocity field of the upstream flow as the main reason for the flame acceleration, while turbulence plays only a supplementary role, if at all, depending on the experimental conditions. Flame acceleration can be influenced by various factors such as tube width, smooth or rough walls, and obstacles in the flame path. The velocity of the flame initiated near the closed end of the channel is controlled by the flow ahead of the flame, which leads to the flame acceleration being compatible with the physical boundary conditions.

After the flame was ignited near the closed end of the tube, the flame front has a near hemispherical shape. Clanet and Searby [52] have shown that the first stage is the formation of the finger shape flame, with the flame tip Z_{tip} accelerating as

$$Z_{tip} \propto \exp(4\Theta U_f t/D), \tag{15.3.1}$$

where $\Theta = \rho_u/\rho_b$ is the density ratio of the unburned gas and combustion products, U_f is the normal velocity of the laminar flame relative to the fresh mixture, D is the tube width. After ignition near the closed end, the volume V of the burned gas increases as

$$\frac{dV}{dt} = \Theta U_f S_f, \tag{15.3.2}$$

where S_f is the total surface area of the flame. Evaluating the volume of the burned matter as $V = \pi D^2 Z_{tip}/4$, and the increase of the flame surface $S_f = \pi D Z_{tip}$, we can rewrite Eq. (15.3.2) as

$$\frac{dZ_{tip}}{dt} = \frac{4\Theta U_f}{D} Z_{tip}. \tag{15.3.3}$$

The solution of this equation is given by formula (15.3.1). This stage elapsed a short time after ignition $t \sim D/2\Theta U_f$ and ends when the flame skirt touches the side walls. According to the experimental measurements [52], this stage elapses about $t \approx 0.1D/U_f$.

At the next stage the burned gas pushes compression waves into the fresh fuel ahead of the flame. Since the flame velocity relative to the fresh fuel is small in comparison with the sound speed, the pressure waves quickly run away from the flame and a quasi-steady regime is established. The continuity equation reads

$$\rho\big(U_{fL} - u\big) = const, \tag{15.3.4}$$

where U_{fL} is the flame velocity in the laboratory frame of reference, and u is the velocity in the upstream flow (the flow ahead of the flame front). The boundary condition at the closed end requires zero velocity of the burned products, $u_b = 0$, so that Eq. (15.3.4) becomes

$$\rho_u(U_{fL} - u) = \rho_b U_{fL}. \tag{15.3.5}$$

Thus, we obtain that the flow velocity ahead of the flame and the flame velocity in the laboratory frame of reference are

$$u = (\Theta - 1)U_f, \; U_{fL} = \Theta U_f. \tag{15.3.6}$$

This result can be also obtained considering the amount of fuel consumed by the propagating flame. The flame propagating with velocity U_f relative to the fresh fuel in a channel of cross section S consumes SU_f amount of fresh fuel per unit time, producing the volume of $S\Theta U_f$ products. The increase in the volume occupied by combustion products $SU_f(\Theta-1)$ induces flow with velocity $u = (\Theta-1)U_f$ ahead of the flame, while the flame moves with the velocity $U_{fL} = U_f + (\Theta - 1)U_f = \Theta U_f$ in the laboratory reference frame.

Consider a simple model of a planar laminar flame propagating in a semi-infinite channel of width D from the closed to the open end. We will be interested in the velocity of the combustion wave, not the local velocity of a given point at the flame front. The velocity of combustion wave is based on the average heat release rate per frontal area of the stretched or wrinkled flame sheet and it determines the flow field ahead of the flame. The stretched flame consumes fresh fuel over a larger surface area, which leads to an increase in the velocity of the combustion wave, while the local velocity of different parts of the flame surface can increase or decrease. For example, the velocity of the central part of the flame front near the tube axis decreases when a tulip shape of the flame develops, while the velocity of the tulip petals remains almost the same. The analytical model does not take into account all the details, but it is intended to shed light on the features of the flame dynamics that are necessary to understand the mechanism of DDT.

Because of the wall friction the velocity in the upstream flow is maximal at the axis and vanishes at the channel side walls. The specific velocity distribution depends on the channel width. In a wide channel, the flow velocity ahead of the flame is plane-parallel over the channel cross section and drops to zero at the channel walls in a thin boundary layer of thickness $\delta_l = 5\sqrt{\nu \cdot X/U_0} \approx X/\sqrt{\text{Re}}$, where X is the characteristic scale of the upstream flow, $\nu$is the kinematic viscosity coefficient, U_0 is the velocity near the axis, $\text{Re} = U_0 X/\nu$ is the Reynolds number, and numerical factor 5 corresponds to the exact solution of H. Blasius [53, 54]. The characteristic scale of the upstream flow along the x-axis is the distance between from the leading shock wave to the flame front, $X = X_{sh} - X_f \approx (6-10)D$. The boundary layer is laminar in the flow behind the leading shock wave and can become turbulent downstream. It is known that the onset of turbulence in the boundary layer requires sufficiently strong disturbances and at least some low level of turbulence in the external (bulk) flow. The critical Reynolds number determining the transition to a turbulent boundary layer is [54]

$$\text{Re}_{x\,crit} = U_0 X/\nu \approx 3 \cdot 10^6. \tag{15.3.7}$$

In particular, turbulence can be caused by instabilities in the main flow for the Reynolds number larger than the critical Reynolds number, $\text{Re}_{cr} = u\delta_l/\nu > 400 - 500$ [53].

In the reference frame co-moving with the precursor shock, the flow of fresh fuel "enters" the tube behind the shock and propagates toward the flame. The thickness of a "young" boundary layer which is formed at the time $t \ll X/U_0$ is

$$\delta \sim \sqrt{\nu t}. \tag{15.3.8}$$

This explains why the measured thickness of the boundary layer (Fig. 15.5) is almost two times less than the boundary layer thickness in the case of a stationary flow.

For highly reactive mixtures used in experimental studies [42, 44], the Reynolds number is large, $\mathrm{Re}_X \sim 10^5$, and the thickness of the boundary layer $\delta_l \sim X/\sqrt{\mathrm{Re}}$ is much less than the channel width D. Since the total amount of flowing gas is constant, the increase in the thickness of the boundary layer, where the velocity is smaller, will cause an acceleration of the central part of the flow. For a steady flow this process continues until the parabolic (Poiseuille) velocity distribution is established. If we consider a 2D problem of the flow between two half-infinite plates separated by a distance D, the boundary layers grow downstream from the entrance as $\delta_l \sim X/\sqrt{\mathrm{Re}} \propto \sqrt{X}$ from both plates. Eventually, the boundary layers merge at some distance from the entrance and further downstream the Poiseuille parabolic velocity distribution is established. The distance X_p at which the Poiseuille flow is established can be estimated by the condition that $2\delta(X_p) \approx D$. Taking the value $\delta_l = 5\sqrt{\nu \cdot X_p/U_0}$ for a stationary flow, we obtain $X_p = D^2U_0/100\nu$ and for the time of establishing the Poiseuille velocity profile $\tau_P \approx D^2/100\nu$. The numerical factor in the denominator was calculated using different methods in [51, 54]. For example, in [51] it was found $\tau_P = D^2/106\nu$. For the experimental studies of DDT in stoichiometric hydrogen–oxygen and ethylene–oxygen mixtures [42–44], the time elapsed between the flame initiation and the detonation onset is about a few milliseconds (Figs. 15.1, 15.2, 15.3, 15.4, 15.5, 15.6, 15.7 and 15.8), whereas the time of establishing the parabolic velocity profile ahead of the advancing flame is much longer, $\tau_P \sim (0.1 - 1)$ sec. This means that in a wide channel the flow ahead of the flame remains nearly uniform in the bulk, with the flow velocity which drops to zero at the tube walls within a thin boundary layer $\delta_l \ll D$.

Each part of the flame front moves with respect to the unreacted mixture in the upstream flow with normal velocity U_f and simultaneously it is carried by the upstream flow with the local velocity of the upstream flow. Thus, the shape of the flame surface is defined by the relative motion of different parts of the flame front. As the flame advances into the nonuniform velocity field, the flame surface increases due to which it stretches within the boundary layer, to some extent trying to repeat the shape of the velocity profile in the upstream flow. This does not mean that the flame front will become planar, it may have a convex shape, which was acquired after ignition when the finger-like shape was formed. However, this means that the flame skirt is mostly stretched backward within the boundary layer. The stretched flame consumes fresh fuel over a larger surface area which results in the increase in the rate of heat release per unit projected flame area. A higher burning velocity results in an enhancement of the flow velocity ahead of the flame, which in turn gives rise to a larger gradient field and enhances the flame stretching, and so on. In this way, a positive feedback coupling is established between the upstream flow and the burning velocity. Within the model of a thin flame the increase of the burning rate is proportional to the relative increase of the flame surface (length in 2D case), which grows linearly in time with accuracy $\delta_l/D \ll 1$. This, in turn, results in a larger gradient of the velocity field upstream. Since the flame surface increases proportionally to the increase of the burning rate, the flame velocity increases exponentially in time

$$U_{fL} = \Theta U_f \cdot \exp\{\alpha \cdot U_f t/D\}, \tag{15.3.9}$$

where α is a numerical factor of the order of unity. The exponential increase of the flame speed in the channel with no-slip walls was first observed in simulations [67].

Formula (15.3.9) for the combustion wave velocity is similar to the exponential increase of the flame velocity obtained in [52], but it arises from the specific structure of the upstream flow with a thin boundary layer and has an accuracy $\delta_l/D \ll 1$. Similar expressions for the exponential growth of the flame velocity was considered in [55–57] assuming a parabolic velocity profile in the upstream flow. However, the parabolic velocity profile (the Poiseuille flow) in the upstream flow is established only in the case of very narrow capillary tubes. In narrow tubes ($D < 1mm$), the parabolic velocity profile in the flow ahead of the flame establishes in the time much less than the time elapsed between the flame initiation and the detonation onset. In this case, the velocity of the flame tip also increases exponentially due to the positive feedback between the accelerating flame and the upstream flow with a slightly different value of the numerical factor α in formula (15.3.9).

The difference in the flow velocity profiles ahead of accelerating flame in a wide and in a narrow channel explains different dynamics of the flame observed during the transition to detonation. In a wide channel, there are three distinctive stages of flame acceleration: the stage of exponential increase of the flame velocity is replaced by the stage with a slower rate of the flame acceleration and the possible formation of a tulip flame, which ends with a sharp increase in the flame speed and an actual transition to detonation (Figs. 15.1, 15.2, 15.3, 15.4, 15.5, 15.6, 15.7, and 15.8). On the contrary, in a sufficiently narrow channel, the parabolic velocity profile is established in a short time, and the stage of exponential increase of the flame speed ends with the transition to detonation without intermediate stage of a slower rate of the flame acceleration.

Taking into account $L_f \approx 0.07\,mm$ and $U_f \approx 5\,m/s$ for ethylene/oxygen mixture at the initial pressure $P_0 = 1\,bar$, one can find that the Poiseuille velocity profile in the flow ahead of the flame is formed almost from the very beginning in the capillary tubes of diameter $D \approx 0.5\,mm$. Experimentally, it can be observed as a monotonous change of the flame velocity before the transition to detonation. This difference in the flame dynamics is clearly seen in the experimentally measured velocity–time dependence of the ethylene–oxygen flame in capillary tubes measured by Wu et al. [48] and shown in Fig. 15.11.

In the case of rough walls, strong disturbances are generated from the very beginning and move downstream forming turbulent wedge. For the value of the wall roughness d, the velocity distribution in the turbulent boundary layer is $U \approx \frac{u'}{\kappa} ln\left(\frac{y}{d}\right)$, where U is the average speed in the bulk stream, u' is the fluctuating velocity and $\kappa = 0.42$ is the von Karman constant. The thickness of the turbulent boundary layer can be estimated as $\delta_t \sim X\frac{u'}{U}$. In this case, the bulk flow can become turbulent at fairly short distances. The commonly used empirical formula for the thickness of a

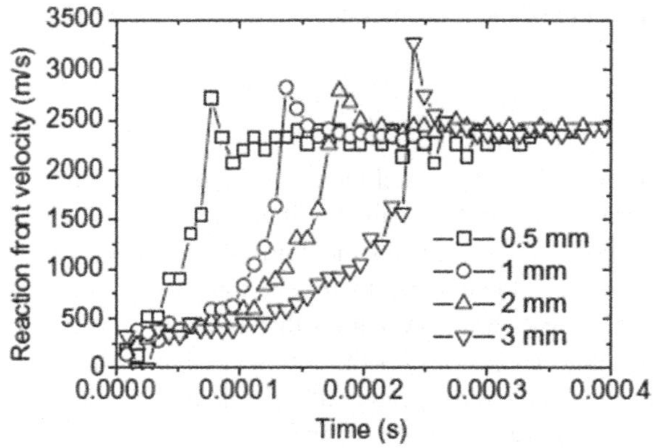

Fig. 15.11 The flame tip velocity evolution in capillary tubes: $D = 0.5, 1, 2, 3$ mm [48]

turbulent boundary layer is well described by the simple Prandtl model

$$\delta_t = 0.16 \cdot \frac{\nu}{U} \mathrm{Re}_x^{6/7}. \tag{15.3.10}$$

15.4 Mechanism of Deflagration-to-Detonation Transition (DDT)

15.4.1 Numerical Simulations of DDT

Experimental studies [42–44] show that the accelerating flame acts as a semipermeable piston, producing compression waves that steepen into weak shock waves. Even the coalescence of these shock waves results in a rather weak shock wave with the Mach number not exceeding $M_{sh} \approx 2 - 2.5$, which is too small to ignite a reaction. The transition to detonation requires a shock wave strong enough to ignite the combustible mixture in a very short time, comparable to the time of passage of particles through the reaction zone behind the shock front.

There are different mechanisms leading to the amplification of the shock wave. The intensity of a shock wave propagating through an inhomogeneous gas of decreasing density (for example, from the center to the surface of a star) can increase unlimitedly [58–61]. The focusing of the shock wave, leading to detonation, was investigated in the design of a tube with a tapered opposite end in [62]. Another possibility is that a nonuniform distribution of temperature or reactivity can lead to spontaneous initiation of detonation due to the reactivity gradient, known as Zel'dovich's gradient mechanism (Chap. 13). A similar concept of Shock-Wave Amplification by Coherent Energy Release (SWACER) was proposed by Lee and co-authors [28, 65, 66]. The SWACER mechanism is based on the principle that the coupling of a chemical reaction with a shock wave can lead to the positive feedback and their mutual amplification if the chemical energy release is coherently synchronized with the shock wave, thereby increasing the shock wave intensity. The authors of [28] wrote that "the acronym SWACER was introduced to emphasize the similarity of the amplification mechanism to the well-known LASER mechanism, which is based on light wave amplification due to coherent energy release by stimulated emission in a resonant cavity".

Because of the complexity of actual chemical kinetics, most of the numerical calculations undertaken in an attempt to understand the DDT mechanism used a simplified one-step Arrhenius model. The results of many years simulations undertaken in an attempt to understand the DDT mechanism using 2D numerical simulations with a one-step chemical model, summarized in review articles [32–34, 38, 39], can be briefly summarize as the following conclusion [32]: "the accelerating flame creates ignition centers ("hot spots") in nearby unreacted material ahead the flame front that lead to detonation formation through Zel'dovich gradient mechanism". Ultimately, it was claimed that the ignition of the spontaneous reaction wave and the Zel'dovich gradient mechanism is a universal mechanism triggering DDT. To justify this approach, it was claimed in [67] that "for practical situations it is more important to have an accurate model of the fluid dynamics coupled to a model for the chemical-energy release that puts the released energy in the 'right' place in the flow at the 'right' time". This was the main trend in DDT studies until it was shown in experiments [43, 44] that the temperature of the mixture ahead of the flame remains too low all the time prior to DDT to cause an exothermic reaction even in a highly reactive mixture such as hydrogen/oxygen and ethylene/oxygen. The measured intensity of the shock waves creating ahead of the flame does not exceed $M_{sh} = 2.3 \div 2.5$, and the maximum temperature ahead of the flame does not exceed $(500 \div 600)\ K$. The ignition time calculated using a detailed chemical model and measured in experiments for stoichiometric hydrogen/oxygen is 10 s at $T = 800\ K$; 0.2 s at $T = 900\ K$, which is by orders of magnitude longer than the time elapsed from the flame ignition to the transition to detonation. Unlike the DDT mechanism suggested in [32], J.H.S. Lee [66] believed that a fast flame at a near CJ deflagration velocity is a key factor in the so-called flame-brush-to-detonation transition via the SWACER mechanism.

The difference between the gradient mechanism and the SWACER mechanism requires a careful analysis of the results obtained in simulations. In practice, the difference is whether DDT is formed ahead of the flame front, in the "hot spot", or its origin is in the flame brush. The key factors required for identifying the mechanism of DDT are the high-resolution simulations and the detailed chemical model, which would guarantee the correct time and place of the chemical energy release and ignition processes. It should also be borne in mind that not only the chemical model, but also a detailed (not oversimplified) transport model is important for correct simulations.

As an example, Fig. 15.12 shows the velocity–time history of the combustion wave in the hydrogen/air mixture during the transition to detonation obtained in the high-resolution simulations ($\Delta = (1.25 \div 10)\ \mu m$) for a simplified transport model and for the full transport model. It should be noted that some authors have argued that

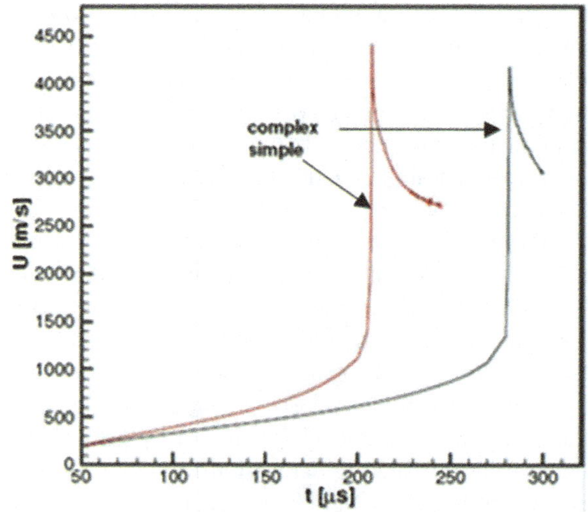

Fig. 15.12 Flame velocities obtained using a simple and complete transport models

the gradient mechanism is the mechanism of DDT without careful analysis, which is a phenomenon known as "adopted over many repetitions".

15.4.2 Flame Acceleration and DDT in 2D Channel with Smooth Walls

In a series of publications [44, 70–73], Liberman et al. analyzed analytically the acceleration of the hydrogen–oxygen flame in a channel with no-slip walls and performed the high-resolution simulations of the multidimensional time-dependent reactive Navier–Stokes equations, with account of compressibility, viscosity, thermal conduction, molecular diffusion, real equation of state, and detailed chemical kinetics for the reactive species. Simulations of the hydrogen–oxygen flame propagating in channels with no-slip walls were performed for channels of different widths, from 0.5 to 10 mm with the minimum computational cell size $\Delta < 10\,m\mu$ (the width of the flame front at $P_0 = 1$ atm is $L_f = 27\,\mu m$). A laminar planar flame was initiated near the left closed end of the 2D and 3D channel and propagates to the right open end.

The velocity–time dependence of the flame and transition to detonation for different initial pressures $P_0 = 0.6,\ 0.8,\ 1.0 bar$ is presented in Fig. 15.13. An overall picture of the flow, shock waves and the transition to detonation at lower pressure is similar to the case $P_0 = 1.0 bar$. The run-up distance of DDT increases with the decrease of the initial pressure in agreement with the experimental observations in Fig. 15.2. The velocity–time dependence of the flame in a channel of width $D = 10\,mm$ at the initial pressure $P_0 = 1\,bar$ is shown in Fig. 15.14. The insert in Fig. 15.14 shows that the flame velocity increases exponentially

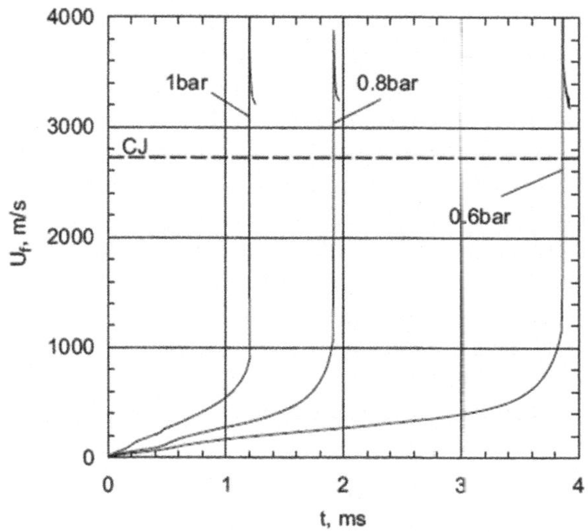

Fig. 15.13 Velocity of H_2/O_2 flame for initial pressures $P_0 = 0.6,\ 0.8,\ 1.0bar$ [70]

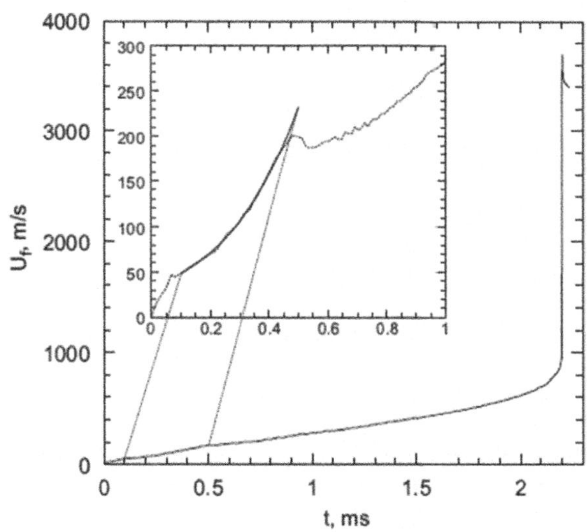

Fig. 15.14 The flame velocity at $P_0 = 1.0bar$. Insert is the flame velocity at the beginning; solid line is approximation by $U_{fL} \propto 32.744 \cdot \exp(3.919 \cdot t/ms)$

before 0.5 ms. The velocity–time dependence before 0.5 ms can be approximated as $U_{fL} \propto 32.744 \cdot \exp(3.919 \cdot t/ms)$ which is consistent with the velocity–time dependence shown in Fig. 15.3 and formula (15.3.9).

During the first stage the pressure wave produced by the accelerating flame steepens into a shock ahead of the flame at the distance about 6–7 widths of the channel.

An overall picture in a wide channel consists of (1) an exponential increase of the flame speed during a short time, (2) the flame continues with a lower rate of acceleration than during the first stage, (3) a very rapid increase of the combustion velocity and the actual transition to detonation.

The temperature, density and pressure computed at t $= 2.2\,ms$, just before the transition to detonation, for $P_0 = 1\,bar$, $D = 10\,mm$, are shown in Fig. 15.15.

For the lower pressure, $P_0 = 0.6bar$, the computed profiles of temperature, density and pressure just before the actual transition to detonation, that starts at $t = 3.85\,ms$, are similar to that shown in Fig. 15.15. The temperature ahead of the flame before the transition to detonation does not exceed 500–550 K, so that the reaction rate is negligible everywhere ahead of the flame front.

The qualitative difference of the flow fields developing ahead of the flame in a wide channel compared with a thin channel is illustrated in Fig. 15.16, computed for $D = 10\,mm$ at $t = 1\,ms$, and in Fig. 15.17 computed for $D = 0.5\,mm$ at $t = 0.1\,ms$. The figures show velocity contours and velocity profiles in the cross section A-A.

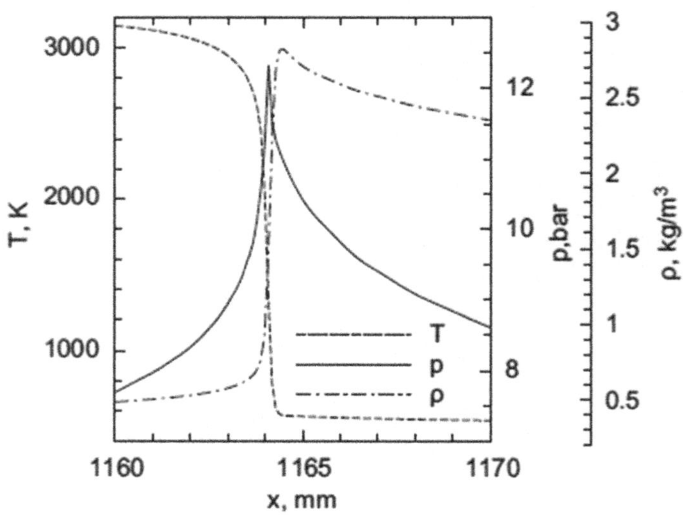

Fig. 15.15 Temperature, density and pressure profiles just before the actual transition to detonation starts at $t = 2.2ms$ for $P_0 = 1\,bar$ [70]

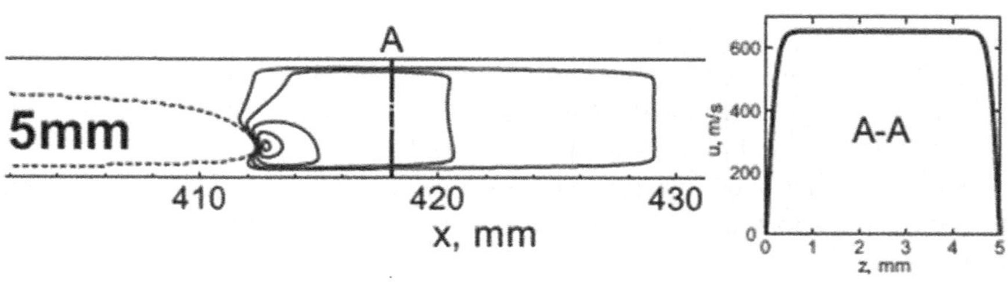

Fig. 15.16 Flow field ahead of the flame front in channel $D = 10\,mm$ at $t = 1\,ms$ (left) and the velocity profile in the cross A-A (right) [71]

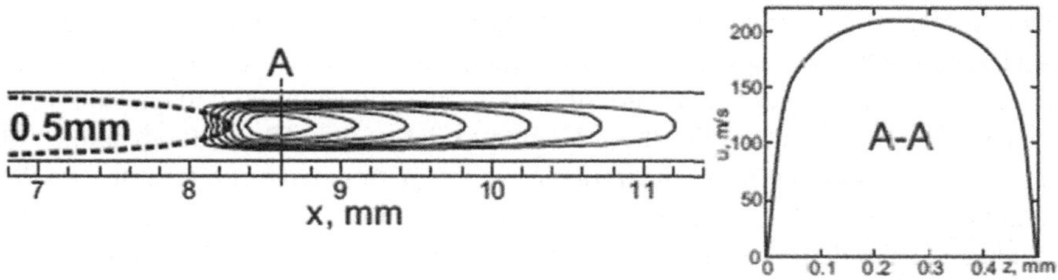

Fig. 15.17 Flow field ahead of the flame front in capillary channel $D = 0.5\,mm$, at $t = 0.1\,ms$ (left), and the velocity profile in the cross A-A (right) [71]

The velocity profiles shown in Figs. 15.14 and 15.15 agree well with the estimates in Sect. 15.3. The parabolic velocity distribution in the flow ahead of the flame is established in a narrow channel, while in a wide channel the flow ahead of the flame is nearly uniform in the bulk with the flow velocity dropping to zero within a thin boundary layer.

Figure 15.18 shows the computed velocity–time dependence of the flame during the flame acceleration and the transition to detonation in channels of widths: 0.5, 1, 5, and 10 mm. For channels wider than 1 mm, the velocity–time dependences demonstrate three distinctive stages of the flame acceleration: a stage of the exponential increase of the flame velocity; the second stage when the rate of the flame acceleration decreases; the sharp increase of the flame velocity and actual transition to detonation.

For a wide channel, the initial stage, which is well approximated by the exponential time dependence shown by the dashed lines in Fig. 15.17, turns into the stage of slower rate of the velocity increase, which is well approximated by a polynomial

Fig. 15.18 The flame velocity in channels of widths: 0.5, 1, 5, and 10 mm [71]

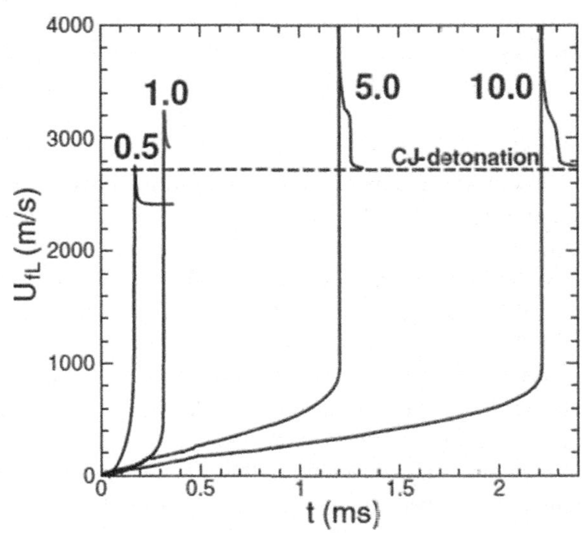

function $U_{Lf} \propto a + bt^{n_1} + ct^{n_2} + ...$ with the exponents $n_i < 1$. On the contrary, if the channel width is less than 1 mm, the exponential increase of the flame velocity (the dashed lines in Fig. 15.17) is not limited in time and an initial deflagration wave accelerates exponentially up to the transition to detonation. Thus, the dynamics of the flame acceleration prior to the transition to detonation agrees quite well with the theoretical analysis of Sect. 15.3. In particular, the numerical factor α in Eq. (15.3.9) calculated in simulations is within the range $1 < \alpha < 3$ depending on the flame front configuration.

The accelerating flame acts as a semipermeable piston producing pressure waves in the flow ahead of the flame. The time and location where the pressure wave steepens into a shock wave are determined by the condition that velocity $u(x, t)$ given by the Riemann solution for a simple traveling wave becomes a multi-valued function [53]. Simple calculations show that for a piston moving with the exponential velocity–time dependence (15.3.9) the pressure wave steepens into the shock at the intermediate point in the pressure wave defined as

$$u = \frac{2a_{s0}}{(\gamma+1)}\left[1 - \gamma(\Theta - 1)\frac{U_{fL}}{a_{s0}}\right]. \tag{15.4.1}$$

The shock wave intensity is $M_{sh} \approx 1.2 \div 1.3$ for $\gamma = 1.4$, and up to the first-order terms in $U_f/a_{s0} \ll 1$, the distance from the flame ($x = X_f$) to the point where the shock formed is

$$X_{sh} \simeq \frac{D}{\alpha(\gamma+1)}\left(\frac{a_{s0}}{U_{Pf}}\right)\left\{\gamma^2 + \gamma + 2 - 2\left[\frac{(\gamma+1)^2}{4} - 1\right]\ln\left(\frac{2a_{s0}}{\Theta(\gamma+1)U_{Pf}}\right)\right\}. \tag{15.4.2}$$

Here, a_{s0} is the sound speed ahead of the flame, and U_{Pf} is the flame velocity in the co-moving with the upstream flow coordinate system. Using Eq. (15.4.2), the distance between the flame ($x = X_f$) and the shock ($x = X_{sh}$) can be estimated as

$$(X_{sh} - X_f) \sim (5 \div 7)\, D. \tag{15.4.3}$$

Using this estimate, the maximum thickness of the boundary layer in the close proximity ahead of the flame is $\delta_l \sim (X_{sh} - X_f)/\sqrt{\text{Re}}$, where the Reynolds number in the upstream flow is $\text{Re}_x = (\Theta - 1)U_f(X_{sh} - X_f)/\nu \approx \Theta U_f D/\nu$. For hydrogen–oxygen flame in the channel of width $D = 10\,mm$, taking $U_f \approx 10\,m/s$, $\Theta = 10$, $\nu = 0.38\,cm^2/s$, we obtain $\text{Re} = 1.1 \cdot 10^5$, and $\delta_l \approx 0.5\,mm$, which is consistent with the results of simulations.

As the flame front stretches along the tube walls, the rear part of the flame skirt moves closer to the wall, and when the distance to the wall becomes about the width of the flame, the rear part of the flame skirt is extinguished. This leads to the reduction of the flame surface (length in the 2D case) and correspondingly to the decrease the flame speed. The characteristic time of the flame surface decrease is of the order $t_T \sim L_f/U_f$. Since this time is small, the flame acceleration during the previous

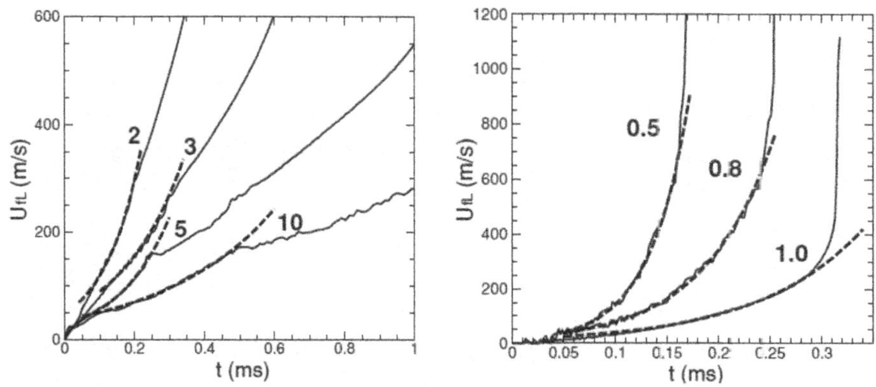

Fig. 15.19 The flame velocity in channels of width $D = 2,\ 3,\ 5,\ 10\,mm$ (left), and the enlargement of the flame velocity for $D = 0.5,\ 0.8,\ 1\,mm$ (right). Dashed lines are exponential approximations of the velocity–time dependence [71]

stage is nearly constant up to the first-order terms in $L_f/D \ll 1$ in series expansion of formula (15.3.9). This means that during the next stage, for $t > t_T$, the flame velocity–time dependence can be approximated as

$$U_{fL} \approx \Theta U_f\left[1 + \beta(t/\tau_f)^n\right],\ where\ 0 < n < 1,\ \tau_f = L_f/U_f, \qquad (15.4.4)$$

which agrees with the velocity–time dependence shown in Fig. 15.19.

It is known [53], that for a piston moving with the velocity–time dependence similar to Eq. (15.4.4), the Riemann solution for the velocity of simple traveling wave $u(x, t)$ is multi-valued function everywhere. This means that the pressure wave steepens and the shock wave ahead of the flame is formed immediately ahead of the flame surface (formally on the flame tip). Specific value of n does not matter, because we are interested only in the location where the compression wave steepens into the shock wave. The shock wave formed in the immediate vicinity ahead of the flame creates a layer of heated and compressed unreacted gas entering the flame front.

Unlike a stationary flow, the accelerating flame is not isobaric, and the pressure increases at approximately the same rate as the flame velocity. At the time when the pressure wave steepens into a shock wave in the immediate vicinity ahead of the flame front, an increased amount of fuel compressed in the shock wave (e.g. $\rho_2/\rho_0 \approx 2 - 2.5$ for $M_{sh} = 1.3 - 1.5$) enters the flame reaction zone. This causes an increase in the energy released in the reaction and can amplify the shock wave, which can lead to a positive feedback coupling between the flame and the shock wave.

The overall picture of the flow, the flame development and the transition to detonation, representing the time sequence of the density gradient computed for the channel of width $D = 10mm$, is shown in Figs. 15.20a, b. Figure 15.20a shows that just before the transition to detonation, a shock wave appears at the flame tip in accordance with the qualitative scenario described above. The shape of the flame

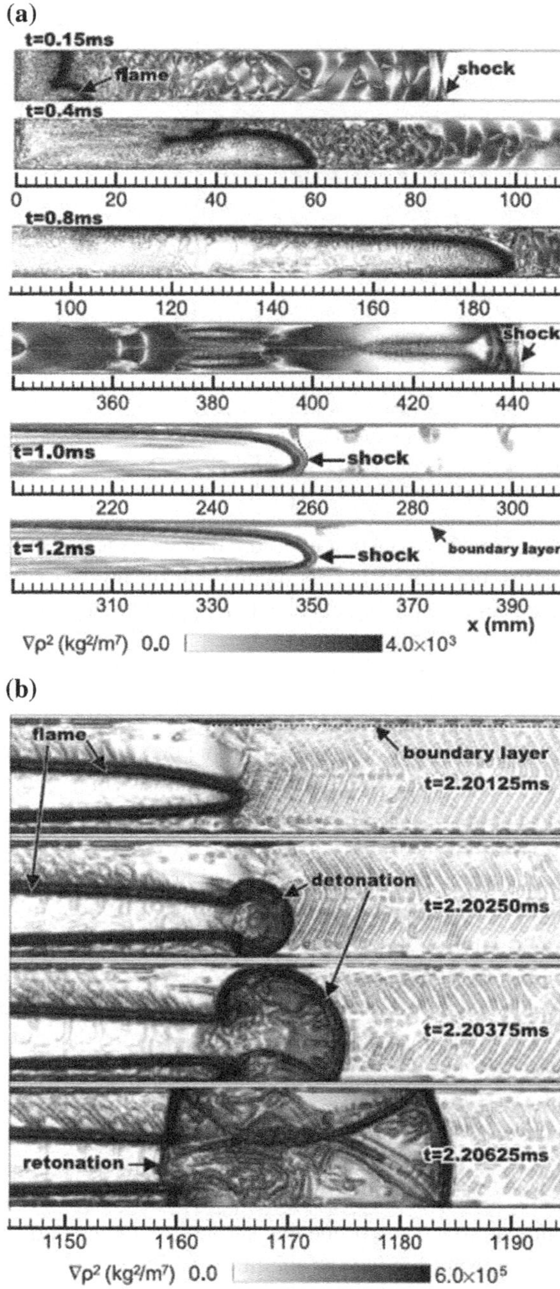

Fig. 15.20 **a** Density gradient at sequential times showing the overall flow development in the channel: $D = 10mm$, $P_0 = 1\,bar$. **b** Density gradient during the last stage of the transition to detonation. DDT starts after 2.2 ms. The detonation and retonation waves are seen in the last frame

front depends on the conditions of its initiation. Usually a well-familiar tulip-shaped front is formed, but if the initial small perturbations are applied, the flame front quickly acquires the form of a single-mode asymmetric hump. Computations show that the overall dynamics of the flame acceleration and the velocity of the combustion wave are almost not affected by the initial small perturbations (less than 1% of the flame velocity). The experimentally observed stochasticity of the process better corresponds to a more realistic situation of a slightly perturbed flame after its initiation. The perturbations tend to destroy the tulip shape, making the front asymmetrical as shown in Fig. 15.20a. The first frame at 0.15 ms shows the perturbed flame in the left end of the computational domain and the compression waves produced by the accelerating flame in the unburned gas, which steepen into the shocks in the upstream flow of unreacted fuel. The upstream flow remains laminar during the whole process until the transition to detonation. A thin boundary layer of thickness 0.3–0.4 mm is seen in Fig. 15.20b. The pressure waves produced by the accelerating flame in the unburned gas steepen into the shocks far ahead of the flame at the distance approximately $(6 \div 7)D$. After $t = 0.8\,ms$, the rate of the flame acceleration decreases, and the pressure waves steepen into the shocks in the immediate vicinity ahead of the flame front as shown in frames 5 and 6 (from the top) in Fig. 15.20a. From 0.8 to 2.2 ms, the shock waves are formed in close proximity ahead of the flame front. They merge and create a layer of compressed and heated unreacted gas attached to the flame front, so the shock wave now "sits" at the flame tip. The precursor shock, which has run far ahead from the flame, is not shown in Fig. 15.20a. After $t = 0.8\,ms$ the flux of unreacted fuel entering the flame front increased significantly, and as a result, a pressure pulse forms at the tip of the flame. At later times the peak of the pressure pulse is located at the flame front and grows exponentially in time. From $t = 2.20125\,ms$ to $t = 2.20250\,ms$ the pressure pulse steepens into a strong shock which is coupled with the flame reaction zone forming overdriven detonation. Figure 15.20b shows the last stage of the transition to detonation. The last frame in Fig. 15.20b shows the overdriven detonation and a blast wave propagating into the combustion products, which is referred to as the retonation wave. After $t = 0.8\,ms$ the pressure peak at the flame tip grows exponentially fast as shown by the dashed line in Fig. 15.21. Effectively, the pressure peaks move together with the flame tip. In the subsequent stage, the fast acceleration of the flame tip is caused by the coupling between the reaction heat release and the exponentially fast growing pressure peak.

The coupling begins just before the transition to detonation when the pressure peak exceeds $P \approx 4\,bar$. From this time, $t = 2\,ms$, and just before the actual transition to detonation begines, $t = 2.2\,ms$, the growth rate of the pressure sharply increases, and the transition to detonation occurs in less than 1 μs. It should be emphasized that during the entire process the temperature ahead of the flame front does not exceed 550 K, so that the rate of chemical reactions ahead of the flame remains negligibly small. The pressure peak reaches 10 bar at 2.2 ms, and it increases above 20 bar at $t = 2.1\,ms$ during the transition to detonation.

The last stage of the actual transition to detonation is shown in Fig. 15.22. Figure 15.22 shows the variations in pressure, temperature and concentration of H-radicals at successive times, from $t = 2.2\,ms$ till $t = 2.20225\,ms$ with the time

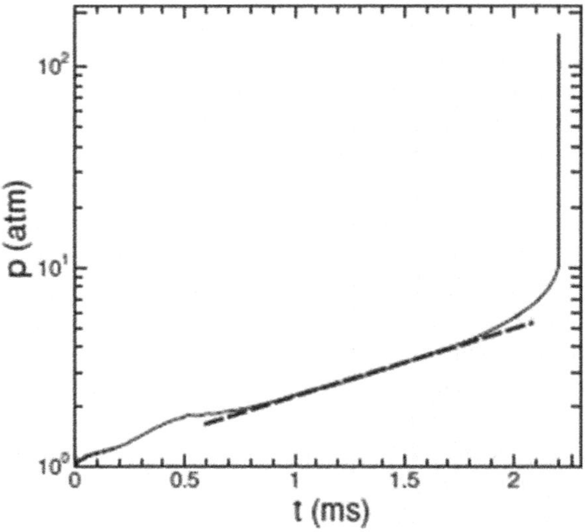

Fig. 15.21 Peak pressure history in the channel of width $D = 10\,mm$. Dashed line is the exponential approximation

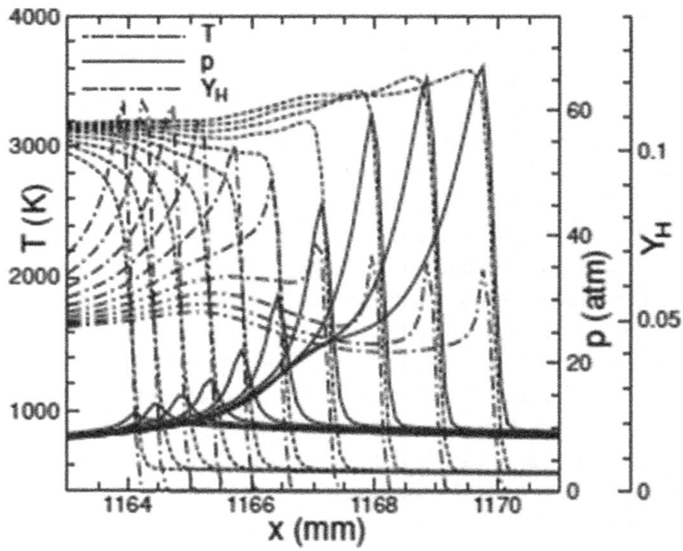

Fig. 15.22 Profiles of pressure, temperature and H-radical concentrations during the actual transition to detonation from $t = 2.2\,ms$ till $t = 2.20225\,ms$; $\Delta t = 0.25\,\mu s$

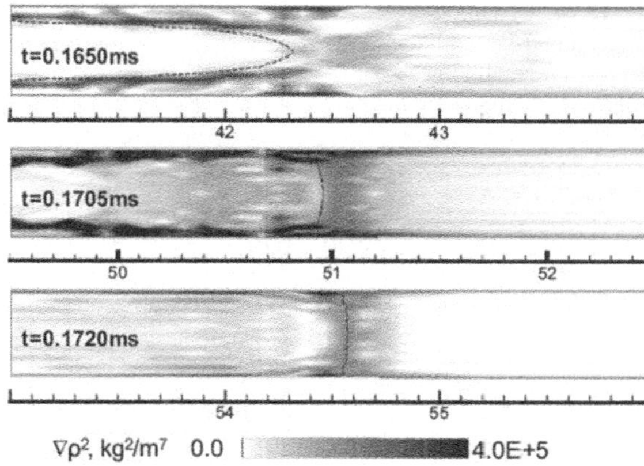

Fig. 15.23 Calculated DDT in the channel: $D = 0.5\,mm$, $P_0 = 1\,bar$

interval 0.25 μs in the channel of width $D = 10mm$, and initial pressure $P_0 = 1\,bar$. The transition to detonation is clearly seen from the steepening of the temperature profile and the increased temperature of the products to the value corresponding to a detonation wave. One can see this also from the specific distribution of H-radicals, which is different for a deflagration and a detonation.

In the deflagration regime, H-radicals appear within the reaction zone, while structure of the detonation wave consists of the well pronounced shock wave with the jump in temperature and pressure following by the reaction zone where H-radicals are born.

In a thin channel, the overall picture of the flame velocity evolution differs from that in a wide channel, but the mechanism of DDT is similar to that in a wide channel. Figure 15.23 shows the evolution of the density gradient at the sequential times for the hydrogen–oxygen flame and the transition to detonation in the channel of width $D = 0.5\,mm$, for the initial pressure $P_0 = 1\,bar$.

The general scenario of DDT in a narrow channel shown in Fig. 15.24 is similar to the scenario in a wide channel, but there are some peculiarities associated with the dynamics of the flame in a narrow channel. In a narrow channel, the flame velocity increases exponentially with time until DDT. Since the H_2/O_2 detonation cell size at 1 atm is larger than the channel width, a steady detonation cannot be properly formed. Therefore, the CJ detonation is formed almost bypassing the stage of an overdriven detonation, with the speed slightly below the Chapman–Jouguet velocity.

15.4.3 Flame Acceleration and DDT in 3D Channels

The first 3D simulations of the hydrogen/oxygen flame acceleration and the deflagration-to-detonation transition were carried out in [73] using a detailed chemical reaction model for the reactive species H_2, O_2, H, O, OH, H_2O, H_2O_2, HO_2. The computational domain was taken in

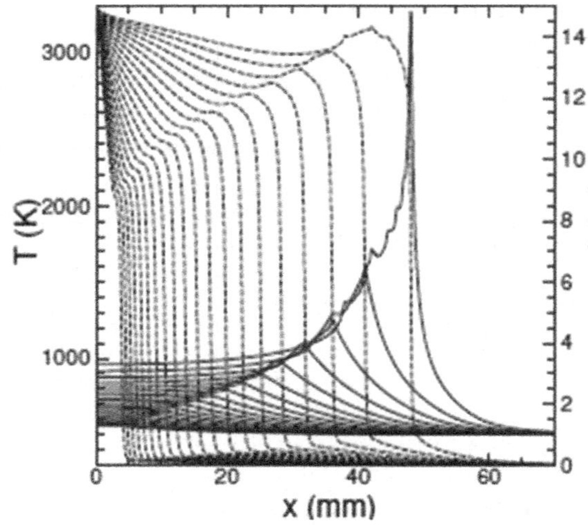

Fig. 15.24 Evolution of temperature (dashed lines) and pressure (solid lines) profiles on the flame tip in the channel $D = 0.5\,mm$; $t = 73.75\,\mu s$ to $t = 168.75\,\mu s$, $\Delta t = 5\,\mu s$

the form of a rectangular channel with a cross section of $10 \times 10\,mm$ and a length of 250 mm with the closed left and open right ends. The flame was ignited by a thin heated layer near the closed end and propagates in the x-axis direction to the opened end.

While in 2D case the preferable direction of the flame propagation is associated with the side walls, and initially planar flame evolves into the tulip-shaped or single-mode flame which leading tips are moving along the walls as in Fig. 15.20a, b, in 3D case the preferential directions for the flame propagation are the edges between the channel side walls. Typical structures of the flame front during the initial stage of the flame propagation are shown in Fig. 15.25 for: (a) initially unperturbed planar flame; (b) initial regular perturbation; (c) initial random (chaotic) perturbation. The type of the perturbations is shown in Fig. 15.15 at the rear end of the channel. Only in the case of non-regular initial perturbations, the flame evolves into an asymmetric shape with tips moving in the bulk far away from the side walls. It is seen that initial perturbations tend to destroy the symmetrical shapes specified by the channel geometry. In case of the initially unperturbed planar flame (Fig. 15.25a) the stretching inside the boundary layers, preferentially at the edges between the channel side walls, causes formation of the tulip-shaped flame. The initial regular perturbations (Fig. 15.25b) result in a single-mode flame shape, and stochastic perturbations (Fig. 15.25c) cause formation of the flame brush with a strongly corrugated flame surface consisting of several leading bulged tips. At the same time, the overall dynamics of flame acceleration remains almost the same regardless of the initial perturbations. Moreover, the 3D flame dynamics is similar to the dynamics obtained in the 2D case, except for the higher acceleration rate associated with the additional degree of freedom.

This is seen from Fig. 15.26, which shows the flame velocities (solid lines) computed for 3D and 2D (dashed-dotted line) channels of the same width $D = 10\,mm$.

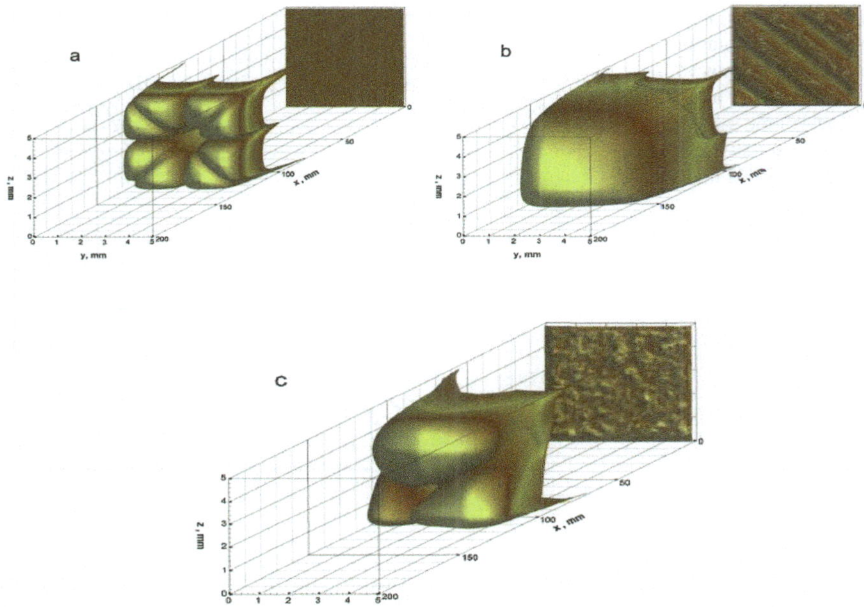

Fig. 15.25 **a**, **b**, **c** Structure of the flame front depending on the initial small perturbations (shown at rear surface of the channel) at $t = 400\mu s$: **a** initially unperturbed planar flame, **b** initial regular perturbations, **c** initial chaotic perturbations [73]

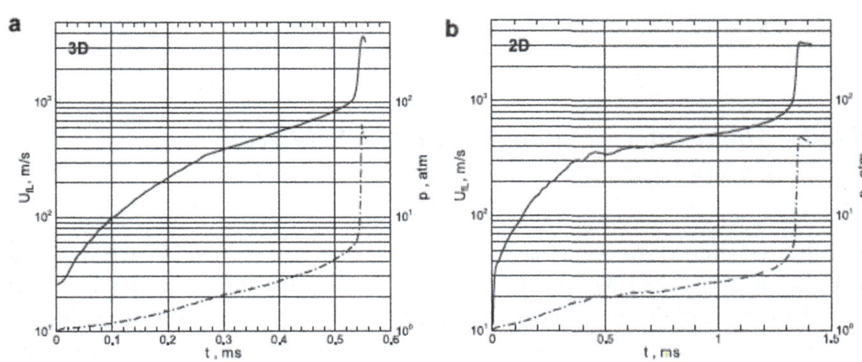

Fig. 15.26 Flame velocity (solid lines) and the pressure peak (dash-dotted lines) time dependences computed for 3D channel (**a**) and for 2D channel (**b**). The width of channels $D = 10\,mm$ in both cases [73]

In both cases, the velocity–time dependences demonstrate the same feature of three distinctive stages of flame acceleration: (0) the initial stage of the flame expansion from the ignition zone; (1) the stage of exponential increase of the flame velocity; (2) the stage when the flame acceleration decreases; (3) the actual transition to detonation. The transition time to detonation in the 3D case is about three times shorter,

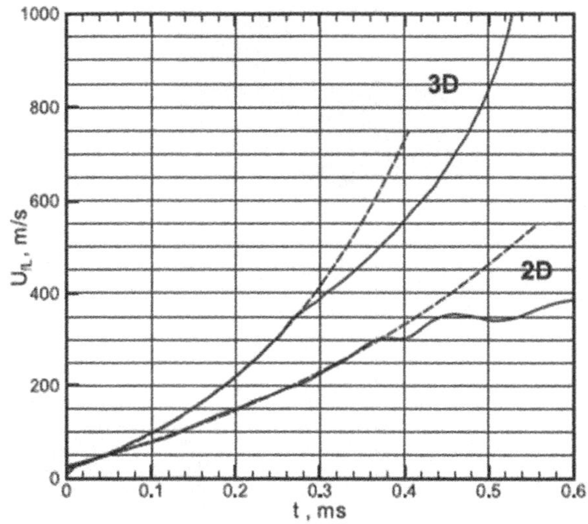

Fig. 15.27 The flame velocity-time evolution during initial stages computed for 3D and 2D channels of the same width. The exponential velocity is shown by dashed lines

mainly due to a faster increase in velocity during the first stage, which leads to more severe conditions and a significant shortening of the second stage, which, in turn, determines the DDT mechanism. Figure 15.27 shows the time evolution of the combustion wave in 3D and 2D channels of the same width. Note that, in the 3D case, the run-up distance and the time before the transition to detonation are about three times shorter compared to the 2D channel of the same width. The pre-detonation time is 0.55 ms for the 3D case and 1.355 ms for the 2D case.

15.4.4 About Interpretation of Shadow/Schlieren Photographs

Since the first attempts to understand the origin and mechanism of DDT, schlieren and shadowgraph techniques have been widely used to visualize the flame propagating in tubes. This is one of the most informative method for studying combustion processes and high-speed flows involving shocks, where the gradients in the refractive index are large, because the method is sensitive to the changes in a refractive index within the material and do not interfere with the flow [81]. Contrary to a shadowgraph method, which measures the second derivative of density, in the schlieren method a knife-edge is placed at the focal point to block about half of the light beam. This results in a set of lighter and darker patches corresponding to positive and negative fluid density gradients in the direction normal to the knife-edge, so that the schlieren method measures the first derivative of density in the direction of the knife-edge, which controls contrast in light intensity distribution on the screen. This method revealed that one of the leading roles for the DDT phenomena belongs to the shock waves, which are forming close to the flame front just before the transition to detonation.

Often, shadow photographs of flame and DDT in smooth-walled tubes are interpreted as "showing the turbulent flame". To capture the features of the flame acceleration and the origin of detonation, one needs accurately resolve both time and space scales which are far beyond the limits of the existing resolution of schlieren method. Even the best up to now temporary resolution of the cameras used in DDT experiments is too low for feasible resolution, which would make possible to resolve the process of DDT in details needed for understanding physical processes during DDT. In most cases, DDT can only be recognized by two sequential snapshots. The first snapshot shows the flow pattern in the vicinity of the flame front. The second one shows already the formed detonation wave propagating at the background of the previous flow pattern.

Usually, interpolation of images of snapshots gives information that the origin of detonation is in the immediate vicinity inside the flame brush, which agrees with the mechanism of DDT described above. However, sometimes the snapshots give impression that the detonation was formed and propagating out from some location ahead the flame. Urtiew and Oppenheim [24] called such scenario "explosion in the explosion" and it was usually associated with the concept of "hot spots" arising in unreacted material. However, since schlieren and shadow photographs give a 2D projection of the original 3D pattern on the planar screen, this may lead to misinterpretation of the process.

Until holographic photography becomes available for combustion experiments, 3D modeling provides us with 3D images of the process. In addition to 2D projections of the calculated 3D images, we can obtain also 2D cross sections of 3D images with streamlines showing the velocity field. As an example, Fig. 15.28 shows sequence of shadow photographs in the cross section XY of the flow during the first stage of flame acceleration obtained using 3D images of 3D simulations and shown in Figs. 15.28a, b at time instants $t = 200\,ms$ and $300\,ms$, correspondingly. The shadowgraph visualization obtained experimentally in [43] for the first stage of flame acceleration is shown in the frame Fig. 15.28c.

It is seen that the "artificial" shadow photographs 15.28a, b are very similar to the real shadow photograph in Fig. 15.28c. The flame reproduced by these shadow photographs looks like a turbulent one, which is usually called "turbulent flame brush". However, the advantage of numerical simulations compared with experimental data is that we can verify whether the flow is turbulent, or this is artifact.

The left images in Fig. 15.29 show sequence of 3D structures of the flame and flow in the channel of square cross section ($10 \times 10\,mm$). The initial irregular small perturbations imposed in the beginning cause the formation of the flame brush, which shortly develops in many well pronounced bulged tips. The compression waves produced by the accelerating flame are seen as slightly corrugated surfaces close ahead the flame front and the planes far ahead the flame front. They steepen into the shock waves far ahead the flame $(6 \div 7)D$ (not shown in Fig. 15.19). The boundary layer is seen as lighter white stripes on the side walls of the channel. Just before the actual transition to detonation the shock wave is formed on the leading bulged tip of the flame and practically "sits" at the tip, and then, within a few microseconds the transition to detonation occurs. The cross sections of each 3D frame in the plane

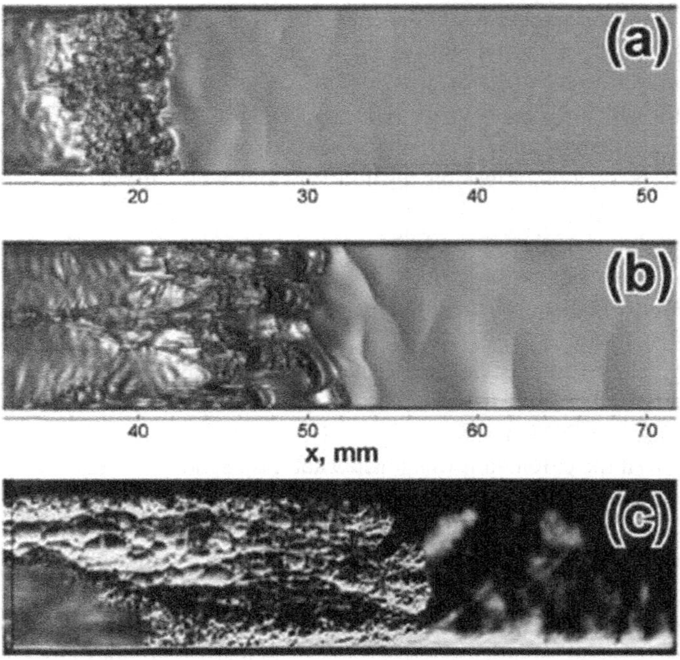

Fig. 15.28 Frames **a** and **b** are obtained using 3D simulations, t = 200 μs and 300 μs correspondingly. Frame **c** is an experimental shadow photo for H_2/O_2 flame [43]

$(x, y, z = 5.0\,mm)$ are shown at the right part of Fig. 15.19. It should be emphasized that the images represented by the cross sections in the plane (x, y) do not correspond to experimental schlieren photographs, but rather similar to the 2D simulation of the process. While the shadow photographs look like a turbulent flame, it is seen from the 3D images and 2D cross sections of Fig. 15.29 that either flame or flow are not turbulent until the detonation formation. The "turbulent" pattern here is the result of the interference of light beams refracted at the different pieces of the cellular flame surface. This example shows that the cellular laminar flame can be visualized and interpreted as a turbulent one using the techniques giving 2D projections of the 3D flow. And it may cause misinterpretation of the studying phenomena.

The shadow visualization of the flow on the second and the last stage may cause similar misinterpretation. Figure 15.30 shows shadow photos for the last stage of the transition to detonation obtained from 3D simulations (a, b) and experimental shadow photos (d, e) for a stoichiometric H_2O_2 flame [44].

Despite the differences between the experimental and numerical setups (in [44], the channel width was 5 cm, and the initial pressure was 0.75 atm), a good qualitative agreement of the flow patterns can be seen. The frame (d) in Fig. 15.30 corresponds to the frame (a), and the frame (b) shows already developing detonation wave as in the frame (e). The information about how the structure of the flame front is transformed into the detonation is lost and cannot be obtained at the experimental shadowgraphs.

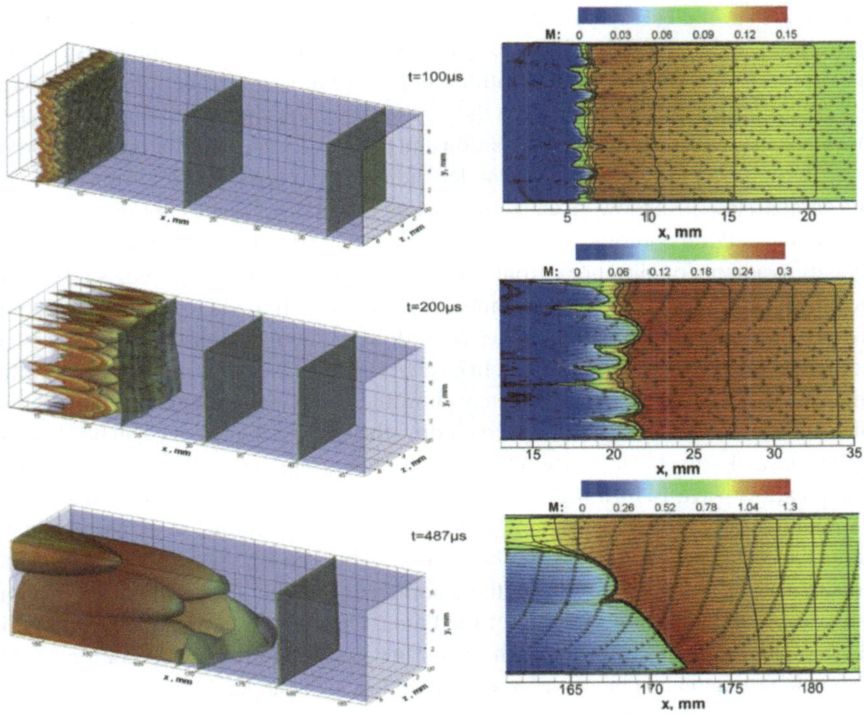

Fig. 15.29 The 3D flame structure and shock waves (green surfaces) at sequential times 100, 200, 487, and 495 µs; the channel width $D = 10\,mm$. Boundary layer is a white strip at the bottom and side walls. Right: 2D cross sections of 3D images in $(x, y, z = 5.0\,mm)$

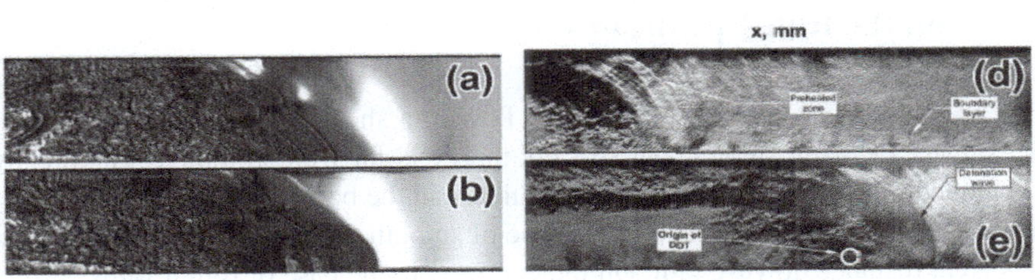

Fig. 15.30 Shadow photos of the flow pattern during the DDT transition taken from 3D simulations at **a** 487 ms, **b** 492 ms. Frames **d** and **e** are the experimentally obtained shadow photos for photo for stoichiometric H_2O_2 flame [44] with time resolution $\Delta t = 0.1\,ms$

On the other hand, the qualitative similarity of the flow patterns confirms that the process in real experiment evolves via the mechanism of the flame–shock wave coupling similar to that described above.

Oppenheim's term "explosion in the explosion" becomes clearer from the analysis of the process shown on Fig. 15.29. First of all, we note that from Fig. 15.29 it is seen that the flame surface at the moment of detonation formation is multidimensional. It consists of several bulge tips that are advanced far ahead compared with neighboring flame bulges. These regions represent leading (tips) edges of the flame front where the pressure pulses arise. On the contrary, shadow photos are projections of all planes intersecting the flame surface along the side walls of the channel. In the diagnostics plane, one can observe the flame position and the shocks intersection points ahead of it, which can be more or less different depending on the taken cross section (see, e.g. the comparison of shadow visualizations using XY and XZ planes in Fig. 15.28a, b). These intersection points are the signals coming to the diagnostics plane out from the density gradients on the shock fronts produced by the leading flame edges. There is no an adequate condition in such points to ignite detonation via local autoignition of the mixture, and the detonation arises according to the mechanism described above in the reaction zone on the flame tip. In case of shadow visualization with laser knife, the pattern in the planar projection may show the detonation arising ahead the flame front since the other part of the flow including the leading flame edge can be clouded by the laser knife. In this case, the shadow photographs give the impression that a detonation wave is forming ahead of the flame front. In fact, as it is seen from 3D simulation, the phenomena of "explosion in the explosion" does not originate from the "hot spot" ahead of the flame front. This can be viewed as a diagnostic artifact caused by the lag between the leading edge of the flame, where detonation occurs, and the remaining surface of the flame. Depending on how the schlieren or shadow photographs were taken, the resulting image and the interpretation of the same phenomenon can be completely different, which can lead to misinterpretation of the physics underlying these phenomena.

15.5 Effect of Wall Roughness and Obstacles on the Run-Up Distance

The transition from laminar to turbulent flow in a shear boundary layer has been the subject of many studies, but still the theory is far from complete. As already mentioned in Sect. 15.3, the onset of turbulence in the boundary layer requires relatively strong perturbations, and in the case of smooth walls, turbulence is expected at a great distance downstream. On the contrary, in the case of rough walls, strong disturbances arise in the boundary layer from the very beginning. In this case, the wrinkling of the flame surface by turbulent eddies increases the total flame area in addition to the flame stretching caused by the nonuniform velocity profile in the upstream flow. Shchelkin [22, 82] has shown that the rate of the flame acceleration is strongly influenced by the roughness of the tube wall due to the development of turbulence in the unburned gas. This observation led him to consider turbulence as the main factor affecting the flame acceleration and leading to the transition to detonation. In order to further enhance turbulence in the flow of fresh mixture ahead of the flame, Shchelkin used wire coils placed on the inner surface of the tube, which are known as Shchelkin spiral and widely used to reduce the detonation run-up distance.

It should be emphasized that Chapman and Wheeler [83] were the first to report that the orifice plates mounted on the inner wall of a tube enhance considerably the flame acceleration in a methane/air mixture.

Velocities of the flame propagating in the hydrogen–oxygen mixture of different initial pressures in the tubes of smooth inner walls ($d = 50\,\mu m$) and very rough walls ($d = 1000\,\mu m$) are shown in Fig. 15.31. The experimental details [42–44] were described in Sect. 15.2. At the initial pressure $P_0 = 0.75\,bar$, the ratio of the run-up distance (X_d) to the tube width decreases almost two times, from $X_d/D \approx 30$ for smooth walls to $X_d/D \approx 15$ for rough walls.

For the initial pressure $P_0 = 0.5\,bar$, the ratio of the run-up distance decreases from $X_d/D \approx 40$ for smooth walls to $X_d/D \approx 30$ for rough walls, and for $P_0 = 0.2\,bar$ it only slightly changes from $X_d/D \approx 56$ for smooth to $X_d/D \approx 47$ for rough walls, correspondingly. The relatively smaller change in X_d/D for rough walls in the latter case is due to the fact that the scales of the main disturbances caused by the wall roughness $d = 1000\,\mu m$ are comparable with the flame thickness ($L_f = 1.4\,mm$) and the cutoff wavelength. In sufficiently long tubes, the interaction of the flow with the tube walls can lead to the formation of a turbulent boundary layer even for tubes with smooth walls. However, the transition to detonation in highly reactive mixtures occurs before the formation of a turbulent boundary layer.

The initial stage of flame acceleration is characterized by the increase of the total flame surface, caused by the flame stretching due to the large-scale velocity gradients in the upstream flow together with the flame front wrinkling by small-scale turbulence generated by obstructed walls of the tube. At later stages the instabilities caused the interaction of the reflected shock wave with the flame, the Richtmyer–Meshkov [84] and Kelvin–Helmholtz [85] instabilities also lead to an additional increase of the flame surface. To lesser extent the increased temperature of the unburned mixture behind the shocks also contributes to the flame acceleration. Eventually, this can continue until the flame velocity reaches the maximum value of the CJ deflagration speed, which is observed in experiments as a choked flame. As soon as the choked

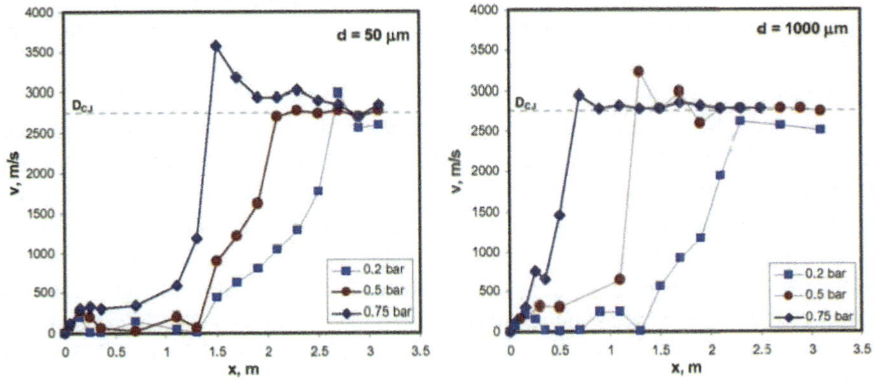

Fig. 15.31 The H_2/O_2 flame velocity in tubes with smooth (left) and rough walls

mode is reached, the flame either continues to propagate in the choked mode with the velocity oscillating around the CJ deflagration speed, or the transition to detonation occurs. There are many experimental and computational studies of the effect of wall roughness, the size of obstacles, their shape and spacing on the run-up distance. These studies are motivated by safety issues, and the difficulties in initiating detonation in pulsed detonation engines [86–112].

Experimental studies [33, 40–42] have shown that the minimum diameter of a tube, $D_{\min}$, for the onset of detonation should be greater than the detonation cell size λ. However, there is no theoretical justification for this assumption, and other experiments give lower values for the minimum diameter. For example, in round tubes the minimum tube diameter is $D_{\min} = \lambda/\pi$ [86, 87, 97, 98].

Although the exact details of autoignition and DDT cannot be resolved experimentally, the methodology can provide an indication of the relative propensity of various combustible mixtures to DDT. Experiments in baffled ducts have shown that the critical value of the flame speed must be reached for the onset of detonation, as suggested earlier by Urtiew and Oppenheim [24]. Earlier studies by Lee et al. [113, 114] have shown that with the help of strong generators of turbulence, including orifices, baffles, and spirals, the flame can be accelerated up to the maximum quasi-stationary value after passing a distance of 30–40 tube diameters, as was observed experimentally in [93]. Experimental studies [99] have shown that the velocity of fast flames in various hydrocarbon mixtures are close to the CJ deflagrations before DDT in agreement with the self-similar model described in Sect. 12.7. In [93], an empirical formula for the run-up distance was proposed, based on the analysis of a large amount of experimental data on the flame speed and acceleration distance of DDT in various combustible mixtures. Such semi-empirical formula makes simplifies the prediction of pressure build-up and shock wave intensities for a given type of fuel and concentration, which can be useful for practical design and safety assessment.

As noted earlier, many DDT studies have used simplified chemical–diffusive models to describe the diffusive transport and chemical reaction. Although some simulations have used powerful mesh refinement techniques, it is unlikely that they actually can resolve fine details in the final stages of DDT, when a fairly high pressure is established before the actual transition to detonation. In this case, a reliable result of simulations (convergence of the solution) may require a very high resolution, $\Delta \sim 0.1\,\mu m$.

A recent high-resolution numerical study of the hydrogen/air DDT in an obstructed channel [109] was focused on the influence of a chemical model on the mechanism of DDT. The authors compared simulations, which used a one-step chemical model designed to reproduce the key features of the flame and detonation, with simulations using a detailed multi-step chemical model. These studies have shown that in simulations with a one-step chemistry, the detonation wave arises from a hot spot ahead of the flame front as the result of a spontaneous wave caused by a reactivity gradient as in [32, 102–106]. On the other hand, in simulations with a detailed chemical model, the necessary conditions for the initiation detonation via the reactivity gradient can hardly be met as long as strong shocks are not formed. In the simulations with a detailed chemical model, the transition to detonation was triggered inside the flame brush in a way similar to that proposed by Liberman et al. [44, 49–73]. In recent experiments by Ssu and Wu [115], with various channel sizes and equivalence ratios, high-speed schlieren visualizations show that a composite reaction-shock cluster structure is always formed in the last stage of flame acceleration before the transition to detonation in ethylene/oxygen mixture and identical structure was observed in propane/oxygen mixtures before the actual transition to detonation in 1 mm $\times$ 1 mm cross section, 750 mm long square channel. The experiments show that transition to detonation occurs shortly after the formation (coupling) of the reaction-shock cluster. Interesting is that in the case of a narrow channel with early formation of the Poiseuille flow, these are oblique shocks reflected from side walls, which form the reaction-shock cluster, and which later merges into a normal precursor shock at the leading edge of the cluster.

15.6 Unconfined Deflagration-To-Detonation Transition

A worst-case scenario of industrial explosions can occur if a deflagration to detonation transition occurs during the explosion. In the case of confined combustions, for example, when a flame propagates in a tube, there are well-established mechanisms, that cause the flame to accelerate and therefore create conditions for the transition to detonation. The situation is different for a freely propagating flame, and the possibility of transition from deflagration to detonation in an unconfined space is a long-standing but still unresolved problem. On the other hand, it is believed that the transition from deflagration to detonation occurs during thermonuclear burning of white dwarfs in Type Ia (SNIa) supernova events. However, thermonuclear burning of white dwarfs occurs in the presence of strong gravity, density that drops to zero on the surface of the star, etc. which are conditions that are very different from those of terrestrial burning. In particular, it can be expected that the role of DL instability is small compared to RT instability due to the strong gravity in white dwarfs. The RT instability can lead to the formation of bubbles at the nonlinear stage, which run from the center of a white dwarf to its surface and may cause turbulence [116].

A freely expanding spherically flame is a usual problem set-up when investigating an unconfined explosion. After ignition, the flame expands from the ignition point for a short time at a constant speed. On the early stage, the flame front is smooth, forming a small ball, while instabilities are suppressed by the strong stretching due to the large curvature of the small flame ball. The evolution of the expanding flame depends on the Lewis number [117]. When the flame grows in size and the stretching intensity weakens, the thermal-diffusive instability can develop cells over the flame front if the Lewis number, $Le < 1$. However, with a further increase in the flame radius, when the flame thickness becomes small in comparison with the average flame radius, the Darrieus–Landau instabilities prevail regardless of the Lewis number. The DL instability of spherical expanding flames leads to the spontaneous formation of cells at the flame front. As the flame further expands, small-scale cells increase in size, the local stretch on the cell surface decreases and the cells themselves becomes unstable, which leads to the cascade development of the cellular structure of the flame surface. The continuous production of new cascades of cells is accompanied by the corresponding increase in the total flame surface compared to a smooth spherical surface. This process leads to the increase in the total consumption rate of the reacting mixture and the flame speed, which, in turn, means the continuous self-acceleration of the flame (see Chap. 11).

Self-acceleration of spherically expanding flames in the large-scale experiments was investigated by Gostintsev et al. [118]. They found that the radius of a spherically expanding flame increases according to the power law expression:

$$R_f(t) = r_0 + At^{\alpha} \approx At^{\alpha}, \tag{15.6.1}$$

where r_0 is radius of the initial flame ball, A and α are constants, with the exponent α in Eq. (15.6.1) estimated as $\alpha = 3/2$. The self-similar regime of the expanding flame had been interpreted as a manifestation of the fractal flame structure with a cascade of cells of different sizes imposed one on another. It was found that the self-similar regime occurs after increasing the flame radius to a critical value estimated as $R_{cr} \approx 1\,m$ for a stoichiometric hydrogen–air flame. In large scale [119–129] and in small-scale experiments at elevated pressures [130–136], the exponent α was found in the range between $\alpha = 3/2$ and $\alpha = 4/3$. In Chap. 11 it was shown that if a fractal flame structure is assumed, then for a spherically expanding flame $\alpha = 3/2$, and for 2D cylindrically expanding flames, $\alpha = 5/4$. The average radial velocity of a spherically expanding flame in the laboratory reference frame is

$$U_{fL} = \frac{dR}{dt} = \alpha A t^{\alpha-1} = \alpha^{1/\alpha} A^{1/\alpha} R^{1-1/\alpha}. \tag{15.6.2}$$

The average velocity of the flow immediately ahead of the spherically expanding flame front is

$$u = \frac{\Theta - 1}{\Theta} U_{fL} \approx \alpha A t^{\alpha-1}. \tag{15.6.3}$$

Eqs. (15.6.2) and (15.6.3) follow that the average velocity of the flame front relative to the fresh mixture is $U_{fu} = U_{fl}/\Theta \approx \alpha A t^{\alpha-1}/\Theta$. Since $\alpha - 1 > 0$, the radial flame velocity increases indefinitely with time, and it can be expected that the transition from deflagration to detonation can occur at a sufficiently large radius, where the flame velocity approaches the CJ deflagration speed and the deflagration mode becomes unstable.

In most large-scale experiments [119–129], hydrogen–air was ignited in the center of a semispherical tent formed by a thin polyethylene plastic film with the radius $\sim 1 \div 3\,m$. The total time of the spherical flame expansion before the destruction of the tent was about 300 ms, and the excess pressure did not exceed 3–4%. The largest unconfined explosion described in literature were conducted in 1983 by the Fraunhofer Institute for Propellants and Explosives [137]. These explosions were initiated inside the hemispherical balloons on the ground level made of thin polyethylene foil and filled with stoichiometric hydrogen–air mixture. The mixture was ignited at the center of the balloon on the ground level. The balloons radii were $R_b = 1.53, 2.88$, 5, and 10 m. The explosion pressure was measured at several distances from the ignition point from 2.0 to 8.0 m and the overpressure was found within 4% of the atmospheric pressure. More detailed information about experiment and estimation of overpressure, velocity, etc. can be found in [135, 136]. Approximately the same $\Delta P/P < 4\%$ overpressure was obtained in LES simulations [120, 138, 139] with the standard k-ε model and the RNG k-ε model. In the Fraunhofer experiments [137], the maximum velocity of the flame occurs between the original radius balloon R_b and $1.5R_b$. The measured flame velocities for hemispheres with radii: 1.53 m, 2.8 m, 5 m, and 10 m were, respectively: 43 m/s, 54 m/s, 60 m/s, and 84 m/s.

There are no reliable estimates for the coefficient A and for the value of the exponent α obtained from the large-scale experiments, while small-scale experiments [134, 135] suggested $\alpha \approx 4/3$. The value $A \approx 1.4 \times 10^4 cm/s^{4/3}$ for hydrogen–air obtained from small-scale experiments is close to the empirical value of A proposed in [119] in the form

$$A \approx 2 \cdot 10^{-3} \Theta^2 U_f^2 / \sqrt{\chi}, \tag{15.6.4}$$

where U_f, Θ, and χ are the laminar flame speed, expansion ratio, and thermal diffusivity, respectively. Taking for hydrogen–air at the atmospheric pressure the exponent $\alpha = 3/2$ and $A \approx 1.4 \times 10^4 cm/s^{3/2}$, and using Eq. (15.6.2), we obtain for the flame velocities measured in [137] for the hemisphere with radii 1.53, 2.8, 5, 10 m the following values: $U_{fL} = 44, 53.8, 64.9, 81.6$ m/s. These values agree well with the experimentally measured values. The flame velocity relative to the fresh mixture is in $\Theta = 8$ times smaller than U_{fL}. Thus, even for the largest unconfined explosion the flame velocity remains much smaller than the sound speed, $a_s = 408\, m/s$. Therefore, the overpressure $\Delta P/P_0 = M^2 \leq 0.04$ is in good agreement with the experimental data. It should be noted that since the average speed of the flame depends on time as $t^{\alpha-1}$, with a positive exponent $(\alpha - 1) < 1$, the accelerating flame, acting as a piston, produces compression waves that steepen into a weak shock close ahead of the flame surface. This means that, in principle, the reaction wave and the shock wave could be coupled with mutual amplification, facilitating the transition to detonation. However, even the maximum velocity of the flame relative to the fresh mixture in experiments [137] is too small, and for such small flame velocities, a weak compression wave has a negligibly small influence on the reaction rate.

It is interesting to estimate the order of magnitude of the spherical flame radius, for which the hydrogen/air flame velocity approach the sound speed, $a_s = 408 m/s$. From the Eq. (15.6.2), we find that the flame radius corresponding to the flame velocity close to the sound speed, $U_{fL} = a_s = 408\, m/s$, is $R_s \approx 1.5\, km$. Using the results of small-scale experiments for hydrogen-air [133] at elevated pressures of 5 atm and 10 atm, we find that the flame velocity reaches the sound speed at $R_s \approx 500\, m$ for $P_0 = 5\, atm$, and $R_s \approx 100\, m$ for $P_0 = 10\, atm$. Experiments with H_2-air on the scale of the order of 1 km in an open atmosphere are unrealistic, and even scales of tens of meters are too large for laboratory experiments, even at $\sim 10\,atm$. Presumably, at such large sizes, other processes that are not usually considered may become important. For example, radiative cooling of combustion products can slowdown and even extinguish combustion when the flame radius approaches $R_s \sim 100\,m$. Therefore, unconfined DDT of a spherically expanding hydrogen/air flame at terrestrial conditions is unlikely.

For mixtures with higher reactivity, such as, e.g. H_2/O_2, C_2H_2/O_2, and at high pressures, the velocity of the outward propagating spherical flame can reach the sound speed at much smaller radii. The critical radius at which the velocity of a hydrogen–oxygen freely expanding flame approaches the speed of sound is $R_s \approx 7.5\,m$ at $P = 1\,atm$ and $R_s \approx 4\,cm$ at $P = 10\,atm$. This means that experiments with a spherical hydrogen–oxygen flame at a pressure of 10 atm or more are possible under laboratory conditions. Unfortunately, only one such experiment is known in the literature [140], carried out by Zel'dovich and Rozlovsky in 1947. The authors reported the experimentally observed transition to detonation of a spherically expanding hydrogen–oxygen flame in a spherical bomb with the inner radius 7.5 cm at the initial pressure $P_0 = 10\,atm$. Since then, there have been no publications reproducing the experiment [140].

It should be emphasized that the speed of the flame does not have to reach the speed of sound at the radius, where the transition to detonation begins. For example, it is seen in Fig. 15.1 that the transition to detonation in H_2–O_2 mixture begins when the flame velocity is approximately one half of the sound speed. Therefore, the obtained estimates for the critical radius of the onset of DDT can be viewed as a conservative approximation. It is unlikely that the gradient mechanism can work in this case, and of interest is the possibility of the flame coupling with the shock wave, leading to their mutual amplification and DDT. The successful coupling of the flame and shock depends on the mixture reactivity and intensity of the shock wave. As it was said earlier since for the velocity–time dependence of a spherical expanding flame, the accelerating flame produces pressure wave, which can steepen into a shock wave in the close vicinity ahead of the flame surface. Therefore, the shock–flame coupling with positive feedback mutual amplification seems plausible. The result of the flame–shock coupling may include the pressure build-up and the increase in the density of fresh mixture. However, the shock wave expanding in the radial direction weakens, and the shock wave–flame coupling must compensate the attenuation of the shock wave. This is unlikely for a hydrogen–air flame but for a hydrogen–oxygen flame, the reaction order is $n \approx 2.7$, so that the increase in pressure behind the shock wave accelerates the flame. For an initial weak shock wave, the flame velocity increases by a factor 1.23 for $M_{sh} \sim 1.3$. More important is the increase in density behind the shock wave, which leads to an increase in the flow of fuel entering the reaction zone. For the same Mach number, the density increases by a factor of ≈ 1.8. Using the experimental results in Fig. 15.1, we can estimate the detonation onset time for the experimental conditions used in [140] as $t_{\mathrm{det}} \sim 0.01ms$, which is much less than the time $t_{sound} \sim 0.3ms$ it takes for a sound wave to travel from the expanding fireball to the inner wall of the bomb. These estimates agree surprisingly well with the experimental results obtained by Zel'dovich and Rozlovsky [140], where the transition to detonation was observed at $R_D \approx 2.5\,cm$, while our simple estimates give $R_D \approx (2-3)\,cm$. Therefore, experiments [140] can be regarded as a reliable example of unconfined DDT in an outward expanding flame.

Interest to the unconfined DDT in the context of outward expanding spherical flames was recently revived in a series of numerical studies [141–143]. These studies are mainly aimed at the parametric study of the transition to detonation, assuming unlimited acceleration of a spherical expanding flame due to the permanent developing folders on the flame surface. Moreover, the degree of folding was treated as a prescribed gradually increasing factor. A similar concept was applied to the parametric study of DDT in thermonuclear explosions of a white-dwarf in SN Ia events [144]. The reliability of these results is not entirely clear. With an increase in the radius of the flame to tens—hundreds of meters, radiation losses become significant and can lead to cooling of combustion products and condensation of water vapor, a significant decrease in pressure inside the fireball, slowdown and even extinguishing the expanding flame.

Even 2D simulations of the problem corresponding to the Zel'dovich–Rozlovskii experiment [140] is a challenging problem, that has not been solved, yet. The main difficulty is that the DDT mechanism must be clearly shown in simulations. Also, it is necessary to verify the convergence of the obtained solution in order to prove that the reaction-shock coupling is real and not an artifact. For the initial pressure 10 atm, the pressure before the transition to detonation increases and can reach $\sim 100\,atm$. Since the width of the hydrogen–oxygen flame at 10 atm is 27 μm, the initial resolution must be at least 2–3 μm, but at the later stages the required the resolution is about $\sim 0.1\mu m$. The maximum resolution used in [145] ranged from 25 to 12.5 μm, which is much coarser for obtaining reliable simulation results. The convergence of numerical simulation for the shock-reaction wave (flame) coupling in the context of the transition to detonation was thoroughly studied in [146]. It was shown that shock-reaction wave coupling observed in coarse-resolution simulations can be an artifact if validation of the solution at much higher resolutions does not prove convergence of the solution.

References

1. M. Kaneshige, J.E. Shepherd, Detonation Database. Explosion Dynamics Laboratory Report FM97–8. California Institute of Technology Pasadena, CA 91125 (1999)
2. E. Mallard, H. Le Chatelier, Sur la vitesse de propagation de l'inflammation dans les mélanges gazeux explosives. Comptes Rendus Acad. Sci. Paris Fr. **93**, 145–148 (1881)
3. E. Mallard, H.L. Le Chatelier, Combustion des melanges gaseux explosifs. Ann. Mines **4**, 379–568 (1883)
4. E. Mallard, H. Le Chatelier, Recherches Experimentales et Theoriques sur la Combustion des Melanges Gaseoux Explosifs. Ann. Mines **8**, 274–568 (1883)
5. M. Berthelot, P. Vieille, Sur les Vitesse de Propagation de Phenomenes Explosifs Dans le Gaz (On the velocity of propagation of explosive processes in gases). Comptes Rendus Acad. Sci. Paris **93**, 18–22 (1881)

6. M. Berthelot, P. Vieille, L'onde explosive. Ann. Chim. Phys. Ser. **5**(28), 289–332 (1883)
7. W.J.M. Rankine, On the thermodynamic theory of waves of finite longitudinal disturbance. Philos. Trans. Roy. Soc. Lond. **160**, 277–288 (1870)
8. H. Hugoniot, Mémoire sur la propagation des mouvements dans les corps et spécialement dans les gaz parfaits (première partie) [Memoir on the propagation of movements in bodies, especially perfect gases (first part)]. (in French) J. École Polytechnique, **57**, 3–97 (1887)
9. H. Hugoniot, Mémoire sur la propagation des mouvements dans les corps et spécialement dans les gaz parfaits (deuxième partie)" [Memoir on the propagation of movements in bodies, especially perfect gases (second part)]. J. E´cole Polytechnique **58**, 1–125 (1889)
10. D.L. Chapman, On the rate of explosion in gases. Philos. Mag. **47**, 90–104 (1899)
11. E. Jouguet, Sur la propagation des réactions chimiques dans les gaz [On the propagation of chemical reactions in gases]. J. de mathematiques Pures et Appliquees **1**, 347–425 (1905). ((in French))
12. E. Jouguet, Sur la propagation des réactions chimiques dans les gaz [On the propagation of chemical reactions in gases], Journal de Mathématiques Pures et Appliquées, series 6 (in French), 2 1906) pp. 5–85
13. Y. B. Zel'dovich, On the theory of propagation of detonation in gaseous systems. Zh. Eksp. Teor. Fiz.10 (1940), pp. 542–568. (English translation: NACA TM 1261, 1950)
14. J. von Neumann, Progress report on "Theory of detonation waves". OSRD. Rept. No. 549, National Defense Research Committee Div. B.1942. Reprinted in the collected works of John von Neumann, ed. by A.J. Taub (Pergamon, 1963)
15. W. Döring, On detonation processes in gases. Ann. Phys. **43**, 421–436 (1943)
16. Lees' Loss prevention in the process industries: hazard identification, assessment and control, ed. by S. Mannan, F.P. Lees (Butterworth-Heinemann Publisher, 2005)
17. W. Breitung, C. Chan, S. Dorofeev, A. Eder, B. Gerland, Flame acceleration and deflagration-to-detonation transition in nuclear safety, State-of-the Art Report, OCDE-Nuclear Safety, NEA/CSNI/R (2000)
18. G.D. Roy, S.M. Frolov, A.A. Borisov, D.W. Netzer, Pulse detonation propulsion: challenges, current status, and future perspective. Prog. Energy Combust. Sci. **30**, 545–672 (2004)
19. P. Mazzali, F. Ropke, S. Benetti, W. Hillebrandt, A common explosion mechanism for type Ia supernovae. Science **315**, 825–837 (2007)
20. W. Hillebrandt, J.C. Niemeyer, Type Ia supernova explosion models. Ann. Rev. Astr. Astrophys. **38**, 191–230 (2000)
21. K.I. Shchelkin, A.S. Sokolik, Detonation in gas mixtures. IV. Effect of 'Chemical Preparation' on detonation origination in gas mixtures, Zh. Fiz. Khim. **10**, 484–488 (1937) (in Russian)
22. K.I. Shchelkin, Influence of tube roughness on the formation and detonation propagation in gas. J. Exp. Theor. Phys. **10**, 823–829 (1940)
23. K.I. Shchelkin, On the transition from a slow combustion to detonation. J. Exp. Theor. Phys. **1953**(24), 1176–1179 (1953)
24. P. Urtiew, A.K. Oppenheim, Experimental observations of the transition to detonation in an explosive gas. Proc. R. Soc. London, Ser. A **295**, 13–28 (1966)
25. J.W. Meyer, P.A. Urtiew, A.K. Oppenheim, On the inadequacy of gasdynamic processes for triggering the transition to detonation. Combust. Flame **14**, 13–20 (1970)
26. A.K. Oppenheim, R.I. Soloukhin, Experiments in gasdynamics of explosions. Annu. Rev. Fluid Mech. **3**, 31–58 (1973)
27. K.I. Shchelkin, Y.K. Troshin, *Gasdynamics of Combustion* (Mono Book Corp. Baltimore, 1965)
28. J.H.S. Lee, I.O. Moen, The mechanism of transition from deflagration to detonation in vapor cloud explosion. Prog. Energy Combust. Sci. **6**, 359–389 (1980)
29. J.F. Clarke, Fast flames waves and detonation. Progr. Energy Combust. Sci. **13**, 241–271 (1989)
30. J.E. Shepherd, J.H.S. Lee, On the transition from deflagration-to-detonation, in *Major Research Topics in Combustion*, ed. by M.Y. Hussaini, A. Kumar, R.G. Voigt (Springer, New York, 1992) , pp. 439–487

31. A.M. Bartenev, B.E. Gelfand, Spontaneous initiation of detonations. Prog. Energy Combust. Sci. **26**, 29–55 (2000)
32. E.S. Oran, V.N. Gamezo, Origins of the deflagration-to-detonation transition in gas-phase combustion. Combust. Flame **148**, 4–47 (2007)
33. G. Ciccarelli, S. Dorofeev, Flame acceleration and transition to detonation in ducts. Prog. Energy Combust. Sci. **34**, 499–550 (2008)
34. S.B. Dorofeev, Flame acceleration and explosion safety applications. Proc. Combust. Inst. **33**, 2161–2175 (2011)
35. D. Bradley, G.A. Chamberlain, D.D. Drysdale, Large vapor cloud explosions, with particular reference to that at Buncefield. Phil. Trans. R. Soc. A **370**, 544–566 (2012)
36. S.M. Frolov, Acceleration of the deflagration-to-detonation transition in gases: from shchelkin to our days. Combust. Expl. Shock Waves **48**, 258–268 (2012)
37. E.S. Oran, F.A. Williams, The physics, chemistry, and dynamics of explosions. Phil. Trans R. Soc. A **379**, 531–799 (2012)
38. E.S. Oran, Understanding explosions – From catastrophic accidents to creation of the universe. Proc. Combust. Inst. **35**, 1–35 (2015)
39. E.S. Oran, G. Chamberlain, A. Pekalski, Mechanisms and occurrence of detonations in vapor cloud explosions. **77**, 100804 (1–37) (2020)
40. M. Kuznetsov, C. Ciccarelli, S. Dorofeev, V. Alekseev, Y. Yankin, T.H. Kim, DDT in methane-air mixtures. Shock Waves **12**, 215–220 (2002)
41. M.S. Kuznetsov, V.I. Alekseev, Y.G. Yankin, S.B. Dorofeev, Slow and fast deflagrations in hydrocarbon-air mixtures. Combust. Sci. Technol. **174**, 157–172 (2002)
42. M. Kuznetsov, V. Alekseev, I. Matsukov, S. Dorofeev, DDT in a smooth tube filled with a hydrogen-oxygen mixture. Shock Waves **14**, 205–215 (2005)
43. M. Kuznetsov, M. Liberman, I. Matsukov, Experimental study of the preheat zone formation and deflagration-to-detonation transition. Combust Sci. Technol. **182**, 1628–1644 (2010)
44. M.A. Liberman, M.F. Ivanov, A.D. Kiverin, M.S. Kuznetsov, A.A. Chukalovsky, T.V. Rakhimova, Deflagration-to-detonation transition in highly reactive combustible mixtures. Acta Astron. **67**, 688–701 (2010)
45. W.C. Reynolds, *The Element Potential Method for Chemical Equilibrium Analysis: Implementation in the Interactive Program STANJAN Version 3* (Stanford University, Palo Alto, California, Dept. of Mechanical Engineering, 1986)
46. D.G. Goodwin, *Cantera User's Guide* (CalTech., Pasadena, CA, 2001)
47. M.-H. Wu, M.P. Burke, S.F. Son, R.A. Yetter, Flame acceleration and the transition to detonation of stoichiometric ethylene/ oxygen in microscale tubes. Proc. Combust. Inst. **31**, 2429–2436 (2007)
48. M.-H. Wu, C-Yu. Wang, Reaction propagation modes in millimeter-scale tubes for ethylene/oxygen mixtures. Proc. Combust. Inst. **33**, 2287–2293 (2011)
49. M.-H. Wu, W.-C. Kuo, Accelerative expansion and DDT of stoichiometric ethylene/oxygen flame rings in micro-gaps. Proc. Combust. Inst. **34**, 2017–2024 (2013)
50. J. Li, P. Zhang, L. Yuan, Z. Pan, Y. Zhu, Flame propagation and detonation initiation distance of ethylene/oxygen in narrow gap. Appl. Therm. Eng. **110**, 1274–1282 (2017)
51. Y.B. Zeldovich, On the theory on transition to detonation in gases. Zh. Tekhn. Fiz. (J. Techn. Phys) **17**, 3–26 (1947) (in Russian)
52. C. Clanet, G. Searby, On the "Tulip flame" phenomenon. Combust. Flame **105**, 225–238 (1996)
53. L.D. Landau, E.M. Lifshitz, *Fluid Mechanics* (Pergamon Press, Oxford, 1989)
54. H. Schlichting, K. Gersten, *Boundary-Layer Theory*, 9th edn. (Springer, Berlin Heidelberg 2017)
55. V. Bychkov, A. Petchenko, V. Akkerman, L.-E. Eriksson, Theory and modeling of accelerating flames in tubes. Phys. Rev. E **72**, 046307 (1–10) (2005)
56. V. Bychkov, V. Akkerman, G. Fru, A. Petchenko, L.E. Eriksson, Flame acceleration in the early stages of burning in tubes. Combust. Flame **150**, 263–267 (2007)

57. D.M. Valiev, V. Bychkov, V. Akkerman, L.-E. Eriksson, Different stages of flame acceleration from slow burning to Chapman-Jouguet deflagration. Phys. Rev. E **80**, 036317 (1–11) (2009)
58. A. Sakurai, Propagation of spherical shock waves in stars. J. Fluid Mech. **1**, 436–453 (1956)
59. G.B. Whitham, On the propagation of shock waves through regions of non-uniform area or flow. J. Fluid Mech. **4**, 337–360 (1958)
60. A. Sakurai, On the problem of a shock wave arriving at the edge of a gas. Commun. Pure Appl. Math. **13**, 353–370 (1960)
61. Y.B. Zel'dovich, Y.P. Raizer, *Physics of Shock Waves and High-Temperature Hydrodynamic Phenomena* (Academic Press, New York San Francisco London 1967)
62. N.N. Smirnov, O.G. Penyazkov, K.L. Sevrouk, V.F. Nikitin, L.I. Stamov, V.V. Tyurenkova, Onset of detonation in hydrogen-air mixtures due to shock wave reflection inside a combustion chamber. Acta Astron. **149**, 77–92 (2018)
63. Ya. B. Zel'dovich, V.B. Librovich, G.M. Makhviladze, G.I. Sivashinsky, On the development of detonation in a non-uniformly preheated gas, Acta Astron. **15**, 313–321 (1970)
64. Ya. B. Zel'dovich, Regime classification of an exothermic reaction with nonuniform initial conditions, Combust. Flame **39**, 211–214 (1980)
65. J.H.S. Lee, R. Knystautas, N. Yoshikawa, Photochemical initiation and gaseous detonations. Acta Astronaut. **5**, 971–982 (1978)
66. J.H.S. Lee, *The Detonation Phenomenon* (Cambridge University Press, Cambridge, UK, 2008)
67. J.D. Ott, E.S. Oran, J.D. Anderson, A mechanism for flame acceleration in narrow tubes. AIAA J. **41**, 1391–1396 (2003)
68. D.A. Kessler, V.N. Gamezo, E.S. Oran, Simulations of flame acceleration and deflagration-to-detonation transitions in methane–air systems. Combust. Flame **157**, 2063–2077 (2010)
69. M. Liberman, A. Kiverin, M. Ivanov, On detonation initiation by a temperature gradient for detailed chemical reaction models. Phys. Lett. A **375**, 1803–1808 (2011)
70. M.F. Ivanov, A.D. Kiverin, M.A. Liberman, Hydrogen-oxygen flame acceleration and transition to detonation in channels with no-slip walls for a detailed chemical reaction model. Phys. Rev. E **83**, 056313 (2011)
71. M. Ivanov, A. Kiverin, M. Liberman, Flame acceleration and DDT of hydrogen-oxygen gaseous mixtures in channels with no-slip walls. Int. J. Hydrogen Energy **36**, 7714–7727 (2011)
72. M.A. Liberman, A.D. Kiverin, M.F. Ivanov, Regimes of chemical reaction waves initiated by nonuniform initial conditions for detailed chemical reaction models, Phys. Rev. E **85**, 056312 (1–11) (2012)
73. M.F. Ivanov, A.D. Kiverin, I.S. Yakovenko, M.A. Liberman, Hydrogen-oxygen flame acceleration and deflagration-to-detonation transition in three-dimensional rectangular channels with no-slip walls. Int. J. Hydrogen Energy **38**, 16427–16440 (2013)
74. R.W. Houim, A. Ozgen, E.S. Oran, The role of spontaneous waves in the deflagration-to-detonation transition in submillimetre channels. Combust. Theor. Model. **20**, 1068–1087 (2016)
75. E. Dzieminska, M. Fukuda, A.K. Hayashi, E. Yamada, Fast flame propagation in hydrogen/oxygen mixture. Combust. Sci. Technol. **184**, 1608–1615 (2012)
76. E. Dzieminska, A. Hayashi, Auto-ignition and ddt driven by shock wave—boundary layer interaction in oxyhydrogen mixture. Int. J. Hydrogen Energy **38**, 4185–4193 (2013)
77. M. Fukuda, E. Dzieminska, A. Hayashi, E. Yamada, N. Tsuboi, Effect of wall conditions on DDT in hydrogen-oxygen mixtures. Shock Waves **23**, 191–200 (2013)
78. T. Machida, M. Asahara, A.K. Hayashi, N. Tsuboi, Three-Dimensional simulation of deflagration-to-detonation transition with a detailed chemical reaction model. Combust. Sci. Technol. **186**, 1758–1773 (2014)
79. C. Wang, C. Qian, J. Liu, M.A. Liberman, Influence of chemical kinetics on detonation initiating by temperature gradients in methane/air. Combust. Flame **197**, 400–415 (2018)
80. M. Liberman, C. Wang, C. Qian, J. Liu, Influence of chemical kinetics on spontaneous waves and detonation initiation in highly reactive and low reactive mixtures. Combust. Theor. Model. **23**, 467–495 (2019)

81. G.S. Settles, *Schlieren and Shadowgraph Techniques: Visualizing Phenomena in Transparent Media* (Springer, Berlin-Heidelberg-New York, 2001)
82. K. Shchelkin, K. Occurrence, of detonation in gases in rough walled tubes. Scv. J. Tech. Phys. **17**, 613–618 (1947)
83. W.R. Chapman, R.V. Wheeler, The propagation of flame in mixtures of methane and air. Part IV. The effect of restrictions in the path of the flame. J. Chem. Soc. **21**, 2139–2147 (1926)
84. L. Massa, P. Jha, Linear analysis of the Richtmyer-Meshkov instability in shock-flame interactions. Phys. Fluids **24**, 056101(1–30) (2012)
85. T. Funada, D.D. Joseph Viscous potential flow analysis of Kelvin-Helmholtz instability in a channel. J. Fluid Mech. **445**, 263–283 (2001)
86. Y. Gao, H.D. Ng, J.H.S. Lee, Minimum tube diameters for steady propagation of gaseous detonations. Shock Waves **24**, 447–454 (2014)
87. Y. Gao, B. Zhang, H.D. Ng, J.H.S. Lee, An experimental investigation of detonation limits in hydrogen-oxygen-argon mixtures. Int. J. Hydrogen Energy **41**, 6076–6083 (2016)
88. E. Schultz, E. Wintenberger, J.E. Shepherd, Investigation of deflagration to detonation transition for application to pulse detonation engine ignition systems, in *Proceedings of the 16th JANNAF Propulsion Symposium*. Chemical Propulsion Information Agency (1999)
89. C.T. Johansen, G. Ciccarelli, Visualization of the unburned gas flow field ahead of an accelerating flame in an obstructed square channel. Combust. Flame **156**, 405–416 (2009)
90. A. Teodorczyk, Scale effects on hydrogen–air fast deflagrations and detonations in small obstructed channels. J. Loss Prev. Process Ind. **21**, 147–153 (2008)
91. A. Teodorczyk, P. Drobniak, A. Dabkowski, Fast turbulent deflagration and DDT of hydrogen–air mixtures in small obstructed channel. Int. J. Hydrogen Energy **34**, 5887–5893 (2009)
92. W. Rudy, M. Kuznetsov, R. Porowski, A. Teodorczyk, J. Grune, K. Sempert, Critical conditions of hydrogen-air detonation in partially confined geometry. Proc. Combust. Inst. **34**, 1965–1972 (2013)
93. M. Silvestrini, B. Genova, G. Parisi, F.L. Trujillo, Flame acceleration and DDT run-up distance for smooth and obstacles filled tubes. J. Loss Prev. Process Ind. **21**, 555–562 (2008)
94. G. Ciccarelli, C.T. Johansen, M. Parravani, The role of shock–flame interactions on flame acceleration in an obstacle laden channel. Combust. Flame **157**, 2125–2136 (2010)
95. R. Porowski, A. Teodorczyk, Experimental study on DDT for hydrogen-methane-air mixtures in tube with obstacles. J. Loss Prev. Process Ind. **26**, 374–379 (2013)
96. M. Cross, G. Ciccarelli, DDT and detonation propagation limits in an obstacle filled tube. J. Loss Preven. Process Indust. **36**, 380–386 (2015)
97. L.-Q. Wang, H.-H. Ma, Z.-W. Shen., D.-G. Chen, Experimental study of DDT in hydrogen-methane-air mixtures in a tube filled with square orifice plates. Process Safety Environ. Protection **116**, 228–234 (2018)
98. W. Rudy, A. Teodorczyk, DDT limits in hydrogen-air mixtures in tube filled with obstacles, in *The 27th International Colloquium on the Dynamics of Explosions and Reactive Systems* 2019 *(The 27th ICDERS)* (Beijing, China, 2019).
99. M. Saif, W. Wang, A. Pekalski, M. Levin, M.I. Radulescu, Chapman-Jouguet deflagrations and their transition to detonation. Proc. Combust. Inst. **36**, 2771–2779 (2017)
100. S. Ishihara, K. Ishii, H. Kataoka, Mechanism of detonation transition from accelerating flames in a channel. Proceed. Combust. Institute **36**, 2753–2759 (2017)
101. A.V. Gaathaug, K. Vaagsaether, D. Bjerketvedt, Experimental and numerical investigation of DDT in hydrogen air behind a single obstacle. Int. J. Hydrogen Energy **37**, 17606–17615 (2012)
102. V. Gamezo, T. Ogawa, E. Oran, Flame acceleration and DDT in channels with obstacles: effect of obstacle spacing. Combust. Flame **155**, 302–315 (2008)
103. T. Ogawa, V. Gamezo, E. Oran, Flame acceleration and transition to detonation in an array of square obstacles. J. Loss Prev. Process Ind. **26**, 355–362 (2013)
104. T. Ogawa, E. Oran, V. Gamezo, Numerical study on flame acceleration and DDT in an inclined array of cylinders using an amr technique. Comput. Fluids **85**, 63–70 (2013)

105. G. Goodwin, R. Houim, E. Oran, Effect of decreasing blockage ratio on DDT in small channels with obstacles. Combust. Flame **173**, 16–26 (2016)
106. G. Goodwin, R. Houim, E. Oran, Shock transition to detonation in channels with obstacles. Proc. Combust. Inst. **36**, 2717–2724 (2017)
107. R.W. Houim, E. Oran, Effect of surface roughness on deflagration-to-detonation transition in submilimeter channels. 26th ICDERS, paper 1067 (Boston, MA, 2017)
108. A.M. Coate, D.L. Mathias, B.J. Cantwell, Numerical investigation of the effect of obstacle shape on deflagration to detonation transition in a hydrogen–air mixture. Combust. Flame **209**, 278–290 (2019)
109. O. Dounia, O. Vermorel, A. Misdariis, T. Poinsot, Influence of kinetics on DDT simulations. Combust. Flame **200**, 1–14 (2019)
110. H. Ma, Z. Xia, W. Gao, C. Zhuo, D. Wang, Numerical simulation of the deflagration-to-detonation transition of iso-octane vapor in an obstacle-filled tube. Int. J. Spray and Combust. Dynamics **10**, 244–259 (2018)
111. Y. Gao, P. Dai, Z. Chen, Numerical studies on autoignition and detonation development from a hot spot in hydrogen/air mixtures. Combust. Theory Model. **24**, 245–261 (2020)
112. H. Xiao, E.S. Oran, Flame acceleration and deflagration-to-detonation transition in hydrogen-air mixture in a channel with an array of obstacles of different shapes. Combust. Flame **220**, 378–393 (2020)
113. J.H.S. Lee, R. Knystautas, A. Freiman, High speed turbulent deflagrations and transition to detonation in H_2-air mixtures. Combust. Flame **56**, 227–239 (1984)
114. J. Chao, J.H.S. Lee, The propagation mechanism of high-speed turbulent deflagrations. Shock Waves **12**, 277–289 (2003)
115. H.-W. Ssu, M.-H. Wu, Formation and characteristics of composite reaction—shock clusters in narrow channels. Proc. Combust. Institute **38**, 3473–3480 (2021)
116. V.V. Bychkov, M.A. Liberman, On the theory of type ia supernova explosion. Astron. Astrophys. **304**, 440–448 (1995)
117. Z. Chen, Y. Ju, Theoretical analysis of the evolution from ignition kernel to flame ball and planar flame. Combust. Theory Modelling **11**, 427–453 (2007)
118. Y.A. Gostintsev, A. Istratov, Y.V. Shulenin, Self-similar propagation of a free turbulent flame in mixed gas mixtures. Combust. Expl. Shock Waves **24**, 563–569 (1988)
119. D. Bradley, T. Cresswell, J. Puttock, Flame acceleration due to flame-induced instabilities in large-scale explosions. Combust. Flame **124**, 551–559 (2001)
120. V. Molkov, D. Makarov, H. Schneider, Hydrogen-air deflagrations in open atmosphere: large eddy simulation analysis of experimental data. Int. J. Hydrogen Energy **32**, 2198–2205 (2007)
121. S.B. Dorofeev, Evaluation of safety distances related to unconfined hydrogen explosions. Int. J. Hydrogen Energy **32**, 2118–2124 (2007)
122. W.K. Kim, T. Mogi, K. Kuwan, R. Dobashi, Self-similar propagation of expanding spherical flames in large scale gas explosions. Proc. Combust. Inst. **35**, 2051–2058 (2015)
123. C.R. Bauwens, J.M. Bergthorson, S.B. Dorofeev, Experimental study of spherical-flame acceleration mechanisms in large-scale propane–air flames. Proc. Combust. Inst. **35**, 2059–2066 (2015)
124. C.R. Bauwens, J.M. Bergthorson, S.B. Dorofeev, On the interaction of the Darrieus-Landau instability with weak initial turbulence. Proc. Combust. Inst. **36**, 2815–2822 (2017)
125. W.K. Kim, T. Mogi, R. Dobashi, Flame acceleration in unconfined hydrogen/air deflagrations using infrared photography. J. Loss Prevent. Proc. **26**, 1501–1505 (2013)
126. W.K. Kim, T. Mogi, K. Kuwana, R. Dobashi, Self-similar propagation of expanding spherical flames in large scale gas explosions. Proc. Combust. Inst. **35**, 2051–2058 (2015)
127. W. Kim, T. Imamura, T. Mogi, R. Dobashi, Experimental investigation on the onset of cellular instabilities and acceleration of expanding spherical flames. Int J Hydrogen Energy **42**, 14821–14828 (2017)
128. W. Kim, T. Endo, T. Mogi, K. Kuwana, R. Dobashi, Wrinkling of large-scale flame in lean propane-air mixture due to cellular instabilities. Combust Sci Technol. **191**, 491–503 (2019)

129. X. Cai, J. Wang, Z. Bian, H. Zhao, H. Dai, Z. Huang, On transition to self-similar acceleration of spherically expanding flames with cellular instabilities. Combust. Flame **215**, 364–375 (2020)
130. D. Bradley, C.G.W. Sheppard, R. Woolley, D.A. Greenhalgh, R.D. Lockett, Combust. Flame **122**, 195–209 (2000)
131. O. Kwon, G. Rozenchan, C.K. Law, Cellular instabilities and self-acceleration of outwardly propagating spherical flames. Proc. Combust. Inst. **29**, 1775–1783 (2002)
132. G. Jomaas, C.K. Law, J.K. Bechtold, On transition to cellularity in expanding spherical flames. J. Fluid Mech **583**, 1–26 (2007)
133. F. Wu, G. Jomaas, C.K. Law, An experimental investigation on self-acceleration of cellular spherical flames. Proc. Combust. Inst. **34**, 937–945 (2013)
134. S. Yang, A. Saha, F. Wu, C.K. Law, Morphology and self-acceleration of expanding laminar flames with flame-front cellular instabilities. Combust. Flame **171**, 112–118 (2016)
135. J. Huo, A. Saha, Z. Ren, C.K. Law, Self-acceleration and global pulsation in hydrodynamically unstable expanding laminar flames. Combust. Flame **194**, 419–425 (2018)
136. X. Cai, J. Wang, H. Zhao, M. Zhang, Z. Huang, Flame morphology and self-acceleration of syngas spherically expanding flames. Int. J. Hydrogen Energy **43**, 17531–17541 (2018)
137. H. Schneider, H. Pförtner, Fraunhofer-ICT Internal Report: PNP-Sichcrheitssofort programm, Prozebgasfreisetzung-Explosion in der Gasfabrik und Auswirkungen von Druckwellen auf das Containment, Dezember 1983
138. L. Pu, X. Shao, Q. Li, Y. Li, A simple and effective approach for evaluating unconfined hydrogen/air cloud explosions. Int. J. Hydrogen Energy **43**, 10193–10204 (2018)
139. I.C. Tolias, A.G. Venetsanos, N. Markatos, C.T. Kiranoudis, CFD evaluation against a large scale unconfined hydrogen deflagration. Int J Hydrogen Energy **42**, 7731–7739 (2017)
140. Y.B. Zeldovich, A.I. Rozlovskii, On the onset of unstable normal burning: transition of a spherical flame to detonation. Dokl. Akad. Nauk SSSR **57**, 365–368 (1947). ((in Russian))
141. L. Kagan, G. Sivashinsky, Transition to detonation of an expanding spherical flame. Combust. Flame **175**, 307–311 (2017)
142. A. Koksharov, V. Bykov, L. Kagan, G. Sivashinsky, Deflagration-to-detonation transition in an unconfined space. Combust. Flame **195**, 163–169 (2018)
143. P.V. Gordon, L. Kagan, G. Sivashinsky, Parametric transition from deflagration to detonation revisited: spherical geometry. Combust. Flame **211**, 405–4015 (2020)
144. P.V. Gordon, L. Kagan, G. Sivashinsky, Parametric transition from deflagration to detonation in stellar medium. Phys. Rev. E **103**, 033106 (1–9) (2021)
145. A. Kiverin, I. Yakovenko, Mechanism of transition to detonation in unconfined volumes. Acta Astr. **176**, 647–652 (2020)
146. C. Qian, C. Wang, J.N. Liu, A. Brandenburg, N.E.L. Haugen, M.A. Liberman, Convergence properties of detonation simulations. Geophys. Astrophys. Fluid Dyn. **114**, 58–76 (2020)

Chapter 16
Effects of Radiation on Particle-Laden Combustion

Interest in the effect of thermal radiation on combustion is mainly associated with explosions in coal mines and other dust-related industries. While studying combustion in gaseous mixture, the radiation of hot combustion products is usually not important, since the radiation absorption length in a gaseous mixture is much larger than typical scales characterizing gaseous combustions. The gas mixture is almost completely transparent to radiation, so radiation heat transfer does not affect the flame dynamics. For example, the average mean free path of photons in the Earth's atmosphere is about hundreds of meters due to the small cross section of Thompson scattering and "bremsstrahlung" cross section, which are about $(10^{-24} \div 10^{-25})\,\mathrm{cm}^2$. Therefore, the contribution of radiation from the hot combustion products to the heat transfer is negligible compared with the gas thermal conduction and can be neglected. Even at elevated pressures the gas mixture can be treated as almost transparent for the radiation. Besides, the intensity of radiation is noticeable only for a very large and optically thick cloud of hot combustion products.

The situation changes drastically in the presence of small particulates of size r_p suspended in the gas mixture, which is typical for explosions in coal mines, chemical industry, forest fire, etc. In this case, the radiation absorption length is almost completely determined by its absorption and scattering on the suspended particles and can be estimated as $L = 1/N_p\sigma_p$, where N_p is the particles number density, and $\sigma_p \approx \pi r_p^2$ is the particles cross section for the radiation absorption. The flux of radiant energy is absorbed by the particles ahead of the flame, the particles are heated by radiation with subsequent re-emission and heat transfer from the surface of the particle to the surrounding unreacted gas phase, so that the temperature of the gas phase lags behind the temperature of the particles. It should also be noted that the presence of even a relatively small concentration of solid particles increases considerably the luminosity of combustion products, so that the radiant flux emitted by the hot combustion products may be approximated by the blackbody radiation. At the same time, if the mass density of suspended particles is small compared to the mass density of the gaseous phase, $\varsigma = m_p N_p/\rho_g \ll 1$, then the effect of particles on the flame dynamics is small.

M. A. Liberman, *Combustion Physics*,
https://doi.org/10.1007/978-3-030-85139-2_16

The role of radiation has been studied experimentally, analytically, and numerically mainly in the context of dust explosions [1–4]. Many studies of dust explosions were aimed at studying the characteristics of mixture ignition by the thermal radiation of the primary flame or by radiation from an external light source [4–16], heating of reacting particles and their devolatilization [17] due to radiation. The radiation-induced ignition requires the release of more energy in the mixture than, for example, ignition from an electric spark. Direct thermal ignition of gaseous combustible mixture the radiation causing a rapid increase in temperature up to 1000 K is possible by focusing a high-power laser radiation and was demonstrated both theoretically and experimentally [4–15]. However, ignition at low power levels is at normal conditions. To ignite a gas mixture with suspended particles as a carrier of radiation energy, the spatial distribution of the particles must be such that most of the radiation is absorbed by particles far ahead of the flame, since the particles must be heated to a relatively high temperature before the flame arrival.

The effect of particles on flame propagation was studied by various groups of authors on the assumption of uniformly dispersed, gravity-free particles with and without regard to radiation heat transfer [18–39]. Essenhigh and Csaba [40] were probably the first who have drawn attention to significant role of the radiation heat transfer for a planar flame propagating in a dust cloud. In early studies, the flame dynamics affected by the radiative preheating was investigated using asymptotic methods and a one-step chemical model with high activation energy. Most of the research has been focused on coal combustion, with two aspects of practical interest: the formation of volatiles due to the thermal decomposition of coal dust and the ignition conditions. The combustible volatiles can react and release energy, which in turn may contribute to the heat-up of the particles, enhance the combustion energy release due to energy feedback mechanism resulting in an explosion [17, 29, 34, 35]. For the coal-fired burners, the coal-dust suspension, and for rocket engines using the solid or fluid fuels as well as for coal-fire mining safety, both the ignition and the flame evolution are of paramount importance. There is a lack of radiation measurements related to explosions in large-scale systems, and the actual level of thermal radiation emitted by the flame remains conjectural. However, recent experiments have shown that the flame propagating in a dust cloud is strongly influenced by thermal radiation. The available data make it possible to estimate the likelihood of combustion initiation due to the forward thermal radiation.

16.1 Flame Velocity in Particle-Laden Mixtures. Effect of Radiation

To reveal the effect of thermal radiation, consider the propagation of a flame through a combustible gas with suspended inert dust particles. This means that we will only consider the effect of additional heat transfer by radiation to cold reactants, which can affect the flame speed. We consider a planar flame propagating in a particle-laden hydrogen–oxygen mixture from the closed to the open end of a duct. The governing equations for the gaseous phase are the 1D, time-dependent, multispecies reactive Navier–Stokes equations including the effects of compressibility, molecular diffusion, thermal conduction, viscosity, gas-particles momentum and energy exchange and detailed chemical kinetics for eight reactive species H_2, O_2, H, O, OH, H_2O, H_2O_2, and HO_2. The governing equations for the gaseous phase are

$$\frac{\partial \rho}{\partial t} + \frac{\partial (\rho u)}{\partial x} = 0, \tag{16.1.1}$$

$$\frac{\partial Y_i}{\partial t} + u\frac{\partial Y_i}{\partial x} = \frac{1}{\rho}\frac{\partial}{\partial x}\left(\rho D_i \frac{\partial Y_i}{\partial x}\right) + \left(\frac{\partial Y_i}{\partial t}\right)_{ch}, \tag{16.1.2}$$

$$\rho\left(\frac{\partial u}{\partial t} + u\frac{\partial u}{\partial x}\right) = -\frac{\partial P}{\partial x} + \frac{\partial \sigma_{xx}}{\partial x} - \rho_p \frac{(u - u_p)}{\tau_{St}}, \tag{16.1.3}$$

$$\begin{aligned}\rho\left(\frac{\partial E}{\partial t} + u\frac{\partial E}{\partial x}\right) &= -\frac{\partial (Pu)}{\partial x} + \frac{\partial}{\partial x}(\sigma_{xx} u) + \frac{\partial}{\partial x}\left(\kappa(T)\frac{\partial T}{\partial x}\right) + \\ &+ \sum_k h_k \left(\frac{\partial}{\partial x}\left(\rho D_k(T)\frac{\partial Y_k}{\partial x}\right)\right) + \rho \sum_k h_k \left(\frac{\partial Y_k}{\partial t}\right)_{ch} \\ &- \rho_p u_p \frac{(u - u_p)}{\tau_{St}} - \rho_p c_{P,p} Q,\end{aligned} \tag{16.1.4}$$

$$P = R_B T n = \left(\sum_i \frac{R_B}{m_i} Y_i\right)\rho T = \rho T \sum_i R_i Y_i, \tag{16.1.5}$$

$$\varepsilon = c_v T + \sum_k \frac{h_k \rho_k}{\rho} = c_v T + \sum_k h_k Y_k, \tag{16.1.6}$$

$$\sigma_{xx} = \frac{4}{3}\mu\left(\frac{\partial u}{\partial x}\right). \tag{16.1.7}$$

Here, we used standard notations: P, ρ, u, are pressure, mass density, and flow velocity of the gas mixture, $Y_i = \rho_i/\rho$—mass fractions of the species, $E = \varepsilon + u^2/2$—total energy density, ε—internal energy density, R_B—universal gas constant, m_i—molar mass of i-species, n—gaseous molar density, σ_{ij}—viscous stress tensor, $c_v = \sum_i c_{vi} Y_i$—constant volume specific heat, c_{vi}—constant volume specific heat of i-species, h_i—enthalpy of formation of i-species, $\kappa(T)$ and $\mu(T)$—coefficients of thermal conductivity and viscosity, $D_i(T)$—diffusion coefficients of i-species, $(\partial Y_i/\partial t)_{ch}$—variation of i-species concentration in chemical reactions, $\rho_p = m_p N_p$—mass density of the suspended particles, N_p—particles number density, u_p—particles velocity, $\tau_{St} = m_p/6\pi\mu r_p$—the Stokes time for the spherical particles of the radius r_p and mass m_p, Q—interphase thermal exchange source, $c_{P,p}$—constant pressure specific heat of the particles material.

Changes in the concentrations of the components Y_i of the gaseous phase due to chemical reactions are determined by the solution of the system of chemical kinetics

$$\frac{dY_i}{dt} = F_i(Y_1, Y_2, \ldots, Y_N, T), \quad i = 1, 2, \ldots, N. \tag{16.1.8}$$

The right-hand parts of Eq. (16.1.8) contain the rates of chemical reactions for the reactive species H_2, O_2, H, O, OH, H_2O, H_2O_2, and HO_2 with subsequent chain branching, production of radicals and energy release. The simulation uses the standard reduced chemical kinetic scheme of the hydrogen/oxygen combustion with elementary Arrhenius-type reactions.

The equations of state for the gas phase and for the combustion products are taken with the temperature dependence of the specific heats and enthalpies of each species borrowed from the JANAF tables and interpolated by the fifth-order polynomials. The viscosity and thermal conductivity coefficients of the mixture were calculated from the gas kinetic theory using the Lennard–Jones potential.

$$\mu = \frac{1}{2}\left[\sum_i \alpha_i \mu_i + \left(\sum_i \frac{\alpha_i}{\mu_i}\right)^{-1}\right], \tag{16.1.9}$$

where $\alpha_i = \frac{n_i}{n}$ is the molar fraction, $\mu_i = \frac{5}{16}\frac{\sqrt{\pi \hat{m}_i kT}}{\pi \Sigma_i^2 \tilde{\Omega}_i^{(2,2)}}$ is the viscosity coefficient of i-species, $\tilde{\Omega}^{(2,2)}$—is the collision integral which is calculated using the Lennard–Jones potential, $\hat{m}_i$ is the molecule mass of the i-th species of the mixture, Σ_i is the effective molecule size. The thermal conductivity coefficient of the mixture is calculated as

$$\kappa = \frac{1}{2}\left[\sum_i \alpha_i \kappa_i + \left(\sum_i \frac{\alpha_i}{\kappa_i}\right)^{-1}\right]. \tag{16.1.10}$$

The coefficient of the heat conduction of i-th species $\kappa_i = \mu_i c_{pi}/\mathrm{Pr}$ can be expressed via the kinematic viscosity μ_i with the Prandtl number, $\mathrm{Pr} \approx 0.75$.

The binary coefficients of diffusion are

$$D_{ij} = \frac{3}{8}\frac{\sqrt{2\pi kT \hat{m}_i \hat{m}_j / (\hat{m}_i + \hat{m}_j)}}{\pi \Sigma_{ij}^2 \tilde{\Omega}^{(1,1)}(T_{ij}^*)} \cdot \frac{1}{\rho}, \tag{16.1.11}$$

where $\Sigma_{ij} = 0,5(\Sigma_i + \Sigma_j)$, $T_{ij}^* = kT/\varepsilon_{ij}^*$, $\varepsilon_{ij}^* = \sqrt{\varepsilon_i^* \varepsilon_j^*}$; $\varepsilon*$ are constants in the expression of the Lennard–Jones potential, and $\tilde{\Omega}_{ij}^{(1,1)}$ is the collision integral similar to $\tilde{\Omega}^{(2,2)}$. The diffusion coefficient of i-th species is

$$D_i = (1 - Y_i)/\sum_{i \neq j} \alpha_i / D_{ij}. \tag{16.1.12}$$

The simplest way to incorporate thermal radiation is the diffusion approximation. In this approximation, the attenuation of the radiant flux is due to the absorption of the incident radiation by particles, followed partly by re-emission of radiation and partly by the transfer of the absorbed radiative energy from particles to the surrounding gas phase. Analytically, the simplest form of this approximation is the Beer–Lambert's law [41]. For simplicity, we assume that radiation scattering is small, $\lambda_{rac} \ll r_p$.

The equations for solid particles are taken in the hydrodynamic approximation. For a small concentration of particles, the interaction between particles is insignificant, and only the Stokes force between the particle and the gas phase can be considered. Then, the equations for the suspended particles are

$$\frac{\partial N_p}{\partial t} + \frac{\partial (N_p u_p)}{\partial x} = 0, \tag{16.1.13}$$

$$\left(\frac{\partial u_p}{\partial t} + u_p \frac{\partial u_p}{\partial x}\right) = \frac{(u - u_p)}{\tau_{St}}, \tag{16.1.14}$$

$$\left(\frac{\partial T_p}{\partial t} + u_p \frac{\partial T_p}{\partial x}\right) = Q - \frac{2\pi r_p^2 N_p}{c_{P,p} \rho_{p0}} \left(4\sigma T_p^4 - q_{\text{rad}}\right). \tag{16.1.15}$$

Here T_p—is temperature of the particles, $2\pi r_p^2 N_p \left(4\sigma T_p^4 - q_{\text{rad}}\right)$ is the thermal radiation energy absorbed and re-emitted by the particles. The energy transferred from the particle surface to the surrounding gas mixture is

$$Q = (T_p - T)/\tau_{pg}, \tag{16.1.16}$$

where $\tau_{pg} = 2r_p^2 c_{P,p} \rho_{p0} / 3\kappa Nu$ is the characteristic time of the energy transfer from the particle surface to surrounding gaseous mixture, $c_{P,p}$ and ρ_{p0} are the specific heat and mass density of the particles, Nu is the Nusselt number [43].

For a 1D problem, the heat transfer equation of thermal radiation in the diffusion approximation has the form

$$\frac{d}{dx}\left(\frac{1}{\chi} \frac{dq_{\text{rad}}}{dx}\right) = -3\chi \left(4\sigma T_p^4 - q_{\text{rad}}\right), \tag{16.1.17}$$

where $\chi = 1/L = \pi r_p^2 N_p$ is the coefficient of radiation absorption and $L = 1/\pi r_p^2 N_p$ is the absorption length. The radiation energy flux emitted by the hot combustion products from the flame surface is assumed to be equal to the blackbody radiative heat source, $q_{\text{rad}}(x = X_f) = \sigma T_b^4$, where $\sigma = 5.6703\ 10^{-8}$ W/m^2 K^4 is the Stefan–Boltzmann constant, and $T_b \approx 3000$ K is temperature of the hot combustion products.

Such formulated problem has been solved for stoichiometric hydrogen/oxygen at initial pressure $P_0 = 1$ atm, assuming a small mass density of particles, $\rho_p = m_p N_p$ in comparison with the gaseous phase density, $\zeta = m_p N_p/\rho \ll 1$, so that there is only one way of momentum coupling of the particles and the gaseous phase.

16.1.1 Flame Propagation Through Uniformly Dispersed Particles

Consider the hydrogen–oxygen flame propagating in the gas mixture with uniformly suspended identical solid spherical particles of radius $r_p \approx 1\,\mu$m, for the particle number densities, $N_p = 5.7 \cdot 10^7, 2.85 \cdot 10^7, 1.4 \cdot 10^7\,\text{cm}^{-3}$, and the material density of particles $\rho_{p,0} = 1\,\text{g/cm}^3$. Assuming that radiation is absorbed and scattered only by particles, this corresponds to the radiation absorption lengths: $L = 1, 2, 4$ cm. The problem set-up is shown schematically in Fig. 16.1. Heating the gas mixture by radiation ahead of the flame front is effective on the scale of the thermal radiation absorption length.

Characteristic time scales of the problem are $\tau_{St} = 2r_p^2 \rho_{p0}/9\nu_g \rho_g \approx 3\,\mu$s, $\tau_{pg} \approx 1\,\mu$s, and the hydrodynamic time scale is $L_f/U_f \approx 20\,\mu$s. Since the characteristic time of energy transfer between the particles and the gas phase, τ_{pg}, is much smaller than the hydrodynamic time scale, the temperatures of the particles and the gas phase are almost equal, $T_p \approx T_g$. Mass loading parameter values are $\varsigma = \rho_p/\rho_g = 0.2$; 0.1 and 0.05 for $L = 1, 2$ and 4 cm, respectively. Because $\varsigma = \rho_p/\rho_g \ll 1$, the momentum coupling of the particles and the gaseous phase is small, and the influence of particles on the flame dynamics is small also. Thus, only the heating of particles by the absorbed radiation and corresponding heating of the gas mixture ahead of the flame will influence the flame dynamics. The presence of neutral solid particles is analogous to diluting a combustible mixture with an inert gas, reducing the adiabatic flame temperature. This effect is also small for $\varsigma \cdot (c_{p,p}/c_{V,g}) \ll 1$.

Since a stationary flow is established during the time of the order L/U_f, it is straightforward to calculate the maximum temperature increase ahead of the flame caused by the radiation heating of the particles. In the reference frame co-moving

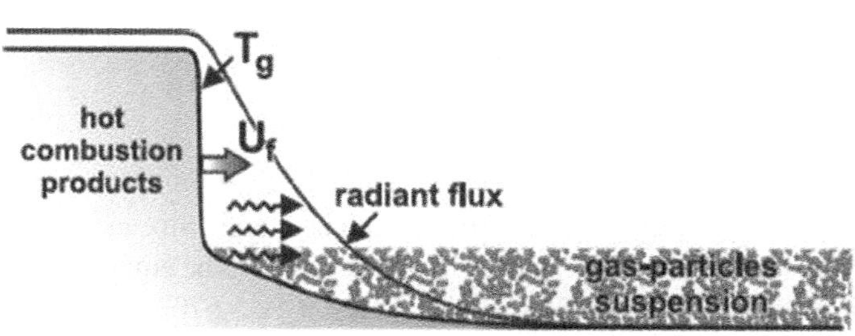

Fig. 16.1 Flame propagation in a cloud of uniformly suspended particles [44]

with the flame, the unburned mixture with suspended particles flows toward the flame with the laminar flame velocity U_f. The thermal radiation is absorbed by the particles, which are located at the distance $x \leq L$ ahead of the flame front, and their temperature can be estimated from the energy balance

$$\rho_p c_{p,p} \frac{dT_p}{dt} = \sigma T_b^4 \exp\left(-\frac{x - U_f t}{L}\right) \pi r_p^2 N_p - \frac{\rho_p c_{p,p}}{\tau_{pg}} (T_p - T). \qquad (16.1.18)$$

The gas phase temperature ahead of the flame increases due to the heat transferred from the particles to the surrounding unburned mixture.

$$dT/dt = (T - T_p)/\tau_{gp} \qquad (16.1.19)$$

For $T \approx T_p$ and $L = 1/\pi r_p^2 N_p$, from Eqs. (16.1.18) and (16.1.19) for gas phase temperature it follows

$$\rho_p c_{p,p} (1 + \xi) \frac{dT}{dt} = \sigma T_b^4 \frac{1}{L} \exp\left(-\frac{x - U_f t}{L}\right) \qquad (16.1.20)$$

where $\xi = \tau_{gp}/\tau_{pg} = c_{V,g}/\varsigma c_{P,p}$.

The time of radiation heating of a Lagrangian particle can be estimated as the time of its arrival at the flame front, $t \approx L/U_f$. Taking this into account, the maximum increase of temperature in the unburned mixture close ahead of the flame is

$$\Delta T = \sigma T_b^4 \frac{1}{U_f} \frac{(1 - e^{-1})}{\rho_p c_p (1 + \xi)} \approx 0.63 \frac{\sigma T_b^4}{(\rho_p c_p + \rho c_{V,g}) U_f} \qquad (16.1.21)$$

From Eq. (16.1.21), it is easy to see that the maximum temperature rise due to radiation heating does not depend on the absorption length of radiation. This is due to the fact that although the local radiant heat flux absorbed by a particle is larger for a shorter absorption length, but in the case of a larger absorption length, the particles absorb the radiant heat flux over a longer time before their arrival to the flame front. This difference in time for shorter and longer absorption lengths compensates the lesser local heating for the case of longer absorption lengths. Taking into account that the adiabatic flame temperature is less than it is in a pure gaseous mixture due to its dilution by the inert solid particles, and that the photons emitted from the flame front originated mainly within the layer of finite thickness near the flame front, the effective temperature of the radiation emitted from the flame surface can be estimated as $T_{b,eff} \approx 2700$ K.

Then, the maximum temperature increase caused by the radiative heating can be estimated as $\Delta T \approx 160$ K, which is in good agreement with the results of numerical simulations [44, 45] shown in Fig. 16.2. It is also seen that the stationary values of the gas temperature ahead of the flame are established during $t \approx L/U_{f0}$, where U_{f0} is the laminar flame velocity in a pure gas mixture.

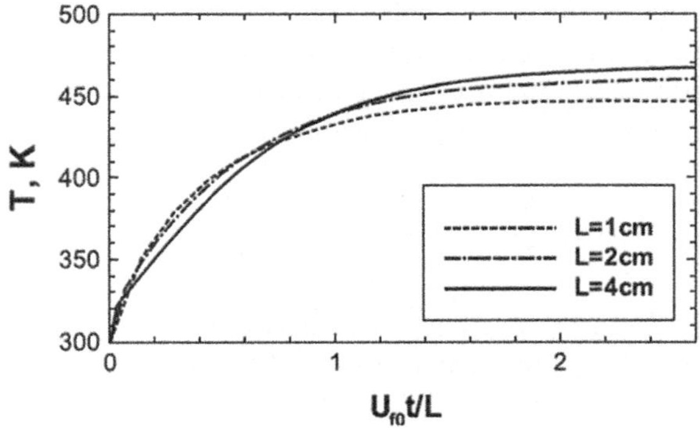

Fig. 16.2 Temporal evolution of the gas temperature during radiative heating at the distance 2 mm ahead the flame for $L = 1, 2$, 4 cm. Time is in units L/U_{f0} [44]

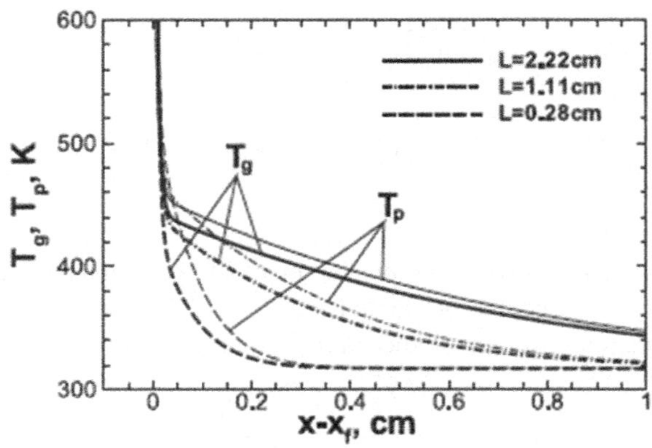

Fig. 16.3 Temperature ahead of the flame for uniformly distributed particles: L = 0.28, 1.11, 2.22 cm. Solid curves—T_g, thin curves—particles' temperature T_p [45]

Figure 16.3 shows obtained in simulations [45] profiles of the gas phase and the particles temperatures, which are established for the radiation absorption lengths: $L = 0.28$, 1.11, and 2.22 cm, after the stationary flow was settled. These radiation absorption lengths correspond to $r_p = 6.0;\ 1.5;\ 0.75\,\mu\text{m}$, $N_p = 3.18 \cdot 10^6;\ 1.27 \cdot 10^7,\ 2.55 \cdot 10^7\,\text{cm}^{-3}$. The characteristic time scales for these cases are $\tau_{St} = 40\,\mu\text{s}$ and $\tau_Q = 30\,\mu\text{s}$.

It should be noted that if the characteristic time of energy transfer from the surface of a particle of the surrounding gas mixture τ_Q is less than the characteristic gas-dynamic time of flame propagation L_f/U_f, then heat transfer between the gas phase and particles occurs quickly, and the temperatures of particles and gas are almost equal.

16.1.2 Flame Structure; Radiation Dominated Regime

From Eq. (16.1.21), it follows that the effect of radiation heating is stronger at a lower laminar flame velocity and at a lower initial density of the gas phase. The increase

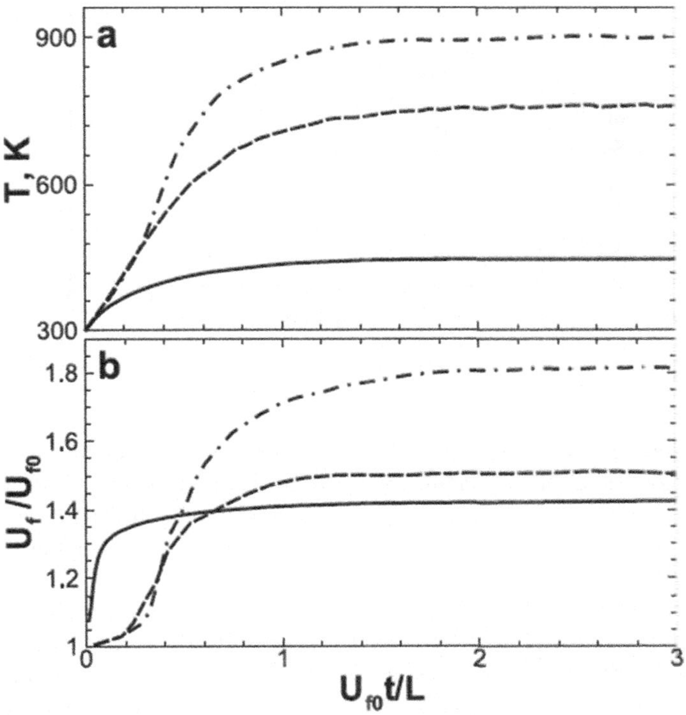

Fig. 16.4 **a** Temperature at 2 mm ahead the flame front; **b** the flame velocity increase. Solid line—$P_0 = 1$ atm, dashed line—$P_0 = 0.3$ atm|, dashed-dotted line—$P_0 = 0.2$ atm. The radiation absorption length is $L = 1$ cm. Units in x-axis is L/U_{f0} [46]

in gas temperature and the corresponding increase in the H_2/O_2 flame propagation velocity at initial pressures of 1.0 atm, 0.3 atm, 0.2 atm is shown in Fig. 16.4a, b. It is interesting to note that in the framework of a one-step chemical model, the flame width decreases with increasing flame speed. The velocity and structure of a laminar flame is determined by the equation of thermal conduction, which in a 1D case is

$$T(x,t) \propto \frac{1}{2\sqrt{\pi \chi_g t}} \exp(-x^2/4\chi_g t), \tag{16.1.22}$$

where $\chi_g = \kappa_g/\rho_g c_{g,p} \approx$ const is coefficient of the thermal diffusivity.

From Eq. (16.1.22), it follows that during time interval Δt the heat propagates at the distance $x \approx \sqrt{4\chi_g \Delta t}$. Taking this into account, an order of magnitude estimates for the speed and the width of the laminar flame are $U_f \approx \sqrt{\chi_g/\tau_R}$, $L_f \approx \sqrt{\chi_g \tau_R}$. These formulas are based on the assumption that the timescale for heat propagation is much greater than the heat release time in the reaction τ_R. In this way, we arrive to the well-known expression relating the width and velocity of a laminar flame, $L_f U_f \approx \chi_g$.

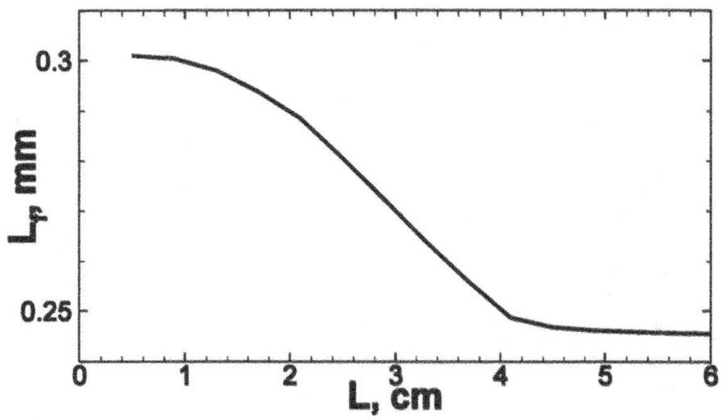

Fig. 16.5 Hydrogen–oxygen flame width versus the radiation absorption length; $P_0 = 1$ atm [44]

On the contrary, an increase of the flame velocity caused by radiative heating leads to an increase of the flame width. Figure 16.5 shows the flame width depending on the radiation absorption length. Large values of radiation absorption length L correspond to pure gas mixture with the flame width $L_f \approx 0.24$ mm for H_2/O_2 mixture at $P_0 = 1$ atm.

According to Eq. (16.1.21), for a flame with a sufficiently small velocity, the maximum temperature of the unburned gas ahead of the flame may exceed the ignition threshold. For a hydrogen–oxygen this is the crossover temperature, when the endothermic reaction stage passes to the fast-exothermic stage. In this case, radiant heat transfer can dominate the conductive heat transfer, and thermal radiation makes a decisive contribution to the entire energy transfer. As an example, the calculated time evolution of the maximum temperature ahead of the flame in a methane–air mixture with suspended particles and the corresponding increase of flame velocity due to the radiative heating for the radiation absorption length $L = 1$ cm are shown in Fig. 16.6 [46]. Because of much smaller velocity of the methane–air flame, the time of radiation heating is much longer than for case of H_2–O_2 or H_2-air flames.

The flame structure changes significantly if radiative heat transfer dominates the convective heat transfer. In the framework of a one-step chemical model, the total enthalpy is constant inside of the flame, which implies similarity in distributions of temperature and concentration of reagent in the flame structure. This conclusion is obtained in the approximation of linear thermal conduction and when the coefficient of thermal diffusivity equals the diffusion coefficient of reagents. On the contrary, if radiative heat transfer is the predominant process, heat propagation is determined by a strongly nonlinear equation of heat conduction, for which the coefficient of radiative thermal conductivity in the diffusion approximation is

$$\kappa_{\mathrm{Rad}} = 16\sigma T^3 L/3. \tag{16.1.23}$$

In this case, the total enthalpy of the unburned mixture and combustion products remains equal but the total enthalpy is not constant inside of the flame structure and there is no similarity of the temperature and the density distributions.

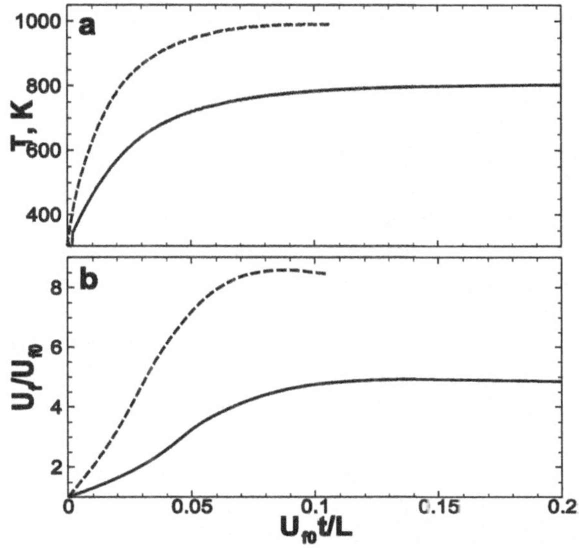

Fig. 16.6 **a** temperature at the distance 2 mm ahead of the methane–air flame; **b** flame velocity. $P_0 = 1$ atm (solid line), $P_0 = 0.5$ atm (dashed line), $L = 1$ cm [45]

Solution of the thermal conduction equation for the nonlinear coefficient of thermal conductivity is much smoother than the solution given by Eq. (16.1.22)

$$T(x) \propto T_0\left(1 - x^2/x_0^2\right)^{1/3} \tag{16.1.24}$$

In the last case the reaction can be ignited within the preheat zone, and the flame thickness by the order of magnitude equals the radiation absorption length, i.e. $L_{\text{f-rad}} \sim L$. The flame velocity can be estimated as $U_{\text{f,rad}} \sim (L/\tau_R) \approx (L/L_f)U_f \gg U_f$, where L_f and U_f are the width and velocity of the laminar flame in a pure gaseous mixture. As soon as combustion has been initiated by the primary particle-laden flame, the radiation of combustion products can generate secondary explosions in the mixture ahead of the flame. Such combustion wave will look like a sequence of thermal explosions and can be accompanied by a strong pressure increase. The necessary condition for the radiation to become dominating mechanism of a flame propagation is that the characteristic time of ignition in the mixture ahead of the flame to be small compared with the time required for the original flame to pass through the radiation preheat zone, $\tau_{\text{ign}}(T_{cr}) \ll L/U_f$. It is unlikely that this condition can be met for a fast flame, such as H_2/O_2 or H_2/air, but it is possible for a slow flame, such as, e.g. methane–air flame, where the process can be further enhanced by the combustion of coal dust volatile. For a noticeable effect, ignition must take place well ahead of the advancing flame, therefore relatively long radiation length scales are needed. Such a scenario is unlikely for the case of uniformly dispersed particles, but it is possible for the non-uniform distribution of suspended particles.

16.2 Nonuniform Distribution of Suspended Particles

An interesting scenario appears in the case of a non-uniform distribution of the particles number density (layered dust cloud), when the mixture is practically transparent to thermal radiation at the layer ahead of the flame, and the radiation is absorbed only far ahead of the flame front as it is shown schematically in Fig. 16.7.

It is a well known that in coal mines, the pressure wave of a weak accidental explosion can disperse dust deposits, leading to the formation of non-uniformly distributed suspended particles forming a layered dust-air cloud. In this case, the time of radiation heating of particles ahead of the flame can be much longer than in the case of uniform distribution of particles, since the radiant flux is not attenuated as long as the flame propagates through gap, which separates the flame and the cloud of particles and which is transparent for radiation.

Following [48], we consider a non-uniform distribution of the particle number density, schematically shown in Fig. 16.8, and assume that the "gap" between the flame and the left boundary of the particles-cloud mixture is transparent for the radiation emitted from the flame surface and that the radiant heat flux is fully absorbed in the layer 3. Particles in the dense cloud far ahead from the advancing flame absorb much more thermal radiant energy before the flame arrival to the left boundary of the denser cloud. Therefore, the temperature of particles and surrounding gas mixture can rise up to the value sufficient for the gas phase ignition before the flame arrival. In such conditions the ignition starts when the temperature exceeds the crossover value, which is for hydrogen/oxygen at 1 atm is about 1050 $\pm$ 50 K and determines the limit where the induction endothermic stage passes into the fast, exothermic stage. An order of magnitude estimate of the time interval needed to rise temperature of the suspended particles up to the crossover temperature is about 1 ms for hydrogen–oxygen mixture. The radiative heating of the mixture in the gas-particle cloud depends on the thermal radiation absorption length, which is also influenced by the gas-particle cloud expansion during the radiative heating. Depending on the steepness of the temperature gradient formed during the heating and expansion of the dust cloud, either a deflagration or a detonation can be ignited via Zel'dovich's gradient mechanism (see Chapter 13.2).

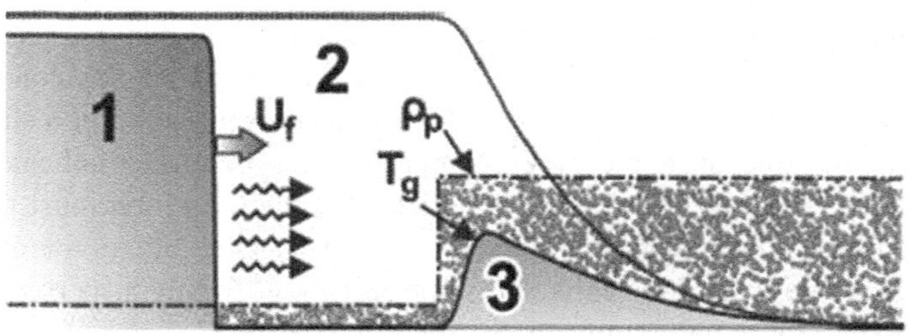

Fig. 16.7 1—combustion products; 2—"transparent gap" with low concentration of particles; 3—high concentration of particles; T_g—temperature of the gas phase [46].

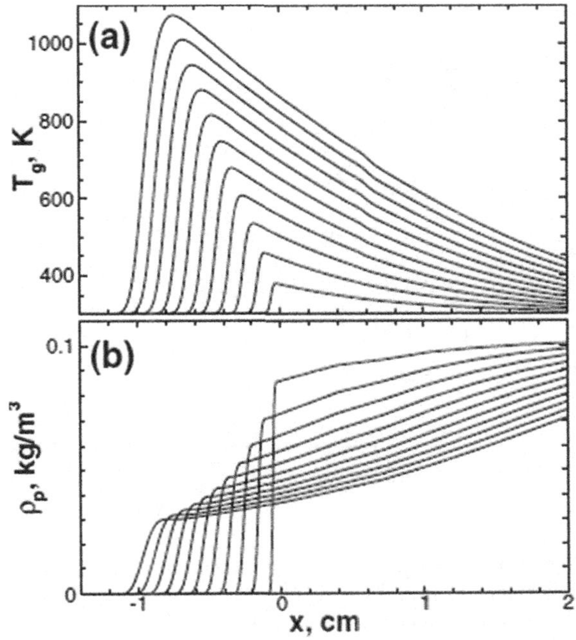

Fig. 16.8 Temporal evolution of the gas phase temperature (**a**) and the mass density of the suspended particles (**b**). Time intervals $\Delta t = 80\,\mu s$, $N_p = 2.5 \cdot 10^7\,cm^{-3}$, $r_p = 1\,\mu m$

Figure 16.8 shows the computed temporal evolution of the gas phase temperature and the particles density profile during the radiation heating, when the maximum temperature raised up to the crossover value and the final temperature gradient is formed at $t_0 \approx 900\,\mu s$. This value agrees with the time needed to rise temperature in the nearest denser layer up to the the crossover temperature by the radiation heating estimated as $\delta t \approx (\rho_p c_p + \rho c_{V,g})\Delta T/\sigma T_b^4$.

The calculations were performed for the initial stepwise density of inert particles of radius $r_p = 1\,\mu m$ with the maximum number density of particles within the gas-particles layer $N_p = 2.5 \cdot 10^7\,cm^{-3}$. The scale of the temperature gradient in Fig. 16.8, which is formed near the left boundary of the dust cloud is $\Delta = (T * -T_0)/|dT/dx| \approx 1\,cm$, which is close to the value of the radiation absorption length $L = 1/\pi r_p^2 N_p \approx 1.2\,cm$. According to the classification of combustion regimes [47] initiated by the initial temperature gradient in hydrogen–oxygen at the initial pressure $P_0 = 1\,atm$, this temperature gradient can ignite a deflagration. The time evolution of the gas phase temperature, shown in Fig. 16.9, shows the development of a spontaneous reaction wave for the same conditions as in Fig. 16.8, and the formation of a deflagration wave. The dashed line in the upper frame in Fig. 16.9 shows the initial profile of the particle number density, while the solid line shows the particles number density profile formed at the instant $t_0 \approx 900\,\mu s$, just prior to the ignition, when the temperature gradient with the maximum temperature $T^* = 1050\,K$ is formed; the first temperature profile in the frame Fig. 16.9b. The profiles in the bottom frame Fig. 16.10b indicate a small variation of pressure during the formation of deflagration wave.

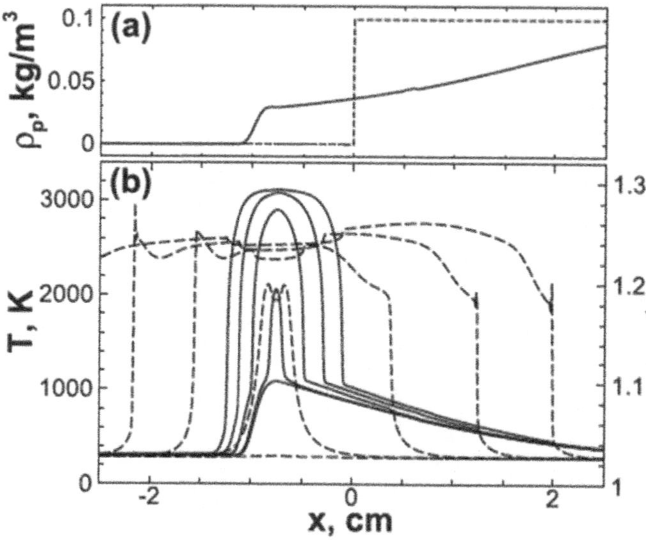

Fig. 16.9 Particle density, gas phase temperature, and pressure profiles during the formation of a deflagration wave in the vicinity of the left boundary of the gas-particles cloud. The profiles are shown starting from $t_0 \approx 890\,\mu s$ with intervals $\Delta t = 10\,\mu s$

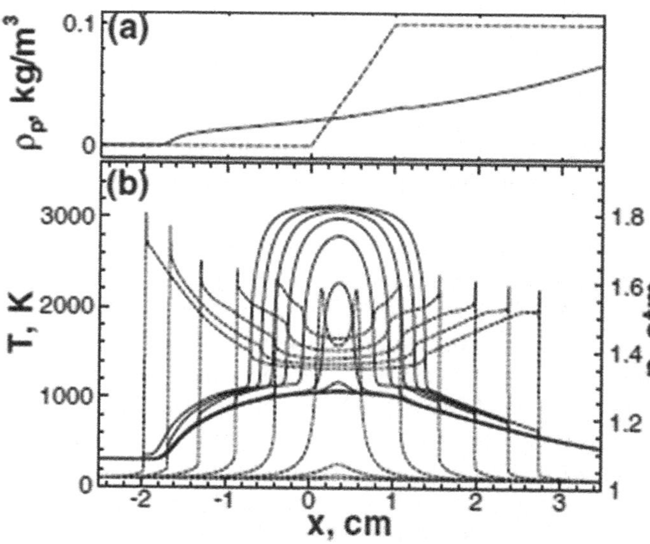

Fig. 16.10 Particle density, gas temperature, and pressure profiles during the formation of fast deflagration wave behind the outrunning shock wave near the left boundary of the gas-particles cloud. The profiles are shown from $t_0 = 1650\,\mu s$, $\Delta t = 4\,\mu s$

It is interesting to note that in a system of periodic layers of optically thick dust clouds separated by optically thin layers, the combustion wave can propagate at a speed much higher than the speed of a laminar flame. Indeed, the radiation of the primary flame causes ignition of a new flame in the nearest denser layer. The radiation of the secondary flame causes ignition of the flame in the next denser layer, etc. As a result of such successive ignition, combustion can propagate with the velocity which is defined by the spacing between the dust layers and can be as high as 1000 m/s. Such scenario of successive ignitions, which looks like the volume thermal explosions can be accompanied by a noticeable pressure increase and may be useful for understanding the origin of the explosions in underground coal mines.

For a dust cloud with a diffuse left interface, a smoother temperature gradient is formed during the radiation heating. In the latter case, the radiation absorption length varies along the diffusive particle-cloud interface resulting in the formation of the temperature profile with a smoother temperature gradient capable to ignite either fast deflagration or detonation.

Two examples of the particle-gas cloud with diffuse interface and the computed temperature profiles caused by the radiative preheating are shown in Figs. 16.10 and 16.11. The upper frame in Fig. 16.10, shows the initial particle density distribution (dashed line) which drops linearly on the scale 1 cm from its maximum value. The diffuse interface (solid line) of the dust cloud is smeared due to the expansion of the dust cloud during the particles and gas phase heating by radiation. The bottom frame, Fig. 16.10, shows the calculated temporal evolution of the gaseous phase

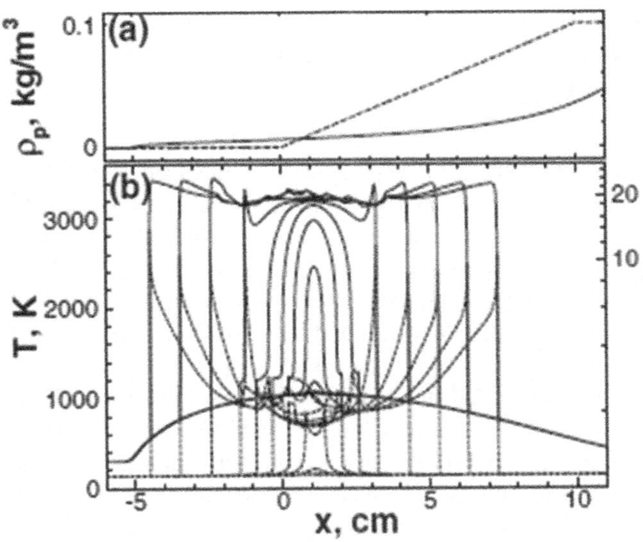

Fig. 16.11 Particle density, gas temperature, and pressure profiles during the detonation formation in the vicinity of the left boundary of the dust cloud near the left boundary of the gas-particles cloud. The profiles are shown from $t_0 = 5015\,\mu s$, $\Delta t = 4\,\mu s$

temperature and pressure profiles during the formation of the fast deflagration behind the outrunning weak shocks in the vicinity of the dust cloud boundary.

Initiation a detonation via Zel'dovich's gradient mechanism requires much shallower temperature gradient. According to [47], the scale of the initially linear temperature gradient in hydrogen/oxygen mixture under normal conditions ($P_0 = 1$ atm) and temperature $T^* = 1050$ K at the top of the gradient is $(T^* - T_0)/\nabla T \approx 20$ cm. The results of simulations for the dust cloud with initial diffuse boundary, where the particles number density drops linearly on the scale of 10 cm from its maximum value $N_p = 2.5 \cdot 10^7$ cm^{-3}, are shown in Fig. 16.11. Here, the upper frame depicts the initial diffuse boundary of the dust cloud, which was smeared due to gas expansion during the thermal radiation heating up to 20 cm. Time evolution of the gaseous temperature and pressure profiles during the detonation formation in the vicinity of the left boundary of the particle-laden H_2/O_2 gas mixture are depicted in the bottom frame of Fig. 16.11.

Dashed line in the upper frame shows the initial density profile, and the solid line shows density profile at the time instant $t_0 = 5015$ μs prior to the ignition of a spontaneous combustion wave triggering the detonation (solid line).

Finally, it should be noted that velocity of the "fast deflagration" initiated by the temperature gradient (Fig. 16.10) decreases as shock waves outrun and it is saturated at the same value as the speed of normal deflagration wave propagating in the particle-laden combustible mixture and accelerated due to the radiant preheating mechanism.

16.3 Turbulent Clustering and Multipoint Radiation Induced Ignition

Historically, large-scale chemical explosions that involved a gas fuel (methane, propane, butane, etc.) mixed with the air were considered as unconfined dust explosions. Admittedly, large-scale coal mine explosions share traits with large vapor cloud incidents and may be considered as either partly or almost fully unconfined dust explosions. In real explosions, it is common that the pressure waves produced by the primary flame disturb dust on the ground ahead of the flame, so a dust explosion occurs when the dust particles are entrained and dispersed in the fuel–air. Conventional flame propagation mechanisms in a deflagration regime cannot cause high overpressures to explain the observed level of damage caused by dust explosions. Since detonation is the only established phenomenon that allows to create a sufficiently rapid burning accompanied by a large pressure jump to be sustained in open areas, there has been a general presumption [48–50] that deflagration-to-detonation transition (DDT) following turbulent flame acceleration is the mechanism explaining the destructive overpressures in dust explosions. Small particles, suspended in a flammable gas mixture ahead of the advancing flame and heated by thermal radiation, can enhance the flame speed and cause ignition of the surrounding fuel–air mixture. In unconfined (large scale) dust explosions, the radiative flux emitted from the advancing flame into the reactants is significantly enhanced by the increased emissivity of the large volume of burned products with hot solid particles. The forward radiative flux, emitted by hot combustion products in large gaseous explosions, is close to the blackbody radiation at the stoichiometric flame temperatures and can make a decisive contribution to the overall energy transport, structure and velocity of flames.

In a turbulent flow with suspended small particles with material density ρ_p greater than that of a gas density ρ_g, turbulent eddies, acting like small centrifuges, expel particles in the area between eddies, where the pressure fluctuations is maximal and the intensity of vortices is minimal. Thus, particles in a turbulent flow are assembled in small clusters with the size of about several Kolmogorov turbulent scales, $\ell_\eta = \ell_0/\mathrm{Re}^{3/4}$. This effect, known as inertial clustering, has been investigated in a number of analytical [51, 52], numerical [53, 54], and experimental [55–58] studies. It was shown in analytical studies [59, 60] and in laboratory experiments [61] that the particle clustering is significantly enhanced in the presence of a mean temperature gradient, so the turbulence is temperature stratified and the turbulent heat flux is not zero. This causes correlations between fluctuations of fluid temperature and velocity, and non-zero correlations between fluctuations of pressure and fluid velocity. As a result, the particle concentration in clusters may rise by a few orders of magnitude, as compared to the mean concentration of evenly dispersed particles.

In the case of uniformly dispersed particles, the intensity of the incident radiation decreases exponentially on the scale of the radiation absorption length, $L_a = 1/\langle\kappa\rangle$, where $\langle\kappa\rangle = \sigma_p \langle \mathrm{N}\rangle$ is the mean particle absorption coefficient, $\sigma_p \approx \pi d_p^2/4$ is the absorption cross section of the particle, d_p is the particle size (diameter), $\langle \mathrm{N}\rangle$ is the mean particle number density. The radiation absorption length for the dust cloud mass densities, $\rho_d \sim 0.01 - 0.05\,\mathrm{kg/m^3}$ typical for dust explosions, is in the range of (1–10) cm for the microns size particles $d_p \sim 1 - 10\,\mu\mathrm{m}$. Therefore, radiation-induced ignition of a fuel–air mixture by the radiatively heated particles is unlikely in case of the evenly dispersed particles. The situation is completely different when the particles are dispersed unevenly, for example, in the form of an optically thick dust layer separated from the flame by a transparent gaseous gap with a sufficiently low concentration of particles in the gap. If this radiation-transparent gap is sufficiently wide, the particles in the dust layer can be exposed to and heated by the radiation from the flame sufficiently long time to become ignition kernels, as it was discussed in Sect. 16.2.

There are two effects related to turbulent clustering of dust particles that affect the heat radiative transfer: (i) clustering of particles leads to the formation of radiation transparent windows between the clusters and (ii) particles inside optically thick clusters are screened from radiation by the particles on the cluster surface, which is equivalent to a decrease in the average number density of particles [62, 64]. If the effect of turbulent clustering considerably increases the radiation absorption length $L_{\mathrm{eff}} \gg L_a$, such that particles are sufficiently heated by the radiation to form ignition kernels in a large volume far ahead of the flame, this can lead to explosion of a large volume of fuel–air ahead of the flame. If the effective absorption length is large enough, then the explosion will be accompanied by a strong shock wave which can promote transition to detonation. The corresponding condition is $L_{\rm eff} \gg c_s \tau_{\rm ign}$, where c_s is the speed of sound in the flow ahead of the flame and $\tau_{\rm ign}$ is the timescale of fuel–air ignition by the radiatively heated ignition kernels.

16.3.1 Radiation Absorption Coefficient and Turbulent Clustering

To investigate how clustering of particles affects the radiation absorption length and how this effect depends on the turbulence parameters, we consider a turbulent flow with suspended particles exposed by a radiative flux. The radiative transfer equation for the radiation intensity $I(\hat{\mathbf{s}}, \mathbf{r})$ in a two-phase flow is (see, e.g. [41, 42]):

$$(\hat{\mathbf{s}} \cdot \nabla)\, I(\hat{\mathbf{s}}, \mathbf{r}) = -\big(\kappa_g(\mathbf{r}) + \kappa_p(\mathbf{r}) + \kappa_s(\mathbf{r})\big) I(\hat{\mathbf{s}}, \mathbf{r}) + \kappa_g I_g + \kappa_p I_p + \frac{\kappa_s}{4\pi} \int_\Omega \Phi(\mathbf{r}, \hat{\mathbf{s}}; \hat{\mathbf{s}}') I(\hat{\mathbf{s}}', \mathbf{r}) d\Omega' \tag{16.3.1}$$

where $\kappa_g(\mathbf{r})$ and $\kappa_p(\mathbf{r})$ are absorption coefficients for the gas phase and particles, $\kappa_s(\mathbf{r})$ is the scattering coefficient, Φ is the scattering phase function, $I_g(\mathbf{r})$ and $I_p(\mathbf{r})$ are the blackbody radiation intensities for the gas phase and for particles, respectively, $\hat{\mathbf{s}} = \mathbf{k}/k$ is the unit vector in the direction of radiation, $\mathbf{k}$ is the wave vector, $d\Omega = \sin\theta \, d\theta \, d\varphi$, and (k, θ, φ) are the spherical coordinates in $\mathbf{k}$ space.

Since the scattering and absorption cross sections for gases at normal conditions are very small, the contribution from the gas phase is negligible in comparison with that of particles. We also take into account that for $\sqrt{\pi}\, d_p \sqrt{\varepsilon}/\lambda_{\rm rad} \gg 1$, where ε is the dielectric permeability of the dust particles, $\lambda_{\rm rad}$ is the radiation wavelength, the scattering coefficient is negligible compared to the absorption coefficient of particles, $|\kappa_p(\mathbf{r})| \gg |\kappa_s(\mathbf{r})|$. Then, Eq. (16.3.1) is reduced to

$$(\hat{\mathbf{s}} \cdot \nabla) I(\hat{\mathbf{s}}, \mathbf{r}) = \kappa(\mathbf{r}) \big(I_b(\mathbf{r}) - I(\hat{\mathbf{s}}, \mathbf{r})\big), \tag{16.3.2}$$

where $\kappa \equiv \kappa_p(\mathbf{r})$ and $I_b \equiv I_{b,p}$ which depends on the local temperature, and for simplicity we have omitted the "p" subscript.

To derive the equation for the effective radiation absorption coefficient with account of particle clustering in a turbulent flow, we use the mean-field approach. Within the framework of this approach, all quantities are decomposed into mean and fluctuating parts: $I = \langle I \rangle + I'$, $I_b = \langle I_b \rangle + I_b'$, and $\kappa = \langle \kappa \rangle + \kappa'$. The fluctuating parts have zero mean values, and angular brackets denote averaging over an ensemble of fluctuations.

The radiation absorption length for evenly dispersed particles is $L_a = 1/\langle \kappa \rangle$, where $\langle \kappa \rangle = \sigma_p \langle N \rangle$ is the mean particle absorption coefficient, $\sigma_p \approx \pi d_p^2/4$, and $\langle N \rangle$ is the mean number density of uniformly dispersed particles. The instantaneous particle number density is $n = \langle N \rangle + n'$, with n' being fluctuations of the particle number density. The instantaneous particle absorption coefficient is $\kappa = \sigma_a n$, so that the fluctuations of the absorption coefficient are $\kappa' = n'\sigma_a = n' \langle \kappa \rangle / \langle N \rangle$. Averaging Eq. (16.3.2) over the ensemble of the particle number density fluctuations, we obtain the equation for the mean radiation intensity $\langle I(\hat{\mathbf{s}}, \mathbf{r}) \rangle$:

$$(\hat{\mathbf{s}} \cdot \nabla)\langle I(\hat{\mathbf{s}}, \mathbf{r})\rangle = \kappa(\langle I_b \rangle - \langle I \rangle) + \langle \kappa' I_b' \rangle - \langle \kappa' I' \rangle \tag{16.3.3}$$

Subtracting Eq. (16.3.3) from Eq. (16.3.2), we obtain the equation for fluctuations of I':

$$(\hat{\mathbf{s}} \cdot \nabla + \overline{\kappa} + \kappa')\, I'(\mathbf{r}, \hat{\mathbf{s}}) = I_{\text{source}}, \tag{16.3.4}$$

where $s = \mathbf{r} \cdot \hat{\mathbf{s}}$, and the source term is

$$I_{\text{source}} = \kappa'(\langle I_b \rangle - \langle I \rangle) - \langle \kappa' I_b' \rangle + \langle \kappa' I' \rangle + (\langle \kappa \rangle + \kappa')\, I_b'. \tag{16.3.5}$$

The solution of Eq. (16.3.5) is

$$I'(\mathbf{r}, \mathbf{s}) = \int_{-\infty}^{+\infty} I_{\text{source}} \exp\left(- \left| \int_{s'}^{s} [\langle \kappa \rangle + \kappa'(s'')] ds'' \right| \right) ds'. \tag{16.3.6}$$

Expanding the exponent under the integral in Eq. (16.3.6) in Taylor series over a small parameter, $\kappa' \ell_\eta / Sc^{1/2} \ll 1$, multiplying Eq. (16.3.6) by κ' and averaging over the ensemble of fluctuations, we obtain using the quasi-linear approach [61] and neglecting by small third and higher moments in fluctuations of κ' due to a small parameter in the theory, $\kappa' \ell_\eta / Sc^{1/2} \ll 1$, as well neglecting by small correlation between the particle number density fluctuations and the gas temperature fluctuations, which implies that $\langle \kappa' I_b' \rangle = 0$, we obtain the one-point correlation function $\langle \kappa' I' \rangle$ in the form:

$$\langle \kappa' I' \rangle = -\langle \kappa \rangle \big(\overline{I} - \overline{I}_b\big) \frac{2\beta J_1}{1 + 2\beta J_2}, \tag{16.3.7}$$

where $Sc = \nu / D$ is the Schmidt number$\nu$is the kinematic viscosity, and D is the coefficient of the Brownian diffusion of the particles, and values J_1 and J_2 in Eq. (16.3.7) are expressed by integrals:

$$J_1 = \int_0^{+\infty} \Phi(Z) \exp(-\beta Z)\, dZ, \tag{16.3.8}$$

$$J_2 = \beta \int_0^{+\infty} \left(\int_0^{Z} \Phi(Z') dZ' \right) \exp(-\beta Z)\, dZ, \tag{16.3.9}$$

where $Z = \left|s - s'\right|/\ell_D$, $\beta = \langle\kappa\rangle\,\ell_D$, $\ell_D = a\,\ell_\eta/Sc^{1/2}$, $\Phi(t, \mathbf{R}) = \langle n'(t, \mathbf{x})n'(t, \mathbf{x}+\mathbf{R})\rangle$ is the two-point instantaneous correlation function of the particle number density fluctuations, the numerical factor $a \gg 1$. Substituting the correlation function $\langle\kappa'\,I'\rangle$, determined by Eq. (16.3.7) into Eq. (16.3.3), we obtain equation for the mean radiation intensity:

$$(\hat{\mathbf{s}}\cdot\nabla)\langle I(\hat{\mathbf{s}}, \mathbf{r})\rangle = \kappa_{eff}(\langle I_b\rangle - \langle I\rangle). \tag{16.3.10}$$

where κ_{eff} is the effective absorption coefficient, which takes into account the particle clustering in a temperature stratified turbulence, is given by expression

$$\kappa_{eff} = \langle\kappa\rangle\left(1 - \frac{2\beta J_1}{1 + 2\beta J_2}\right). \tag{16.3.11}$$

16.3.2 Effect of Turbulent Clustering on Radiation Heat Transfer

The integrals J_1 and J_2 in formula (16.3.11) can be calculated using the normalized two-point correlation function of the particle number density fluctuations, $\Phi(R) = \langle n'(\mathbf{r})n'(\mathbf{r}+\mathbf{R})\rangle/\langle N\rangle^2$. The two-point correlation function that accounts for particle clustering in temperature stratified turbulence was derived in [61, 62], where a reader can find details of calculations. The two-point correlation function can be written in the following form:

$$\Phi(R) = (n_{cl}/\langle N\rangle)^2 \;\; for\;\; 0 \le R \le \ell_D, \tag{16.3.12a}$$

$$\Phi(R) = \left(\frac{n_{cl}}{\langle N\rangle}\right)^2\left(\frac{R}{\ell_D}\right)^{-\mu}\cos\left(\alpha\ln\frac{R}{\ell_D}\right) \quad \text{for } \ell_D \le R < \infty. \tag{16.3.12b}$$

The two-point correlation function (16.3.12a and 16.3.12b) determines the maximum increase of the particle number density inside a cluster, where $\ell_D = a\,\ell_\eta/Sc^{1/2}$, $R = \mathbf{R}\cdot\hat{\mathbf{s}}$, and

$$\mu = \frac{1}{2\,(1 + 3\sigma_T)}\left[3 - \sigma_T + \frac{20\,\sigma_v(1 + \sigma_T)}{1 + \sigma_v}\right], \tag{16.3.13}$$

$$\alpha = \frac{3\pi\,(1 + \sigma_T)}{(1 + 3\sigma_T)\,\ln Sc}. \tag{16.3.14}$$

Here, $\sigma_T = (\sigma_{T0}^2 + \sigma_v^2)^{1/2}$ is the degree of compressibility of the turbulent diffusion tensor, and $\sigma_{T0} = \sigma_T(St = 0)$ is the degree of compressibility of the particle velocity field:

$$\sigma_v = \frac{8St^2K^2}{3(1+bSt^2K^2)}, \quad K^2 = \left[1 + \mathrm{Re}\cdot\left(\frac{L_*\nabla\langle T\rangle}{\langle T\rangle}\right)^2\right]^{1/2}, \tag{16.3.15}$$

where $Re = \ell_0 u_0/\nu$ is the Reynolds number, $\langle T\rangle$ is the mean fluid temperature, the length $L_* = c_s^2 \tau_\eta^{3/2}/9\nu^{1/2}$ and a constant b determines the value of the parameter σ_v in the limit $St^2K^2 \gg 1$, c_s is the speed of sound.

The particle inertia is characterized by the Stokes number, $St = \tau_p/\tau_\eta$, where $\tau_p = m_p/(3\pi\,\rho\,\nu\,d_p)$ is the particle Stokes time, $\tau_\eta = \tau_0/\mathrm{Re}^{1/2}$ is the Kolmogorov turbulent turnover time, $\tau_0 = \ell_0/u_0$ and u_0 are the characteristic time and velocity at the turbulent integral scale, ℓ_0, respectively. The size of a cluster is given by $\ell_D = a\ell_\eta/Sc^{1/2}$. The value of the numerical factor $a \approx 5-10$ corresponds to the values obtained in the laboratory experiments [61] and the atmospheric measurements [62, 66]. The parameter b enters in the expression for the degree of compressibility of particle velocity field, σ_v. Its value is $\sigma_v = 8/3b$ in the limit of large temperature gradients, $St^2K^2 \gg 1$. Since the degree of compressibility of particle velocity field $\sigma_v \leq 1$, the value of the coefficient b must be $b \geq 8/3$. On the other hand, when b is very large ($b \gg 1$), $\sigma_v \to 0$, i.e. the role of stratification of turbulence or particle inertia is negligible, which implies the absence of the turbulent clustering. Therefore, it is reasonable to choose the value of b in the range $5 \leq b \leq 10$. In fact, the analysis shows that the final result for the effective radiation penetration length, $L_{\mathrm{eff}} = 1/\kappa_{\mathrm{eff}}$, is not very sensitive to the particular value of b.

In the analytical study, we can neglect the third and higher moments in κ' in the correlation function $\langle\kappa' I'\rangle$, due to a small parameter $\kappa'\ell_\eta/Sc^{1/2} \ll 1$. This condition is satisfied since for inertial particles, $Sc = \nu/D \gg 1$. However, to study the turbulent clustering of particles we consider higher moments in the particle number density fluctuations, which contribute to the effective penetration length of radiation. The higher moments in κ' (i.e. in fluctuations of particle number density n') are important only within the particle clusters, while outside the clusters the higher moments in κ' are much smaller than the second moments.

The growth of the particle number density inside the cluster is limited by nonlinear effects. An increase in the particle number density inside the cluster can be limited by the depletion of particles in the surrounding flow caused by their accumulation inside the cluster. It was shown in [59, 60] that the maximum number density of particles inside the cluster is limited by the depletion of particles outside the clusters, so that

$$\frac{n_{cl}}{\langle N\rangle} = \left(1 + \frac{e\mu}{\pi} Sc^{\mu/2} \ln Sc\right)^{1/2}. \tag{16.3.16}$$

Another effect, that restricts the growth of the particle number density inside the cluster, is related to a momentum coupling of particles and turbulent flow, when kinetic energies of particles and turbulence become close. Numerical modeling shows that this effect becomes important when the mass loading parameter (the ratio of the

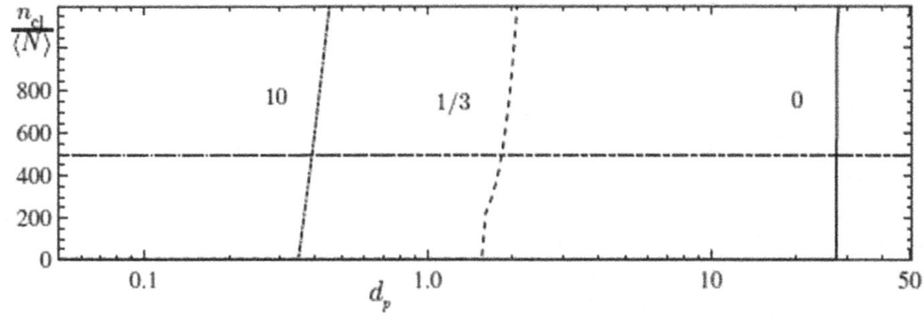

Fig. 16.12 The particle number density, $n_{cl}/\langle N \rangle$, within the cluster normalized by the mean particle number density versus the particle diameter d_p for different temperature gradients $|\nabla\langle T \rangle| = 10$ K/m (dashed-dotted), 1/3 K/m (dashed) and 0 K/m (solid). The inclined lines are related to the saturation mechanism due to the particle depletion effect. The horizontal line corresponds to the second saturation mechanism caused by the momentum coupling of particles and turbulent flow with the mass loading parameter $m_p n_{cl}/\rho \approx 1/2$, from [63]

particle mass density inside the cluster to the gas density) is of the order of unity, $m_p n_{cl}/\rho \sim 1$.

Using Eqs. (16.3.12)–(16.3.15) with account that $\beta = \langle \kappa \rangle \ell_D \ll 1$, the integrals J_1 and J_2 can be presented in the following form:

$$J_1 = J_2 = \left(\frac{n_{cl}}{\langle N \rangle}\right)^2 \left[1 + \frac{\mu - 1}{(\mu - 1)^2 + \alpha^2}\right] \tag{16.3.17}$$

Since the main contributions to the integrals J_1 and J_2 are from the size of the cluster, $\ell_D = a\ell_\eta/Sc^{1/2} \propto \nu^{3/4}$, these integrals depend on the kinematic viscosity ν.

The calculated maximum particle number density within the cluster normalized by the mean particle number density $n_{cl}/\langle N \rangle$ versus the particle diameter d_p for different temperature gradients $|\nabla\langle T \rangle| = 10$ K/m; 1/3 K/m; 0 K/m is shown in Fig. 16.12. The inclined lines in the figure are related to the first saturation mechanism due to the depletion of particles outside the cluster. The horizontal line corresponds to the second saturation mechanism caused by the momentum coupling of particles and turbulent flow with the mass loading parameter $m_p n_{cl}/\rho \approx 1/2$.

16.3.3 Turbulent Clustering: Effective Radiation Absorption Length

Substituting Eqs. (16.3.17) and (16.3.11) for the effective absorption coefficient κ_{eff}, we obtain the expression for the effective absorption (penetration) length of radiation, $L_{\rm eff} \equiv 1/\kappa_{\rm eff}$. The effective absorption length radiation normalized to the radiation absorption length, $L_a = 1/\sigma_p \langle {\rm N} \rangle$ for uniformly dispersed particles, is

$$L_{\text{eff}}/L_a = 1 + \frac{2a}{Sc^{1/2}} \left(\frac{n_{cl}}{\langle N \rangle} \right)^2 \left(\frac{\ell_\eta}{L_a} \right) \left(1 + \frac{\mu - 1}{(\mu - 1)^2 + \alpha^2} \right). \tag{16.3.18}$$

As mentioned earlier, there are two physical effects associated with particle clustering that affect the radiation transfer: (i) the formation of transparent for radiation windows between particle clusters due to the particle depletion and (ii) the effect of radiation screening of particles inside the cluster in the case of optically thick clusters. Formula (16.3.18) considers only the collective effect of turbulent clustering on the radiation transfer—the effect of the formation of transparent for radiation windows between particle clusters, the screening effect of optically thick clusters cannot be calculated in the present model. Therefore, the obtained results can be viewed as a lower limit for the increase in the penetration length of radiation, L_{eff}, whereas the overall effect may be much stronger.

We first consider turbulence in the absence of the mean temperature gradient (isothermal turbulence). For isothermal turbulence, the particle clustering instability occurs only for relatively large dust particles, when the particle diameter $d_p > 25\,\mu\text{m}$ [51, 52, 59].

The calculated ratio L_{eff}/L_a depending on the particle size is shown in the top panel in Fig. 16.13 and depending on the Stokes number (bottom panel) for turbulent clustering of particles at different Reynolds numbers and for the following parameters of the turbulent flow (methane–air): $\nu = 0.2\,\text{cm}^2/s$, $c_s = 450\,\text{cm/s}$, $n_{cl}/\langle N \rangle = 500$, $\sigma_{T0} = 1/2$. It is seen that the effective penetration length of radiation in isothermal turbulence increases due to the turbulent clustering at least by one order of magnitude for the particles with $d_p > 25\,\mu\text{m}$.

16.3.4 Radiation-Induced Multipoint Secondary Explosions

According to Eq. (16.3.18) the ratio L_{eff}/L_a depends strongly on $(n_{cl}/\langle N \rangle)^2$. It was shown in [59, 60] that for temperature stratified turbulence the ratio $n_{cl}/\langle N \rangle$ can increase up to three orders of magnitude. In addition, according to [59] the rate of formation of particle clusters for a temperature stratified turbulence is much larger (by almost an order of magnitude) than for isothermal turbulence. On the contrary, in the case of isothermal turbulence the rate of particle clustering is rather low. Consequently, under isothermal turbulence, the density of particles in clusters is unlikely to increase significantly. In this case, clusters either do not survive for a long time, or they are absorbed by the advancing flame long before the density of particles in clusters can increase significantly.

An inhomogeneous temperature distribution and the formation of temperature-stratified turbulence ahead of the advancing flame are typical in dust explosions. Therefore, it is natural to assume that in dust explosions the gaseous mixture temperature in the turbulent flow ahead of the primary flame is non-uniform. As a representative example, we choose the mean temperature gradient to be in the range 0.5–3 K/m. Figure 16.14a shows the ratio L_{eff}/L_a depending on particle size and the

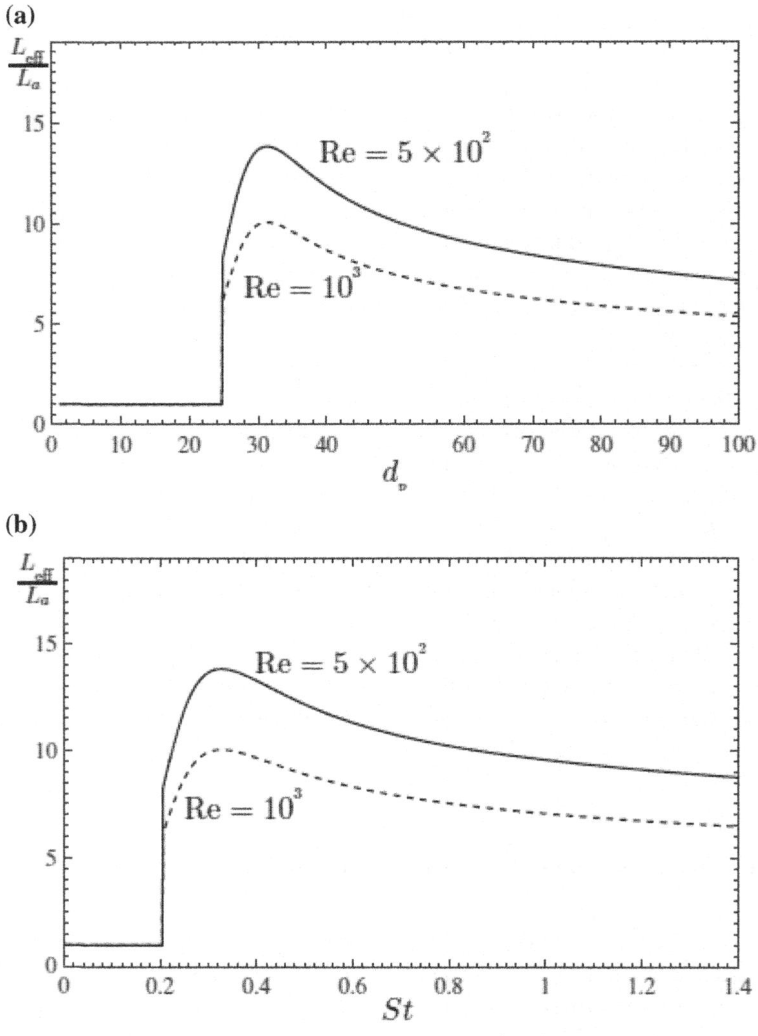

Fig. 16.13 The ratio L_{eff}/L_a versus particle size (top panel) and Stokes number (bottom panel) for the particle clustering in isothermal turbulence with different Reynolds numbers: $\mathrm{Re} = 5 \cdot 10^2$ (solid curve) and $\mathrm{Re} = 10^3$ (dashed curve), from [63]

dependence on the Stokes number in Fig. 16.14b for different values of the mean temperature gradient. In these figures, the following parameters typical for turbulent flow in unconfined dust explosions were used: $\mathrm{Re} = 5 \cdot 10^4$, $\ell_0 = 1\,\mathrm{m}$, $u_0 = 1\,\mathrm{m/s}$, $\nu = 0.2\,\mathrm{cm/s}^2$, $c_s = 450\,\mathrm{m/s}$, $n_{cl}/\langle N \rangle = 500$, $\sigma_{T0} = 1/2$.

As seen in Fig. 16.14, the increase in the effective radiation penetration length for temperature stratified turbulence is much larger than for isothermal turbulence: the value of L_{eff} increases up to 2–3 orders of magnitude compared with L_a.

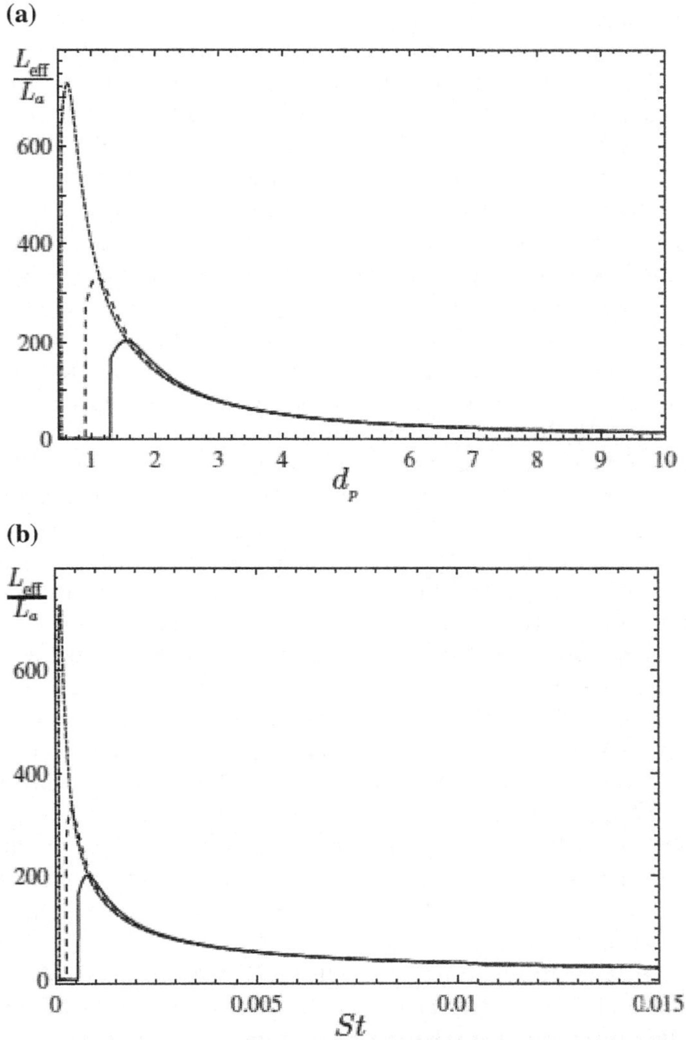

Fig. 16.14 **a** The ratio L_{eff}/L_a depending on the particle sizes for the mean temperature gradients: 0.5 K/m (solid), 1 K/m (dashed), 3 K/m (dashed-dotted). The particle diameter is in microns, from [63]. **b** L_{eff}/L_a vs the Stokes number for the conditions of (**a**), from [63]

During the early stage of dust explosions, the accidentally ignited subsonic flame accelerates due to hydrodynamic instabilities and interaction with the turbulent flow. The pressure waves, produced by the accelerating flame, run away enhancing turbulence ahead of the flame. With an increase of the primary flame surface and the flame velocity, the parameters of the turbulent flow ahead of the flame, u_0, ℓ_0, Re, St, $\nabla\langle T\rangle$, change continuously, and dust particles in a turbulent flow are assembled into clusters with the characteristic time of the order of milliseconds. According to our analysis, the turbulent clustering of particles in the temperature stratified turbulent flow ahead of the primary flame may increase the radiation penetration length up to 2–3 orders. This effect ensures that clusters of particles are exposed to and heated by the radiation from the primary flame for a sufficiently long time to become ignition kernels in a large volume ahead of the flame. A multi-point radiation-induced ignition of the fuel–air mixture around the clusters effectively increases the total area of the flame, so the distance that each piece of the combustion must travel to completely burn out the fuel is significantly reduced. It results in a strong increase of the effective combustion speed, defined as the rate of reactant consumption of a given volume. If, for example, the radiation absorption length for uniformly dispersed particles with mass density 0.03 kg/m^3 is in the range of several centimeters, then the effect of clustering of dust particles can increase the radiation absorption length up to $L_{\mathrm{eff}} = 10-20$ m. In this case, the clusters of particles will be heated by radiation at distances up to 10–20 m ahead of the primary flame.

It should be emphasized that Moore and Weinberg were the first [65, 66] who proposed a radiation-based ignition, assuming that radiation from hot combustion products in large gaseous explosions can cause multi-point ignition ahead of the flame due to ignition of the gaseous mixture by radiatively heated suspended particles. Later, they [67] also demonstrated ignition of a propane–air mixture using a uniform radiation flux of intensity up to 400 kW/m^2 from a CO_2 laser.

To implement radiation-induced ignition, it is necessary to consider two different time scales. The first is the time of the primary flame arrival at the radiation induced ignition location, which must be longer than the time required for the irradiated suspended particles to be heated enough to cause ignition at this location. The experimentally measured [13, 68] ignition time of the fuel–air heated by the radiatively particles is 100 ms for 10 μm nonreactive particles. The level of thermal radiation of $S \approx 300-400\,\mathrm{kW/m^2}$ radiated from hot combustion products in dust explosions is sufficient to rise the temperature of particles by $\Delta T_p \approx 1000$ K during $\tau_T \approx d_p\, \rho_p\, c_p\, \Delta T_p/S < 10$ ms, where c_p is the particle specific heat. The interphase (particle-gas) energy exchange time is $\tau_{pg} = d_p^2\, \rho_p\, c_p/6\, \lambda\, Nu \le 1$ ms, where $\lambda = \chi \rho_g c_p$ is the coefficient of gas phase thermal conduction, and $Nu \approx 2$ is the Nusselt number [43]. It is known that for ignition to occur, a volume of fuel–oxygen mixture of the order of the flame thickness must be heated above the ignition temperature in order to ignite a self-sustained combustion. This means that for an isolated particle it takes a relatively long time, $t_\chi \sim L_f^2/\chi \simeq 500$ ms for methane/air mixture at normal conditions. For typical parameters of dust explosions, the Kolmogorov viscous time scale is $\tau_\eta \sim 5 \times 10^{-3}$ s, and of the same order is the lifetime of turbulent clusters. Since the lifetime of clusters is larger than the ignition time, the estimates in previous section describe satisfactory the origin and mechanism of the secondary explosion.

Numerical simulations of fuel–air ignition by a radiatively heated cluster of particles show that the ignition time scales decrease approximately as $1/n_{cl}^{1/3}$, with the increase of the particle number density n_{cl} in the cluster. In this case, the radiation-heated particles ignite the fuel–air mixture in a much shorter time, since the gas mixture is heated by thermal conductivity at a short distance corresponding to the distance $\Delta r \approx 1/n_{cl}^{1/3}$ between the particles in the cluster. Analysis shows that the ignition time of the mixture by clusters of radiation-heated particles is about 10 ms. Thus, the effective propagating rate of combustion caused by the multipoint radiation induced ignition can be estimated as $L_{\text{eff}}/\tau_{\text{ign}} \sim 10^3$ m/s. Such rate of the "secondary explosion" corresponds to the intensity of a shock wave with Mach number $Ma = 2{-}3$, producing overpressures of 8–10 atm.

The multipoint explosions caused by thermal radiation induced ignition of small particles in a turbulent flow was demonstrated experimentally by Li and Lindstedt [69] using a near infrared radiation of the CW laser as the radiation source with very fine acetylene black particles of size $d_p \approx 0.035$ μm used as targets. The measured ignition time was found in the range of 8–10 ms. The experimental study [69] has shown that the impact of thermal radiation in large vapor cloud explosions can cause the formation of secondary explosions, as it was proposed in theoretical studies [62, 63].

16.3.5 Impact of Radiation in Vapor Cloud and Dust Explosions

In the majority of large-scale vapor cloud explosions (VCEs), there is clear forensic evidence that a severe explosion propagated into open uncongested areas. This is a feature of all large-scale vapor cloud incidents for which detailed primary evidence is available, e.g. the 2005 Buncefield fuel storage depot explosion. Unconfined dust explosions and VCEs start with a primary accidentally ignited flame propagating in the deflagration regime. In the majority of dust explosions, the primary flame becomes turbulent due to obstacles such as bushes, trees, buildings in the flame propagation path, and this can enhance the flame propagation speed significantly. Large-scale accidental industrial explosions have disastrous consequences for human life and property. When such explosions occur, industry carries out detailed investigations to determine the root cause of the accident. This information is then used to design facilities or create procedures that will minimize the chance of recurrence. Currently, most prevention and safety methods rely on empirical correlations. Although the technology for preventing dust explosions has progressed considerably, recurring accidents in the mining industry and fuel storage facilities call for a fundamentally improved understanding of the mechanism of flame propagation and the origin of high overpressures in these explosions. Advances in knowledge and understanding of the large-scale VCEs and dust explosions can be found in numerous review papers and books [48–50, 70–77].

Most if not all discussions on the mechanism of large explosions, VCEs or Dust Explosions are based on studies of confined DDT. VCEs, however, should be considered as unconfined, even if there is ground below. When analyzing catastrophic large-scale explosions, the biggest problem for explosion scientists and process safety engineers is an explanation of how the explosion developed and how the associated overpressures could give rise to some specific localized damage recorded by the original incident investigators. The answers to these questions are closely related to understanding the fundamental combustion problems: what are the mechanisms of flame acceleration and what is the mechanism of the deflagration-to-detonation transition (DTT) in the worst-case scenario. The current conventional explanation is that a developed detonation following flame acceleration which is caused by turbulence developing in the gaseous mixture ahead of the flame, followed by the flame turbulization. These conclusions were made on the basis of explosive tests in a partially confined space, as well as laboratory experimental studies of DDT in channels with obstacles and corresponding predominantly 2D numerical calculations. However, turbulence and a turbulent flame acceleration observed in simulations is essentially influenced by the flame interaction with the channel walls. The impact of obstacles or mechanical obstructions in the case of a confined flame is considerably different from the case of an unconfined flame. Applying it to trees, bushes, and buildings is taking the model beyond its validation range. Therefore, the models obtained from numerical simulations and laboratory experiments should be used with great care as a predictive tool for large-scale unconfined explosions, and any conclusion would be speculative at best. The process is scale dependent, and therefore small laboratory scale studies may not reveal important features of the phenomenon. In particular, it was shown in [78] that the impact of thermal radiation losses, which is usually considered to be negligible either in laboratory scale experimental studies or simulations, can become significant with flame acceleration and corresponding pressure increase, having noticeable impact on the gas dynamics and run-up distance. An important point is that DDT mechanisms have microscopic scales, which are many orders of magnitude smaller than the scale of explosions. This disparity in scales creates serious and almost insurmountable difficulties for numerical simulations of a real large-scale explosion. In general, any mechanism of DDT consists of the coupling of a shock wave and a reaction wave with subsequent mutual amplification. In the case of flame propagation in a pipe, the coupling of the flame with the shock wave occurs due to the peculiarity of the flame dynamics in the tube, when a shock wave is formed in the immediate vicinity of the flame or when the flame is accelerated in the pipe so that it can catch up with the leading shock wave. It should be noted that in the latter case the flame speed does not have to reach the speed of sound in the mixture, or about half of the Chapman–Jouguet detonation velocity, as is often believed. The condition for the flame coupling with the shock wave requires that the flame move at a speed equal to the flow velocity behind the shock wave.

In the case of unconfined explosions, the accelerating turbulent flame generates pressure waves that eventually steepen into a shock wave ahead of the flame. However, it is not clear how the coupling between the flame (reaction wave) and the shock occurs. The large-scale experiments have shown the importance of certain congestion. For instance, no DDT occurred in the congestion generated by pine trees, but it occurred in the congestion generated by deciduous trees in a propane–air mixture. Nevertheless, the Buncefield Major Incident Investigation Board concluded that the explosion involved a deflagration to detonation transition, with the detonation propagating through the whole vapor cloud area. As a result, much effort has been expended on the study of flame acceleration and the ability of an accelerating flame to induce the transition to detonation. Conventional explosion prediction methods cannot explain how the explosion developed and how the associated overpressures could give rise to some specific localized damage recorded by the incident investigators. A multsipoint radiation induced ignition may be a potential scenario that could explain the high localized overpressures and development a detonation caused by the coupling between the reflected shock produced in the secondary, radiation-induced explosion and the primary flame.

2D numerical simulations were performed in [79–81] to study the influence of thermal radiation on the propagation and structure of layered coal dust explosions. The simulations used global irreversible reactions for devolatilization, char burning, and methane combustion. It was found that the effect of radiation heat transfer on the structure and propagation of coal dust layered explosions can have different effects for different scenarios depending on the volume fraction of the dust layer, as well as the modeling choice of the chemical reaction model, the radiation absorption and scattering models for particles and for gas.

References

1. R.K. Eckhoff, *Dust Explosions in the Process Industries*, 3rd ed. (Gulf Professional Publishing, 2003).
2. R.K. Eckhoff, Current status and expected future trends in dust explosion research. J. Loss Prev. Process Ind. **18**, 225–237 (2005)
3. C. Proust, A few fundamental aspects about ignition and flame propagation in dust clouds. J. Loss Prev. Process Ind. **19**, 104–120 (2006)
4. R.K. Eckhoff, Ignition of combustible dust clouds by strong capacitive electric sparks of short discharge time. Zeitschrift Fur Physikalische Chemie **231**, 1683–1707 (2017)
5. D.R. Ballal, Ignition and flame quenching of quiescent dust clouds. Proceed. Royal Soc. London A **369**, 479–500 (1980)
6. T.F. Smith, K.-H. Byun, L.-D. Chen, Effects of radiative and conductive transfer on thermal ignition. Combust. Flame **73**, 67–74 (1988)
7. S.W. Baek, Ignition of combustible gases by radiative heating of inert particles. Combust. Flame **97**, 418–422 (1994)
8. S.W. Baek, Ignition of particle suspensions in slab geometry. Combust. Flame **81**, 366–377 (1990)
9. R. Blouquin, G. Joulin, Ignition of an absorbing/emitting medium by an imposed radiant heat flux. Combust. Sci. Technol. **113**, 109–135 (1996)
10. P.C. Hills, D.K. Zhang, P.J. Samson, T.F. Wall, Laser ignition of combustible gases by radiative heating of small particles. Combust. Flame **91**, 399–412 (1992)
11. T. Sloane, Energy requirement for spherical ignitions in methane–air mixtures at different equivalence ratio. Combust. Sci. Techn. **73**, 351–365 (1990)
12. Ch. Proust, Laser ignition of dust clouds. J. Phys. IV **12**, 79–88 (2002)
13. F. Beyrau, M.A. Hadjipanayis, R.P. Lindstedt, Ignition of fuel/air mixtures by radiatively heated particles. Proc. Combust. Inst. **34**, 2065–2072 (2013)
14. J. Ma, C. Zhang, C. Wang, W. Han, Ignition of methane/air mixtures with inert and reactive particles heated by radiation. Fuel **278**, 118309 (1–7) (2020)
15. N.N. Smirnov, V.F. Nikitin, J.C. Legros, Ignition and combustion of turbulized dust-air mixtures. Combust. Flame **123**, 46–67 (2000)

16. E.K. Addai, D. Gabel, U. Krause, Experimental investigation on the minimum ignition temperature of hybrid mixtures of dusts and gases or solvents. J. Hazardous Mater. **301**, 314–326 (2016)
17. S. Choi, C.H. Kruger, Modeling coal particle behavior under simultaneous devolatilization and combustion. Combust. Flame **61**, 131–144 (1985)
18. G. Joulin, B. Deshaies, On radiation-affected flame propagation in gaseous mixtures seeded with inert particles. Combust. Sci. Techn. **47**, 299–315 (1986)
19. G. Joulin, B. Deshaies, Radiative transfer as a propagation mechanism for rich flames of reactive suspensions. SIAM J. Appl. Math. **46**, 561–581 (1986)
20. G. Joulin, M. Eudier, Radiation-dominated propagation and extinction of slow, particle-laden gaseous flames. Symp. (Int.) Combust. **22**, 1579–1585 (1989)
21. K. Seshadari, A.L. Berlad, V. Tangirala, The structure of premixed particle-cloud flames. Combust. Flame **89**, 333–342 (1992)
22. S.W. Baek, C. Lee, Heat transfer in a radiating medium between flame and fuel surface. Combust. Flame **75**, 153–163 (1989)
23. R. Blouquin, G. Joulin, Radiation-affected hydrodynamic instability of particle-laden flames. Combust. Sci. Technol. **110–111**, 341–359 (1995)
24. R. Blouquin, G. Joulin, On the influence of gradient transports upon the burning velocities of dust-flames. Combust. Sci. Technol. **115**, 355–367 (1996)
25. R. Blouquin, P. Cambray, G. Joulin, Radiation-affected dynamics of unconfined spherical flames in particle laden premixtures. Combust. Sci. Technol. **104**, 55–73 (1995)
26. R. Blouquin, G. Joulin, Y. Merhariz, Combustion regimes of particle-laden gaseous flames: influences of radiation, molecular transports, kinetic-quenching, stoichiometry. Combust. Theory Modelling **1**, 217–242 (1997)
27. J.L. Chen, R. Dobashi, T. Hirano, Mechanisms of flame propagation through combustible particle clouds. J. Loss Prev. Process Ind. **9**, 225–229 (1996)
28. F.N. Egolfopoulos, C.S. Campbell, Dynamics and structure of dusty reacting flows: Inert particles in strained, laminar, premixed flames. Combust. Flame **117**, 206–226 (1999)
29. S.R. Rockwell, A.S. Rangwala, Influence of coal dust on premixed turbulent methane–air flames. Combust. Flame **160**, 635–640 (2013)
30. C.R. Kaplan, S.W. Baek, E.S. Oran, J.L. Ellzey, Dynamics of a strongly radiating unsteady ethylene jet diffusion flame. Combust. Flame **96**, 1–21 (1994)
31. S.W. Baek, J.H. Park, C.E. Choi, Investigation of droplet combustion with nongray gas radiation effects. Combust. Sci. Technol. **142**, 55–79 (1999)
32. J.H. Sun, R. Dobashi, T. Hirano, Structure of flames propagating through aluminum particles cloud and combustion process of particles. J. Loss Prev. Process Ind. **19**, 769–773 (2000)
33. L. Qiao, J. Xu, Detailed numerical simulations of flame propagation in coal-dust clouds. Combust. Theory Model. **16**, 747–773 (2012)
34. W. Cao, W. Gao, J. Liang, S. Xu, F. Pan, Flame-propagation behavior and a dynamic model for the thermal-radiation effects in coal-dust explosions. J. Loss Prev. Process Ind. **29**, 65–71 (2014)
35. W. Gao, T. Mogi, J. Yu, X. Yan, J. Sun, R. Dobashi, Flame propagation mechanisms in dust explosions. J. Loss Prev. Process Ind. **36**, 186–194 (2015)
36. K. Hüttenbrenner, H. Kern, S. Tomasch, H. Raupenstrauch, Effects of inert particulate additives on ignition and flame propagation in dust-air mixtures. Combust. Explos. Shock Waves **56**, 427–434 (2020)
37. M. Bidabadi, S. Zadsirjan, S.A. Mostafavi, Radiation heat transfer in transient dust cloud flame propagation. J. Loss Prevent. Process Ind. **26**, 862–868 (2013)
38. R. Dobashi, K. Senda, Detailed analysis of flame propagation during dust explosions by UV band observations. J. Loss Prevent. Process Ind. **19**, 149–153 (2006)
39. A.D. Benedetto, A.G. Agreda, O. Dufaud, I. Khalili, R. Sanchirico, N. Cuervo, L. Perrin, P. Russo, Flame propagation of dust and gas-air mixtures in a tube, in *Proceedings of 7th Mediterranean Combustion Symposium, Sardinia, Italy* (2011)

40. R.H. Essenhigh, J. Csaba, The thermal radiation theory for plane flame propagation in coal dust clouds. Symp. Int. Combust. **9**, 111–125 (1963)
41. R. Siegel, J. Howell, *Thermal Radiation Heat Transfer* (Taylor & Francis, New York, New York, 2002)
42. Ya.B. Zel'dovich, Yu.P. Raizer, *Physics of Shock waves and High-Temperature Hydrodynamic Phenonema* (Academic Press, New-York-London, 1966)
43. A. Acrivos, T.D. Taylor, Heat and mass transfer from single spheres in Stokes flow. Phys. Fluids. **5**, 387–394 (1962)
44. M.A. Liberman, M.F. Ivanov, A.D. Kiverin, Radiation heat transfer in particle-laden gaseous flame: flame acceleration and triggering detonation. Acta Astronaut. **115**, 82–93 (2015)
45. M.F. Ivanov, A.D. Kiverin, M.A. Liberman, Ignition of deflagration and detonation ahead of the flame due to radiative preheating of suspended micro particles. Combust. Flame **162**, 3612–3621 (2015)
46. M.A. Liberman, M.F. Ivanov, A.D. Kiverin, Effects of thermal radiation heat transfer on flame acceleration and transition to detonation in particle-cloud hydrogen flames. J. Loss Prev. Process Ind. **38**, 176–186 (2015)
47. M.A. Liberman, A.D. Kiverin. M.F. Ivanov, Regimes of chemical reaction waves initiated by nonuniform initial conditions for detailed chemical reaction models. Phys. Rev. E **85**, 056312 (1–12) (2012)
48. D. Bradley, G. Chamberlain, D. Drysdale, Large vapor cloud explosions, with particular reference to that at Buncefield. Phil. Trans. R. Soc. A **370**, 544–566 (2012)
49. A. Pekalski, J. Puttock, S. Chynoweth, Deflagration to detonation transition in a vapor cloud explosion in open but congested space: large scale test. J. Loss Prev. Process Ind. **36**, 365–370 (2015)
50. E.S. Oran, Understanding explosions – from catastrophic accidents to the creation of the universe. Proc. Combust. Inst. **35**, 1–35 (2015)
51. T. Elperin, N. Kleeorin, I. Rogachevskii, Self-excitation of fluctuations of inertial particle concentration in turbulent fluid flow. Phys. Rev. Lett. **77**, 5373–5376 (1996)
52. T. Elperin, N. Kleeorin, M.A. Liberman, V.S. L'vov, I. Rogachevskii, Clustering of aerosols in atmospheric turbulent flow. Environm. Fluid Mech. **7**, 173–193 (2007)
53. J. Bec, L. Biferale, M. Cencini, A. Lanotte, S. Musacchio, F. Toschi, Heavy particle concentration in turbulence at dissipative and inertial scales. Phys. Rev. Lett. **98**, 084502 (1–4) (2007)
54. F. Toschi, E. Bodenschatz, Lagrangian properties of particles in turbulence. Ann. Rev. Fluid Mech. **41**, 375–404 (2009)
55. Z. Warhaft, Why we need experiments at high Reynolds numbers. Fluid Dyn. Res. **41**, 021401 (1–10) (2009)
56. H. Xu, E. Bodenschatz, Motion of inertial particles with size larger than Kolmogorov scale in turbulent flows. Physica D **237**, 2095–2100 (2008)
57. S. Balachandar, J.K. Eaton, Turbulent dispersed multiphase flow. Ann. Rev. Fluid Mech. **42**, 111–133 (2010)
58. E.-W. Saw, G.P. Bewley, E. Bodenschatz, S. Sankar Ray, J. Bec, Extreme fluctuations of the relative velocities between droplets in turbulent airflow. Phys. Fluids **26**, 111702 (2014)
59. T. Elperin, N. Kleeorin, M. Liberman, I. Rogachevskii, Tangling clustering instability for small particles in temperature stratified turbulence. Phys. Fluids **25**, 085104 (1–18) (2013)
60. T. Elperin, N. Kleeorin, B. Krasovitov, M. Kulmala, M. Liberman, I. Rogachevskii, S. Zilitinkevich, Acceleration of raindrop formation due to the tangling-clustering instability in a turbulent stratified atmosphere. Phys. Rev. E **92**, 013012 (1–11) (2015)
61. A. Eidelman, T. Elperin, N. Kleeorin, B. Melnik, I. Rogachevskii, Tangling clustering of inertial particles in stably stratified turbulence. Phys. Rev. E **81**, 056313 (1–10) (2010)
62. M. Liberman, N. Kleeorin, I. Rogachevskii, N.-E.L. Haugen, Mechanism of unconfined dust explosions: turbulent clustering and radiation-induced ignition. Phys. Rev. E **95**, 051101 (R1–5) (2017)

63. M. Liberman, N. Kleeorin, I. Rogachevskii, N.E.L. Haugen, Multipoint radiation induced ignition of dust explosions: turbulent clustering of particles and increased transparency. Combust. Theory Model. **22**, 1084–1102 (2018)
64. M. Sofiev, V. Sofieva, T. Elperin, N. Kleeorin, I. Rogachevskii, S. Zilitinskevich, Turbulent diffusion and turbulent thermal diffusion of aerosols in stratified atmospheric flows. J. Geophys. Res. **114**, D18209 (1–19) (2009)
65. S. Moore, F. Weinberg, High propagation rates of explosions in large volumes of gaseous mixtures. Nature **290**, 39–40 (1981)
66. S.R. Moore, F.J. Weinberg, A study of the role of radiative ignition in the propagation of large explosions. Proc. R. Soc. A: Math. Phys. Eng. Sci. **385**, 373–387 (1983)
67. S.R. Moore, F.J. Weinberg, Further studies of the role of radiative ignition in the propagation of large explosions. Proc. R. Soc. A: Math. Phys. Eng. Sci. **409**, 1–20 (1987)
68. F. Beyrau, M. Hadjipanayis, R. Lindstedt, Time-resolved temperature measurements for inert and reactive particles in explosive atmospheres. Proc. Combust. Inst. **35**, 2067–2074 (2015)
69. T. Li, R.P. Lindstedt, Thermal radiation induced ignition of multipoint turbulent explosions. Process Saf. Environ. Prot. **107**, 108–121 (2017)
70. R.K. Eckhoff, *Dust Explosions in the Process Industries: Identification, Assessment and Control of Dust Hazards*, 3rd edn. (Gulf Professional Publishing, Houston, TX, 2003)
71. R.K. Eckhoff, Understanding dust explosions. The role of powder science and technology. J. Loss Prevent. Proc. Ind. **22**, 105–116 (2009)
72. T. Abbasi, S. Abbasi, Dust explosions: cases, causes, consequences, and control. J. Hazard. Mater. **140**, 7–44 (2007)
73. G. Atkinson, L. Cusco, Buncefield: a violent, episodic vapour cloud explosion. Proc. Safety Environ. Protection **89**, 360–370 (2011)
74. S. Kundu, J. Zanganeh, B. Moghtaderi, A review on understanding explosions from methane-air mixture. J. Loss Prev. Process Ind. **40**, 507–523 (2016)
75. G. Atkinson, E. Cowpe, J. Halliday, D. Painter, A review of very large vapour cloud explosions: cloud formation and explosion severity. J. Loss Prev. Process Ind. **48**, 367–375 (2017)
76. G.O. Thomas, Buncefield: a possible alternative multi-stage route to localised overpressure generation. Process Saf. Environ. Prot. **119**, 253–260 (2018)
77. E.S. Oran, G. Chamberlain, A. Pekalski, Mechanisms and occurrence of detonations in vapor cloud explosions. Prog. Energy Combust. Sci. **77**, 100804 (1–37) (2020)
78. G. Krishnamoorthy, L.N. Mulenga, Impact of radiative losses on flame acceleration and deflagration to detonation transition of lean hydrogen-air mixtures in a macro-channel with obstacles. Fluids **3**, 104 (1–14) (2018).
79. R.W. Houim, E.S. Oran, Numerical simulation of dilute and dense layered coal-dust explosions. Proc. Combust. Inst. **35**, 2083–2090 (2015)
80. R.W. Houim, E.S. Oran, Structure and flame speed of dilute and dense layered coal-dust explosions. J. Loss Prev. Process Ind. **36**, 214–222 (2015)
81. S. Guhathakurta, R.W. Houim, Influence of thermal radiation on layered dust explosions. J. Loss Prev. Process Ind. (2021) in press. https://doi.org/10.1016/j.jlp.2021.104509

Chapter 17
Astrophysical Combustion

The thermonuclear burning of stars such as our Sun and the explosions of white dwarfs during the birth of a supernova is a unique astrophysical laboratory where scientists can study combustion processes in unusual conditions not available in terrestrial laboratories. It is generally accepted that the different supernovae (SN) are associated with different physical mechanisms of the explosion. Type Ia (SN Ia) supernovae are explained by the thermonuclear explosion of a carbon–oxygen white dwarf; SN II are associated with the gravitational collapse of a massive star. Supernovae SN Ia are bright stellar explosions that occur when roughly one solar mass of stellar material, composed primarily of carbon and oxygen, is burned up. A type Ia supernova is a thermonuclear explosion of a carbon–oxygen white dwarf, that gained enough mass as a result of the mass accretion from a binary companion to gravitationally compress the core, so that the core temperature and density rise enough to trigger thermonuclear reactions. Ignition occurs when the core density reaches high values, of the order of $2 \times 10^9\,\mathrm{g/cm^3}$, so that neutrino losses in the convective "Urca" process cannot balance the energy produced by thermonuclear reactions. Supernovae are characterized by a sudden increase in brightness, with a decrease in various shapes and durations after maximum luminosity, which lasts for several weeks, followed by an exponential decrease that can last for several years. The total electromagnetic radiation obtained from the light curve is $\sim 10^{49}$ erg, while the maximum luminosity is $\sim 10^{10} L_{\odot}$. Here and below $L_{\odot}$, $M_{\odot}$ means luminosity and mass of the Sun. The kinetic energy, estimated from the expansion velocity of the ejecta, which is of the order of $u_{\mathrm{eject}} \sim (5-10) \times 10^3\,\mathrm{km/s}$, corresponds to the energy $\sim 10^{51}$ erg. Such amount of energy can be obtained either as a result of the gravitational collapse of the electron degenerate stellar core or from the incineration of a carbon–oxygen core. The combustion and explosions of SN Ia have long been the subject of intense research, but the details of the combustion that lead to the observed spectrum and luminosity of SN Ia are still not completely understood. It is now widely believed that the most likely precursor of Type Ia supernovae is the electron-degenerate White Dwarf stars. It is believed that carbon/oxygen white dwarfs are produced by stars of moderate mass, less than about $8M_{\odot}$. From the very beginning,

M. A. Liberman, *Combustion Physics*,
https://doi.org/10.1007/978-3-030-85139-2_17

they burn hydrogen, when the hydrogen in the central core is burned out, a helium core is formed. In the range of moderate masses, thermonuclear burning of helium will create a core composed of roughly equal parts of carbon and oxygen nuclei and the Fermi degenerate electrons. The light curves of SN Ia have unique property that allows to use them as distance indicators, but to improve these measurements, theoretical and numerical models of SN Ia are needed, which describe details of the burning and explosion and explain the observed characteristics of SN Ia. Supernovae are classified according to their spectrum at maximum light and by their light curve. If there are no hydrogen lines, they are classified as type I (SN I), and if a noticeable Si II absorption line is visible in the spectrum around 6100 Å, they are called SN Ia. A more detailed classification consists of nine different types of SN Ia.

17.1 Compact Objects: The Birth and the Death of Stars

We briefly describe how white dwarfs, which belong to the family of *Compact objects* along with neutron stars and black holes, are formed during stellar evolution. How these stars born, live, and die? Many details that are beyond the scope of this book can be found in textbooks and reviews on stellar evolution and related issues in astrophysics [1–12].

It is a general believe that the Universe and all elementary particles, neutrons, protons, electrons, neutrinos, and others, were formed from a hot soup of quarks during the very early stage after the Big Bang on the Planck time scale $\sim 10^{-43}$ s. During the next stage of the Universe expansion and cooling the first light elements, hydrogen and helium were formed. At the early stage of the expansion of the Universe, the background density of matter was too low for the nucleosynthesis of heavier elements that are later formed in thermonuclear reactions. The stars and galaxies are formed from a cloud of interstellar gas, mostly hydrogen, through its gravitational contraction and subsequent heating. A star formation implies that the total energy of a hydrogen atom in a gravitational field of an interstellar gas cloud is negative

$$E_H \sim \frac{m_H u^2}{2} - g\frac{m_H M_{cl}}{R} < 0, \tag{17.1.1}$$

where $M_{cl} \sim \rho_{cl} R^3$ is the total mass of the gas cloud. Assuming that the cloud density is constant, $\rho_{cl} \leq 10^{-22}\,\mathrm{g/cm^3}$, and $m_H u^2/2k_B = T \sim 100\,\mathrm{K}$, we obtain that the cloud size (radius) is greater than the Jeans radius

$$R_J \approx \left(\frac{\pi k_B T}{m_H \rho_{cl} g}\right)^{1/2}, \tag{17.1.2}$$

and the mass of the gravitationally contracting cloud is

$$M_J \approx 5 \times 10^{-10} \left(\frac{T^3}{\rho_{cl}} \right)^{1/2} M_\odot \sim 5 \times 10^4 M_\odot. \tag{17.1.3}$$

In course of the gravitational collapse such huge interstellar cloud disintegrates to form thousands of protostars.

In astronomy, the newly formed stars are classified using the *main sequence* diagram according to stellar brightness. The initial mass of the star formed during gravitational contraction can vary depending on local conditions within the gas/dust cloud. The star will begin to burn when the temperature near its center becomes high enough to ignite the thermonuclear combustion of hydrogen. This imposes low and upper limits on stellar masses. The mass of the star must be large enough for the core temperature to rise up to $T \geq 10^7$ K before the pressure of the degenerate electrons stops the gravitational contraction and heating. This gives the low mass limit

$$M_{\min} \sim 0.05 M_\odot. \tag{17.1.4}$$

The upper limit of stellar masses follows from the condition that the radiation pressure of hydrogen burning in the star center does not exceed the gravitational forces holding the matter on the star surface. Otherwise, if the luminosity exceeds the Eddington limit, $L_{\text{Edd}} \approx 1.3 \times 10^{38} (M/M_\odot)$ erg/s, matter will be expelled from the outer surface of the star. This condition gives the upper mass limit

$$M_{\max} \sim 40 M_\odot. \tag{17.1.5}$$

A lifetime of stars is determined by the amount of hydrogen that can be burned in the star core and consists of successive stages of thermonuclear burning: the first is hydrogen, then, hydrogen ashes—helium, etc. Each stage of thermonuclear combustion releases thermal energy and temporarily interrupts the previous stage of gravitational contraction. In low mass stars the temperature in the core of the star cannot become high enough to ignite C^{12} and O^{16}, which are ashes from helium fusion, and the last stage is established by hydrostatic equilibrium between gravitational compression and pressure of the Fermi degenerate electron liquid.

Compact objects (white dwarfs, neutron stars, and black holes) which are the end products of stellar evolution, are "born" when normal stars "die", that is, when they burn up available for them nuclear fuels. The primary factor determining whether a star ends up as a white dwarf, neutron star, or black hole is the star's mass. White dwarfs originate from light stars with masses of about $M_\odot$, which has already burnt hydrogen and helium. As the WD accumulates sufficient materials for gravitational contraction, the central density of the WD increases and the cooling caused by neutrino emission became less than the heat of the compression and carbon fusion. From this stage the main heat losses will be due to the convection. This may continue a century, until the temperature rises to the point where convection cannot stop the ignition of the thermonuclear burning of C^{12} and O^{16}.

All three types of compact object differ from normal stars in two fundamental ways. First, since they cannot ignite thermonuclear reaction in the ashes left from the previous stage, they cannot support themselves against gravitational collapse by a thermal pressure. White dwarfs stop gravitational collapse by the pressure of degenerate electron liquid, while neutron stars support themselves by the pressure of degenerate neutrons. The predominance of neutrons in their interior follows the mutual elimination of electrons and protons by inverse beta-decay, which are held together by self-gravity at densities comparable to nuclear values. Black holes are completely collapsed stars, that is, stars that could not withstand gravitational collapse and therefore collapsed to singularities. Another feature that distinguishes compact objects from ordinary stars is their small size. Compact objects have much smaller radii than ordinary stars of comparable mass and therefore much stronger gravitational forces on their surface. Since nothing, including light, can escape from black holes, they are "black" to any observer. On the contrary, neutron stars can be observed as pulsating radio sources ("pulsars") and as gas-accreting, periodic X-ray sources ("X-ray pulsars"). White dwarfs can be observed directly in optical telescopes during their long cooling time. Because of their small radii, luminous white dwarfs have much higher effective temperatures radiating away their thermal energy.

The existence of a mass limit for white dwarfs can be understood as follows. Let a star of radius R consists of N fermions (baryons or electrons) so that the number density of fermions is $n \sim N/R^3$. At very high densities when the average volume per atom becomes less than the volume of the atom itself, $a^3 \sim N/V$, the atoms lose their individual electrons, and the matter is transformed into a high-density plasma consisting of electrons and nuclei. At high density, when the Fermi energy exceeds the energy of the Coulomb interaction between electrons and nuclei, $\varepsilon_F = \frac{\hbar^2}{2m}\left(\frac{N}{V}\right)^{2/3} \gg Ze^2/a$, and if temperature is not too high, the electrons form a degenerate Fermi liquid. The Coulomb interaction between electrons and nuclei is negligible if

$$n = \frac{N}{V} \gg \left(\frac{e^2 m}{\hbar^2}\right)^3 Z^2 \tag{17.1.6}$$

The Fermi energy increases with the increasing n, and when it exceeds $m_e c^2$, we must use relativistic expression for energy. From the uncertainty principle the momentum of a fermion is $\sim \hbar n^{1/3}$, and the Fermi energy of the electron is $\hbar c N^{1/3}/R$. An equilibrium corresponds to the minimum of the Fermi energy of electrons plus gravitational energy

$$\varepsilon = \varepsilon_F + \varepsilon_G = \hbar c \frac{N^{1/3}}{R} - \frac{g N m_n m_e}{R} \tag{17.1.7}$$

The total energy is negative at large N so that the minimum is determined by the condition $\varepsilon = 0$, which gives the maximum number of baryons

$$N_{\max} = \left(\frac{\hbar c}{Gm_b^2}\right)^{3/2} \sim 2 \times 10^{57}. \tag{17.1.8}$$

The corresponding maximum mass of a white dwarf star, known as the Chandrasekhar limit is

$$M_{\max} \sim m_b N_{\max} = 1.67 \times 10^{-24} \text{ g } \times 2 \times 10^{57} \approx 3.3 \times 10^{33} \text{ g} \approx 1.5 M_{\odot} \tag{17.1.9}$$

The equilibrium radius of a star depends on whether the maximum value $M_{\mathrm{Ch}} = M_{\max}$ is achieved due to the balance of gravity and pressure of degenerate electrons or baryons:

$$R_{\mathrm{Ch}} \leq \frac{\hbar}{mc}\left(\frac{\hbar c}{gm_b^2}\right)^{1/2} \approx \begin{cases} 5 \times 10^8 \text{ cm} \sim 10^{-2} R_{\odot}, & m = m_e \\ 3 \times 10^5 \text{ cm} \sim 10^{-5} R_{\odot}, & m = m_n \end{cases} \tag{17.1.10}$$

Not too small stars with the initial mass $M \sim M_{\odot}$ go through the stages of hydrogen and helium combustion and end up their life as white dwarfs. Stars with the initial mass $M < 4M_{\odot}$ are likely to eject excess the amount of mass $M - M_{\mathrm{Ch}}$, which is spent on forming a planetary nebula. It is believed that neutron stars and black holes originate from more massive stars, $M > 4M_{\odot}$, for which the pressure of degenerate electrons cannot stop gravitational collapse. In the process of gravitational contraction, massive stars go through all stages of thermonuclear combustion of elements for which thermonuclear reactions are exothermic, starting from hydrogen and until the stage when only iron peak elements, ^{56}Fe, remain, for which the reaction is endothermic. Table 17.1 shows the masses and sizes of compact objects expressed in the mass and size of the Sun.

White dwarfs (WDs) no longer burn thermonuclear fuel since light fusion fuels such as hydrogen, helium, and lithium have already been burned out in previous stages of stellar evolution, and neither the density nor the temperature of the WD core is high enough to ignite remaining carbon or oxygen, which are products of the previous stage. Instead, toward the end of its life, a white dwarf will shed its outer thin layers as its core shrinks down. Radius of a white dwarf star with a mass of about the mass of the Sun is close to that of the Earth, $\approx 10^{-2} R_{\odot}$. Without the

Table 17.1 Compact objects

Object	Mass	Radius	Mean density (g/cm^3)
Sun	$M_{\odot} = 1.99 \times 10^{33}$ g	$R_{\odot} = 6.96 \times 10^{10}$ cm	1
White dwarf	$\leq 1.4 M_{\odot}$	$\sim 10^{-2} R_{\odot}$	$\geq 10^7$
Neutron star	$\sim (1-3) M_{\odot}$	$\sim 10^{-5} R_{\odot}$	$\leq 10^{13}$
Black hole	Arbitrary	$2GM/c^2$	$\sim M/R^3$

supply of thermonuclear energy and having a small radiating surface, a single white dwarf cools down slowly, gradually fading away over a long time.

It is believed that SN Ia is most likely to appear as the result of a thermonuclear explosion of carbon–oxygen remnant cores of a medium-size white dwarf that have exhausted their hydrogen and helium fuels [10–14]. The classification of SN Ia is based on spectroscopic features [15]: the absence of hydrogen absorption lines, which distinguishing them from Type II supernovae, the presence of strong silicon lines and intermediate-mass elements (Si, S, Ca, etc.) in their optical spectra. Burning of these supernovae produces much of the Fe-group elements found in the galaxy. One of the most striking properties of SN Ia is the spectrophotometric homogeneity exhibited by a large fraction of them. These properties, together with the fact that they appear in all kind of galaxies, have led to the conclusion that the most plausible scenario is a thermonuclear explosion of a CO white dwarf near the Chandrasekhar's mass in a close binary system. Unfortunately, up to now it has not been possible to identify any of pre-explosion candidates. Meantime, the regularity, brightness, and extraordinary uniform luminosity of the type Ia supernovae allow to used them as distance indicators by cosmologists, which led to the discovery of the accelerated expansion, changed our understanding of the universe geometry to a nearly flat universe that is expanding at an accelerated rate and the constraints for the dark energy equation of state parameter [16–19]. This discovery implies that a negative pressure compensates the gravitational force, dominating the dynamical evolution of the universe. On the other hand, observations of the relict cosmic background radiation are compatible with a flat geometry, while the observations of galaxy clusters and the measurement of the redshifts of galaxies with account of the anisotropy of cosmic microwave background are in favor a low-density universe requiring the total mass-density parameter $\Omega \approx 0.3$. These independent methods are consistent with the universe composed of approximately 30% matter and 70% dark energy, corresponding to the universe models with the cosmological constant introduced by Albert Einstein. Alternative models without the cosmological constant contribution would mean that either the measurements or interpretations of the observations of high-redshift SN Ia are not correct. That is why accurate modeling of SN Ia explosions and better understanding the mechanism of explosions is of paramount importance. The uniformity of the type Ia supernovae luminosity is usually associated with the formation mechanism of SN Ia in a binary system, where WD accretes additional mass from its companion star until it approaches the Chandrasekhar limit, $\approx 1.5 M_{\odot}$. At this point, the pressure of the degenerate electrons no longer supports the white dwarf against its own gravitational contraction. When the center of the white dwarf becomes sufficiently hot and dense due to compression, an explosion is triggered by the ignition of carbon fusion.

The evolution of a star is determined not only by the sources of energy that drive the luminosity but also by processes by which energy is transported to the star surface, where it is lost by radiation. There is a clear distinction between the core of the white dwarf, where the electrons are degenerate and the heat transport is dominated by electron thermal conduction, and the thin, nondegenerate outer envelope, in which heat transports by convection and/or radiative diffusion. The outer layer is thin and carries little mass, but it plays an important role in the WD cooling, since energy transfer and loss are determined by the outer layer.

The observations show that during thermonuclear combustion most of the white dwarf is converted into the iron-group elements. In combustion products, Ni^{56} decays into Co^{56} and then Fe^{56} and these beta decays produce enough light during several weeks to outshine the entire host galaxy and be seen billions of light-years away. An important feature of the SNIa spectrum is the presence of lines corresponding to intermediate mass elements: such as silicon, calcium, and sulfur. The presence of intermediate mass elements in the SNIa spectrum imposes strict limitations on the possible regimes of white dwarf combustion. Another feature of the SNIa is a material ejected from the surface of the star at a speed from 10.000 to 12.000 km/s, which is measured by the shift of the Doppler lines.

17.2 The Mysterious Stars: Numerical Models of SN Ia Explosion

Different theoretical scenarios consider how the SN Ia explosion is triggered. The single degenerate scenario assumes that a carbon–oxygen white dwarf accretes matter from a non-degenerate companion that can be a normal hydrogen star (main sequence, red giant, etc.) or a helium star and explodes when it reaches the Chandrasekhar's mass or when enough helium has been accumulated on the top even if the mass is below the Chandrasekhar's mass. The double degenerate scenario assumes two white dwarfs being close enough. If the separation of both white dwarfs is less than $\sim 3R_{\odot}$, the system loses angular momentum due to gravitational wave emission, the two stars shrink, the less massive star is completely disrupted, forming a thick disk around the more massive star, and the centrifugal force of the rotation acts as a finetuning mechanism of the accretion process in the evolution of the accreting WD. The WD-WD collision scenario assumes that the collision of two white dwarfs immediately causes ignition. These collisions can occur for example in the dense central region of a galaxy. Various numerical models have been developed that more or less successfully describe thermonuclear combustion of type Ia supernovae. The reader can find detailed discussions in excellent reviews that published regularly and summarize the progress in numerical modeling of SN Ia [8–12, 20–24].

Due to the uncertainties and complexity of the WD models, numerical studies of the SN Ia explosion often rely on ad hoc assumptions, choosing the initial conditions as simple as possible. Many models assume that the explosion remains spherically symmetric, though this assumption is questionable. Numerical models must satisfy all the constraints given by the observed SN Ia spectra, light curves, and nucleosynthesis. In particular, the initial stage of deflagration is necessary in order to ensure the production of large quantities of intermediate mass elements in addition to the iron group elements [20–22]. Another constrain is associated with the thermonuclear destruction of the white dwarf and the ejection of nuclear combustion products during the explosion.

After half a century of intensive studies of the SN Ia explosion, starting from the pioneering work of Arnett [25], still there is no consensus on the dynamics of thermonuclear combustion waves in SN Ia events. All the 3D versions of the pure deflagration models show deficiency of intermediate-mass elements, leave too much unburned carbon and oxygen close to the center, too low mass of ^{56}Ni ejected, and too low final kinetic energies compared to the observed value $\sim 10^{51}$ erg [26–28]. In this respect the underproduction of intermediate mass elements found in pure deflagration models of Type Ia supernovae is a long-standing problem. These elements must be synthesized somewhere below densities $\rho \approx (3-5) \times 10^7$ g/cm^3 in an expanding medium. Nevertheless, the resolution achieved in the best multidimensional models carried out to date is far from adequate to solve unambiguously this problem. Besides, the deflagration model cannot explain the observed mass ejection form the WD surface. In the case of pure deflagration, the star expands at a speed exceeding the flame velocity so that the temperature drops drastically and all burning terminates. The only feasible successful regime is a mixed one, when the flame starts with deflagration in the high-density matter where the expansion coefficient is small. Then it somehow accelerates and, at a radius characterized by a certain critical density, transits to detonation mode. The detonation burns the outer layers, eliminating the outer unburned carbon that makes unacceptable the pure deflagration models. This model successfully explains the parameters of the explosion at the expense of one tuning parameter, the critical density ρ_{cr}. From comparison of simulations with observations, it is generally accepted that $\rho_{cr} \sim 2 \times 10^7$ g/cm^3 (e.g. [10, 12, 26–28]).

The initial ignition in a detonation mode is disfavored because it does not produce sufficient amount of intermediate mass elements, such as Mg, Si, S, and Ca [10, 25–28]. Besides much less energy is required for the initial ignition of the deflagration mode than for the ignition of detonation. Therefore, it is generally accepted that the explosion begins as a deflagration, which may become strongly turbulent and lasts either until the end of the explosion in pure deflagration models or in the scenario the deflagration-to-detonation transition (delayed detonation), deflagration is supposed to spontaneously turn into detonation at late stages after the flame has entered the distributed burning regime [22, 29–34]. Assuming that collisions between the convective elements can efficiently convert their kinetic energy into thermal energy, this results in temperature fluctuations $\Delta T/T_0 \approx 0.01$ for the background temperature $T_0 \approx 5 \times 10^8$ K. Given adequate physical conditions, these fluctuations could grow to finally trigger the nuclear runaway. Typical SN Ia also have a clear correlation, known as Phillips relationship [35], between their peak luminosity, the rate of decline after maximum light, and the color at maximum. The observed diversity in terms of luminosity ranges, light curve shape, color, etc. implies that SN Ia are presumably the end products of various evolutionary pathways. There are various progenitor scenarios of exploding C + O white dwarfs leading a Chandrasekhar mass (M_{Ch}), a sub-M_{Ch} mass and a super-M_{Ch} mass white dwarfs [13, 14, 36, 37].

On the other hand, attempts to overcome the shortage of deflagrations models using the delayed detonation paradigm face other challenges. The density at which the transition to detonation takes place has to be estimated in advance because the physics the deflagration-to-detonation transition is not fully understood. In general, there are three ways how the initially ignited deflagration can transit to detonation: (i) the deflagration-to detonation (DDT) can occur via a spontaneous reaction wave, the Zel'dovich gradient mechanism; (ii) DDT can occur via catching and coupling of the accelerating flame and shock wave, the SWASER mechanism; (iii) DDT can occur if a flame is accelerated in a high-intensity turbulent flow so that the turbulent flame velocity reaches the Chapman-Jouguet deflagration speed. The last possibility was numerically studied in [38–41].

In the DDT paradigm, the duration of the deflagration phase determines the degree of expansion of the star, which is critical for the composition of the material synthesized during the explosion. It is widely agreed that a detonation wave alone does not produce sufficient amounts of intermediate mass elements. If the entire star burns in the detonation mode, it incinerates the entire star down to the Fe-peak elements, which contradicts observations. A sufficient amount of intermediate mass elements can be formed if a significant part of the star burns by a detonation at low densities. Therefore, the observed spectrum of SN Ia can be explained if the white dwarf was pre-expanded before the ignition of a detonation. The desirable pre-expansion may occur if part of the star was consumed by deflagration before a detonation was triggered. A good agreement with observations was found in the numerical simulations where the detonation regime was artificially "turned on" after the flame has consumed about 10% of the white dwarf mass, providing the necessary expansion of the star. In simulations [42–44] the transition to detonation was artificially prescribed once the deflagration wave reaches a certain density threshold. However, an explosion model without tuning parameters, which allows to calculate the critical density of the transition to detonation from "first principles" was not developed. The problem is so complex that direct simulations are not feasible, and the only way to gain insight into the conditions of runaway is by combining analytical models with multidimensional simulations. The need for such models follows from the huge variety of scales of the problem. At high densities, the width of a laminar nuclear flame is $\sim(10^{-1}-10^{-3})$ cm while the radius of the white dwarf is $\sim 10^{8}$ cm. Even with adaptive mesh refinement, simulations of an entire star cannot simultaneously resolve nuclear flames. An open and still debated remains the question: what is the "sufficient" resolution in numerical simulations?

The model of delayed detonation has become very popular, but the mechanism of the DDT for the SN Ia conditions still remains unclear. It was assumed that the turbulent flame causes detonation when the speed of the flame approaches the CJ deflagration speed. However, it is known from laboratory studies that when the turbulent flame velocity approaches the CJ deflagration speed, combustion transits in the chocking mode not necessary to detonation. For the transition to detonation, either the formation of a shock wave with its subsequent coupling with the reaction zone with their mutual amplification is necessary, or the formation of the suitable gradient of reactivity is needed. This may be either extended high temperature zone of nonuniform induction time, where the transition may occur through the Zel'dovich gradient mechanism, or a relatively strong shock wave coupling with the reaction zone.

Röpke and Niemeyer [45] explored the range of explosions that lead to the DDT in the outer shells at $R \sim 1.7 \times 10^8$ cm after huge expansion caused by subsonic combustion initiated in several hot bubbles. They used a physically motivated criterion to switch on the DDT, when the combustion entered the distributed regime, in which mixing of burnt and fresh fuel by turbulence at the scale of the flame width might trigger a detonation. Although there is no justification for this criterion and for all others currently used, it has led to consistent results for the final kinetic energy and mass of synthesized intermediate-mass and Fe-group elements, and to almost complete depletion of carbon in the center. In three-dimensional simulations of delayed detonations in white dwarfs, Bravo and García-Senz [46] used the condition to switch-on the DDT, which is similar to the condition that is usually assumed in one-dimensional simulations. In their model, DDT was triggered when the mean density at the flame surface was $\rho \sim (2-3) \times 10^7\,\text{g/cm}^3$. They also verified the sensitivity of their results to the initial geometrical configuration of the flame by computing a delayed detonation model in which the deflagrative phase began at the center of the white dwarf compared with another model in which it began in 30 hot bubbles, as in the deflagration model [47]. Whether a turbulent flame is followed by a delayed detonation during the expansion or after many pulses, what is the mechanism of the DDT, and how is a turbulent flame developed still not completely clear. Some authors [48, 49] suggested that the mechanism of DDT is the Zel'dovich gradient mechanism, but this was not confirmed in simulations.

Until the mid-1990s, most SN Ia explosions models were one-dimensional spherically symmetric models that could not account for important aspects of combustion, the instability of a propagating flame, the development of turbulence, off-center ignition, etc. Nevertheless, one-dimensional models still represent the only reasonable approach to combining hydrodynamics with calculations of the detailed thermonuclear reactions. It was emphasized by Hillebrandt and Niemeyer [10] that a realistic simulation must be multidimensional, but multidimensional numerical studies can only investigate a very small fraction of the available parameter space. Therefore, a failure or success of a particular model to explain certain observations should not be considered conclusive. The most critical aspect in modeling WD explosions is the huge diversity of spatial and time scales, which are not feasible and cannot be resolved in multidimensional simulations. This means that combustion fronts are considered as discontinuity surfaces, and, therefore, nuclear reactions are not explicitly represented, as well as the mechanism of energy release and the flame propagation. Correct modeling of turbulence in Type Ia supernovae is also limited by the extremely large difference in spatial scales. The Reynolds number on $\ell_0 \sim 10^7$ cm is $\text{Re} \sim 10^{14}$, and the Kolmogorov scale is $\ell_\eta = \ell_0/\text{Re}^{3/4} \sim 10^{-4}$ cm, which is much smaller than the flame thickness. Similar problem is time scales, $\tau_\eta = \tau_0/\text{Re}^{1/2}$. This means that small turbulent eddies also cannot be resolved. It is usually assumed that turbulence is strong and within the inertial range of the turbulent cascade the velocity fluctuations decrease according to the Kolmogorov-Obukhov law $u'_\lambda \propto (\varepsilon\lambda)^{1/3}$. Depending on the fuel density, the laminar flame speed is less or comparable to turbulent velocity fluctuations. Therefore, turbulent eddies ultimately penetrate the internal flame structure and mix fuel and ashes. Then, laminar mode of the flame transits to the mode of "distributed reaction regime". Many authors [28, 49–56] suggest that when the turbulence becomes sufficiently strong, $U_t > U_f$, the turbulent flame speed becomes independent of the laminar flame velocity and, therefore, of the microphysics of burning and diffusion. In this case, the laminar flame width becomes comparable to the size of the largest turbulent eddies, which means that turbulence dominates the transport processes and the flame enters a state of the "distributed reaction zone" also called "well-stirred reactor regime". It is assumed that in this case the temperature of the fuel ahead of the flame raises so much that DDT occurs according to the Zel'dovich reactivity gradient mechanism [57]. Since the change in the properties of a laminar flame depends strongly on the fuel density, it is assumed that the transition to detonation occurs near the star surface, at $\rho \sim (10^6 - 10^7)\ \mathrm{g/cm^3}$. However, this estimate is only an order of magnitude, and the assumption that the transition to detonation occurs through the Zel'dovich gradient mechanism. Neither flame nor detonation can be resolved in stellar scale simulations because no computer is capable of resolving this scale range. Therefore, it is necessary to resort to statistical or scale approximations for those length scales that are not properly resolved.

In a series of three-dimensional models [58–61], deflagration ignition was used in the form of a set of spherical ignition spots that began to burn simultaneously (multipoint ignition) and the ignition geometry was treated as a free parameter. It was found that in the case of a large number of ignition points, the flame is always efficiently ignited centrally. Such central ignition with many nuclei of instability is the only way that allows for purely deflagration explosions that reach the brightness of normal SN Ia. On the contrary, in the case of a small number of ignition spots, the outermost ignition spot tends to dominate the burning, resulting in the one-sided deflagration in most of simulations. This is consistent with pre-ignition simulations [62] which predict bipolar flows in the case of a one-sided off-center ignition. On the other hand, according to [60], rotation can suppress such bipolar flows and increase the probability of central ignition.

By using one or two free parameters to tune the 3D modeling of SN Ia, a good agreement with observations can be obtained. However, there are too many unknown key properties of white dwarfs that make it currently impossible to create a predictable model of the type Ia supernova explosion. Among them are turbulence, which probably developed during a long convective period before the flame ignition, and the interaction of turbulence with the flame. In addition, simulations [63] have recently shown that the initial magnetic field exceeding $10^9 G$ (the Larmor radius is much smaller than the flame thickness) tends to decrease the flame velocity and, accordingly, suppress the development of RT instabilities in the early stages of the explosion. Such strong magnetic fields were not present in the conventional scenario of the WD explosion although they can be created by the mechanism of the magnetic turbulent dynamo.

Many authors suggest that when the flame reaches a state of distributed burning, when the ratio of turbulent intensity to the laminar flame speed exceeds some unknown threshold, the fuel and ashes are mixed and the temperature and size of this mixture became high and suitable to produce the reactivity gradient required for the transition to detonation [38, 39, 48, 49, 64, 65]. However, since according to [66] the distributed burning mode does not guarantee the formation of a reactivity gradient, more stringent requirements were proposed in [44] for achieving conditions required for DDT.

Since the grid sizes used in the full-star simulations is by orders of magnitude larger than the width of combustion wave, which therefore must be considered as a discontinuity surface, the simulations must include some subgrid model(s) to calculate the flame speed. An excellent discussion of possible RT-type subgrid models is by Hicks and Rosner [67] and Hicks [68]. Generally, there are two different approaches depending on the effect that dominates the flame speed: the RT instability of the flame front and turbulence, which sets the flame speed, or the flame speed is established due to the interaction with RT-generated turbulence. One way is to consider the flame moving through pre-existing turbulence created by burning bubbles rising to the star surface ahead of the flame front, or assuming that the initial turbulence have been created by a strong pre-ignition convection [38, 39, 65, 69]. Another possible scenario is to consider the effect of turbulence created by the carbon flame on the trailing oxygen flame [70, 71] depending on the effect that dominates the flame velocity.

17.3 Ignition and the Early Stage of White Dwarfs Explosion

Ignition occurs when the temperature in the WD core increases as a result of gravitational contraction so that the reaction time becomes shorter than that of the cooling processes that take energy from the hot reacting center. These are neutrino losses or thermal conductivity, radiative heat transfer or convective energy flux from a hot core to cold outer layers. Among them convection is the fastest cooling process. The time of the carbon thermonuclear reaction at the white dwarf becomes shorter than the convection time when the temperature at the star center exceeds $56.89\,\text{keV} \approx 6.6 \times 10^8\,\text{K}$.

17.3.1 Ignition of Self-accelerating Reaction in White Dwarfs

As the white dwarf grows close to the Chandrasekhar mass limit, $M_{\text{Ch}} \approx 1.4 M_{\odot}$, the energy budget near the core is governed by plasma neutrino losses and compressional heating. The neutrino losses increase with the increase of central density until the latter reaches approximately $\rho_c \approx 2 \times 10^9\,\text{g/cm}^3$ [20]. At this point, the plasmon creation is strongly suppressed while electron screening of nuclear reactions enhances the energy generation rate until it exceeds the neutrino losses.

Let us consider the conditions necessary for initiating a self-sustained thermonuclear reaction near the white dwarf center [72]. The rate of the thermonuclear reaction depends strongly on the initial temperature [73]. The initial temperature in the core of a the single-degenerate WD ($M_{\mathrm{Ch}} = 1.4 M\odot$) just before the ignition is about $(2 \div 4) \times 10^8$ K. The ignition temperature depends on the matter density, and for $\rho \sim 10^9\,\mathrm{g/cm^3}$ it is $T \sim 10^8$ K [74]. At the moment of the thermal run-away, the typical central density of a white dwarf is about $2.9 \times 10^9\,\mathrm{g/cm^3}$. In white dwarfs a flame is ignited by a hot pocket of size $\delta_f \sim \sqrt{\chi \tau_R} \sim 0.3\,\mathrm{m} \ll R_{WD}$, where χ is the thermal diffusivity and τ_R is the time scale of nuclear burning. On the contrary, a hot pocket of size $L_{Det} \sim a_s \overline{\tau}_R \sim 10^3$ km is required to initiate detonation, which is comparable to the size of the hot core of the star. Therefore, in supernova explosions, the deflagration mode is always the first to start. The kinetic equation for the fraction a of the fuel is

$$\frac{da}{dt} = -\frac{a^2}{\tau_0} \exp(-\sqrt[3]{E_a/T}), \tag{17.3.1}$$

where τ_0 is the constant of time dimension, which depends slightly on temperature, $E_a = 5.9 \times 10^{14}$ K is the activation energy for the carbon reaction. The energy balance equation for the energy release Q per unit mass in the reaction is

$$\frac{dH(T)}{dt} = -Q\frac{da}{dt}, \tag{17.3.2}$$

where the enthalpy $H(T)$ per unit mass for the degenerate matter of white dwarfs can be expressed as a function of temperature for the isobaric process [75]

$$H = \frac{4P}{\rho} = 3.1\frac{\hbar c}{\rho}(\rho Z/m)^{4/3}\left[1 + 2.06\left(\frac{m}{\rho Z}\right)^{2/3}\left(\frac{k_B T}{\hbar c}\right)^2\right], \tag{17.3.3}$$

where m is the average mass, eZ is the average electrical charge of ions, k_B is the Boltzmann constant, and $k_B = 1.38 \times 10^{-23}$ J/K. Since, from the beginning the process is isobaric and, neglecting by energy losses, Eqs. (17.3.1)–(17.3.3) can be combined into the equation for temperature

$$\left(\frac{\partial H}{\partial T}\right)_P \frac{d(k_B T)}{dt} = \frac{Q}{\tau_0}\left[1 + \left(\frac{H_c - H}{Q}\right)^2\right] \exp\left(-\sqrt[3]{E_a/k_B T}\right), \tag{17.3.4}$$

where subscript "c" denotes the initial enthalpy value at the white dwarf center.

From Eq. (17.3.4) it follows that due to the exponential dependence of the reaction rate at a high activation energy $\sqrt[3]{E_a/k_BT} \gg 1$, even a small increase in temperature, $(T - T_c) \ll T_c\sqrt[3]{k_BT_c/E_a}$, will lead to a significant increase in the reaction rate. This means that the characteristic reaction time is determined by the time of the first slow induction stage, when the temperature exceeds the initial value by $\Delta T \approx 3T_c\sqrt[3]{kT_c/E_a}$. The characteristic reaction time obtained using Eqs. (17.3.3) and (17.3.4) is

$$\tau_i \simeq 28\tau_0\sqrt[3]{kT_c/E_a})\frac{(kT_c)^2}{\rho Q\hbar c}(\rho Z/m)^{2/3}\exp(\sqrt[3]{E_a/kT_c}). \tag{17.3.5}$$

The induction stage is followed by the explosion stage. The characteristic explosion time τ_{exp} obtained from Eq. (17.3.4) is much shorter than the induction time:

$$\tau_{\text{exp}} \ll \tau_i\sqrt[3]{kT_c/E_a}. \tag{17.3.6}$$

An increase in temperature by $\Delta T \approx 3T_c\sqrt[3]{kT_c/E_a})$ during the induction stage causes a decrease in density near the center of the white dwarf

$$\Delta\rho = -\left(\frac{\partial\rho}{\partial T}\right)_P 3T_c\sqrt[3]{kT_c/E_a}). \tag{17.3.7}$$

Convection occurs because $(\partial\rho/\partial T)_P < 0$, and it can be roughly described as bubbles of lower density combustion products rising from the center to the surface of the star.

The equation of motion of a bubble of lower density, $\rho = \rho_c - \Delta\rho$, surrounded by the substance of unburned fuel of higher density ρ_c, can be written in the form

$$\ddot{\zeta} = \frac{4\pi}{3}\frac{\Delta\rho}{\rho_c}\Gamma\rho_c\zeta, \tag{17.3.8}$$

where $\Gamma = 6.67 \times 10^{-12}\,\text{cm}^3/\text{gs}^2$ is the gravitational constant and ζ is the displacement of the bubble from the center of the star. The characteristic convection time is

$$\tau_{\text{conv}} = \left[-4\pi\Gamma T_c\sqrt[3]{kT_c/E_a}\left(\frac{\partial\rho}{\partial T}\right)_P\right]^{-1/2} \tag{17.3.9}$$

Convection distributes the energy released in the reaction from the hot white dwarf center to the cold outer layers and can prevent a self-accelerating reaction (thermal run-away) at the center of the star, and it probably responsible for the development of turbulence prior to thermal run-away. The thermal run-away begins when the induction time of the reaction becomes shorter than the convection time $\tau_i \leq \tau_{\text{conv}}$.

Using Eqs. (17.3.3), (17.3.5), and (17.3.9), the condition of the thermal run-away at the star center is

$$\tau_i(T_c) \leq \left[12\pi \Gamma \rho_c \sqrt[3]{kT_c/E_a} \left(\frac{m}{\rho_c Z} \right)^{2/3} \left(\frac{kT_c}{\hbar c} \right)^2 \right]^{-1/2}. \tag{17.3.10}$$

From Eqs. (17.3.5) and (17.3.10) it follows that for the $0.5\,C + 0.5\,O$ white dwarf with the density at the star center $\rho_c = 3 \times 10^9\ \mathrm{g/cm^3}$, the temperature of the thermal run-away at the star center is $T_c = 0.66 \times 10^9$ K in agreement with results obtained in [74], and the time of the thermal run-away is $\tau_r \approx \tau_{\mathrm{conv}} \approx (3 \div 4)$ s.

The development of convection establishes an iso-entropic profile of the average temperature $\overline{T}$ of a large mass body in equilibrium (see [75])

$$\overline{T} = \overline{T}_c \left(1 - \frac{1}{2} \frac{R^2}{R_{WD}^2} \right), \tag{17.3.11}$$

where $\overline{T}$ and $\overline{T}_c$ are averaged over the surface $R =$ const, and the characteristic length scale of the white dwarf follows from the equilibrium condition of a large mass body:

$$R_{WD} \approx 0.38 \sqrt{\frac{\hbar c}{\Gamma \rho_c^{2/3} m_p^{4/3}}} \tag{17.3.12}$$

Using Eqs. (17.3.11) and (17.3.12) we find that for the central density of a white dwarf $\rho_c = 2.9 \times 10^9\ \mathrm{g/cm^3}$, the characteristic length scale of the average temperature profile in the white dwarf is $R_{WD} = 600$ km.

The average induction time $\overline{\tau}_i = \overline{\tau_i(T)}$ can be estimated using Eqs. (17.3.5) and (17.3.11). If $\overline{\tau}_i$ is less than the time of energy transfer by convection, then the self-accelerating reaction can occur in the form of a spontaneous explosion on the scale of about $R_c = \sqrt{3}\left(k\overline{T}_c/E_a\right)^{1/6} R_{WD} \approx 100$ km. On the other hand, local temperature deviations from the average distribution given by Eq. (17.3.11) are inevitable due to the multidimensional nature of convection. Consequently, the local induction time can be much shorter than the average value of the induction time in the central core of the white dwarf. We can conclude that the self-accelerating reaction in the central part of the white dwarf can be ignited by a small and essentially local increase in temperature and probably occurs in the form of multipoint ignition within the white dwarf core.

17.3.2 Flame Ignition in the White Dwarf Core

Consider the ignition of a nuclear flame that propagates in the highly degenerate matter of white dwarfs. The condition necessary for the ignition of the flame means the possibility that the self-sustaining reaction can propagate due to the transfer of heat by electron thermal conduction or radiative heat transfer. If ignition occurs, the burning becomes localized with a sharp boundary between the fuel and combustion products. As it was shown in Chap. 10, the flame propagation in a strong gravitational field can be completely controlled by the bubbles of light burned matter that are rising out of the star center. The dynamics of rising bubbles will be considered in the following sections.

The minimal size of the high temperature reacting volume which is sufficient to ignite a flame in white dwarfs can be found using the Zel'dovich–Frank-Kamenetskii theory. Let us consider a sphere of radius R_0 filled with a thermonuclear fuel. Let the temperature of the fuel at the surface of the sphere equals to the average temperature $\overline{T}_c$ of the surrounding matter. The ignition of the flame means that the energy released during the induction stage of the reaction is higher than the energy transferred by thermal conduction from the reaction zone to the surrounding fresh fuel. With an increase in the radius of the sphere, the effect of heat losses due to thermal conductivity weakens, and a flame can be ignited. The critical (minimum) radius of a sphere R^* at which a flame can be ignited is the maximum radius at which a stationary temperature profile is possible. At the induction stage, the reduction in the fuel fraction is small and the fuel consumption can be neglected. The stationary temperature profile for a sphere of radius R_0 is determined by the equation

$$\frac{\chi}{r^2}\frac{d}{dr}\left(r^2\frac{dT}{dr}\right)+3\sqrt[3]{k\overline{T}_c/E_a}\frac{\overline{T}_c}{\tau_r}\exp\left(\sqrt[3]{E_a/k\overline{T}_c}-\sqrt[3]{E_a/kT_c}\right)=0, \quad (17.3.13)$$

where χ is thermal diffusivity and the boundary condition is

$$T(r=R_0)=\overline{T}_c. \quad (17.3.14)$$

Using dimensionless variables

$$\xi = r/\sqrt{2\chi\tau_r}\ and\ \vartheta = \sqrt[3]{E_a/k\overline{T}_c}\cdot(T-\overline{T}_c)/3\overline{T}_c,$$

the problem is reduced to

$$\frac{d}{d\xi}\left(\xi^2\frac{d\vartheta}{d\xi}\right)+2\xi^2\exp\vartheta=0,\ \vartheta(\xi=\xi_0)=0. \quad (17.3.15)$$

The minimum radius of the sphere $R^* = \xi_0^*\sqrt{2\chi\tau_r}$ when the flame is ignited is determined by the condition that the solution to Eq. (17.3.15) does not exist for

$\xi_0 > \xi_0^*$. The numerical solution of the problem gives $\xi_0^* = 1.29$ for the flame ignition inside a sphere.

Using the numerical value of the electron thermal diffusivity we find that a flame can be ignited near the center of the white dwarf in a volume of size $R = 1.8\sqrt{\chi\tau_r} \approx 30$ cm, which is negligible compared to the size of the hot central core, $R_c \approx 100$ km. Therefore, the flame is likely ignited by the local thermal run-away at many points near the white dwarf center [60, 62]. The time required to ignite a flame in the central core of white dwarfs is much shorter than the average induction time in the center of the star because even a small increase in the local initial temperature, for example, by 5%, leads to a decrease in the local induction time by about five times.

The condition necessary for the ignition of detonation can be obtained considering spontaneous ignitionv by the nonuniform temperature distribution $T = T(r)$ near the center of a white dwarf. According to the Zel'dovich gradient mechanism a spontaneous reaction propagates with the velocity

$$u_{sp} = \left(\frac{d\tau_r}{dT}\right)^{-1} \frac{\nabla T}{(\nabla T)^2}. \tag{17.3.16}$$

For the iso-entropy distribution of the average temperature near the white dwarf center, Eq. (17.3.11), the average velocity of the spontaneous reaction wave is

$$\overline{u}_{sp} = 3\sqrt[3]{\overline{T}_c/E_a}\,\frac{R_{WD}^2}{R\overline{\tau}_r}. \tag{17.3.17}$$

At the center of the star ($R = 0$), the velocity of the spontaneous wave is infinite, which formally means the infinite thickness of the wave front, $\delta_{exp} = u_{sp}\tau_{exp}$, where τ_{exp} is the characteristic time of the explosion. The infinite thickness implies that the spontaneous wave of the explosion is not formed yet. Formation of the spontaneous explosion wave near the white dwarf center occurs when the thickness of the wave front becomes less than the size of the ignition area, $\delta_{exp} = u_{sp}\tau_{exp} < R$. This means that the front of the spontaneous explosion wave is formed at the radius

$$R > R_{WD}\sqrt{3\tau_{exp}/\overline{\tau}_r}(k\overline{T}_c/E_a)^{1/6} \tag{17.3.18}$$

The initial velocity of the spontaneous wave decreases as the wave propagates along the temperature gradient. The resulting combustion mode depends on the steepness of the temperature gradient and on the initial velocity of the spontaneous wave. In analytical analysis minimal condition for initiating a detonation cannot be obtained, but can be obtained in numerical simulations. The ignition problem has been the topic of a number of numerical studies [47, 60, 62]. Three-dimensional simulations of deflagration ignition, leading to several bubbles of burned matter scattered around the central region of the white dwarf, demonstrated a scenario qualitatively similar to the present analytical model, including the characteristic size of deflagration ignition, assuming a 5% of the background temperature fluctuation from the initial temperature profile $T_{max} = 7.5 \times 10^8$ K. High resolution ($\Delta = 2.17$ km) 3D simulations of

Type Ia supernovae using the adaptive mesh refinement [62] have shown that ignition is likely to occur off-center with a probable range from 40 to 75 km. At the same time, it was shown that the ignition of the first flames due to temperature fluctuations occurs at a size less than or near $R = 1.8\sqrt{\chi \tau_r} \sim 1$ m, which is thousands of times less than the best resolution available in simulations [47, 62].

17.4 Effect of the Flame Instabilities

Niemeyer and Woosley [49] emphasized that if a type Ia supernova explosion starts in the deflagration mode, then, in order to agree with observations, its velocity should be significantly higher than the laminar flame velocity obtained in multidimensional simulations, for example [10, 28, 51, 69, 76–80]. The effect of DL instability and the possibility for speeding up the deflagration wave: turbulent combustion and/or multipoint ignition were discussed in [80].

17.4.1 Dynamics of the Flame for Arbitrary Equation of State

The thermonuclear flame in a white dwarf is subject to two types of instabilities: the hydrodynamic Darrieus–Landau (DL) instability and the Rayleigh-Taylor (RT) instability. Which one dominates depends on the value of the dimensionless parameter—the Froude number: $Fr = U_f^2/gL_f$. According to simulations [81] the width of the deflagration wave at the central core of the white dwarf is $L_f \sim 10^{-2}$–10^{-3} cm, and the Froude number is $Fr \approx U_f^2/gL_f \sim 10^5$. The Darrieus–Landau instability of the carbon/oxygen flame propagating outward from the center of a white dwarf was studied theoretically in [82, 83] using the equation of state of the degenerate electrons near the white dwarf center. The effect of the DL instability on the flame speed was studied using two-dimensional direct numerical simulations by Bell et al. [84].

First, we derive an equation for the velocity of a deflagration wave propagating in a medium with an arbitrary equation of state [85]. The system of equations describing the flame propagation is

$$\frac{\partial \rho}{\partial t} + \nabla(\rho \mathbf{u}) = 0, \tag{17.4.1}$$

$$\rho \frac{\partial \mathbf{u}}{\partial t} + (\rho \mathbf{u} \nabla)\mathbf{u} + \nabla P = \rho \mathbf{g}, \tag{17.4.2}$$

$$\rho \frac{\partial H}{\partial t} + \rho \mathbf{u} \nabla H = \nabla(\kappa \nabla T) + \frac{\rho Q}{\tau_R} a^n \exp\left(-\sqrt[3]{E_a/kT}\right), \tag{17.4.3}$$

$$\frac{\partial a}{\partial t} + \mathbf{u}\nabla a = -\frac{a^n}{\tau_R} \exp\left\{-\sqrt[3]{E_a/kT}\right\}. \tag{17.4.4}$$

Since the time scale and length scale of hydrodynamic instability are much larger than the time scale and the length scale of the reaction zone, the development of instabilities in the Wentzel-Kramers-Brillouin (WKB) approximation can be considered as a quasi-stationary process compared to the fast rate of the reaction. In WKB approximation the concentration can be excluded from Eqs. (17.4.3) and (17.4.4), and we obtain

$$\rho\frac{\partial H}{\partial t} + \rho\mathbf{u}\nabla H = \nabla(\kappa\nabla T) + \frac{\rho_2 Q}{\tau_R} W(T), \tag{17.4.5}$$

where the dimensionless function $W(T)$ vanishes everywhere except the reaction zone of width $\Delta_R \approx 3L_f\sqrt[3]{kT_2/E_a}$; labels 1 and 2 used for variables ahead and behind of the flame. The specific structure of the function $W(T)$ depends on the reaction order n, and for $n = 1$ it is

$$W(T) = \exp\left(-2\sqrt[3]{E_a/kT}\right)\left\{\exp\left(\sqrt[3]{E_a/kT}\right) - \exp\sqrt[3]{E_a/kT_2}\right\}. \tag{17.4.6}$$

A stationary solution of Eqs. (17.4.1), (17.4.2), and (17.4.5) is the planar flame front propagating along z-axis. In the reference frame co-moving with the flame, the stationary solution consists of the upstream flow of the fresh fuel, the transition region of burning and heat release in the reaction, and the homogenous downstream flow of the combustion products. For the incompressible flow, $\rho u = const$ and $P = const$, and the velocity of the flame is determined as the eigenvalue of the equation for enthalpy (17.4.5). We introduce dimensionless variables: $\xi = z/L_f$, $\tau = tu_2/L_f$, $r = \rho/\rho_2$, $\Theta = T/T_2$, $h = H/H_2$, $\chi = \kappa/\kappa_2$, $G = 1/Fr_2 = gL_f/u_2^2$, and $\Lambda = QL_f/(H_2 u_2 \tau_R)$, $L_f = \kappa_2 T_2/(H_2 u_2 \rho_2)$, and Λ is the eigenvalue of the equation of heat transfer

$$\frac{d}{d\xi}\left(\chi\frac{d\Theta}{d\xi}\right) - \frac{dh}{d\xi} + \Lambda W(\Theta) = 0, \tag{17.4.7}$$

with the boundary conditions $\Theta = \Theta_1$ at $\xi \to -\infty$ and $\Theta = 1$ at $\xi \to +\infty$.

Outside the reaction zone $W(\Theta) = 0$ and Eq. (17.4.7) reduces to

$$\chi\frac{d\Theta}{d\xi} = h - h_1. \tag{17.4.8}$$

An approximate analytical solution of Eq. (17.4.8) can be obtained following to the Zel'dovich–Frank-Kamenetskii method. The value of Λ obtained in this way gives the velocity of the flame as the function of the energy released in the reaction

$$\Lambda \approx 0.5(1-h_1)^2 \left(\int_{\Theta_1}^{1} W(\Theta) d\Theta \right)^{-1}. \tag{17.4.9}$$

17.4.2 Hydrodynamic Instability of the Flame in White Dwarfs

Consider the perturbations of a planar and stationary flame, taken in the form

$$\tilde{\varphi}(\xi, \zeta, \tau) = \tilde{\varphi}(\xi) \exp(iK\zeta + S\tau), \tag{17.4.10}$$

where $\zeta = x/L_f$ is the dimensionless coordinate perpendicular to the direction of the flame propagation (z-axis), $K = kL_f$ is the dimensionless wave number, and $S = \sigma L_f/u_2$ is the dimensionless growth rate of instability. Introducing the dimensionless perturbations of the mass flux $\tilde{j} = (\rho\tilde{u}_z + \tilde{\rho}u)/\rho_2 u_2$, the transverse velocity $\tilde{v} = i\tilde{u}/u_2$, and the momentum flux $\tilde{\Pi} = (\tilde{P} + 2\rho u\tilde{u}_z + \tilde{\rho}u^2)/\rho_2 u_2^2$, we write down the linearized Eqs. (17.4.1), (17.4.2), and (17.4.5) in the dimensionless variables

$$\frac{d\tilde{j}}{d\xi} = -S\frac{dr}{d\Theta}\tilde{\Theta} - Kr\tilde{v}, \tag{17.4.11}$$

$$\frac{d\tilde{v}}{d\xi} = -Sr\tilde{v} + K\Pi - 2K\tilde{j}/r - \frac{K}{r^2}\frac{dr}{d\Theta}\tilde{\Theta}, \tag{17.4.12}$$

$$\frac{d\tilde{\Pi}}{d\xi} = -S\tilde{j} - K\tilde{v} + G\frac{dr}{d\Theta}\tilde{\Theta}. \tag{17.4.13}$$

In addition, we introduce the operator, defined as

$$\hat{F}(\tilde{\Theta}) = \frac{d^2(\chi\tilde{\Theta})}{d\xi^2} - \frac{d}{d\xi}\left(\frac{dh}{d\Theta}\tilde{\Theta}\right) + \Lambda\frac{dW}{d\Theta}\tilde{\Theta} = \tilde{j}\frac{dh}{d\xi} + Sr\frac{dh}{d\Theta}\tilde{\Theta} + \chi K^2\tilde{\Theta} \tag{17.4.14}$$

It is easy to verify that

$$\hat{F}(d\Theta/d\xi) = 0$$

The boundary conditions for the problem (17.4.11)–(17.4.14) require that perturbations vanish at infinity ahead and behind the flame front, which can be expressed in the form of algebraic equations at the coordinates $\xi = \xi_1$ and $\xi = \xi_2$ taken far enough in the upstream and downstream flows where the flows can be considered as homogeneous.

We know that a flame propagating in the gas mixture in a gravitational field is unstable against the long-wavelength perturbations, $K \ll 1$, while the short-wavelength perturbations, $K \geq 1$, are stabilized if gravitational acceleration is not too large. According to [81] this is typical situation for the core of carbon–oxygen white dwarfs. Therefore, we will use the same method as in Chap. 5 to solve Eqs. (17.4.11)–(17.4.14).

The solution of Eqs. (17.4.11)–(17.4.14) in homogeneous flows upstream and downstream gives the following boundary conditions for the variables $\tilde{j}$, $\tilde{v}$, $\tilde{\Pi}$:

$$r_1\tilde{v}_1 = -\tilde{j}_1,\ r_1\tilde{\Pi}_1 = (1 - r_1 S/K)\tilde{j}_1,\ \text{for}\ \xi \to -\infty, \tag{17.4.15}$$

$$\tilde{v}_2 + \tilde{\Pi}_2 - (2 + S/K)\tilde{j}_2 = 0,\ \text{for}\ \xi \to +\infty. \tag{17.4.16}$$

The solution of Eqs. (17.4.11)–(17.4.13) up the first-order to terms in $K \ll 1$ is

$$\tilde{\Theta} = \phi\frac{d\Theta}{d\xi}, \tag{17.4.17}$$

$$\tilde{j} = \tilde{j}_1 + \phi S(r_1 - r), \tag{17.4.18}$$

$$\tilde{v} = \tilde{v}_1 - \phi K\frac{(r_1 - r)}{rr_1}, \tag{17.4.19}$$

$$\tilde{\Pi} = \tilde{\Pi}_1 - \phi(r_1 - r)g. \tag{17.4.20}$$

The coupling equation for the amplitudes ϕ and $\tilde{j}_1$ with accuracy of the first-order terms in $K \ll 1$ follows from Eq. (17.4.14)

$$\tilde{j}_1 = -\phi S r_1. \tag{17.4.21}$$

Substituting (17.4.18)–(17.4.21) into boundary conditions (17.4.15)–(17.4.16), we obtain the instability growth rate within first-order terms in $K \ll 1$

$$S_0 = \sqrt{\frac{r_1^2 + r_1 - 1}{r_1(r_1 + 1)^2}K^2 + \frac{r_1 - 1}{r_1 + 1}GK} - \frac{K}{r_1 + 1}. \tag{17.4.22}$$

It should be noted that the structure of the Eq. (17.4.22) is similar to the Eq. (5.3.4) in Chap. 5 for the growth rate of instability of the infinitely thin flame front. The growth rate (17.4.22) does not depend on the particular equation of state. This dependence enters the expression for S_0 through the velocity of the unperturbed flame.

Within the model of a discontinuous flame front, it follows from Eq. (17.4.22) that the growth rate of perturbations tends to infinity for short-wavelengths $K \to \infty$. To obtain the growth rate of the fastest perturbations, it is necessary to consider the finite thickness of the flame front. For this purpose, we have to calculate the next terms in the expansion of the growth rate in powers of $K \ll 1$. The expressions for $\tilde{j}$, $\tilde{v}$, $\tilde{\Pi}$ to within the next order terms in $K \ll 1$ are given by integrals of the Eqs. (17.4.11)–(17.4.13):

$$\tilde{j} = \tilde{j}_1 + \phi S(r_1 - r) - K \int_{\xi_1}^{\xi} r\tilde{v}_0 d\eta, \tag{17.4.23}$$

$$\tilde{v} = \tilde{v}_1 - \phi K \frac{(r_1 - r)}{r r_1} - S \int_{\xi_1}^{\xi} r\tilde{v} d\eta + K \int_{\xi_1}^{\xi} (\tilde{\Pi}_0 - 2\tilde{j}_0 r^{-1}) d\eta, \tag{17.4.24}$$

$$\tilde{\Pi} = \tilde{\Pi}_1 - \phi(r_1 - r)G - \int_{\xi_1}^{\xi} (S\tilde{j}_0 + K\tilde{v}_0) d\eta. \tag{17.4.25}$$

Here $\tilde{j}_0$, $\tilde{v}_0$, $\tilde{\Pi}_0$ are the first-order terms of the expansion of $\tilde{j}$, $\tilde{v}$, $\tilde{\Pi}$, defined by Eqs. (17.4.18)–(17.4.20). Substituting Eq. (17.4.23) into Eq. (17.4.14) and transforming the operator $\hat{F}$ into Hermitian form, we find the coupling condition for the amplitudes ϕ and $\tilde{j}_1$ up to terms of the next order in $K \ll 1$:

$$\tilde{j}_1(1 - K\xi) + \phi\left[Sr_1 + K^2\langle f\rangle\right] = 0, \tag{17.4.26}$$

where

$$f = \chi\frac{d\Theta}{dh} + \xi + (1 + r_1 S/K)(I - r\xi/r_1),$$

$$I = \int_{\xi_1}^{\xi} \eta\frac{dr}{d\eta} r_1^{-1} d\eta, \; \langle f\rangle = \frac{\int_{-\infty}^{+\infty} f\psi^2 d\xi}{\int_{-\infty}^{+\infty} \psi^2 d\xi},$$

$$\psi = \sqrt{\chi\frac{dh}{d\xi}\frac{d\Theta}{d\xi}} \exp\left(-\frac{1}{2}\int_{\xi_1}^{\xi} \chi^{-1}\frac{dh}{d\Theta} d\eta\right).$$

Finally, substituting Eqs. (17.4.23)–(17.4.26) in the boundary conditions (17.4.15) and (17.4.16) and using Eq. (17.4.8), we obtain the following expression for the instability growth rate of the flame-front of finite thickness

$$S = S_0 - \frac{1}{2}\frac{K^3\Phi}{K + (1 + r_1)S_0}, \tag{17.4.27}$$

where S_0 is the growth rate of the zero-thickness flame instability, defined by Eq. (17.4.22), and the function Φ is

$$\Phi = \frac{1}{1-h_1}\left(r_1(1+h_1)\frac{S_0^2}{K} + 2(1+r_1h_1)\frac{S_0}{K} + 2h_1 - (1-h_1)r_1\frac{G}{K}\right)\times$$
$$\int\limits_{\Theta_1}^{1}\frac{r_1 - r}{r_1}\frac{\chi d\Theta}{h - h_1} + \int\limits_{\Theta_1}^{1}\frac{r_1 - r}{rr_1}\frac{\chi d\Theta}{h - h_1} +$$
$$+\frac{2(1+S_0/K)(1+r_1S_0/K)}{1-h_1}\int\limits_{\Theta_1}^{1}\left[1 - \frac{h(r_1 - r)}{r_1(h - h_1)}\right]\chi d\Theta \tag{17.4.28}$$

In the derivation of the instability growth rate (17.4.27) we did not specify the equation of state of the burning matter, $r = r(\Theta)$, $h = h(\Theta)$, the growth rate was obtained for an arbitrary type of thermal conductivity and for arbitrary form of the energy release $W(\Theta)$. The dependence of the instability growth rate on the particular equation of state is given by the function Φ in Eq. (17.4.28).

The obtained analytical solution is in good agreement with the numerical solution of the spectral problem for a flame in a gas mixture, considered in previous chapters. The instability growth rate as a function of the wave number has a maximum, and it vanishes for long wavelength perturbations in comparison with the flame thickness, if the acceleration G is not too large. In a general case of an arbitrary equation of state the expression for the wavelength of the fastest growing perturbation is rather complicated, but in some particular cases interesting for the flame stability in white dwarfs it can be obtained in an explicit analytical form.

17.4.3 Flame Speed in White Dwarfs and the DL Instability

We consider the flame speed enhancement caused by the development of the flame instability in the central part of the white dwarf. The equation of state for the degenerate relativistic electrons (17.3.3) can be written as

$$H = 4P/\rho, \tag{17.4.29}$$

$$P = 0.77\hbar c\left(\frac{\rho Z}{m_i}\right)^{4/3}\left[1 + 2.06\left(\frac{m_i}{\rho Z}\right)^{2/3}\left(\frac{kT}{\hbar c}\right)^2\right]. \tag{17.4.30}$$

While the equation for enthalpy (17.4.29) is valid for a large body with high density and any temperature, Eq. (17.4.30) holds only for highly degenerate matter, when the temperature is less than the Fermi energy

$$\left(\frac{m_i}{\rho Z}\right)^{2/3}\left(\frac{kT}{\hbar c}\right)^2 \ll 1. \qquad (17.4.31)$$

This condition is valid in the WD core, where thermal conduction is determined by the electron thermal conduction with the thermal conduction coefficient [81, 86–88]

$$\kappa = \frac{0.81\hbar c^2 k^2}{e^4 Z \Lambda_{ei}}\left(\frac{\rho Z}{m_i}\right)^{1/3} T, \qquad (17.4.32)$$

where $\Lambda_{ei} \sim 1$ is the Coulomb logarithm. Using the explicit form of the equation of state (17.4.29 and 17.4.30), and expression (17.4.32), the instability growth rate of the flame propagating through the white dwarf central core was calculated in [82, 83]

$$\sigma = \sigma_0 - r_1^2 u_1 L_f k^2 \frac{I_1 u_1 k + I_2 \sigma_0}{r_1 u_1 k + \sigma_0(1 + r_1)}. \qquad (17.4.33)$$

The instability growth rate of the infinitely thin flame is

$$S_0 = \sqrt{\frac{r_1^2 + r_1 - 1}{(r_1 + 1)^2} r_1 u_1^2 k^2 + \frac{r_1 - 1}{r_1 + 1} gk} - \frac{r_1 u_1 k}{r_1 + 1}, \quad r_1 = \rho_1/\rho_2. \qquad (17.4.34)$$

The coefficients in formula (17.4.33) are

$$I_1 = \frac{3r_1^{3/4} - 2r_1 - 1}{6(r_1^{4/3} - 1)} + \frac{r_1 + 1}{12(r_1 - 1)(r_1^{4/3} + 1)}\left[3r_1^{8/3} - 3r_1^{6/3} + 4r_1^{4/3} + r_1^{2/3} + 1\right],$$

$$I_2 = \frac{r_1^{3/4}}{2(r_1 - 1)} - \frac{r_1^{2/3} - 1}{6(r_1^{4/3} + 1)}\left[3r_1^{4/3} + 2r_1^{2/3} + 1\right].$$

The growth rate of the fastest perturbation mode is determined by the DL instability if

$$\frac{r_1 - 1}{r_1 + 1}\frac{g}{u_1^2 k_{\max}} \ll 1, \ \textit{or}\ Fr \approx \frac{u_1^2 k_{\max}}{g} \gg 1. \qquad (17.4.35)$$

The growth rate of the fastest perturbation is determined by the Rayleigh-Taylor instability in the opposite case, if

$$\frac{r_1 - 1}{r_1 + 1} \frac{g}{u_1^2 k_{max}} \gg 1, \; or \; Fr \approx \frac{u_1^2 k_{max}}{g} \ll 1. \tag{17.4.36}$$

Since the gravitational acceleration g is proportional to the distance from the center of the star, condition (17.4.35) holds near the star center, where the instability growth rate is defined by the hydrodynamic instability. Thus, near the center of the WD core the growth rate of perturbations with the wave numbers close to the maximum of the dispersion relation (17.4.33) is

$$\sigma = \Sigma_0 u_1 k (1 - k\lambda_{max}/4\pi), \tag{17.4.37}$$

where Σ_0 is the growth rate of the Darrieus–Landau instability of the infinitely thin flame-front, $\Sigma_0 = \frac{r_1}{r_1+1}\left(\sqrt{r_1 + 1 - 1/r_1} - 1\right)$, and the wavelength of the fastest perturbation is

$$\lambda_{max} = \frac{4\pi r_1^2 L_f}{\Sigma_0} \frac{I_1 + \Sigma_0 I_2}{(r_1 + 1)\Sigma_0 + r_1}. \tag{17.4.38}$$

The maximum growth rate is

$$\sigma_{max} = \frac{\Sigma_0^2 u_1}{4 r_1^2 L_f} \frac{(r_1 + 1)\Sigma_0 + r_1}{I_1 + \Sigma_0 I_2}. \tag{17.4.39}$$

At the center of the carbon–oxygen white dwarf $\rho \approx 3 \times 10^9$ g/cm^3 and $r_1 \approx 1.15$, and the wavelength of the fastest perturbations is $\lambda_{max} \approx 813\, L_f$. At the outer layers of the carbon white dwarf, where the effect of gravity becomes essential, $\rho < 5 \times 10^7$ g/cm^3, $r_1 \approx 1.4$, the wavelength of the fastest perturbations is also large compared with the thickness of the flame, $\lambda_{max} \approx 193\, L_f$. The relative value of the instability growth rate compared with the inverse time scale of the steady flame, $\sigma_{max} L_f / u_1$, increases with the increase of the densities ratio r_1. The density ratio r_1 in turn increases with the decrease of the density of the burning matter of the white dwarf [81] because the matter with a lower density is less degenerate and more sensitive to the temperature change in the flame. Consequently, the flame is more unstable in the outer layers of the white dwarf where the density is lower.

It should be emphasized that the fastest growing perturbations correspond to the wavelength much larger than the width of the flame. The critical value of the Froude number is reached when the gravitational acceleration becomes large enough in the outer layer of the star at large distances from the center of the star ($g \propto \Gamma R \rho_c$)

$$Fr_{cr} \approx \frac{u_1^2 (L_f/\lambda_{max})}{g L_f} \leq 1. \tag{17.4.40}$$

The effect of gravity also increases with the increase of the flame width and the decrease of the flame velocity. As the density of the white dwarf decreases, the flame speed decreases sharply while the flame width increases. For example, in the outer layers of the white dwarf the laminar flame speed and width are $U_f = 7.84 \times 10^3$ cm/s and $L_f = 0.54$ cm at the density $\rho = 1.5 \times 10^7$ g/cm^3, and $U_f = 2.97 \times 10^3$ cm/s and $U_f = 1.04 \times 10^3$ cm/s, and $L_f = 1.9$ cm and $L_f = 5.6$ cm at the density 10^7 g/cm^3 and 6.6×10 g/cm^3, correspondingly. According to [81] with the increase of the oxygen fraction in the white dwarf, the flame width increases and the flame velocity decreases, so the role of gravity becomes important at higher densities closer to the star center, presumably, at $\rho < 5 \times 10^7$ g/cm^3 for 0.5C – 0.5O white dwarfs.

In the context of SN Ia models, the nonlinear DL instability was studied in [47] using the model Frankel equation. The increase in the flame velocity considered in [47] is associated with fractal structure of a spherically expanding flame. But as it was discussed in Sect. 17.3.1, it is most likely that the explosion in white dwarfs is initiated by multipoint ignitions, which exclude spherically symmetric flame. The 2D simulations in [76] used a one-step model for the reaction rate, and in [79] the DL instability for the low-density C/O flame in a white dwarf was studied using 3D simulations. Both of these studies concluded that the stabilization mechanism precludes strong acceleration of the flame. Consequently, the increase in the flame velocity observed in these simulations is quite small, within a few percent, which is consistent with the theoretical estimate above. A much stronger effect is expected due to the flame turbulization. Another important consequence is the formation of bubbles of burned products floating up to the stellar surface during the nonlinear stage of the RT instability in the outer less dense layers of white dwarfs. These bubbles, in turn, participate in the development of turbulence.

17.5 Thermal-Diffusion Instabilities of the Flame in White Dwarfs

17.5.1 A Steady Flame in White Dwarfs

Thermal-diffusion instability of the flames in white dwarfs is another possible mechanism that can affect the propagation of flames in white dwarfs [89]. The main results of a theoretical study [89] of thermal-diffusion instability leading to pulsations of the flame velocity in white dwarfs were confirmed in one-dimensional simulations [90]. Thermal-diffusion instability and pulsations of the flame velocity were studied using a one-dimensional simulation of a spherically expanding flame in [91], where pulsation regimes were observed in simulations for a simplified one-step reaction model and high Zel'dovich numbers. However, the flame velocity differs strongly in more sophisticated simulations using a detailed network of nuclear reactions.

Since the main contribution to thermal conductivity is made by the thermal conductivity of a relativistic degenerate electron liquid with thermal-diffusivity $\chi_e = k_e/(\rho c_p)$, the Lewis number at the star's core is $Le = D/\chi \sim 10^{-5}$–10^{-7}, where $D \sim \ell_i a_s$ is the ion diffusion coefficient, ℓ_i is the ion mean free path and a_s is the sound speed. For the kinematic viscosity coefficient $\nu = \mu/\rho \sim \ell_i a_s$, the Prandtl number can be estimated as $\mathrm{Pr} = \nu/\chi \sim 10^{-4}$. In fact, both estimates of the Lewis number and the Prandtl number are a consequence of the difference in the velocities of electrons and ions in thermal equilibrium.

The thickness and velocity of the deflagration wave can be estimated using a simple dimensional analysis. The order of magnitude of the distance over which heat is transferred by the thermal conductivity is $\delta_f \sim \sqrt{\chi \tau_R}$, where $\chi = \kappa/\rho_f C_P$. During the characteristic reaction time τ_R, the flame travels a distance of the order of δ_f so that the velocity of the flame is $U_f \sim \delta_f/\tau_R = \sqrt{\chi/\tau_R}$. Although these are rough estimates, but they agree by the order of magnitude with results obtained in simulations [81]: $\delta_f \approx 1.0 \times 10^{-6}\,\mathrm{cm}$, $U_f \approx 1.3 \times 10^8\,\mathrm{cm/s}$, at density $\rho \approx 2 \times 10^9\,\mathrm{g/cm^3}$.

The values of the Lewis number and Prandtl number depend on whether the explosion of the star is caused by thermonuclear reactions of burning carbon or oxygen. Strictly speaking, one should consider a detailed nuclear reaction network [10–12], which begins with the most energetic branch of carbon reaction, followed by oxygen reactions, etc. However, in [91] it was shown that in this case the effective Zel'dovich number is smaller than the critical one, and therefore the flame is stable against pulsations.

It was shown in Chap. 5 that two conditions are necessary for the development of thermal–pulsation instability. First, the thermal-diffusion must be larger than the mass diffusion. This condition is satisfied in pre-supernovae, where $Le \gg 1$. The second condition is that the Zel'dovich number, $Ze = E_a/kT_b > Ze_{cr} \gg 1$, which characterizes the temperature dependence of the heat release in the reaction. Thermal instability occurs if the Zel'dovich number is higher than a critical value, $Ze_{cr} = 4 + 2\sqrt{5} \approx 8.47$. For a simplified model of a one-step carbon reaction ($^{12}\mathrm{C} + {}^{12}\mathrm{C} \rightarrow {}^{24}\mathrm{Mg}$) with a small temperature increase $\Theta = T_b/T_u$ in the reaction zone, the Zel'dovich number can be taken in the form

$$Ze = \sqrt[3]{E_a/kT_b}(\Theta - 1)/3\Theta. \tag{17.5.1}$$

The propagation of a one-dimensional planar subsonic deflagration is described by the equations

$$\frac{\partial \rho}{\partial t} + \frac{\partial}{\partial z}(\rho u) = 0, \tag{17.5.2}$$

$$\frac{\partial}{\partial t}(\rho u) + \frac{\partial}{\partial z}\left(\rho u^2 + P\right) = 0, \tag{17.5.3}$$

$$\rho \frac{\partial H}{\partial t} + \rho u \frac{\partial H}{\partial z} = \frac{\partial}{\partial z}\left(\kappa \frac{\partial T}{\partial z}\right) + \rho Q \dot{\mathcal{R}}. \tag{17.5.4}$$

We need to specify the rate of thermonuclear reaction $\dot{\mathcal{R}}$ and the energy release per unit mass, which is $Q \approx 5.6 \times 10^{17}\,\mathrm{erg/g}$ for carbon reaction and $Q \approx 4.9 \times$

10^{17} erg/g for oxygen reaction [81, 92]. The fastest reaction in the burning of the carbon–oxygen white dwarfs is the reaction $^{12}C+^{12}C$. Therefore, if the initial carbon mass fraction in a white dwarf is not small, the flame velocity is mainly determined by the rate of carbon reaction

$$\dot{\mathcal{R}} \equiv -\frac{da}{dt} = -\left(\frac{\partial a}{\partial t} + u\frac{\partial a}{\partial z}\right) = \frac{a^2}{\tau_R}\exp\left(-\sqrt[3]{E_a/kT}\right). \tag{17.5.5}$$

The activation energy for the carbon thermonuclear reaction is $E_a = 84^3 \times 10^9$ K, and τ_R is a coefficient with dimension of time, which depends on temperature [81, 92]. If the initial carbon fraction is small, then the reaction governing the flame propagation is the oxygen–oxygen reaction. The reaction rate has the same form as in Eq. (17.5.5), but with the activation energy $E_a = 136^3 \times 10^9$ K.

For simplicity, we consider only the carbon–carbon reaction. This initial reaction will be followed by the oxygen reaction after the environment temperature increases due to the energy released in the carbon reaction. This is typical situation confirmed by many numerical simulations of white dwarfs burning [23, 24]. The intermediate mass elements, such as Si, S, Ca, are produced in the reactions following the reactions with the formation Fe-peak elements. They are slow and can be neglected if we are interested in the thermal instability of the flame. The equation of state for the relativistic degenerate electrons is given by Eq. (17.3.3), and the thermal conduction in a white dwarf is dominated by the electron thermal conduction, with the coefficient of the relativistic electron thermal conduction given by expression (17.4.32) [92, 93].

Because of the strong temperature dependence of the reaction rate (17.5.5), we can neglect by spontaneous reactions in the initially cold fuel. Then, Eqs. (17.5.2)–(17.5.5) have a stationary solution for a planar flame propagating with the velocity u_1. In the coordinate system co-moving with the flame, the stationary solution consists of the homogenous upstream flow of a fresh fuel, the transition zone where reaction and heating take place, and the homogenous downstream flow. The first integrals of Eqs. (17.5.2)–(17.5.5) for the incompressible flow are

$$\rho_1 u_1 = \rho_2 u_2, \tag{17.5.6}$$

$$P = P_1 = P_2. \tag{17.5.7}$$

This means that the density and enthalpy in the incompressible flow depend only on temperature, $\rho = \rho(T)$ and $H = H(T)$. The temperature profile in the transition zone is determined by the equation of thermal conduction and the energy released in the reaction. For the stationary flow, it is defined by equations

$$\rho_1 u_1 \frac{\partial H}{\partial z} = \frac{\partial}{\partial z}\left(\kappa\frac{\partial T}{\partial z}\right) + \rho Q a^2 \exp\left(-\sqrt[3]{E_a/kT}\right)/\tau_R, \tag{17.5.8}$$

$$\rho_1 u_1 \frac{da}{dz} = -\rho a^2 \exp\left(-\sqrt[3]{E_a/kT}\right)/\tau_R. \tag{17.5.9}$$

The boundary conditions for Eqs. (17.5.8) and (17.5.9) are

$$T = T_1, \quad a = a_1 \quad \text{for } z \to -\infty, \tag{17.5.10}$$

$$dT/dz = 0, \quad a = 0 \quad \text{for } z \to \infty. \tag{17.5.11}$$

For the formulated problem the flame velocity is an eigenvalue of the solution of Eqs. (17.5.8) and (17.5.9) with the boundary conditions (17.5.10) and (17.5.10).

The first integral of Eqs. (17.5.8)–(17.5.11) is

$$\kappa \frac{\partial T}{\partial z} - \rho_1 u_1 (H - H_2 + Qa) = 0. \tag{17.5.12}$$

Taking into account the boundary conditions, we obtain

$$H_2 - H_1 = Qa_1. \tag{17.5.13}$$

Using (17.5.5) and Eqs. (17.5.7) and (17.5.13), we obtain expressions for the density and temperature of the combustion products:

$$\rho_2 = \rho_1 \left(1 + 0.325 \left(\frac{m}{\rho_1 Z}\right)^{1/3} \frac{a_1 Qm}{\hbar c Z}\right)^{-1}. \tag{17.5.14}$$

$$kT_2 = 0.4\sqrt{(\hbar c a_1 Qm/Z)}(\rho_2 Z/m)^{1/6}. \tag{17.5.15}$$

Note that due to the degenerate equation of state of the white dwarf material, the density of the burned products ρ_2 differs only slightly from the initial density ρ_1 of the fresh fuel. On the contrary, the temperature of the burned products T_2 is much higher than the initial temperature of the fresh fuel.

Due to very large activation energy, which in dimensionless units can be written as $\mathfrak{E} \simeq \sqrt[3]{(E_a/kT_2)} \gg 1$, the nonzero reaction rate is only in the narrow reaction zone, where temperature is close to the final temperature, i.e. $(T_2 - T) \sim T_2/\mathfrak{E}$. It is seen from Eqs. (17.5.14) and (17.5.15) that the reaction time, which determines the flame speed, is determined only by the final temperature T_2, which, in turn, depends on the initial fuel density ρ_1 and is almost independent of the initial temperature T_1. The rates of other reactions following the reactions of carbon and oxygen, are also determined by the final temperature T_2, which, as a result, influences the nuclear content of the combustion products. If the temperature is sufficiently low, then the intermediate-mass elements will not have time to be burned and transformed into the Fe-peak elements. That is why the flame can produce the Fe-peak elements at the center of the white dwarf, but in the outer layers of lower density, $\rho_1 < (5 \div 7) \times 10^7\ \text{g/cm}^3$, there will many intermediate-mass elements in the final combustion products [10–12, 27]. The flame velocity can be obtained using the Zel'dovich–Frank-Kamenetskii (ZFK) method. Out of the reaction zone $\dot{\mathcal{R}} \approx 0$, the temperature profile is determined by equation

$$\kappa \frac{dT}{dz} = \rho_1 u_1 (H - H_1). \tag{17.5.16}$$

The ZFK solution of the eigenvalue problem for a flame in a gas mixture under terrestrial conditions was obtained for conditions when there is a finite mass diffusion and negligible convective heat transfer compared to thermal conduction in the reaction zone. This condition can be formulated as an inequality

$$\Delta \frac{dT}{dz} \gg T_2 - T, \tag{17.5.17}$$

where $\Delta = \kappa_2 T_2 / \rho_1 u_1 a_1 Q$ is the flame thickness, which is much larger than the thickness of the reaction zone $\Delta_R \approx 3(kT_2/E_a)^{1/3} \Delta$ due to $kT_2/E_a \ll 1$. It should be emphasized that inequality (17.5.17) holds only in the reaction zone, and the ZFK method is applicable either in the case of a finite mass diffusion or for the first-order reaction in the fuel with zero mass diffusion. In the case of a flame propagating in white dwarfs the reaction is of the second-order and the Lewis number is $Le = 10^{-4} \div 10^{-7}$, which means that the mass diffusion is negligible. Then, from Eq. (17.5.9) it follows that the asymptotic behavior of the concentration behind the flame-front, for $z \to \infty$, is $a \sim 1/z$. Therefore, the temperature behind the flame-front approaches the final value at $z \to \infty$ as $T_2 - T \sim 1/z$ and $dT/dz \sim 1/z^2$. This means that the inequality (17.5.17) does not hold, and we must use another method to solve the problem.

To do this we rewrite Eqs. (17.5.8)–(17.5.11) using dimensionless variables for the scaled concentration $\alpha = a/a_1$, the dimensionless temperature $\vartheta = \mathfrak{E}(T_2 - T)/T_2$, and the dimensionless eigenvalue

$$\Lambda = \frac{3\kappa_2 T_2}{\rho_2 u_2^2 \tau_R Q} \sqrt[3]{\frac{T_2}{E_a}} \exp\left(-\sqrt[3]{T_2/E_a}\right). \tag{17.5.18}$$

In the following calculations a parameter will be used, $q \equiv a_1 Q T_2^{-1} (dH/dT)_2^{-1}$, the numerical value of which for a highly degenerate matter is 0.5 with a good accuracy. In the reaction zone of the thickness $\Delta_r = \Delta/\mathfrak{E} \ll \Delta$, Eqs. (17.5.8)–(17.5.11), written in dimensionless variables can be reduced to the unique first-order equation for dimensionless variables

$$\left(\alpha - \frac{\vartheta}{q\mathfrak{E}}\right) \frac{d\alpha}{d\vartheta} = \Lambda \alpha^2 \exp(-\vartheta). \tag{17.5.19}$$

The boundary conditions in dimensionless variables are

$$\alpha = 0 \quad \text{for } \vartheta = 0, \quad \text{and } \alpha = 1 \quad \text{for } 1 \ll \vartheta \ll \mathfrak{E}, \quad \text{where } \exp(-\vartheta) \approx 0. \tag{17.5.20}$$

Equation (17.5.19) is the Abel equation of the second kind. The solution of this equation cannot be expressed in terms of elementary functions. Nevertheless, we can obtain the approximate eigenvalue of the problem (17.5.19) and (17.5.20) by matching the asymptotic solutions of Eq. (17.5.19) for the unburned fuel and the burned combustion products.

In the upstream flow of the unburned fuel ($\alpha \approx 1 \gg \vartheta/q\mathfrak{E}$) the asymptotic solution of Eq. (17.5.19) is

$$\ln\alpha = -\Lambda\exp(-\vartheta), \quad \frac{d\alpha}{d\vartheta} = \Lambda\alpha\exp(-\vartheta). \tag{17.5.21}$$

In the downstream flow for the combustion products ($\alpha \to 0$), the asymptotic solution of Eq. (17.5.19) is

$$\alpha = \vartheta/q\mathfrak{E}, \quad \frac{d\alpha}{d\vartheta} = 1/q\mathfrak{E}. \tag{17.5.22}$$

Matching the asymptotic solutions for α and $d\alpha/d\vartheta$, we obtain the eigenvalue

$$\Lambda = (\exp\vartheta^*)/\vartheta^*, \tag{17.5.23}$$

where $\vartheta^* = 1/\ln[q\mathfrak{E}(nq\mathfrak{E}....)]$. It is easy to verify that the eigenvalue Λ depends only slightly from the dimensionless activation energy. For example, $\Lambda \approx 4.6$ for $q\mathfrak{E} \approx 9$ and $\Lambda \approx 5.3$ for $q\mathfrak{E} \approx 15$, typical for oxygen flames. The flame velocity is determined by Eq. (17.5.18) and the calculated values of Λ. The calculated velocity of the thermonuclear flame in a white dwarf is

$$\mathrm{u}_1 = \sqrt{\frac{3\kappa_2\rho_2\mathrm{T}_2}{\Lambda\rho_1^2\tau_R\mathrm{Q}}\left(\frac{\mathrm{T}_2}{\mathrm{E_a}}\right)^{1/6}}\exp\left(-\frac{1}{2}\sqrt[3]{\mathrm{E_a}/\mathrm{kT}_2}\right). \tag{17.5.24}$$

This formula for the flame velocity differs by factor $\sqrt{\Lambda} \approx 2.2$ from the formula for the velocity of a flame supported by chemical reactions in a gas mixture. Using formula (17.5.15) for the temperature of combustion products, we obtain from Eq. (17.5.24) for the velocity of the carbon flame near the center of the white dwarf, where $\rho \approx 10^9\,\mathrm{g/cm^3}$, the value $u_1 \approx 69\,\mathrm{km/s}$, which differs by 25% from the value $u_1 \approx 93\,\mathrm{km/s}$ of the flame velocity obtained in simulations [81] using the exact equation of state and the detailed network of 130 nuclear reactions. However, for smaller densities ($\rho \sim 10^7\,\mathrm{g/cm^3}$) the analytical solution for the flame velocity differs considerably from the velocities obtained in [81]. Presumably, the discrepancy between the analytical solution for the flame velocity and the flame velocity obtained in numerical simulations is because the temperature of the combustion products in the analytical approach is calculated using a simplified equation of state (17.3.3). It should be noted that the flame speed is highly sensitive to the temperature of the combustion products: a change in the temperature of the combustion products even by 30% leads to a twofold change in the flame speed. Using a more accurate but much more complex equation of state gives much better agreement between the analytically calculated flame velocity and the velocity obtained in numerical simulations. From Eqs. (17.5.14), (17.5.15), and (17.5.24) it follows that the flame velocity is larger in the center of the star and decreases in the outer layers with a lower density for both carbon and oxygen white dwarfs. Besides, in agreement with numerical simulations, the velocity of the carbon flame is larger than the velocity of oxygen flames in the layers of the same density, which is consistent with the results of simulations [81].

17.5.2 Thermal-Diffusion Instability and Flame Pulsation in White Dwarfs

In Chap. 5, it was shown that a flame propagating in a combustible gas with negligible small mass diffusion is unstable with respect to one-dimensional perturbations. This instability is associated with a strong dependence of the flame speed on the temperature in the reaction zone, which is expressed through a large value of the Zel'dovich number. At a high activation energy, $3(kT_2/E_a)^{1/3} \ll 1$, the thickness of the reaction zone, $\Delta_R \approx 3(kT_2/E_a)^{1/3}\Delta$, is much smaller than the thickness of the flame, Δ, which is of the order of the characteristic scale of thermal conduction. In this case, the inner reaction zone can be treated as a discontinuity surface inside the flame-front. This is a weak discontinuity, where the temperature is continuous, but there is a jump of the temperature derivative caused by the energy release inside the reaction zone. This model differs from the flame model, in which the flame is considered as a surface of discontinuity and all hydrodynamic variables: speed, temperature, and density are discontinuous due to the energy release in the reaction zone. On the contrary, the present model corresponds to the continuous temperature profile because the thermal conductivity is explicitly included and the flame thickness is assumed to be finite. A similar model was used to study the flame-front stability and pulsations in solid propellants [94, 95].

Considering the reaction zone as a weak discontinuous surface, the equation of thermal conduction in the coordinate system co-moving with the stationary flame can be written as

$$\rho_1 \frac{\partial H}{\partial t} + \rho_1 u_1 \frac{\partial H}{\partial z} = \frac{\partial}{\partial z}\left(\kappa \frac{\partial T}{\partial z}\right). \qquad (17.5.25)$$

The boundary conditions at the discontinuous reaction front $\zeta = \zeta(t)$ are

$$T_{\zeta+0} = T_{\zeta-0}, \tag{17.5.26}$$

$$\left(\kappa\frac{\partial T}{\partial z}\right)_{\zeta+0} - \left(\kappa\frac{\partial T}{\partial z}\right)_{\zeta-0} = -\rho_1 u_1 (H_2 - H_1)\exp\left[\frac{1}{2}\left(\sqrt[3]{\frac{E_a}{kT_2}} - \sqrt[3]{\frac{E_a}{kT_\zeta}}\right)\right]. \tag{17.5.27}$$

Here T_ζ is the temperature at the discontinuity surface $\zeta = \zeta(t)$. The equation for the reaction front can be written using Eq. (17.5.24) as follows:

$$\frac{\partial\zeta}{\partial t} = -u_1\left\{\exp\left[\frac{1}{2}\left(\sqrt[3]{E_a/kT_2} - \sqrt[3]{E_a/kT_\zeta}\right)\right] - 1\right\}. \tag{17.5.28}$$

It is seen that the stationary solution of Eqs. (17.5.25)–(17.5.28) coincides with the solution outside of the reaction zone, obtained in Sect. 17.5.1.

It can be shown [95] (see Chap. 5, Sect. 5.4) that discontinuity model underestimates the growth rate of the thermal instability by approximately 50% if the flame parameters are not close to the stability limit. On the contrary, if the flame parameters are close to the stability limit, then the discrepancy between the discontinuity model and the exact solution is small [95]. Below we will show that the parameters of a thermonuclear flame in white dwarfs are close to the stability limit, and therefore the discontinuity model is well applicable to the case of white dwarfs. The model also assumes that the density of the burning matter is weakly dependent on temperature. The last condition is acceptable since the equation of state for a flame in white dwarfs refers to a highly degenerate matter.

We denote one-dimensional perturbation of any variables φ of the stationary solution by $\tilde{\varphi}$, and we will seek a solution in the following form

$$\tilde{\varphi}(z,t) = \tilde{\varphi}(z)\exp(\sigma t), \tag{17.5.29}$$

where σ is the growth rate of instability. From the equation of state (17.3.3) and the expression for the thermal conduction (17.4.32), we obtain

$$\kappa\tilde{T} = 0.5\kappa T_2\tilde{H}/(H_2 - H_1). \tag{17.5.30}$$

The linearized Eqs. (17.5.25)–(17.5.28) with account (17.5.30) are

$$\frac{\sigma}{u_1}\tilde{H} + \frac{d\tilde{H}}{dz} = \frac{\Delta}{2}\frac{d^2\tilde{H}}{dz^2}, \tag{17.5.31}$$

$$\tilde{H}_{\zeta-0} + 2(H_2 - H_1)\zeta/\Delta = \tilde{H}_{\zeta+0}, \tag{17.5.32}$$

$$\frac{\Delta}{2}\left(\frac{d\tilde{H}}{dz}\right)_{\zeta-0}-\frac{\Delta}{2}\left(\frac{d\tilde{H}}{dz}\right)_{\zeta+0}+\frac{2(H_2-H_1)\zeta}{\Delta}=\frac{\mathfrak{E}}{4}\tilde{H}_{\zeta+0}, \tag{17.5.33}$$

$$\sigma(H_2-H_1)\zeta/\Delta=-\frac{\mathfrak{E}}{4}\tilde{H}_{\zeta+0}. \tag{17.5.34}$$

Since perturbations must vanish at infinity, Eq. (17.5.31) follows that out of the reaction zone

$$\tilde{H}(z)=\tilde{H}_{\zeta\pm0}\exp(\mu_{\pm}z), \tag{17.5.35}$$

where

$$\mu_{\pm}=\frac{1}{\Delta}\left(1\pm\sqrt{1+2\sigma\Delta/u_1}\right). \tag{17.5.36}$$

The equation for the instability growth rate follows from the consistency condition of Eqs. (17.5.31)–(17.5.34):

$$(2\sigma\Delta/u_1)^2-2\sigma\Delta/u_1(\mathfrak{E}^2/16-\mathfrak{E}-1)+\mathfrak{E}=0. \tag{17.5.37}$$

The thermonuclear flame in the degenerate matter of a white dwarf is unstable if $\mathrm{Re}(\sigma)>0$. This condition leads to

$$\mathfrak{E}=\frac{1}{3}\sqrt[3]{E_a/kT_2}>\mathfrak{E}_{cr}=8+4\sqrt{5}\approx 17. \tag{17.5.38}$$

Note that for the equation of state of an ideal gas, the critical Zel'dovich number is $Ze_{cr}=4+2\sqrt{5}=8.47$, while for the equation of state of strongly degenerate matter it is about two times larger.

If the flame propagates due to the carbon reactions with the activation energy $84^3\times 10^9$ K, then according to (17.5.38) the flame becomes unstable if the temperature of the combustion products is below than $T_{cr}\approx 4.5\times10^9$ K. According to Eq. (17.5.15), this corresponds to the flame propagating in the white dwarf layer with the density smaller than $\rho_c\approx 5\times10^8\,\mathrm{g/cm^3}$. In the case of a small initial carbon fraction, the thermonuclear flame in a white dwarf propagates due to carbon–oxygen and oxygen–oxygen reactions. The activation energy of these reactions is larger, and therefore the flame will be more unstable. If the flame propagates due to oxygen reactions, for which the activation energy is $136^3\times10^9$ K, then instability occurs if the temperature of the combustion products is less than $\approx 19\times10^9$ K. Using Eq. (17.5.15) for the final temperature, we find that the oxygen flame is unstable for layers with a density up to $\rho_c\approx 10^{10}\,\mathrm{g/cm^3}$, which is even greater than that assumed in the center of the white dwarf. If the parameters of the flame are close to the stability limit given by Eq. (17.5.38), then the imaginary part of the instability growth rate does not vanish

$$\mathrm{Im}(\sigma) \approx \frac{\sqrt{3}u_1}{2\Delta}\left(\frac{E_a}{kT_2}\right)^{1/6}. \quad (17.5.39)$$

This means that the development of the flame instability leads to the pulsating regime of a thermonuclear flame with the pulsation frequency equals to $\mathrm{Im}(\sigma)$. If the dimensionless activation energy slightly exceeds the critical value, $(\mathfrak{E} - \mathfrak{E}_{cr})/\mathfrak{E}_{cr} \ll 1$, then the amplitude of pulsations is proportional to $\sqrt{(\mathfrak{E}) - \mathfrak{E}_{cr})/\mathfrak{E}_{cr}}$, as it follows from the basic principles of the stability theory. With accuracy up to the first terms in the expansion of Eq. (17.5.28) into power series of $(\mathfrak{E} - \mathfrak{E}_{cr})/\mathfrak{E}_{cr} \ll 1$, we obtain for the average velocity of the pulsating flame-front

$$u_p = u_1[1-\beta(\mathfrak{E} - \mathfrak{E}_{cr})/\mathfrak{E}_{cr}]. \quad (17.5.40)$$

The coefficient β in Eq. (17.5.40) can be obtained if we consider all terms of the first-order in $(\mathfrak{E} - \mathfrak{E}_{cr})/\mathfrak{E}_{cr}$ in Eqs. (17.5.25)–(17.5.28). Such calculations have been done in [94] for the chemical flame propagating in the mixture with the equation of state of an ideal gas, and it was found $\beta \approx 0.25$. Because of similarity of the discontinuity model of the thermonuclear flame and the chemical flame, the calculations of [94] are valid also in our case. Using Eq. (17.5.40), we obtain that the average velocity of the pulsating carbon thermonuclear flame in a white dwarf is slightly, by 3%-4%, reduced compared with the velocity of the steady flame in the white dwarf layer with density $\rho \approx 5 \times 10^7\ \mathrm{g/cm^3}$. The difference between the velocity of a steady oxygen flame and the average velocity of pulsating flame can be larger. In this case the velocity reduces by approximately 12% at high densities, $\rho \approx 3 \times 10^9\ \mathrm{g/cm^3}$ near the star center, and even more, up to 20–25% at the outer layers of density $\rho \approx 5 \times 10^7\ \mathrm{g/cm^3}$.

17.6 Instability of Thermonuclear Detonation in White Dwarfs

Note that the flame speed in the deflagration mode can be obtained with good accuracy by linear interpolation of the results of one-dimensional calculations with a large amount of physical processes [81]. But even with adaptive mesh refinement, the available resolution used in multidimensional simulations of supernovae, is too coarse to determine the regime of burning, and the choice of the combustion mode inevitably becomes an artifact of a particular simulation (see [23, 24]). It was found that the delayed detonation model can satisfactorily explain the observed features of the SNIA explosion if the white dwarf was pre-expanded before the burning turned into detonation. This preliminary expansion can be achieved if part of the star is burnt out in the deflagrationv mode. The models that best reproduce observations are those in which the burning begins as a deflagration during which the star expands and then followed by a detonation [11, 22, 26–34, 38, 39]. The "delayed detonation" models produce an ejecta with sufficient amount of the intermediate mass elements and iron-group elements to reasonably match observations of typical SNIa. In the DDT paradigm, the duration of the deflagration phase sets the amount of expansion of the star, which is critical to the composition of material synthesized in the explosion. Various approaches have been used by different groups to simulate supernova explosions using the delayed detonation paradigm [11, 22, 26–34, 38, 39, 46, 47, 96–99]. However, simulations that successfully reproduce observations usually use an artificially ("by hands") introduced transition to detonation after the deflagration wave has reached some threshold density, $\rho \approx (1 \div 5) \times 10^7\ \mathrm{g/cm^3}$, or after about 10–20% of the white dwarf mass was burned by a deflagration wave. Some studies indicate that the transition occurs when a distributed combustion regime is established, provided that the turbulence intensity is sufficiently high [43, 100]. However, since the resolution available for 2D or 3D simulations of SN Ia is too low to resolve the physical processes leading to the transition to detonation, the mechanism of the deflagration to detonation transition remains unclear.

Another criterion based on the instability of planar detonations with respect to one-dimensional pulsations was considered in [101]. It was shown that the detonation wave in the degenerate C-O material is unstable and should decay at high densities, $\rho > 2 \times 10^7\ \mathrm{g/cm^3}$, but becomes stable at lower densities $\rho < 2 \times 10^7\ \mathrm{g/cm^3}$ typical for pre-expanded white dwarfs.

17.6.1 *Pulsating Instability of Detonation*

Since the time of the classical works of Shchelkin [102], Seidel [103], and Erpenbeck [104], it is known that with a strong temperature dependence of the rate of a chemical reaction, a detonation wave is unstable with respect to small one-dimensional pulsations of the detonation front. Seidel and Zel'dovich [105], and later Lee and Stewart [106] showed that pulsating instabilities can lead to self-extinguishing of detonation so that under certain conditions a supersonic combustion regime becomes impossible. In particular, He and Lee [107] have shown that the detonation instability leads to the self-quenching, if there is an unstable perturbation mode with zero imaginary part of the growth rate. By this reason the effect of the detonation self-quenching can be predicted even by solving the linear problem of the detonation stability, without direct non-linear numerical simulations.

Pulsating regime of the detonation propagating at high densities of a white dwarf has been observed in numerical simulations [108–111]. Since the fastest initial stage of white dwarf burning plays a decisive role for the detonation stability, in [101] was considered only the carbon–carbon reaction $^{12}\mathrm{C} + {}^{12}\mathrm{C} \rightarrow {}^{24}\mathrm{Mg}^*$, which is the fastest reaction in the white dwarf explosion. The kinetic equation for the carbon–carbon reaction is

$$-\frac{1}{\alpha^2}\frac{d\alpha}{dt} = \frac{1}{\tau_r}\exp\left(-E_a/T^{1/3}\right). \qquad (17.6.1)$$

Here α is the fuel concentration, $E_a = 85165K^{1/3}$ is the activation energy, and τ_r is a numerical factor, which weakly depends on temperature. Note that Coulomb corrections are not important at high densities, $\rho > 2 \times 10^7$ g/cm^3, where the Fermi momentum of the electrons $p_{Fi} \approx 10^{-16}$ g cm/s exceeds $m_e c = 2.7 \times 10^{-17}$ g cm/s, and the Fermi energy of the electrons $\varepsilon_{Fi} \sim 10^{-6}$ erg exceeds the temperature of the cold fuel and the temperature in the detonation wave. Thus, the equation of state of the degenerate ultrarelativistic electrons can be used in high-density layers of a white dwarf. However, for densities $\rho < 10^7$ g/cm^3, one should also consider the contribution of other factors to the equation of state, such as radiation, Coulomb screening, and electron-positron pairs. Also, at such densities, electrons cannot be considered as an ultrarelativistic degenerate gas. The carbon–oxygen background can be treated as a Boltzmann gas, since the Fermi energy of the ions is less than $\varepsilon_{Fi} \sim 10^{-8}$ erg. Since the temperature of the nuclei, which determines the reaction rate, is equal to the temperature of electrons, a one-component hydrodynamic equation can be used.

The equation of state of the white dwarf matter is determined by electrons (subscript e), positrons (subscript p), and photons (subscript ph). The thermodynamic potential per unit volume Ω, the internal energy per unit mass, ε, and the electrical charge densities can be written as

$$\Omega = \Omega_e + \Omega_p + \Omega_{ph}, \tag{17.6.2}$$

$$\varepsilon = \varepsilon_e + \varepsilon_p + \varepsilon_{ph}, \tag{17.6.3}$$

$$n_e - n_p = n. \tag{17.6.4}$$

Because the chemical potential of the photons is zero, $\mu_{ph} = 0$, in chemical equilibrium the chemical potentials of the electrons and positrons are $\mu_e + \mu_p = 0$.

It is convenient to introduce the following designations for integrals used in the equation of state:

$$I(\Gamma, \Sigma) = \int_0^\infty \frac{x^2}{1 + \exp\left(\sqrt{\Sigma^2 + x^2} - \Gamma\right)} dx, \tag{17.6.5}$$

$$J(\Gamma, \Sigma) = \int_0^\infty \frac{x^4}{1 + \exp\left(\sqrt{\Sigma^2 + x^2} - \Gamma\right)} \frac{1}{\sqrt{\Sigma^2 + x^2}} dx, \tag{17.6.6}$$

$$K(\Gamma, \Sigma) = \int_0^\infty \frac{x^2 \sqrt{\Sigma^2 + x^2}}{1 + \exp\left(\sqrt{\Sigma^2 + x^2} - \Gamma\right)} dx. \tag{17.6.7}$$

Using these designations, Eqs. (17.6.2)–(17.6.4) take the form

$$P = -\frac{\Omega}{V} = \frac{1}{3\pi^2}\frac{T^4}{(\hbar c)^3}\left[J(\Gamma_e, \Sigma) + J(\Gamma_p, \Sigma)\right] + \frac{4\sigma}{3c}T^4, \tag{17.6.8}$$

$$\rho\varepsilon = \frac{T^4}{\pi^2(\hbar c)^3}\left[K(\Gamma_e, \Sigma) + K(\Gamma_p, \Sigma)\right] + \frac{4\sigma}{c}T^4, \tag{17.6.9}$$

$$n = \frac{1}{\pi^2}\left(\frac{T}{\hbar c}\right)^3\left[I(\Gamma_e, \Sigma) - I(\Gamma_p, \Sigma)\right]. \tag{17.6.10}$$

Here $\Gamma = \mu T$ is the scaled chemical potential, $\Sigma = mc^2/T$ is the inverse scaled temperature, σ is the Stefan-Boltzmann constant, and the density $\rho = 2nm_a$, where m_a is the atomic mass unit. The Euler equations describing dynamics of a detonation wave with the reaction rate (17.6.1), the equation of state given by Eqs. (17.6.8)–(17.6.10), and the boundary conditions at the leading shock of the detonation specify the problem of detonation dynamics and stability.

The stationary solution of the Euler equations corresponds to a plane steady detonation wave, which consists of a planar stationary shock wave followed by the reaction zone. The shock compresses and heats initially cold fuel igniting the thermonuclear reaction, while expansion of the burning matter in the reaction zone plays the role of a piston supporting a steady propagation of the leading shock. In the reference frame co-moving with the leading shock the steady solution is

$$J \equiv \rho v = J_0, \tag{17.6.11}$$

$$\Pi \equiv P + \rho v^2 = \Pi_0, \tag{17.6.12}$$

$$H \equiv \rho v\left(\frac{\rho\varepsilon + P}{\rho} + \frac{v^2}{2} + \alpha q\right) = H_0, \tag{17.6.13}$$

where q is the energy release of the reaction per unit mass of the fuel. Equations (17.6.11)–(17.6.13) determine all quantities in the detonation wave if the state of the thermonuclear fuel ahead of the shock (P_0, ρ_0) and the velocity of the detonation with respect to the fuel $D_0 = v_0$ are known.

In the range of the white dwarf densities $3 \times 10^9 > \rho > 10^7$ g/cm^3, the fuel temperature $T \approx (1-5) \times 10^8$ K is much lower than the Fermi energy of the degenerate electron gas. Therefore, the initial pressure of the fresh fuel P_0 is a function of the fuel density ρ_0 only. The relations between the fuel density, pressure, and internal energy can be presented in the following parametric form:

$$\frac{\rho_0}{2m_a} = \left(\frac{m_e c}{\hbar}\right)^3 \frac{1}{3\pi^2}\sinh^3\frac{\xi}{4}, \tag{17.6.14}$$

$$P_0 = \frac{m_e^4 c^5}{32\pi^2\hbar^3}\left(\frac{1}{3}\sinh\xi - \frac{8}{3}\sinh\frac{\xi}{2} + \xi\right), \tag{17.6.15}$$

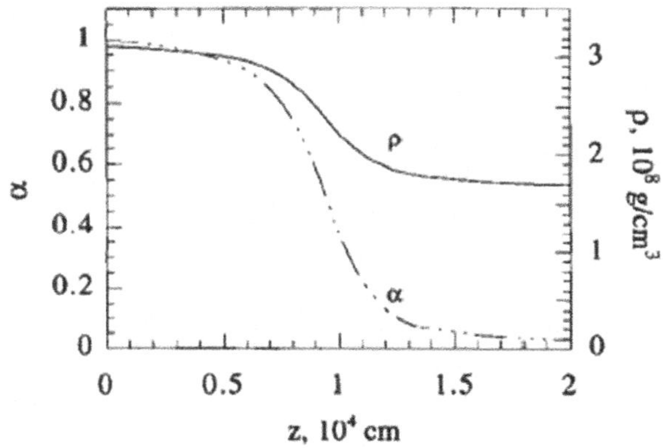

Fig. 17.1 A steady CJ detonation wave profiles at the density $\rho \sim 10^8$ g/cm^3 [101]

$$\varepsilon_0\rho_0 = \frac{m_e^4 c^5}{32\pi^2\hbar^3}(\sinh\xi - \xi), \tag{17.6.16}$$

where the parameter $\xi = 4\text{arc}\sinh(p_{Fi}/mc)$ varies from $\xi \approx 0.3$ for $\rho = 10^6$ g/cm^3 and $\xi \approx 2$ for $\rho = 3 \times 10^9$ g/cm^3.

The detonation velocity is restricted from below by the velocity of CJ detonation, which is determined only by the energy release in the thermonuclear reaction and by the equation of state of burning matter. The profiles of the hydrodynamic variables behind the leading shock (at $z > 0$) in a stationary detonation are determined by integrals (17.6.11)–(17.6.13) together with Eq. (17.6.1), which a stationary solution has the form

$$-\frac{1}{\alpha^2}v\frac{d\alpha}{dz} = \frac{1}{\tau_r}\exp\left(-E_a/T^{1/3}\right). \tag{17.6.17}$$

The profiles of the hydrodynamic variables for the stationary detonation calculated in [101] are shown in Fig. 17.1.

It is seen that the reaction region consists of two zones: the induction zone, where the thermonuclear burning is slow, and the reaction zone, where most of the fuel is burned. The reaction zone is followed by the downstream flow where combustion ends. In the case of strong dependence of the reaction rate on temperature, the width of the reaction zone is much less than the thickness of the induction zone so that the total width of detonation is determined by the thickness of the induction zone. Thus, the width of the detonation is mainly determined by the width of the induction zone, which can be estimated as $v_1\tau_{1/2}$, where v_1 is the fuel velocity immediately behind the leading shock wave and $\tau_{1/2}$ is the half reaction time, that is, the time required to burn half of the fuel. The half reaction time for a steady plane thermonuclear detonation in white dwarfs is shown in Fig. 17.2 versus the density of the white dwarf. Its value decreases significantly with an increase in the density of the white dwarf matter. The structure of the reaction zone strongly depends on the effective

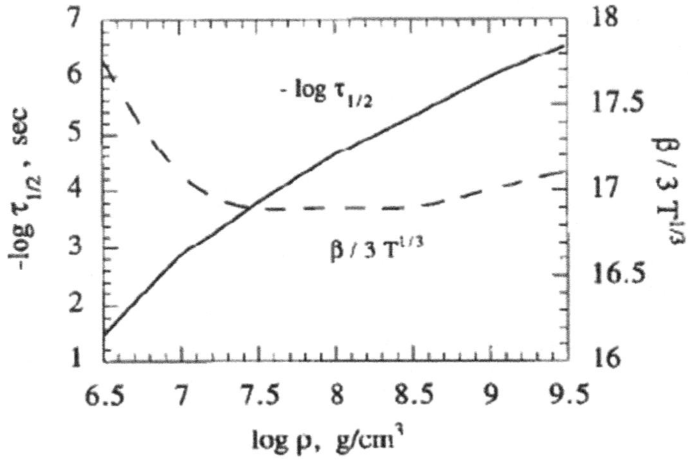

Fig. 17.2 Half reaction time and the scaled effective activation energy of the reaction $\beta/3T_1^{1/3}$ versus density of the white dwarf [101]

activation energy of the reaction, scaled by the temperature immediately behind the leading shock, $\beta/(3T_1^{1/3})$. In particular, the half reaction time changes mainly due to the change in the effective activation energy, $\tau_{1/2} \sim \exp(\beta/3T_1^{1/3})$. As a result, the effective activation energy determines the ratio of the induction zone thickness and the thickness of the effective reaction zone, as well as stability of the stationary detonation.

Since the typical temperature of the fuel in the detonation wave in a white dwarf is about 5×10^9 K, the characteristic values of the scaled effective activation energy are rather high: $\beta/3T_1^{1/3} \approx 20$. Because of this, the thickness of the active reaction zone is much smaller than the induction zone thickness. It is known that steady chemical detonation waves with such structure are always unstable [102–105]. Therefore, it can be expected that thermonuclear detonation will also be unstable. It should be emphasized that the stability of detonation depends on the effective activation energy $\beta/3T_1^{1/3}$, but not on the reaction time $\tau_{1/2}$ [102, 103]. It can be said that the reaction time provides only the characteristic time scale of instability and, therefore, it can be eliminated in a dimensionless analysis of the stability problem. For this reason, small inaccuracies in the temperature calculation immediately behind the leading shock wave are not amplified by the exponential function $\tau_{1/2} \sim \beta/3T_1^{1/3}$.

Investigation of the detonation stability implies solving the linearized Euler equations with the boundary conditions specified by the Hugoniot relations for the leading shock wave and the kinetic Eq. (17.6.1). In the reference frame of the leading shock wave, small perturbations are taken in the form

$$\delta\phi = \phi'(z + \Delta) - \phi(z),$$

where $\Delta = \delta z$ is a small shift of the shock wave, the prime denotes the perturbed quantity. Since the unperturbed solution does not depend upon time, small perturbations can be taken in the form

$$\delta\phi(z,t) = \delta\phi(z)\exp(\sigma t), \tag{17.6.18}$$

where σ is the instability growth rate, and since we are interested in unstable solutions, we assume $\mathrm{Re}\sigma > 0$. Then, the system of linearized equations becomes

$$\frac{d\delta J}{dz} + \sigma\left(\delta\rho - \Delta\frac{\partial\rho}{\partial z}\right) = 0, \tag{17.6.19}$$

$$\frac{d\delta\Pi}{dz} + \sigma\delta J = 0, \tag{17.6.20}$$

$$\begin{aligned} &\frac{d\delta H}{dz} + \sigma\left[\delta P + \delta(\rho\varepsilon) + J\delta v + \left(\frac{v^2}{2} + \alpha q\right)\delta\rho + q\rho\delta\alpha\right] \\ &\quad - \sigma\Delta\left[\frac{\partial P}{\partial z} + \frac{\partial(\rho\varepsilon)}{\partial z} + J\frac{\partial v}{\partial z} + \left(\frac{v^2}{2} + \alpha q\right)\frac{\partial\rho}{\partial z} + q\rho\frac{\partial\alpha}{\partial z}\right] = 0 \end{aligned}, \tag{17.6.21}$$

$$\frac{d\delta\alpha}{dz} + \sigma\frac{\delta\alpha}{v} - \frac{1}{v}\left[\left(\alpha^2\frac{\delta v}{v} - 2\alpha\delta\alpha\right)R(T) - \alpha^2\frac{dR(T)}{dT}\delta T\right] = 0, \tag{17.6.22}$$

where the following expressions were used for the perturbed mass flux, the momentum flux and the energy flux:

$$\delta J = \rho\delta v + v\delta\rho, \tag{17.6.23}$$

$$\delta\Pi = \delta P + 2J\delta v + v^2\delta\rho, \tag{17.6.24}$$

$$\begin{aligned} \delta H =& v\delta P + v\delta(\rho\varepsilon) + \left(\frac{v^2}{2} + v\alpha q\right)\delta\rho + \\ &+ \left(P + \rho\varepsilon + \frac{3}{2}\rho v^2 + \rho\alpha q\right)\delta v + Jq\delta\alpha \end{aligned}. \tag{17.6.25}$$

The quantities $\delta\rho$, δP, and $\delta\varepsilon$ are related via the linearized equation of state. In the case of the CJ detonation the determinant of the system of Eqs. (17.6.23)–(17.6.25) tends to zero as $\alpha \to 0$. This implies that one of the independent modes of Eqs. (17.6.19)–(17.6.22) grows faster than exponential growth at $z \to \infty$.

Boundary conditions at the leading shock wave follow from the linearized Hugoniot relations for the leading shock wave

$$\delta J = \sigma\Delta\rho|_0^1, \tag{17.6.26}$$

$$\delta\Pi = 0, \tag{17.6.27}$$

$$\delta H = \sigma\Delta\left(\rho\varepsilon + \frac{1}{2}\rho v^2 + \rho q\right)\bigg|_0^1, \tag{17.6.28}$$

$$\delta\alpha = 0. \tag{17.6.29}$$

The last condition, Eq. (17.6.29), means that there is no burning ahead of the leading shock wave, which is considered as a discontinuity surface.

The boundary condition in the downstream flow implies that the solution of the linear problem is regular at $z \to \infty$ [106]. To specify this condition, we need to find the asymptotic solution of the linearized hydrodynamic Eqs. (17.6.19)–(17.6.22) in the uniform downstream flow of the burnt matter at $z \to \infty$, which is a superposition of four independent linear modes of Eqs. (17.6.19)–(17.6.22). Two of them correspond to the sound waves propagating downstream (from the detonation wave) and upstream (to the detonation wave). The other two modes are related to perturbations of entropy and fuel concentration. Of these four modes, only the amplitude of the sound mode propagating upstream from the burnt matter to the detonation front increases as $z \to \infty$. Therefore, for the regular solution of the linear problem, the amplitude of this mode must be taken equal to zero in the flow of the burnt matter; this is called the "radiation condition". The entropy mode, the concentration mode, and the sound mode propagating from the detonation front vanish at infinity so that their amplitudes remain unspecified by the boundary condition in the flow of the burnt matter.

For the numerical solution of the stability problem the "radiation condition" may be formulated as a condition that the perturbation velocities are zero at some point far away from the detonation front, $\delta v = 0$ for large values of z. The point where this condition holds should be taken sufficiently far from the reaction zone, much farther than the typical instability length scale $v_2/\mathrm{Re}\sigma$, where v_2 is the velocity of the burnt matter. The choice of this point depends also on the accuracy requirements of the numerical solution. For a usual laboratory detonation this condition may be interpreted as the condition at the piston surface. If the piston is placed sufficiently far from the detonation front, it does not influence the detonation stability.

17.6.2 Thermonuclear Detonations. Spectrum of Instabilities

For the thermonuclear detonation in white dwarfs the main propagation regime is the CJ detonation. The CJ detonation velocity D is calculated as a smallest velocity for which Eqs. (17.6.11)–(17.6.13) could be solved for $\alpha = 0$. The obtained values of the CJ-detonation velocity are in good agreement with the results of simulation [112]. For an ideal gas the CJ detonation velocity can be determined analytically, and in [101] it was used as a test to evaluate the accuracy of the numerical solution.

Comparison of the CJ-velocity calculated analytically differs by about 5% from the numerical value for an ideal gas. Once the detonation velocity is known, the steady solution can be determined from Eqs. (17.6.11)–(17.6.13) and tabulated.

The spectral problem has been solved in [101] using the shooting method. Since the equation of state does not depend on the fuel concentration, Eqs. (17.6.19)–(17.6.22) can be written versus the reaction coordinate:

$$\frac{d\delta J}{d\alpha} = \sigma \hat{\Re} \rho, \tag{17.6.30}$$

$$\frac{d\delta \Pi}{d\alpha} = \sigma \hat{\Re} J, \tag{17.6.31}$$

$$\frac{d\delta H}{d\alpha} = \sigma \left[\hat{\Re} P + \hat{\Re} e + J \hat{\Re} v + \left(\frac{v^2}{2} + \alpha q \right) \hat{\Re} \rho + q\rho \hat{\Re} \alpha \right], \tag{17.6.32}$$

$$\frac{d\delta\alpha}{d\alpha} = \frac{\sigma}{v} \hat{\Re} \alpha - \frac{2\delta}{v} - \frac{2\delta\alpha}{v} + \frac{1}{R}\frac{dR}{dT}\delta T, \tag{17.6.33}$$

where the operator $\hat{\Re}$ is defined as

$$\hat{\Re} = \sigma \frac{d}{d\alpha} - \frac{dz}{d\alpha}\delta. \tag{17.6.34}$$

Integration of Eqs. (17.6.30)–(17.6.33) is carried out from the leading shock, where the boundary conditions are given by (17.6.26)–(17.6.29) and then solving equation $\delta v = 0$ in the downstream flow.

Functions (17.6.5)–(17.6.7) are tabulated in the two-dimensional $(\Gamma - \Sigma)$ plane. Then, using formulas (17.6.8)–(17.6.10) and the boundary conditions (17.6.11)–(17.6.13) for the steady solution, the dependence of all thermodynamic quantities and their derivatives upon α was obtained. Using the obtained stationary solution, the system of Eqs. (17.6.30)–(17.6.33) was numerically integrated from $\alpha = 1$ to $\alpha = \alpha_f > 0$ using the fourth-order Runge–Kutta method.

The roots σ of the equation $\delta v(\sigma, \alpha_f) = 0$ were calculated in [101] using the modified one-dimensional Newton's method in the rectangular region of the complex plane ("carpet search"). When the spectrum for a given α_f is found the convergence of the spectrum at $\alpha_f \to 0$ must be investigated. For an ideal polytropic gas, calculations show that the value $\alpha_f = 0.1$ already provides reasonable accuracy. Therefore, it is reasonable to set $\alpha_f = 0.1$ in all calculations of the unstable spectrum of detonation in the white dwarfs.

The numerical method used in [101] was tested for the case of an ideal gas. The eigenvalue problem of the detonation stability has been solved in [101] for different values of the overdrive factor $f = (D/D_{CJ})^2$. The results obtained for an ideal gas fuel with the adiabatic exponent $\gamma = 1.2$, dimensionless energy release

$q\rho_0/P_0 = 50$, and activation energy $E_a/T_0 = 50$ were in good agreement with calculations [104, 106, 107].

Direct numerical simulations of the detonation instability [107] show that the unstable mode with the smallest frequency, $\mathrm{Im}(\sigma)$, plays a decisive role in the nonlinear stage of detonation instability. For the CJ-detonation, the frequency and the growth rate of this mode are determined by the activation energy of the reaction. With an increase in the activation energy, the growth rate $\mathrm{Re}(\sigma)$ of the mode increases, while its frequency tends to zero until it vanishes at sufficiently high activation energies $E_a/T_1 > 7.4$. According to simulations [107] of the nonlinear stage of detonation instability, the detonation is self-quenching for these activation energies. For the critical value of the activation energy $E_a/T_1 = 7.4$ the instability growth rate $\mathrm{Re}(\sigma)$ of the zero frequency mode ($\mathrm{Im}(\sigma) = 0$) is close to the growth rates of another unstable modes, but it becomes much higher than the growth rate of another modes with further increase of the activation energy.

As it was said, the only independent parameter of the stability problem of the plane stationary CJ detonations in the carbon thermonuclear fuel is the initial density of the degenerate matter. The numerical solution to the stability of detonation, formulated in the previous section, was used in [101] to study the stability of thermonuclear detonation for densities in the range $10^6 < \rho < 3 \times 10^9$ g/cm^3. Figure 17.3a, b shows the dependence of the growth rates of unstable modes on their frequencies. The growth rate and frequencies of the detonation instability are shown in the dimensionless units, scaled with the half reaction time: $\sigma_{1/2} = \sigma\tau_{1/2}$.

It is seen in Fig. 17.3a that at high densities, $\rho \sim 10^8$–10^9 g/cm^3, the stationary detonation is unstable with respect to a large number of perturbation modes with different growth rates and different frequencies. The total number of the unstable modes can be about 20. Since the detonation dynamics is determined mainly by the unstable modes of low frequencies, we can consider only modes with $\mathrm{Im}\{\sigma\tau_{1/2}\} < 10$. With a decrease in the density, the total number of unstable modes becomes smaller and the scaled rates of their instabilities decrease. In addition, at lower densities, the maximum growth rate is achieved at lower frequencies of the

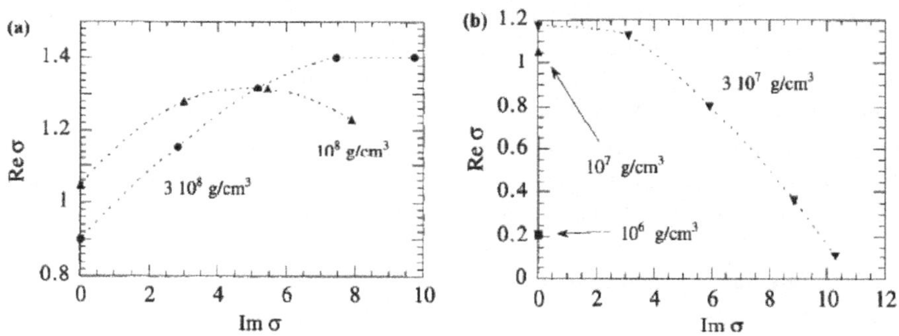

Fig. 17.3 Unstable spectrum of the CJ-detonation in a white dwarf at various densities. Here $\sigma_{1/2} = \sigma\tau_{1/2}$ is the complex frequency σ scaled with the half reaction time [101]

unstable modes. It is seen that at $\rho = 3 \times 10^8\,\mathrm{g/cm^3}$ the scaled maximal growth rate $\mathrm{Re}\{\sigma\tau_{1/2}\} = 1.4$ corresponds to the fifth mode with $\mathrm{Im}\{\sigma\tau_{1/2}\} = 9.7$, while for a slightly lower density $\rho = 10^8\,\mathrm{g/cm^3}$, the third mode with the scaled frequency $\mathrm{Im}\{\sigma\tau_{1/2}\} = 5.4$ has the maximal growth rate $\mathrm{Re}\{\sigma\tau_{1/2}\} = 1.3$. This trend becomes more pronounced at lower densities, as shown in Fig. 17.3b. For $\rho = 3 \times 10^7\,\mathrm{g/cm^3}$, the maximum growth rate $\mathrm{Re}\{\sigma\tau_{1/2}\} = 1.2$ is for the mode of zero frequency, while for $\rho = 10^7\,\mathrm{g/cm^3}$ the mode of zero frequency remains the only unstable mode.

Thus, the general trend is that the lower the density of the degenerate fuel, the more stable the thermonuclear detonation. For the thermonuclear detonations in white dwarfs the temperature behind the leading shock increases with the decrease of the fuel density (see Fig. 17.2). Therefore, the scaled effective activation energy $\beta/3\sqrt[3]{T}$ decreases and the detonation becomes more stable. The unstable root with zero imaginary part leaves the real axis at the fuel density $\rho \approx 1.6 \times 10^6\,\mathrm{g/cm^3}$, which agrees quantitatively with results of direct numerical simulations [109], where the instability of an overdriven thermonuclear detonation in pure carbon fuel was observed at the density $\rho \approx 2.3 \times 10^7\,\mathrm{g/cm^3}$ and higher. More accurate estimate shows that the pure real root of the unstable spectrum leaves the real axis at the density $\rho = 2.1 \times 10^6\,\mathrm{g/cm^3}$, which agrees with the stability limit $\rho \approx 2 \times 10^7\,\mathrm{g/cm^3}$ obtained in simulations [108]. Self-quenching of detonation at densities above $\rho \approx 2 \times 10^7\,\mathrm{g/cm^3}$ was not observed in the simulations in [108], which can probably be explained by the coarse resolution, which is inevitably used in 2D and 3D simulations of white dwarf explosions.

From the numerical simulations [107] follows that in the case of zero frequency mode, the intensity of the leading pulsating shock wave falls to a low level, insufficient to ignite the fuel behind the shock. The solution of the eigenvalue problem of the detonation stability in white dwarfs shows that the spectrum of detonation instability has a zero-frequency mode at fuel densities, $\rho \geq 2 \times 10^7\,\mathrm{g/cm^3}$. As a result, it can be expected that, due to self-quenching, thermonuclear detonation cannot propagate at densities higher than $2 \times 10^7\,\mathrm{g/cm^3}$.

The "delayed detonation" paradigm, which aims to explain the mass ejection and the balance between intermediate mass and iron group elements observed by SN Ia, suggests that the transition to detonation occurs through a specific DDT mechanism. However, there are scenarios of the transition from deflagration to detonation, which have not been properly studied. For example, in the scenarios of a spherically expanding flames, the increase of the flame velocity propagating in decreasing density was not considered. In a sense, this effect is analogous to the known unlimited increase in the intensity of a shock wave propagating in the direction of decreasing density. The initially ignited flame in the deflagration mode is accelerated due to various processes and therefore generates pressure waves which can steepen into weak shock waves running toward the star surface ahead of the flame. The self-similar solution to the problem of a shock wave propagating to the star surface in a power law decreasing density has been obtained by Gandelman and Frank-Kamentskii [113], Sakurai [114], and Whitham [115]. If the density decreases as $\rho \propto X^\delta$, where $X = R - r$ is the coordinate measured from the star surface and δ is a constant, the temperature and velocity in the shock wave behave as

$$T \sim u^2 \sim X^{-2(1-\alpha)/\alpha}, \quad P \sim X^{\delta-2(1-\alpha)/\alpha}, \quad u \sim X^{-(1-\alpha)/\alpha}, \tag{17.6.35}$$

where the exponent $\alpha < 1$ depends on the values of δ and the adiabatic constant γ. For example, for $\delta = 2$ and $\gamma = 7/5$, $\alpha = 0.718$; the temperature and pressure increase as $T \propto 1/X^{0.718}$ and $P \sim 1/X^{1.28}$, when the distance to the surface decreases, $X \to 0$.

It is likely that the intensity of the shock wave can increase enough to ignite a detonation. The energy deficit known for "pure deflagration" models can probably be compensated by very strong shock waves emerging at the star surface.

The scenario of a white dwarf explosion in SN Ia events, based on the amplification of shock waves propagating along an inhomogeneous density decreasing toward the stellar surface, was used in the simulations by Charignon and Chièze [116]. They showed that when the initial deflagration wave propagates at the velocity 75 km/s in the central core of a C + O white dwarf of mass $1.43M_{\odot}$, it generates weak ($M \sim 0.003$) pressure waves, and these pressure waves will intensify as they travel along the density gradient toward the stellar surface and can trigger detonation in the outer layers. This scenario looks promising, but unfortunately it remains unclear where exactly the shock triggers a detonation and how the expansion of the star influences the location where the detonation is triggered. In the context of SN Ia, further investigations of this scenario would be desirable.

17.7 Rayleigh–Taylor Instabilities, Bubbles and Turbulence

The burning of a white dwarf (WD), most likely, first begins in the deflagration mode during a local thermal runaway in many chaotically distributed ignition kernels within 50–100 km from the center of the star134. As shown in previous section, much more energy is required to ignite a detonation, and even if a detonation ignites, it will be unstable in a high-density core and decay into shock wave and deflagration.

Most of the multidimensional simulations of the WD explosion predominantly assumed that SN Ia events took place in a non-rotating progenitor. However, recent studies [117–121] have questioned this hypothesis, arguing that the same accretion process that creates the WD mass also transfers angular momentum from the disk to the accreting star. The observed values of the peak brightness, and light curves of SN Ia sensitively depend on the ignition conditions, the density at the moment of carbon ignition, the deflagration velocity, the density at which deflagration transits to detonation, and the rate of the star rotation. We will consider only the key processes that affect the scenarios of SN Ia explosions. It should be emphasized that there are many uncertainties regarding the possible progenitors of SN Ia, therefore regarding ignition. Therefore, despite the great progress made in recent years, all simulations inevitably rely on models with some free parameters in order to obtain agreement with observations. Extensive observational studies of SN Ia have shown that, despite their relative homogeneity, there are several distinct subclasses, and the observed heterogeneity may require different progenitors and/or explosion mechanisms. With regard to the use of SN Ia as standard candles, the uniformity of SN Ia events and their high luminosity at maximum are a fundamental requirement. In this context, the relation between maximum luminosity and another observable property requires more careful study. The single-degenerate Chandrasekhar-mass scenario, in which the white dwarf companion is a nondegenerate star, is still generally considered the preferred progenitor of SN Ia. In this scenario, the explosion is triggered when the WD mass approaches the Chandrasekar value ($M_{\rm Ch} \approx 1.5 M_{\odot}$) at the expense of mass accretion from its companion star.

17.7.1 Nonlinear Stage of the RT Instability

Subsonic deflagration, which begins near the center of the white dwarf, is subject to various instabilities, Darrieus–Landau (DL), thermal-diffusion, Rayleigh-Taylor (RT), and Kelvin–Helmholtz (KH) instabilities. The Rayleigh–Taylor instability plays a decisive role when the gravity acceleration is directed opposite to the density gradient. Arising under various physical conditions, from laboratory experiments with the laser-driven inertial fusion or magnetic confinement of high-temperature plasma to gravitational confinement of rising bubbles of light ashes in SN Ia explosions, the RT instability exhibits similar features in its evolution. The mixing and turbulence caused by the RT instability has been the subject of many theoretical and numerical studies. There are still many unresolved problems concerning the nonlinear stage of the RT instability, see e.g. reviews [122–126].

As discussed earlier, it is likely that the initial ignition occurs in the form of multipoint ignition of deflagration in the vicinity of about 50 km from the center of the star. A popular scenario of the carbon–oxygen white dwarf explosion is that the primary ignited deflagration accelerates and transfers to distributed regime due to the Rayleigh–Taylor and Kelvin–Helmholtz instabilities, which probably can create proper conditions for the transition to detonation somewhere in the outer layers of the star, at densities around $10^7\ \mathrm{g/cm^3}$.

Since the flame propagates in a strong gravitational field, the RT instability can become the dominant process determining the fate of combustion. While initial turbulence is caused by the development of large-scale convection due to the unstable configuration of the high temperature core [122, 127], RT-induced turbulence is believed the dominant flame acceleration mechanism at densities around $(1-3) \times 10^7\ \mathrm{g/cm^3}$. In 2D simulations [67, 68, 128], it was found that the velocity of a turbulent flame obeys the scaling law $\sqrt{g\,\Delta}$ for small Froude numbers, where Δ is the unresolved RT-turbulent driving scale equal to one or two computational cell widths. In other words, the turbulent flame speed U_t is associated with the velocity $U_t \propto \sqrt{gR_b}$ of bubbles of the size $R_b \sim \Delta$ that are formed during the nonlinear stage of RT instability. Similar conclusion was obtained in [97, 129], where 3D simulations of were performed for only the outer layers of $0.5\mathrm{C} + 0.5\mathrm{O}$ fuel with density (0.5−1.5) × 10^7 g/cm^3. For the conditions used in [97] the Froude number was $Fr \simeq 5 \times 10^{-3}$, the thickness of the flame itself mattered before the transition to the distributed regime, and the flame velocity was determined by the velocity of raising bubbles. Interesting results concerning morphology of the flame dynamics controlled by the RT instability were obtained in recent fully resolved 3D simulations [130].

17.7.2 Multipoint Ignition and Bubbles in SN Ia Explosions

Obviously, a multipoint ignition violates the spherical symmetry of white dwarf explosions. Let us consider a small spherical bubble of radius $R > R^* = \xi_0^* \sqrt{2\chi\tau_r}$, which is formed from one of the ignited flame kernels near the center of the star. The bubble surface is the front of the outward expanding flame separating the burned products form the fresh fuel outside the bubble. The burning localized on the bubble surface can be considered as a discontinuity. Since the density of the combustion products is less than the density of the fuel, $\Theta = \rho_u/\rho_b = 1.21 - 1.7$, light bubbles emerge from the center to the surface of the star. During this time: (1) the radius of the bubble increases; (2) all bubbles come out to the surface of the star; (3) the flame instability front wrinkles the bubble surface at scales much smaller than the bubble radius. According to [81] the deflagration velocity near the center of the 0.5C+0.5O white dwarf is $U_f \approx 36\,\text{km/s}$. The bubbles formed of the ignited kernels leave the star center during the time of the order of the average induction time at the WD center, which can be estimated using Eq. (17.3.8) as

$$\tau_b \approx \frac{1}{\sqrt{(4\pi/3)(\Theta - 1)\Gamma\rho_c}}. \tag{17.7.1}$$

During this time, the bubble rises to the star surface, and at the same time the radius of the bubble increases as $R_b = U_f \tau_i \approx 10\,\text{km}$, remaining much smaller the central core of WD, $R_c = \sqrt{3}\,(k\overline{T}_c/E_a)^{1/6} R_{WD} \approx 100\,\text{km}$, and the amount of fuel burned inside the bubbles,

$$M_b \approx \frac{4\pi}{3}\rho_b (U_f \tau_i)^3, \tag{17.7.2}$$

is negligible compared to the central core mass $\sim 24\pi\rho_c \sqrt{k\overline{T}_c/E_a R_{WD}^3}$. This means that most of the central core remains unburned and is likely to be burned up later by turbulent flames [41].

The velocity of rising bubbles is different for open cylindrical bubbles and for closed spherical bubbles: $u_{2b} = 0.49\sqrt{g\lambda}$ and $u_{3b} = 0.67\sqrt{gR_b}$. In derivation of these formulas [123, 131135], the density of the matter inside the bubble was assumed to be negligible, and the gravitational acceleration was assumed to be uniform. Estimating the gravitational acceleration in SN Ia at the radius R as

$$g(R) \approx \Gamma(4\pi/3)\rho_c R, \tag{17.7.3}$$

the bubble velocity in [72] was estimated as

$$u_b \approx 0.67 R_b \sqrt{\frac{4\pi}{3}\frac{\Theta - 1}{\Theta}\Gamma\rho_c} \approx 0.67 R_b/\tau_b. \tag{17.7.4}$$

It should be emphasized that the velocity of the rising bubbles becomes greater than the flame speed when the bubble radius increases up to the value $R_b \geq 1.5 U_f \tau_b$. In fact, this condition confirms that only an insignificant part of the central core burns out in the rising bubbles.

It was suggested in Refs. [72, 82, 132, 133] that the scenario of rising bubbles consuming negligible fraction of the hot core as they move toward the stellar surface can explain the development of turbulence ahead of the subsequently ignited deflagration in the stellar core and explain the development conditions for the transition to detonation according to the Zel'dovich gradient mechanism.

The scenario of multipoint ignition and rising bubbles, proposed in [72, 133], obviously leads to violation of the spherically symmetric explosions of SN Ia, in view of chaotic distribution of the ignition kernels, which are progenitors of the rising bubbles. At present, it is difficult to say which model of the SN Ia explosion better explains the observations. The fast-rotating white dwarfs which have accreted a large angular momentum can be detectable sources of gravitational waves, but they may not reach central carbon ignition even if they grow beyond the Chandrasekhar limit of $1.4 M_{\odot}$. The reduced effective gravity in the equatorial direction caused by centrifugal forces changes the structure of the white dwarf progenitor before the explosion and considerably changes the scenario of rising bubbles. Another possibility of explosion can be gravitational confined detonation models [119–121]. Three-dimensional simulations by Fink et al. [121] for the first time makes it possible to sequentially select observables for models of differentially rotating carbon–oxygen white dwarfs, which can lead to more realistic modeling results.

References

1. R. Kippenhahn, A. Weigert, *Stellar Structure and Evolution* (Springer, Berlin, 1990)
2. D. Prialnik, *An Introduction to the Theory of Stellar Structure and Evolution* (Cambridge University Press, Cambridge, 2000)
3. S. Shore, N. Steven, *The Tapestry of Modern Astrophysics* (Wiley, Hoboken, 2003)
4. C.J. Hansen, S.D. Kawaler, V. Trimble, *Stellar Interiors: Physical Principles, Structure, and Evolution*, 2nd edn. (Springer, New York, 2004)
5. D. Arnett, *Supernovae and Nucleosynthesis* (Princeton University Press, Princeton, 1996)
6. S. Chandrasekhar, *An Introduction to the Study of Stellar Structure* (Dover, New York, 1967)

7. M.S. Longair, *The Cosmic Century: A History of Astrophysics and Cosmology* (Cambridge University Press, 2006)
8. J.C. Wheeler, Astrophysical explosions: from solar flares to cosmic gamma-ray bursts. Philos. Trans. R. Soc. A **370**, 774–799 (2012). See also Ref. [46]
9. D. Maoz, F. Mannucci, G. Nelemans, Observational clues to the progenitors of Type Ia supernovae. Annu. Rev. Astron. Astrophys. **52**, 107–170 (2014)
10. W. Hillebrandt, J.C. Niemeyer, Type Ia supernova explosion models. Annu. Rev. Astron. Astrophys. **38**, 191–230 (2000)
11. W. Hillebrandt, M. Kromer, F.K. Röpke, A.J. Ruiter, Towards an understanding of Type Ia supernovae from a synthesis of theory and observations. Front. Phys. **82**, 116–143 (2013)
12. F.K. Röpke, Combustion in thermonuclear supernovae explosions, in *Handbook of Supernovae* (Springer, 2017)
13. K. Maeda, Y. Terada, Progenitors of type Ia supernovae. Int. J. Mod. Phys. D **25**(1–22), 1630024 (2016)
14. M. Livio, P. Mazzali, On the progenitors of type Ia supernovae. Phys. Rep. **736**, 1–23 (2018)
15. A.V. Filippenko, Optical spectra of supernovae. Annu. Rev. Astron. Astrophys. **35**, 309–355 (1997)
16. A.G. Riess, A.V. Filippenko, P. Challis et al., Observational evidence from supernovae for an accelerating universe and a cosmological constant. ApJ. **116**, 1009–1038 (1998)
17. S. Perlmutter, G. Aldering, G. Goldhaber et al., Measurements of Ω and Λ from 42 high-redshift supernovae. ApJ. **517**, 565–586 (1999)
18. P. Garnavich, Discovery of cosmic acceleration, in *Handbook of Supernovae* (Springer, 2017)
19. S. Blondin, T.M. Davis, K. Krisciunas et al., Time dilation in type Ia supernova spectra at high redshift. Astrophys. J. **682**(2), 724–736 (2008)
20. S.E. Woosley, T.A. Weaver, The physics of supernova explosions. Annu. Rev. Astron. Astrophys. **24**, 205–253 (1986)
21. A. Rest, D. Scolnic, R.J. Foley et al., Cosmological constraints from measurements of type Ia supernovae discovered during the first 1.5 yr of the pan-STARRS1 survey. Astrophys. J. **795**, 44–78 (2014)
22. L. Dessart, S. Blondin, D.J. Hillier, A. Khokhlov, Constraints on the explosion mechanism and progenitors of Type Ia supernovae. Mon. Not. R. Astron. Soc. **441**, 532–550 (2014)
23. F.K. Röpke, S.A. Sim, Models for type Ia supernovae and related astrophysical transients. Space Sci. Rev. **214**(1–17), 72 (2018)
24. A.G. Nouri, P. Givi, D. Livescu, Modeling and simulation of turbulent nuclear flames in Type Ia supernovae. Prog. Aerosp. Sci. **108**, 156–179 (2019)
25. W.D. Arnett, A possible model of supernovae: detonation of ^{12}C. Astrophys. Space Sci. **5**, 180–212 (1969)
26. S.E. Woosley, Type I supernovae: carbon deflagration and detonation, in *Supernovae* (Springer, 1990)
27. K. Nomoto, F.K. Thielemann, J.C. Wheeler, Explosive nucleosynthesis and type I supernovae. Astrophys. J. **279**, L23–L26 (1984)
28. M. Fink, M. Kromer, I.R. Seitenzahl, F. Ciaraldi-Schoolmann, F.K. Röpke, S.A. Sim, R. Pakmor, A.J. Ruiter, W. Hillebrandt, Three-dimensional pure deflagration models with nucleosynthesis and synthetic observables for type Ia supernovae. Mon. Not. R. Astron. Soc. **438**(2), 1762–1783 (2013)
29. S.I. Blinnikov, A.M. Khokhlov, Development of detonations in degenerate stars. Sov. Astron. Lett. **12**, 131–134 (1986) (Translation: Pisma v Astron. Zhurnal **12**, 318–324 (1986))
30. S.I. Blinnikov, A.M. Khokhlov, Stage of spontaneous flame propagation in supernovae. Sov. Astron. Lett. **13**, 364–366 (1987)
31. A.M. Khokhlov, Delayed detonation model for type Ia supernovae. Astron. Astrophys. **245**, 114–128 (1991)
32. H. Yamaoka, K. Nomoto, T. Shigeyama, F.K. Thielemann, Late detonation models for the type Ia supernovae SN 1991T and SN 1990N. Astrophys. J. **393**, L55–L58 (1992)

33. P. Höflich, A. Khokhlov, J.C. Wheeler, M.M. Phillips, N.B. Suntzeff, M. Hamuy, Maximum brightness and pos-tmaximum decline of light curves of type supernovae Ia: a comparison of theory and observations. Astrophys. J. **472**(2), L81–L84 (1996)
34. V.N. Gamezo, A.M. Khokhlov, E.S. Oran, Three-dimensional delayed-detonation model of type Ia supernovae. Astrophys. J. **623**, 337–346 (2005)
35. M.M. Phillips, The absolute magnitudes of type Ia supernovae. Astrophys. J. **413**, L105–L108 (1993)
36. D. Branch, S. van den Bergh, Spectroscopic differences between supernovae of type Ia in early-type and in late-type galaxies. Astrophys. J. **105**, 2231–2235 (1993)
37. D. Branch, J.C. Wheeler, *Supernova Explosions* (Springer, Berlin, Heidelberg, 2017)
38. A.Y. Poludnenko, E.S. Oran, The interaction of high-speed turbulence with flames: turbulent flame speed. Combust. Flame **158**, 301–326 (2011)
39. A.Y. Poludnenko, E.S. Oran, The interaction of high-speed turbulence with flames: global properties and internal flame structure. Combust. Flame **157**, 995–1011 (2011)
40. C.A.Z. Towery, A.Y. Poludnenko, P.E. Hamlington, Detonation initiation by compressible turbulence thermodynamic fluctuations. Combust. Flame **213**, 172–183 (2020)
41. A.Y. Poludnenko, J. Chambers, K. Ahmed, V.N. Gamezo, B.D. Taylor, A unified mechanism for unconfined deflagration-to-detonation transition in terrestrial chemical systems and type Ia supernovae. Science **366**(588), 1–12 (2019)
42. A.M. Lisewski, W. Hillebrandt, S.E. Woosley, Constraints on the delayed transition to detonation in Type Ia supernovae. Astrophys. J. **538**, 831–836 (2000)
43. F.K. Röpke, Flame-driven deflagration-to-detonation transitions in type Ia supernovae? Astrophys. J. **668**, 1103–1108 (2007)
44. S.E. Woosley, A.R. Kerstein, V. Sankaran, A.J. Aspden, F.K. Röpke, Type Ia supernovae: calculations of turbulent flames using the linear eddy model. Astrophys. J. **704**, 255–273 (2009)
45. F.K. Röpke, J.C. Niemeyer, Delayed detonations in full-star models of type Ia supernova explosions. Astron. Astrophys. **464**, 683–686 (2007)
46. E. Bravo, D. García-Senz, A three-dimensional picture of the delayed-detonation model of type Ia supernovae. Astron. Astrophys. **478**, 843–853 (2008)
47. D. García-Senz, E. Bravo, Type Ia supernova models arising from different distributions of igniting points. Astron. Astrophys. **430**, 585–602 (2005)
48. A.M. Khokhlov, E.S. Oran, J.C. Wheeler, Deflagration to detonation transition in thermonuclear supernovae. Astrophys. J. **478**, 678–688 (1997)
49. J.C. Niemeyer, S.E. Woosley, The thermonuclear explosion of Chandrasekar mass white dwarfs. Astrophys. J. **475**, 740–753 (1997)
50. S.E. Woosley, S. Wunsch, M. Kuhlen, Carbon ignition in Type Ia supernovae: an analytic model. Astrophys. J. **607**, 921–930 (2004)
51. C.N. Augustine, D.E. Willcox, J. Brooks, D.M. Townsley, A.C. Calder, SN Ia explosions from hybrid carbon–oxygen–neon white dwarf progenitors that have mixed during cooling. Astrophys. J. **887**(1–11), 188 (2019)
52. D.M. Townsley, B.J. Miles, F.X. Timmes, A.C. Calder, E.F. Brown, A tracer method for computing Type Ia supernova yields: burning model calibration, reconstruction of thickened flames, and verification for planar detonations. Astrophys. J. **225**(1–28), 3 (2016)
53. A.C. Calder, B.K. Krueger, A.P. Jackson, D.E. Willcox, B.J. Miles, D.M. Townsley, Cosmic Chandlery with thermonuclear supernovae. J. Phys. Conf. Ser. **837**(012005), 1–14 (2017)
54. A.C. Calder, D.E. Willcox, C.J. DeGrendele, D. Shangase, M. Zingale, D.M. Townsley, Thermonuclear (Type Ia) supernovae and progenitor evolution. J. Phys.: Conf. Ser. **1225**(1–10), 012002 (2019)
55. D.M. Townsley, A.C. Calder, S.M. Asida, I. Seitenzahl, F. Peng, N. Vladimirova, D.Q. Lamb, J. Truran, Flame evolution during Type Ia supernovae and the deflagration phase in the gravitationally confined detonation scenario. Astrophys. J. **668**, 1118–1124 (2007)
56. D.M. Townsley, B.J. Miles, K.J. Shen, D. Kasen, Double detonations with thin, modestly enriched helium layers can make normal Type Ia supernovae. Astrophys. J. Lett. **878**(L38), 6 (2019)

57. E. Brooker, T. Plewa, D. Fenn, Type Ia supernovae deflagration-to-detonation transition explosions powered by the Zel'dovich reactivity gradient mechanism. MNRAS **501**, L23–L27 (2021)
58. I.R. Seitenzahl, F. Ciaraldi-Schoolmann, F.K. Röpke, M. Fink, W. Hillebrandt, M. Kromer, R. Pakmor, A.J. Ruiter, S.A. Sim, S. Taubenberger, Three-dimensional delayed-detonation models with nucleosynthesis for Type Ia supernovae. Mon. Not. R. Astron. Soc. **429**, 1156–1172 (2013)
59. M. Fink, M. Kromer, I.R. Seitenzahl1, F. Ciaraldi-Schoolmann, F.K. Röpke, S.A. Sim, R. Pakmor, A.J. Ruiter, W. Hillebrandt, Three-dimensional pure deflagration models with nucleosynthesis and synthetic observables for Type Ia supernovae. Mon. Not. R. Astron. Soc. **438**, 1762–1783 (2014)
60. F.K. Röpke, W. Hillebrandt, J.C. Niemeyer, S.E. Woosley, Multi-spot ignition in type Ia supernova models. Astron. Astrophys. **448**, 1–14 (2006)
61. H. Ma, S.E. Woosley, C.M. Malone, A. Almgren, J. Bell, Carbon deflagration in Type Ia supernova: I. Centrally ignited models. Astrophys. J. **771**(1–16), 58 (2013)
62. A. Nonaka, A.J. Aspden, M. Zingale, A.S. Almgren, J.B. Bell, S.E. Woosley, High-resolution simulations of convection preceding ignition in Type Ia supernovae using adaptive mesh refinement. Astrophys. J. **745**(1–22), 73 (2012)
63. B. Hristov, D.C. Collins, P. Hoeflich, C.A. Weatherford, T.R. Diamond, Magnetohydrodynamical effects on nuclear deflagration fronts in Type Ia supernovae. Astrophys. J. **858**(1–12), 13 (2018)
64. I. Golombek, J.C. Niemeyer, A model for multidimensional delayed detonations in SN Ia explosions. A&A **438**, 611–616 (2005)
65. A.J. Aspden, J.B. Bell, M.S. Day, S.E. Woosley, M. Zingale, Turbulence-flame interactions in Type Ia supernovae. Astrophys. J. **689**, 1173–1185 (2008)
66. S.E. Woosley, Type Ia supernova: burning and detonation in the distributed regime. Astrophys. J. **668**, 1109–1117 (2007)
67. E.P. Hicks, R. Rosner, Gravitationally unstable flames: Rayleigh-Taylor stretching versus turbulent wrinkling. Astrophys. J. **771**(1–12), 135 (2013)
68. E.P. Hicks, Rayleigh-Taylor unstable flames—fast or faster? Astrophys. J. **803**(1–25), 72 (2015)
69. A.J. Aspden, J.B. Bell, S. Dong, S.E. Woosley, Burning thermals in Type Ia supernovae. Astrophys. J. **738**, 94–107 (2011)
70. S.E. Woosley, A.R. Kerstein, A.J. Aspden, Flames in type Ia supernova: deflagration-detonation transition in the oxygen-burning flame. Astrophys. J. **734**(1–5), 37 (2011)
71. A.J. Aspden, J.B. Bell, S.E. Woosley, Turbulent oxygen flames in Type Ia supernovae. Astrophys. J. **730**(1–8), 144 (2011)
72. V.V. Bychkov, M.A. Liberman, On the theory of Type Ia supernova explosion. Astron. Astrophys. **304**, 440–448 (1995)
73. W. Fowler, G. Caughlan, B. Zimmerman, Thermonuclear reaction rates, II. Annu. Rev. Astron. Astrophys. **13**, 69–112 (1975)
74. A.Y. Potekhin, G. Chabrier, Thermonuclear fusion in dense stars—electron screening, conductive cooling, and magnetic field effects. Astron. Astrophys. **538**(1–9), A115 (2012)
75. L.D. Landau, E.M. Lifshitz, *Course of Theoretical Physics. Vol. 5, Statistical Physics, Part 1* (Pergamon Press, Oxford, 1980)
76. J.C. Niemeyer, W. Hillebrandt, Turbulent nuclear flames in Type Ia supernovae. Astrophys. J. **452**, 769–778 (1995)
77. A.M. Khokhlov, Propagation of turbulent flames in supernovae. Astrophys. J. **449**, 695–713 (1995)
78. A. Khokhlov, Flame modeling in supernovae. Astrophys. J. **419**, L77–L80 (1993)
79. A. Khokhlov, Supernovae deflagrations in three dimensions. Astrophys. J. **424**, L115–L117 (1994)
80. S.L. Blinnikov, P.V. Sasorov, Landau-Darrieus instability and the fractal dimension of flame fronts. Phys. Rev. E **53**(1–15), 4827 (1996)

81. F.X. Timmes, S.E. Woosley, The conductive propagation of nuclear flames. I. Degenerate C + O and O + Ne + Mg white dwarfs. Astrophys. J. **396**, 649–667 (1992)
82. V.V. Bychkov, M.A. Liberman, Hydrodynamic instabilities of the flame front in white dwarfs. Astron. Astrophys. **302**, 727–734 (1995)
83. V.V. Bychkov, M.A. Liberman, Flame instabilities and models of white dwarf burning, in *Thermonuclear Supernovae*, ed. by P. Ruiz-Lapuente, R. Canal, J. Isern (Kluwer, Dordrecht, 1996), pp. 379–388. *Conference on Thermonuclear Supernovae*, NATO Institute for Advanced Science, Aiguablava, Spain, 20–30 June 1995
84. J.B. Bell, M.S. Day, C.A. Rendleman, S.E. Woosley, M. Zingale, Direct numerical simulations of Type Ia supernovae flames. I. The Landau-Darrieus instability. Astrophys. J. **606**, 1029–1038 (2004)
85. V.V. Bychkov, M.A. Liberman, On the dynamics of a curved deflagration front. Zh. Eksp. Teor. Fiz. **111**, 514–527 (1997) (English Translation: JETP **84**, 281–288 (1997))
86. B. Paczynski, Evolution of single stars. V. Carbon ignition in population I stars. Acta Astron. **21**, 271–288 (1971)
87. T.D. Lee, Hydrogen content and energy-productive mechanism of white dwarfs. Astrophys. J. **111**, 625–640 (1950)
88. R.E. Marshak, Wave mechanical scattering at particle collisions. Ann. N. Y. Acad. Sci. **41**, 49–67 (1941)
89. V.V. Bychkov, M.A. Liberman, Thermal instability and pulsations of the flame front in white dwarfs. Astrophys. J. **451**, 711–716 (1995)
90. S.I. Glazyrin, S.I. Blinnikov, A. Dolgov, Flame fronts in Type Ia supernovae and their pulsational stability. MNRAS **433**, 2840–2849 (2013)
91. G. Xing, Y. Zhao, M. Modestov, C. Zhou, Y. Gao, C.K. Law, Thermal-diffusional instability in white dwarf flames: regimes of flame pulsation. Astrophys. J. **841**(1–8), 21 (2017)
92. S.E. Woosley, Type I supernovae: carbon deflagration and detonation, in *Supernovae* (Springer, 1990), pp. 182–212
93. D.G. Yakovlev, V.A. Urpin, Thermal and electrical conductivity in white dwarfs and neutron stars. Sov. Astron. **24**, 303–310 (1980)
94. B.J. Matkovsky, G.I. Sivashinsky, Propagation of a pulsating reaction front in solid fuel combustion. SIAM J. Appl. Math. **35**, 465–478 (1978)
95. V.V. Bychkov, M.A. Liberman, On the stability of solid propellant combustion. Phys. Rev. Lett. **73**, 1998–2000 (1994)
96. J.C. Niemeyer, Can deflagration-detonation transitions occur in Type Ia supernovae? Astrophys. J. **523**, L57–L60 (1999)
97. J.B. Bell, M.S. Day, C.A. Rendleman, S.E. Woosley, M. Zingale, Direct numerical simulations of Type Ia supernovae flames. II. The Rayleigh-Taylor instability. Astrophys. J. **608**, 883–906 (2004)
98. F.K. Röpke, W. Hillebrandt, W. Schmidt, J.C. Niemeyer, S.I. Blinnikov, P.A. Mazzali, A three-dimensional deflagration model for Type Ia supernovae compared with observations. Astrophys. J. **668**, 1132–1139 (2007)
99. E. Bravo, D. García-Senz, Beyond the bubble catastrophe of Type Ia supernovae: pulsating reverse detonation models. Astrophys. J. **642**, L157–L160 (2006)
100. W. Schmidt, F. Ciaraldi-Schoolmann, J.C. Niemeyer, F.K. Röpke, W. Hillebrandt, Turbulence in a three-dimensional deflagration model for Type Ia supernovae. II. Intermittency and the deflagration-to-detonation transition probability. ApJ. **710**, 1683–1693 (2010)
101. S.A. Kriminski, V.V. Bychkov, M.A. Liberman, On the stability of thermonuclear detonation in supernovae events. New Astron. **3**, 363–377 (1998)
102. K.I. Shchelkin, Two cases of unstable combustion. Zh. Eksp. Teor. Fiz. **36**, 600–606 (1959) (English Translation: K.I. Shchelkin, Sov. Phys.-JETP **36**, 416–420 (1959))
103. R.M. Zaidel, Stability of detonation waves in gas mixtures. Dokl. Akad. Nauk SSSR, **136**, 1142–1145 (1961) (English Translation: R.M. Zaidel, Proc. Acad. Sci. USSR, Phys. Chem. Sec. **136**, 167 (1961))

104. J.J. Erpenbeck, Stability of steady-state equilibrium detonations. Phys. Fluids **5**, 604–614 (1962)
105. R.M. Zaidel, Ya.B. Zel'dovich, Одномерная неустойчивость и затухание детонации ПМТФ **6**, 59–65 (1964) (One-dimensional instability and quenching of detonation. Prikl. Mech. Tekh. Fia. **6**, 59 (1964))
106. H.I. Lee, D.S. Stewart, Calculation of linear detonation instability: one-dimensional instability of plane detonation. J. Fluid Mech. **216**, 103–132 (1990)
107. L. He, J.H.S. Lee, The dynamical limit of one-dimensional detonations. Phys. Fluids **7**, 1151–1158 (1995)
108. A.M. Khokhlov, Stability of detonations in supernovae. Astrophys. J. **419**, 200–206 (1993)
109. A.V. Koldoba, E.V. Tarasova, V.M. Chechetkin, Instability of the detonation wave in a thermonuclear supernova model. Astron. Lett. **20**, 377–381 (1994)
110. I. Domınguez, A. Khokhlov, Incomplete carbon-oxygen detonation in Type Ia supernovae. Astrophys. J. **730**(1–15), 87 (2011)
111. F. Hoyle, W.A. Fowler, Nucleosynthesis in supernovae. Astrophys. J. **132**, 565–590 (1960)
112. A.M. Khokhlov, Hugoniot adiabats for the waves of combustion in degenerate matter of supernovae. Astrophys. Space Sci. **149**, 91–106 (1988)
113. G.M. Gandelman, D.A. Frank-Kamentskii, Выход ударной волны на поверхность звезды. ДАН СССР **107**, 811–817 (1956) (in Russian) (English Translation: Emergence a shock wave at a stellar surface. Proc. Russ. Acad. Sci. **1**, 223–226 (1956))
114. A. Sakurai, On the problem of a shock wave arriving to the edge of a gas. Commun. Pure Appl. Math. **XIII**, 353–370 (1960)
115. G.B. Whitham, On the propagation of shock waves through regions of non-uniform area or flow. J. Fluid Mech. **4**, 337–360 (1958)
116. C. Charignon, J.-P. Chièze, Deflagration to detonation transition by amplification of acoustic waves in type Ia supernovae. Astron. Astrophys. **550**(1–9), A105 (2013)
117. J.M.M. Pfannes, J.C. Niemeyer, W. Schmidt, C. Klingenberg, Thermonuclear explosions of rapidly rotating white dwarfs—I. Deflagrations. Astron. Astrophys. **509**(1–12), A74 (2010)
118. J.M.M. Pfannes, J.C. Niemeyer, W. Schmidt, Thermonuclear explosions of rapidly rotating white dwarfs—II. Detonations. Astron. Astrophys. **509**(A75), 1–6 (2010)
119. D. García-Senz, R.M. Cabezón, I. Domínguez, Surface and core detonations in rotating white dwarfs. Astrophys. J. **862**(1–16), 27 (2018)
120. D. García-Senz, R.M. Cabezón, I. Domínguez, F.K. Thielemann, Type Ia supernovae: can Coriolis force break the symmetry of the gravitational confined detonation explosion mechanism? Astrophys. J. **819**(1–13), 132 (2016)
121. M. Fink, M. Kromer, W. Hillebrandt, F.K. Röpke, R. Pakmor, I.R. Seitenzahl, S.A. Sim, Thermonuclear explosions of rapidly differentially rotating white dwarfs: candidates for superluminous Type Ia supernovae? A&A **618**(A124), 1–12 (2018)
122. S. Chandrasekhar, *Hydrodynamic and Hydromagnetic Stability*, 3rd edn. (Dover, New York, NY, 1981)
123. H.J. Kull, Theory of the Rayleigh-Taylor instability. Phys. Rep. **206**, 197–325 (1991)
124. S.I. Abarzhi, Review of theoretical modelling approaches of Rayleigh-Taylor instabilities and turbulent mixing. Philos. Trans. A Math. Phys. Eng. Sci. **368**, 1809–1828 (2010)
125. S.I. Anisimov, R.P. Drake, S. Gauthier, E.E. Meshkov, S.I. Abarzhi, What is certain and what is not so certain in our knowledge of Rayleigh-Taylor mixing? Philos. Trans. A Math. Phys. Eng. Sci. **371**(1–13), 20130266 (2013)
126. S.I. Abarzhi, On fundamentals of Rayleigh-Taylor turbulent mixing. Europhys. Lett. **91**(1–5), 35001 (2010)
127. M.V. Popov, Convective instability in Type Ia supernova explosion model. Gravit. Cosmol. **11**, 177–182 (2005)
128. J. Zhang, O.E. Bronson Messer, A.M. Khokhlov, T. Plewa, On the evolution of thermonuclear flames on large scales. Astrophys. J. **656**, 347–365 (2007)
129. M. Zingale, S.E. Woosley, C.A. Rendleman, M.S. Day, J.B. Bell, Three-dimensional numerical simulations of Rayleigh-Taylor unstable flames in type Ia supernovae. Astrophys. J. **632**, 1021–1034 (2005)

130. E.P. Hicks, Rayleigh-Taylor unstable flames at higher Reynolds number. Mon. Not. R. Astron. Soc. **489**, 36–51 (2019)
131. D. Layzer, On the Instability of superposed fluids in a gravitational field. Astrophys. J. **122**, 1–12 (1955)
132. V. Bychkov, M.V. Popov, A. Oparin, L. Stenflo, V.M. Chechetkin, Dynamics of bubbles in supernovae and turbulent vortices. Astron. Rep. **50**, 298–311 (2006)
133. V.V. Bychkov, M.A. Liberman, Self-consistent theory of white dwarf burning in the supernova Ia events. Astrophys. Space Sci. **233**, 287–292 (1995)
134. F.K. Röpke, W. Hillebrandt, The distributed burning regime in type Ia supernova models. A&A **429**, L29–L32 (2005)
135. R.M. Davies, G.I. Taylor, The mechanics of large bubbles rising through extended liquids and through liquids in tubes. Proc. R. Soc. A **200**, 375–390 (1950)

Chapter 18
Ablation Fronts in Inertial Confinement Fusion

18.1 Inertial Confinement Fusion

The Rayleigh–Taylor instability (RTI) develops at the interface between fluids with different densities, which are accelerated in the direction opposite to the density gradient [1, 2]. In ordinary life, the manifestation of the Rayleigh–Taylor instability is water pouring out of an overturned glass. It is worth noting that atmospheric pressure (1 kg/cm^2) is more than enough to hold water in an overturned glass. If we cover the glass with a sheet of paper and, while carefully holding the paper, turn the glass over, water will not flow out of the overturned glass when we stop supporting the sheet of paper. This simple experiment demonstrates suppression of RT instabilities due to the small elasticity of the sheet of paper, which is sufficient to suppress small disturbances at the water surface.

The Rayleigh–Taylor instability plays an important role in many engineering and scientific problems. In particular, the RT instability limits the spherically symmetric compression of targets in inertial confinement fusion (ICF) [3–8], and the cylindrically symmetric compression of the Z-pinch plasma [9–11]. There is now a considerable volume of literature and review articles on the subject [12–17].

Inertial confinement fusion (ICF) is a scheme in which a deuterium–tritium target must be highly compressed and heated to trigger the DT fusion reaction. The required high density can be achieved by compression a spherical "target", irradiated with powerful laser, ion, or X-ray beams. A high pressure created due to ablation of the capsule surface causes strong compression and heating of the target. As a result, a cold deuterium–tritium shell is accelerated radially inward under almost adiabatic compression [18, 19]. The DT reaction is considered because the cross section for the DT reaction is about two orders of magnitude larger than for other fusion reactions. In the fusion reaction, the nuclei of deuterium and tritium form helium (alfa particle) and neutron with 3.5 meV energy released in helium and 14.1 meV in neutron

$$D + T \to {}^4He(3.5\,\text{Mev}) + n(14.1\,\text{Mev}). \quad (18.1.1)$$

M. A. Liberman, *Combustion Physics*,
https://doi.org/10.1007/978-3-030-85139-2_18

For the fusion reaction to occur, the deuterium–tritium fuel must be confined in a small high-density hot spot to overcome the Coulomb repulsion of the nuclei and to allow thermonuclear fusion to be initiated in a small region of the fuel ("ignition"). The confinement time of the plasma ball of radius R_f is $\tau_{\text{conf}} \approx R_f/a_s$, where a_s is the speed of sound. The confinement time can also be defined considering that at the stagnation point, $\dot{R} = 0$, so that the confinement time $\tau_{\text{conf}} \sim \sqrt{R/\ddot{R}}$. Since the only force acting on the thin shell of mass M_{sh}, is the hot-spot pressure P_{hs}, the equation of motion of the shell is $M_{\text{sh}}\ddot{R} = 4\pi R P_{\text{hs}}$, and we obtain

$$\tau_{\text{conf}} \approx \sqrt{\frac{M_{\text{sh}}}{4\pi P_{\text{hs}} R_s}}. \tag{18.1.2}$$

Assuming an infinitesimally thin shell and that all kinetic energy is converted into internal energy, we obtain the previous estimate, $\tau_{\text{conf}} \approx R_f/a_s$.

The confinement time must be greater or at least comparable to the fusion reaction time, $\tau_{\text{fusion}} = 1/n\langle\sigma v\rangle$, that is, the probability of reaction per unit time per unit number density, $n = \rho_f/m_f$. The condition that the density and radius of the hot spot are large enough to self-heat the fuel and to produce sufficient energy for the reaction propagation, so that a considerable fraction of the fusion α-particles deposit their energy within the hot spot volume, can be expressed as [18]

$$\rho_f R_f > 0.2 - 0.5\,\text{g/cm}^2,\ for\ T_f \sim 10\,\text{KeV}. \tag{18.1.3}$$

The condition for sustained operation, known as Lawson criterion [19], requires that the fusion power output exceeds the rate of thermal energy losses. At typical fusion temperatures, and plasma densities, the main volumetric power loss is the bremsstrahlung radiation, $W_b = 5.34 \times 10^{-24} n^2 T^{1/2} erg/\text{s cm}^3$, while the fusion volumetric power is $W_{DT} \approx 0.25 n^2 \langle\sigma v\rangle Q_{DT}$, where $Q_{DT} = 17.59\,\text{Mev}$. Using expressions for the energy losses the Lawson-type criterion can be written as [5]

$$\rho_f R_f T_f > \frac{1.1\sqrt{T_f}}{1 - 3.47 T_f^{-3/2}} \sqrt{\rho_f/\rho_{DT}}\ \text{g}/rmcm^2. \tag{18.1.4}$$

The criterion (18.1.4) can be written through the hot-spot confinement time τ_{conf} as

$$P_{\text{hs}} \tau_{\text{conf}} > \frac{24}{\varepsilon_\alpha S(T_i)}, \tag{18.1.5}$$

where P_{hs} is the hot-spot pressure in Mbar, τ_{conf} is in seconds, $\varepsilon_\alpha = 3.5\,\text{Mev}$ is the alpha-particle energy, and $S(T_i)$ is a function of the ion temperature T_i at the center of the hot spot [20].

In ICF schemes, the energy supplied by the driver can be applied either directly or indirectly to the outer layer of the fuel pellet, or to a shell containing the fuel, that is made up of an ablator material (e.g. beryllium) that explodes outwardly when heated. The dynamics of the ICF implosion is similar to combustion in a deflagration mode. The energy of the laser or X-ray radiation is absorbed at the critical surface, where the radiation frequency ω is equal to the electron plasma frequency $\omega_p = \sqrt{4\pi e^2 n_e/m_e}$. Here, e, m_e, M_i, $n_e = \rho/M_i$ are the electrical charge, electron and ion masses, and electron number density, respectively. Correspondingly, the density at the critical surface in the ablation flow is

$$\rho_c = \omega^2 m_e M_i / 4\pi e^2. \tag{18.1.6}$$

A schematic of the direct drive spherical implosion is sketched in Fig. 18.1. Laser radiation is absorbed at a critical density (dotted circle), and heat is transferred to the shell surface due to electron thermal conductivity, causing it to be ablated in the form of a high-temperature plasma expanding outward. This outward explosion of the surface layer forces the shell containing the fuel to accelerate inward to conserve momentum (a rocket effect). During this stage the outer surface of the shell is RT unstable. During the deceleration of the collapsing shell, the kinetic energy of the shell is converted into the internal energy of the compressed fuel. During this stage, when the pressure of the compressed fuel decelerates the imploding denser shell, the inner surface of the shell is RT unstable.

High compression of the fuel can be achieved in the isentropic process. According to the first law of thermodynamics the internal energy $dE = TdS - PdV$ can be increased either due to transfer of heat at the constant volume $dQ = TdS$, or by doing work $dE = -PdV$ in an isentropic compression. Ideally, an adiabatic process $P/\rho^\gamma = const$ can lead to unlimited compression to an arbitrary high density. Before stagnation, the adiabatic approximation for the hot spot is valid, because the thermal conduction losses are "recycled" by ablation back into the hot spot. Bremsstrahlung radiation losses do not significantly affect the adiabatic scaling law since the rate of radiation cooling of the hot spot is lower than the rate of $dE = -PdV$ work of the converging shell.

Fig. 18.1 Schematic of the ICF direct drive spherical target implosion

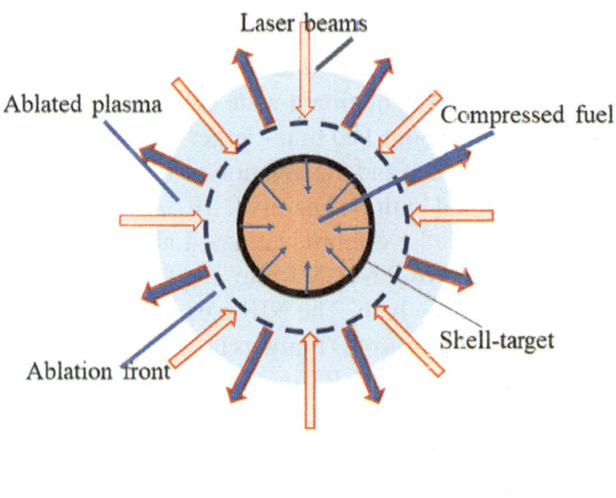

A sudden increase in the radiation power at the beginning initiates the ablation and implosion, but at the same time sends shock waves into the center of the fuel inside the shell. Contrary to the adiabatic compression, the maximum density increase in a shock wave is limited by $\rho/\rho_0 = (\gamma + 1)/(\gamma - 1)$. Nearly isentropic compression can be achieved by the temporal increase in the pressure, creating a properly shaped sequence of superimposing shocks, which merge at the same time. The implosion of a hollow spherical shell, with initial thickness $\Delta R_0 \ll R_0$, has great advantages over a full sphere implosion [21]. In this case, the high value of the implosion velocity can be reached using a lower driving pressure over a larger accelerated distance, since the pressure required to accelerate a shell of constant mass to velocity u_{imp} is $P \propto (u_{imp}^2 \Delta R_0)/R_0$. Another advantage of a hollow spherical shell is that the shocks passing a thin spherical shell almost do not affect the nearly adiabatic convergence of the shell.

There are two major concepts of ICF: (i) a direct drive implosion, in which the driver energy in the form of laser radiation or particle beams directly irradiates the fuel capsule, (ii) an indirect drive implosion, in which the laser irradiates the inner walls of a hollow chamber ("hohlraum"). In the latter case, intense laser beams enter the hohlraum through the holes in the end plates and deposit their energy on the inner walls of the hohlraum, which are covered by a high Z material (e.g. gold). This energy is converted to X-rays by the hot gold plasma. The capsule of fuels, placed in the center of the hohlraum, receives a uniform flux of X-ray energy over its entire surface. The X-rays are continually absorbed and reemitted by the wall material, which results in establishing a radiation equilibrium inside the hohlraum with a uniform effective temperature about 300 eV. Except for the loss at the laser radiation entrance holes, the emitted X-ray energy is essentially trapped within the hohlraum. The most important function of hohlraum is to provide uniform illumination of the spherical capsule. The advantage of the hohlraum concept is that in this case the target is illuminated much more uniformly compared to the direct drive method, and this reduces the development of RT instability. The disadvantage is that part of the original energy of the laser beam is spent on the heating of hohlraum walls. It should be noted that in the case of a direct drive, the temperature of laser-ablated plasma is higher than the temperature of the X-ray ablated plasma in the case of indirect drive, since the critical density in the case of direct drive (laser wavelength $\lambda \sim 1\,\mu\text{m}$) is about two orders of magnitude lower than in the case of an indirect drive. Therefore, in the direct drive ablation a wide conduction layer with temperature about several KeV, which is about 100 times larger than it is in indirect drive, is formed between the critical surface and the ablation front.

The National Ignition Facility (NIF) is the largest laser system operating nowadays at the Lawrence Livermore National Laboratory. It consists of 192 lasers delivering radiation of up to 2.6–3 MJ into a cylindrical gold hohlraum, which converts the energy into X-ray bath. The NIF uses indirect drive concept, and it was designed to demonstrate ignition of DT targets in ICF. The NIF was designed for high gain, but it is sensitive to the ablation RT instability, ultimately resulting in hydrodynamic mixing [22, 23]. Numerical simulations show that much of the damage to the implosion occurs during the early stage, when the capsule is accelerating [24].

18.2 Rayleigh–Taylor Instability; Linear Stage, Discontinuity Model

For simplicity, we consider the RTI problem for the case of ablatively accelerated planar thin foils, which is the first step toward understanding the features of RT instability in regimes relevant to ICF. A review of experimental studies of the ablative RT instability in planar geometry is presented in [25].

The surface of the accelerated foil irradiated by laser light is unstable during the acceleration phase because the higher density foil is accelerated by the lower density ablated plasma. The RT instability develops in the transition layer between the ablation zone and the accelerated denser part of the target, where the density gradient and the hydrodynamic acceleration have the opposite directions. In the reference frame of the ablation front, the effective gravity g is directed from the dense cold foil to the lighter plasma cloud. At the linear stage of the instability, perturbations of the interface grow exponentially with the growth rate $\sigma = \sqrt{A_t gk}$, where $A_t = (\rho_1 - \rho_2)/(\rho_1 + \rho_2)$ is the Atwood number, ρ_2 and ρ_1 are the densities of the heavy and light fluids, respectively, and k is the perturbation wave number [26]. Figure 18.2 shows a schematic sketch of the unstable ablation flow.

Despite the obvious similarities between laser ablation and deflagration waves, there are also features of laser ablation. Since the plasma expands from the critical surface into vacuum, a rarefaction wave is formed. Due to the high electron thermal conduction of the high-temperature plasma, the head of the rarefaction wave propagates with the isothermal, $a_{sT} = \sqrt{P/\rho}$, rather than adiabatic sound speed $a_s = \sqrt{\gamma P/\rho}$. This means that the laser ablation wave is similar to the Chapman–Jouguet deflagration, but, in contrast to the CJ-deflagration, propagates at an isothermal speed. The velocity U_a of the ablation front depends on the intensity of laser radiation absorbed at the critical surface. In the case of a stationary flow, the continuity of the mass flux implies $U_a = U_c\rho_c/\rho_a$. Due to its similarity to deflagration, the ablation foil can be unstable with respect to DL and RT instabilities. Numerous experiments have demonstrated the RT instability of the ablatively accelerated thin foils [27–31]. However, a decrease in the growth rate of the RT instability and even complete stabilization of short-wavelength instabilities, predicted earlier

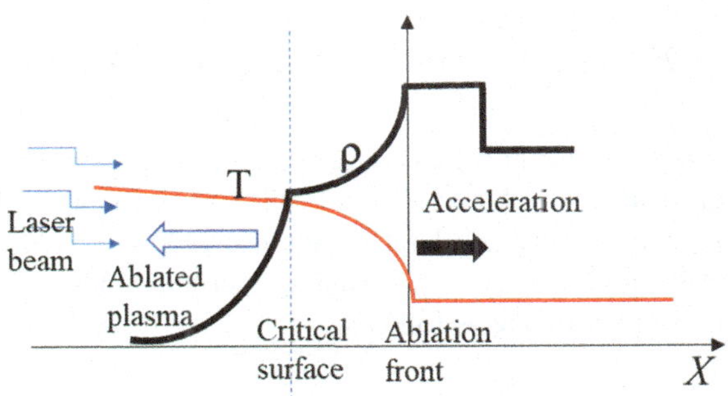

Fig. 18.2 Schematic density and temperature profiles in the ablation wave

in [18], was discovered. Takabe and Mima [32], based on numerical experiments, argued that thermal conductivity can reduce the growth rate of short-wavelength instabilities.

The first solution of the linear stability problem was obtained numerically by Takabe et al. for spherical geometry [33, 34]. A self-consistent linear stability analysis of an ablating plasma was numerically investigated by Takabe et al. [35] as an eigenvalue problem, assuming a stationary ablative plasma. The results of [35] are consistent with the empirical formula for the growth rate of RT instability

$$\sigma = \alpha\sqrt{kg} - \beta k U_a, \tag{18.2.1}$$

where $\alpha \approx 0.9$ and $\beta \approx 3.1$ are the fitting numerical factors, which depend weakly on the equilibrium parameters. It was shown in [35] that formula (18.2.1) with $\alpha = 0.9$ and $\beta = 3.0$ is in excellent agreement with the LASNEX simulations. A slightly different form of Eq. (18.2.1), proposed in [25], considers a finite density gradient, $L = \rho/|\nabla\rho|$

$$\sigma = \alpha\sqrt{kg/(1+kL)} - \beta k U_a. \tag{18.2.2}$$

The stabilizing mechanism causing the suppression of short-wavelength modes with $k > \alpha^2 g/\beta^2 U_a^2$ was associated with the gas flow U_a across the ablation front.

One of the first attempts to explain theoretically the reduced growth rate of the RT instability of the ablation front was undertaken by Bodner [36], who considered the ablation front as a discontinuity surface. Obviously, in the model of the discontinuity surface, the problem cannot be completely determined by the conservation equations, since the velocity of the ablation front is unknown. An additional closure equation is needed to match the number of unknowns with the number of equations. In an effort to overcome the lack of matching conditions, Bodner [36] and a number of subsequent studies [37–39] have proposed various additional conditions that have led to different solutions. The problem of the boundary conditions required for the existence a solution in the form of a discontinuity surface was discussed in Chap. 5 and [40].

18.2.1 RT Instability of Ablatively Accelerated Plasma

A self-consistent solution of the stability problem of a deflagration wave in a gravitational field, obtained in [41, 42], was applied to the problem of RT instability of an ablatively accelerated target in [43, 44]. Hydrodynamic equations of mass, momentum and energy, including thermal conduction, diffusion and absorption of laser radiation are as follows:

$$\frac{\partial\rho}{\partial t} + \nabla\cdot(\rho\mathbf{u}) = 0, \tag{18.2.1.1}$$

$$\rho\frac{\partial \mathbf{u}}{\partial t} + \rho(\mathbf{u}\cdot\nabla)\mathbf{u} + \nabla P = \rho\mathbf{g}, \tag{18.2.1.2}$$

$$\rho c_p\frac{\partial T}{\partial t} + \rho c_p(\mathbf{u}\cdot\nabla T) - \frac{\partial P}{\partial t} - (\mathbf{u}\cdot\nabla P) - \nabla\cdot(\kappa\nabla T) = -\mathbf{n}\cdot\nabla I. \tag{18.2.1.3}$$

Here, $P = c_p\rho T(\gamma-1)\gamma$, $\mathbf{n}$ is a unit vector in the direction of laser radiation, and the absorption of laser light with intensity I is considered in the Wentzel–Kramers–Brillouin (WKB) approximation

$$\mathbf{n}\cdot\nabla I = -I_0 \exp\left(-\int_{\infty}^{z}\Omega dl\right), \tag{18.2.1.4}$$

where the integral is taken along the path of the laser radiation. Since the length scale of the absorption region of the laser radiation is of the order of the electron mean free path, $L_a \approx \chi_{el}/U_{el}$, ($\chi_{el}$ is electron thermal diffusivity) the corresponding term in the equation for energy transfer (18.2.1.3) can be taken the form of δ-function. Strictly speaking, a flow of the high-temperature plasma is not a uniform, but expands in the form of an isothermal rarefaction wave. This means that velocity at the critical surface is equal to the isothermal sound velocity, and the compressibility effects may be important. In general, this complicates the analytical study of the problem. Still, the problem can be considerably simplified assuming that the flow is subsonic and uniform.

In the WKB approximation the function of the energy release can be written in the dimensionless form:

$$c_p\rho_2 u_2 T_2\frac{W(\rho,T)}{L} \equiv -I_0\,\Omega\exp\left(-\int_{\infty}^{z}\Omega dl\right), \tag{18.2.1.5}$$

where L is the dimension of length, and the subscript 2 labels the variables on the surface of the energy release. Instead of the δ function in the expression for $W(\rho,T)$ it is convenient to choose $W(\rho,T)$ in the following form:

$$W(\rho,T) = \begin{cases}(\rho/\rho_2 - 1)\exp(-E\rho/\rho_2), & \rho > \rho_2\\ 0, & \rho < \rho_2\end{cases}, \tag{18.2.1.6}$$

where the large parameter $E \gg 1$ characterizes the ratio of the thermal conduction scale to the thickness of the energy release zone (the critical surface). Such formulated, the problem is similar to the problem of the flame stability.

Considering a plane stationary solution in the approximation of a small Mach number, we have the following integrals of Eqs. (18.2.1.1)–(18.2.1.2)

$$\rho u = \rho_2 u_2, \tag{18.2.1.7}$$

$$P = P_2. \tag{18.2.1.8}$$

Introducing dimensionless variables, $\rho_2/\rho = u/u_2 = T/T_2 = \Theta$, $\xi = z/\Delta_2$, where $\Delta_2 = \kappa_2/c_p\rho_2 u_2$ is the thickness of the ablation region determined by the electron thermal conduction, we obtain from (18.2.1.3) the dimensionless equation, which defines the dimensionless temperature as the eigenfunction of the eigenvalue problem

$$\frac{d}{d\xi}\left(\chi(\Theta)\frac{d\Theta}{d\xi}\right) - \frac{d\Theta}{d\xi} + \Lambda W(\Theta) = 0, \tag{18.2.1.9}$$

where $\chi(\Theta) = \kappa/\kappa_2 = \Theta^{5/2}$ is the dimensionless coefficient of the electron thermal conduction. The solution to the Eq. (18.2.1.9) with boundary condition $\Theta = \Theta_1$ for $\xi \to -\infty$ and $\Theta = 1$ for $\xi \to \infty$ can be obtained using the same method as was used to obtain a temperature profile in a deflagration wave (Chap. 5). In particular, for the function $W(\rho, T)$ in the form (18.2.1.6), the eigenvalue is

$$\Lambda_E = 0.5(1-\Theta_1)^2\left(\int_{\Theta_1}^{1} W(\Theta)d\Theta\right)^{-1} = 0.5E^2(1-\Theta_1)^2\exp(E). \tag{18.2.1.10}$$

Small perturbations to the stationary solution are taken in the form

$$\tilde{\varphi}(x, z, t) = \tilde{\varphi}(z)\exp(\sigma t + ikx). \tag{18.2.1.11}$$

The z-axis is taken in the direction of propagation of the ablation wave, and the x-axis is taken along the ablation front. For isobaric flow ($M^2 \ll 1$), linearized Eqs. (18.2.1.1)–(18.2.1.3) have the form

$$\frac{d\tilde{j}}{d\xi} = S\frac{\tilde{\Theta}}{\Theta^2} - K\frac{\tilde{v}}{\Theta}, \tag{18.2.1.12}$$

$$\frac{d\tilde{v}}{d\xi} = -S\frac{\tilde{v}}{\Theta} - K\tilde{\Pi} - 2K\Theta\tilde{j} - K\tilde{\Theta}, \tag{18.2.1.13}$$

$$\frac{d\tilde{\Pi}}{d\xi} = -S\tilde{j} - K\tilde{v} - \frac{\tilde{\Theta}}{Fr\Theta^2}, \tag{18.2.1.14}$$

$$\frac{d^2}{d\xi^2}(\chi\tilde{\Theta}) - \frac{d\tilde{\Theta}}{d\xi} + \Lambda\frac{dW}{d\Theta}\tilde{\Theta} = \tilde{j}\frac{d\Theta}{d\xi} + S\frac{\tilde{\Theta}}{\Theta} + \chi K^2\tilde{\Theta}, \tag{18.2.1.15}$$

where

$$\tilde{\Theta} = \frac{\tilde{T}}{T_2},\ \tilde{v} = i\frac{\tilde{u}_x}{u_2},\ \tilde{j} = \frac{\tilde{\rho}u + \rho\tilde{u}_z}{\rho_2 u_2},\ \tilde{\Pi} = \frac{\tilde{P} + \tilde{\rho}u^2 + 2\rho u\tilde{u}_z}{\rho_2 u_2^2}, \tag{18.2.1.16}$$

$$K = k\Delta_2,\ S = \sigma\Delta_2/u_2. \tag{18.2.1.17}$$

The boundary conditions for Eqs. (18.2.1.12)–(18.2.1.15) require that all perturbed quantities vanish at uniform flows at infinity, $\xi \to \pm\infty$. The flow can be considered as uniform at $\xi = \xi_1$ and $\xi = \xi_2$, if

$$\left|\frac{d(\ln\Theta)}{d\xi}\right| \ll \min\{\Theta, K\}\ for\ \xi = \xi_1, \xi_2. \tag{18.2.1.18}$$

The problem is considerably simplified assuming that the flow is subsonic everywhere. In this case, the boundary condition at the critical surface follows from the condition that there are no sound perturbations propagating in the direction to the ablation front, and there is no sound perturbation coming from infinity.

In a uniform flow, for $\xi < \xi_1$ and $\xi > \xi_2$, the solution of the linearized equations can be expressed as a superposition of decaying eigenmodes

$$\tilde{\varphi}(\xi) = \tilde{\varphi}\exp(\mu\xi), \tag{18.2.1.19}$$

where $\mu > 0$ for $\xi < \xi_1$, and $\mu < 0$ for $\xi > \xi_2$. For $\xi < \xi_1$, there are two such modes ahead of the ablation zone. This is acoustic mode.

$$\tilde{\varphi} = \tilde{\varphi}_s,\ \mu = K > 0, \tag{18.2.1.20}$$

and thermal mode

$$\tilde{\varphi} = \tilde{\varphi}_T,\ \mu = \mu_T = 0.5\chi_1^{-1} + \sqrt{0.25\chi_1^{-2} + S\Theta_1^{-1}\chi_1^{-1} + K^2} > 0. \tag{18.2.1.21}$$

Behind the ablation zone, for $\xi > \xi_2$ there are three modes: acoustic mode

$$\tilde{\varphi} = \tilde{\varphi}_a,\ \mu = -K, \tag{18.2.1.22}$$

vorticity mode

$$\tilde{\varphi} = \tilde{\varphi}_v,\ \mu = -S, \tag{18.2.1.23}$$

and energy mode

$$\tilde{\varphi} = \tilde{\varphi}_W, \ \mu = \mu_W = 0.5 - \sqrt{0.25 + S + K^2 - \Lambda \left(\frac{dW}{d\Theta} \right)_2} < 0. \qquad (18.2.1.24)$$

The boundary conditions for Eqs. (18.2.1.12)–(18.2.1.15) are that the solution at the points $\xi = \xi_1$ and $\xi = \xi_2$ is represented by the superposition of the eigenfunctions (18.2.1.20–18.2.1.21), and (18.2.1.22–18.2.1.24), respectively:

$$\tilde{\varphi}(\xi_1) = \tilde{\varphi}_s + \tilde{\varphi}_T, \qquad (18.2.1.25)$$

$$\tilde{\varphi}(\xi_2) = \tilde{\varphi}_a + \tilde{\varphi}_v + \tilde{\varphi}_W. \qquad (18.2.1.26)$$

There are three algebraic equations for the components of functions $\tilde{\varphi}(\xi_1)$ and two equations for the components of functions $\tilde{\varphi}(\xi_2)$. The system (18.2.1.12)–(18.2.1.15) together with the boundary conditions (18.2.1.25) and (18.2.1.26) completely specify the eigenvalue problem for the stability of a planar steady ablation wave.

Figure 18.3 shows the instability growth rate σ/σ_0 as a function of the dimensionless wave number $u_1^2 k/g$ for different values of the Froude number $Fr = u_2^2/g\Delta_2$ and for the same ratio of the initial and final temperatures Θ_1 obtained in [44]. It can be seen that the dispersion relation changes significantly with variation of $u_1^2 k/g$ and Θ_1 only for the long-wavelength perturbations. For the short-wavelength perturbations, the instability growth rate depends slightly on the Froude number and the temperatures ratio.

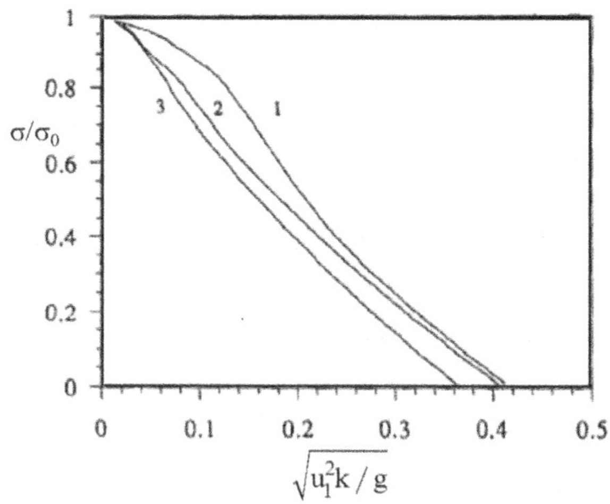

Fig. 18.3 The scaled instability growth rate σ/σ_0 versus $\sqrt{ku_1^2/g}$; (1) $Fr = 1$; (2) $Fr = 0.25$; (3) $Fr = 0.11$; the temperatures ratio $\Theta_1 = 0.1$, from [44]

18.2.2 Discontinuity Model: Long-Wavelength Perturbations

Perturbations of long wavelengths, which are much larger than the thickness of the ablation region $K = k\Delta_2 \ll 1$, are not important since the growth rate of these perturbations is small. At the same time this is the long-wavelength perturbations that can be associated with the discontinuity model used in [41, 44] and in many subsequent publications on the RT instability in approximation of incompressible fluids [45–55]. The solution to this problem is similar to the solution of the stability problem of a planar flame front, but with the electron thermal conduction instead of a gas thermal conduction.

In the limit of long-wavelength perturbations, different modes vary at different length scales. Since the thermal modes (18.2.1.21) and (18.2.1.24) decay rapidly as functions of the distance from the ablation region, the boundary conditions imposed on the perturbed hydrodynamic variables $\tilde{j}$, $\tilde{v}$, and $\tilde{\Pi}$ depend only on the structure of hydrodynamic modes (18.2.1.20), (18.2.1.22), and (18.2.1.23). Thus, we obtain

$$\tilde{v}_1 = -\Theta_1 \tilde{j}_1,\ \tilde{\Pi}_1 = (\Theta_1 - S/K)\tilde{j}_1\ for\ \xi = \xi_1(\Theta_1),\ \Theta_1 = T_1/T_2 \tag{18.2.2.1}$$

$$\tilde{v}_2 + \tilde{\Pi}_2 - (CS/K + 2)\tilde{j} = 0,\ for\ \xi = \xi_2(\Theta_2 = 1). \tag{18.2.2.2}$$

It can be shown in [41, 42] that the solution of Eqs. (18.2.1.12)–(18.2.1.15) with the boundary conditions (18.2.1.25) and (18.2.1.26) up to terms of zero order in $K \ll 1$ is

$$\tilde{\Theta} = \zeta_T \frac{d\Theta}{d\xi}, \tag{18.2.2.3}$$

$$\tilde{j} - \tilde{j}_1 = S\zeta_T \frac{\Theta - \Theta_1}{\Theta\Theta_1}, \tag{18.2.2.4}$$

$$\tilde{v} - \tilde{v}_1 = -K\zeta_T(\Theta - \Theta_1), \tag{18.2.2.5}$$

$$\tilde{\Pi} - \tilde{\Pi}_1 = -\frac{\zeta_T}{Fr} \frac{\Theta - \Theta_1}{\Theta\Theta_1}, \tag{18.2.2.6}$$

where ζ_T is the amplitude of the ablation front perturbation. By integrating the equation of thermal conduction, we obtain the relation between the amplitudes ζ_T and $\tilde{j}_1$

$$S/\Theta_1 + \tilde{j}_1/\zeta_T = 0. \tag{18.2.2.7}$$

In dimensional variables, Eq. (18.2.2.7) takes the following form:

$$\tilde{u}_{z1} - \frac{\partial \zeta}{\partial t} = 0, \quad (18.2.2.8)$$

where $\zeta = \zeta_T \Delta_2$ is the perturbation of the ablation front of zero thickness. This equation is the same additional condition, which has been used by Landau and Darrieus to solve the stability problem of a flame front of zero thickness.

The consistency condition for Eqs. (18.2.2.4)–(18.2.2.7) at the downstream flow $\xi = \xi_2$, together with the conditions (18.2.2.1) and (18.2.2.2), gives the dispersion relation for the growth rate of perturbations

$$S_0 = \frac{\sigma \Delta_2}{u_2} = \sqrt{\frac{1 + \Theta_1 - \Theta_1^2}{(1+\Theta)^2}\Theta_1 K^2 + \frac{1-\Theta_1}{1+\Theta_1}\frac{K}{Fr}} - \frac{\Theta_1}{\Theta_1 + 1}K. \quad (18.2.2.9)$$

The growth rate (18.2.2.9) monotonous increases with the increase of the wave number K. A slight reduction in the RT growth rate associated with smooth gradients can be obtained as the first-order term in the power series expansion in $K = k\Delta_2 \ll 1$. The integrals of Eqs. (18.2.1.12)–(18.2.1.15) with accuracy up to terms $\sim K^2$ are

$$\tilde{j} = \tilde{j}_1 + S\zeta_T \frac{\Theta - \Theta_1}{\Theta\Theta_1} - K \int_{\xi_1}^{\xi} \Theta^{-1}\tilde{v}_0 d\eta, \quad (18.2.2.10)$$

$$\tilde{v} = -\Theta \tilde{j}_1 - K\zeta_T(\Theta - \Theta_1) - S \int_{\xi_1}^{\xi} \Theta^{-1}\tilde{v}_0 d\eta + K \int_{\xi_1}^{\xi} \tilde{\Pi}_0 d\eta - 2K \int_{\xi_1}^{\xi} \Theta \tilde{j}_0 d\eta, \quad (18.2.2.11)$$

$$\tilde{\Pi} = \left(\Theta_1 - \frac{S}{K}\right)\tilde{j}_1 - \frac{\zeta_T}{Fr}\frac{\Theta - \Theta_1}{\Theta\Theta_1} - S \int_{\xi_1}^{\xi} \tilde{j}_0 d\eta - K \int_{\xi_1}^{\xi} \tilde{v}_0 d\eta. \quad (18.2.2.12)$$

Here $\tilde{j}_0$, $\tilde{v}_0$, and $\tilde{\Pi}_0$ are the zeroth-order terms in the expansion of $\tilde{j}$, $\tilde{v}$, and $\tilde{\Pi}$ in powers of K, defined by (18.2.2.4)–(18.2.2.6), and the boundary conditions (18.2.2.1). After simple but somewhat cumbersome calculations, we can obtain the following relation between the amplitudes $\tilde{j}_1$ and ζ_T accurate up to the linear terms in K^2

$$(1 - K\xi_1)\tilde{j}_1 + (S/\Theta_1 + K\langle f \rangle)\zeta_T = 0, \quad (18.2.2.13)$$

where

$$\langle f\rangle = \int_{-\infty}^{\infty} f(\xi)\tilde{\psi}_0^2\chi^{-1}d\xi \Big/ \int_{-\infty}^{\infty} \tilde{\psi}_0^2\chi^{-1}d\xi$$

$$f(\xi) = \chi + \xi - \left(\frac{S}{K} + \Theta_1\right)\left(\frac{\xi}{\Theta} + \int_{-\infty}^{\infty} \Theta^{-2}\eta\frac{d\Theta}{d\eta}d\eta\right), \tag{18.2.2.14}$$

$$\tilde{\psi}_0 = \zeta_T\chi\frac{d\Theta}{d\xi}\exp -\frac{1}{2}\int_{\xi^*}^{\xi} \chi^{-1}d\eta. \tag{18.2.2.15}$$

Finally, from Eqs. (18.2.2.10)–(18.2.2.12) taken at the point $\xi = \xi_2$ together with Eq. (18.2.2.13) and boundary conditions, the dispersion relation with the same accuracy can be written as

$$S = S_0\left(1 - \frac{\Theta_1\Phi(K)K^3}{2S_0^2(1+\Theta_1+\Theta_1 K/S_0)}\right), \tag{18.2.2.16}$$

where S_0 is defined by Eq. (18.2.2.9), and the function $\Phi(K)$ is

$$\Phi(K) = 2\left(1+\frac{S_0}{K}\right)\langle f\rangle + \left(\frac{S_0^2}{K^2} + \frac{2S_0}{K} + \frac{1}{KFr} + 2\Theta_1\right)\times$$
$$\times \int_{-\infty}^{\infty} \Theta^{-2}\xi\frac{d\Theta}{d\xi}d\xi - \int_{-\infty}^{\infty} \xi\frac{d\Theta}{d\xi}d\xi. \tag{18.2.2.17}$$

The second term in (18.2.2.16) describes the reduction in the RT growth rate due to the ablation flow. For a small thickness of the energy release zone, the Eq. (18.2.1.9) outside the energy release zone is

$$\chi\frac{d\Theta}{d\xi} = \Theta - \Theta_1. \tag{18.2.2.18}$$

Using (18.2.2.18) to estimate the integrals in (18.2.2.17), we obtain for $\Phi(K)$

$$\Phi(K) = \frac{1+\Theta_1}{\Theta_1(1-\Theta_1)}\left(\frac{S_0^2}{K^2} + \frac{4\Theta_1}{(1+\Theta_1)}\frac{S_0}{K} + \frac{2\Theta_1^2}{1+\Theta_1} - \frac{1-\Theta_1}{(1+\Theta_1)KFr}\right)\times$$
$$\times\left[\int_{\Theta_1}^{1} \Theta^{-1}\chi d\Theta\right] + \int_{\Theta_1}^{1} \chi d\Theta \tag{18.2.2.19}$$

It should be noted that expression (18.2.2.19) was obtained for an arbitrary type of thermal conductivity and does not depend on the specific form of the function $W(\rho, T)$. Returning to dimensional variables, the increment of RT instabilities take the form

$$\sigma = \sigma_0\left(1 - k\Delta_2 \frac{u_1^2 k^2 \Phi(k)}{2\Theta_1\sigma_0^2(1+\Theta_1+u_1k/\sigma_0)}\right), \qquad (18.2.2.20)$$

where

$$\sigma_0 = \sqrt{\frac{1+\Theta_1-\Theta_1^2}{(1+\Theta)^2}u_1^2k^2 + \frac{1-\Theta_1}{1+\Theta_1}gk} - \frac{u_1k}{\Theta_1+1}. \qquad (18.2.2.21)$$

In dimensional variables, the function $\Phi(K)$ for the case of the electron thermal conduction is

$$\Phi(K) = \frac{2\Theta_1(1+\Theta_1)(1-\Theta_1^{5/2})}{5(1-\Theta_1)} \times$$
$$\times\left(\frac{\sigma_0^2}{u_1^2k^2} + \frac{4\sigma_0}{(1+\Theta_1)u_1k} + \frac{2}{1+\Theta_1} - \frac{(1-\Theta_1)g}{(1+\Theta_1)u_1^2k}\right) + \frac{2}{7}(1-\Theta_1^{7/2}). \qquad (18.2.2.22)$$

In the limit of large acceleration, $Fr = u_1^2k/g \ll 1$, the increment of RT instabilities (18.2.2.20) can be written as

$$\sigma = \sqrt{A_t gk} - bu_1k, \qquad (18.2.2.23)$$

where the Atwood number $A_t = (1-\Theta_1)/(1+\Theta_1) = (\rho_1-\rho_2)/(\rho_1+\rho_2)$, and

$$b = \frac{2(1-\Theta_1^{5/2})}{5(1-\Theta_1^2)}\Delta_2 k. \qquad (18.2.2.24)$$

The obtained solution (18.2.2.23) for the RT instability growth rate is an exact solution to the problem for the long-wavelength perturbations, $\Delta_2 k \ll 1$, and for the case of incompressible fluids. The high accuracy of the solution is illustrated in Fig. 18.4, where curve 1 is the numerical solution of the spectral problem, and curve 2 is the analytical solution (18.2.2.23) for $\Theta_1 = 0.1$ and $Fr = 10$.

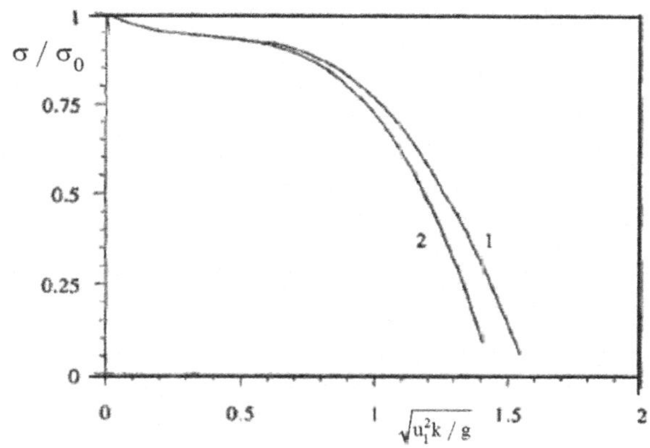

Fig. 18.4 Results of the numerical solution of the spectral problem (curve 1) and the analytical solution Eq. (18.2.2.20) (curve 2) for $\Theta_1 = 0.1$, $Fr = 10$, from [44]

18.2.3 Discontinuity Model: Short Wavelength Perturbations

At high acceleration, the Froude number is small, $Fr \ll 1$, and the wavelength of the most dangerous perturbations is much less than the total thickness of the ablation region $k\Delta_2 \gg 1$, but is usually large in comparison with the scale of thermal conduction in cold layers of the target $k\Delta_1 \ll 1$, where $\Delta_1 = \Theta^{5/2}\Delta_2$. It is possible to estimate the "cutoff" wave number K_g, at which the growth rate of perturbations tends to zero in the limit of a short wavelength. In this limit, the perturbations are located out of the energy release zone, and instead Eq. (18.2.1.15) we have

$$\frac{d^2}{d\xi^2}(\chi\tilde{\Theta}) - \frac{d\tilde{\Theta}}{d\xi} - \chi K_g^2\tilde{\Theta} = \tilde{j}\frac{d\Theta}{d\xi}. \tag{18.2.3.1}$$

For short-wavelength perturbations, the term $\chi K_g^2\tilde{\Theta}$ in Eq. (18.2.3.1) is not small, as it was in the limit of long wavelength. The perturbations of temperature decay downstream as $\exp(-K_g\xi)$, and can be used the following estimate

$$\tilde{\Theta}_g = \zeta_g \frac{d\Theta}{d\xi}\exp(-K_g\xi), \tag{18.2.3.2}$$

where ζ_g is the amplitude. Equation (18.2.3.1) with estimate (18.2.3.2) takes the form

$$\hat{F}_g[\tilde{\Theta}] = \tilde{j}\frac{d\Theta}{d\xi} + K_g\tilde{\Theta}, \tag{18.2.3.3}$$

where the operator $\hat{F}_g$ satisfies the condition

$$\hat{F}_g[\tilde{\Theta}_g] = 0. \tag{18.2.3.4}$$

Introducing a new variable

$$\tilde{\Theta} = \frac{\tilde{\psi}}{\chi} \exp\left(\frac{1}{2} \int\limits_{\xi^*}^{\xi} \chi^{-1} d\eta \right),$$

the operator $\hat{F}_g$ can be reduced to the Hermitian form relative to the variable $\tilde{\psi}$. Multiplying Eq. (18.2.3.3) by $\tilde{\psi}_g$, corresponding to $\tilde{\Theta}_g$, and integrating, we obtain

$$\int\limits_{-\infty}^{\infty} \left(\tilde{j} \frac{d\Theta}{d\xi} + K_g \tilde{\Theta} \right) \exp(-K_g \xi) d\xi = 0. \tag{18.2.3.5}$$

From Eqs. (18.2.1.12)–(18.2.1.15), we obtain the equation for the perturbation mode corresponding to the sound wave φ_s propagating toward the cold matter, $\xi \to -\infty$:

$$\frac{d}{d\xi}\left(\tilde{\upsilon} + \tilde{\Pi} - 2\Theta\tilde{j}\right) - K_g\left(\tilde{\upsilon} + \tilde{\Pi} - 2\Theta\tilde{j}\right) = -2\tilde{j}\frac{d\Theta}{d\xi} - K_g\tilde{\Theta} - \frac{\tilde{\Theta}}{\Theta^2 Fr}. \tag{18.2.3.6}$$

Multiplying (18.2.3.6) by $\exp(-K_g\xi)$, integrating, and using (18.2.3.5) and boundary conditions (18.2.1.25)–(18.2.1.26), we obtain

$$-2\Theta \tilde{j}_s + K_g \int\limits_{-\infty}^{\infty} \tilde{\Theta} \exp(-K_g \xi) d\xi = \frac{1}{Fr} \int\limits_{-\infty}^{\infty} \tilde{\Theta}\Theta^{-2} \exp(-K_g \xi) d\xi, \tag{18.2.3.7}$$

where $\tilde{j}_s$ is the amplitude of the sound mode φ_s in the upstream flow.

If the solution for the temperature perturbation $\tilde{\Theta}$ and the relation between the amplitude of this solution and the amplitude of the sound mode $\tilde{j}_s$ are known, then Eq. (18.2.3.7) inexplicitly determines the dependence of the cutoff wave number K_g on the dimensionless acceleration $1/Fr$. Estimating the temperature perturbation using (18.2.3.2) and the amplitude of sound mode using (18.2.3.5), we find

$$\tilde{j}_s = -K_g \zeta_g. \tag{18.2.3.8}$$

Finally, after a simple algebra using Eqs. (18.2.3.2), (18.2.3.7), and (18.2.3.8), we come to the following equation for the cutoff wave number K_g

$$Fr\, K_g \left(\int\limits_{\Theta_1}^{1} \exp(-2K_g \xi) d\Theta + 2\Theta_1 \right) = \int\limits_{\Theta_1}^{1} \Theta^{-2} \exp(-2K_g \xi) d\Theta. \tag{18.2.3.9}$$

From Eq. (18.2.2.18), which is obtained from the energy equation (18.2.1.9) outside the narrow zone of energy release, it follows that the function $\xi(\Theta)$ grows faster than $\Theta^{5/2}$, while the singularity of $\xi(\Theta)$ for $\Theta \to \Theta_1$ is excluded from the integrals in (18.2.3.9) for $K_g \chi_1 < 0.5$, and the function $\exp(-K_g \xi)$ can be approximated as $\exp(-K_g \xi) \approx 1$ for $\Theta < \Theta_K \equiv K_g^{-2/5}$, and $\exp(-K_g \xi) \approx 0$ for $\Theta > \Theta_K$. The temperature Θ_K corresponds to the length scale of the thermal conduction equals the length scale of perturbations. The numerical accuracy of this approximation for the integral in (18.2.3.9) is better than 5%. Then, the equation for the cutoff wave number can be written as

$$K_g \Theta_1^2 Fr = \frac{\Theta_1}{\Theta_K} \frac{\Theta_K - \Theta_1}{\Theta_K + \Theta_1}, \tag{18.2.3.10}$$

where $\Theta_K \equiv K_g^{-2/5}$. It should be noted that the function

$$f_k(\Theta_1 K_g^{2/5}) = \Theta_1 K_g^{2/5} \frac{1 - \Theta_1 K_g^{2/5}}{1 + \Theta_1 K_g^{2/5}} \tag{18.2.3.11}$$

changes only slightly over a wide range of K_g. For example, in the interval $0.01 < k_g \Delta_1 < 0.5$, where the wave number changes by two orders of magnitude, $f_k(0.1) \approx 0.082$, $f_k(0.7) \approx 0.12$. The maximum of f_k is $Max\{f_k\} = 0.17$.

When the acceleration changes over a wide range of values, the cutoff wave number of the instability is determined by the formula

$$k_g u_1^2 / g = (\Delta_1 k_g)^{2/5} \frac{1 - (\Delta_1 k_g)^{2/5}}{1 + (\Delta_1 k_g)^{2/5}} \approx 0.13. \tag{18.2.3.12}$$

Formula (18.2.3.12) is in good agreement with the results of the numerical solution of the spectral problem performed for short-wavelength perturbations shown in Fig. 18.3, and with the results of 2D simulations of the RT instability of ablatively accelerated targets [35]. A similar formula for the cutoff wavelength was also obtained in [45–49].

Finally, using Eqs. (18.2.2.23) and (18.2.3.12), the growth rate of instabilities can be written in the form

$$\sigma = \sqrt{A_t g k} - b u_1 k, \tag{18.2.3.13}$$

where according to formula (18.2.3.12) $b \approx 3$. This can be interpreted as a decrease in the growth rate of the instability RT due to the ablative mass flow for perturbations with wavelengths that are small compared to the thickness of the ablation region, $k\Delta_2 \gg 1$ but not too small, so that $k\Delta_1 = k\Delta_2 \Theta_1^{5/2} \ll 1$. This is the interval of the fastest perturbations for typical regimes of the ablatively accelerated thin foil.

In the limiting case of very large acceleration, $g\Delta_1/u_1^2 = \sqrt{\Theta_1}/Fr \gg 1$, the wavelength of fastest perturbations is small compared to the length scale of thermal conduction in the cold layers $k\Delta_1 \gg 1$. In this limit, perturbations are localized in a thin zone providing the maximum growth rate at the given wave number, and the amplitudes of the perturbations change on the length scale comparable to the thickness of the zone. Therefore, the derivatives in the left-hand side of Eqs. (18.2.1.12)–(18.2.1.15) are small in comparison with terms in the right-hand side and Eqs. (18.2.1.12–18.2.1.15) become a system of algebraic equations. The consistency condition of this system of equations gives the instability growth rate in the form

$$S = \max_{\Theta}\left\{\sqrt{\frac{1}{\Theta\, Fr}\frac{d\Theta}{d\xi} + \frac{\chi^2\Theta^2K^4}{4}} - \frac{\chi\Theta K^2}{2}\right\}. \tag{18.2.3.14}$$

Let us consider the zone of the shortest length scale of the flow $\Theta/(d\Theta/d\xi)$ as the zone providing the maximum growth rate. Outside the narrow zone of the energy release according to Eq. (18.2.2.18) we have

$$\chi\frac{d\Theta}{d\xi} = \Theta - \Theta_1, \tag{18.2.3.15}$$

which leads to the following estimates:

$$\Theta_m = \frac{7}{5}\Theta_1, \text{ and } \Theta_m^{-1}\left(\frac{d\Theta}{d\xi}\right)_m = \frac{2}{5}\left(\frac{5}{7}\right)^{7/2}\Theta_1^{-5/2}. \tag{18.2.3.16}$$

Using these estimates, the growth rate can be written as

$$S = \sqrt{\frac{0.12}{Fr}\Theta_1^{-5/2} + 2.56\Theta_1^7K^4} - 1.6\Theta_1^{7/2}K^2. \tag{18.2.3.17}$$

In dimensional variables, formula (18.2.3.17) has the form

$$\sigma = \sqrt{\frac{0.12g}{\Delta_1} + 2.56(\chi_1k^2)} - 1.6\chi_1k^2. \tag{18.2.3.18}$$

It is seen that the growth rate (18.2.3.18) is less than the growth rate $\sigma_0 = \sqrt{0.12g/\Delta_1}$, which corresponds to the RT instability of a stratified gas at rest with the density profile determined by Eqs. (18.2.1.7), (18.2.1.8) and (18.2.3.15).

Thus, under the assumption of an incompressible flow, it follows from Eq. (18.2.3.13) that the wavelength of the fastest perturbations is

$$\lambda_{\max} = 2\pi/k_{\max} = 72\pi u_1^2/g, \tag{18.2.3.19}$$

and the corresponding growth rate is

$$\sigma_0 = g/12u_1. \tag{18.2.3.20}$$

If P_a is the ablation pressure, then the acceleration of a thin foil of thickness d can be estimated as $g = P_a/(\rho_1 d)$. The structure of a steady-state ablation flow and the effect of high thermal conduction was considered in [56–58]. According to [56], the dependences between the parameters of the steady-state plane ablation flow and the laser radiation flux I_0 are as follows: $I_0 = 4\rho_2 u_2^3$, and $P_a = 2\rho_2 u_2^2$, and the critical density $\rho_2 \equiv \rho_c$ is determined by the laser radiation wavelength λ_L

$$\rho_2 = \frac{\pi m_i m_e c^2}{Ze^2\lambda_L^2}. \tag{18.2.3.21}$$

For the fastest growing and correspondingly, the most dangerous perturbations, we obtain

$$\lambda_{\max} = 36\pi d\rho_2/\rho_1, \; \sigma_{\max} = \frac{u_2}{6d} = \frac{I_0^{1/3}}{6d(4\rho_2)^{1/3}}. \tag{18.2.3.22}$$

It is seen that the wavelength of the most dangerous perturbations does not depend on the energy flux of laser radiation, but increases with decreasing radiation wavelength as λ_L^2. The growth rate increases with the increase of the laser radiation flux, $\sigma_{maz} \propto I_0^{1/3}\lambda_L^{2/3}$. It follows from (18.2.3.22) that the growth rate of the most dangerous perturbations with the wavelength in the range $k_{\max}\Delta_2 \gg 1$ and $k_{\max}\Delta_1 \ll 1$ depends on the intensity and the frequency of the laser radiation, but it is independent of the width of the ablation wave, i.e. independent of the thermal conduction.

Expression (18.2.2.23) for the growth rate of perturbations together with expression (18.2.2.24) for the coefficient b gives a fairly good analytical description of the linear stage of RT instability, and it is in good agreement with 2D modeling [35] and with available experimental results. This opens up the possibility of choosing the optimal regime of the target acceleration by choosing the appropriate material and driver parameters. At the same time, it should be emphasized that solution (18.2.2.23) is valid for the relatively long wavelengths of perturbations, $\Delta_2 k \ll 1$, and it was obtained for the case of an incompressible fluid. We also recall that an ablatively accelerated target, we are mostly interested in the case of small Froude numbers, when the RT instability dominates.

18.3 Effect of Compressibility

The structure of the ablation flow shown schematically in Fig. 18.2 consists of a narrow region of energy release at the critical surface, from which the laser energy is transferred by thermal conduction into the cold high-density plasma layers leading to the ablation flow. The transitional region between the accelerated high-density target and the expanding low-density plasma should be considered as a front of heating that propagates into a cold high-density plasma. The expanding plasma forms a rarefaction wave which moves with the velocity of the isothermal sound speed, which means that effect of compressibility can be important.

18.3.1 Structure of Ablation Wave

In the case of a compressible fluid, Eq. (18.2.1.7) remains valid, but Eq. (18.2.1.8) does not, since it is a consequence of the incompressibility condition. In the reference frame where the ablation front is at rest, the planar stationary ablation flow is determined by the equations of conservation of mass, momentum, and the energy transfer

$$\rho_c u_c = \rho_2 u_2, \tag{18.3.1.1}$$

$$P_c + \rho_c u_c^2 = P_2 + \rho_2 u_2^2, \tag{18.3.1.2}$$

$$\frac{d}{dz}\left\{\rho_c u_c\left(C_p T + \frac{u^2}{2}\right) - \kappa\frac{dT}{dz}\right\} = W(\rho, T). \tag{18.3.1.3}$$

In [56] it was noted that due to the high ($\kappa = \kappa_c T^{5/2}$) thermal conductivity of hot plasma, the head of the rarefaction wave moves with the isothermal speed of sound $u_c = a_{sT} = \sqrt{P/\rho}$, while beyond the critical surface the plasma expands outwards.

A steady ablative flow is mainly determined by the laser wavelength and the absorbed energy flux at the critical surface. The absorbed energy flux at the critical surface is

$$I_0 \approx \rho_c T_c^{3/2} \approx \rho_c a_{sc}^3. \tag{18.3.1.4}$$

Then, using (18.3.1.1) the ablation wave speed can be written as

$$U_a \equiv u_2 = \frac{\rho_c}{\rho_a}\left(\frac{I_0}{\rho_c}\right)^{1/3}. \tag{18.3.1.5}$$

From (18.3.1.5) it follows that the dynamic pressure on the ablation surface $\rho_a U_a^2$ is small compared with $\rho_c u_c^2$ as $\rho_c/\rho_a \ll 1$, and Eq. (18.3.1.2) can be written as

$$P_a \approx 2\rho_c a_{sc}^2 = 2\rho_c T_c = 2\rho_c \left(\frac{I_0}{\rho_c}\right)^{2/3}. \qquad (18.3.1.6)$$

The estimate by order of magnitude is $P_a \propto (I_0/10^{13}\,\mathrm{W/cm^2})^{2/3}$Mbar, which is in good agreement with the experimentally measured laser ablation pressure in [56]. The acceleration of the planar foil of the thickness L_{foil} can be estimated as $g \approx P_a/\rho_a L_{\mathrm{foil}}$. Then, using (18.3.1.4)–(18.3.1.6) the estimate the Froude number is

$$Fr = \frac{U_a^2}{gL_{\mathrm{foil}}} \approx \frac{\rho_c}{\rho_a} \ll 1. \qquad (18.3.1.7)$$

This means that in cases of interest for inertial thermonuclear fusion, the RT instability always prevails. Introducing dimensional variables in the same way as in Sect. 18.2.1, we find from Eq. (18.3.1.1) the relation between the dimensionless density and velocity, which is convenient to write for two points: at the critical surface and at the ablation front

$$\rho_a/\rho_c = u_c/u_a = \Theta_{ac}. \qquad (18.3.1.8)$$

The relation between temperature and density differs from the case of an incompressible flow. For the dimensionless temperature from Eqs. (18.3.1.1) and (18.3.1.2) follows

$$\frac{T}{T_c} = (1 + \gamma M_c^2)\frac{\rho_c}{\rho} - \gamma M_c^2 \left(\frac{\rho_c}{\rho}\right)^2, \qquad (18.3.1.9)$$

where $M_c = M_T/\sqrt{\gamma}$, and $M_T = u/a_{sT}$ is the isothermal Mach number, which is equal to unity at the critical surface. While the ratio of temperatures is determined by the expansion ratio, $T_2/T_1 = \rho_1/\rho_2 = \Theta$, in the case of compressible fluid, these values differ considerably in the case of compressible flow. For example, from (18.3.1.9) the temperature at the ablation front is

$$T_a = [(2\Theta_{ac} - 1)/\Theta_{ac}^2]T_c. \qquad (18.3.1.10)$$

For typical direct drive experiments the energy flux of the Nd-glass laser is $I_0 \approx 10^{13} - 10^{15}\,\mathrm{W/cm^2}$, the laser wavelength is $\lambda_L = 1.05\,\mu\mathrm{m}$, and the corresponding electron number density at the critical surface (vertical dashed line in Fig. 18.5), where the laser radiation is absorbed, is $n_e = 10^{21} cm^{-3}$, which is about two orders of magnitude smaller than the initial electron number density of the ablated target ($\sim 5 \times 10^{22}\,cm^{-3}$). The same is true for the mass density at the critical surface, $\rho_a/\rho_c \sim$

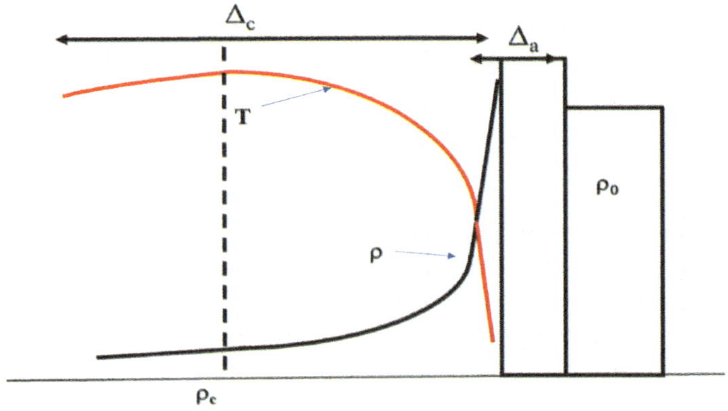

Fig. 18.5 Schematic temperature and density profiles in the ablation wave

10^2. Equation (18.3.1.10) implies a similar relationship between the temperatures at the ablation front and at the critical surface

$$T_c \approx \Theta_{ac} T_a \sim 100 T_a. \tag{18.3.1.11}$$

The large difference between the temperatures T_c and T_a also leads to a significant difference between the length scales characterizing the structure of the ablation wave, which is determined by the thermal conductivity, $\chi_{c,a} = \kappa_{c,a}/\rho_{c,a} C_p u_{c,a}$ in the hot region, $\Delta_c = \chi_c/u_c$, and in the cold ablation flow, $\Delta_a = \chi_a/u_a$. Due to the large ratio, $\Delta_c/\Delta_a = (T_c/T_a)^{5/2} \sim 10^5$, the internal structure of the ablation wave, schematically shown in Fig. 18.5, consists of a wide high-temperature region with relatively small gradients and a narrow region near the ablation wave front characterized by large gradients.

Before analyzing the effect of compressibility, it is necessary to make several remarks about the similarities and differences between the structure and instabilities of deflagration and ablation waves. It should be noted that the release of energy in the deflagration wave occurs within the flame front, which is characterized by the greatest gradients. On the contrary, in the ablation wave, energy is released in a thin layer at the critical surface, while the largest gradients are at the ablation front, which is separated from the critical surface (Fig. 18.5). Since the Darrieus–Landau instability is inherent in any wave that converts the density ahead of the wave front into the "products" of lower density behind the wave front, ablation waves must also be subject to DL instability. The above differences between deflagration and ablation structures inevitably lead to differences in the development of instabilities.

In general, the mathematical formulation of the stability problem for an ablation wave is similar to that for a flame propagating upwards in a gravitational field. Since the velocity of the fastest flame is much smaller than the speed of sound, the pressure remains constant with high accuracy up to small terms in $M^2 \ll 1$, and $\rho_2 = \rho_0 T_0/T_2$. Another difference is that the thermal conductivity of the combustible gas depends weakly on temperature, and therefore the temperature and density vary smoothly within the width of the flame. The situation is different with ablation waves (see Fig. 18.5), where the absorbed laser radiation creates a fully ionized plasma with high temperature ($\sim 1 \div 10$ KeV), high pressure (> 1 Mbar) and the thermal conductivity, which is strongly dependent on temperature. The most pronounced difference from a flame is a large difference in the scales of temperature and density profiles as is seen in Fig. 18.5. The velocity of ablation wave, $U_a = (\rho_c/\rho_a)U_c$, is about two orders of magnitude less than the isothermal sound speed at the critical surface, $U_c = \sqrt{P_c/\rho_c}$. At the same time, the temperature changes on the scales Δ_c, which is an order of magnitude larger than the scale Δ_a on which the density changes, $\Delta_c/\Delta_a \sim (T_c/T_a)^{5/2}$.

Instabilities of ablation fronts for all modes with wavelengths shorter than the total thickness of the thermal wave (Δ_c) but larger than the conduction length in the cold material (Δ_a) were studied and compared with the flame instabilities using the quasi-isobaric approximation in [52–55]. The main distinguishing feature of ablation fronts is associated with a large difference in length scales, which leads to certain functional relationships that distinguish them from those for deflagrations. At moderate accelerations (in fact, for $Fr \gg 1$), the results obtained for the flame are more or less analogous to the results obtained for the ablation wave. Assuming that stationary uniform profiles can exist in the vicinity of the critical surface, the Darrieus–Landau instability can be present in the ablation front for long-wavelength perturbations with a wavelength exceeding the critical distance Δ_c. At the same time, according to Eq. (18.3.1.7), ablation waves for experimental conditions typical for ICF are characterized by $Fr \ll 1$, when the RT instability dominates. It is possible that the DL instability manifests itself at the very beginning of the initial stage, when $Fr > 1$. Strictly speaking, the unperturbed flow in laser ablation is not stationary. The unsteadiness of ablation flows is well known and is described by self-similar solutions with nonlinear thermal conduction [59–64], which were used in numerical studies of the effect of compressibility on the hydrodynamic stability of ablation fronts in order to clarify the characteristics of more realistic hydrodynamic properties of such flows. The reliability of approaches using low Mach numbers, which were used to model laser-driven ablation flows and their hydrodynamic stability, was found to be insufficient for describing ablative waves with fast expansions. A numerical study of the hydrodynamic instability of the radiation-driven ablation flow with respect to 3D linear perturbations was carried out in [64], while the physics of ablation was modeled by a self-similar solution with nonlinear radiative heat transport.

18.3.2 Growth Rate of RT and DL Instabilities for a Finite Mach Numbers

In most analytical studies of the effect of compressibility on the growth rate of DL and RT instabilities, the discontinuous ablation front model was used [52, 67–73]. The influence of compressibility on the growth rate of the Darrieus–Landau instability of a deflagration was investigated numerically in [65, 66], taking into account the continuous structure of the deflagration front in a compressible flow and, as a consequence, eliminating the deficit of the boundary conditions. Due to the lack of in the number of boundary conditions inherent in the discontinuous model, some ad hoc additional matching condition at the ablation front was used in [52, 67–70].

The method developed in [41, 42] and used in the previous sections is the simplest way to solve the stability problem of the ablation wave in the case of a finite Mach number. Small perturbations to the stationary solution $\varphi(z)$ are taken in the form $\varphi(x, z, t) = \varphi(z) + \tilde{\varphi}(z)\exp(\sigma t + ikx)$. The system of linearized Eqs. (18.2.1.1–18.2.1.3) for a compressible flow differs from linearized Eqs. (18.2.1.12–18.2.1.15) for an incompressible flow. The linearized Eqs. (18.2.1.1)–(18.2.1.3) for compressible flow are as follows:

$$\sigma\tilde{\rho} + \frac{\partial}{\partial z}(\tilde{\rho}u_z + \rho\tilde{u}_z) + ik\rho\tilde{u}_x = 0, \tag{18.3.2.1}$$

$$\sigma\tilde{\rho}\tilde{u}_x + \rho u_z\frac{\partial\tilde{u}_x}{\partial z} + ik\tilde{P} = 0, \tag{18.3.2.2}$$

$$\sigma\rho\tilde{u}_z + \rho u_z\frac{\partial\tilde{u}_z}{\partial z} + \frac{\partial u_z}{\partial z}(\tilde{\rho}u_z + \rho\tilde{u}_z) + \frac{\partial\tilde{P}}{\partial z} = 0, \tag{18.3.2.3}$$

$$\tilde{P} = \frac{\gamma - 1}{\gamma}\left(C_P\rho\tilde{T} + \tilde{\rho}T\right), \tag{18.3.2.4}$$

$$\begin{aligned}
&\sigma\left(\rho C_v\tilde{T} + \rho u_z\tilde{u}_z - \tilde{\rho}RT\right) + (\tilde{\rho}u_z + \rho\tilde{u}_z)\frac{\partial}{\partial z}\left(C_P T + \frac{u_z^2}{2}\right) + \\
&+ \rho u_z\frac{\partial}{\partial z}\left(C_P\tilde{T} + u_z\tilde{u}_z\right) + \\
&+ \frac{\kappa_c}{T_c^{5/2}}\left\{k^2T^{5/2}\tilde{T} - \frac{\partial}{\partial z}\left(\frac{5}{2}\frac{\partial T}{\partial z}\tilde{T} + T\frac{\partial\tilde{T}}{\partial z}\right)\right\} = \frac{\partial W}{\partial\rho}\tilde{\rho} + \frac{\partial W}{\partial T}\tilde{T}
\end{aligned} \tag{18.3.2.5}$$

The boundary conditions for perturbations require vanishing all perturbations on the cold surface of the accelerated foil and in the outgoing flow far from the critical surface. An analytical solution can be obtained using additional simplification. For example, in [52] it was assumed that coefficients in Eqs. (18.3.2.1)–(18.3.2.5) are constant in the uniform flows on both sides from the ablation wave, while perturbations exponentially decay, $\tilde{\varphi}(x, z, t) = +\tilde{\varphi}(0)\exp(\sigma t + \mu z + ikx)$, where $\mu > 0$, in the direction of target, $z \rightarrow -\infty$, and $\mu < 0$ downstream from the critical surface, at $z \rightarrow +\infty$. Another simplification used in [71] assumes a strongly localized energy source, which means that the right-hand part in Eq. (18.3.2.5) equals zero.

Considering a discontinuity model, the solution of Eqs. (18.3.2.1) – (18.3.2.5) in homogeneous flows should be matched at the perturbed front $\zeta(z,t) = \zeta\exp(\sigma t + ikx)$. Using continuity equations for mass and momentum, we obtain

$$\rho_1\left(\tilde{u}_{z1} - \frac{\partial\zeta}{\partial t}\right) + \tilde{\rho}_1 u_{z1} = \rho_2\left(\tilde{u}_{z2} - \frac{\partial\zeta}{\partial t}\right) + \tilde{\rho}_2 u_{z2}, \tag{18.3.2.6}$$

$$\tilde{u}_{z1} + u_{z1}\frac{\partial\zeta}{\partial x} = \tilde{u}_{z2} + u_{z2}\frac{\partial\zeta}{\partial x}, \tag{18.3.2.7}$$

$$\tilde{P}_1 + \tilde{\rho}_1 u_{z1}^2 + 2\rho_1 u_{z1}\left(\tilde{u}_{z1} - \frac{\partial\zeta}{\partial t}\right) = \tilde{P}_2 + \tilde{\rho}_2 u_{z2}^2 + 2\rho_2 u_{z2}\left(\tilde{u}_{z2} - \frac{\partial\zeta}{\partial t}\right). \tag{18.3.2.8}$$

Subscripts 1 and 2 refer to the values taken at the ablation front, i.e. u_{z1} is the ablation front velocity, and at the critical surface, i.e. $u_{z2} \equiv U_c$. If the discontinuity model is considered, then it remains to solve only Eqs. (18.3.2.6)–(18.3.2.8). The energy equation, which determines the internal structure of the wave, is not included in the model. This means that an additional condition is required, since for 4 unknown modes: two sound waves (one on both sides of the discontinuity surface), one entropy wave in the outgoing flow, plus the perturbation of the front, there are only 3 Eqs. (18.3.2.6)–(18.3.2.8). In the case of a flame and under the assumption of an incompressible gas, the additional condition was the Landau condition, according to which the flame velocity in the direction normal to the flame surface remains unchanged. If $\zeta(z,t)$ is a perturbation of the flame front, $\tilde{u}_z - \partial\zeta/\partial t = 0$ (see Chap. 5). However, without an "unjustified" assumption, the additional condition can be obtained only considering the internal structure of a deflagration or the ablation wave. If we express the perturbations through the perturbation of the front, in the form

$$\tilde{\rho} = -\zeta\frac{\partial\rho}{\partial z}, \quad \tilde{T} = -\zeta\frac{\partial T}{\partial z}, \quad \text{etc.}, \tag{18.3.2.9}$$

then the additional condition in the case of compressible flow will be

$$\rho_1\left(\tilde{u}_{z1} - \frac{\partial\zeta}{\partial t}\right) + \tilde{\rho}_1 u_{z1} = 0. \tag{18.3.2.10}$$

In the limit of an incompressible flow, the condition (18.3.2.10) reduces to the Landau condition, $\rho_1(\tilde{u}_{z1} - \partial\zeta/\partial t) = 0$. It is seen that the effect of compressibility is due to the additional term $\tilde{\rho}_1 u_{z1}$ in the Landau condition, which vanishes in the case of an incompressible flow. If the energy release is taken in the form of δ-function, the condition (18.3.2.10) does not depend on the mechanism of energy release. Written in the form $\rho_a(\tilde{u}_{za} - \partial\zeta/\partial t) + \tilde{\rho}_a U_a = 0$, this condition can be interpreted as an unchanged perturbation of the mass flux instead of unchanged normal flame velocity in the Landau condition in the case of incompressible fluids. Note, that u_{z1} is the velocity of the ablation front, which was previously designated as $u_{z1} \equiv U_a$.

In general, the system of Eqs. (18.3.2.1)–(18.3.2.5) can be solved only numerically [71, 72]. It is convenient to use dimensionless variables, $\xi = z/\Delta_c$, $u = u_z/u_c$, $r = \rho/\rho_c$, $\Theta = T/T_c$, where $\Delta_c = \chi_c/u_c$ is the width of the ablation front near the critical surface, $\chi_c = \kappa_c/\rho_c C_P$. Introducing dimensionless perturbations, as in [41, 42]:

$$\tilde{\Theta} = \frac{\tilde{T}}{T_2},\ \tilde{v} = i\frac{\tilde{u}_x}{u_2},\ \tilde{j} = \frac{\tilde{\rho}u_z + \rho\tilde{u}_z}{\rho_2 u_2},\ \tilde{\Pi} = \frac{\tilde{P} + \tilde{\rho}u_z^2 + 2\rho u_z \tilde{u}_z}{\rho_2 u_2^2}, \tag{18.2.2.11}$$

and an auxiliary variable $\Phi = \Theta^{5/2}\frac{\partial\Theta}{\partial\xi}$, allows to present a second-order Eq. (18.3.2.5) in the form of two first-order equations. Then, the problem is reduced to the system of five linearized equations for five unknowns. For perturbations $\tilde{\varphi} = \tilde{\varphi}_i \exp(\mu z)$, this is an eigenvalue problem with an eigenvalue μ and eigenvectors $\tilde{\varphi}_i$ defined by the 5×5 matrix $\mathfrak{M}_{ij}$ for the coefficients of the linearized equations

$$\det\left|\mathfrak{M}_{ij} - \mu\delta_{ij}\right| = 0. \tag{18.2.2.12}$$

For an incompressible flow we come to the results obtained in Chap. 5. In the case of a compressible flow, the solution can only be obtained numerically, using different methods, for example, as used in [73]. When searching for a solution in the form $\sigma = \Gamma(M, \Theta)kU_a$, it was found that the numerical solutions of the dispersion equation strongly depend on $\Gamma(M, \Theta)$. The factor $\Gamma(M, \Theta)$ obtained numerically in [73] is almost two times larger than predicted in [52, 67–71] using the Landau condition as an additional condition.

However, it should be noted that the assumption of a uniform flow behind the critical surface is a serious drawback of the model. In reality, the rarefaction wave behind the critical surface makes the flow highly inhomogeneous Also, the assumption of a uniform flow from the opposite side of the ablation wave implicitly means that the thickness of the accelerated target (foil) is much larger than the scale Δ_c.

18.3.3 WKB Model: Stabilization of RT Instability by Convection

A self-consistent analytical model based on the Wentzel–Kramers–Brillouin (WKB) approximation was developed by Bud'ko and Liberman [74, 75] to investigate the stabilization of the Rayleigh–Taylor instability due to the combined action of inhomogeneous convective mass flow and stratified profiles. In this approach, the eigenvalue problem for the instability growth rate is reduced to solving a system of algebraic equations with coefficients depending on unperturbed variables.

We consider 2D perturbations in the form $\Psi(t, x, z) = \Psi(z)\exp(\sigma t + ikx)$, with a wave number k and growth rate σ; the z-axis is directed along the target acceleration and unperturbed gradients of density, temperature, etc. For simplicity, we will consider an ablatively accelerated planar foil. Since the energy of laser radiation is transferred from the critical surface to the accelerated target faster than the characteristic time scales of hydrodynamic motion, an almost stationary self-similar dynamics is formed after the initial shock reaches the back side of the foil [59, 76]. In the case of a nonstationary unperturbed flow with the characteristic time scale of evolution τ, the value of $\sigma(t)$ is considered as the instantaneous growth rate. The linearized hydrodynamic equations are

$$\sigma\tilde{\rho} + \frac{\partial}{\partial z}(\tilde{\rho}u_z + \rho\tilde{u}_z) + ik\rho\tilde{u}_x = 0, \tag{18.3.3.1}$$

$$\sigma\rho\tilde{u}_x + \rho u_z\frac{\partial\tilde{u}_x}{\partial z} + ik\tilde{P} = 0, \tag{18.3.3.2}$$

$$\rho\left(\sigma\tilde{u}_x + \frac{\partial(u_z\tilde{u}_z)}{\partial z}\right) + \frac{\partial\tilde{P}}{\partial z} - \tilde{\rho}\left(g - u_z\frac{\partial u_z}{\partial z}\right) = 0. \tag{18.3.3.3}$$

$$\sigma\tilde{P} + \gamma\tilde{P}\frac{\partial u_z}{\partial z} + \gamma P\left(\frac{\partial\tilde{u}_z}{\partial z} + ik\tilde{u}_x\right) + u_z\frac{\partial\tilde{P}}{\partial z} + \rho g\tilde{u}_x = 0. \tag{18.3.3.4}$$

Assuming that the characteristic Péclet number is large,

$$Pe(k) = \frac{2\pi}{k}\frac{\gamma P}{T}\frac{u}{\kappa} \gg 1, \tag{18.3.3.5}$$

we can neglect by the terms with perturbed electron heat flux in Eq. (18.3.3.4). Here, $\kappa = 3 \times 10^4 T^{3/2}[\mathrm{eV}]/\mathrm{n}\,[\mathrm{cm}^{-3}]$ is the electron thermal conductivity. Condition (18.3.3.5) implies a limit from below on the perturbation wavelength λ:

$$\lambda\,[cm] = \frac{2\pi}{k} \gg \frac{T^{5/2}[\mathrm{eV}]}{(1+\overline{Z})n(10^{20}\,[\mathrm{cm}^{-3}]\,u\,[\mathrm{cm/s}]}, \tag{18.3.3.6}$$

where $\overline{Z}$ is the average charge of ions.

The growth rate is the eigenvalue of the boundary value problem for Eqs. (18.3.3.1)–(18.3.3.4). The boundary conditions are that all perturbations vanish in the regions of homogeneous flows at $z \to \pm\infty$. In the WKB model, eigenfunctions are localized in the unstable transition layer and the eigenvalues depend weakly on the specific form of the boundary conditions.

Consider first the RT instability of an accelerated plasma in a co-moving reference frame without a convective mass flow. After simple algebra Eqs. (18.3.3.1)–(18.3.3.4) can be reduced to the second-order equation for the z component of the perturbation velocity:

$$\frac{d^2}{dz^2}\left[\frac{\rho\sigma^2\gamma P}{\rho\sigma^2+k^2\gamma P}\left(\frac{d}{dz}\tilde{u}_z\right)\right]- \\ -\left[\rho\sigma^2+g\frac{d\rho}{dz}-\frac{kg\rho^2}{\rho\sigma^2+k^2\gamma P}-\frac{d}{dz}\left(\frac{\sigma^2\rho^2 g}{\rho\sigma^2+k^2\gamma P}\right)\right]\tilde{u}_z=0. \qquad (18.3.3.7)$$

In the WKB approximation, we assume that the wavelengths of the perturbations are small compared to the characteristic scale of stratification, $L \sim a_s^2/g$, where $a_s = \sqrt{\gamma P/\rho}$. Then, Eq. (18.3.3.7) is reduced to

$$\frac{d^2\tilde{u}_z}{dz^2}-k^2\left[1-\frac{g\alpha}{\sigma^2}-\left(\frac{g}{\sigma a_s}\right)^2\right]\tilde{u}_z=0, \qquad (18.3.3.8)$$

where $\alpha(z) = -d(\ln\rho(z))/dz$. Except for the last term in square brackets, Eq. (18.3.3.8) is related to incompressible perturbations in a stratified plasma [77]. It is readily seen that Eq. (18.3.3.8) is identical to the Schrödinger equation for the quantum mechanical problem of stationary states of a particle with energy $E = 0$ in the 1D potential $U(z) = k^2\{1-g\alpha/\sigma^2-(g/\sigma a_s)^2\}$. The parameters of the stationary localized states determine the eigenvalues spectrum $\sigma = \sigma_n$. Expanding the potential $U(z)$ in the vicinity of its minimum at $z = z_m$ into the Taylor series, and assuming $d^2U(z_m)dz^2 \neq 0$, we obtain the eigenvalues spectrum σ_n using the well-known formula for the energy levels of a harmonic oscillator [78]. With accuracy up to the zeroth-order terms in $1/(kL/2\pi) \ll 1$, we obtain

$$\sigma_n^2=\left(g\alpha+\frac{g^2}{a_s^2}\right)(z_m)-\frac{2n+1}{4k}\sqrt{-2\frac{d^2(g\alpha+g^2/a_s^2)}{dz^2}(z_m)}, \qquad (18.3.3.9)$$

where $n = 0, 1, 2, \ldots$ The expression for the spectrum (18.3.3.9) is valid for $n \ll kL/2\pi$. The corresponding eigenfunctions can be expressed via the Hermitian polynomials $H_n(z)$:

$$\tilde{u}_z = \exp(-\omega_n z^2/2)H_n(\omega_n^{1/2}z)$$

$$\omega_n=\sqrt{-\frac{k^2}{2\sigma_n^2}\frac{d^2(g\alpha+g^2/a_s^2)}{dz^2}(z_m)}.$$

In the zeroth approximation $1/(kL/2\pi) \ll 1$, the splitting of the eigenvalues σ_n, in the spectrum (18.3.3.9) can be neglected, so that for the maximal growth rate we obtain $\sigma_{\max} = \sqrt{g\alpha + g^2/a_s^2}$. The eigenvalues spectrum for large $n \gg 2\pi kL$ corresponds to the Bohr–Sommerfeld quantization rule [78]:

$$k\int_{z_1}^{z_2}\sqrt{\frac{g\alpha}{\sigma_n^2}+\frac{g^2}{\sigma_n^2 a_s^2}-1}\,dz = \pi\left(n+\frac{1}{2}\right), \tag{18.3.3.10}$$

where z_1 and z_2 corresponds to the points determined by the condition $U(z) = 0$. It is interesting to note, that for an incompressible fluid with the unperturbed density profile $\rho(z) \propto \exp(-\alpha_0 z)$ and constant acceleration $g = g_0$, Eq. (18.3.3.10) gives

$$\sigma_n = \sqrt{\frac{g_0\alpha_0}{1+[\pi(n+1/2)/kL]^2}},$$

which is in the limit $n \gg kl/2\pi$ reduced to the solution obtained by Mikaelian [79].

Let us now consider the RT increments for the steady-state mass flow in a stratified accelerated plasma. The solutions of the linearized Eqs. (18.3.3.1) – (18.3.3.4) in the WKB approximation will be taken in the form

$$\Psi(z) = \sum_i \psi_i \exp\left(k\int^{z}\varphi_i(z')dz'\right), \tag{18.3.3.11}$$

where $\Psi = \tilde{u}_z, \tilde{u}_x, \tilde{\rho}, \tilde{P}$, and the perturbation amplitudes, $\psi_i(z)$, as well as the phase functions, $\varphi_i(z)$, are slowly varying functions of the coordinate z in comparison with the exponential terms. Substitution of perturbations in the form (18.3.3.11) into Eqs. (18.3.3.1) – (18.3.3.4) leads to four coupled equations for the amplitudes

$$\left(\sigma+\frac{du_z}{dz}+u_z k\varphi\right)\tilde{\rho}+\left(\frac{\partial\rho}{\partial z}+\rho k\varphi\right)\tilde{u}_z + ik\rho\tilde{u}_x = 0, \tag{18.3.3.12}$$

$$\rho(\sigma+u_z k\varphi)\tilde{u}_x + ik\tilde{P} = 0, \tag{18.3.3.13}$$

$$\rho\left(\sigma+\frac{\partial u_z}{\partial z}+ku_z\varphi\right)\tilde{u}_z + k\varphi\tilde{P} - \left(g-u_z\frac{\partial u_z}{\partial z}\right)\tilde{\rho} = 0, \tag{18.3.3.14}$$

$$\left(\sigma+\gamma\frac{\partial u_z}{\partial z}+ku_z\varphi\right)\tilde{P} + i\gamma Pk\tilde{u}_x + (\rho g+\gamma Pk\varphi)\tilde{u}_z = 0, \tag{18.3.3.15}$$

where g is determined by the unperturbed pressure gradient, and the term $u_z du_z/dz$ in Eq. (18.3.3.14), which is small compared to g, can be omitted hereinafter. By setting the determinant of the homogeneous system (18.3.3.12)–(18.3.3.15) equal to zero, we obtain the characteristic equation for the phase functions φ_i:

$$D_4(\varphi) \equiv (\Sigma+\varphi)^2\left((\Sigma+\varphi)^2+\frac{1-\varphi^2}{M^2}\right)- \frac{\alpha^2}{k^2 Fr}\left[(\Sigma+\varphi)\left(\Sigma+\varphi-(\gamma-1)\frac{\alpha_u}{\alpha}\right)+\frac{1}{M^2}+\frac{1}{Fr}\right]=0. \quad (18.3.3.16)$$

Here $\Sigma(z) = \sigma/uk$ is the dimensionless growth rate, $\alpha_u = d(\ln u(z))/dz$ is the steepness of the velocity profile, $Fr = u^2/gL$ is the local Froude number, and $M(z) = u_z/a_s$ is the Mach number.

Depending on the relative values of the dimensionless parameters, different regimes of plasma dynamics and stability are possible. Assuming that the flows in a dense slab and in the ablating plasma beyond the unstable transition layer are uniform, the term in square brackets in Eq. (18.3.3.16) vanishes, and the solution for the modes in a uniform flow is

$$\varphi_{1,2} = \frac{\Sigma M^2 \pm \sqrt{1+M^2(\Sigma^2-1)}}{1-M^2},\ \varphi_{3,4} = -\Sigma. \quad (18.3.3.17)$$

The roots $\varphi_{1,2}$ represent sonic waves, while the roots $\varphi_{3,4}$ correspond to entropy and vorticity modes. In accordance with typical experimental conditions, we assume a subsonic flow in a dense plate ($z \to -\infty$), while the ablation flow at $z \to +\infty$ can be subsonic or supersonic, depending on the laser intensity and wavelength [80–82]. The asymptotic behavior of the localized perturbation in the dense slab at $z \to -\infty$ implies the sonic solution $\Psi(z) = A\exp(k\varphi_1 z)$, $z \to -\infty$, $\mathrm{Re}(\varphi_1) > 0$, and the coupled vorticity, entropy, and sonic waves in the downstream flow, $\Psi(z) = \sum_{i=2}^{4} \psi_i \exp(k\varphi_i z)$, $z \to +\infty$, $\mathrm{Re}(\varphi_i) < 0$. Location of the roots φ_i of Eq. (18.3.3.16) versus the profile coordinate z in the complex plane φ, depends on the parameters of the problem. The behavior of the roots in the subsonic flow regime is shown in Figs. 18.6, 18.7, 18.8 and 18.9. There can be no more than two roots simultaneously in the half-plane $\mathrm{Re}(\varphi) > 0$. At certain values of z, the roots can intersect and form the multiple ones. We conclude that in order for the perturbation to be localized, there must be points on the profile at which a multiple root of the Eq. (18.3.3.16) is formed by the intersection of the root $\varphi_1(z)$ with some other root $\varphi_i(z)$ ($i = 2-4$) in the half-plane $\mathrm{Re}(\varphi) > 0$. The multiple root can be formed if the value of the parameter σ is sufficiently small: $\sigma < \sigma_{\max}(k)$, see Figs. 18.6 and 18.8. Greater values of $\sigma > \sigma_{\max}(k)$ correspond to the pattern shown in Figs. 18.7 and 18.9 without the intersection points. The coordinates z_*, z_{**} of these points and the corresponding values of φ_*, Σ_* are obtained from two coupled equations.

$$D_4[\varphi(z_*)] = 0 \text{ and } \frac{d(D_4)}{d\varphi} = 0. \quad (18.3.3.18)$$

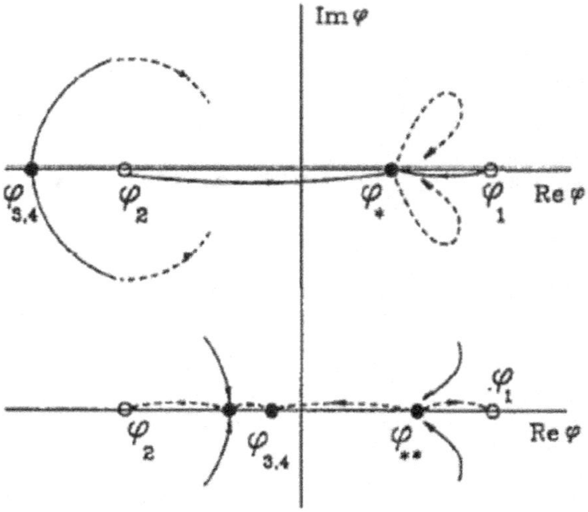

Fig. 18.6 Location of the roots $\varphi_i(z)$ $(i = 2-4)$ of Eq. (18.3.3.16) in the complex plane φ for subsonic ablation flow, $ku_z(z \to -\infty) < \sigma_{\max} < ku_z(z \to +\infty)$, $\sigma < \sigma_{\max}(k)$. Solid circles represent the modes in the homogeneous flows, dashed circles -the multiple roots. Solution of Eqs. (18.3.3.12)–(18.3.3.15) in the downstream homogeneous flow corresponds to the vorticity mode. Figures from upper to lower represent the roots $\varphi_i(z)$ paths with the profiles in the unperturbed flow changing continuously from the upstream flow [the upper figure) to the downstream flow (the lower figure)

Fig. 18.7 The same as in Fig. 18.6, but for $\sigma > \sigma_{\max}(k)$. There is no intersection point in the half-plane $\mathrm{Re}(\varphi) > 0$

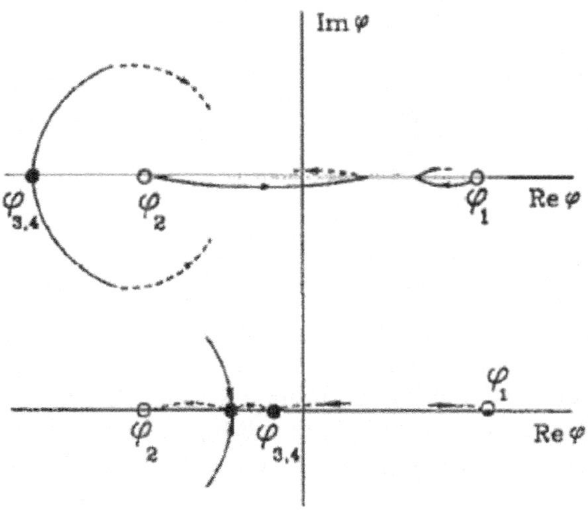

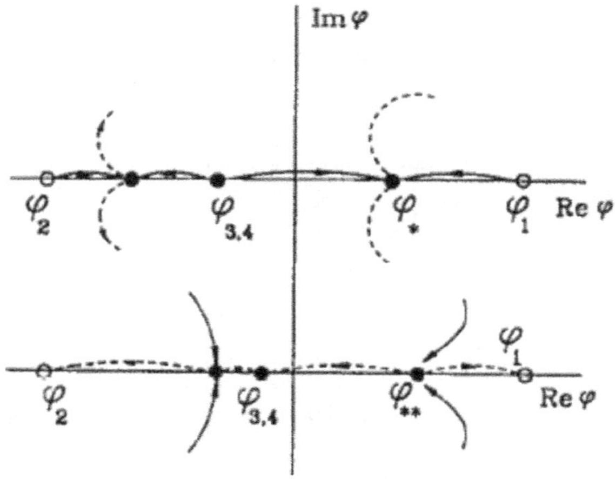

Fig. 18.8 The same as in Fig. 18.7, for $\sigma < \sigma_{\max}(k) < ku(z \to -\infty)$. The solution of Eqs. (18.3.3.12) – (18.3.3.15) in the downstream homogeneous flow corresponds to the sonic mode

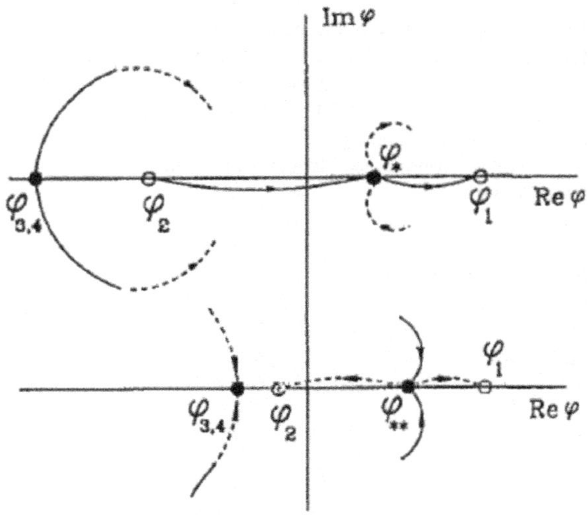

Fig. 18.9 The same as in Fig. 18.6 for $ku(z \to -\infty) < ku(z \to +\infty)$

Since the relative difference between the eigenvalues, σ_n and $\sigma_{\max}$ is the first-order term in the small parameter $(k/2\pi\alpha)^{-1}$, the eigenvalues splitting at small n in the zeroth-order approximation is negligible. Thus, neglecting the first-order terms, we can write $\sigma_k \approx \sigma_{\max}(k)$ with $\sigma_{\max}$ corresponding to the case, where there is the only intersection point, and the roots move alone the axis $\mathrm{Im}\varphi = 0$ in the half-plane $\mathrm{Re}\varphi > 0$. The eigenfunction corresponding to the maximum eigenvalue in the zeroth-order approximation in the parameter $(k/2\pi\alpha)^{-1}$ is

$$\Psi(z) = \exp\left(k\int_{z_*}^{z} \varphi_+(z')dz'\right), \; z > z_*, \; \Psi(z) = \exp\left(k\int_{z_*}^{z} \varphi_-(z')dz'\right), \; z < z_*,$$

where φ_+ is the root of the characteristic Eq. (18.3.3.16), moving along the axis $\text{Im}\varphi = 0$ from the point φ_* to the negative values of φ at $z > z_*$, and φ_- is the root moving at $z < z_*$ to the point $\varphi = 1$. The function $\Psi(z)$ is localized at the point z_0, where the root $\varphi_+(z)$ changes its sign passing through the point $\varphi = 0$.

From Eq. (18.3.3.16), it follows that the coordinate z_0 of this point is obtained from the equation

$$\Sigma^2(z_0)\left(\Sigma^2(z_0) + \frac{1}{M^2(z_0)}\right) - \frac{G(z_0)}{M^2(z_0)} = 0, \tag{18.3.3.19}$$

where the dimensionless growth rate is $\Sigma(z) = \sigma/uk$, and the dimensionless acceleration is $G(z) = 4g\alpha(z)/(uk)^2 = 4\alpha^2/k^2 Fr$. Solution of Eq. (18.3.3.19) must satisfy the condition $z_0 > z_*$.

The root φ_- represents the sonic wave, which is the only possible solution in the upstream flow. The root φ_+ can be associated with the sonic solution in the downstream ablation flow or with the vorticity mode, depending on the eigenvalue $\varphi_{\max}(k)$ as well the plasma parameters in the ablation zone. In Fig. 18.6, coupling of the sonic wave with the vorticity mode is illustrated for the subsonic flow regime. In this case the pressure and density perturbations vanish in the downstream flow and only vortex velocity perturbations left. This limit is consistent with the vortex shedding stabilization mechanism discussed by Manheimer and Colombant [80] for the discontinuity model. If the condition $\sigma > ku(+\infty)$ is valid, the sonic waves coupling shown in Fig. 18.8 takes place, which implies nonzero pressure and density disturbances as well as potential velocity.

The approach based on the WKB approximation is self-consistent if the local Froude number is small

$$Fr(z) \ll 1. \tag{18.3.3.20}$$

Assuming that the length scales of density and velocity are comparable, we can neglect the first term in square brackets in Eq. (18.3.3.16) regardless of the Mach number value, and from Eqs. (18.3.3.16) and (18.3.3.18), after a rather cumbersome algebra, we obtain

$$\begin{aligned} &M^2\Sigma_*^8 + (1-4M^2)\Sigma_*^6 - 3\,[1-2M^2+(1-20M^2-8M^4]\times \\ &\quad G/12)]\,\Sigma_*^4 + [3-4M^2+(1-M^2)(5+4M^2)\,G]\,\Sigma_*^2 - \\ &\quad (1-M^2)[1-(1-M^2)G]^2 = 0, \end{aligned} \tag{18.3.3.21}$$

$$\varphi_* = \frac{-(1-4M^2)\Sigma_* + \sqrt{8(1-M^2)+(1+8M^2)\Sigma_*^2}}{4(1-M^2)}, \tag{18.3.3.21a}$$

where $G(z)$ denotes the dimensionless acceleration: $G(z) = 4\alpha^2(1+M^2/Fr)/k^2 Fr$. Inspection of the roots in the half-plane $\text{Re}\varphi > 0$ and formation of the multiple roots implies that the solution of the equation

$$\Sigma(z_*) = \Sigma_*(z_*) \tag{18.3.3.22}$$

exists, where Σ_* is a real and positive root of Eq. (18.3.3.21). Since we are looking for the maximal eigenvalue for a given k, corresponding to the unique root's intersection point then, in addition to Eq. (18.3.3.22), the following conditions must be met:

$$\frac{d}{dz}\Sigma(z_*) = \frac{d}{dz}\Sigma_*(z_*). \tag{18.3.3.23}$$

The remaining complex conjugate roots of Eq. (18.3.3.21) correspond to the complex solutions z_* of Eq. (18.3.3.22).

Dividing Eq. (18.3.3.23) by Eq. (18.3.3.22) and taking into account the explicit form of $\Sigma(z)$, we can exclude the eigenvalue σ from this system and obtain the equation for the coordinate of the intersection point

$$\frac{d}{dz}\ln[\Sigma_*(z_*)] = -\frac{d}{dz}\ln[u(z_*)]. \tag{18.3.3.24}$$

The growth rate (maximal eigenvalue) then will be

$$\sigma = ku(z_*)\Sigma_*(z_*). \tag{18.3.3.25}$$

Equations (18.3.3.24) and (18.3.3.25) give the solution to the problem: the growth rate is obtained as an implicit function of the wave number and the structure of unperturbed profiles.

First, we consider the low-k asymptotic, assuming $\frac{k^2}{\alpha^2} \ll \frac{M^2}{Fr}\left(1 + \frac{M^2}{Fr}\right)$.

The approximate solution of Eqs. (18.3.3.21) and (18.3.3.24) is

$$\Sigma_* = \frac{3 - \sqrt{(1+8M^2}}{2}\left(\frac{1 + 2M^2 + \sqrt{1+8M^2}}{8M^2(1-M^2)}G\right)^{1/4}$$

$$\frac{d}{dz}\left[\frac{3 - \sqrt{1+8M^2}}{2}\left(\frac{1 + 2M^2 + \sqrt{1+8M^2}}{8M^2(1-M^2)}G\right)^{1/4} u\right]_{z=z_*} = 0. \tag{18.3.3.26}$$

Assuming

$$u_z \ll a_{sT},\ 2\pi/k \ll a_{sT}^2/g,\ \sigma \ll ka_{sT}, \tag{18.3.3.27}$$

where $a_{sT} = \sqrt{P/\rho}$ is the isothermal sound speed, we obtain an intermediate asymptotic for small Mach numbers $M(z)$. For the perturbation wave numbers in the range

$$\frac{\sqrt{g\alpha}}{a_{sT}} \ll \frac{k}{2\pi} \ll \frac{\sqrt{g\alpha}}{u_z}, \tag{18.3.3.28}$$

Equations (18.3.3.21) and (18.3.3.24) yield

$$\Sigma_* = \frac{\sqrt{G}}{2}, \quad \frac{d}{dz}\left[\frac{\alpha^2 u^2}{Fr}\left(1 + \frac{M^2}{Fr}\right)\right]_{z=z_*} = 0. \tag{18.3.3.29}$$

As $G \propto 1/k^2$ the growth rate σ does not depend on wave number k in the limit (18.3.3.28). In particular, if M^2 is small compared to the Froude number, the saturated RT growth rate in the static plasma with smooth density gradient, given by Eq. (18.3.3.9) is reproduced.

Next, we consider the perturbation with wavelengths close to the total stabilization limit. As a solution of Eq. (18.3.3.21), the value of the function Σ_* in the intersection point of roots φ_i depends on k as a parameter. It can be shown that this dependence is a monotonic decreasing. Since in the homogeneous upstream and downstream flows we have $\Sigma_* = -1 < 0$, the necessary and sufficient condition for the positive growth rate $\sigma(k)$ to be the zeroth-order solution of Eq. (18.3.3.22) is $\max\{\Sigma_*(z)\} > 0$. From Eq. (18.3.3.21), it follows that this inequality implies

$$\max\left\{\frac{4(1-M^2)g}{k^2 u_z^2}\left(\alpha + \frac{M^2 g}{u_z^2}\right)\right\} > 1. \tag{18.3.3.30}$$

Thus, for any given hydrodynamic profiles $g(z)$, $u_z(z)$, $\rho(z)$, $T(z)$, the RT growth rate vanishes at the cutoff wave number

$$k = k_0 = \sqrt{\max\left\{\frac{4(1-M^2)g}{u_z^2}\left(\alpha + \frac{M^2 g}{u_z^2}\right)\right\}}. \tag{18.3.3.31}$$

Equation (18.3.3.31) provides the scaling for the cutoff wave number at various flow parameters. If the target is irradiated by a long-wavelength laser radiation with sufficiently high intensity, the inequality $M^2 \ll Fr \ll 1$ is valid in the ablating plasma, and the stabilization of the RT modes occurs at wavenumber

$$k_0 \sim \frac{1}{u_z}\sqrt{\frac{g}{L}} \sim \frac{\alpha}{\sqrt{Fr}}. \tag{18.3.3.32}$$

In the limit of small Froude numbers, $Fr \ll M^2 \sim 1$, typical for a short-wavelength laser radiation, the scaling is

$$k_0 \sim \frac{gM}{u_z^2} \sim \frac{\alpha M}{Fr}, \tag{18.3.3.33}$$

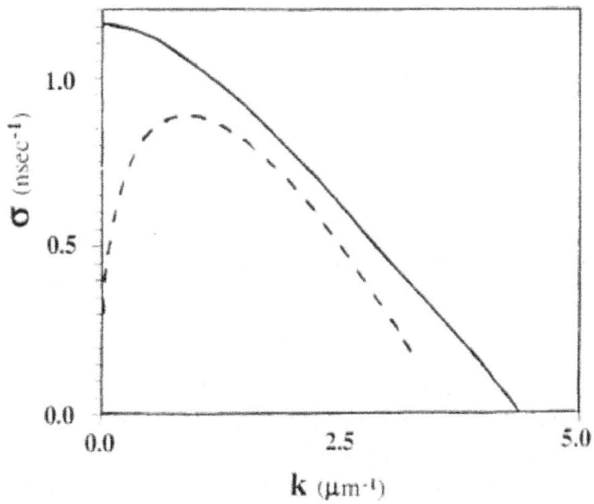

Fig. 18.10 The RT growth rates corresponding to the profiles in Fig. 18.11 versus wave number obtained numerically (dashed curve) and from the WKB model (solid curve) [75]

which agrees qualitatively with the well-known Takabe formula (18.2.1). Note also that both conditions (18.3.3.32) and (18.3.3.33) demonstrate that condition (18.3.3.20), that is $Fr(z) \ll 1$, coincides with the condition of applicability of the WKB model, $k/2\pi \gg \alpha = -d(\ln\rho)/dz$.

For parameters in the range determined by inequalities (18.3.3.28), which correspond to the essentially subsonic flow, the solution to Eq. (18.3.3.21) is

$$\varphi_* = \sqrt{1 - \frac{G^{1/3}}{(s+1)^{2/3} + (s-1)^{2/3} + G^{1/3}/3}}, \quad \Sigma_* = \frac{1 - 2\varphi_*^2}{\varphi_*}, \tag{18.3.3.34}$$

where $s = \sqrt{1 + G/27}$. The maximum eigenvalue σ is associated with the single tangent point of the functions $\Sigma(z)$ and $\Sigma_*(z)$. Using the explicit form of the functions $\Sigma(z)$ and $\Sigma_*(z)$, we obtain from Eq. (18.3.3.24)

$$\frac{dG}{dz} = \frac{1 - 2\varphi_*^2}{1 + 2\varphi_*^2}\left[12(1 - 2\varphi_*^2)^2\right]\frac{d\ln(u_z)}{dz}, \tag{18.3.3.35}$$

where $\varphi_* = \varphi_*[G(z_*)]$ is given by Eq. (18.3.3.21a). If $G_{\max} < 1$, then $\Sigma_*(z) < 0$ and positive eigenvalues do not exist in agreement with Eq. (18.3.3.30). This is the exact result in the quasi-classical approximation and for given profiles of the unperturbed flow. The corresponding wave number k_0, for which the growth rate vanishes, is

$$k_0 = \max\sqrt{(-4gu_z^{-2}d\ln\rho/dz)}.$$

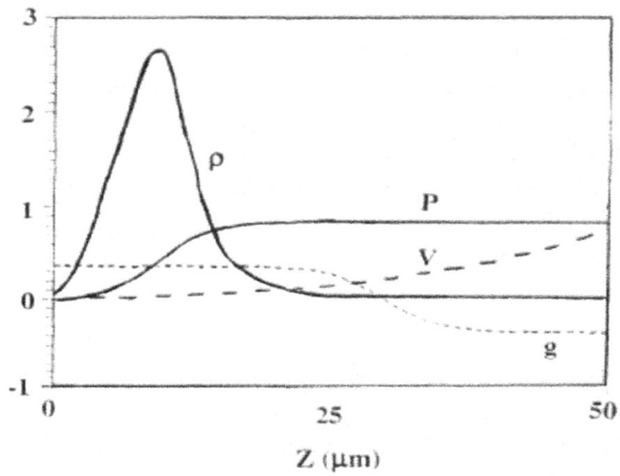

Fig. 18.11 Steady-state smoothed unperturbed profiles: ρ [g/$rmcm^3$], P [Mbar], u_z[10^6 cm/s], 10 μm Al planar target, Nd-glass laser, $\lambda_0 = 1.05\,\mu$m, $I = 10^{13}$ W/cm^2

Figure 18.10 shows growth rates of the RT instability in an aluminum foil of thickness 10 μm exposed to the Nd-glass laser radiation ($\lambda_0 = 1.05\,\mu$m, intensity $I = 10^{13}$ W/cm^2) versus wave number k, obtained using the WKB analytical model [75] and calculated numerically as the eigenvalues of the boundary value problem for Eqs. (18.3.3.1) – (18.3.3.4). The smoothed unperturbed plasma profiles in the coordinate system co-moving with the center of mass of the accelerated target, calculated using the 1D Lagrangian code "IMPULS" with the real equation of state, are shown in Fig. 18.11. The dashed line in Fig. 18.10 represents the growth rate obtained numerically, and the solid line corresponds to the WKB solution with the unperturbed hydrodynamic profiles in Fig. 18.11 used for calculation the coefficients in Eqs. (18.3.3.24)–(18.3.3.25).

18.4 Nonlinear Rayleigh–Taylor Instability in the Laser Ablation

The growth of RT instability can be considered as consisting of three different stages. At the linear stage, when the perturbation amplitude is small, it grows exponentially with time. When the amplitude has grown and exceeded the size of the order of 0.1 λ, the RTI goes into a nonlinear stage, when significant deviations from the linear theory are observed. During this second stage, the light fluid rises into the heavy fluid in the form of bubbles, and the heavy fluid falls almost freely into the light fluid in the form of spikes. At the next stage, the nonlinearity becomes much stronger, and the Kelvin–Helmholtz instability may develop due to the difference in velocities at the interface between the light liquid rolled up in the form of a mushroom and the heavy fluid in the spikes. At this final stage, instabilities evolve into turbulent mixing of light and heavy fluids.

Despite the numerous efforts undertaken to study the nonlinear stage of RTI, there is still no complete theory. The nonlinear stage is usually described as stationary rising gas bubbles of lower density and an almost freely falling spikes of a heavier density. The theory of rising bubbles was considered in Chap. 10, and we briefly recall the main results. From the quantities characterizing the problem, bubble radius and acceleration, only one combination with dimensionality of velocity can be made, that is

$$U_b = \alpha\sqrt{A_t g R}, \tag{18.4.1}$$

where α is a dimensionless numerical factor of the order of unity, and the bubble radius is associated with the wave number, $k = 2\pi/R$. The value of the numerical factor α depends on the specific geometry, being different for open or closed, 2D or 3D bubbles [83–87]. The theoretical prediction for the velocity of 2D and 3D bubbles obtained in Chap. 10 is

$$U_{2D} = \sqrt{A_t g R/3\pi(A_t + 1)}, \tag{18.4.2}$$

and

$$U_{3D} = \sqrt{A_t g R/\pi(A_t + 1)}. \tag{18.4.3}$$

These values are in reasonable agreement with those obtained using the Layzer model [88], as well as with numerical solutions and experimental results. It should be noted that the condition that the velocity of rising bubble exceeds the velocity of the ablation front, $U_b \sim \sqrt{gR} \gg U_a$, coincides with the criterion for the dominant role of gravity, $Fr = U_a^2/gR \ll 1$.

Various authors have done a great work on different aspects of the Layzer model. To study the evolution of bubbles and ridges in a nonlinear mode, Sanz et al. [84] developed a fully nonlinear 2D-sharp-boundary model (SBM) applicable to ablation fronts for large Froude numbers when the cutoff of the unstable spectrum occurs for long-wavelength perturbations, $k\Delta_a \gg 1$. They also demonstrated saturation of the long-wavelength single-mode amplitude and showed that the ablation mass flow reduces the bubble velocity. Clavin and Williams [86] studied analytically the asymptotic spike evolution for the idealized model of rising with constant velocity bubbles separated by freely falling spikes. Based on the assumption that the asymptotic flow near the tip of the spike is close to that due to free fall, they obtained asymptotically power laws in time for the change in curvature and acceleration of the tip. It was found that the analytical theory developed in Ref. [86] is in good qualitative agreement with simulations [87].

In contrast to the result obtained in [84], the numerical simulations in [89], which were performed for a small Froude number, showed that the bubble velocity in the ablation-accelerated target significantly exceeds the bubble velocity obtained in the classical theory [88] without ablation, $U_b \sim \sqrt{A_t g k}$. This may seem surprising, since, as was shown in the previous sections, the convective ablation flow plays a stabilizing role at the linear stage, and in early nonlinear regimes [83–85]. Betti and Sanz [89] argued that this acceleration is driven by the accumulation of vorticities inside the bubble as a result of mass ablation and convection of vorticities from the ablation front. During a highly nonlinear phase, vorticities dominate the flow field inside the bubble. The accretion of vorticities inside the bubble develops when the peak-to-valley amplitude is of the order of the wavelength. At this point, mass ablation predominates at the spikes with respect to the apex of the bubble, where the plasma is relatively cold. They also suggested that during the highly nonlinear phase the lateral ablation in spikes induces a transversal component to the velocity, thus generating vortices, which penetrate into the bubbles filled with the lower density plasma. As the bubble rises, a lower density plasma, produced by the lateral mass ablation and carrying a large vorticity $\omega \sim kU_a\Theta$, is sucked into the bubble. At this stage, the bubble velocity no longer follows the predictions of the potential flow theory: the increase in the bubble buoyancy can be interpreted as caused by the centrifugal forces, which are directed against the acceleration near the bubble top. In the case of 2D bubbles, estimating the bubble radius as $R \approx \lambda/2 \approx 3/k$, and assuming rigid rotation near the bubble boundary [89] with an angular frequency $\Omega = \omega_0/2$, we obtain for the total force per unit volume acting on the bubble fluid element in the vicinity of the bubble tip

$$\mathbf{F} = \rho_2(g - \omega_0^2 R/4). \tag{18.4.4}$$

Note that the centrifugal force $(\rho_2\omega_0^2 R/4)$ is directed along the positive direction opposite to the gravitational acceleration g. Consequently, the buoyancy force of the bubble increases from $(\rho_1 - \rho_2)g$ and becomes $(\rho_1 - \rho_2)g + \omega_0^2 R/4$. Accordingly, the classical velocity of a 2D bubble $U_b^{2D} = \sqrt{g(\Theta - 1)/3k\Theta}$, determined by the Layzer formula [88], will be modified as

$$U_b^{2D} = \sqrt{\frac{g(\Theta - 1)}{3k\Theta} + \frac{\omega_0^2}{\Theta 4k^2}}. \tag{18.4.5}$$

Similarly to (18.4.5), a semi-analytic formula for the velocity of a 3D bubble with a prescribed vorticity ω_0 was derived and verified using 3D simulations by Yan et al. [90]

$$U_b^{3D} = \sqrt{\frac{g(\Theta - 1)}{k\Theta} + \frac{\omega_0^2}{\Theta 4k^2}}. \tag{18.4.6}$$

Note that the main contribution to the bubble velocity in the case $Fr = U_a^2/gR \ll 1$ is made by the centrifugal term leading to the increase of the bubble velocity in comparison with the classical value.

It should be emphasized that potential flow models, such as those used in Chap. 10 or in [83], assumed a parabolic bubble shape in order to obtain a solution for the flow in the vicinity of the bubble top. Numerical simulations [91] have confirmed that the potential flow models are accurate for the Atwood number $A_t \sim 1$, which is the case of ablatively accelerated targets, where $\Theta = \rho_1/\rho_2 \gg 1$. However, as A_t decreases, the accuracy may deteriorate significantly at a later time due to the onset of the vortex motion.

Mikaelian [92] pointed out an important drawback of analytical models, demonstrating that they cannot correctly describe spikes for $A_t < 1$. In subsequent articles [93–95], Mikaelian proposed an analytic model for the amplitudes and mixing widths of bubble and spike in the linear, nonlinear, and turbulent regimes. This approach is suitable for constant or relatively slow time-varying acceleration and for arbitrary density ratios. For the linear stage, the equation for the amplitude of the interface separating two liquids in the constant gravity field is

$$\ddot{\eta} - gkA_t\eta = 0. \tag{18.4.7}$$

The linear solution of Eq. (18.4.7) for the Rayleigh–Taylor instability at the constant acceleration g is

$$\eta(t) = \eta_0 \cosh(\sigma t), \tag{18.4.8}$$

where $\sigma = \sqrt{gkA_t}$. The linear solution of Eq. (18.4.7) for the Richtmyer–Meshkov instability [96, 97] for the case $g(t) = \Delta u\,\delta(t)$, where Δu is the jump velocity acquired by the interface after passing the shock, $\delta(t)$ is the Dirac delta function, is

$$\eta(t) = \eta_0 + \Delta u k A \eta_0 t. \tag{18.4.9}$$

For some particular acceleration depending on time, for example, $g(t) = g_0(1 + \alpha t^2)$, Eq. (18.4.7) can be reduced to the Schrödinger equation, and the solution can be obtained by solving the initial value problem instead of the eigenvalue problem [94]. Assuming a constant density, Eq. (18.4.7) can be reduced to the following equation for $\theta(t) \equiv \exp[k(\eta - \eta_0)]$:

$$\ddot{\theta}_L - gk_LA_L\theta_L = 0, \tag{18.4.10}$$

where $k_L \equiv k(A_t + 1)/(A_t + 2)$, $A_L \equiv 2A_t/(A_t + 2)$. Since Eq. (18.4.10) has the same form as the linear Eq. (18.4.7), the nonlinear solution to Eq. (18.4.10) is the logarithm of the linear solution

$$\eta = \eta_0 + \frac{1}{k_L} \ln \theta_L. \tag{18.4.11}$$

Another way is to use the WKB approximation [94] to solve Eq. (18.4.10). Introducing $s \equiv \int_0^t \sqrt{g(t)}dt$, Eq. (18.4.10) can be written in the form

$$\frac{d^2\theta_L}{ds^2} - k_L A_L \theta_L + \frac{1}{2g^2}\frac{dg}{dt}\frac{d\theta_L}{dt} = 0. \tag{18.4.12}$$

If the last term in (18.4.12) can be neglected, then the solution of Eq. (18.4.12) with boundary conditions $\theta_L(0) = 1$ and $\dot{\theta}_L(0) = 0$ has the form

$$\eta = \eta_0 + \frac{1}{k_L}\ln\left[\cosh\left(s\sqrt{k_L A_L}\right)\right]. \tag{18.4.13}$$

This nonlinear solution is formally valid for arbitrary $g(t)$, and therefore in [94] is called the scale solution. For the bubble velocity, we obtain

$$\dot{\eta} = \sqrt{gA_L/k_L}\tanh\left(s\sqrt{k_L A_L}\right) \rightarrow_{t\rightarrow\infty} \sqrt{gA_L/k_L}. \tag{18.4.14}$$

For the RM instability ($g(t) = \Delta u\,\delta(t)$), Eq. (18.4.7) in the WKB approximation became $d^2\eta/ds^2 - kA_t\eta = 0$ with the solution

$$\eta(t) = \eta_0 \cosh\left(s\sqrt{kA}\right). \tag{18.4.15}$$

The nonlinear regime begins either at large initial amplitudes, $\eta_0 k > 1$, or if amplitudes have grown, so that, $\eta k > 1$. In [95] Mikaelian used the Layzer model [88], in particular, Eq. (55) from [88] in the form:

$$(1 + 2\tilde{\eta}_2)\ddot{\eta} + k\dot{\eta}^2 + 2g\tilde{\eta}_2 = 0, \tag{18.4.15}$$

where $\tilde{\eta}_2$ is the curvature function, which is defined as

$$\tilde{\eta}_2 \equiv \eta_2/k = \frac{1}{6}\left[1 + 3(\eta_0 k - 1)e^{-3k(\eta-\eta_0)}\right]. \tag{18.4.16}$$

Unlike the expression $\tilde{\eta}_2 \equiv \eta_2/k = \left[1 - e^{-3k\eta}\right]$ used in [88] for the case $\eta_0 = 0$, the expression (18.4.16) for $\tilde{\eta}_2$ allows an arbitrary η_0. Introducing a new variable $s \equiv \int_0^t \sqrt{g(t)}dt$, and assuming $\tilde{\eta}_2 \approx$ const., one can obtain an explicit analytical solution

$$\eta(t) = \eta_0 + \left(\frac{1 + 2\tilde{\eta}_2}{k}\right)\ln\left[\cosh\left(sd\sqrt{k}\right)\right], \tag{18.4.17}$$

where $d^2 = -2\tilde{\eta}_2/(1 + 2\tilde{\eta}_2)^2$. This solution gives and expression for the bubble amplitude as a function of $s \equiv \int_0^t \sqrt{g(t)}dt$; for a constant acceleration $s = \sqrt{gt}$.

If we consider the formation of spikes, we should choose negative η instead of positive η_2 for bubbles. However, for spikes η_2 cannot be constant, but according to Eq. (18.4.16) for large negative η it tends to $+\infty$. The structure of the spike at the nonlinear stage is greatly complicated by the formation of a mushroom with the subsequent development of the Kelvin–Helmholtz instability. The growth rate of the Kelvin–Helmholtz instability is equal to $\sigma_{KH} \propto \frac{\sqrt{\Theta}}{\Theta+1} k(\Delta u)_{1-2}$, where $(\Delta u)_{1-2} \equiv u_1 - u_2$ is the relative velocity of heavy and light liquids. Since the density ratio for typical ablation experiments is large, $\Theta = \rho_1/\rho_2 \gg 1$, a noticeable growth rate of the Kelvin–Helmholtz instability can be expected only at large values of k, so that one can expect the development of only small-scale turbulence in the layer between heavy and light liquids. In addition, the flow at this stage can no longer be considered as potential, and a simple analytical solution, such as that obtained by Layzer [88], is not available even for the bubble motion [89], not to mention spikes and turbulent mixing. Instead, Mikaelian and Olson [98] proposed an interesting semi-phenomenological approach.

The highly nonlinear RTI stage and turbulent mixing caused by the development of the Richtmyer–Meshkov instability are rather challenging problems. In order to explain the results of numerical and experimental studies, it is necessary to accurately consider all terms associated with the propagation of the ablation front, including the production of vorticities. An analytical solution to this problem is unlikely, therefore, numerical modeling and experiments have become the main methods at the present time. The results of 2D and 3D numerical simulations of the RT instability for plane foils accelerated by laser ablation, as well as for spherically symmetric ablative compression, which were performed in [99–104], are fairly well described by expressions (18.2.1) and (18.2.2) for linear stages of the RT instability and in good agreement with the experimental data [82, 105, 106].

References

1. L. Rayleigh, Investigations of the character of the equilibrium of an incompressible heavy fluid of variable density. Proc. Lond. Math. Soc. **14**, 170–177 (1883)
2. G.I. Taylor, The instability of liquid surfaces when accelerated in a direction perpendicular to their planes. I. Proc. R. Soc. Lond. A **201**(1950), 192–196 (1950)
3. J.D. Lindl, P. Amendt, R.L. Berger, S.G. Glendinning, S.H. Glenzer, S.W. Haan, R.L. Kauffman, O.L. Landen, L.J. Suter, The physics basis for ignition using indirect drive targets on the National Ignition Facility. Phys. Plasmas **11**, 339–491 (2004); J. Lindl, O. Landen, J. Edwards, Ed. Moses, and NIC Team, Review of the National Ignition Campaign 2009–2012. Phys. Plasmas **21**, 020501(1–72) (2014)
4. M.K. Matzen, M.A. Sweeney, R.G. Adams, et al., Pulsed-power-driven high energy-density physics and inertial confinement fusion research. Phys. Plasmas **12**, 055502(1–16) (2005)
5. S. Atzeni, J. Meyer-ter-Vehn, *The Physics of Inertial Fusion: Beam Plasma Interaction Hydrodynamics* (Oxford University Press, Oxford, Hot Dense Matter, 2004)
6. D. Besnard, The megajoule laser program—ignition at hand. Eur. Phys. J. D **44**, 207–213 (2007)

7. R.P. Drake, Perspectives of high energy density physics. Phys. Plasmas **16**, 055501(1–7) (2009)
8. S.E. Bodner, D.G. Colombant, J.H. Gardneret et al., Direct-drive laser fusion: status and prospects. Phys. Plasmas **5**, 1901–1918 (1998)
9. B.A. Remington, R.P. Drake, D.D. Ryutov, Experimental astrophysics with high power lasers and Z-pinches. Rev. Mod. Phys. **78**, 755–807 (2006)
10. M.A. Liberman, J.S. DeGroot, A. Toor, R.B. Spielman, *Physics of High-Density Z-Pinch Plasmas* (Springer, New York, Berlin, 1998)
11. D.D. Ryutov, M.S. Derzon, M.K. Matzen, The physics of fast Z pinches. Rev. Mod. Phys. **72**, 167–223 (2000)
12. H. Kull, Theory of the Rayleigh-Taylor instability. Phys Rep. **206**, 197–325 (1991)
13. N. Inogamov, The role of Rayleigh-Taylor and Richtmyer-Meshkov instabilities in astrophysics: an introduction. Astrophys. Space Phys. **10**, 1–335 (1999)
14. S.I. Abarzhi, Review of theoretical modelling approaches of Rayleigh-Taylor instabilities and turbulent mixing. Phil. Trans. R. Soc. **368**, 1809–1828 (2010)
15. S.I. Anisimov, R. Drake, S. Gauthier, E.E. Meshkov, S.I. Abarzhi, What is certain and what is not so certain in our knowledge of Rayleigh–Taylor mixing? Phil. Trans. R. Soc. A **371**, 20130266(1–16) (2013)
16. Y. Zhou, Rayleigh-Taylor and Richtmyer-Meshkov instability induced flow, turbulence, and mixing. I. Phys. Rep. **720**, 1–136 (2017)
17. Y. Zhou, Rayleigh-Taylor and Richtmyer-Meshkov instability induced flow, turbulence, and mixing. II. Phys. Rep. **723**, 1–160 (2017)
18. J. Nuckolls, L. Wood, A. Thiessen, G. Zimmerman, Laser compression of matter to super-high densities: thermonuclear (CTR) applications. Nature (London) **289**, 139–142 (1972)
19. J.D. Lawson, Some criteria for a power producing thermonuclear reactor. Proc. Phys. Soc. B **70**, 6–10 (1957)
20. R. Betti, P.Y. Chang, B.K. Spears, K.S. Anderson, J. Edwards, M. Fatenejad, J.D. Lindl, R.L. McCrory, R. Nora, D. Shvarts, Thermonuclear ignition in inertial confinement fusion and comparison with magnetic confinement. Phys. Plasmas **17**, 058102(1–10) (2010)
21. R.E. Kidder, Laser-driven compression of hollow shells: power requirements and stability limitations. Nucl. Fusion **16**, 3–14 (1976)
22. Laser Indirect Drive input to NNSA 2020 Report. Preprint LLNL-TR-810573
23. O.A. Hurricane, D.A. Callahan, D.T. Casey DT, et al., Fuel gain exceeding unity in an inertially confined fusion implosion. Nature **506**, 343–347 (2014)
24. O.A. Hurricane, P.T. Springer, P.K. Patel, D.A. Callahan, et al., Approaching a burning plasma on the NIF. Phys. Plasmas **28**, 052704(1–12) (2019)
25. J.D. Kilkenny, S.G. Glendinning, S.W. Haan, B.A. Hammel, J.D. Lindl, D. Munro, B.A. Remington, S.V. Weber, J.P. Knauer, C.P. Verdon, A review of the ablative stabilization of the Rayleigh-Taylor instability in regimes relevant to inertial confinement fusion. Phys. Plasmas **1**, 1379–1389 (1994)
26. L.D. Landau, E.M. Lifshitz, *Fluid Mechanics*, 2nd edn. (Butterworth-Heinemann, Oxford, 1987)
27. J. Grun, M.H. Emery, C.K. Manka, T.N. Lee, E.A. McLean, A. Mostovych, J. Stamper, S. Bodner, S.P. Obenshain, B.H. Ripin, Rayleigh-Taylor instability growth rates in targets accelerated with a laser beam smoothed by induced spatial incoherence. Phys. Rev. Lett. **58**, 2672–2676 (1987)
28. S.E. Bodner, M.H. Emery, J.H. Gardner, The Rayleigh-Taylor instability in direct-drive laser fusion. Plasma Phys. Controll. Fusion **29**, 1333–1342 (1987)
29. M. Deselberger, O. Willi, M. Savage, M.J. Lamb, Measurement of the Rayleigh-Taylor instability in targets driven by optically smoothed laser beams. Phys Rev. Lett. **65**, 2997–3002 (1990)
30. S.G. Glendinning, S.N. Dixit, B.A. Hammel et al., Measurements of laser-speckle-induced perturbations in laser-driven foils. Phys. Rev. E **54**, 4473–4475 (1996)

31. B.A. Remington, S.V. Weber, M.M. Marinak, S.W. Haan, J.D. Kilkenny, R.J. Wallace, G. Dimonte, Single-mode and multimode Rayleigh-Taylor experiments on Nova. Phys. Plasmas **2**, 241–255 (1995)
32. H. Takabe, K. Mima, Effects of thermal conduction and compressibility on Rayleigh-Taylor instability. J. Phys. Soc. Jpn. **48**, 1793–1794 (1980)
33. H. Takabe, L. Montierth, R.L. Morse, Self-consistent eigenvalue analysis of Rayleigh-Taylor instability in an ablating plasma. Phys. Fluids **26**, 2299–2307 (1983)
34. H. Takabe, K. Mima, L. Montierth, R.L. Morse, Self-consistent growth rate of the Rayleigh-Taylor instability in an ablatively accelerating plasma. Phys. Fluids **28**, 3676–3682 (1985)
35. M. Tabak, D.H. Munro, J.D. Lindl, Hydrodynamic stability and the direct drive approach to laser fusion. Phys. Fluids B **2**, 1007–1014 (1990)
36. S.E. Bodner, Rayleigh-Taylor Instability and Laser-Pellet Fusion. Phys. Rev. Lett. **33**, 761–765 (1974)
37. J.H. Gardner, S.E. Bodner, J.P. Boris, M.H. Emery, M.J. Fritts, M.J. Herbst, Analysis of stability and symmetry implications for ICF, in *Laser Interaction and Related Plasma Phenomena,* ed. by H. Hora, G.H. Miley (Springer, Boston, MA, 1984)
38. H. Kull, S. Anisimov, Ablative stabilization in the incompressible Rayleigh-Taylor instability. Phys. Fluids **29**, 2067–2675 (1986)
39. D. Book, Convective stabilization of the Rayleigh-Taylor instability with self-consistent treatment of the boundary conditions. Plasma Phys. Controll. Fusion **34**, 737–748 (1992)
40. M. Liberman, A. Velikovich, *Physics of Shock Waves in Gases and Plasmas* (Springer, Berlin, New York, 1985)
41. M.A. Liberman, V.V. Bychkov, S.M. Golberg, On the stability of a flame in the gravitational field. Sov. Phys. JETP **77**, 227–236 (1993)
42. M.A. Liberman, V.V. Bychkov, S.M. Golberg, D. Book, Stability of a planar flame front in the slow-combustion regime. Phys. Rev. E **49**, 445–457 (1994)
43. V.V. Bychkov, S.M. Golberg, M.A. Liberman, Rayleigh-Taylor instability of combustion and laser produced ablation fronts. Phys. Fluids B **5**, 3822–3824 (1993)
44. V.V. Bychkov, S.M. Golberg, M.A. Liberman, Self-consistent model of the Rayleigh-Taylor instability in ablatively accelerated laser plasma. Phys. Plasmas **1**, 2976–2986 (1994)
45. J. Sanz, Self-consistent analytical model of the Rayleigh-Taylor instability in inertial confinement fusion. Phys. Rev. Lett. **73**, 2700–2703 (1994)
46. R. Betti, V.N. Goncharov, R.L. McCrory, C.P. Verdon, Self-consistent cut-off wave number of the ablative Rayleigh-Taylor instability. Phys. Plasmas **2**, 3844–3851 (1995)
47. J. Sanz, Self-consistent analytical model of the Rayleigh-Taylor instability in inertial confinement fusion. Phys. Rev. E **53**, 4026–4045 (1996)
48. V. Goncharov, R. Betti, R. McCrory, C. Verdon, Self-consistent stability analysis of ablation fronts with small Froude numbers. Phys. Plasmas **3**, 4665–4676 (1996)
49. V.N. Goncharov, R. Betti, R.L. McCrory, P. Sorotokin, C.P. Verdon, Self-consistent stability analysis of ablation fronts with large Froude numbers. Phys. Plasmas **3**, 1402–1414 (1996)
50. A.R. Piriz, J. Sanz, L.F. Ibanez, Rayleigh-Taylor instability of steady ablation fronts: the discontinuity model revisited. Phys. Plasmas **4**, 1117–1126 (1997)
51. R. Betti, V. Goncharov, R. McCrory, C. Verdon, Growth rate of the ablative Rayleigh-Taylor instability in inertial confinement fusion. Phys. Plasmas **5**, 1446–1454 (1998)
52. A. Piriz, R. Portugues, Landau-Darrieus instability in an ablation front. Phys. Plasmas **10**, 2449–2456 (2003)
53. P. Clavin, L. Masse, Instabilities of ablation fronts in inertial confinement fusion: a comparison with flames. Phys. Plasmas **11**, 690–704 (2004)
54. P. Clavin, L. Masse, F.A. Williams, Comparison of flame-front instabilities with instabilities of ablation fronts in inertial confinement fusion. Combust. Sci. Tech. **177**, 979–989 (2005)

55. J. Sanz, L. Masse, P. Clavin, The linear Darrieus-Landau and Rayleigh-Taylor instabilities in inertial confinement fusion revisited. Phys. Plasmas **13**, 102702(1–12) (2006)
56. W. Manheimer, D. Colombant, J. Gardner, Steady-state planar ablative flow. Phys Fluids **25**, 1644–1652 (1982)
57. F. Felber, Steady-state model of a flat laser-driven target. Phys. Rev. Lett. **39**, 84–87 (1977)
58. R. Fabbro, C. Max, E. Fabre, Planar laser-driven ablation: effect of inhibited electron thermal conduction. Phys Fluids **28**, 1463–1481 (1985)
59. M.A. Liberman, A.L. Velikovich, Self-similar spherical expansion of a laser plasma or detonation products into a low-density ambient gas. Phys. Fluids B **1**, 1271–1276 (1989)
60. M. Murakami, T. Sakaiya, J. Sanz, Self-similar ablative flow of nonstationary accelerating foil due to nonlinear heat conduction. Phys. Plasmas **14**, 022707(1–7) (2007)
61. C. Boudesocque-Dubois, S. Gauthier, J.-M. Clarisse, Self-similar solutions of unsteady ablation flow in inertial confinement fusion. J. Fluid Mech. **603**, 151–178 (2008)
62. J.-M. Clarisse, C. Boudesocque-Dubois, S. Gauthier, Linear perturbation response of self-similar ablative flows relevant to inertial confinement fusion. J. Fluid Mech. **609**, 1–48 (2008)
63. J.-M. Clarisse, J.-L. Pfister, S. Gauthier, C. Boudesocque-Dubois, A hydrodynamic analysis of self-similar radiative ablation flows. J. Fluid Mech. **848**, 219–255 (2018)
64. V. Grégoire, J.-M. Clarisse, A. Couairon, Investigation of supersonic heat-conductivity hyperbolic waves in radiative ablation flows. Phys. Rev. E **101**, 043215(1–15) (2020)
65. O. Yu. Travnikov, M.A. Liberman, V.V. Bychkov, Stability of a planar flame front in a compressible flow. Phys. Fluids **9**, 3935–3937 (1997)
66. O.Yu. Travnikov, V.V. Bychkov, M. A. Liberman, Influence of compressibility on propagation of curved flames. Phys. Fluids **11**, 2657–2666 (1999)
67. S. Kadowaki, Instability of a deflagration wave propagating with finite Mach number. Phys. Fluids **7**, 220–222 (1995)
68. L. He, Analysis of compressibility effects on Darrieus-Landau instability of deflagration wave. Europhys. Lett. **49**, 576–583 (2000)
69. A. Piriz, Hydrodynamic instability of ablation fronts in inertial confined fusion. Phys. Plasmas **8**, 997–1002 (2001)
70. A. Piriz, Compressibility effects on the Rayleigh-Taylor instability of an ablation front. Phys. Plasmas **8**, 5268–5276 (2001)
71. A.R. Piriz, N.A. Tahir, Physics of ablative Rayleigh–Taylor and Landau–Darrieus instabilities. New J. Phys. **15**, 015013(1–14) (2013)
72. V. Bychkov, M. Modestov, M. Marklund, The Darrieus–Landau instability in fast deflagration and laser ablation. Phys. Plasmas **15**, 032702(1–9) (2008)
73. M. Modestov, V. Bychkov, D. Valiev, M. Marklund, Growth rate and the cutoff wavelength of the Darrieus-Landau instability in laser ablation. Phys Rev E **80**, 046403(1–9) (2009)
74. A.B. Bud'ko, M.A. Liberman, Suppression of the Rayleigh-Taylor instability by convection in ablatively accelerated laser target. Phys. Rev. Lett. **68**, 178–181 (1992)
75. A.B. Bud'ko, M.A. Liberman, Stabilization of the Rayleigh-Taylor instability by convection in smooth density gradient: WKB analysis. Phys. Fluids B **4**, 3499–3507 (1992)
76. L.H. Montierth, R.L. Morse, Stationary flow model of ablatively imploded spherical shells. Phys. Fluids B **2**, 353–370 (1990)
77. S. Chandrasekar, *Hydrodynamics and Hydrodynamic Stability* (Oxford University Press, London, 1961)
78. L.D. Landau and E.M. Lifshitz, Course of theoretical physics, in *Quantum Mechanics,* vol. III (Pergamon Press, New York, 1987)
79. K.O. Mikaelian, Rayleigh-Taylor instabilities in stratified fluids. Phys. Rev. A **26**, 2140–2158 (1982)
80. W. Manheimer, D. Colombant, Slab model for Rayleigh-Taylor stabilization by vortex shedding, compressibility, thermal conduction, and ablation. Phys. Fluids **27**, 983–993 (1984)

81. R. S. Craxton, K.S. Anderson, T.R. Boehly, et al., Direct-drive inertial confinement fusion: a review. Phys. Plasmas **22**, 110501(1–153) (2015)
82. V.A. Smalyuk, C.R. Weber, O.L. Landen, et al., Review of hydrodynamic instability experiments in inertially confined fusion implosions on National Ignition Facility. Plasma Phys. Control. Fusion **62**, 014007(1–9) (2020)
83. V.N. Goncharov, Analytical model of nonlinear, single-mode, classical Rayleigh-Taylor Instability at arbitrary atwood numbers. Phys. Rev. Lett. **88**, 134502(1–4) (2002)
84. J. Sanz, J. Ramirez, R. Ramis, R. Betti, R. Town, Nonlinear theory of the ablative Rayleigh-Taylor instability. Phys. Rev. Lett. **89**, 195002(1–4) (2002)
85. K.O. Mikaelian, Explicit expressions for the evolution of single-mode Rayleigh-Taylor and Richtmyer-Meshkov instabilities at arbitrary Atwood numbers. Phys. Rev. E **67**, 026319(1–7) (2003)
86. P. Clavin, F. Williams, Asymptotic spike evolution in Rayleigh-Taylor instability. J Fluid Mech. **525**, 105–113 (2005)
87. L. Duchemin, C. Josserand, P. Clavin, Asymptotic behavior of the Rayleigh-Taylor instability. Phys. Rev. Lett. **94**, 224501(1–4) (2005)
88. D. Layzer, On the instability of superposed fluids in a gravitational field. Astrophys. J. **122**, 1–12 (1955)
89. R. Betti, J. Sanz, Bubble acceleration in the ablative Rayleigh-Taylor instability. Phys Rev. Lett. **97**, 205002(1–4) (2006)
90. R. Yan, R. Betti, J. Sanz, H. Aluie, B. Liu, A. Frank, Three-dimensional single-mode nonlinear ablative Rayleigh-Taylor instability. Phys. Plasmas **23**, 022701(1–6) (2016)
91. H. Liang, Q. X. Li, B. C. Shi, Z. H. Chai, Lattice Boltzmann simulation of three-dimensional Rayleigh–Taylor instability. Phys. Rev. E **93**, 033113(1–16) (2016)
92. K.O. Mikaelian, Limitations and failures of the Layzer model for hydrodynamic instabilities. Phys. Rev. E **78**, 015303(R) (2008)
93. K.O. Mikaelian, Nonlinear hydrodynamic interface instabilities driven by time-dependent accelerations. Phys. Rev. E **79**, 065303 (2009)
94. K.O. Mikaelian, Analytic approach to nonlinear hydrodynamic instabilities driven by time-dependent accelerations. Phys. Rev. E **81**, 016325 (2010)
95. K.O. Mikaelian, Solution to Rayleigh-Taylor instabilities: bubbles, spikes, and their scalings. Phys. Rev. E **89**, 053009(1–7) (2014)
96. R.D. Richtmyer, Taylor instability in shock acceleration of compressible fluids. Commun. Pure Appl. Math. **13**, 297–319 (1960)
97. E.E. Meshkov, Instability of the interface of two gases accelerated by a shock wave. Fluid Dyn. **4**, 101–104 (1969)
98. K.O. Mikaelian, B.J. Olson, On modeling Richtmyer-Meshkov turbulent mixing widths. Physica D **402**, 132243(1–6) (2020)
99. A. Casner, V.A. Smalyuk, L. Masse, I. Igumenshchev, S. Liberatore, L. Jacquet, C. Chicanne, P. Loiseau, O. Poujade, D.K. Bradley, H.S. Park, B.A. Remington, Designs for highly nonlinear ablative Rayleigh-Taylor experiments on the National Ignition Facility. Phys. Plasmas **19**, 082708(1–10) (2012)
100. L.F. Wang, W.H. Ye, W.Y. Zhang, X.T. He, Numerical investigation of nonlinear ablative single-mode Rayleigh–Taylor instability in the presence of preheating. Phys. Scripta T **155**, 014018 785(1–8) (2013)
101. W.H. Liu, L.F. Wang, W.H. Ye, X.T. He, Temporal evolution of bubble tip velocity in classical Rayleigh-Taylor instability at arbitrary Atwood numbers. Phys. Plasma **20**, 062101(1–11) (2013)
102. H.G. Lee, J. Kim, Numerical simulation of the three-dimensional Rayleigh-Taylor instability. Comput. Math. Appl. **66**, 1466–1474 (2013)
103. W. Liu, C. Yu, H. Jiang, X. Li, Bell-Plessett effect on harmonic evolution of spherical Rayleigh-Taylor instability in weakly nonlinear scheme for arbitrary Atwood numbers. Phys. Plasmas **24**, 022102 (2017)

104. J.W. Bates, A.J. Schmitt, M. Karasik, S.T. Zalesak, Numerical simulations of the ablative Rayleigh-Taylor instability in planar inertial-confinement-fusion targets using the FastRad3D code. Phys. Plasma **23**, 122701(1–12) (2016)
105. V.A. Smalyuk, C.R. Weber, H.F. Robey, et al., Hydrodynamic instability growth of three-dimensional modulations in radiation-driven implosions with "low-foot" and "high-foot" drives at the National Ignition Facility. Phys. Plasmas **24**, 042706(1–11) (2017)
106. L.F. Wang, J.F. Wu, W.H. Ye, J.Q. Dong, Z.H. Fang, G. Jia, Z.Y. Xie, X.G. Huang, S.Z. Fu, S.Y. Zou, Y.K. Ding, W.Y. Zhang, X.T. He, Nonlinear ablative Rayleigh–Taylor growth experiments on Shenguang–II. Phys. Plasmas **27**, 072703(1–6) (2020)

Appendix A
Conversion Formulas and Constants

Prefixes; Units

10^{-3}—milli [m], 10^{-6}—micro [μ], 10^{-9}—nano [n], 10^{-12}—pico [p], 10^{-15}—femto [f], 10^{3}—kilo [k], 10^{6}—mega [M], 10^{9}—giga [G], 10^{12}—tera [T], 10^{15}—peta [P].

Length units: 1 km $= 10^3$ m $= 10^5$ cm; 1 cm $= 10$ mm $= 10^4$ μm $= 10^7$ nm $= 10^8$ Å.

Volume units: 1 $m^3 = 10^3$ L $= 10^6$ $cm^3 = 10^9$ mm^3.

Mass units: 1 ton $= 10^3$ kg, 1 kg $= 10^3$ g $= 10^6$ mg $= 10^9$ μg.

Energy units:

1 cal $= 4.1868$ J $= 4.1868 \cdot 10^7$ erg $= 2.6126 \cdot 10^{19}$ eV

1 J $=$ 1 N m $=$ 1 W s $= 10^7$ erg $= 6.2419 \cdot 10^{11}$ eV

1 eV $= 11600$ K $= 6.021 \cdot 10^{-12}$ erg $= 6.021 \cdot 10^{-19}$ J $= 3.827 \cdot 10^{-20}$ cal.

Temperature units: T [K] $=$ T [°C] $+ 273.15$.

Density units: 1 kg/$m^3 = 10^{-3}$ g/cm^3.

Pressure units:

1 atm $= 1.0133$ bar $= 1.0133 \cdot 10^5$ N/$m^2 = 1.0133 \cdot 10^6$ dyn/$cm^2 = 760$ Torr $=$ 760 mmHg
1 bar $= 10^5$ N/$m^2 = 10^5$ Pa.

Force units: 1 kg $= 9.806$ N $= 9.806 \cdot 10^5$ dyn.

M. A. Liberman, *Combustion Physics*,
https://doi.org/10.1007/978-3-030-85139-2

Fundamental Constants

Universal gas constant: R_B = 8.3144 J/g - mol K = 8.3144 kJ/kmol K = 1.987 cal/g - mole K = $8.3144 \cdot 10^7$ erg/mol K.

Boltzmann constant: $k = R_B/N_A = 1.38 \cdot 10^{-23}$ J/K.

Avogadro number:

$N_A = 6.022 \cdot 10^{23}$ molecules/g - mol. Number of molecules in 1 cm^3 at pressure 1 atm: $n = 2.686 \cdot 10^{19}\,cm^{-3}$.

Volume of 1 mol at normal conditions: $V_0 = 2.2414 \cdot 10^4\,cm^3/mol$.

Stefan-Boltzmann constant:

$\sigma = \pi^2 k^4/60\hbar^3 c^2 = 5.670 \cdot 10^{-8}\,W/m^2\,K^4 = 5.670 \cdot 10^{-5}\,erg/cm^2\,s\,K^4 = 1.355 \cdot 10^{-12}\,cal/cm^2\,s\,K^4$.

Atomic mass unit:

$m_a = 1.660 \cdot 10^{-24}\,g = 1.660 \cdot 10^{-27}\,kg$

$m_p = 1.67264 \cdot 10^{-24}\,g$, $m_n = 1.67495 \cdot 10^{-24}\,g$, $m_e = 9.109534 \cdot 10^{-28}\,g$.

Speed of light: $c = 2.997 \cdot 10^{10}$ cm/s $= 2.997 \cdot 10^8$ m/s.

Electron charge: $e = 1.602189 \cdot 10^{-19}$coulombs $= 10^{-20}$cgsm units.

Max Plank constant:

$h = 6.626 \cdot 10^{-27}$ erg s $= 6.626 \cdot 10^{-34}$ J s, $\hbar = h/2\pi = 1.054 \cdot 10^{-27}$ erg s.

Density, thermal conductivity, viscosity

Density of some gases at$T = 0\,°C$, $P = 760$ mmHgρ [kg/m^3]

Gas	Air	H_2	O_2	C_2H_2	CH_4	C_2H_4	C_2H_6	C_3H_8	CO	CO_2	NO
ρ [kg/m^3]	1.29	0.089	1.429	1.170	0.716	1.260	1.356	2.003	1.250	1.976	1.34

Thermal conduction of some gases at$T = 0\,°C$, $P = 760$ mmHg: $\lambda\left[10^{-6}\frac{cal}{cm\,s\,K}\right]$

Gas	Air	H_2	O_2	C_2H_2	CH_4	C_2H_4	C_2H_6	C_3H_8	CO	CO_2	NO
λ [10^{-6} cal/cm s K]	57.5	401	57.5	44	69.6	41.5	43.5	35.5	51.1	34.6	56.4

Coefficient viscosity of some gases at $T = 0\,°C$, $P = 760\,mmHg$: $\eta\left[10^{-8}\,\frac{kg}{m\,s}\right]$

Gas	Air	H_2	O_2	C_2H_2	CH_4	C_2H_4	C_2H_6	C_3H_8	CO	CO_2	NO
$\eta\,[10^{-8}\,kg/m\,s]$	1720	866	1993	955	1028	855	1223	1390	1660	1463	1800

Temperature dependence is approximated by Sutherland's formula $\eta_T = \eta_0\sqrt{\frac{T}{T_0}}\frac{1+a/T_0}{1+b/T}$.

Appendix B
Combustion Characteristics of Some Mixtures

Hydrogen/Oxygen, H_2/O_2, $T_0 = 300$ K, $P_0 = 1$ Bar[**1.01325**]

Adiabatic Temperature:	$T_b = 3072$ K [3072.819653549975]
Expansion coefficient:	$\Theta = 10$ [10.242732178499917]
Flame Thickness [cm]:	$L_f = 0.027$ cm [0.02609975030387686]
Laminar Flame velocity [m/s]:	$U_f = 9.4$ m/s [9.398938589729063]
Density [kg/m^3]:	$\rho = 0.48$ kg/m^3 $= 4.8 \cdot 10^{-4}$ g/cm^3 [0.487877]
Viscosity [kg/(m * s)]:	$\mu = 1.8 \cdot 10^{-5}$ kg/m s $= 1.8 \cdot 10^{-4}$ g/cm s; $\nu = 0.38$ cm^2/s
Thermal conductivity [J/(m * K * s):	$\kappa = 9.8 \cdot 10^{-2}$ J/m K s $=$ $9.8 \cdot 10^3$ erg/cm s K
Adiabatic constant $C_P/C_V = 1.4$:	$\gamma = 1.4$ [1.4013930689762282]
Sound speed [m/s]:	$a_s = 539.5$ m/s [539.4892686580549]
Specific heat:	$C_V = 1.725 \cdot 10^7$ erg/g K, $C_P = 2.417 \cdot 10^7$ erg/g K
Thermal diffusivity:	$n\chi = \kappa/\rho C_p = 0.85$ cm^2/s

Detonation characteristics of hydrogen–oxygen mixture at 293 K

Initial pressure	Chapman-Jouguet velocity	Chapman-Jouguet pressure	Chapman-Jouguet temperature	Detonation cell width
P (bar)	D_{CJ} (m/s)	P_{CJ} (bar)	T_{CJ} (K)	λ (mm)
0.2	2755	3.7	3395	8.5
0.5	2805	9.5	3556	3.2

(continued)

M. A. Liberman, *Combustion Physics*,
https://doi.org/10.1007/978-3-030-85139-2

(continued)

Initial pressure	Chapman-Jouguet velocity	Chapman-Jouguet pressure	Chapman-Jouguet temperature	Detonation cell width
0.6	2815	11.5	3590	2.7
0.75	2827	14.4	3631	2.1

Hydrogen/air ($2H_2$, O_2, 3.76 N_2), $T_0 = 300$ K, $P_0 = 1$ Bar

Adiabatic Temperature:	$T_b = 2350$ K [2350.118565471589]
Expansion coefficient:	$\Theta = 7.8$ [7.833728551571964]
Flame Thickness [cm]:	$L_f = 0.035$ cm [0.035649272246440217]
Laminar Flame velocity [m/s]:	$U_f = 2.36$ m/s [2.3610670072685855]
Density [kg/m^3]:	$\rho = 0.85\,kg/m^3 = 8.5 \cdot 10^{-4}\,g/cm^3$ [0.849]
Viscosity [kg/(m * s)]:	$\mu = 1.83 \cdot 10^{-5}$ kg/m s $= 1.83 \cdot 10^{-4}$ g/cm s; $\nu = 0.22\,cm^2/s$
Thermal conductivity [J/(m * K * s)]:	$\kappa = 5.4 \cdot 10^{-2}$ J/m K s $= 5.4 \cdot 10^3$ erg/cm s K, $\chi = \kappa/\rho C_p = 0.64\,cm^2/s$
Adiabatic constant $C_P/C_V = 1.4$:	$\gamma = 1.4$ [1.4008]
Sound speed [m/s]:	$a_s = 408.77$ m/s [408.7779]
Specific heat [erg/g/K]:	$C_V = 9.9 \cdot 10^6$ erg/g/K$C_P = 1.389 \cdot 10^7$ erg/g/K

Ethylene/oxygen ($C_2H_4/3O_2$), $T_0 = 300$ K, $P_0 = 1$ Bar

Adiabatic Temperature:	$T_b = 3172$ K [3172.7428]
Expansion coefficient:	$\Theta = 10.6$ [10.57580943]
Flame Thickness [cm]:	$L_f = 0.0067$ cm [0.006738065]
Laminar Flame velocity [m/s]:	$U_f = 4.86$ m/s [4.86148]
Density [kg/m^3]:	$\rho = 1.26\,kg/m^3 = 1.26 \cdot 10^{-3}\,g/cm^3$ [0.849]
Viscosity [kg/(m * s)]:	$\mu = 1.88 \cdot 10^{-5}$ kg/m s $= 1.88 \cdot 10^{-4}$ g/cm s, $\nu = 0.15\,cm^2/s$
Thermal conductivity [J/(m * K * s):	$\kappa = 2.67 \cdot 10^{-2}$ J/m K s $= 2.67 \cdot 10^3$ erg/cm s K
Adiabatic constant $C_P/C_V = 1.4$:	$\gamma = 1.339$ [1.3395]
Sound speed [m/s]:	$a_s = 328.2$ m/s [328.2317]
Specific heat [erg/g/K]:	$C_V \approx 7.89 \cdot 10^6$ erg/g/K$C_P \approx 1.085 \cdot 10^7$ erg/g/K

See Table B.1.

Table B.1 Combustion characteristics of ethylene-oxygen mixtures at 293 K

Initial pressure	Density	Sonic speed in reactants	Sonic speed in products	Adiabatic combustion temperature	Adiabatic combustion pressure	Expansion rate	Laminar flame speed	Laminar flame thickness	Overall reaction order	Activation energy
P_0 (bar)	P (kg/m^3)	c_r (m/s)	c_p (m/s)	T_b, K (K)	P_{icc} (bar)	Θ	S_L (m/s)	L_f (mm)	n	E_a (K)
0.05	0.064	325	1173	3053	0.77	14.1	5.28	1.540	1.99	14,220
0.075	0.095	325	1170	3014	1.17	14.0	5.29	1.025	1.98	14,210
0.1	0.127	325	1166	2986	1.58	13.9	5.28	0.762	1.98	14,950
0.12	0.153	325	1164	2975	1.90	13.9	5.20	0.615	1.97	15,370
0.15	0.191	325	1162	2968	2.40	13.8	5.20	0.506	1.96	15,940
0.2	0.255	325	1165	2977	3.24	13.9	5.23	0.379	1.96	16,640
0.25	0.318	325	1166	2993	4.08	13.9	5.15	0.304	1.95	17,210
0.3	0.382	325	1167	3011	4.93	14.0	5.12	0.255	1.94	17,560

Methane/air (CH_4/air)

Adiabatic Temperature:	$T_b = 2231$ K
Expansion coefficient:	$\Theta = 10.6$ [10.57580943]
Flame Thickness [cm]:	$L_f = 451\ \mu m = 0.0451$ cm
Laminar Flame velocity [m/s]:	$U_f = 0.3803$ m/s $= 38$ cm/s
Density [kg/m^3]:	$\rho_{CH_4/air} = 1.1 \cdot 10^{-3}$ g/cm^3

Index

M. A. Liberman, *Combustion Physics*,
https://doi.org/10.1007/978-3-030-85139-2

E

F

S

Stress tensor, 95, 158, 355, 467

T

The manufacturer's authorised representative in the EU is Springer Nature Customer Service Centre GmbH, Europaplatz 3, 69115 Heidelberg, Germany. If you have any concerns regarding our products, please contact ProductSafety@springernature.com

Printed and bound by CPI Group (UK) Ltd, Croydon, CR0 4YY
30/09/2026
02225288-0018